Lecture Notes in Computer Science 7334

Commenced Publication in 1973
Founding and Former Series Editors:
Gerhard Goos, Juris Hartmanis, and Jan van Leeuwen

Editorial Board

David Hutchison
Lancaster University, UK

Takeo Kanade
Carnegie Mellon University, Pittsburgh, PA, USA

Josef Kittler
University of Surrey, Guildford, UK

Jon M. Kleinberg
Cornell University, Ithaca, NY, USA

Alfred Kobsa
University of California, Irvine, CA, USA

Friedemann Mattern
ETH Zurich, Switzerland

John C. Mitchell
Stanford University, CA, USA

Moni Naor
Weizmann Institute of Science, Rehovot, Israel

Oscar Nierstrasz
University of Bern, Switzerland

C. Pandu Rangan
Indian Institute of Technology, Madras, India

Bernhard Steffen
TU Dortmund University, Germany

Madhu Sudan
Microsoft Research, Cambridge, MA, USA

Demetri Terzopoulos
University of California, Los Angeles, CA, USA

Doug Tygar
University of California, Berkeley, CA, USA

Gerhard Weikum
Max Planck Institute for Informatics, Saarbruecken, Germany

Beniamino Murgante   Osvaldo Gervasi
Sanjay Misra   Nadia Nedjah
Ana Maria A.C. Rocha   David Taniar
Bernady O. Apduhan (Eds.)

Computational Science and Its Applications – ICCSA 2012

12th International Conference
Salvador de Bahia, Brazil, June 18-21, 2012
Proceedings, Part II

Springer

Volume Editors

Beniamino Murgante
University of Basilicata, Potenza, Italy, E-mail: beniamino.murgante@unibas.it

Osvaldo Gervasi
University of Perugia, Italy, E-mail: osvaldo@unipg.it

Sanjay Misra
Federal University of Technology, Minna, Nigeria, E-mail: smisra@futminna.edu.ng

Nadia Nedjah
State University of Rio de Janeiro, Brazil, E-mail: nadia@eng.uerj.br

Ana Maria A. C. Rocha
University of Minho, Portugal, E-mail: arocha@dps.uminho.pt

David Taniar
Monash University, Clayton, VIC, Australia, E-mail: david.taniar@infotech.monash.edu.au

Bernady O. Apduhan
Kyushu Sangyo University, Fukuoka, Japan, E-mail: bob@is.kyusan-u.ac.jp

ISSN 0302-9743　　　　　　　　　　e-ISSN 1611-3349
ISBN 978-3-642-31074-4　　　　　　ISBN 978-3-642-31075-1 (eBook)
DOI 10.1007/978-3-642-31075-1
Springer Heidelberg Dordrecht London New York

Library of Congress Control Number: 2012939389

CR Subject Classification (1998): C.2.4, C.2, H.4, F.2, H.3, D.2, F.1, H.5, H.2.8, K.6.5, I.3

LNCS Sublibrary: SL 1 – Theoretical Computer Science and General Issues

© Springer-Verlag Berlin Heidelberg 2012
This work is subject to copyright. All rights are reserved, whether the whole or part of the material is concerned, specifically the rights of translation, reprinting, re-use of illustrations, recitation, broadcasting, reproduction on microfilms or in any other way, and storage in data banks. Duplication of this publication or parts thereof is permitted only under the provisions of the German Copyright Law of September 9, 1965, in its current version, and permission for use must always be obtained from Springer. Violations are liable to prosecution under the German Copyright Law.
The use of general descriptive names, registered names, trademarks, etc. in this publication does not imply, even in the absence of a specific statement, that such names are exempt from the relevant protective laws and regulations and therefore free for general use.

Typesetting: Camera-ready by author, data conversion by Scientific Publishing Services, Chennai, India

Printed on acid-free paper

Springer is part of Springer Science+Business Media (www.springer.com)

Preface

This four-part volume (LNCS 7333-7336) contains a collection of research papers from the 12th International Conference on Computational Science and Its Applications (ICCSA 2012) held in Salvador de Bahia, Brazil, during June 18–21, 2012. ICCSA is one of the successful international conferences in the field of computational sciences, and this year for the first time in the history of the ICCSA conference series it was held in South America. Previously the ICCSA conference series have been held in Santander, Spain (2011), Fukuoka, Japan (2010), Suwon, Korea (2009), Perugia, Italy (2008), Kuala Lumpur, Malaysia (2007), Glasgow, UK (2006), Singapore (2005), Assisi, Italy (2004), Montreal, Canada (2003), (as ICCS) Amsterdam, The Netherlands (2002), and San Francisco, USA (2001).

The computational science community has enthusiastically embraced the successive editions of ICCSA, thus contributing to making ICCSA a focal meeting point for those interested in innovative, cutting-edge research about the latest and most exciting developments in the field. We are grateful to all those who have contributed to the ICCSA conference series.

ICCSA 2012 would not have been made possible without the valuable contribution of many people. We would like to thank all session organizers for their diligent work, which further enhanced the conference level, and all reviewers for their expertise and generous effort, which led to a very high quality event with excellent papers and presentations. We specially recognize the contribution of the Program Committee and local Organizing Committee members for their tremendous support and for making this congress a very successful event. We would like to sincerely thank our keynote speakers, who willingly accepted our invitation and shared their expertise.

We also thank our publisher, Springer, for accepting to publish the proceedings and for their kind assistance and cooperation during the editing process.

Finally, we thank all authors for their submissions and all conference attendants for making ICCSA 2012 truly an excellent forum on computational science, facilitating the exchange of ideas, fostering new collaborations and shaping the future of this exciting field. Last, but certainly not least, we wish to thank our readers for their interest in this volume. We really hope you find in these pages interesting material and fruitful ideas for your future work.

We cordially invite you to visit the ICCSA website—http://www.iccsa.org—where you can find relevant information about this interesting and exciting event.

June 2012

Osvaldo Gervasi
David Taniar

Organization

ICCSA 2012 was organized by Universidade Federal da Bahia (Brazil), Universidade Federal do Recôncavo da Bahia (Brazil), Universidade Estadual de Feira de Santana (Brazil), University of Perugia (Italy), University of Basilicata (Italy), Monash University (Australia), and Kyushu Sangyo University (Japan).

Honorary General Chairs

Antonio Laganà	University of Perugia, Italy
Norio Shiratori	Tohoku University, Japan
Kenneth C.J. Tan	Qontix, UK

General Chairs

Osvaldo Gervasi	University of Perugia, Italy
David Taniar	Monash University, Australia

Program Committee Chairs

Bernady O. Apduhan	Kyushu Sangyo University, Japan
Beniamino Murgante	University of Basilicata, Italy

Workshop and Session Organizing Chairs

Beniamino Murgante	University of Basilicata, Italy

Local Organizing Committee

Frederico V. Prudente	Universidade Federal da Bahia, Brazil (Chair)
Mirco Ragni	Universidade Estadual de Feira de Santana, Brazil
Ana Carla P. Bitencourt	Universidade Federal do Recôncavo da Bahia, Brazil
Cassio Pigozzo	Universidade Federal da Bahia, Brazil
Angelo Duarde	Universidade Estadual de Feira de Santana, Brazil
Marcos E. Barreto	Universidade Federal da Bahia, Brazil
José Garcia V. Miranda	Universidade Federal da Bahia, Brazil

International Liaison Chairs

Jemal Abawajy	Deakin University, Australia
Marina L. Gavrilova	University of Calgary, Canada
Robert C.H. Hsu	Chung Hua University, Taiwan
Tai-Hoon Kim	Hannam University, Korea
Andrés Iglesias	University of Cantabria, Spain
Takashi Naka	Kyushu Sangyo University, Japan
Rafael D.C. Santos	National Institute for Space Research, Brazil

Workshop Organizers

Advances in High-Performance Algorithms and Applications (AHPAA 2012)

Massimo Cafaro	University of Salento, Italy
Giovanni Aloisio	University of Salento, Italy

Advances in Web-Based Learning (AWBL 2012)

Mustafa Murat Inceoglu	Ege University, Turkey

Bio-inspired Computing and Applications (BIOCA 2012)

Nadia Nedjah	State University of Rio de Janeiro, Brazil
Luiza de Macedo Mourell	State University of Rio de Janeiro, Brazil

Computer-Aided Modeling, Simulation, and Analysis (CAMSA 2012)

Jie Shen	University of Michigan, USA
Yuqing Song	Tianjing University of Technology and Education, China

Cloud Computing and Its Applications (CCA 2012)

Jemal Abawajy	University of Deakin, Australia
Osvaldo Gervasi	University of Perugia, Italy

Computational Geometry and Applications (CGA 2012)

Marina L. Gavrilova	University of Calgary, Canada

Chemistry and Materials Sciences and Technologies (CMST 2012)

Antonio Laganà   University of Perugia, Italy

Cities, Technologies and Planning (CTP 2012)

Giuseppe Borruso   University of Trieste, Italy
Beniamino Murgante   University of Basilicata, Italy

Computational Tools and Techniques for Citizen Science and Scientific Outreach (CTTCS 2012)

Rafael Santos   National Institute for Space Research, Brazil
Jordan Raddickand   Johns Hopkins University, USA
Ani Thakar   Johns Hopkins University, USA

Econometrics and Multidimensional Evaluation in the Urban Environment (EMEUE 2012)

Carmelo M. Torre   Polytechnic of Bari, Italy
Maria Cerreta   Università Federico II of Naples, Italy
Paola Perchinunno   University of Bari, Italy

Future Information System Technologies and Applications (FISTA 2012)

Bernady O. Apduhan   Kyushu Sangyo University, Japan

Geographical Analysis, Urban Modeling, Spatial Statistics (GEOG-AN-MOD 2012)

Stefania Bertazzon   University of Calgary, Canada
Giuseppe Borruso   University of Trieste, Italy
Beniamino Murgante   University of Basilicata, Italy

International Workshop on Biomathematics, Bioinformatics and Biostatistics (IBBB 2012)

Unal Ufuktepe   Izmir University of Economics, Turkey
Andrés Iglesias   University of Cantabria, Spain

International Workshop on Collective Evolutionary Systems (IWCES 2012)

Alfredo Milani — University of Perugia, Italy
Clement Leung — Hong Kong Baptist University, Hong Kong

Mobile Communications (MC 2012)

Hyunseung Choo — Sungkyunkwan University, Korea

Mobile Computing, Sensing, and Actuation for Cyber Physical Systems (MSA4CPS 2012)

Moonseong Kim — Korean intellectual Property Office, Korea
Saad Qaisar — NUST School of Electrical Engineering and Computer Science, Pakistan

Optimization Techniques and Applications (OTA 2012)

Ana Maria Rocha — University of Minho, Portugal

Parallel and Mobile Computing in Future Networks (PMCFUN 2012)

Al-Sakib Khan Pathan — International Islamic University Malaysia, Malaysia

PULSES - Transitions and Nonlinear Phenomena (PULSES 2012)

Carlo Cattani — University of Salerno, Italy
Ming Li — East China Normal University, China
Shengyong Chen — Zhejiang University of Technology, China

Quantum Mechanics: Computational Strategies and Applications (QMCSA 2012)

Mirco Ragni — Universidad Federal de Bahia, Brazil
Frederico Vasconcellos Prudente — Universidad Federal de Bahia, Brazil
Angelo Marconi Maniero — Universidad Federal de Bahia, Brazil
Ana Carla Peixoto Bitencourt — Universidade Federal do Reconcavo da Bahia, Brazil

Remote Sensing Data Analysis, Modeling, Interpretation and Applications: From a Global View to a Local Analysis (RS 2012)

Rosa Lasaponara Institute of Methodologies for Environmental Analysis, National Research Council, Italy
Nicola Masini Archaeological and Mconumental Heritage Institute, National Research Council, Italy

Soft Computing and Data Engineering (SCDE 2012)

Mustafa Matt Deris Universiti Tun Hussein Onn Malaysia, Malaysia
Tutut Herawan Universitas Ahmad Dahlan, Indonesia

Software Engineering Processes and Applications (SEPA 2012)

Sanjay Misra Federal University of Technology Minna, Nigeria

Software Quality (SQ 2012)

Sanjay Misra Federal University of Technology Minna, Nigeria

Security and Privacy in Computational Sciences (SPCS 2012)

Arijit Ukil Tata Consultancy Services, India

Tools and Techniques in Software Development Processes (TTSDP 2012)

Sanjay Misra Federal University of Technology Minna, Nigeria

Virtual Reality and Its Applications (VRA 2012)

Osvaldo Gervasi University of Perugia, Italy
Andrès Iglesias University of Cantabria, Spain

Wireless and Ad-Hoc Networking (WADNet 2012)

Jongchan Lee Kunsan National University, Korea
Sangjoon Park Kunsan National University, Korea

Program Committee

Jemal Abawajy Daekin University, Australia
Kenny Adamson University of Ulster, UK
Filipe Alvelos University of Minho, Portugal
Paula Amaral Universidade Nova de Lisboa, Portugal
Hartmut Asche University of Potsdam, Germany
Md. Abul Kalam Azad University of Minho, Portugal
Michela Bertolotto University College Dublin, Ireland
Sandro Bimonte CEMAGREF, TSCF, France
Rod Blais University of Calgary, Canada
Ivan Blecic University of Sassari, Italy
Giuseppe Borruso University of Trieste, Italy
Alfredo Buttari CNRS-IRIT, France
Yves Caniou Lyon University, France
José A. Cardoso e Cunha Universidade Nova de Lisboa, Portugal
Leocadio G. Casado University of Almeria, Spain
Carlo Cattani University of Salerno, Italy
Mete Celik Erciyes University, Turkey
Alexander Chemeris National Technical University of Ukraine "KPI", Ukraine
Min Young Chung Sungkyunkwan University, Korea
Gilberto Corso Pereira Federal University of Bahia, Brazil
M. Fernanda Costa University of Minho, Portugal
Gaspar Cunha University of Minho, Portugal
Carla Dal Sasso Freitas Universidade Federal do Rio Grande do Sul, Brazil
Pradesh Debba The Council for Scientific and Industrial Research (CSIR), South Africa
Frank Devai London South Bank University, UK
Rodolphe Devillers Memorial University of Newfoundland, Canada
Prabu Dorairaj NetApp, India/USA
M. Irene Falcao University of Minho, Portugal
Cherry Liu Fang U.S. DOE Ames Laboratory, USA
Edite M.G.P. Fernandes University of Minho, Portugal
Jose-Jesus Fernandez National Centre for Biotechnology, CSIS, Spain
Maria Antonia Forjaz University of Minho, Portugal
Maria Celia Furtado Rocha PRODEB–PósCultura/UFBA, Brazil
Akemi Galvez University of Cantabria, Spain
Paulino Jose Garcia Nieto University of Oviedo, Spain
Marina Gavrilova University of Calgary, Canada

Jerome Gensel	LSR-IMAG, France
Maria Giaoutzi	National Technical University, Athens, Greece
Andrzej M. Goscinski	Deakin University, Australia
Alex Hagen-Zanker	University of Cambridge, UK
Malgorzata Hanzl	Technical University of Lodz, Poland
Shanmugasundaram Hariharan	B.S. Abdur Rahman University, India
Eligius M.T. Hendrix	University of Malaga/Wageningen University, Spain/The Netherlands
Hisamoto Hiyoshi	Gunma University, Japan
Fermin Huarte	University of Barcelona, Spain
Andres Iglesias	University of Cantabria, Spain
Mustafa Inceoglu	EGE University, Turkey
Peter Jimack	University of Leeds, UK
Qun Jin	Waseda University, Japan
Farid Karimipour	Vienna University of Technology, Austria
Baris Kazar	Oracle Corp., USA
DongSeong Kim	University of Canterbury, New Zealand
Taihoon Kim	Hannam University, Korea
Ivana Kolingerova	University of West Bohemia, Czech Republic
Dieter Kranzlmueller	LMU and LRZ Munich, Germany
Antonio Laganà	University of Perugia, Italy
Rosa Lasaponara	National Research Council, Italy
Maurizio Lazzari	National Research Council, Italy
Cheng Siong Lee	Monash University, Australia
Sangyoun Lee	Yonsei University, Korea
Jongchan Lee	Kunsan National University, Korea
Clement Leung	Hong Kong Baptist University, Hong Kong
Chendong Li	University of Connecticut, USA
Gang Li	Deakin University, Australia
Ming Li	East China Normal University, China
Fang Liu	AMES Laboratories, USA
Xin Liu	University of Calgary, Canada
Savino Longo	University of Bari, Italy
Tinghuai Ma	NanJing University of Information Science and Technology, China
Sergio Maffioletti	University of Zurich, Switzerland
Ernesto Marcheggiani	Katholieke Universiteit Leuven, Belgium
Antonino Marvuglia	Research Centre Henri Tudor, Luxembourg
Nicola Masini	National Research Council, Italy
Nirvana Meratnia	University of Twente, The Netherlands
Alfredo Milani	University of Perugia, Italy
Sanjay Misra	Federal University of Technology Minna, Nigeria
Giuseppe Modica	University of Reggio Calabria, Italy

José Luis Montaña	University of Cantabria, Spain
Beniamino Murgante	University of Basilicata, Italy
Jiri Nedoma	Academy of Sciences of the Czech Republic, Czech Republic
Laszlo Neumann	University of Girona, Spain
Kok-Leong Ong	Deakin University, Australia
Belen Palop	Universidad de Valladolid, Spain
Marcin Paprzycki	Polish Academy of Sciences, Poland
Eric Pardede	La Trobe University, Australia
Kwangjin Park	Wonkwang University, Korea
Ana Isabel Pereira	Polytechnic Institute of Braganca, Portugal
Maurizio Pollino	Italian National Agency for New Technologies, Energy and Sustainable Economic Development, Italy
Alenka Poplin	University of Hamburg, Germany
Vidyasagar Potdar	Curtin University of Technology, Australia
David C. Prosperi	Florida Atlantic University, USA
Wenny Rahayu	La Trobe University, Australia
Jerzy Respondek	Silesian University of Technology Poland
Ana Maria A.C. Rocha	University of Minho, Portugal
Humberto Rocha	INESC-Coimbra, Portugal
Alexey Rodionov	Institute of Computational Mathematics and Mathematical Geophysics, Russia
Cristina S. Rodrigues	University of Minho, Portugal
Octavio Roncero	CSIC, Spain
Maytham Safar	Kuwait University, Kuwait
Haiduke Sarafian	The Pennsylvania State University, USA
Qi Shi	Liverpool John Moores University, UK
Dale Shires	U.S. Army Research Laboratory, USA
Takuo Suganuma	Tohoku University, Japan
Ana Paula Teixeira	University of Tras-os-Montes and Alto Douro, Portugal
Senhorinha Teixeira	University of Minho, Portugal
Parimala Thulasiraman	University of Manitoba, Canada
Carmelo Torre	Polytechnic of Bari, Italy
Javier Martinez Torres	Centro Universitario de la Defensa Zaragoza, Spain
Giuseppe A. Trunfio	University of Sassari, Italy
Unal Ufuktepe	Izmir University of Economics, Turkey
Mario Valle	Swiss National Supercomputing Centre, Switzerland
Pablo Vanegas	University of Cuenca, Equador
Piero Giorgio Verdini	INFN Pisa and CERN, Italy
Marco Vizzari	University of Perugia, Italy
Koichi Wada	University of Tsukuba, Japan

Krzysztof Walkowiak   Wroclaw University of Technology, Poland
Robert Weibel   University of Zurich, Switzerland
Roland Wismüller   Universität Siegen, Germany
Mudasser Wyne   SOET National University, USA
Chung-Huang Yang   National Kaohsiung Normal University, Taiwan
Xin-She Yang   National Physical Laboratory, UK
Salim Zabir   France Telecom Japan Co., Japan
Albert Y. Zomaya   University of Sydney, Australia

Sponsoring Organizations

ICCSA 2012 would not have been possible without tremendous support of many organizations and institutions, for which all organizers and participants of ICCSA 2012 express their sincere gratitude:

Universidade Federal da Bahia, Brazil
(http://www.ufba.br)

Universidade Federal do Recôncavo da Bahia, Brazil
(http://www.ufrb.edu.br)

Universidade Estadual de Feira de Santana, Brazil
(http://www.uefs.br)

University of Perugia, Italy
(http://www.unipg.it)

University of Basilicata, Italy
(http://www.unibas.it)

XVI   Organization

 Monash University, Australia
(http://monash.edu)

 Kyushu Sangyo University, Japan
(www.kyusan-u.ac.jp)

 Brazilian Computer Society
(www.sbc.org.br)

 Coordenação de Aperfeiçoamento de Pessoal de
Nível Superior (CAPES), Brazil
(http://www.capes.gov.br)

 National Council for Scientific and
Technological Development (CNPq), Brazil
(http://www.cnpq.br)

  Fundação de Amparo à Pesquisa do Estado
da Bahia (FAPESB), Brazil
(http://www.fapesb.ba.gov.br)

Table of Contents – Part II

Workshop on Econometrics and Multidimensional Evaluation in the Urban Environment (EMEUE 2012)

Knowledge and Innovation in Manufacturing Sector: The Case of Wedding Dresses in Southern Italy 1
Annunziata de Felice, Isabella Martucci, and Dario A. Schirone

Marketing Strategies: Support and Enhancement of Core Business 17
Dario Antonio Schirone and Germano Torkan

The Rational Quantification of Social Housing: An Operative Research Model 27
Gianluigi De Mare, Antonio Nesticò, and Francesco Tajani

Simulation of Users Decision in Transport Mode Choice Using Neuro-Fuzzy Approach 44
Mauro Dell'Orco and Michele Ottomanelli

Multidimensional Spatial Decision-Making Process: Local Shared Values in Action 54
Maria Cerreta, Simona Panaro, and Daniele Cannatella

A Proposal for a Stepwise Fuzzy Regression: An Application to the Italian University System 71
Francesco Campobasso and Annarita Fanizzi

Cluster Analysis for Strategic Management: A Case Study of IKEA 88
Paola Perchinunno and Dario Antonio Schirone

Clustering for the Localization of Degraded Urban Areas.............. 102
Silvestro Montrone and Paola Perchinunno

A BEP Analysis of Energy Supply for Sustainable Urban Microgrids ... 116
Pasquale Balena, Giovanna Mangialardi, and Carmelo Maria Torre

The Effect of Infrastructural Works on Urban Property Values: The *asse attrezzato* in Pescara, Italy............................... 128
Sebastiano Carbonara

Prospect of Integrate Monitoring: A Multidimensional Approach 144
Marco Selicato, Carmelo Maria Torre, and Giovanni La Trofa

The Use of Ahp in a Multiactor Evaluation for Urban Development
Programs: A Case Study 157
  Luigi Fusco Girard and Carmelo Maria Torre

Assessing Urban Transformations: A SDSS for the Master Plan of
Castel Capuano, Naples. 168
  Maria Cerreta and Pasquale De Toro

Workshop on Geographical Analysis, Urban Modeling, Spatial Statistics (Geo–An–Mod 2012)

Computational Context to Promote Geographic Information Systems
toward Human-Centric Perspectives 181
  Luis Paulo da Silva Carvalho and Paulo Caetano da Silva

Voronoi-Based Curve Reconstruction: Issues and Solutions 194
  Mehran Ghandehari and Farid Karimipour

Geovisualization and Geostatistics: A Concept for the Numerical and
Visual Analysis of Geographic Mass Data 208
  Julia Gonschorek and Lucia Tyrallová

Spatio-Explorative Analysis and Its Benefits for a GIS-integrated
Automated Feature Identification 220
  Lucia Tyrallová and Julia Gonschorek

Peer Selection in P2P Service Overlays Using Geographical Location
Criteria. 234
  Adriano Fiorese, Paulo Simões, and Fernando Boavida

Models for Spatial Interaction Data: Computation and Interpretation
of Accessibility 249
  Morton E. O'Kelly

Am I Safe in My Home? Fear of Crime Analyzed with Spatial Statistics
Methods in a Central European City 263
  Daniel Lederer

Developing a GIS Based Decision Support System for Resource
Allocation in Earthquake Search and Rescue Operation 275
  Abolfazl Rasekh and Ali Reza Vafaeinezhad

Concepts, Compass and Computation: Models for Directional
Part-Whole Relationships 286
  Gaurav Singh, Rolf A. de By, and Ivana Ivánová

SIGHabitar – Business Intelligence Based Approach for the
Development of Land Information Systems: The Multipurpose
Technical Cadastre of Ouro Preto, Brazil............................. 302
   *João Tácio C. Silva, José Francisco V. Rezende, Érika Fidêncio,
Tarick Melo, Brayan Neves, and
Joubert C. Lima e Tiago G.S. Carneiro*

Rehabilitation and Reconstruction of Asphalts Pavement Decision
Making Based on Rough Set Theory.................................. 316
   Shaaban M. Shaaban and Hossam A. Nabwey

Cartographic Circuits Inside GIS Environment for the Construction of
the Landscape Sensitivity Map in the Case of Cremona 331
   Pier Luigi Paolillo, Umberto Baresi, and Roberto Bisceglie

Cloud Classification in JPEG-compressed Remote Sensing Data
(LANDSAT 7/ETM+) 347
   Erik Borg, Bernd Fichtelmann, and Hartmut Asche

A Probabilistic Rough Set Approach for Water Reservoirs Site Location
Decision Making 358
   Shaaban M. Shaaban and Hossam A. Nabwey

Definition and Analysis of New Agricultural Farm Energetic Indicators
Using Spatial OLAP 373
   *Sandro Bimonte, Kamal Boulil, Jean-Pierre Chanet, and
Marilys Pradel*

Validating a Smartphone-Based Pedestrian Navigation System
Prototype: An Informal Eye-Tracking Pilot Test 386
   Mario Kluge and Hartmut Asche

Open Access to Historical Atlas: Sources of Information and Services
for Landscape Analysis in an SDI Framework 397
   *Raffaella Brumana, Daniela Oreni, Branka Cuca,
Anna Rampini, and Monica Pepe*

From Concept to Implementation: Web-Based Cartographic
Visualisation with CartoService 414
   Hartmut Asche and Rita Engemaier

Multiagent Systems for the Governance of Spatial Environments:
Some Modelling Approaches 425
   Domenico Camarda

A Data Fusion System for Spatial Data Mining, Analysis and
Improvement 439
   Silvija Stankute and Hartmut Asche

Dealing with Multiple Source Spatio-temporal Data in Urban Dynamics
Analysis 450
  João Peixoto and Adriano Moreira

Public Decision Processes: The Interaction Space Supporting Planner's
Activity 466
  Giuseppe B. Las Casas, Lucia Tilio, and Alexis Tsoukiàs

Selection and Scheduling Problem in Continuous Time
with Pairwise-Interdependencies 481
  Ivan Blecic, Arnaldo Cecchini, and Giuseppe A. Trunfio

Parallel Simulation of Urban Dynamics on the GPU 492
  Ivan Blecic, Arnaldo Cecchini, and Giuseppe A. Trunfio

Geolocalization as Wayfinding and User Experience Support in Cultural
Heritage Locations 508
  Letizia Bollini and Roberto Falcone

Climate Alteration in the Metropolitan Area of Bari: Temperatures
and Relationship with Characters of Urban Context 517
  Pierangela Loconte, Claudia Ceppi, Giorgia Lubisco,
  Francesco Mancini, Claudia Piscitelli, and Francesco Selicato

Study of Sustainability of Renewable Energy Sources through GIS
Analysis Techniques 532
  Emanuela Caiaffa, Alessandro Marucci, and Maurizio Pollino

The Comparative Analysis of Urban Development in Two Geographic
Regions: The State of Rio De Janeiro and the Campania Region 548
  Massimiliano Bencardino, Ilaria Greco, and Pitter Reis Ladeira

Land-Use Dynamics at the Micro Level: Constructing and Analyzing
Historical Datasets for the Portuguese Census Tracts 565
  António M. Rodrigues, Teresa Santos, Raquel Faria de Deus, and
  Dulce Pimentel

Using Hydrodynamic Modeling for Estimating Flooding and Water
Depths in Grand Bay, Alabama 578
  Vladimir J. Alarcon and William H. McAnally

Comparison of Two Hydrodynamic Models of Weeks Bay, Alabama   589
  Vladimir J. Alarcon, William H. McAnally, and Surendra Pathak

Connections between Urban Structure and Urban Heat Island
Generation: An Analysis trough Remote Sensing and GIS 599
  Marialuce Stanganelli and Marco Soravia

Taking the Leap: From Disparate Data to a Fully Interactive SEIS for
the Maltese Islands... 609
  Saviour Formosa, Elaine Sciberras, and Janice Formosa Pace

Analyzing the Central Business District: The Case of Sassari in the
Sardinia Island 624
  Silvia Battino, Giuseppe Borruso, and Carlo Donato

That's ReDO: Ontologies and Regional Development Planning......... 640
  Francesco Scorza, Giuseppe B. Las Casas, and Beniamino Murgante

A Landscape Complex Values Map: Integration among *Soft* Values and
Hard Values in a Spatial Decision Support System 653
  Maria Cerreta and Roberta Mele

Analyzing Migration Phenomena with Spatial Autocorrelation
Techniques 670
  Beniamino Murgante and Giuseppe Borruso

From Urban Labs in the City to Urban Labs on the Web.............. 686
  Viviana Lanza, Lucia Tilio, Antonello Azzato,
  Giuseppe B. Las Casas, and Piergiuseppe Pontrandolfi

General Track on Geometric Modelling, Graphics and Visualization

Bilayer Segmentation Augmented with Future Evidence............... 699
  Silvio Ricardo Rodrigues Sanches, Valdinei Freire da Silva, and
  Romero Tori

A Viewer-dependent Tensor Field Visualization Using Multiresolution
and Particle Tracing.. 712
  José Luiz Ribeiro de Souza Filho, Marcelo Caniato Renhe,
  Marcelo Bernardes Vieira, and Gildo de Almeida Leonel

Abnormal Gastric Cell Segmentation Based on Shape Using
Morphological Operations 728
  Noor Elaiza Abdul Khalid, Nurnabilah Samsudin, and
  Rathiah Hashim

A Bio-inspired System for Boundary Detection in Color Natural
Scenes 739
  Karin S. Komati, Evandro O.T. Salles, and Mario Sarcinelli-Filho

Author Index 753

Knowledge and Innovation in Manufacturing Sector: The Case of Wedding Dresses in Southern Italy

Annunziata de Felice[1], Isabella Martucci[2], and Dario Antonio Schirone[2]

[1] Department Jonico, University of Bari, P.zza Umberto I, 70121 Bari, Italy
annunziata.defelice@lex.uniba.it
[2] Department of Mediterranean Societies of Bari, P.zza Umberto I,
70121 Bari, Italy
i.martucci@lex.uniba.it, darioschirone@libero.it

Abstract. The neoclassical production function assumes that economic growth depends on exogenous factors of production centred on capital, labour and technology. However, residual variables, notably social capabilities and knowledge, are neglected. This study seeks to highlight that, in fact, they are key variables for understanding the economic growth and recent structural changes of an industrial cluster, both in technical and organizational terms. In this work, the peculiarity of knowledge and in particular of tacit knowledge form a crucial element in the social capabilities that are associated with enlarging knowledge learning processes and network diffusion. The aim of this research is to analyse the key role that knowledge and innovations play in the local wedding system of Bari in Puglia. They are the decisive factors in the survival of firms in a global market for the creation of competitive advantage and provide a basis for continuous innovation. The relationship between innovation and knowledge is discussed in the theoretical part of the paper, while the empirical aspect remains based upon results of consumer and producer surveys. The objective is to show how innovation, including demand-driven, can influence companies' behaviours.

Keywords: Innovation, Knowledge, Social Capabilities, Wedding Dresses.

1 Introduction

Unlike the neoclassical production function, where economic growth depends exclusively on traditional resources such as capital, labor and technology – the latter an exogenous datum - the cognitive approach also considers residual variables - social capabilities and knowledge. They are in fact the key variables for understanding the recent structural changes and economic growth of an industrial cluster. In this work, the peculiarities of knowledge include social capabilities or social abilities, since they help enlarge knowledge learning processes and promote network diffusion. The former depends on the degree of cumulativeness, and appropriability, which represents the capacity of new knowledge to generate yet more knowledge and innovation. The higher the degree of appropriability of knowledge, the lower the capacity of diffusion in a cluster and, consequently, its growth.

From a theoretical prospective, this work analyses the different typologies of knowledge, its creation and diffusion. Through an empirical analysis conducted on a sample of firms and consumers in the field of wedding dresses, we show how the innovative capacity, generated by knowledge and by social interactions, causes changes in the competitive dynamics of the sector.

2   The Key Role of Knowledge in the Innovation

Changes in production technology have, over time, been considered independent factors of production function; they are also time-dependent, associated with investment and, therefore, with capital accumulation. These can be the manifestation of an expansion phase of the production cycle, which is positively correlated with an increase in demand or, on the contrary, they can grow into an anti-cyclical function or be connected to non-economic variables. In any case, we cannot deny the crucial role that technical progress has had in the economic development "that is a distinct phenomenon, entirely foreign to what may be observed in the circular flow or in the tendency towards equilibrium. It's spontaneous and discontinuous change in the channels of the flow, disturbance of equilibrium which forever alters and displaces the equilibrium state previously existing"[1]. According to Schumpeter, the changes that lead to development are the introduction of a new good or a new production method, not necessarily based on a scientific discovery; but consisting, for example, in the determination of a new form of marketing or in the access to a new supply source of raw or semi-finished materials or, still, in a new industrial organization. This technological change concerns not only the firms and the adopters, respectively as technology sellers and buyers, but also public institutions; each contributed to the technological change with their experience, hence the expectations.

Schumpeter [2] pointed out the difference between invention and innovation, being the latter dependent from the former, and identified the different innovation route taken by small and large enterprises. In 19th century capitalism, the single entrepreneur gains market shares thanks to his own ability to innovate products, production processes and marketing strategies. This is possible only in a market without barriers to entry and, therefore, with a structural configuration in which perfect information and production fragmentation prevail.

Otherwise, the distinction between invention and innovation makes it clear that if research contributes to innovation, the value derivable from it cannot be ensured without the entrepreneur's initiative. Consequently, if the enterprise allows its competitors to prevail in the innovation of products/processes or marketing, it neither creates more added value for the consumers nor enjoys competitive advantages.

So, if we believe that business is the engine of innovation, it is not feasible for it to invest in R&D if technical progress is a public good characterized by non-rivalry and non-exclusiveness.

The neoclassical models of development [3] regard technical progress as an exogenous parameter that prevents the zero growth rate. In this way, any economic system, coherently with its institutional specificities, can keep its own path to growth and converge to equilibrium. Any attempt to exceed the limits of this theory must keep in mind that the technical progress is endogenously produced but not appropriable [4]. According to Arrow, productive activity produces experience that in the long run becomes knowledge and, therefore, learning useable by everyone without supporting any costs. The existence of economies external to the single enterprise but internal to the industrial area, already singled out by Marshall to justify the validity of the hypothesis of perfect competition in long run, are employed by Arrow to explain that the connection between running productivity and capital goods production - which happened in past and which can be estimated through the learning bend - where you can get increasing returns to scale, notwithstanding the decreasing return of capital factor.

Moreover, is should be kept in mind that human capital, like any physical capital, can be accumulated in time to produce "self-sustained" growth [5a,5b]. Technical progress is an activity whose purpose is to develop knowledge that, once incorporated into the physical capital, determines increasing returns [6]. The three fundamental hypotheses that the theory of the endogenous growth is built on and that are singled out by Romer are: 1) technical progress is endogenous to the economical system, because it's generated by market incitement; 2) technical progress is the main factor to accumulate and to grow; 3) the production units should include between fixed costs the expenses for innovation

The process leading to modern economic systems owes its existence to great discoveries and, therefore, to the inventions that, once translated into innovation by enterprises, create the evolutionary process that leads to economic growth characterized by the progress of some industries and the regress of others [7]; [8]. Consequently, it can be stated that there is a connection between economic growth and the evolution of the industrial structure [9]. This has been analyzed, according to different paradigms, by Bain [10], Coase [11] and Williamson [12], among others.

The enterprise's evolution, the choice of more complex legal forms, the separation between ownership and management, and the consequent or existing modification of the market's structure, all favor large enterprises, which, in turn, take control of market shares and raise strong barriers. In this context, innovation is produced by the large enterprises, whose research activity produces innovation which is distributed according to precise industrial policy choices, leaving little room for the individual entrepreneur.

There are two main types of innovation, one regarding SMEs operating in the traditional sectors - defined as Mark I - and another – defined as Mark II – which regards large enterprises in advanced sectors. They are connected to the technology used [13]. In fact, if innovation depends on the level, variety and pervasiveness of knowledge, the then effectiveness of innovation and its ability to give monopoly to the firms will be positively proportional to the level of knowledge appropriability, and negatively proportional to the degree of externality within the industrial sector. Time, as we know, decreases monopoly power, allowing the potential for imitation. Consequently, the firm requires constant innovative actions to increase its competitiveness and maintain its market share. A process of knowledge accumulation that produces innovation is much needed.

Knowledge accumulation within the firm creates innovation strictly connected to acquired or acquirable skills. The well-funded research laboratories of large enterprises make it possible to elaborate innovative projects. In a specific industrial sector, knowledge accumulation depends on the low grade of knowledge appropriability and on the large extent inter-firm diffusion. This is the key for strengthening knowledge commutability in a local context, where the firms can take advantage of localized externalities and geographic proximity.

Innovation activity depends, therefore, on knowledge that can be classified as context-specific, tacit, complex and independent [14]. In fact, the more changeable the knowledge, the higher the possibility to share it thanks to personal interactions. On the contrary, codified knowledge uses standard communication means like patents, licenses, and so on.

This means that a higher degree of knowledge appropriability is positively correlated with a higher degree of monopoly power, a higher concentration of production activities, and little possibility of knowledge diffusion, which can be regarded as a private property of the firm.

Vice versa, a low grade of knowledge appropriability results in a higher fragmentation of the production activities, and in a higher imitation capacity and higher possibility of creating a continuous process of knowledge diffusion with a consequent increase in innovation. The latter differs from the former in that there is a good chance for the entry of new innovative entrepreneurs who, making use of the skills available in the sectors in which they operate, allow the whole production system to keep the market shares steady and to get new ones.

Knowledge is a "good" that cannot be considered either public or individual; it is more of a collective good in that it can be accumulated, transmitted and can generate innovation according to the capacity of cooperation between people, firms and countries. In fact, if knowledge remains the patrimony of a few, who would benefit from it more than others, differences in growth and development would grow wider. Hence, it is important that every country – whether economically advanced or developing - employ Schumpeter's model of innovation and learning. In fact, only when we recognize the importance of life-long learning can we carry out endogenous growth and self-sustained/sustainable development, which are universal medium-long term targets. Each invention helped the increase in the production of goods and services to improve the welfare of those people who benefited from them, overcoming the borders of production and allowing access to points outside these borders by means of international and interregional markets.

In addition to codified knowledge, which results from the genius of the few, we also have to consider tacit knowledge [15], which is based on experience and can give rise to innovation by inductive methods. The combination of codified and tacit knowledge leads to localized knowledge [16a]; [16b]; [17], which is not easily imitable and is characteristic of industrial districts.

Over time, companies that operate in industrial districts accumulate experience of production techniques, learn from theirs and others' mistakes, interact with suppliers and customers and share the information collected, all of which enables them to increase yield and using known techniques. In the neoclassical model exogenous changes in production and utility functions cause changes in the behaviour of the operator, but not in the structure of their preferences. When the preferences and technologies are endogenous, the social interactions simultaneously modify and complete the ones of the market [18]. In other words, each firm and each consumer change his behaviour because of their interactions with one another. This makes easier the access to external knowledge, which generates new technological knowledge and then innovation. As a matter of fact, if we think that knowledge is at the same time input and output, cannot be ignored the crucial role of the external knowledge in generating the technological one.

3 From Personal to Social, Firm and Cluster Knowledge

It is important to take a more in-depth look at the evolution and transformation of knowledge from personal to social, firm or organizational and cluster or inter-organizational knowledge. According to Nelson & Winter [13], Nonaka & Takeuchi [19], Grant [20], Spender [21] and Howells [22] it is not only individuals that are able to create knowledge. It is necessary to distinguish between individual or personal knowledge, and social or group knowledge, or what Metcalfe and Ramlogan [call "understanding", organizational and inter-organizational knowledge. However, organizational knowledge cannot exist without individuals. Initially, individual knowledge is private, it is in the mind of the individual, and is difficult to transfer because it derives from perceptions, memory, inferences and experience allied with reason [23].

The same object can be seen through different prisms of personal knowledge; it depends on the conceptual system [24]. When individuals interact within the same geographical or local space or context, using a common language, personal knowledge is augmented and becomes interdependent. It becomes social knowledge that is collective, and is derived from individual interactions through formal or informal meetings.

People socialize within political, economic and social organizations, and exchange ideas, information and experience. Members of organizations share their knowledge, for example to improve their individual standing within the organization, or inflate their role [25]. A firm, in terms of its internal structure and its organization, can be considered to be a social and/or knowledge system [26], within which workers exchange ideas, opinions, information, experience and knowledge. The importance of knowledge in the structural organization of the firm was described by Marshall [27], which proposes an analogy between the functional process of the mind and the structural organization of the firm. Both involve different levels: the decision-making level where decisions are codified and transferred to the executive level. At this point knowledge is transformed into routines that can be adapted to different situations. External factors, coming from the demand and suppliers, generate further knowledge, which can make the enterprise's actions more flexible to the market fluctuations. It is necessary to specify that firms and organizations have no self-knowledge in the direct sense. Knowledge can be shared within the firm among customers and managers,

managers and employees, managers and buyers, who all help to create new knowledge. The organization and the internal structure of the firm form a social system. According to Barney [28a]; [28b], organizational knowledge can be considered a competitive advantage, and makes an important contribute to the firm's success by acting as a strategy formulation. So amongst firms, it is information - not knowledge - that is transferred, enabling each firm to retain its competitive advantage. Knowledge becomes available through publication, patents, informal networks, trade and goods. It is possible for knowledge to be transferred between partners that are part of an inter-firm cooperative arrangement or a strategic alliance [29].

Cooperation can reduce competition and networking can enable organizations to access complementary resources and capabilities. Multinational firms and outsourcing agreements are other important ways of transferring knowledge. But it is still difficult to transfer knowledge between firms [30a]; [30b]; [30c], and especially tacit knowledge, because one partner might choose to protect a particular innovation or competitive advantage. In any given geographical area, the relationship between firms and the circulation of workers facilitate the exchange of information and transform organizational/firm knowledge into inter-organizational or cluster knowledge. In fact, when firms are spatially concentrated, knowledge externalities will be more frequent and intensive [31]. Thus concentrations of firms or spatial proximity have power and influence over knowledge interactions [22]. However, these will be dependent on the types of relationships that are established in terms of horizontal and /or vertical integration.

In a district, social networks form more easily and stimulate transfers of information and inputs, which generate new knowledge to an extent determined by the abilities of the firm or social capabilities.

We can consider the firm to be the result of the organization of the production process connected to a mental one that establishes procedures, which, when codified, are transformed into individual knowledge that can be utilized, revised and adapted to different situations [32]. If we also include Marshall's idea of the relationship between the mind and the organization - in other words, between mental process and organizational structure - the system becomes a place where capacities are stimulated, scientific potential is used and technology is developed. More specifically, in any society, we need to consider the steps based on knowledge and on agent relations as they are at the core of any growth process.

Consequently, the key determinants of the birth and growth of an industrial cluster can be found in the historical and cognitive reasons explained by the economic, social, cultural and institutional relations that characterize a population in a specific territorial context. Geographic proximity alone is not sufficient to generate learning and knowledge [33]; [34].

4 The Italian and Southern Italian Economy: Some Notes

If the crisis has hit the developed economies, the Italian one, already slowing down for some time, due to the unsatisfactory performance of industrial production, the fall in investments, the lack of ability to innovate and, consequently, a loss in competitiveness and a fall in the productivity of manufacturers, received a real jolt.

In 2007, Italian GDP grew by 1.5%. In 2008 this was reduced by 1.3%, household consumption fell by 0.8% and gross fixed capital investments by 4%. The situation worsened in 2009 with GDP recording a fall of 5%, consumption -1.7% and investment falling 12.1% [35a]. In the second half of 2009, the economic picture improved with increasing business and household confidence. The year 2010 ended with a GDP growth of 1.3%, which, however, corresponds to a modest recovery of that lost during the crisis. The first months of 2011 lent hope to a recovery, but the financial strains in Greece, Spain and more recently in Portugal are still shaking the markets and the real economy still seems to be slowing.

During 2010 the added value increased in all sectors and in particular in industry excluding construction, growing by 4.8%, thanks to the favorable trend in international trade.

It is clear that the crisis has made its effects felt in the labor market. Between April 2008 and March 2010 resident employment fell by 815,000 units, returning to the levels of early 2006. The number of people looking for work grew by about half a million, due to the contraction in people being taken on and, although only in part, as a result of layoffs. The unemployment rate, rising steadily from the first half of 2007, in March of 2010 reached 8.8%, returning to the levels of 2001 and, in the age group 20-34 years, rose 2%. The rise in unemployment appears less than expected, as there is a more limited labor supply, due to the exit from the market of those discouraged from seeking work. If these people are taken into account together with the hours of state-subsidized employment, the unemployment rate would be 10.6%. After a reduction of 110,000 in 2008, in 2009 those inactive and of working age grew by 329,000 (+2.3%), an increase in absolute terms greater than that recorded for the unemployed. Nevertheless, hourly labor costs grew by 2.3%, adding to the dramatic drop in productivity per hour, resulting in a further acceleration in labor costs per unit of product.

Within this scenario, the southern economy, already in trouble and always lagging behind in terms of growth appears to have slowed down even more. If we dwell for a moment on the data, we can see that the gap between the two geographical areas widens even more, precisely in terms of employment and unemployment. In 2009, in the South, 194,000 people (145,000 men and 49,000 women), of which 125,000 were young people aged 15 - 29 years, lost their jobs and, in addition, the remaining workers were less protected. In fact, against the 186,000 jobs lost in the North, 438,000 people benefited from state-subsidized employment, while in the South of the more than 200,000 employees less than 96,000 were protected in this way [35b].

If we take a quick look at the economy of Apulia, it can be noted that in 2009 the regional GDP decreased between 5 and 6% and employment dropped by 3.8%, at a faster pace than the average for the entire southern sector, while the unemployment rate stood at 12.6% with an increase in both the use of state-subsidized employment, and in the inactivity rate [36a].

By examining the area data [36b] for 2010, we note that in the South, the GDP grew by 0.2%, set against a fall of 4.9% in 2009, but well below the 1.7% growth recorded in the Centre-North. Within the South, the region achieving the greatest growth was the Abruzzi, while in Campania, Molise and Apulia GDP decreased.

By analyzing, in particular Apulia's export data, this shows that during the past decade, this grew by less than the other Italian regions which, in turn, carried little weight in world trade terms. Against a decline in exports [35c] of 25% in Italy,

Apulia, in February 2008, (which saw a peak in Apulian exports) and July 2009, the minimum point, witnessed a drop of -28%, which translates to a +28% if you look at the data up to December 2010.

This data, both negative and positive, can be ascribed to the trends in exports of capital goods and intermediates, while non-durable consumer goods contributed to the recovery to the tune of 26.4%. The sectors that in 2010 recorded an increase in sales abroad, for about one quarter of the total, included chemical and pharmaceuticals, steel and mechanical, while almost 35% was contributed by foodstuffs, leather, leather-working and footwear. The main export markets for Apulia are EU countries, with Germany at the top, followed by France. Markets in the BRIC countries represent a very small share of total Apulian exports and during the worst of the crisis, there was a significant decrease in sales (-26.3%), and in the timid recovery which took place in the course of 2010, the increase was only 2.2%.

At the national level the clothing sector registered a rise, thanks to an improvement in consumer spending which is recorded in Apulia in the sector of formal and ceremonial clothing, which is the subject of our empirical analysis.

5 The Case of Wedding Dresses: The Results of Survey

5.1 Characteristics of Supply and Demand

The empirical investigation was conducted for the province of Bari, the regional capital of the Apulia region in southern Italy, through the administration of questionnaires to consumers and producers in the wedding clothing sector. The choice of the sector resulted from its having taken over the leading position in the textile and clothing industry in the Apulia region, with the presence of c. 7,000 businesses and 38,000 active employees [37]. In particular, the province of Bari, produces 60-70% of the wedding dresses that are certified as being Made in Italy.

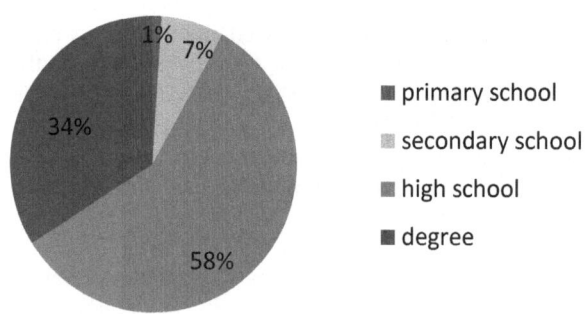

Fig. 1. Consumers' level of education

80% of consumers, of a sample of 250 respondents, are female with 66% aged 25 - 34 years, 58% have a good level education, having earned at least a high school diploma (Fig. 1).

More than 50% of the consumers surveyed belong to the income bracket (Fig. 2) of between 10,000 and 25,000 Euros per annum (p.a) with only 5% having an annual income of over 50,000 Euros p.a.

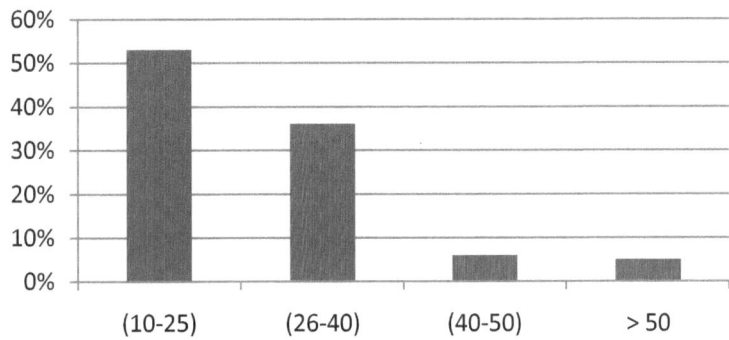

Fig. 2. Income class of consumers interviewed

By analyzing the answers regarding the total expenses for the wedding, one can note that more than 50% of the respondents have a budget of between 10,000 and 30,000 Euros available (Fig.3).

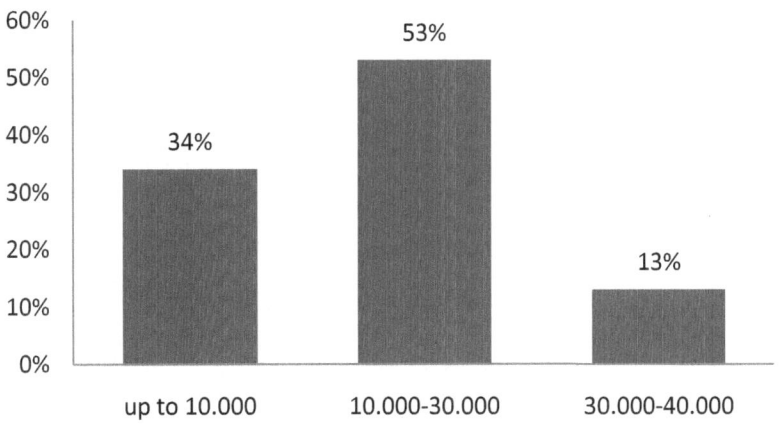

Fig. 3. Total expenditure budget for the marriage

For the wedding dress those interviewed expected to spend at least 3,000 Euros and expressed a preference for a non-designer dress (69%) to be purchased from a local retailer (64%).

Of the 54 companies in the sector under examination [38], 40 of them were interviewed during events or by direct contact. Of these, 36% have their operational headquarters in the municipality of Bari and 37% in Putignano, an area specializing in the production of wedding and ceremonial dress, the remaining 27% being located in other areas of the province of Bari. The survey shows that 20% of the enterprises in question were started in the 1960s and 40% in the 1980s and that they are mostly individual businesses and, to a lesser extent, capital companies.

An examination of the size class shows that 40% of the firms have no permanent but only seasonal employees, 30% are micro-enterprises with an average number of 5.1employees, 20% have an average of 12.5 employees and only 10% have an average of 30.6 permanent employees. Analysis of the questionnaires also revealed that the number of employees has been reduced considerably over the period 2001-2011, dropping, at least in the cluster in question, from an average of 35.6 to 7.1. The lower number of first marriages in Italy – dropping from 392,000 in 1972 to 197,000 in 2009 [39] – as well as the current economic crisis and the adoption of new technologies which improve worker productivity, may explain the contraction in employment levels for the cluster as a whole in which, when staff reach pensionable age, they are not replaced. The number of customers is more than 50 for 77% of companies, and among these, some are working for 180-200 annually even while 3% have a numerically lower, but selected number of customers (1-3 or 4-10). The customer data derived from the responses to the questionnaires may, in our view, also include companies that commission the full package of clothes onto which they then put their own label with an exclusivity clause. In this case, some of the microenterprises produce both independently and for third parties. The companies that make up the sample under investigation have between 11 and 50 suppliers.

In about 30% of companies surveyed the distribution of the finished product is done through wholesalers, or directly to retailers, both Italian and foreign, while some companies do not have their own points of sale. The companies do not relocate their production activities abroad, but limit this to the marketing, approximately 17% of which is achieved through the signing of contracts with agents, importers and distributors. In a few cases (c. 3.3%) agreements are signed with third parties.

5.2 Knowledge, Social Capabilities and Innovation

The analysis conducted shows that local companies, often small or very small in size, do not cooperate with each other, limiting themselves to exchanges of information on suppliers or distributors to reduce transport costs, while no innovative knowledge or market information is exchanged. The intensity of cooperative relationships with suppliers, both local and national, is very important including those in Milan for embroidery and lace and those in Como for silk and fabrics as well as international ones, especially lace in France. With the suppliers they do engage in the exchange of knowledge with regard to new materials and fashion trends. Relations with the majority of external buyers are established at industry events reserved for professionals working in the sector. Participation in local events reaches 80%, with slightly less than 7% also participating in trade fairs and fashion shows that take place both in Europe and the United States and Latin America

90% of the companies surveyed pay particular attention to product innovation (Fig.4) that in small companies is carried out through the presence of an average of 2.3 employees for research and design, a number that rises to an average of 3 units in micro-enterprises.

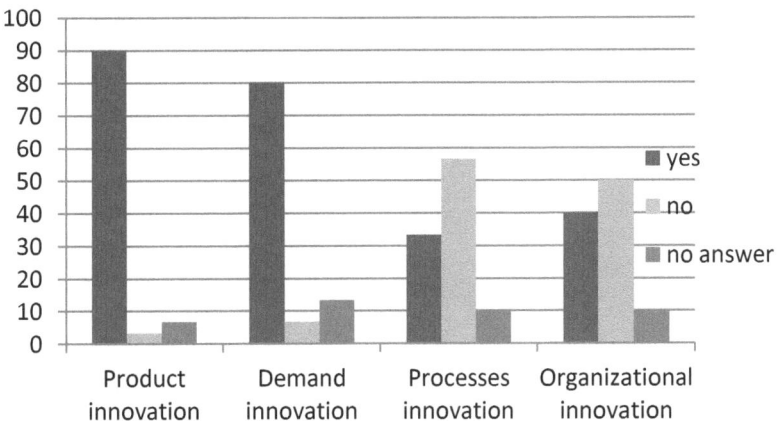

Fig. 4. Levels of innovative activity in enterprises

This activity is carried out in-house in 87% of cases and only about 7% in collaboration with other companies. Product innovation is closely linked to the fabric used in the groom's dress with 100% of consumers preferring a classical style and not the cut of the suit as is more often the case for the bride, for example, with a study of the structure of the bodice. In most cases (80%), innovative activity is induced by the requests or preferences of the bride-to-be, and only in 18% of cases for a casual dress, 36% for a modern one and predominantly, equivalent to 46 %, for a classic look, almost always enriched with a detail that makes it unique and different for the wearer.

The innovation process (Fig. 4), carried out by 33.3% of companies, culminates in the purchase of new machinery and the use of CAD (Computer Aided Design) or CAM (Computer Aided Manufacturing), in order to improve production efficiency. 40% of companies surveyed are also careful to innovate with regard to their organization and marketing through websites that for 73% of the companies makes market access easier, facilitating contact with consumers.

Attention is also devoted to the process of staff training which for 10% of respondents translates into a cost of between 5,000 and 10,000 Euros and 20% in skills costing up to 5,000 Euros.

The survey shows that, despite the great attention paid to innovation, spending on R&D and Design, in 2011, for 20% of the companies, does not exceed 5,000 Euros (Table 1), reaching more than 20,000 Euros for 14% of firms and in only a few cases, especially s.r.l (ltd.) companies, is there a higher than average size class with higher spending as much as 50,000 or 90,000 Euros.

Table 1. Expenditure for innovation and training (2011)

Budget	Expenditure for innovation and training		
	Expenditure on R&D and Design (%)	Expenditure in Learning (%)	Expenditure on skills (%)
Up to 5,000	20.0	7.0	20.0
From 5,000 to 10,000	0.0	10.0	2.0
From 10,000 to 20,000	3.3	4.0	3.3
More than 20,000	14.0	0.0	0.0
N/A	63.0	79.0	75.0
Total	100.0	100.0	100.0

Source: investigation data

In 35% of cases the prototypes are processed and manufactured using CAD or CAM, and in other cases are drawn by hand. The companies also carry out experiments with regard to processes in 17% of cases and the product in 48%.

The lively innovative performance, met with in this traditional sector, reflects the importance that is attributed to the exchange of information and knowledge, seen both from the side of the consumer and that of the manufacturer. Extrapolating from the set of consumers surveyed, a subset of more than 50% shows that the various purchasing decisions are influenced by the experience of others and the confidence derived from knowing this induces the same in the product itself. It seems worth noting that this subset of consumers has a level of education that is higher, compared to the that noted overall, by two percentage points - from 58 to 60% - in the case of high school graduates and more than 4 percentage points - 34 to 38% - in that of graduates

It should also be noted that these consumers, distancing themselves from the overall group by three percentage points and reaching 38% spend more than 3,000 Euros for the purchase of the dress, but the structure of the preferences remains unchanged.

5.3 The Local Wedding Production System

An analysis of the results of this local production system (Fig.5) does not seem to have the characteristics of the cases identified in the literature [40];[41];[42].

The exchange of knowledge between companies that are direct competitors producing goods that are close substitutes but differentiated in terms of quality and design in order to meet the needs of consumers is, in fact, absent. The knowledge is transmitted only within the same company, while with their competitors the exchange of information on upstream or downstream phases of the process is limited. It is worth noting that in most of the micro-enterprises that make up the system the meetings between the workers and owners, including stylists and designers are, by and large, very informal with only the larger firms organizing formal meetings with designers and stylists from outside the company, usually to define a collection of clothes.

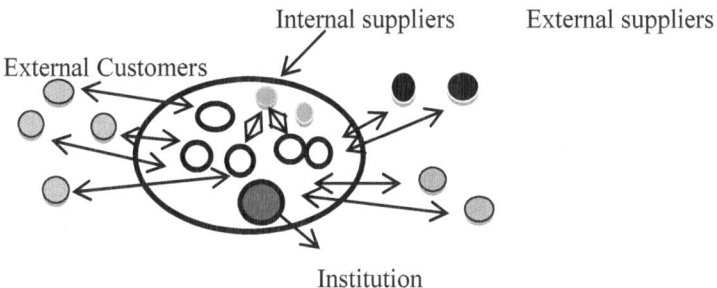

Fig. 5. Local wedding production system

One can identify a transmission of knowledge in dealing with suppliers, especially those that are non-local, where the relationship is one of a cooperative activity with the contracted suppliers. The latter, in fact, producing complementary goods (shoes, accessories, jewelry, flowers, confetti) or offering services, including those of a wedding planner, increase the demand for the product. Great importance is attached (in 93% of cases) to information relating to innovative knowledge and the market, obtained mostly from customers (70%) and suppliers (66.7%), both internal and external to the system. Knowledge is also acquired by consulting specialist journals (43.3%) and in some cases through access to the internet to study the techniques used directly through tutorials (ateliers videos). Similarly relevant for 40% of businesses are commercial relations with agents, both Italian and foreign.

The companies aspire to create synergy with institutions, which, however, "ignore" them, as well as with associations and consortia, which are not particularly present, seeking to fill these gaps by participating in trade fairs and industry events at a level that is compatible with the costs that this sort of exposure entails.

6 Conclusion

The analysis demonstrates that in the field of wedding dresses, external relations are the primary source of knowledge, which is transformed into innovation through social interactions. These, however, are engaged in within each firm as well as between the entrepreneur and his collaborators, including suppliers and designers. In the sector being analyzed, social interaction, involving all stakeholders, leading to the creation of an industrial atmosphere of Marshallian memory that combines tacit and codified knowledge, allowing cooperation albeit with competitors, seems to be lacking.

In our view, the external knowledge that is translated into knowledge and technological innovation, being limited to a single company, does not induce competitive dynamics across the industry. In other words, it would seem that the ability to organize and create the system both locally and internationally is still at an embryonic stage. This deficiency could be overcome if the synergy with the institutions, sought by many of the companies surveyed, were to be realized.

Effective public governance would facilitate the establishment of relations between the external environment and the enterprises themselves, which are vital for economic growth. The gulf between the regions of central and northern Italy and those in the south could also be filled with the competitive advantages, consisting of superior human, social and institutional capital.

Production activities, not only in the sector concerned but covering the entire manufacturing sector in Apulia, can improve their competitiveness domestically and internationally only by strengthening the formation of networks and the creation of knowledge.

References

1. Schumpeter, J.A.: Teoria dello sviluppo economico. Sansoni ed., Firenze (1971)
2. Schumpeter, J.A.: Il processo capitalistico. Cicli economici. Bollati Boringhieri, Torino (1977)
3. Solow, R.M.: Technical Change and the Aggregate Production Function. Review of Economic and Statistics 39(3), 312–320 (1957)
4. Arrow, K.J.: The economic implication of learning by doing. Review of Economic Studies 29(3), 155–173 (1962)
5a. Lucas, R.E.: On the Mechanism of Economic Development. Journal of Monetary Economics 22(1), 3–42 (1988)
5b. Lucas, R.E.: Making a Miracle. Econometrica 61(2), 251–272 (1993)
6. Romer, P.M.: Endogenous Technological Change. Journal of Political Economy 98(5), 71–103 (1990)
7. Marshall, A.: Principles of Economic. MacMillan & Co., London (1890)
8. Schumpeter, J.A.: Business Cycle. Porcupine Press, Philadelphia (1939)
9. Kuznets, S.: Secular Movements in Production and Prices. Houghton Mifflin, Boston (1930)
10. Bain, J.: Barriers to New Competition. Harvard University Press, Cambridge (1956)
11. Coase, R.: The Nature of the Firm. Economica 4(16), 386–405 (1937)
12. Williamson, O.: Markets and Hierarchies: Analysis and Antitrust Implications. The Free Press, NY (1975)
13. Nelson, R.R., Winter, S.G.: An Evolutionary Theory of Economic Change. Harvard University Press, Cambridge (1982)
14. Winter, S.: Knowledge and Competence as Strategic Assets. In: Teece, D.J. (ed.) The Competitive Challenge. Ballinger, Cambridge (1987)
15. Polanyi, M.: Personal Knowledge. Towards a Post-Critical Philosophy. Routledge & Kegan, London (1958)
16a. Antonelli, C.: L'economia dell'innovazione. Cambiamento tecnologico e dinamica industriale. Laterza ed., Bari (1995)
16b. Antonelli, C.: a cura di, Conoscenza tecnologica. Nuovi paradigmi dell'innovazione e specificità italiana. Fondazione Agnelli, Torino (1999)

17. Metcalfe, S.J.: L'innovazione come problema europeo: vecchie e nuove prospettive della divisione del lavoro nel processo innovativo. In: Antonelli, C. (ed.) a cura di, Conoscenza Tecnologica. Nuovi Paradigmi dell'innovazione e Specificità Italiana. Fondazione Agnelli, Torino (1999)
18. Hanush, H., Pyka, A.: Principles of Neo-Schumpeterian economics. Cambridge Journal of Economics 31(2), 275–289 (2007)
19. Nonaka, I., Takeuchi, H.: The knowledge – creating company. Oxford University Press (1995)
20. Grant, R.M.: Toward a knowledge-based theory of the firm. Strategic Management Journal 17, 109–122 (1996) (Winter Special Issue)
21. Spender, J.C.: Organizational knowledge, learning and memory: three concepts in search of a theory. Journal of Organizational Change Management 9(1), 63–78 (1996)
22. Howells, J.: Tacit knowledge, innovation and economic geography. Urban Studies 39(5-6), 871–884 (2002)
23. Metcalfe, J.S., Ramlogan, R.: Limits to the Economy of Knowledge and Knowledge of the Economy. Futures 37, 655–674 (2005)
24. Putnam, R.D.: The Prosperous community: social capital and public life. The American Prospect 4(13), 11–18 (1993)
25. Cappellin, R.: Le reti di conoscenza e di innovazione e le politiche di sviluppo regionale. In: Mazzola, F., Maggioni, F.M. (eds.) Crescita Regionale e Urbana nel Mercato Globale: Modelli, Politiche e Processi di Valutazione, pp. 200–224. Franco Angeli, Milano (2001)
26. Tsoukas, H.: The Firm as a Distributed Knowledge System: A Constructionist Approach. Strategic Management Journal 17, 11–25 (1996) (Winter Special Issue)
27. Marshall, A.: Ye machine. Macmillan, London (1867)
28a. Barney, J.B.: Organizational culture: can it be a source of sustained competitive advantage? Academy of Management Review 11(3), 656–665 (1986)
28b. Barney, J.B.: Firm resources and Sustained Competitive Advantage. Journal of Management 17(1), 99–120 (1991)
29. Collins, J., Hitt, M.: Leveraging Tacit Knowledge in Alliances: the Importance of Relational Capabilities to Build and Leverage Relational Capital. Journal of Engineering & Technology Management 23(3), 147–167 (2006)
30a. Kogut, B., Zander, U.: Knowledge of firm, combinative capabilities and the replication of technology. Organization Science 3(3), 383–397 (1992)
30b. Kogut, B., Zander, U.: What firms do? Coordination, Identity and Learning, Organization Science 7(5), 502–518 (1996)
30c. Kogut, B., Zander, U.: Knowledge of the firm, combinative capabilities, and the replication of technology. In: Foss, N.J. (ed.) Resources Firms and Strategies, pp. 306–326. Oxford University Press (1997)
31. Krugman, P.R.: Geography and Trade. Leuven University Press (1991)
32. Rizzello, O.S.: L'economia della mente. Università Laterza Economia, Roma (1997)
33. Maskell, P., Malmberg, A.: Localized Learning and Industrial Competitiveness. Cambridge Journal of Economics 23(2), 167–185 (1999)
34. Amin, A., Cohendet, P.: Architectures of knowledge. Firms, capabilities and communities. Oxford University Press, Oxford (2004)
35a. Banca d'Italia, Relazione annuale sul 2009, Roma (2010)

35b. Banca d'Italia, L'economia della Puglia nel 2009, Bari (2010)
35c. Banca d'Italia, L'economia della Puglia nel 2010, Bari (2011)
36a. SVIMEZ, Rapporto 2009 sull'economia del Mezzogiorno, Roma (2010)
36b. SVIMEZ, Rapporto 2010 sull'economia del Mezzogiorno, Roma (2011)
37. Osservatorio Nazionale dei Distretti Industriali, II Rapporto, Roma (2010)
38. INFOIMPRESE.it, Dalle Camere di Commercio l'archivio di tutte le imprese italiane (2011)
39. ISTAT, Il Matrimonio in Italia, Roma (2010)
40. Becattini, G.: Dal Distretto industriale allo sviluppo locale. Bollati Boringhieri, Torino (2000)
41. Markusen, A.: Sticky Place in Slippery Space: a Typology of Industrial Districts. Economic Geography 72(3), 293–313 (1996)
42. Guerrieri, P., Pietrobelli, C.: Industrial Districts' Evolution and Technological Regimes: Italy and Taiwan. Technovation 24(11), 899–914 (2004)

Marketing Strategies: Support and Enhancement of Core Business

Dario Antonio Schirone and Germano Torkan

Department for the Study of Mediterranean Societies, University of Bari, Bari, Italy
darioschirone@libero.it,
germano.torkan@gmail.com

Abstract. In theory, the use of atypical marketing techniques holds the potential to support commerce. One such unconventional approach is demonstrated in the example of a specialty furniture company. The company implemented a food-marketing policy as collateral service to the core business The strategic importance of this policy was demonstrated through the resulting boost in sales. The empirical evidence was gathered through measurement and analysis of consumer expenditure. This was monitored under various purchasing conditions. Analysis presents the dimensions resulting from such a successful strategy, showing how the most effective promotional technique is carried out within a single store.

Keywords: Traffic driver, food, marketing mix, promotion.

1 Introduction

In this age of global information sharing and broadening individual knowledge, computers and communication networks are capable of merging the local and the global. The role of the consumer in the market has undergone a profound transformation. No longer the simple end-user of a product or service, the consumer has become an atypical stakeholder. The company must now approach the consumer in a new way and with enhanced content. In accepting that the client can be the object of a raid by the advertising companies, it is undeniable that his contractual power has increased [1]. This is proved in several ways. On the one hand, greater knowledge and the commitment of the individual in the process of "thinning" information asymmetries supports this increase in power. While it can also be proven by unwittingly involving the consumer.

Actions of rival companies in the market determine a strong review of the competitive paradigm, in which there is a need to gain a dominant position according to an innovative method. "The need to remain competitive and to intercept a growing target of customers drives companies to modify their strategies, in order to become not only suppliers of products, but also - and increasingly - of services " [2] . This goal can be reached by focusing, not so much on the product itself, but on shifting the focus onto the way the product is presented to the consumer. This therefore shows that the monopoly of a new market is basically tied to a company's ability to manage an innovative relationship with its customers [3]. The competition is no longer strictly tied to a logic of the 'traditional' offer, but requires a strategic plan in terms of pre and post-sales. This means a "premium price corresponding to an effective higher value recognized by customers" [4]. In this sense, Montemerlo's insight [5] is shared according to the outcome of a competitive strategy depending "on the harmony between the three macro variables," including the product system, the competitive system, and corporate resources.

This intuition demonstrates the need to combine the company's internal and external elements according to the principles of coherence and creativity [6]. This strategy assumes some relevance on the way to saturation in mature industries; where the collateral-offered services are inherent to the main product. However, it is more interesting to wonder whether this insight translates into other market situations which are not necessarily affected by saturation. It is also important to understand whether the spread of a product on the market is enhanced by the availability of collateral services closely related to it, or by taking advantage of other atypical ways to drum up business. The case study seeks to answer this question if compared to the core business of the company.

2 Food as Business Strategy

The megastore analyzed contains multiple self-branded eateries. An objective was established through the analysis of sales trend within the furniture megastore. The goal is to calculate whether the offer of the restaurant service, collateral compared to the business mission, can provide a driving force capable of increasing sales. The alternative could be to find out that the impact of food collateral is physiological or is at least minimal compared to the original overall intent. It must be determined if this can be described as an elaboration of the highly evolved 'promotional' component of the marketing mix. Although the traditional approach is intended to be closely linked to the product offered on the market, unavoidably recalling to mind that generically, a promotion "identifies the set of promotional tools variously directed to create the image of the enterprise" [7]. For this reason, it may suggest that this commitment is not bound simply to the advertisement of a single product or service. But it also represents an unusual method to promote the company system in its entirety. So conceived, the individuals purchases are strongly influenced by external factors, such as the perception that the consumer has of the bidder. This perception is considered as a process of gathering and interpretation of the information from the external, which translates into purchasing decisions [8].

It is also important to note that in this particular case, as well as considering the promotional component of the aforementioned activity, one has to deal with a strategic business area (SBA); which is by its very nature, a "sub company distinguished by its own specific mission in terms of products offered and markets served" [9]. One cannot exclude that the analysis could lead to considering the possibility that the parallel activity can evolve independently and could even emancipate itself from its ancillary role with respect to the core business. Nevertheless, the analysis must take into account another considerable aspect. Megastores are increasingly developing a configuration similar to that of traditional malls. They are no longer simply places of traditional business negotiations, but are becoming an attraction that is a form of entertainment for the whole family. If it is true, it could be inappropriate to consider these places as potential competitors of locations traditionally perceived as places of entertainment. On the other hand, one cannot deny that they leverage the same psychological mechanisms upon the consumer, perpetuating certain behaviors even in these megastores, which deeply affects the propensity to buy.

This scenario is not only intended to measure the extent of the contribution of this food-marketing strategy on the total sales, but to see if it is even possible to formalize this strategy through a methodology ad-hoc. The concept of 'customer loyalty' itself deserves further analysis. Could the use of the food as a promotional tool be counted as an implement for cementing the supplier-consumer relationship?

3 The Survey

3.1 Methodology

The questionnaire was administered in November 2010 over seven consecutive days. Seven hundred customers of an important commercial leader in the furniture industry located in Bari, Italy were surveyed. There are multiple catering areas in this store, including restaurants, bars, and fast food joints. After the surprising socio-demographic characteristics of the sample, the study will seek to monitor the behavior of the persons concerned through the analysis of the average receipt of those who made purchases of furnishings. The second step is to see if the propensity to consume for those who did use the catering service differs from those who did not partake. The possible differences in shopping habits between the occasional customers and loyal customers, who hold a rewards cards will also be inspected.

Interviews were conducted using a randon sample, which was formed by systematic sampling [10], following a few basic instructions:

- The interviewee was approached in the final stage of the buying process, next to the main;
- Respondents were chosen regardless of gender, but necessarily of adult age, excluding children and teenagers;
- Employees of the mall were excluded from the survey;
- The choice of the interviewees was made with a predetermined frequency (one in every ten visitors met), i.e. through a mathematical recurrent process instead[11];
- The dynamics of choice were not affected in any way by physiognomic sympathies;
- Spontaneous applications have not been considered.

3.2 Socio-Demographic Analysis of the Sample

The analysis conducted shows that 78% of participants surveyed were female (Figure1).

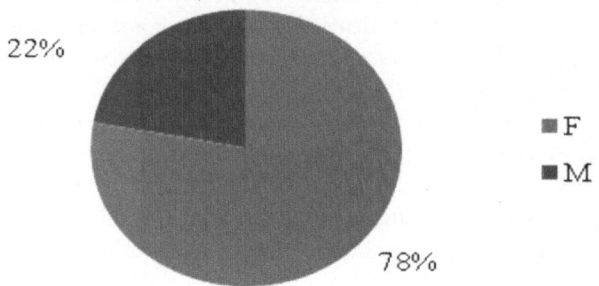

Fig. 1. Gender of respondents

The family group size is at least doubled on the weekends at about 4.2 individuals per receipt. Whereas on working days it remains around 2,39 individuals (Table 1). This figure confirms that the megastore has become a preferred form of weekend family entertainment.

Table 1. Group size

Off-peak days (Mon thru Fri)	2,39
Peak day Saturday	2,87
Peak day Sunday	4,19

This concept is additionally supported by the age demographics of clients interviewed. As one can see from the table below, all those sampled form a relatively equal representation of ages of 25 through 54 years old (Table.2).

Out of the total of respondents, 64% are permanent employees of either a private enterprise or public administration.

The income bracket most represented was that of the medium-to-low income class, as evidenced by the following figure:

Table 2. Breakdown of the age demographic

18 - 24 years	6%
25 - 34 years	23%
35 - 44 years	31%
45 - 54 years	20%
55 - 64 years	18%
Over 65 years	4%

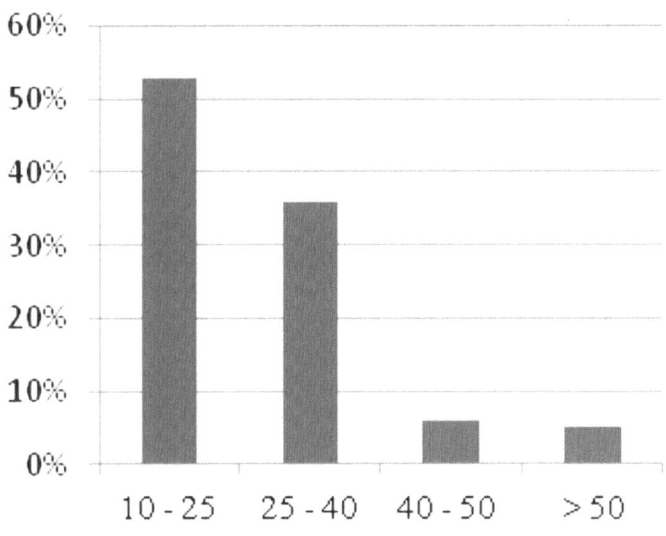

Fig. 2. Income classes

Indeed, more than half of the sample has a gross income of between 10,000 and 25,000 Euros a year. While only 5% are in the high income bracket at an average of more than 50,000 Euros yearly. Consequently, the target of the megastore is reflected in its heterogeneous audience.

3.3 How Food Influences Purchasing Habits

Among the 700 customers surveyed, about 70% asserts having rested and used the catering service. The remaining 30% of customers self-limited to only making purchases in the megastore (Figure 3):

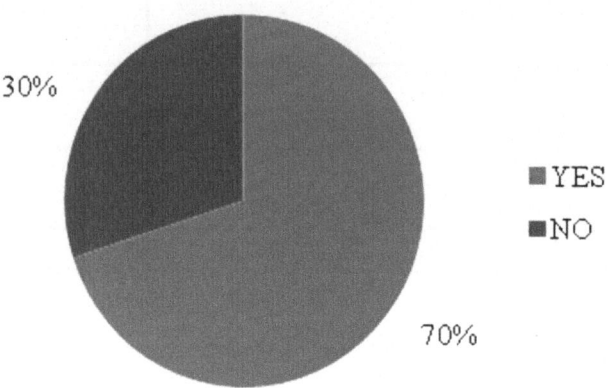

Fig. 3. Composition score: Food / No Food

As evident by the chart below, the total expenditure on furniture and home accessories made by those who had previously rested at the refreshment areas is four times greater than those who had not. The amount of in-store expenditure is shown as the sum of receipts recorded at the bank. Those who purchased at the refreshment areas spent an average of 58,282 Euros; while those who did not averaged a 13,496 Euros bill (Figure 4).

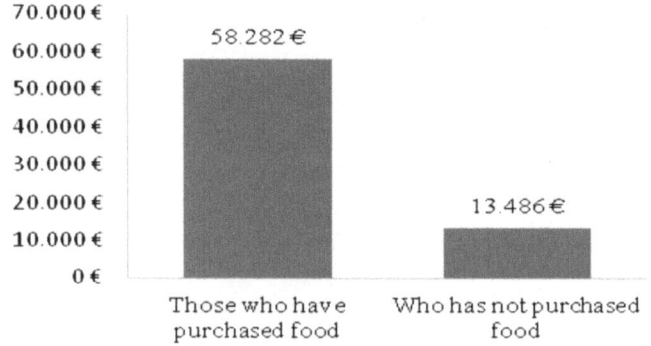

Fig. 4. Overall in-store spending

It is important to note that because of the greater weight of the purchases made by customers in the refreshment areas, the data must be made unbiased and given a unit of measure that is genuinely representative of the driving effect that food exerts on the average receipt of furniture and furnishings. In this regard, the average receipt will be calculated between two categories (Figure 5):

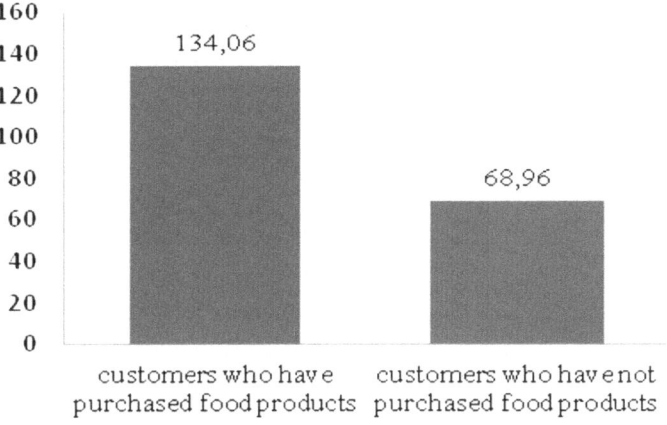

Fig. 5. Average receipt in store (Food/No Food)

A further demographic division functional to the inquiry appears between the customer fidelity card holders and those are just occasional shoppers (Figure 6). Because of this delineation, it is interesting to note the distribution of the accumulated expenditures between loyal customers and those who bought occasionally in refreshment areas.

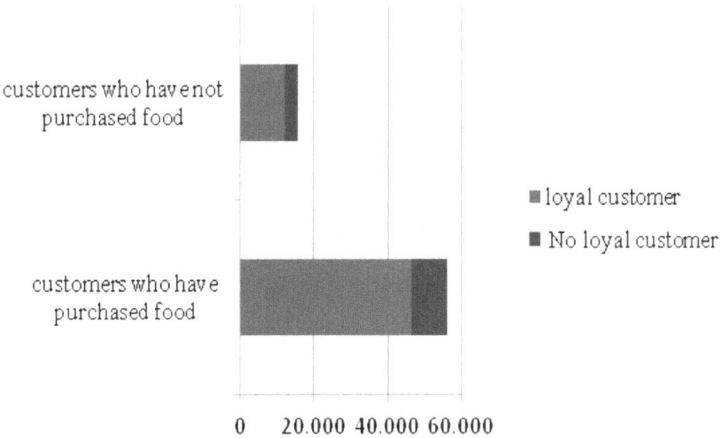

Fig. 6. Spending accumulation: Food / No Food and Loyal customers / Occasional customers

While analyzing those who have purchased at the refreshment areas; one learns that the loyal customer generates spends an average of 145.72 Euros, which is 31% higher than a "non-loyal customer". Occasional shoppers spend about 99.72 Euros on average (Figure 7).

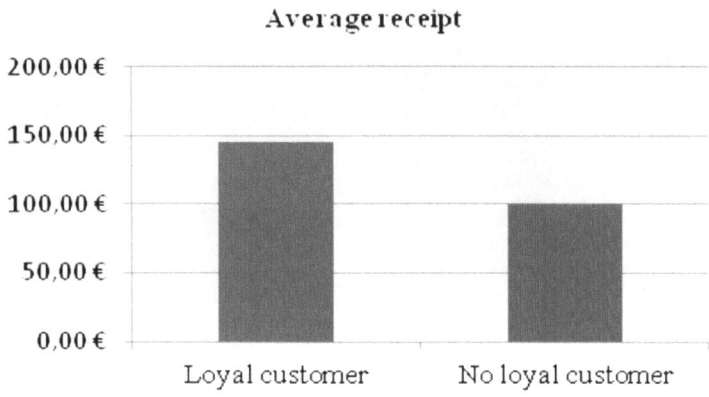

Fig. 7. Average spending in Euros (Loyal customer v. Occasional customer)

Given that half of those interviewed were served at the refreshment areas also own a loyalty card; one may record an increase of 15% on the average store receipt so as to offset the price change cause by the card (Figure 8).

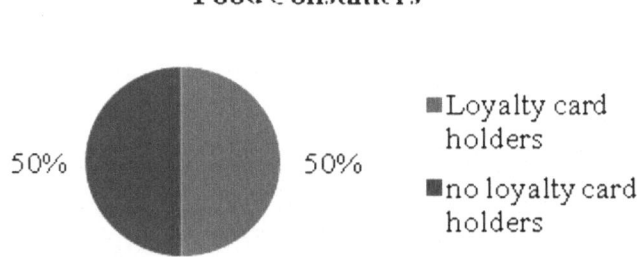

Fig. 8. Customers who have eaten in the dining areas

The correlation in the data of loyalty and attendance of the refreshment areas certainly highlights the average spending of loyal *Food consumer* customers, resulting in a higher propensity to purchase than with the typical buyer. Indeed, the increase found in the average spent of this type of consumer is about one sixth above average, suggesting promotion in this direction. The company's goal is linked to its ability to increase the purchasing power of its customers through both loyalty and participation in the *Food store experience*.

4 Conclusions

The analysis clearly highlights the strategic value of promotional incentives. Maintaining multiple dining areas within a megastore is highly attractive for family groups. During the peak weekend days, the purchasing group doubles, attesting to the attractiveness of having varied products and services offered to the public. Such circumstances go beyond the mere commercial vision of the traditional store. So conceived, the store becomes a center of attraction, targeting not just a single type of customer, but also widening the target demographic by both age and income.

The use of food as a promotional incentive augments consumer demand by affecting the dynamics of the consumer mindset, creating the need to buy. Measuring average spending of customers who have previously used the catering service, shows that the break assumes an additional purpose. Taking this dining break serves to subvert the idea of a frenetic 'race to buy'; the highly sought after perceptive and sensorial stimuli, both improving the shopping experience as well as stimulating the birth of the consumer needs. The latter allows the consumer to regain his sense of rationality that was temporarily inhibited by impulsive purchasing. The quality and type of purchase increases both the supply and demand.

The empirical data produces a significant effect; the evolution of promotional incentives from instrumental and supporting activities, into core business practices and assets of an independent company. The role food plays is not limited to simply strengthening the starting business, it creates a self-supporting business both identified and supported by customers. Establishing that the company's target demographic is, without a doubt, the 'loyal client' came as an unexpected result. While simultaneously protagonist to promotional drivers, this type of customer guarantees better performance with repeat business for the company. It is even encouraged to target this new idea of consumer as a replacement for the traditional customer demographic because of the exclusivity of its features.

References

1. Cohen, J.B., Pham, M.T., Andrade, E.B.: The nature and role of affect in consumer behavior. In: Haugtvedt, C., Kardes, F., Herr, P. (eds.) Handbook of Consumer Psychology, pp. 297–348. Erlbaum, Mahwah (2008)
2. Pace, R., Schirone, D.A.: Over 65, un nuovo target per chi fa marketing. Neodemos, Firenze (2011)
3. Aerni, P., Scholderer, J., Ermen, D.: How would Swiss consumers decide if they had freedom of choice? Evidence from a field study with organic, conventional and GM corn bread. Food Policy 36, 830–838 (2011)

4. Spano, F.M.: Il vantaggio competitivo di differenziazione. In: Dipartimento di Economia Aziendale (eds.) Lezione di Economia Aziendale, pp. 460–467. Giappichelli, Torino (1996)
5. Montemerlo, D.: Il sistema competitivo: la struttura del sistema competitivo, il modello della concorrenza allargata. In: Airoldi, G., Brunetti, G., Coda, V. (eds.) Corso di Economia Aziendale, pp. 322–328. Il Mulino, Bologna (2005)
6. Giaccari, F.: Attività economica e le aziende. In: Di Cagno, N., Adamo, S., Giaccari, F. (eds.) Lezioni di Economia Aziendale, pp. 11–38. Cacucci, Bari (2003)
7. Giaccari, F.: Le funzioni e i processi aziendali. In: Di Cagno, N., Adamo, S., Giaccari, F. (eds.) Lezioni di Economia Aziendale, pp. 135–162. Cacucci, Bari (2003)
8. Pellicelli, A.: Il Marketing: segmentazione e selezione del target. In: Dipartimento di Economia Aziendale (eds.) Lezione di Economia Aziendale, pp. 86–87. Giappichelli, Torino (1996)
9. Parolini, C.: Un modello di sintesi: le determinanti dei risultati reddituali e patrimoniali. In: Airoldi, G., Brunetti, G., Coda, V. (eds.) Corso di Economia Aziendale, pp. 573–602. Il Mulino, Bologna (2005)
10. Girone, G., Salvemini, T.: Lezioni di Statistica. Cacucci, Bari (2004)
11. Robert, C.P., Casella, G.: Monte Carlo Statistical Methods. Springer, New York (2004)

The Rational Quantification of Social Housing
An Operative Research Model

Gianluigi De Mare, Antonio Nesticò, and Francesco Tajani

Università degli Studi di Salerno, Dipartimento di Ingegneria Civile
Via Ponte don Melillo - 84084 - Fisciano (SA), Italy
{gdemare,anestico,ftajani}@unisa.it

Abstract. This work has addressed the issue of the urban redevelopment of brownfield sites. It has developed an evaluation model for the quantification of the social housing component that the private investor must make in favor of the public administration. The protocol is based on finding a balance between the interests of the parties involved in urban redevelopment. The algorithm is constructed by borrowing techniques from Operations Research Linear Programming. The objective function is to maximize the social housing quota. The constraints formalize the utility functions of the two contracting parties (public and private). The result is a simple to use mathematical process, which can be implemented in any local context, in order to make choices consistent with the potential of the local housing markets. The algorithm developed is applied to a case located in a municipality of the Agro-Nocerino-Sarnese (SA). The output returned enhances the rationality of the model, capable of contributing to the feasibility of the redevelopment of disused areas.

Keywords: brownfield sites, housing market, linear programming, social housing, urban redevelopment, public utility, private utility.

1 Introduction

In the current macroeconomic situation, the re-launching of the Italian economy is based on: a) a strict fiscal policy that should help to contrast the growth of public debt; b) the enactment of a package of measures to stimulate investment through administrative simplification. The structure of the national economy is at the centre of the current difficulties, i.e. small and medium-sized businesses hit by declining demand, competition from foreign low-cost production systems and the restrictions on the flow of financing from banks, with this, consequently, resulting in a growing number of bankruptcies. This phenomenon, in addition to creating more unemployment, causes an increase in urban brownfield sites that need to be redeveloped and assigned a new use.

The possibility to regenerate urban brownfield sites is particularly noticeable when the structures are located in central areas. The transformation from a productive use to

---

[*] This paper is to be attributed in equal parts to the three authors.

either a residential, commercial or tertiary one[1] is therefore more appealing. Most of these sites were built at the beginning of the last century, situated close to city centres, at a time when there was a need for the strict regulation of the urban development of cities. Around these industries, new centres were built, characterized by a mix of uses, leading to a particular form of urban sprawl. This is not only widespread in Italy but also in other European countries, where urban centres have developed over time sometimes becoming forms of urban sprawl.

The need for urban reorganisation can be associated to the pressing demand for housing, with one of the basic aims being to try and limit land consumption (so-called *soil sealing*[2]). In this perspective, there has been a persistent need for national and regional regulations in recent years. The measures are aimed, on the one hand, at simplifying and accelerating the administrative procedures for the approval of conversion of use of buildings projects, recognising a rewarding volume in the implementation phase of new structures. While on the other, to meet the housing needs of disadvantaged social classes, by making changes to the construction of social housing.

2   Aim of the Study

In current literature, the term *brownfields* is used and includes all those "abandoned industrial areas, potentially contaminated that require conversion"[3]. Brownfields may also be defined as those redeveloped properties and whose reuse may be complicated by the presence – potential or real – of hazardous substances, pollutants or contaminants. The uncertainty related to the quantification of the costs of the reclamation of "potentially contaminated" or "contaminated"[4] due to, for the public, the difficulty of providing adequate compensation – new urban destinations, transfer of development rights, etc. – that make the areas more attractive for investors

---

[1] "In most cases [...] the redevelopment deals with areas, once on the outskirts, that have become internal to urban centres with high market values related to positional annuity phenomena. [...] These are areas that were initially outside the city limits, subsequently incorporated over the years due to expansion, and now occupy strategic positions, often with the aim of converting them to new uses. Up until recently, they were considered a problem, as places of a high concentration of social problems, degraded areas that are now considered as a resource by both the public and private sectors" [17].

[2] The term *soil sealing* is used to indicate the degradation process specifically related to the transformation of urban land. It is caused by the covering of land with materials that make it waterproof and is generally irreversible or hardly ever reversible.

[3] [21].

[4] "In accordance to D.Lgs. 152/2006, Title V of Part IV, so that a potentially contaminated site can be defined as such, it must have already exceeded the table limits, whereas a contaminated site can only be defined so, when, following the execution of a characterization plan, it has exceeded the risk threshold, which are site-specific and defined on a case by case basis upon application of an environmental health risk analysis" [1].

determine the impossibility to estimate a priori the convenience margins of urban transformation interventions[5].

The situation of disused industrial areas is different due to the fact that the reclamation costs are not required or easily estimated[6]. It is therefore worthwhile to distinguish between two different types of brownfield sites: 1) land for which reclamation costs are quantifiable and contained, it is possible to determine a margin of economic convenience (by a private investor) in the transformation, 2) actual brownfields or, in other words, "contaminated" and "potentially contaminated" sites for which it is difficult to reduce the uncertainty inherent in estimating the reclamation costs.

With reference to the first category, taking into account private needs and public interventions, this paper aims to develop a model that can establish the social housing quota required to ensure the convenience for both parties involved of the proposed redevelopment. It should be noted that for several decades the use of models has spread as an investigation and study technique of urban and metropolitan planning. In some European cases, the use of these techniques is usually entrusted to the ability and willingness of innovative designers and projects. In other contexts, such as the English one, it is sanctioned by the law, so much so that some peripheral organizations of the State publish manuals on the use of models by the directors of local communities.

The model is herein constructed by adapting techniques from Operations Research Linear Programming. The objective function seeks the maximization of the social housing quota through a resolution software[7] that uses the Simplex method. The constraints formalize the utility functions of the two contracting parties (public and private). The result is a simple to use mathematical process, which can be implemented in any local context. The algorithm developed is applied to a case study located in a municipality of the Agro-Nocerino-Sarnese (SA). The returned output shows the appropriateness of the model tested for the activation of the redevelopment of derelict areas. Obviously, the model presented makes specific reference to the national regulatory landscape. Nevertheless, small changes can be made to its mathematical structure, so as to allow it to correspond to the laws of other countries. In this sense, the model seems to be generally valid.

---

[5] "[...] It is possible that the high market value of these areas and the huge amounts that frequently must be supported in order to make them suitable, considerably influence transaction costs, so as to make any intervention by private investors inconvenient. Neither an increase of the buildable land index – to offset the extra costs – may go beyond the limits imposed by the absorption capacity of the building by the local property market or the design and zoning choices, set out by the regulations for the "intervention area" [17].

[6] It is worth considering brownfield sites where there are buildings with asbestos roofing. In these cases, it is possible to develop a valid evaluation of the reclamation costs that the contractor will incur in the conversion process.

[7] LPSolve IDE – Versione 5.5.2.0.

3 Social Housing

EU countries are characterized by the heterogeneity of their national housing situations and policies. The concept of social housing tends to vary from one country to another. The CECODHAS – European Committee for the promotion of housing rights – defines social housing as "housing solutions for those families whose needs can not be satisfied with market conditions and for which there are rules of allocation". The country with the largest area of social rent is the Netherlands, with a percentage equal to 35%. The northern and western European countries tend to have a social sector that is larger than the Mediterranean countries. In the countries under forms of transition in Eastern Europe, what was a vast public housing sector has substantially dissolved with the fall of communist regimes and the mass privatization which followed.

In Italy social housing consists of the set of housing services, to safeguard social cohesion, aimed at satisfying the primary housing needs for those who are unable to access, for economic reasons, adequate housing in terms of health, safety and dignity.

Under the Ministerial Decree dated 22 April 2008[8], the classification of social housing on the basis of the management mode and owner profile is as follows:

a) social housing in public ownership. In this case, the private owner of the redeveloped land builds social housing that is then donated to the Local Authority. The management mode is regulated by regional regulations;
b) social housing of private property which can be sold immediately. The criteria for determining the sale price of social housing are set by the Regional Authorities, as a result of the agreement with the regional ANCI;
c) social housing of private property, used as permanently or temporarily rented (at least eight years). The rent is set by the Regional Authority, in consultation with the regional ANCI, "according to the different economic capacities of the beneficiaries, the composition of the household and housing characteristics"[9]. In any case, the set rent cannot be greater than that resulting from the fixed-price values resulting from the local territorial agreements[10]. For temporary social housing, the selling price at the end of the rental period is defined by the Regional Authority, in accordance with the Regional ANCI.

Between types b) and c), the most convenient for the private owner is b), which provides for the immediate sale of housing, thus making it possible to quickly recuperate the investment cost. From a public perspective category c) is preferential. In fact, the management mode of "controlled" rental is a solution that benefits those families who cannot afford to either buy – even if the price is "controlled" – or rent accommodation on the free market.

---

[8] [14].
[9] Ministerial Decree of 22 April 2008, Art. 2, co. 2.
[10] See: Art. 2, co. 3, L. 431/1998.

4 The Private Benefits Related to the Redevelopment of a Disused Area

The financial benefit for the owner to redevelop the area is directly proportional to the density and marketability of the intended uses consented. While, it is negatively affected by the percentage of social housing to ensure. Since, ordinarily, the owner of the disused area is not the investor who wishes to develop the land, the utility he is eligible for depends on the value that results from the change in use.

Under a disciplinary profile, the market value of a buildable area is influenced by a direct process, noting the recent transactions in similar areas in the reference market, or through an indirect process, using the transformation value as the process one.

It is evident that the owner of a disused area deems the redevelopment project economically convenient if the market value of the property related to the transformation is at least equal to the current market value generated by the actual use of the property. It is worthwhile noting that abandoned buildings often have an earnings value that results from the renting of volumes for any purposes other than the original one. The *limit value* which leads the owner to carry out the transformation intervention is, in fact, generally greater than the market value of the area on the basis of the current use. It is worth considering the extent of the difference between the market value of the area and the limit value which increases the appeal of investments in the sector. It is therefore plausible that a reasonable limit value is an expression of the mortgage value which the owner ordinarily expects from the transformation.

As previously mentioned, the social housing quota represents a reduction in the total private utility of the transformation. The induced loss is diversifiable due to the type of social housing built.

Taking into consideration the categories of social housing described in section 3, it is possible to deduce that: 1) the social housing quota donated by the private investor – type a) – makes up for a missed revenue equal to the selling price on the free market of the gross surface areas (SLP) transferred to public ownership, 2) the social housing quota that can be immediately sold – type b) – sets a minus value equal to the difference between the selling price on the free market and the "controlled" selling price. and 3) the social housing quota used for renting – type c) – generates a contraction of private benefits that, in the case of temporary accommodation, including two quotas: i) the initial accumulation of the difference between the free market fee and the agreed fee, ii) the difference between the controlled selling price of social housing at the end of the lease period, with both being estimated at the end of the lease.

The mathematical formal of private utility ($U_{private}$) deriving from the redevelopment of the disused area is the following:

$$U_{private} = V_{m,trasf} - K_{sh} \qquad (1)$$

in relation to the three conditions

$$\begin{cases} V_{m,trasf} - K_{sh} \geq V_{\lim it} \\ V_{\lim it} \geq V_{current} \\ V_{current} = c_p \cdot R \end{cases} \qquad (2)$$

Obviously, the impact on the utility for the owner deriving from the *social housing* is proportional to the values of the same through an appropriate exchange coefficient.

Table 1 reports the terms in expressions (1) and (2). It is understood that the term $V_{m,trasf}$ does not reconcile the damage caused to the owner by the social housing quota. The cost associated with social housing (K_{sh}) obviously depends on the three aforementioned types of social housing. It also depends on the selling price of social housing (p_s), the controlled rent (Ca_s), the number of years of temporary renting (n), the discount rate for the private owner (r). This sample is used for the items that make up the cost of type c)social housing.

$$K_{sh} = f(x_1, x_2, x_3, p_s, Ca_s, n, r) \tag{3}$$

Revenues from the sale of the finished product (R) include the reduction of value generated by the predictions of social housing. This term is a function of K_{sh}, the intended uses (*dest*), the total construction amount realized (*quant*) and the selling prices of the buildings on the open market (p).

$$R = f(K_{sh}, dest, quant, p) \tag{4}$$

Table 1. Private utility parameters

$V_{m,trasf}$	market value of the disused area due to the effect of the possible transformation
V_{limit}	limit value for the owner until the transformation is deemed economically convenient
$V_{current}$	market value generated by the current use of the property
K_{sh}	cost for the owner of the developed area by the social housing quota
c_p	mortgage coefficient for the owner of the land
R	revenues from the sale of the finished product
p_s	"social" selling price, i.e. controlled
Ca_s	annual unitary "social" renting price, i.e. controlled
n	number of years of temporary renting of social housing
r	discount rate for the private investor
dest	intended uses
quant	total construction amount realized
p	selling price on the open market of the building

5 The Public Utility Resulting from the Transformation of an Abandoned Area

Within a public context, the transformation programmes of degraded or disused areas generate positive economic, social, urban and environmental effects. There are numerous contributions in current literature on methods to determine the economic value from the redevelopment and conversion of brownfields[11], using *contingent*

---

[11] See [21]: "In recent decades, many European cities are facing the problem of disused land, often former industrial sites".

evaluation [12] or *hedonic price*[13] techniques. In this paper, the complex under the name of *Benefits from redevelopment* will be defined.

---

[12] See amongst others [11]: In the paper, the authors implement the *stated discrete choice experiment* (DCE) for the estimation of the value of the environmental improvements resulting from urban redevelopment interventions. The analysis refers to the territory of Seaham, a small town in the North East of England (21,000 inhabitants) and was carried out in January of 2010. The amenities considered were: open spaces, redevelopment of disused properties, outdoor recreational facilities, street cleaning, cycle lanes and walking paths. An econometric Logit model was used for the analysis of the *stated preferences*. 106 respondents were asked what they would be willing to pay in terms of annual additional municipal taxes for an improvement (or realization) of each of the amenities listed above. The cost would cover the costs of annual maintenance to ensure the efficiency of each of the considered amenities. The following results were obtained: among the reference amenities, the recovery of disused property and improvement of street cleaning were preferred by the respondents. For all the amenities considered, there was a willingness to pay per family per year about £ 70. The authors showed that the value attributed to new urban standards – such as the amenities used for the study – depends on the type and quantity of the existing standards in the area of interest.

[13] Damodaran A. [5], Des Rosiers, F., M. Theriault [8] believe that the hedonic approach is the most reliable method to assess properties with environmental externalities, as it can reveal the perceptions of buyers on the amenities in the surrounding area. A different opinion is held by Dolan, P. and Metcalfe, R. [10]. The authors stated that "property prices may not reflect the value of the environmental amenities due to information asymmetries. Potential buyers may see the process of urban redevelopment as a sign of the presence of negative externalities". In current literature, there are numerous applications of the hedonic pricing method for assessing the contribution of urban standards (trees, green area facilities, etc..) on the market value of the homes in the surrounding area. Morancho, A.B. [16], using the *hedonic price*, estimated that every 100 metre increase in green areas results in a decrease of about €1800 in the market value of a home. Morales, D.J. [15] by means of regression, determines that the presence of trees weighs 6% on the market value of a home. Martin, C.W., May R.C. and Appel, D.N. [13] estimate that the value of trees weighs between 13% and 19% of the market values of homes in the area of interest. Bolitzer, B. and Netustil N.R. [4], applying the hedonic price, show how to estimate the value of open spaces (parks, natural areas, golf courses), "the net effect of nearby open spaces is theoretically uncertain because the positive externalities associated with proximity as a view or proximity to recreational facilities may be underestimated by negative factors (traffic congestion and noise)". The research was carried out in Portland, Oregon, a large metropolitan area in the United States. The analysis showed that a home located within 1,500 feet of an open space is sold at $ 2,105 more than a house outside the radius of 1,500 feet. Furthermore, the authors estimate that each additional acre of open space increases the market value of a house by $ 28.33. The authors estimate that a dwelling located within 1,500 feet of a 20-acre open space is sold at $ 2,670 more than a home outside this radius. By specifying the type of open space, with a public park within 1,500feet of a home leads to an increase in value of $2,262, while proximity to a golf course increases the market value by about $ 3,400. An interesting application is made by Tyrvainen L., Miettinen A. [23]. The authors use the hedonic price to estimate the value of the forest seen as positive externalities for the surrounding properties. The analysis shows that the increase of a kilometre away from forest areas leads to a decrease of 5.9% of the market value of homes. Homes with views of the forests are 4.9% more expensive than homes without any view but with other similar characteristics.

A further and more specific public benefit resulting from the reuse of existing buildings is represented by the lower land consumption that comes from reutilization interventions. The phenomenon of *soil sealing* is particularly felt in over-concreted cities, where the constant population increase is accompanied by a reduction of natural areas. Conservation of land is a quantifiable benefit to the community due to the lack of mitigation costs of the direct and indirect effects generated by the irreversible sealing of the territory.

From an exclusively financial point of view, the budget related to the redevelopment of disused sites reconciles, positively, four quotas: 1) the contribution of the fee paid for the transformation; 2) the value of the donated social housing, equal to the agreed sale price multiplied by the SLP of social housing; 3) the saving of the realisation costs of social housing assigned for temporary renting; 4) the market value of the realised urban standards for the functional transformation of the disused area. Negatively, the unlimited accumulation of management costs of acquired urban standards for the redevelopment of disused areas.

The term that completes the definition of public utility is the differential of municipal taxes provided for in the situation "with" urban redevelopment and those collected in the context "without" the transformation (ΔIMU). Component for which the sign of the contribution to the estimate of the public utility is a function of the specific case to which the model is applied. The public utility (U_{public}) associated to the redevelopment of the disused area can be formalized in the following:

$$U_{public} = B_{requal} + B_{land} + C_c + p_s \cdot x_1 + k \cdot x_3 + V_{st} - K_{gest} \pm \Delta IMU \qquad (5)$$

Table 2 reports the terms in equation (5).

Table 2. Public utility parameters

B_{requal}	economic benefits deriving from the urban requalification of the disused area
B_{land}	economic benefits deriving from minimum use of the land
C_c	concession fee
p_s	"social" selling price, i.e. controlled
x_1	SLP of donated social housing
K	unitary construction cost of buildings
x_3	SLP of rent controlled social housing
V_{st}	market value of the achieved urban standards
K_{gest}	unlimited accumulation of the annual management costs of the achieved urban standards
ΔIMU	difference in municipal tax determined by the urban development

6 The Model

Having discussed the utility functions of the two entities (private and public) involved in the urban transformation of a disused area, a model that returns the optimal combination of the types of social housing a), b) and c) described in paragraph 3 has been defined. The result of the *social housing* quota must allow for the maximization of economic benefits of both the private owner and public entities mentioned in paragraphs 4 and 5. A linear programming model can therefore be created, that summarizes the variables in appropriate mathematical formalisms in order to determine the predetermined objective function and analysis constraints.

The algorithms in Table 3 model the concepts presented in the previous paragraphs. Table 4 reports the items that make the algorithm resolutive and the terms that define K_{sh}.

Table 3. Explanation of the model

	x_1
variables	x_2
	x_3
objective function	$\max(x_1 + x_2 + x_3)$
constraints	$V_{m,trasf} - K_{sh} \geq V_{\lim it}$
	$V_{\lim it} = c_p \cdot R \geq V_{current}$
	$V_{m,trasf} - K_{sh} = B_{requal} + B_{land} + C_c + p_s \cdot x_1 + k \cdot x_3 + V_{st} - K_{gest} \pm \Delta IMU$
	$x_i \geq 0$, $i = 1, 2, 3$

Table 4. Parameters of the proposed model

x_1	SLP of donated social housing
x_2	SLP of social housing at an agreed selling price
x_3	SLP of rent controlled social housing
$V_{m,trasf}$	market value of the disused area influenced by possibility of redevelopment
K_{sh}	cost for the owner of the developed area by the social housing quota
V_{limit}	limit value for the owner until judging the redevelopment to be economically convenient
c_p	mortgage coefficient for the owner of the area
R	revenues from the sale of the finished product
$V_{current}$	market value generated by the current use of the property
B_{requal}	economic benefits deriving from the urban requalification of the disused area
B_{land}	economic benefits deriving from minimum use of the land
C_c	concession fee
p_s	"social" selling price, i.e. controlled
k	unitary construction cost of buildings

Table 4. (*Continued*)

V_{st}	market value of the achieved urban standards
K_{gest}	unlimited accumulation of the annual management costs of the achieved urban standards
ΔIMU	difference in municipal tax determined by the urban development
Ca_s	annual unitary "social" rent, i.e. controlled
n	number of years of temporary renting of social housing
r	private discount rate

7 The Case Study

The model has been applied to a disused abandoned industrial area located in the Campania region known as the Agro-Nocerino-Sarnese (SA). If the housing demand, the rationalization of land use and redevelopment of existing property are typical problems of Campania, the area of the Agro-Nocerino-Sarnese is characterized by the widespread presence of brownfield sites near the city centre, with the need for a functional transformation.

The case considered is represented by a former industrial area located near the historical centre, in a residential area. The land covers 12,000 m². The built land, which includes offices, warehouses, houses, locker rooms and sheds, is equal to 7,500 m² covering a total volume of 60,000 m³. The assets are currently in fair condition, with most generating an income.

In order to implement the algorithm developed, a housing redevelopment of the area has been assumed, with the demolition of existing volumes and the construction of new structures with a change of intended use to residential and commercial property, notwithstanding the urban provisions set out by the municipal PRG and within the constraints set out by D.D. 1444/1968. The project aims to improve land use as well as the architectural quality of the area. In particular, it includes the creation of a link road between the two municipal roads and a large square with public parking. Every building in the project has five floors that are above ground (ground floor shops and accommodation on the four upper floors). The overall volume is 32,900 m³. The area for urban standards covers 5,923 m², divided into a public park (1,550 m²), parking (1,473 m²) and a square (2,900 m²). The project includes the construction of 62 underground garages (commercial area = 2,782 m²), 13 shops (retail area = 1,311 m²) and 62 homes (commercial area = 8,663 m²).

Table 5 summarises the items of income and costs required to estimate the market value of the area on the basis of its construction potential. Table 6 reports the values of the parameters required for the application of the model[14].

---

[14] The data used for the application of the model reflect the real trends of the territory in which the area under study is located.

Table 5. Summary of the revenues and private costs of the urban redevelopment project without social housing

REVENUE		Parameter	Quantity	U.M.	TOTAL
Residential		2.500	8.663	m^2	21.658.000
Commercial		2.800	1.311	m^2	3.671.000
Underground Parking		30.000	62	n	1.860.000

TOTAL REVENUE (A)	27.189.000

COSTS	Incidence				TOTAL
GENERAL COSTS					
Technical expenses	6%				699.000
General expenses	3%				349.000
Commercialization expenses	2%				543.780
Total General Costs					**1.591.780**
Concession fee		Parameter	quantity	U.M.	TOTAL
		%	8%		931.360
Total					**931.360**
CONSTRUCTION COSTS		Parameter	quantity	U.M.	TOTAL
Demolition buildings		10	60.000	m^3	600.000
Waste disposal		37	5.000	m^3	186.000
Buildings		283	32.900	m^3	9.311.000
Underground parking		15.000	62	n	930.000
Urban furniture of square above underground parking		50	4.000	m^2	200.000
Realization standards		70	5.923	m^2	415.000
Total Construction costs					**11.642.000**
TOTAL COST					**14.165.140**
Fees		Parameter	quantity	U.M.	TOTAL
		%	6%		849.908
Total					**849.908**
Profit for investor		Parameter	quantity	U.M.	TOTAL
		%	20% of A		5.437.800
Total					**5.437.800**

TOTAL COST	20.452.848

The following assumptions are valid:

1) type c) social housing is in a temporary lease for at least 8 years and then sold at controlled prices;
2) the private discount rate (r) is equal to 7%. This rate is estimated with reference to the average annual yield of ten-year BTPs (about 6%),

adjusted for expected inflation (2%) and considering a rate assumed to be representative of the risk premium for property investments (3%);
3) the annual operating costs of achieved urban standards (K_{gest}), the economic benefits associated with redevelopment (B_{requal}) and the minimum soil consumption (B_{land}), the market value of achieved urban standards (V_{st}) and the municipal tax differential (ΔIMU) compensate each other; this assumption is ordinarily subject to the benefit of the public;
4) the advantage in the budget of public utility is assumed equal to half of the private sector. This hypothesis stems from assumption 3). This simplification, as mentioned, determines an underestimate of the actual public utility;
5) the private utility is assumed equal to the exchange rate in the ordinary common reference.

Table 7 reports the expressions of the model with the numerical values in Table 6.

Table 8 summarizes the items of income and expense necessary to estimate the market value of the area considering the social housing component.

Table 6. Parameter values of the model for the case study

x_1	SLP of donated social housing
x_2	SLP of social housing sold at an agreed price
x_3	SLP of controlled rent social housing
$V_{m,trasf}$	27.189.000 – 20.452.848 = 6.736.152 €
K_{sh}	$f(x_1, x_2, x_3, p_s, Ca_s, n, r)$
c_p	20%
R	$f(K_{sh}, dest, quant, p)$
$V_{current}$	3.500.000 €
C_c	8% × 11.642.000 = 931.360 €
K_{gest}	$= B_{requal} + B_{land} + V_{st} \pm \Delta IMU$
p_s[15]	1.800 €/m²
K	850 €/m²
Ca_s[16]	35 €/m²
n	8

---

[15] In Campania, the main reference for the "social" price is law n. 7 of 14 January 2009. This Decree establishes the price of social housing in relation to the implementation cost and a set of parameters.

[16] In applying the model, the unitary annual controller rent is reduced by 25% due to the costs that the owner has during the eight year lease.

Table 6. (*Continued*)

r	7%
dest	residential, commercial, underground parking
quant	8.663 m², 1.311 m², 62 box
p	2.500 €/m², 2.800 €/m², 30.000 €

Table 7. Application of the model to the case study

variables	x_1
	x_2
	x_3
objective function	$max\ (x_1 + x_2 + x_3)$
constraints	$6.736.152 - K_{sh}(x_1, x_2, x_3, p_s, Ca_s, n, r) \geq 20\% \cdot R(K_{sh}, dest, quant, p)$
	$20\% \cdot R(K_{sh}, dest, quant, p) \geq 3.500.000$
	$\frac{1}{2} \cdot [6.736.152 - K_{sh}(x_1, x_2, x_3, p_s, Ca_s, n, r)] = 931.360 + 1.800 \cdot x_1 + 850 \cdot x_3$
	$x_i \geq 0 \qquad i = 1, 2, 3$

Table 8. Summary of revenues and costs for a private urban redevelopment project with a social housing component

REVENUE	Parameter	Quantity	U.M.	TOTAL
Residential	2.500	7.479	m²	18.695.000
Social Housing	1.800	1.184	m²	719.005
Commercial	2.800	1.311	m²	3.671.000
Underground Parking	30.000	62	n	1.860.000

TOTAL REVENUE (A)				24.945.005

COSTS	Incidence			TOTAL
GENERAL COSTS				
Technical Fees	6%			699.000
General Expenses	3%			349.000
Commercialization Fees	2%			498.900
Total General Costs				1.546.900
Concession fee		Parameter	quantity	U.M.	TOTAL
		%	8%		931.360
Total				931.360
CONSTRUCTION COSTS		Parameter	quantity	U.M.	TOTAL
Demolition buildings		10	60.000	m³	600.000

Table 8. (*Continued*)

Waste disposal		37	5.000	m^3		186.000
Buildings		283	32.900	m^3		9.311.000
Underground parking		15.000	62	n		930.000
Urban furniture of square above underground parking		50	4.000	m^2		200.000
Realization standards		70	5.923	m^2		415.000
Total Construction costs						11.642.000
TOTAL COST						14.120.260
Fees		Parameter	quantity	U.M.		TOTAL
		%	6%			847.216
Total						847.216
Profit for investor		Parameter	quantity	U.M.		TOTAL
		%	20% di A			4.989.001
Total						4.989.001

TOTAL COST	19.956.477

The solution to the problem of maximizing the total social housing quota ($x_1+x_2+x_3$), in accordance with the ratio equal to half the profitability between public and private, determines the following results:

$x_1 = 588$ m^2,
$x_2 = 0$ m^2,
$x_3 = 596$ m^2,

for a total of 1,184 m^2, equal to 13.67% of the total surface area intended for residential housing in the intervention. As expected, the variable x_2 vanishes, as it reduces private utility without increasing the public one. In fact, it is entirely absent in the formal expression of the latter. Besides, in the public logic, between the x_1 and x_2, it is preferable that the x_1 component passes to complete ownership of the Administration, as it serves the same purpose of x_2 while leaving more flexibility to use the same public entity. Thus, it might decide to use it for disadvantaged social classes and then rent it even an even lower rent or sell it at a token, symbolic price.

Obviously, the model also lends itself to a more articulated use. For example, the public may request to increase the availability of social housing due to the contingent condition. In fact, in general terms, the Administration should check in advance the demand for social housing and then distribute it proportionally through various redevelopment interventions, or similar, which can accommodate social housing units. The case study could have been intended a specific quota, non-absorbable by adopting the optimal first derivative vector. Thus, requiring a greater consistency, for example for facilitated sale, which will be the case of introducing an x_2 quota, which is less expensive than x_1 for the private investor. It is worth noting, in this case, that by adding a further 588 m^2 of x_2 (for similarity to x_1), the profitability of the private investor drops to 19.03%, while for the public entity it remains at 10.17%. In total, the percentage of social housing, compared to the open housing market, rose to 20.45%. Obviously, in this case, the private investor loses about 1% profit, amounting to € 250,000.

In general terms, the entire protocol described provides a very attractive solution to the public entity, due to the Administration making a great deal of money through the construction and sale of housing that has been donated as well as through the savings in the construction of social housing, while delivering the services demanded by citizens. It could, alternatively, even give up some of the profit by transforming it into housing. It would consequently take advantage of a business mechanism that involves owners and managers, and therefore be able to guarantee a high efficiency of the intervention. The Administration would also refrain from transforming the money collected into social housing, thus avoiding the related business risks.

The hypothesis of increasing x_2 could also be supported by cancelling x_1, given the cost that x_1 implies for the private investor and the substantial similarity of results between x_1 and x_2. It is also true that the elimination of x_1 implies that the whole process remains in the hands of the private investor, while the quota of x_1 involves the transfer of ownership of a part of the property directly to the public entity. In contrast, the solution with $x_1 = 0$ reduces the management burden for the Administration that merely acts as a check on the procedures implemented by the private sector. Thus, assuming $x_1 = 0$, the following vector is assumed:

$x_1 = 0$ m^2,
$x_2 = 1,000$ m^2,
$x_3 = 1,200$ m^2,

for a total of 2,200 m^2, equal to 25.40% of the total surface area intended for residential housing in the intervention. With a private return essentially stable at 19.97% and a public one that is reduced to 7.82%.

Financially, the public loses about two percentage points, but increases by about 1,000 m^2, the availability of social housing. The administration has lost the quota of property that could be allocated to the most extreme social functions. The result is generally consistent. In fact, the Administration waives a significant part of the profit (x_1 quota) but increases the supply to the demand for social housing.

8 Conclusions

The model developed and applied to a case study makes it possible to find a compromise between the parties involved in the redevelopment of a disused area. The balance is sought between the parties by trying to maximize the amount of social housing that the private investor must guarantee to the public entity. The Operational Research algorithms developed elicit the utility functions of the private owner of the property and the public administration. The implementation of the model allows to make some considerations that illustrate the rationality.

1. It is evident that the assumptions adopted in the calculation of the public utility in which the management costs of the urban standards are offset by economic benefits of redevelopment, from the minor land loss and the value of these standards is a simplification for the benefit of communities. In fact, an increase of public utility, connected to the greater value of the positive terms in question in relation to the single negative one, corresponds to a reduction in the social housing quota that provides a balance of overall income. In contrast, the assumption made can greatly simplify the computational phase of the parameters useful in implementing the model. Obviously, the results in a more detailed specification of the items, in order to obtain a higher reliability of the algorithm.
2. The results of the model are not binding in an absolute sense, the quotas allocated to the various categories of social housing can be easily changed, while maintaining a substantial compliance with the predetermined yield ($U_{public} = \frac{1}{2} U_{private}$).
3. The most significant result of the model, which enhances its operational value, is the fact that it takes into account the characteristics of the territory in the planning phase. The social housing quota is not established by a table and spread throughout the region (a concept used by the current laws in effect on the housing plan in Campania and on Expressions of interest), but the percentage attributed to each intervention stems from a prior market analysis of the area and the items to be included in the model. It is evident that the application of the model to the centre of Naples and Salerno, where market values are high, generates a significantly higher proportion of social housing than in the provinces. Thus, apodictic determinations contained in the standards mentioned in the Campania Region, are verifiable by the logic of the local market, different for different areas and essential to determine what resources are available to replace the shortage of public funds.

References

1. ARPA Campania: Relazione sullo stato dell'ambiente in Campania 2009. Consorzio STA, Napoli (2009)
2. AUDIS: Carta della Rigenerazione Urbana (2008)
3. Barberis, R.: Consumo di suolo e qualità dei suoli urbani. ARPA Piemonte
4. Bolitzer, B., Netustil, N.R.: The impact of open spaces on property values in Portland, Oregon. Journal of Environmental Management 59 (2000)
5. Damodaran, A.: Valuation approaches and metrics: a survey of the theory and evidence. Stern School of Business (2006), http://pages.stern.nyu.edu
6. De Mare, G.: La valutazione della qualità nella riqualificazione urbana. Un modello multicriteriale per le scelte partecipate. Edizioni Graffiti, Napoli (2004)
7. De Mare, G., Manganelli, B.: La programmazione lineare per la selezione dei progetti di riqualificazione urbana. In: Atti del XXXII Incontro di Studio Ce. S.E.T., Venezia (2002)
8. Des Rosiers, F., Theriault, M.: Mass appraisal, hedonic price modelling and urban externalities: understanding property value shaping processes. In: Advances in Mass Appraisal Methods Seminar. Delft University of Technology (2006)

9. Di Gennaro, A., Malucelli, F., Filippi, N., Guandalini, B.: Dinamiche di uso dei suoli: analisi per l'Emilia Romagna, tra il 1850 e il 2003. Territori 1. Editrice Compositori, Bologna (2010)
10. Dolan, P., Metcalfe, R.: Comparing preferences and subjective well-being for valuing non-market goods. Centre for Economic Performance Discussion Paper 890. London School of Economics (2008)
11. Lanz, B., Provins, A.: Valuing local environmental amenity: using discrete choice experiments to control for the spatial scope of improvements. CEPE working paper 79 (2011)
12. Manganelli, B., Tajani, F.: Modelli di stima nel mercato immobiliare. L'utilizzazione della programmazione lineare. Valori e Valutazioni, vol. 3. DEI, Roma (2009)
13. Martin, C.W., Maggio, R.C., Appel, D.N.: The contributory value for trees to residential property in the Austin, Texas Metropolitan Area. Journal of Arboriculture 15 (1989)
14. Ministero delle Infrastrutture: Definizione di alloggio sociale ai fini dell'esenzione dall'obbligo di notifica degli aiuti di Stato, ai sensi degli articoli 87 e 88 del Trattato istitutivo della Comunità europea (2008)
15. Morales, D.J.: The contribution of trees to residential property value. Journal of Arboriculture 7 (1980)
16. Morancho, A.B.: A hedonic valuation of urban green areas. Landscape and Urban Planning 66 (2003)
17. Morano, P.: Un modello di perequazione urbanistico estimativo. Edizioni Graffiti, Napoli (1998)
18. Morano, P., Nesticò, A.: Un'applicazione della programmazione lineare discreta alla definizione dei programmi di investimento. In: Aestimum. 50. Firenze University Press (2007)
19. Orlando, L.: Il Sole 24 ORE (September 14, 2011)
20. Regione Campania: Linee Guida per la Programmazione in materia di Edilizia Residenziale Pubblica e fondi fitto, di cui alla legge 431/98 (2008)
21. Riganti, P., Alberini, A., Longo, A.: Public Preferences for Land uses' changes: valuing urban re generation projects at the Venice Arsenale. In: 45th Congress of the European Regional Science Association (2005)
22. Sforza, A.: Modelli e metodi della Ricerca Operativa. Edizioni Scientifiche Italiane, Napoli (2005)
23. Tyrvainen, L., Miettinen, A.: Property Prices and Urban Forest Amenities. Journal of Environmental Economics and Management 39 (2000)

Simulation of Users Decision in Transport Mode Choice Using Neuro-Fuzzy Approach

Mauro Dell'Orco and Michele Ottomanelli

Technical University of Bari, Bari, Italy
{dellorco,m.ottomanelli}@poliba.it

Abstract. In this paper, soft computing and artificial intelligence techniques have been used to define a model for simulating users' decisional process in a transportation system. Through this framework, the variables involved are expressed by approximate or linguistic values, like in the humans' reasoning way, in order to forecast users' mode choice behavior. The model has been specified and calibrated using a set of real life data. Results appear good in comparison with those obtained by a classical random utility based model calibrated with the same data, and the methodology seems promising also in case of different applications in the field of choice behavior simulation.

1 Introduction and Background

Recently, great attention has been addressed to new paradigms and theories in order to model uncertainty laying in transportation problem and, in particular, in transportation systems users choice behaviour. Such an interest is due to the concept of uncertainty that is closely related to the travellers information issue as well as to the vagueness, subjectivity and imprecision of many variables and data relevant to traffic and transportation problems as well as to socio-economic elements relevant to individuals.

In this environment, fuzzy set theory and related techniques seems to be a consistent paradigm in order to model traffic and transportation problems characterized by ambiguity, subjectivity and uncertainty.

Literature reviews show (1, 2, 3) how many issues of transportation science have been covered by fuzzy sets based applications starting from the pioneer work by Pappis and Mamdami (5) in the field of traffic light control.

In particular, also the interesting issue of representing the decision process of hu-man being acting as transportation user/decision maker has been faced. Namely, fuzzy logic has been used to model the natural approximate reasoning of users when they have to choice among transport alternatives, such as departure time, trip destination, mode of transport, route, parking etc.(i.e. when decision of transport users has to be simulated).

Most of the effort has been devoting to the problem of route choice modelling by means of rule-based Fuzzy Logic Inference System (FIS) constituted of a set of IF ... THEN rules in which input and output variables are fuzzy sets. This approach allows analysts to handle both approximate reasoning and managing/control users attitudes by means of linguistic attributes. For example, a rule for representing user route choice decision can be modelled as follow:

```
IF Travel Time on route 1 is very short AND Travel Time
on route 2 is long THEN preference for route 1 is strong
```

where "very short", "long" and "strong" are fuzzy sets representative user perception of travel cost on the given routes.

The first attempt to apply the fuzzy logic inference to transport users' behaviour simulation was focused on route choice modelling and presented by Teodorović and Kikuchi (4). Fuzzy Logic based model in the field of route choice was also presented by Akiyama and Kawahara (6). The simulation of route choice drivers' behaviour with information provision have been proposed later in (7) and (8) also using a Neuro-Fuzzy approach (9).

Fuzzy Sets have been used in this field also by Possibility Theory approach as described in Henn and Ottomanelli (10) or Dell'Orco et al. (11)

For a wider discussion of the use of fuzzy logic in transportation engineering and relevant problems the reader could refer to the reference work by Teodorović (3).

Most of the research show how Fuzzy Logic is a powerful approach to represents the natural approximate reasoning of human decision makers (i.e. system users). Studies on transport mode choice in the field of Fuzzy Logic are often open to some criticisms because of the absence of comparisons with classical and well established behavioural models and because of the scale of the study.

This paper presents a model to simulate the users behaviour when facing a mode choice problem in a transportation system at real size scale and propose a comparison with the well-known Multinomial Logit mode choice model.

The proposed model is mainly based on two mathematical framework: the Subtractive Clustering (15) and the Adaptive Neural Network-based Fuzzy Inference System (ANFIS) (16).

Through this approach, the value of the utility perceived by users in a modal choice process is not calculated according to classical random utility models: utility value – in our case, more correctly, Modal Preference – that user associates to each modal alternative is estimated by a Fuzzy Inference System (FIS), simulating user's approximate reasoning when comparing different transport modes.

It is well-known that in a fuzzy system the specification of FIS characteristics, as number and shape of Membership Functions (MF), is a discretionary phase; instead, in this work, such characteristics have been determined by clustering the observed data and then optimizing the MFs' shape through an Adaptive Neural Network (ANN).

The data we used in this paper arise from a survey carried out to specify and calibrate a system of travel demand models for the city of Salerno, a medium-sized town in South-Western side of Italy (14) based on the well-established random utility theory (namely, Multinomial Logit).

The performance of the proposed model has been evaluated through comparisons with observed real data and with the results obtained by a Random Utility based Logit model (13), calibrated using the same dataset.

2 The Proposed Model

In this study, within a more general framework of transportation demand models, the modal choice model has been considered. Four transportation modes (i.e. car, motorbike, bus, on foot) have been taken into account; for each modal alternative, a Fuzzy Inference System (FIS) has been set up in order to determine the preference value of a user with specific socio-economic characteristics, according to the perceived values of attributes relevant to that alternative. Each FIS has a set of attributes as input, and a single element – the *modal preference* - as output. Starting from input-output data, a *Subtractive Clustering* algorithm has been used, to identify subsets of data having similarity relationships within a larger set and to carry out a concise representation of the system behaviour. These subsets are characterized by a radius and a centre.

Given n elements in a M-dimensional space, the potential P_i of each element – that is, a M-dimensional vector - is defined as:

$$P_i = \sum_{j=1, i \neq j}^{n} \exp\left(-\alpha \|x_i - x_j\|^2\right) \tag{1}$$

with

$$\alpha = \frac{4}{r_\alpha^2} \tag{2}$$

where

- r_α is the radius of the cluster, fixed by the analyst;
- $\|x_i - x_j\|$ is the norm of two generic vectors x_i ed x_j.

The following recursive procedure has been used to select cluster centres:

- STEP 1 the potential of all vectors is calculated, and the vector having the highest potential is assumed as centre of the first cluster;
- STEP 2, the procedure does not take any more this vector into account, and continues calculating again the potential of remaining vectors The vector having the highest potential is selected as centre of another cluster;
- STEP 3, if the difference between the potentials of last two centres is less than a fixed threshold, the procedure ends. Otherwise, goes to STEP 2.

In this way, at the end of the procedure, all vectors are put into c clusters: the larger the chosen radius is, the less the number of clusters and the less precise the model.

Each vector x consists of:

- a set of components y forming the input, and
- a component z, forming the output; for the vectors centre of cluster, these components will be indicated by y^* e z^*, while the centre itself will be indicated by x_r^* (r = 1,..c).

The centres of clusters can be considered as a rule of a fuzzy inference system; then, the degree to which each rule is satisfied can be defined as:

$$\mu_r = \exp\left(-\alpha \|y - y_r\|^2\right) \quad (3)$$

While the output is given by:

$$z = \frac{\sum_{r=1}^{c} \mu_r z_r^*}{\sum_{r=1}^{c} \mu_r} \quad (4)$$

The equations (3) and (4) are equivalent to fuzzy rules such as:

IF y_1 is A_1 AND y_2 is A_2 AND... y_k is A_k THEN z is B (5)

where A_1, A_2, .. A_k e B are Membership Functions (MF), determined through the following equations:

$$A_r(q) = \exp\left[-\alpha(q - y_r^*)^2\right] \quad (6)$$

$$B_r = z_r^* \quad (7)$$

All the MFs' have then been calibrated training the Adaptive Neural Network corresponding to the FIS obtained by clustering (Fig. 1).

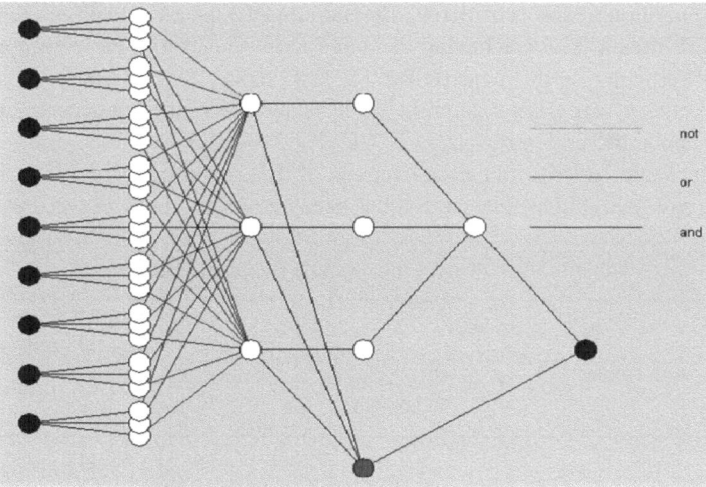

Fig. 1. Adaptive Neural Network of the FIS

An artificial neural network (ANN) is an analytical tool, able to learn and simulate the behavior of a real system, working along with it during the training phase, and modifying its own characteristics in order to minimize the errors.

In the ANFIS procedure, all characteristics of the ANN, like the number of input-output nodes, depend on the FIS they refer to.

Given a certain number of input–output pairs, the error ε is the difference between the calculated output value (Y*) and the real output value, the so-called 'pattern' Y.

$$\varepsilon = |Y^* - Y| \qquad (8)$$

During the training phase, the neural network calibrates the MFs' parameters, such as, for example, mean and variance for a Gaussian membership function.

3 Model Specification and Calibration

A modal choice model has been specified and calibrated. The input vectors are made of 361 interviews to transportation users in the city of Salerno, a middle-sized city in the South-Western side of Italy, and refer to home-work trips during the morning rush hours. For more details about the survey, the reader can refer to the report by Elasis-CSST (14) in which Logit mode choice model calibration is described.

The validation of the proposed model has been carried out using 50 vectors as *hold out sample*, drawn randomly from the starting sample.

The FIS has eight input variables and one output. The input variables are referred to socio-economics attributes relevant to the user and to the transport system attributes: age, income, number of motorbikes and cars owned, generalized costs for the four transportation modes considered. The output is the user's modal preference.

Each FIS uses a database having the same input and modal preference as unique output. In particular, in the database for the "car" mode, the preference value is 1 for the modal choice "car" (last column in Table 1) and 0 for other choices. Likewise, for the "bus" mode the preference value is 1 for the modal choice "bus" and 0 for other choices. In Table 1 there is a partial example of the database for the FIS "car" mode; the fourth row is highlighted because it has been chosen for the *hold out sample*.

Table 1. Some items of the database for the FIS "car" mode

User ID	Age	Income	No. of Bikes	No. of Cars	Generalized cost				Mode Preference
					Motorbike	Car	Walk	Bus	
1	42	5	0	1	502.50	997.58	3483.33	6395.00	0
2	37	4	1	2	2191.00	3135.35	14970.41	9216.50	1
3	31	3	1	1	408.50	779.06	2807.78	3184.00	0
4	29	5	1	1	756.50	1406.12	5127.36	7763.00	0
5	23	1	1	2	1549.50	2023.74	10563.47	4780.00	0

In the presented model, for the clusters radius r_α, values between 0.85 and 1.15 have been used. They are rather large values, determining 3 clusters: consequently, also the inference engine will have 3 rules and 3 membership functions for each input.

A *trial-and-error* procedure has been used for the specification of the model: several tests with different numbers of MFs have been carried out, considering that a larger number of MFs makes the FIS more precise, but increases the number of variables.

In this way, through clustering, a first result is obtained in terms of FIS. Each FIS has an error, calculated as difference between observed (real) values and values of modal preference estimated by the FIS.

In the presented model, for the clusters radius $r\alpha$, values between 0.85 and 1.15 have been used. They are rather large values, determining 3 clusters: consequently, also the inference engine will have 3 rules and 3 membership functions for each input.

A trial-and-error procedure has been used for the specification of the model: several tests with different numbers of MFs have been carried out, considering that a larger number of MFs makes the FIS more precise, but increases the number of variables.

In this way, through clustering, a first result is obtained in terms of FIS. Each FIS has an error, calculated as difference between observed (real) values and values of modal preference estimated by the FIS.

Afterwards, a calibration has been carried out through four neural networks, one for each FIS, and an error reduction of about 25%, with respect to the results obtained by only the clustering procedure, has been found.

As example, in Fig. 2 are reproduced the MF for the input "Car Generalized Cost" and the FIS for the "car" mode (Fig. 3), obtained through clustering and subsequently calibrated through the ANFIS.

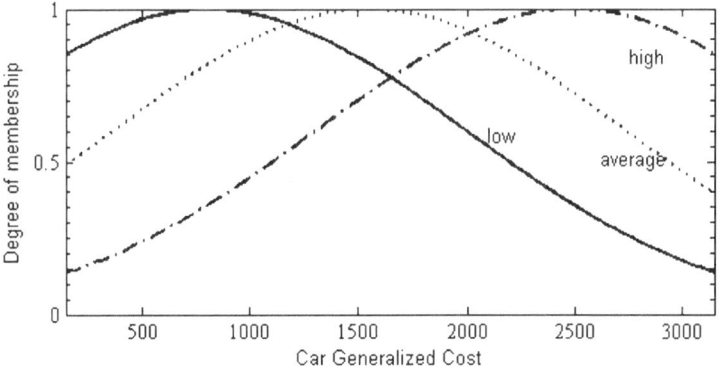

Fig. 2. Membership Functions for the input variable "Car Generalized Cost"

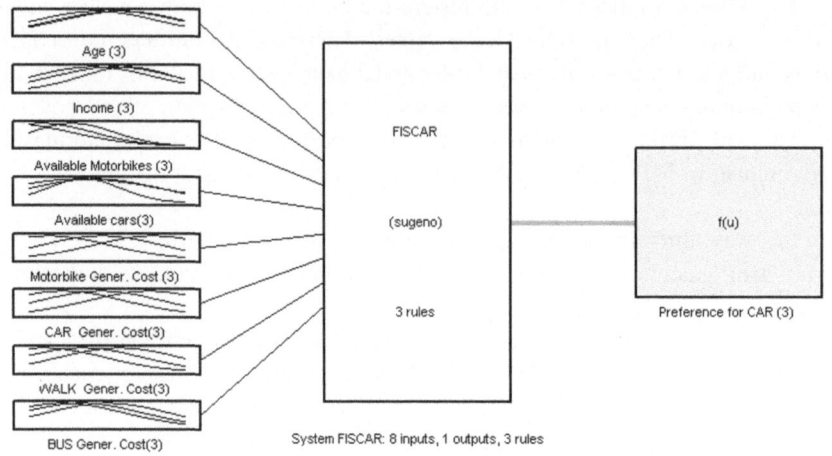

Fig. 3. FIS for Car Mode Choice

The three rules embedded in the FIS relevant to the car mode choice are the following:

1. IF (Age is high) AND(Income is high) AND(Available Motorbikes is Low) AND (Available Cars is high) AND (Motorbike Gen. Cost is average) AND(Car Gen. Cost is average) AND (Walk Gen. Cost is average) AND (Bus Gen. Cost is low) THEN (Preference for CAR is Low);
2. IF (Age is average)AND(Income is high) AND (Available Motorbikes is average)AND(Available Cars is low) AND (Motorbike Gen. Cost is low) AND(Car Gen. Cost is low) AND (Walk Gen. Cost is low) AND (Bus Gen. Cost is average) THEN (Preference for CAR is average);
3. If (Age is low)AND(Income is low) AND (Available Motorbikes is high)AND(Available Cars is average) AND (Motorbike Gen. Cost is high) AND(Car Gen. Cost is high) AND (Walk Gen. Cost is high) AND (Bus Gen. Cost is high) THEN (Preference for CAR is high);

In order to reduce the number of input values and therefore make the system training simpler, we used the generalized cost instead of separated costs for monetary and time costs. Then, the time costs have been transformed to monetary ones through the Value-of-Time (VoT). Using only one value for VoT, disregarding the income, has been found more convenient, in terms of model errors: probably different values of VoT reflected subjective choices that did not allow, in the training phase, establishing an objective structure of relationships between input and output.

Validation of the model has been carried out on the hold out sample formed of 50 data drawn randomly from the initial sample. In Table 2 some of the obtained results are reproduced. From the second to the fifth column the preference index is reported for each transportation mode and for each user.

In the proposed model, the user chooses the mode with the highest preference, shown in the "FIS" column.

Table 2. Model validation

User ID	Pref. Car	Pref. Bus	Pref. Bike	Pref. Walk	FIS Output	Observed preference	Logit Output
1	0.226	0.424	0.622	0.433	Bike	Walk	Car
2	0.424	0.070	0.084	0.446	Walk	Car	Car
3	0.267	0.108	0.204	0.694	Walk	Car	Walk
4	0	0	0.212	0.655	Walk	Bike	Car
5	0.360	0	0.274	0.593	Walk	Walk	Car
6	0.639	0.164	0.161	0.043	Car	Bus	Car
7	0.605	0.065	0.186	0.584	Car	Walk	Car
8	0.328	0.101	0.141	0.613	Walk	Walk	Car
9	0.549	0.076	0.291	0.806	Walk	Walk	Walk
10	1	0	0	0	Car	Car	Car
11	0.752	0	0	0.061	Car	Car	Car
12	0.692	0	0.127	0.160	Car	Car	Car

The choices simulated by the proposed model have been compared to those obtained by a Logit model, and reported in the "Logit" column in Table 2.

From the comparison appears that the percentage of correct results is 64% for the Logit model and 68% for the proposed model. Although the values are rather close to each other, it is worth noting that the Logit model has been calibrated on about 2000 phone interviews, as mentioned in the CSST's report, while the proposed model was defined using 361 values only. Of course, the results obtained by the ANFIS procedure can be further improved using a larger data set.

Moreover, the comparison between the two approaches shows that the Logit model tends to overestimate the preference of the "car" mode, while no overestimation is shown by the proposed one.

4 Conclusions

In this study, a model for the simulation of transportation mode choice by users in a urban area has been carried out. The user choice behaviour model was built by using a fuzzy inference engine ; through this framework, the variables involved in users' decisional process are expressed by approximate or linguistic values (i.e. *low, average, high*), closer to human reasoning way and to the imprecision-uncertainty lying in available data.

From this point of view, the model appears rather realistic and simple, because of the small number of rules (three) used in the FIS; in fact, human decision maker chooses more likely on the basis of a limited set of rules, rather than on the basis of a complicated system of mathematical elaborations.

The model was applied to a medium-sized Italian town and the obtained results appears good, even in comparison with those obtained by a random utility based model, and capable of further improvements

The value of the simulation performance is very interesting also with respect to the amount of data used in the two approaches. Further analysis will be carried out such as cross-validation of the model.

References

1. Avineri, E.: Soft Computing Applications in Traffic and Transport Systems: A Review. In: Hoffmann, F., Köppen, M., Klawonn, F., Roy, R. (eds.) Soft Computing: Methodologies and Applications, pp. 17–25. Springer, Heidelberg (2005)
2. Teodorović, D., Vukadinović, K.: Traffic Control and Transport Planning: A Fuzzy Sets and Neural Networks Approach. Kluwer Academic Publishers, Boston (1998)
3. Teodorović, D.: Fuzzy Logic Systems for Transportation Engineering: The State of the Art. Transportation Research-Part A 33, 337–364 (1999)
4. Teodorović, D., Kikuchi, S.: Transportation route choice model using fuzzy inference technique. In: 1st International Symposium on Uncertainty Modeling and Analysis, pp. 140–145. IEEE Press, USA (1990)
5. Pappis, C., Mamdani, E.: A fuzzy logic controller to a traffic junction. IEEE Trans. on Systems, Man and Cybernet. 7(10), 707–717 (1977)
6. Akiyama, T., Kawahara, T.: Traffic assignment model with fuzzy travel time information. In: Proceedings of the 9th Mini-EURO Conference "Fuzzy sets in Traffic and Transport Systems", Budva, Yugoslavia, September 17-19 (1997)
7. Lotan, T.: Modelling route choice behaviour in the presence of information using concepts of fuzzy sets theory and approximate reasoning. Thesis (PhD). MIT, Boston, MA (1992)
8. Lotan, T., Koutsopoulos, H.N.: Approximate reasoning models for route choice behavior in the presence of information. In: Daganzo, C.F. (ed.) Transportation and Traffic Theory, pp. 71–88. Elsevier, Amsterdam (1993)
9. Koutsopoulos, H.N., Lotan, T., Yang, Q.: A driving simulator and its application for modeling route choice in the presence of information. Transportation Research Part C: Emerging Technologies 2, 91–107 (1994)
10. Henn, V., Ottomanelli, M.: Handling uncertainty in route choice models: From probabilistic to possibilistic approaches. European J. of Operational Research 175, 1526–1538 (2006)
11. Ottomanelli, M., Dell'Orco, M., Sassanelli, D.: Modelling parking choice behaviour using Possibility Theory. Transp. Planning and Technology 34, 647–667 (2011)
12. Cascetta, E., Nuzzolo, A., Velardi, V.: A System of mathematical models for the evaluation of integrated traffic planning and control policies. In: Workshop on "Integration Problems in Urban Transportation Planning and Management Systems", Ottobre, Capri, Italy, pp. 28–29 (1993)

13. Cascetta, E.: Transportation systems engineering: theory and methods. Kluwer Academic Publishers, Dordrecht (2001)
14. CSST: Un sistema di supporto alle decisioni per la gestione della mobilità e per la pianificazione dei trasporti urbani, Report no. 26 (1993) (in Italian)
15. Chiu, S.L.: A cluster estimation method with extension to fuzzy model identification. In: Proc. of the 3rd IEEE World Congress on Computational Intelligence, pp. 1240–1245. IEEE Press (1994)
16. Jang, J.S.R.: ANFIS: Adaptive-Network-Based Fuzzy Inference System. IEEE Transaction on System, Man and Cybernetics 23, 665–685 (1993)
17. Mendel, J.M.: Uncertain Rule-Based Fuzzy Logic Systems. Prentice Hall, Upple Saddle River (2001)
18. Zimmermann, H.J.: Fuzzy Set Theory and its Applications. Kluwer Academic Publishers (1991)

Multidimensional Spatial Decision-Making Process: Local Shared Values in Action

Maria Cerreta, Simona Panaro, and Daniele Cannatella

Department of Conservation of Architectural and Environmental Heritage,
University of Naples Federico II, via Roma 402, 80132 Naples, Italy
cerreta@unina.it, {panaro.simona,cannatelladaniele}@gmail.com

Abstract. This paper is about developing a methodological framework for a multidimensional spatial decision-making process oriented to the identification of a territorial transformation strategy reflecting shared values. Through the empirical investigation in an operative case study, the Avellino-Rocchetta S. Antonio railway line, in the South of Italy, an integrated evaluative approach implemented in a SDSS can make us go beyond space and hierarchical limits. Taking into account the different multidimensional components of decision-making process, making clear the weights and recognizing the different priorities, fit and situated strategies have been identified, according to an interactive and dynamic dialogue among expertise and local communities.

Keywords: Spatial Decision Support System (SDSS), Spatial indicators, Multi-Criteria Analysis (MCA), Analytic Hierarchy Process (AHP), Soft Systems Methodology (SSM).

1 Introduction

The concept of planning-evaluation proposed by Lichfield [1] makes explicit the close interaction and reciprocal framing of evaluation and planning: evaluation can be conceived as deeply embedded in planning, affecting planning, and evolving with it. Indeed, the evolution of evaluation methods reflects their evolving relationship with the planning process and the way in which they interact with the diversity and multiplicity of domains and values. To identify an evaluative framework able to integrate different purposes and multidimensional values within the decision-making processes not only means focusing on the environmental, social and economic effects of different options, but also considering the nature of the stakes, selecting priorities and values in a multidimensional perspective. It is crucial to structure complex decision-making processes oriented to an integrated planning, that can support the selection, the monitoring and the management of different resources, and the interaction among decision-makers, decision-takers, stakeholders and local community [2]. It is essential to adopt "explorative" approaches, open to plurality and dialogue among the different expertise involved [3], taking into account that facing the complexity of interacting perspectives, interests, and preferences [4] means to identify a dynamic decision-making process, where *integration* represents the crucial point. An integrated approach to planning-evaluation involves many institutional and non-institutional stakeholders with divergent and conflicting values and mandates, with a high complexity of issues and interdependencies.

According to the above perspective, the concept of *knowledge generation* [5] is essential to build integrated strategies, where *socialization* (tacit with tacit: sharing experiences to create new tacit knowledge, observing other participants, brainstorming without criticism), *externalization* (tacit with explicit: articulating tacit knowledge explicitly, writing it down, creating metaphors, indicators and models), *combination* (explicit with explicit: manipulating explicit knowledge by sorting, adding, combining, looking to best practices) and *internalization* (explicit with tacit: learning by doing, developing shared mental models, goal based training) [6] [7] represent the main phases and the four key modes of knowledge conversion.

Through a process of knowledge generation iteratively acting in all four modes of knowledge conversion, interplaying between tacit knowledge end explicit knowledge, and by experiencing the four knowledge conversion modes, planners can develop a shared explicit language and use it to develop integrated strategies [8] [5].

This approach to knowledge management to support strategy-making is also consistent with the epistemological structure of "post-normal science" developed by Funtowicz and Ravetz [9], considering two crucial aspects: uncertainty and conflict of values. According to post-normal science, to recognize the importance of difference implies a different way to address complex systems and to face complexity means to take into account the self-organization chances, non-linear dynamics, non-continuous behaviours of complex systems and participated decision-making processes. This means to broaden the field of decision-makers and to involve new social actors in order to create an "extended community", able to elaborate new solutions [10]. The approach of post-normal science forces decision-makers and planners to find solutions not only coming from the "expert knowledge", but also legitimated by "common knowledge", including uncertainty as part of the decision problem, and considering solutions based not only on exact scientific data (*hard data*), but also on public decisions, shared by the community (*soft data*). Indeed, facing and/or solving complex problems depends on the capability to consider them under different points of view, and to manage uncertainty, filling the gap between experts and community.

A key tool able to support the decision-making process, especially when uncertainty, complexity and values of different social groups are many, different and conflicting [11] is represented by the "integrated evaluations". Not only do they consider the inputs of data expressing the impacts of different solutions, but also are "open" to a wide public participation, so that they can offer more information for the evaluation itself and, in addition, can make the decision-making processes and the results more acceptable [12] [13]. Participation becomes essential not only to examine and evaluate choices on social, ethic, political, economic, environmental levels, but also to legitimate choices and make them acceptable for the community itself. Integrated evaluations are an ongoing process both, iterative and interactive, multi-disciplinary and participative, able to recognize the relevance of technical indeterminacy and value multiplicity.

In this view, it is important to combine different approaches in the same framework, integrating different evaluation tools [14], considering the possibility of combining Multi-Criteria Analysis and Multi-Groups Analysis with Geographical Information Systems (GIS), Internet Technology, Spatial Decision Support Systems, Cellular Automata Models, etc. In particular, integration of differing evaluation models with GIS [15] becomes important in the construction of a Spatial Decision Support System: a variety of territorial information (social, economic and environmental) may be easily combined and related to the characteristics of the different options of territorial use, facilitating the construction of appropriate indicators and improving impacts forecasting, leading up to a preference priority list of the various options. Integration among Multi-Criteria Analysis, Multi-Group Analysis and GIS may be useful when there are strong conflicts, in which the role of local actors, their relations and objectives may be considered as a structuring element in the process of information construction and values identification, in a spatial and dynamic evaluative model [16] [17] [18] [19] [20] [21] [22].

The paper explores potential of a Spatial Decision Support System (SDSS) in the field of land-use planning, considering both the technical knowledge of the decision-making problem and the lay knowledge of the local community for the construction of shared transformation choices. Through the empirical investigation in an operative case study, the Avellino-Rocchetta S. Antonio railway line, in the South of Italy, an integrated evaluative approach implemented in a SDSS can facilitate the interaction among the different multidimensional components (tangible and intangible, *hard* and *soft*), making clear weights and priorities, and defining a "strategic map", expression of a an interactive and dynamic dialogue among expertise and local communities, able to provide a value added for decision support processes in complex evaluation problems, and useful to overcome sectorial perspectives diffused in transport planning field.

2 GIS-Based Decision-Making Process

Lately, theoretical research and new technologies have improved the identification and implementation of integrated approaches for building planning strategies and actions. GIS-based Multi-Criteria Decision Analysis (GIS-MCDA) can be considered an interesting expression of a process able to transform and combine geographical data and value judgments (decision-makers' preferences and uncertainties) in order to obtain appropriate and useful information for decision-making process [23]. Spatial Multi-Criteria Analysis is a relatively new but growing research field, that is still developing with further improvements of GIS [15] [24]. Examples for the application and new approaches of spatial MCA are illustrated in many recent papers [25] [26] [27] [28] [29] [30] [31]. It has been argued that the GIS-MCDA systems can potentially enhance group decision-making processes by providing a flexible problem-solving framework where participants can explore, understand and redefine a decision problem [32] [33] [34]. Central to GIS-MCDA is the concept of decision rules or evaluation algorithms. A decision (or combination) rule can be defined as a procedure that enables the decision-maker to rank and select one or more alternatives from a set of available ones [35] [15]. Over the last 20 years numerous multicriteria decision rules have been implemented in the GIS environment, including the weighted linear combination (WLC) or weighted summation/Boolean overlay methods [36], the Analytic Hierarchy Process (AHP) [37], the ideal/reference point methods and the outranking methods [38]. The benefits of combining MCDA and GIS have been widely discussed in the literature [38]: in general terms, GIS enables the computation of spatial criteria and/or indicators, whereas MCDA is used to group these criteria and/or indicators into a suitability index, which is assigned to each mapping unit.

In the case study of the Avellino-Rocchetta S.Antonio railway line, using GIS and AHP allowed to test the possibility of interfacing spatial components with two different types of data: *hard data*, more easily expressed in a GIS environment, and *soft data*, often related to intangible data, and therefore more complex to communicate through a spatial definition. This has led to a series of arguments that aimed primarily to give a spatial representation to intangible actions, and secondly to create indexes that could relate *hard data* with *soft data* and that, at the same time, can be represented in GIS environment. The innovative aspect of the methodological approached structured is related to the spatial representation of soft data, able to describe the soft values specific of the territory, considered relevant as the hard data for the identification of a suitable transformation strategy. Indeed, the ultimate goal of this study was therefore to create a *strategic map* that synthesizes the hypothesis of transformation proposed by stakeholders through soft data, taking into account of the potential and critical aspects expressed by hard data. The strategic map helps to understand local preferences in relation to the potential of the area and the infrastructure in question, identifying a common vision able to reduce the different conflicts and to combine objective observation and subjective perception.

3   Avellino-Rocchetta S. Antonio Railway Line[1]: Towards a Situated Strategy

Avellino-Rocchetta S. Antonio railway line, crossing the inland areas of Campania, Basilicata and Puglia in the Southern of Italy, closed in December 2010. The methodological approach, structured in order to take into account the needs of affected communities, tried to find wide-shared transformation strategies. It followed four main stages (fig. 1):

- construction of the cognitive framework, with the collection of hard data and soft data; where hard data have been analyzed through fishbone diagrams and support the selection of indicators of main relevant systems; while soft data have been identified starting from an Institutional Analysis and the related stakeholders map;

---

[1] Avellino-Rocchetta S. Antonio case study has been carried out within the elaboration of the degree thesis in Architecture of arch. Simona Panaro, tutor prof. Maria Cerreta, University of Naples Federico II, October 2011.

- representation of relevant systems, which means the elaboration of indicators and spatial indicators by GIS maps; but also the exploration of preferences with some tools proper of Soft System Methodology as in-depth interviews, structured by the CATWOE approach and decoded through the Rich Pictures;
- identification of transformation scenarios through the combined application of AHP and GIS, defining a decision tree consisting of goal, criteria, subcriteria, indicators and spatial indicators;
- elaboration of a strategic map, where the results of spatial assessment support the identification of a ranking of actions, considering both the relevance of actions and the relevance of areas.

The methodological approach (fig. 1), using tools and techniques linked to each other, has led to outline a structured knowledge of the area, collecting both *hard data* and *soft data*, expression of *hard values* and *soft values*, elaborated according to a multidimensional and incremental approach in order to identify the most sensitive areas to a scenario for sustainable transformations.

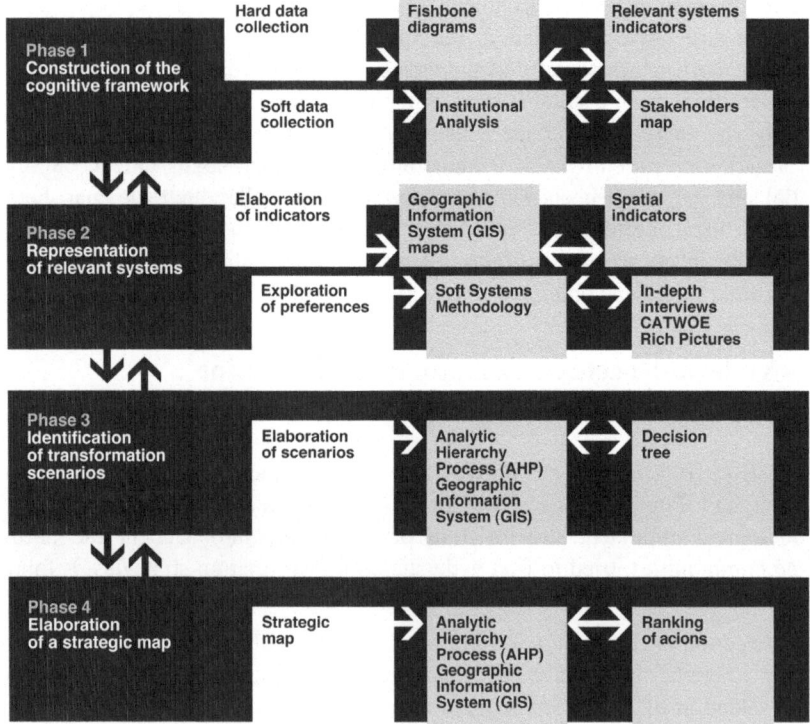

Fig. 1. Methodological framework: phases, methods, results

According to the above perspective, it has been explored the interaction among:

- the methodology of the survey and representation of soft data, typical of Soft System Methodology [39];
- the Analytic Hierarchy Process [40] [41], a multi-criteria evaluation method, in its application to ArcGIS 9.2 software [42];
- the GIS, as a tool to process spatial data.

The result of this process is a *strategic map*, expression of the preferable transformations ranking, where the priority actions are localized. This operation took into account the constraints, resources and values of the area analyzed, as a result of knowledge generation process, where tangible and intangible values can be combined.

3.1 A Bottom-Up Approach for the Construction of Spatial Indicators

The Avellino-Rocchetta S. Antonio is a secondary railway line, which came into operation in 1895, at that time, only link for 43 municipalities (35 of the Campania, 7 of the Basilicata and 1 of the Puglia) (fig. 2). The railway layout is very articulated as respect to the territory orography and the presence of three rivers (Sabato, Calore, Ofanto). Today, the line is closed and only two terminal stations are active of the 33 original. The depopulation of the Southern Apennines, the structural difficulties of the line, the completion of the so-called Ofantina road represent some of the causes that have contributed to exclude the railway line from local public transport. In recent years, the Regional Plan of the metropolitan network has reported the line to the attention of the Campania region institutions. But the results of the feasibility study, on its inclusion in the regional network, not have been encouraging and the recent fund cuts to public transport have determined its closure in December 2010. Despite the many critical aspects, in many weekends of 2011 some local associations have experienced the use of the railway as a tourist line in order to show that it can be relevant for some sectors of the local economy (food and wine, tourism, and landscape).

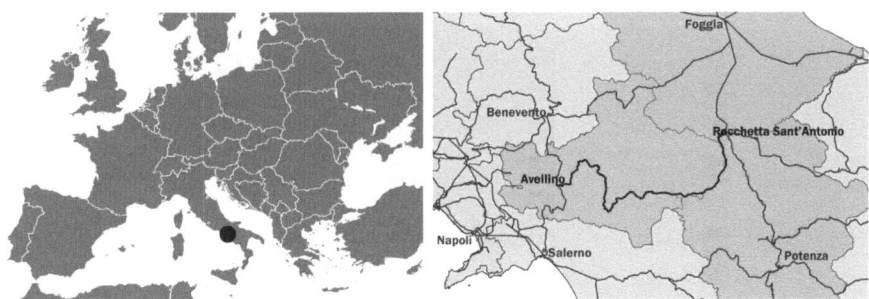

Fig. 2. The study area: the Avellino-Rocchetta S. Antonio railway line

Starting from these considerations, hard data were identified to describe the environmental, social and economic relevant issues. Through the Institutional

Analysis [43], the main stakeholders were also selected that over time have been, or that might become, the reference for the implementation of new strategies for developing the territory and the railway line. It was therefore structured the map of different stakeholders according to three main groups: *promoters* (Campania Region, Avellino Province, Mountain Community Terminio Cervialto, Municipalities of Avellino, Montella, Nusco, Cairano, Rapone), *operators* (environmental associations, touristic promotion association, association of merchants and artisans, association for railway promotion, Trenitalia) and *experts* (president of Sustainable mobility, professor of urban planning, professor of transport planning, professor of technical construction, professor of restoration, professor of rural development, citizens of involved municipalities). Soft Systems Methodology [39] [44] was used at this stage, in order to investigate the local complexities and to identify and represent *soft data*. In particular, the CATWOE technique has been used to structure the interview submitted to stakeholders and the Rich Picture to represent the issues emerged (preferred transformation scenarios, constraints, obstacles and resources) (fig. 3).

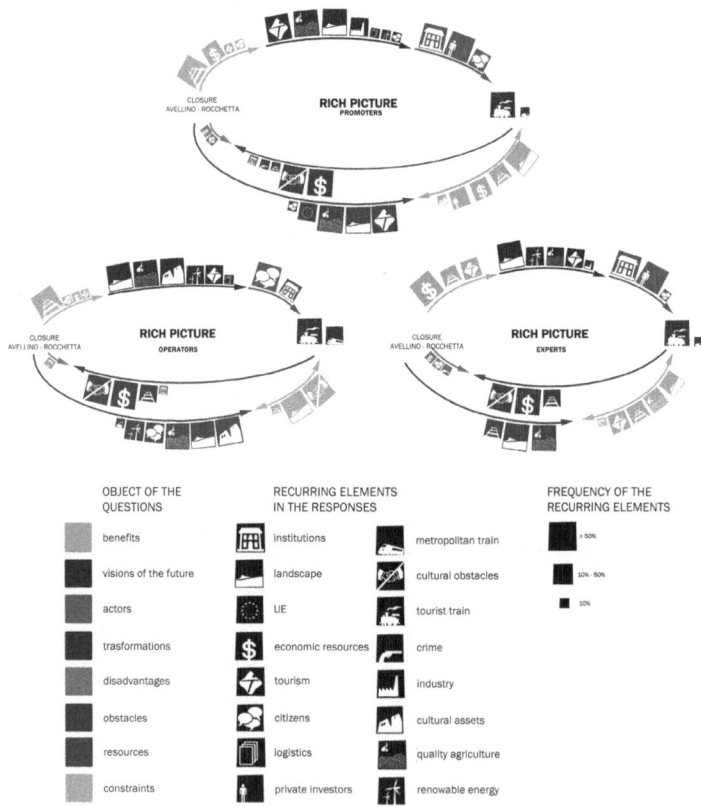

Fig. 3. Rich Pictures: representation of the main issues emerging from interviews

The Rich Picture was built from the detection of the frequency of recurring elements in the responses of interviews. These elements were then put in a logical sequence that can explain the relationship between the issues raised and therefore better define preferred transformation scenarios. From the Rich Pictures it can be seen that some identity characteristics of the area (orography, landscape, food and wine) are perceived by all categories of respondents as the potentials for new development models. The re-use of railway infrastructure represents an advantage due to the economic and cultural renewal of existing assets and, at the same time, a constraint for future transformations. The lack of funds from the government is, for respondents, the greatest obstacle to any transformation mainly because the local economy, despite the fact that it is largely based on industry, is not accustomed to the joint investments of the small private enterprises. The greatest obstacles are therefore of a cultural and financial nature. Through Rich Pictures, the significant alternatives were expressed according to the point of view of the different stakeholders and the causes of conflict, gradually expanding the degree of knowledge of the area and refining the selection of indicators for the evaluation. It has passed gradually from the selection of data to the identification of values, *hard* and *soft*, which characterize the study area.

In particular, the selection of *hard data* has been driven by *soft data* emerged as significant; from these the appropriate indicators were developed and the related spatial indicators and indexes. This was a tricky phase in the data processing. The construction of the indexes included a progressive refinement of the information, combining the main relevant indicators. For example, from the interviews elaboration it is clear the importance of the definition of an index that measures the cultural vitality of local governments, especially for the scenario of tourist transformation of the railway line. This index was then calculated by taking into account the available data relative to the number, type and season of the events within a calendar year:

$$Ey + Et*0,5 + Es*0,5 = Vc. \qquad (1)$$

where:

Ey = number of Events per year
Et = number of Events varied by type
Es = number of seasons where there is at least one Event
Vc = index of cultural Vitality

Attributing to this index 5 degrees of relevance (from extreme to null), it was possible to compare the cultural vitality of the municipalities concerned and build the GIS map representing the specific differences (Table 1). The map (fig. 4) shows that the cultural vitality not only depends on the policy importance of the municipalities. Indeed, some small villages have the same cultural vitality as Avellino town, the main city of the Province. In the smallest centres, it seems that cultural vitality spreads "infecting" neighbouring municipalities.

Table 1. The index of cultural vitality

Reference Range	Degree of relevance
>12	Extreme
From 8 to 12	High
From 5 to 8	Medium
From 3 to 5	Little
From 0 to 3	Null

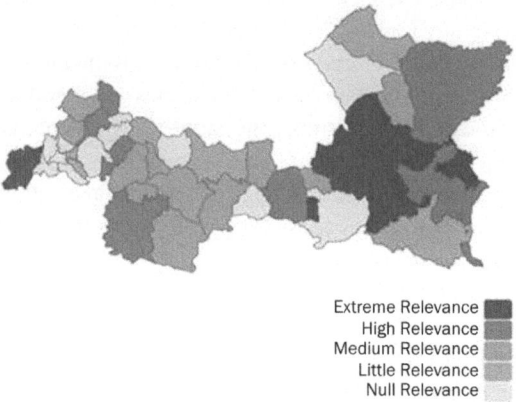

Fig. 4. The spatial representation of the index of cultural vitality

Table 2. Organization of data according to the hierarchical logic of the AHP

Criteria	Sub-criteria	Spatial indicators
Social infrastructure	Population	Residents 2001 - Variation 1991 - Residents over 64 - Level of education
	Occupation	Work force - Youth unemployment - Employed agriculture and industry
	Intangible heritage	Cultural vitality
Economic infrastructure	Railway network	Transformability railway - Influence areas stations
	Local mobility	Mobility attracted - Bus per day
	Economy - production	Agricultural enterprises - Employees agriculture, industry and services - Governance
Environmental infrastructure	Hydrosphere - Biosphere	Environmental resources
	Geosphere	Productions of quality - Seismic risk - Environmental risks
	Cultural heritage	Valuable elements

Multidimensional Spatial Decision-Making Process: Local Shared Values in Action 63

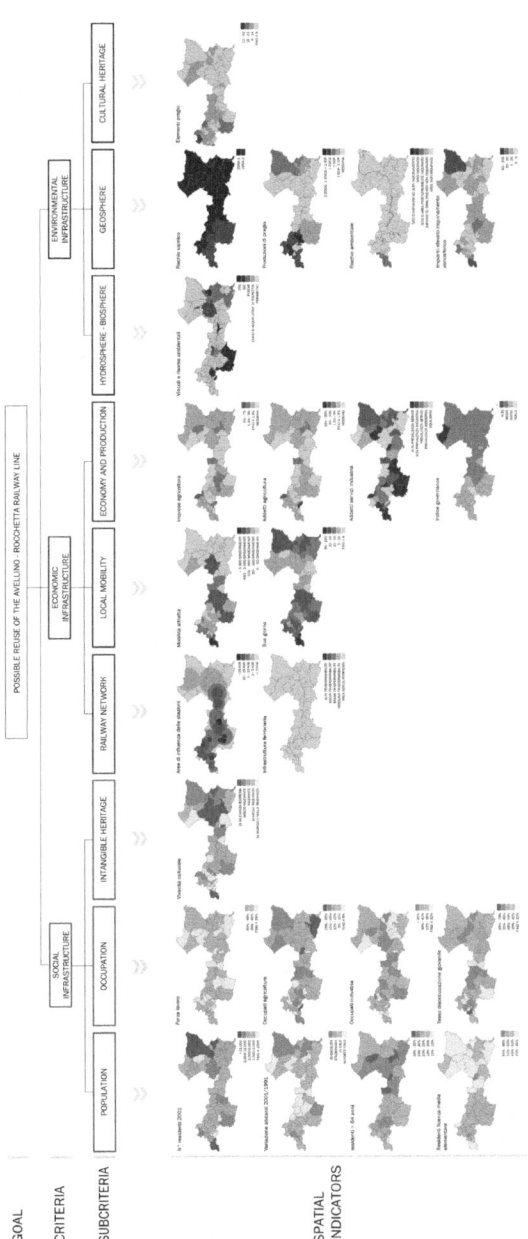

Fig. 5. The decision tree according to AHP approach

In general terms, the elaboration of the indices was done by selecting the families of indicators able to explain, and then to evaluate the significant elements of the preferred alternative for stakeholders. For the development of indexes, the basic indicators have been transformed into spatial indicators.

Then these were organized according to the hierarchical framework provided by the AHP method. In the specific case, three main categories have been identified: social infrastructure, economic infrastructure and environmental infrastructure. They represent the evaluation criteria, and each of them, in turn, was divided into environmental issues (sub-criteria) and finally each environmental issue has been expressed through the related spatial indicators or indexes. A database of information and, at the same time, of values, has been structured according to the hierarchical logic of the AHP method (Table 2) (fig. 5).

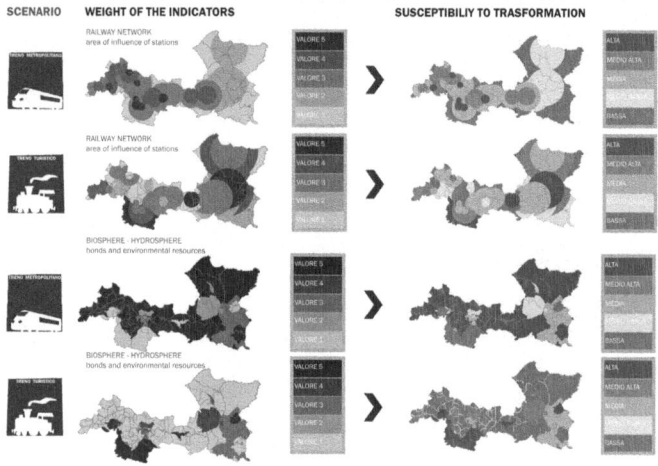

Fig. 6. Attribution of weight to the spatial indicators according to the scenario goal

It should be noted that the desire to integrate the social, economic and environmental characteristics, typical of a process of integrated assessment, has met with the difficulty in obtaining reliable data. The location of the study area, the border of three regions has required a homogenization of data for spatial or temporal coverage. In the absence of precise localization, the indices have been reported to the municipal boundaries or to the appropriate areas of influence, which allow an approximation to reality.

3.2 Alternative Scenarios for a Strategic Map

The preferred scenarios, elaborated by the interaction with lay knowledge, were two: the railway line as a *metropolitan train* and the railway line as a *tourist train*. The scenario of the metropolitan train refers to the local aspiration to have a competitive network infrastructure and the Avellino-Rocchetta S. Antonio railway line, as the only line of the Irpinia, is perceived as a resource to be preserved and strengthened. The scenario of the tourist train sees the line as non-competitive for the public transport, but as an alternative to enjoy the landscape and environment: the train is an opportunity to improve local economies. Each of them represented the goal of the hierarchical decision model as respect to which it was assigned a value to each of the indicators, with reference to a five point scale. This makes it possible to identify the different weight that each of the spatial indicators takes compared to the two scenarios (fig. 6).

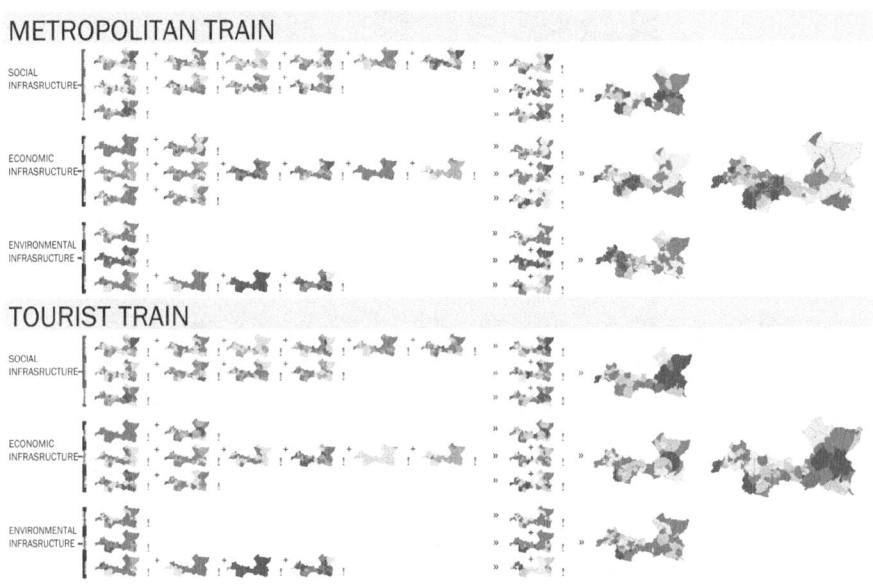

Fig. 7. Evaluation of scenarios emerged: the metropolitan train and the tourist train

Then we proceeded to the pairwise comparison between all indicators of a single environmental theme (sub-criteria), using the "Saaty scale". The resulting maps identify a ranking of areas of the greatest interest for that sub-criteria. In the same way, pairwise comparison was replicated for each level of the hierarchy: among sub-criteria with respect to the upper criterion and, finally, among the criteria as respect to objective (scenario) (fig. 7). In order to obtain a greater readability of the results it has been adopted a chromatic scale to distinguish the areas characterized by a high susceptibility to transformation (green) from those with low susceptibility to transformation (red). Therefore, for each scenario it was possible to identify an order of preferability of suitable areas for transformation, spatially represented (fig. 7).

Throughout the evaluation process, weights were given taking into account the issues that emerged from the interviews, according to preferences and values identified as significant by stakeholders.

Fig. 8. Localization of the priority actions: the strategic map

3.3 From Values to Actions

The comparison between the results shows that the most peripheral areas are insignificant for both scenarios. Given the technical limits, the railway line is not attractive to areas that have a metropolitan suction.

For the tourist scenario, however, are relevant the areas that define a continuous path involving municipalities and communities crossed by the railway line. The tourist scenario shows that the railway line reopening is feasible where existing infrastructure is compatible for the naturalistic and cultural fruition, especially along the course of Ofanto river. The *train of water* can be characterized by some poles of attraction: the stations of the wine (Montemiletto-Taurasi), the stations of the landscape (Montella and Monticchio) and the arrival-departure station of the services pole (Avellino).

At this primary use it could be added to, only for the first part of the line, the local public transport, including some opportunities to revitalize the railway line, though this is not decisive for its reopening.

At the end of the decision-making process it was developed a strategic map (fig. 8), where the priority actions emerged have been localized in areas with higher susceptibility to transformation.

This operation show the constraints and resources of the territory already spelled out in the GIS maps. In this way the intangible preferences, turned into actions, were compared to the spatial dimension, contributing to transparency and awareness of the possible choices.

4 Conclusions

The SDSS built during the case study may be considered as a basis to increase the level of integration of local and expert knowledge in a more extensive participatory GIS process, oriented to involve different expertise and to improve the completeness of hard and soft data and the consistency of the whole evaluation process. By means of this decision-making process, local communities can contribute actively to the implementation and updating of GIS data and improving evaluation of alternative strategies and actions. In addition to supplying the experts with valuable information for a better comprehension of the territory, they are also made much more aware of the characteristics and values of their own contexts.

In the above perspective, the use of Multi-Criteria Analysis has a privileged role as decision-making tool. Indeed, through the hierarchical construction of decisional objectives it was easy to involve the local community and different experts and obtain a shared elaboration of scenarios, strategies and actions. This has contributed to the creation of a richer and complex knowledge framework of the territory and a bottom-up construction of transformation ideas. Indeed, the different maps obtained from the GIS and AHP interaction were the expression of multidimensional interface about the meaning and the role of the different evaluation criteria, contributing together to the scenarios definition. They help recognizing their technical effectiveness and, at the same time, improving the transparency of evaluation process, building the decision able to reflect the different needs and expectations. The construction of knowledge framework and the knowledge management have revealed the need to structure a constant interaction between hard data and soft data, including objective indicators and subjective points of view, highlighting the most relevant issues or those particularly critical. For example, the inability to converge on common objectives represented, according to the stakeholders, is a serious stop to the enhancement of activities development and land management. Therefore, special attention was paid to the development of indicators that would allow to assess the impacts of actions that can affect this cultural obstacle. In addition, the combined use of methods and techniques made it possible to tackle a complex decision-problem, characterized by many variables and a high level of uncertainty, in an incremental evaluation process, characterized by continuous feedback and constant interaction. In this way, assessment has become a fundamental part of planning choices elaboration, and can be seen as a preventive check of the territorial sustainability and, at the same time, a tool to stimulate the identification of alternative solutions in the spatial decision-making process.

References

1. Lichfield, N.: Community Impact Evaluation. UCL Press, London (1996)
2. Cerreta, M.: Thinking Through Complex Values. In: Cerreta, M., Concilio, G., Monno, V. (eds.) Making Strategies in Spatial Planning, pp. 381–404. Springer, Dordrecht (2010)
3. Fusco Girard, L., Cerreta, M., De Toro, P., Forte, F.: The Human Sustainable City: Values, Approaches and Evaluative Tools. In: Deakin, M., Mitchell, G., Nijkamp, P., Vreeker, R. (eds.) Sustainable Urban Development. The Environmental Assessment Methods, vol. 2, pp. 65–93. Routledge, London (2007)
4. Wiek, A., Walter, A.: A Transdisciplinary Approach for Formalized Integrated Planning and Decision-Making in Complex Systems. European Journal of Operational Research 197(1), 360–370 (2009)
5. Te Brömmelstroet, M., Bertolini, L.: Integrating Land Use and Transport Knowledge in Strategy-making. Transportation 37(1), 85–104 (2010)
6. Nonaka, I., Takeuchi, H.: The Knowledge-creating Company: How Japanese Companies Create the Dynamics of Innovation. Oxford University Press, New York (1995)
7. Nonaka, I., von Krogh, G., Voelpel, S.: Organizational Knowledge Creation Theory: Evolutionary Paths and Future Advances. Organization Studies 27(8), 1179–1208 (2006)
8. Healey, P.: Urban Complexity and Spatial Strategies: Towards a Relational Planning for our Times. Routledge, London (2007)
9. Funtowicz, S., Ravetz, J.R.: Science for the Post-Normal Age. Futures 25, 568–582 (1993)
10. Funtowicz, S., Ravetz, J.R.: Emergent Complex Systems. Futures 26, 739–755 (1994)
11. van der Sluijs, J.P.: A Way Out of the Credibility Crisis of Models Used in Integrated Environmental Assessment. Futures 34, 133–146 (2002)
12. Golub, A.L.: Decision Analysis: An Integrated Approach. John Wiley & Sons Inc., New York (1997)
13. Munda, G.: Social Multi-Criteria Evaluation for a Sustainable Economy. Springer, Heidelberg (2008)
14. Finnveden, G., Nilsonn, M., Johansonn, J., Personn, Å., Moberg, Å., Carlsonn, T.: Strategic Environmental Assessment Methodologies – Application within the Energy Sector. Environmental Impact Assessment Review 23, 91–123 (2003)
15. Malczewski, J.: GIS and Multicriteria Decision Analysis. John Wiley & Sons Inc., New York (1999)
16. Al-Shalabi, M.A., Bin Mansor, S., Bin Ahmed, N., Shiriff, R.: GIS Based Multicriteria Approaches to Housing Site Suitability Assessment. In: XXIII FIG Congress, Shaping the Change, Munich, Germany, October 8-13, pp. 1–17 (2006)
17. Joerin, F., Musy, A.: Land Management with GIS and Multicriteria Analysis. International Transactions in Operational Research 7, 67–78 (2000)
18. Nekhay, O., Arriaza, M., Guzmán-Álvarez, J.R.: Spatial Analysis of the Suitability of Olive Plantations for Wildlife Habitat Restoration. Computers and Electronics in Agriculture 65, 49–64 (2009)
19. Şener, Ş., Şener, E., Nas, B., Karagüzel, R.: Combining AHP with GIS for Landfill Site Selection: A case Study in the Lake Beys Ehircatchment Area (Konya, Turkey). Waste Management 30, 2037–2046 (2010)
20. Thirumalaivasan, D., Karmegam, M., Venugopal, K.: AHP-DRASTIC: Software for Specific Aquifer Vulnerability Assessment using DRASTIC Model and GIS. Environmental Modelling & Software 18, 645–656 (2003)
21. Vizzari, M.: Spatial Modelling of Potential Landscape Quality. Applied Geography 31, 108–118 (2011)

22. Boroushaki, S., Malczewski, J.: Implementing an Extension of the Analytical Hierarchy Process Using Ordered Weighted Averaging Operators with Fuzzy Quantifiers in ArcGIS. Computers & Geosciences 34(4), 399–410 (2008)
23. Malczewski, J.: GIS-based Multicriteria Decision Analysis: A Survey of the Literature. International Journal of Geographic Information Science 20(7), 703–726 (2006)
24. Malczewski, J., Chapman, T., Flegel, C., Walters, D., Shrubsol, D., Healy, M.A.: GIS-Multicriteria Evaluation with Ordered Weighted Averaging (OWA): Case Study of Developing Watershed Management Strategies. Environmental Planning A 35(10), 1769–1784 (2003)
25. Simonovic, S.P., Nirupama, N.: A Spatial Multi-objective Decision-making under Uncertainty for Water Resources Management. Journal of Hydroinform 7(2), 117–133 (2005)
26. Strager, M.P., Rosenberger, R.S.: Incorporating Stakeholder Preferences for Land Conservation: Weights and Measures in Spatial MCA. Ecological Economics 57(13), 627–639 (2006)
27. Aceves-Quesada, F., Díaz-Salgado, J., López-Blanco, J.: Vulnerability Assessment in a Volcanic Risk Evaluation in Central Mexico through a Multi-criteria-GIS Approach. Natural Hazards 40(2), 339–356 (2006)
28. Geneletti, D., van Duren, I.: Protected Area Zoning for Conservation and Use: A Combination of Spatial Multicriteria and Multiobjective Evaluation. Landscape and Urban Planning 85, 97–110 (2008)
29. Coutinho-Rodrigues, J., Simão, A., Henggeler Antunes, C.: A GIS-based Multicriteria Spatial Decision Support System for Planning Urban Infrastructures. Decision Support Systems 51, 720–726 (2011)
30. Kordi, M., Brandt, S.A.: Effects of Increasing Fuzziness on Analytic Hierarchy Process for Spatial Multicriteria Decision Analysis. Computers, Environment and Urban Systems 36, 43–53 (2012)
31. Feick, R.D., Hall, G.B.: Consensus Building in a Multiparticipant Spatial Decision Support System. URISA Journal 11(2), 17–23 (1999)
32. Jankowski, P., Nyerges, T.: Geographic Information Systems for Group Decision-making: Towards a Participatory Geographic Information Science. Taylor & Francis, New York (2001)
33. Kyem, P.A.K.: On Intractable Conflicts Participatory GIS Applications: the Search for Consensus Amidst Competing Claims and Institutional Demands. Annals of the Association of American Geographers 94(1), 37–57 (2004)
34. Starr, M.K., Zeleny, M.: MCDM-State And Future of the Arts. In: Starr, M.K., Zeleny, M. (eds.) Multiple Criteria Decision-Making, pp. 5–29. North-Holland, Amsterdam (1977)
35. Carver, S.J.: Integrating Multi-Criteria Evaluation with Geographical Information Systems. International Journal of Geographic Information System 5(3), 321–339 (1991)
36. Banai, R.: Fuzziness in Geographic Information Systems: Contributions from the Analytic Hierarchy Process. International Journal of Geographic Information System 7(4), 315–329 (1993)
37. Joerin, F., Theriault, M., Musy, A.: Using GIS and Outranking Multicriteria Analysis for Land-use Suitability assessment. International Journal of Geographic Information Science 15(2), 153–174 (2001)
38. Checkland, P.: Soft System Methodology. In: Mingers, J., Rosenhead, J. (eds.) Rational Analysis for a Problematic World Revisited: Problem Structuring Methods for Complexity, Uncertainty and Conflict, pp. 61–89. John Wiley & Sons Inc., Chichester (2001)

39. Saaty, T.L.: The Analytical Hierarchy Process. McGraw Hill, New York (1980)
40. Saaty, T.L., Peniwati, K.: Group Decision Making: Drawing out and Reconciling Differences. RWS Publications, Pittsburgh (2007)
41. Marinoni, O.: Implementation of Analytic Hierarchy Process with VBA in ArcGIS. Computational Geosciences 30, 637–646 (2004)
42. De Marchi, B., Funtowicz, S.O., Lo Cascio, S., Munda, G.: Combining Participative and Institutional Approaches with Multi-Criteria Evaluation. An Empirical Study for Water Issue in Troina, Sicily. Ecological Economics 34(2), 267–282 (2000)
43. Checkland, P.: Systems Thinking, Systems Practice - Soft Systems Methodology - a 30-year Retrospective. John Wiley & Sons Inc., Chichester (1999)

A Proposal for a Stepwise Fuzzy Regression: An Application to the Italian University System[*]

Francesco Campobasso and Annarita Fanizzi

Department of Economics and Mathematics, University of Bari,
Via C. Rosalba 53, 70100 Bari, Italy
{fracampo,a.fanizzi}@dss.uniba.it

Abstract. Fuzzy regression techniques can be used to fit fuzzy data into a regression analysis. Diamond treated the case of a simple least square model introducing a metrics into the space of triangular fuzzy numbers; in this paper we propose a stepwise procedure to select independent variables in a multivariate model.

At each iteration we introduce into the equation the variable which is less correlated with the already present ones and, at the same time, significantly explains the total sum of the squares of the estimated model; in any case a variable, whose explanatory contribution is subrogated by the combination of those later introduced, can be eliminated until the end of the iterations.

The goodness of the proposed selection procedure is reviewed in the evaluative context of the Italian university system. In our country educational offer has been recently enriched of innovative services, such as those directed to information for students and, more specifically, to their input or output guidance; as an example, teaching regulations recently allow students to gain a training experience directly in workplaces. In the perspective of monitoring more closely the innovative services offered by universities, we evaluate the effectiveness of the activated internships through the opinion (itself fuzzy) expressed by students on many aspects concerning them.

Keywords: Multivariate fuzzy regression, stepwise selection procedure, Italian university system, evaluation of internship activities.

1 Introduction

Private companies and public institutions periodically commission market researches and opinion polls in order to know how customers assess the consumed products or services and to identify which aspects determine their choices. In these surveys judgements are often expressed through common linguistic terms, such as ordered categories from "Unquestionably unsatisfied" to "Unquestionably satisfied".

---

[*] The contribution is the result of joint reflections by the authors, with the following contributions attributed to F. Campobasso (chapters 2, 4 and 5), and to A. Fanizzi (chapters 1 and 3).

An appropriate way to manage customers' responses, which represent verbal labels of sets with vague and uncertain borders, is provided by using fuzzy numbers.

In this paper we deepen the problems of defining a multivariate regression model [1], specifically proposing a stepwise procedure to select of the most significant independent variables.

In each step a single variable is included between the starting ones, according to two basic criteria: its originality with respect to the other variables already included in the model and its explanatory contribution to the model. The procedure provides the possibility of eliminating at each iteration variables already included in the model, whose explanatory contribution is subrogated by the combination of the later introduced variables.

As an application case of the proposed procedure we evaluate the effectiveness of the internships activated by Italian universities, detecting those factors which contribute more than others to the corresponding satisfaction expressed by the interviewed graduates in a sample national survey specifically conducted by consortium *AlmaLaurea* in 2008.

2 The Fuzzy Least Square Regression Model

2.1 Introduction to Triangular Fuzzy Numbers

The fuzzy theory allows the intrinsic complexity of phenomenons under investigation to be adequately taken into account (see for example [2-4]).

A *triangular fuzzy number* $\tilde{X} = (x, x_L, x_R)_T$ for the variable X is characterized by a function $\mu_{\tilde{X}} : X \to [0,1]$, like the one represented in Fig. 1, that expresses the *membership degree* of any possible value of X to $\tilde{X}$ [5].

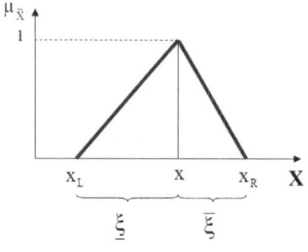

Fig. 1. Representation of a triangular fuzzy number

The x value is considered the core of the fuzzy number, while $\underline{\xi} = x - x_L$ and $\overline{\xi} = x_R - x$ are considered the left spread and the right spread respectively. Note that x belongs to $\tilde{X}$ with the highest degree (equal to 1), while the other values included between the left extreme x_L and the right extreme x_R belong to $\tilde{X}$ with a gradually lower degree.

The set of triangular fuzzy numbers is closed under addition: given two triangular fuzzy numbers $\tilde{X}=(x,x_L,x_R)_T$ and $\tilde{Y}=(y,y_L,y_R)_T$, their sum $\tilde{Z}$ is still a triangular fuzzy number $\tilde{Z}=\tilde{X}+\tilde{Y}=(x+y,x_L+y_L,x_R+y_R)_T$. Moreover the opposite of a triangular fuzzy number $\tilde{X}=(x,x_L,x_R)_T$ is $-\tilde{X}=(-x,-x_R,-x_L)_T$.

It follows that, given n fuzzy numbers $\tilde{X}_i = (x_i, x_{Li}, x_{Ri})_T$, $i = 1, 2, .., n$, their average is $\overline{X} = \frac{\sum \tilde{X}_i}{n} = \left(\frac{\sum x_i}{n}, \frac{\sum x_{Li}}{n}, \frac{\sum x_{Ri}}{n} \right)_T$.

Diamond [6] introduced a metrics into the space of triangular fuzzy numbers; according to this metrics, the squared distance between $\tilde{X}$ and $\tilde{Y}$ is

$$d(\tilde{X},\tilde{Y})^2 = d((x,x_L,x_R)_T,(y,y_L,y_R)_T)^2 = (x-y)^2 + (x_L-y_L)^2 + (x_R-y_R)^2.$$

The latter definition allowed the same author to derive the expression of the estimated coefficient in a simple regression model.

The adequacy of such a model in many contexts, even different from the opinion polls, has been repeatedly demonstrated (for example see [7]).

2.2 The Multiple Regression Model Including a Fuzzy Asymmetric Intercept

Recently [8] we generalized this estimation procedure to the case of a multiple regression model with a fuzzy asymmetric intercept, which seems more appropriate than the non-fuzzy one as it expresses the average value of the dependent variable (which is also fuzzy) when the independent variables equal zero.

Assuming to regress a dependent variable $\tilde{Y}_i = (y_i, y_{Li}, y_{Ri})_T$ on k independent variables $\tilde{X}_{ij} = (x_{ij}, x_{Lij}, x_{Rij})_T$ in a set of n units, the linear regression model with a fuzzy asymmetric intercept is given by

$$\tilde{Y}_i^* = \tilde{A} + \sum b_j \tilde{X}_{ij}, \qquad i = 1, 2, ..., n; \quad j = 1, 2, ..., k; \quad a, b_j \in \mathbf{R}$$

(with * denoting the theoretical value) where $\tilde{A} = (a,a_L,a_R)_T$, $a_L = a - \underline{\gamma}$, $a_R = a + \overline{\gamma}$ and $\underline{\gamma}$, $\overline{\gamma} > 0$. The estimates of the corresponding parameters a, $\underline{\gamma}$, $\overline{\gamma}$, b_1, ..., b_j, ..., and b_k, are determined by minimizing, with respect to them, the sum of Diamond's squared distances between theoretical and empirical values of the dependent variable

$$\sum d(\tilde{Y}_i, \tilde{A} + \sum b_j \tilde{X}_{ij})^2.$$

The function to minimize assumes different expressions according to the signs of the regression coefficients.

In matricial terms, the estimates of the fuzzy regression coefficients are given by

$$\beta = (\mathbf{X'X + X_L'X_L + X_R'X_R})^{-1} (\mathbf{X'y + X_L'y_L + X_R'y_R}),$$

where:

y = [y_i] is the n-dimensional vector of centers of the dependent variable;

y_L = [y_{Li}] and **y_R** = [y_{Ri}] are the n-dimensional vectors of the lower extremes and the upper extremes respectively of the dependent variable;

X is the n×(k+3) matrix formed by the vector **1**, k vectors of the cores of the independent variables and two vectors **0**;

X_L is the n×(k+3) matrix formed by the vector **1**, k vectors of the lower bounds of the independent variables and the vectors **-1** and **0**;

X_R is the n×(k+3) matrix formed by the vector **1**, k vectors of the upper bounds of the independent variables and the vectors **0** and **1**;

β is the vector (a, b_1, ..., b_j, ..., b_k, $\underline{\gamma}$, $\overline{\gamma}$)'.

The total sum of squares (TotSS) of the dependent variable can be decomposed according to Diamond's metrics [9]. In particular we obtained the expressions of two components, the regression (RegSS) and the residual (ResSS) sums of squares, like in the OLS estimation procedure for classical variables.

Conversely, in a model with a not asymmetric fuzzy intercept we obtained [10] the expressions of two additional components, besides the regression and the residual sums of squares, which arise from the diversity between theoretical and empirical values of the average fuzzy dependent variable (unlike in the OLS estimation procedure for classical variables). In such a case an increase in the regression sum of squares does not necessarily imply a better fit to observed data: this is because the theoretical average value, from which the regression sum of squares is calculated, may be very different from the empirical one. Only a decrease in the residual sum of squares necessarily implies a better fit to observed data.

2.3 A Fuzzy Model Fit Index

We have just demonstrated that the total sum of squares of the dependent variable consists only of two addends (the regression and the residual sums of squares), when the intercept is fuzzy asymmetric.

In order to assess the goodness of fit of the regression model, we propose the following index, for simplicity called Fuzzy Fit Index (FFI):

$$\text{FFI} = 1 - \frac{\text{Res SS}}{\text{Tot SS}}.$$

The more this index is next to 1, the smaller the residual sum of squares is and the better the model fits the observed data.

In order to compare models that explain the same dependent variable by means of a different number of independent variables, it is appropriate to refer to an index that takes into account the corresponding degrees of freedom (closely linked at this number). As in the classic model, an increase in FFI does not necessarily mean that the new independent variable contributes significantly to explain $\tilde{Y}$; any excess in measuring the fit of the model can be corrected by deflating FFI for a term which increases with the number k of independent variables included in the equation.

The proposed version of the adjusted FFI is

$$\overline{\text{FFI}} = 1 - \left(\frac{\text{ResSS}}{\text{TotSS}} \cdot \frac{n-1}{n-k-1} \right)$$

which increases only if the increase in FFI (i.e. in the regression sum of squares) exceeds the penalty induced by having one more independent variable in the model, and decreases otherwise.

3 A Stepwise Procedure to Select Independent Variables

The selection of the most significant independent variables presents greater difficulties from a computational point of view in a fuzzy regression model than in the classic one.

Recently we proposed [11] a *stepwise forward* identification procedure, which inserts at each iteration a variable in the regression equation among the p starting ones, according to two fundamental criteria: the significance of its contribution, measured by the relative increase of the total sum of squares in the dependent variable, and its originality, i.e. the ability to introduce information into the equation which other variables have not already introduced (assessed in terms of correlation with the latter).

In this work, we introduce a *stepwise* procedure which enables us to find the optimal combination of the independent variables not only by including only one of them at a time, but also by eliminating in each iteration that variable, whose explanatory contribution is subrogated by the combination of the other ones included after it was.

This procedure drastically reduces the number of models to be estimated in order to identify the best one among them. Let's examine how it works in detail, starting from the forward selection.

After identifying an initial simple regression model (in which $\tilde{X}_{(1)}$ presents the highest correlation with $\tilde{Y}$), in each successive iteration we select the variable less correlated with those already present, provided that it significantly explains the total sum of squares of the model. In other words, focusing on the q.th step, $\tilde{X}_{(q)}$ is candidate to be also included in the equation if its contribution is original with respect to the previous q-1 variables.

Such a contribution is evaluated by measuring the so called tolerance $T_q = 1 - \overline{\text{FFI}}_{q;1,2,\ldots,q-1}$, in which $\overline{\text{FFI}}_{q;1,2,\ldots,q-1}$ represents the share of variability of $\tilde{X}_{(q)}$ explained by $\tilde{X}_{(1)}, \tilde{X}_{(2)}, \ldots, \tilde{X}_{(q-1)}$. The tolerance ranges between 0 and 1, depending on the degree of linear correlation between $\tilde{X}_{(q)}$ and the other variables; therefore, only if T_q exceeds a threshold identified between 0 and 1, $\tilde{X}_{(q)}$ is candidate to become part of the model.

Note that a high value of the threshold allows us to select very original variables, but it can also stop the process since from the initial steps; on the contrary, a low value allows us to select a greater number of variables, although more correlated to each other. In any case, if none of the variables not yet included in the equation proves significantly its originality, the selection process would stop.

The opportunity of actually introducing the selected variable $\tilde{X}_{(q)}$ into the equation is now evaluated in terms of its explanatory contribution. In particular such a contribution is measured as the increase in the adjusted Fuzzy Fit Index of the model due to the entry submitted for consideration, i.e. as $\overline{FFI}_{y;1,2,...,q}$ - $\overline{FFI}_{y;1,2,...,q-1}$ (where the two terms of the subtraction represent the proportion of the sum of squares of $\tilde{Y}$ explained by the model including and not including $\tilde{X}_{(q)}$ respectively).

The selected variable ends up being introduced into the equation if the correspondent increase in the adjusted Fuzzy Fit Index is higher than an arbitrary threshold value. The higher such an arbitrary value is, the easier the procedure inhibits the entry of new independent variables, whose explanatory contribution is not that relevant.

If the explanatory contribution of $\tilde{X}_{(q)}$ is not significant, we pass to consider the inclusion of the remaining candidate variables on the basis of the same criterion. In any case, the selection procedure stops when none of them contributes significantly to explain the sum of squares of $\tilde{Y}$.

The proposed procedure provides also the possibility of eliminating in each iteration one of the variables already included in the equation. For example, once $\tilde{X}_{(q)}$ is inserted, the explanatory contribution of every $\tilde{X}_{(i)}$ (i = 1, 2, ..q-1) is still valued as the reduction of the adjusted Fuzzy Fit Index caused by the elimination of $\tilde{X}_{(i)}$ from the model, i.e. as $\overline{FFI}_{y;1,2,...,q}$ - $\overline{FFI}_{y;1,2,...,q (-i)}$ (where the two terms of the subtraction represent the proportion of the sum of squares of $\tilde{Y}$ explained by the model excluding and not excluding $\tilde{X}_{(i)}$ respectively).

The variable which shows the smallest reduction is excluded from the equation if such a reduction does not exceed an arbitrary threshold value. This would happen if the explanatory contribution of the variable to be discarded was subrogated by the combination of the independent variables introduced after it was.

In conclusion it is worth noting that the threshold value of the variation in the adjusted Fuzzy Fit Index could be lower in the case of the forward selection, where priority is given to the role of tolerance, than in the case of the backward selection. This allows us to include into the equation a greater number of variables which otherwise would play no role, maintaining the opportunity to evaluate their exclusion at a later time.

The just exposed procedure has been developed through the *MatLab Editor*. In particular, providing the matrices of the centers, the left extremes and the right extremes both of the dependent variable and of the independent variables as input parameters, the implemented function actives the described selection procedure and provides the estimated models at each iteration together with the adjusted FFI indices.

The correspondent algorithm uses recursively the FLS function, defined in the following way:

```
function [BA,res,regr,tot,FFI] = FLS(x,xR,xL,y,yR,yL,p)
    num=size(x);
    q=num(1,2)-1;
    n=num(1,1);
    numcomb=(2^p)-1;
    num= 0:numcomb;
    data=dec2bin(num,p);
    numcomb=2^q;
    comb=data(1:numcomb,p-q+1:end);
    BA=zeros(q+3,1);
    res=0;
    regr=0;
    tot=0;
    FFI=0;
    uno=ones(n,1);
    zer=zeros(n,1);
    z=uno;
    w=uno;
    i=0;
    while (i<numcomb)
        i=i+1;
        j=1;
        while (j<=q)
          if comb(i,j)== '0'
            z=[z,xR(:,j+1)];
            w=[w,xL(:,j+1)];
          else
            z=[z,xL(:,j+1)];
            w=[w,xR(:,j+1)];
          end;
          j=j+1;
        end;
        X=[x,zer, zer];
        Z=[z,-uno,zer];
        W=[w,zer, uno];
        a=eig(Z'*Z+X'*X+W'*W);
        if(a>0)
           b=inv( Z'*Z + W'*W + X'*X )*( Z'*yL + W'*yR + X'*y )
           count=0;
           j=1;
           l=1;
           while (j<=q)
              l=j+1;
              if (b(l,1)<0 && comb(i,j)=='0') || (b(l,1)>=0 && comb(i,j)=='1')
                count=count+1;
```

```
            end;
            j=j+1;
         end;
         if ((count==q) && (b(q+1,1)>0) && (b(q+2,1)>0))
            BA = [BA(:,:),b];
            yth=X*b;
            ythL=Z*b;
            ythR=W*b;
            devregr=(ythL-(uno'*ythL*uno)/n)'*(ythL-(uno'*ythL*uno)/n)+(yth-
(uno'*yth*uno)/n)'*(yth-(uno'*yth*uno)/n)+(ythR-(uno'*ythR*uno)/n)'*(ythR-
(uno'*ythR*uno)/n);
            devres=(yL-ythL)'*(yL-ythL)+(y-yth)'*(y-yth)+(yR-ythR)'*(yR-ythR);
            devtot=(yL-(uno'*yL*uno)/n)'*(yL-(uno'*yL*uno)/n)+(y-
(uno'*y*uno)/n)'*(y-(uno'*y*uno)/n)+(yR-(uno'*yR*uno)/n)'*(yR-(uno'*yR*uno)/n);
            res = [res,devres];
            regr= [regr,devregr];
            tot= [tot,devtot];
            FFI= [FFI,(devregr/devtot)];
         end;
      end;
      z=ones(n,1);
      w=ones(n,1);
   end;
```

The FLS function provides the matrix of admissible solutions, arising from the possible combinations of the signs relative to the q variables taken into account, and identifies the optimal one among them, after checking the consistency of the signs of the estimated model with its predetermined assumptions for each solution.

As previously described, we determine the tolerance of each variable not yet introduced in the model in order to identify the more original candidate one:

```
(...)
while(s<=q)
        y_q=comp1(:,s+2);
        yL_q=comp1L(:,s+2);
        yR_q=comp1R(:,s+2);
        [BA,res,regr,tot,FR2] = FLS(x,xR,xL,y_q,yR_q,yL_q,p);
        h=size(BA);
        if(h(1,2)>1)
            BA = BA(:,2:end);
            res = res(:,2:end);
            FR2= FR2(:,2:end);
            [devres,imin] = min(res);
            v_devregr_devtot(s) = FR2(:,imin);
            v_toll(s)= (1-v_devregr_devtot(s));
        else
```

```
            v_devregr_devtot_agg(s)=0;
            v_devregr_devtot(s)=0;
            v_toll(s)=0;
        end;
        s=s+1;
    end;
(...)
```

Among all the analyzed variables, we select the one with the highest tolerance; if its originality is significant, then we estimate the regression model after including such a variable and we evaluate the significance of its explanatory contribution.

Once a new variable entered the model, we verify that the explanatory contribution of the variables introduced in previous steps is still significant:

```
(...)
while(s<q)
    if(s>1)
        x_p=[x(:,1:s),x(:,s+2:end)];
        xL_p=[xL(:,1:s),xL(:,s+2:end)];
        xR_p=[xR(:,1:s),xR(:,s+2:end)];
    else
        x_p=[x(:,1),x(:,3:end)];
        xL_p=[xL(:,1),xL(:,3:end)];
        xR_p=[xR(:,1),xR(:,3:end)];
    end;
    BA=zeros(q+3,1);
    res=0;
    regr=0;
    tot=0;
    FR2=0;
    [BA,res,regr,tot,FR2] = FLS(x_p,xR_p,xL_p,y,yR,yL,p);
    h=size(BA);
    if(h(1,2)>1)
        BA = BA(:,2:end);
        res = res(:,2:end);
        FR2= FR2(:,2:end);
        [devres,imin] = min(res);
        m_sol = [m_sol,BA(:,imin)];
        v_devregr_devtot(s) = FR2(:,imin);
        v_devregr_devtot_agg(s)=1-((1-v_devregr_devtot(s))*(n-1)/(n-q));
        v_diff(s)= (FR2_agg(:,end)-v_devregr_devtot_agg(s));
    else
        v_devregr_devtot_agg(s)=0;
        v_devregr_devtot(s)=0;
        m_sol = [m_sol,BA];
        v_diff(s)=99;
```

```
      end;
      s=s+1;
    end;
(...)
```

Among all the analyzed variables, we select the one which presents the lowest explanatory contribution. The latter will be removed from the model if such a contribution is found to be not significant.

4 An application Case: The Effectiveness of the Activated Internships in Graduates' Opinion

The university system can fully respond to its responsibility of promoting development only by establishing a connection with manufacturing companies and other institutions of the territory in which it operates. In this perspective, the growth of human capital cannot be reduced to mere relationship between teacher and student established in a classroom, but needs new educational models, alternative to the pure transfer of theoretical knowledge.

In an effort to encourage the organization of learning communities, university regulations allow their students to gain a training experience directly in operational structures. The aim pursued by such regulations is to engage students to adapt in various contexts and to solve operational problems that will occur more frequently during their working life; moreover the direct experimentation of the theoretical skills stimulates the academic system to recognize the practical needs of the productive system, encouraging the development of new skills.

The utility of internship activities actually organized can be measured on the basis of objective data, such as by monitoring how many graduates have pursued the goal of job placement as a result of the gained experience, but also on the basis of subjective data, such as by analyzing the assessments of satisfaction expressed by each of them in relation to that experience.

Since the satisfaction depends not only on the quality of internships, but also on many other factors (the most important of which is the local context within which the activity is carried out), the analysis of the collected questionnaires is conducted separately depending on whether the interviewees have a degree in a university based in Northeast, Northwest, Centre, South and Islands of our country. In each of these four areas we regress the overall satisfaction for the concluded internship with respect to the satisfaction expressed for the following nine aspects concerning it: organization, guidance activities, clearness of formative goals, tutor's helpfulness and competence, autonomy in carrying out the assigned tasks, utility for personal training, prestige of the company, ability to convey something useful to trainees.

The opinions on internship are collected by the *AlmaLaurea* consortium in a sample survey conducted specifically between 2 and 23 April 2008 (i.e. between 16 and 28 months after graduation). The reference population is represented by 61,347 graduates in calendar year 2006 who said they supported an internship approved by the University, of which 58,904 were potentially reached by e-mail.

The survey, carried out by subjecting 58,904 respondents to a questionnaire prepared for this purpose, is concluded with a response rate of 42.8 percent. Note that the valid interviews were weighted by the Consortium according to specific characteristics of the reference population correlated with the studied phenomenon, such as the course of study, the University and the Faculty of enrollment, gender and so on, in order to avoid as far as possible that the sample is distorted; in particular the answers provided by every graduate interviewed were multiplied by a weighting factor equal to the ratio between the theoretical proportion of the joint distribution (note) of the aforementioned features found in the reference population and the proportion found in the corresponding category.

The missing responses to individual questions considered in this study are very small in percentage terms and, therefore, are not subject to further investigations.

We consider only the curricular internships, structured in learning (and not working) activities, among the ones carried out by Italian students who graduated in 2006; according to an estimate prepared by the consortium, they represent 75.9% of all curricular internships (about three out of four).

Respondents express their opinion on a ordinal scale formed by the following categories: "Unquestionably unsatisfied", "More unsatisfied than satisfied", "More satisfied than unsatisfied", "Unquestionably satisfied". As both Chiandotto and Gola [12] and Lalla, Ferrari and Pirotti [13] proposed a quantification of such an ordinal scale, assigning respectively 2, 5, 7 and 10 to the above mentioned categories, we treat "Unquestionably unsatisfied" as the triangular fuzzy number "about 2", "More unsatisfied than satisfied" as "about 5", "More unsatisfied than satisfied" as "about 7" and "Unquestionably satisfied" as "about 10". In so doing, rather than establish an unrealistic one-to-one correspondence between verbal terms and numerical values, we associate to each verbal term an interval represented by a neighborhood of the chosen value, the amplitude of which varies depending on the intensity of the expressed judgment [14].

The use of triangular membership functions allows to allocate to each value included in the range a degree of representativeness inversely proportional to its distance from the core of the fuzzy number. It is chosen, in particular, to match the two extreme (intermediate) categories, which describe a net (vagus) belief from respondents, triangular fuzzy numbers characterized by rather low (extended) spreads. They are analytically represented in the figure below (Fig. 2).

Note that the membership functions of the two extreme categories differ from each other because they must be triangular in shape and must be centered respectively on the values 2 and 10 instead of on the extremes of the interval (1, 10).

Once defined triangular fuzzy numbers for the collected opinions, we candidate to enter the model the variable which shows the highest tolerance, provided that the latter is above 70% (i.e. the multiple correlation coefficient with the other variables of the model is not more than 30%); the boundary imposed to the tolerance is identified as a result of specific attempts experienced, that is on the basis of simple heuristic considerations: higher (lower) values of the tolerance inhibit (facilitate) the entry of new variables in the model. Moreover the candidate variable is not included in the equation if its explanatory contribution is less than 1%.

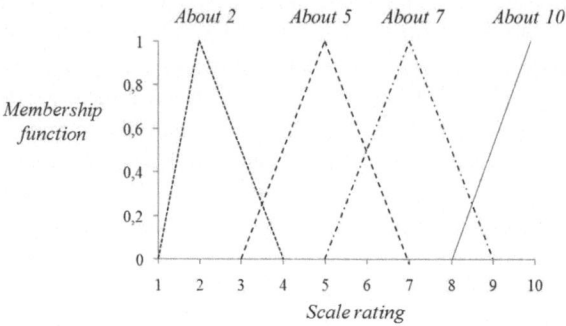

Fig. 2. Fuzzy numbers used to express the four response categories

At each step the procedure removes the least significant variable, provided that it helps to explain the variance of the model for less than 2%. The different boundary imposed to the explanatory contribution of variables in exit with respect to those in entrance is justified by the fact that, in selecting the latter (which can always come out even after), the procedure gives greater weight to their correlation with the rest of the variables already present.

A conjoint exam of Fig. 4 and Table 1 shows that the procedure, implemented in order to identify among all the available variables those most relevant, generates at each step an ever-increasing $\overline{FFI}$, except for the case in which a variable, whose contribution has been subrogated by other ones included in the model in successive steps, is excluded. Actually the $\overline{FFI}$ relative to the model estimated after the removal of a variable is reduced, although only a few percent, but increases later in the iteration when a new variable is inserted (in particular it also increases with respect to the model prior to such a removal, i.e. the model with the same number of variables).

The last model estimated by the stepwise procedure, in each of the four areas (Table 1) taken into account, is obtained by removing an explanatory variable, whose contribution was clearly not significant; this obviously causes a reduction, even if minimum, of $\overline{FFI}$ compared to the previous step.

Ultimately the choice of the preferred model among the last ones identified by the procedure depends on the analyst: in such an evaluation the criterion of parsimony can justify the loss of a few percentage points in terms of goodness of fit.

The obtained results lead us to affirm that the utility for personal training impacts decisively on the overall satisfaction in each of the four territorial divisions (note that such an utility explains alone over 63% of the overall satisfaction perceived by the graduates from Southern Italy).

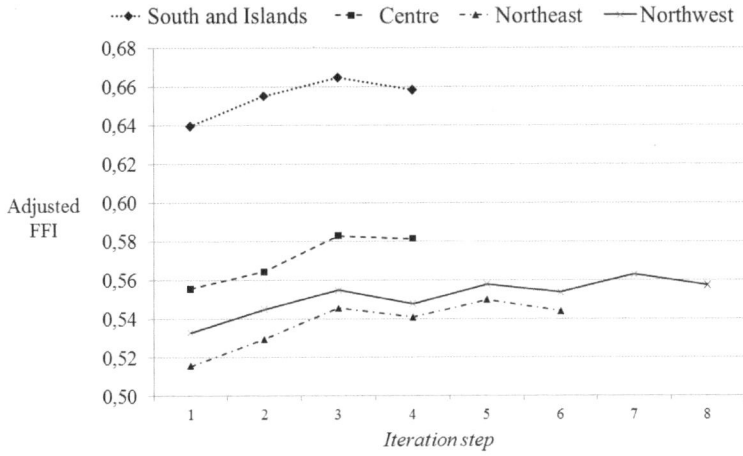

Fig. 3. Values of the adjusted $\overline{FFI}$ in each iteration of the estimated models

In addition the clearness of formative goals is significantly relevant to the trainees from the Centre and South Italy and remains relevant for the other two areas till the last iteration.

The organization of the internship reveals instead in the two areas of Northern Italy; finally it should be noted that, during the selection process, the following variables are never found to be relevant: guidance activities, tutor's helpfulness, prestige of the company, ability to convey something useful to trainees.

The validity of the proposed *stepwise* selection procedure is confirmed by comparison with the estimation of all the possible models, i.e. the models which consider every combination of the 9 variables taken into account (Table 2). In fact, regardless of territoriality, in less than 8% of the estimated models the $\overline{FFI}$ index exceeds that of the best model identified by the stepwise procedure.

In particular, among the models estimated for the South, the highest value of $\overline{FFI}$ is equal to 0.668 (just slightly larger than that identified by the stepwise procedure) and the correspondent model includes two variables in addition to those identified by our procedure: the organization of the internship and the assigned tutor's helpfulness.

Similarly, among the models estimated for the North-East, the one with the highest $\overline{FFI}$ includes in addition the guidance activities and the ability to convey something useful to trainees; on the contrary, considering the models estimated for the Centre and for the North-West with the highest $\overline{FFI}$, instead, there is not a strong correspondence with the variables selected by the iterative procedure proposed by us.

It should be emphasized that, in general, the variables identified by the proposed method are indeed the most influential variables in the model with the highest $\overline{FFI}$ (reported in Table 2). As already mentioned, the choice between the best fit to data and the parsimony of the model depends on the analyst.

Table 1. Estimated coefficients of the variables selected in each step of the iterative procedure, by territorial division

SOUTH AND ISLANDS

Intercept	(0.463, 0.246, 0.226)	(0.266, 0.117, 0.126)	(0.044, 0.042, 0.053)	(0.335, 0.166, 0.092)	
Utility for personal training	0.857	0.723	0.737	0.769	
Autonomy	-	0.151	0.105	-	
Clearness of formative goals	-	-	0.131	0.169	
R²	0.639	0.655	0.665	0.638	

CENTRE

Intercept	(1.059, 0.519, 0.267)	(0.515, 0.182, 0.186)	(0.286, 0.097, 4.105)	(0.135, 0.143, 0.125)
Utility for personal training	0.826	0.777	0.711	0.725
Autonomy	-	0.118	0.053	-
Clearness of formative goals	-	-	0.176	0.194
R²	0.556	0.565	0.583	0.582

NORTHEAST

Intercept	(1.373, 0.362, 0.204)	(0.905, 0.316, 0.224)	(0.247, 0.123, 0.146)	(0.097, 0.204, 0.167)	
Utility for personal training	0.799	0.734	0.680	0.705	
Autonomy	-	0.147	0.091	-	
Clearness of formative goals	-	-	0.159	0.185	0.696
Organization	-	-	-	0.137	
R²	0.516	0.530	0.546	0.541	0.548

NORTHWEST

Intercept	(0.314, 0.346, 0.176)	(0.423, 0.318, 0.219)	(0.144, 0.116, 0.177)	(0.368, 0.222, 0.265)	(0.393, 0.142, 0.195)	(0.694, 0.204, 0.206)	(0.594, 0.152, 0.153) (0.697, 0.217, 0.168)
Utility for personal training	0.794	0.738	0.696	0.736	0.661	0.698	0.674 0.664
Autonomy	-	0.148	0.112	-	-	-	- -
Trust	-	-	0.125	0.150	0.690	-	- -
Clearness of formative goals	-	-	-	-	0.140	0.150	0.094 0.198
Organization	-	-	-	-	-	-	0.138 0.197
R²	0.533	0.545	0.548	0.554	0.558	0.555	0.560 0.557

Table 2. Estimated coefficients of the variables selected in the models with the highest FFI by territorial division

	SOUTH AND ISLANDS	CENTRE	NORTHEAST	NORTHWEST
Intercept	(-0.054, 0.013, 0.028)	(0.039, 0.015, 0.002)	(-0.010, 0.016, 0.018)	(0.013, 0.017, 0.016)
Utility for personal training	0.721	0.693	0.648	0.638
Autonomy	0.104	-	0.068	-
Organization	0.087	-	0.123	0.116
Clearness of formative goals	0.066	0.167	0.076	0.047
Tutor's helpfulness	0.010	0.03	0.002	-
Ability to convey something useful to trainees	-	0.097	0.071	0.106
Tutor's competence	-	-	-	0.083
FFI	0.668	0.590	0.556	0.572

5   Concluding Remarks

In this work we propose a stepwise procedure to select independent variables in a multivariate fuzzy regression model with a fuzzy asymmetric intercept. Such a procedure, developed through the *MatLab Editor*, drastically reduces the number of models to be estimated in order to identify the best one among them.

As an applicative case, we evaluate the effectiveness of the internships activated by Italian universities, detecting those factors which contribute more than others to the corresponding satisfaction expressed by the interviewed graduates.

Since the satisfaction depends not only on the quality of internships, but also on many other factors (the most important of which is the local context within which the activity is carried out), the analysis of the collected questionnaires is conducted separately depending on whether the interviewees have a degree in a university based in Northeast, Northwest, Central, South and Islands of our country. In each of these four areas we regress the overall satisfaction for the concluded internship with respect to the satisfaction expressed for the following nine aspects collected by the consortium *AlmaLaurea* in a sample survey. Respondents express their opinion on a ordinal scale formed by four categories that we treat as the triangular fuzzy number. In so doing, we associated to each verbal term an interval represented by a neighborhood of the chosen value, the amplitude of which varies depending on the intensity of the expressed judgment.

The procedure, implemented in order to identify among all the available variables those most relevant, generates at each step an ever-increasing goodness of fit measure, except for the case in which a variable, whose contribution has been subrogated by other ones included in the model in successive steps, is excluded. Actually the goodness of fit measure relative to the model estimated after the removal of a variable is reduced, although only a few percent, but increases later in the iteration when a new variable is inserted (in particular it also increases with respect to the model prior to such a removal, i.e. the model with the same number of variables).

The obtained results lead us to affirm that the utility for personal training impacts decisively on the overall satisfaction in each of the four territorial divisions, the clearness of formative goals is significantly relevant to the trainees from the Centre and South Italy (and remains relevant for the other two areas till the last iteration), the organization of the internship reveals instead in the two areas of Northern Italy.

The validity of the proposed *stepwise* selection procedure is confirmed by comparison with the estimation of all the possible models, i.e. the models which consider every combination of the variables taken into account. In fact, regardless of territoriality, in less than 8% of them the goodness of fit measure exceeds that of the best model identified by the stepwise procedure.

It should be emphasized that, in general, the variables identified by the proposed method are indeed the most influential variables in the model with the highest goodness of fit measure. However the choice between the best fit to data and the parsimony of the model depends on the analyst.

References

1. Kao, C., Chyu, C.L.: Least-squares estimates in fuzzy regression analysis. European Journal of Operational Research 148, 426–435 (2003)
2. Campobasso, F., Fanizzi, A., Perchinunno, P.: Homogenous Urban Poverty *Clusters* within the City of Bari. In: Gervasi, O., Murgante, B., Laganà, A., Taniar, D., Mun, Y., Gavrilova, M.L. (eds.) ICCSA 2008, Part I. LNCS, vol. 5072, pp. 232–244. Springer, Heidelberg (2008)
3. Montrone, S., Perchinunno, P.: Di giuro, A., Torre, C.M., Rotondo, F.: Identification of hot spot of social and housing difficulty in urban areas. In: ICCSA 2011. LNCS, vol.176, pp. 57-78. Springer (2011)
4. Perchinunno, P., Rotondo, F., Torre, C.M.: A Multivariate Fuzzy Analysis for the Regeneration of Urban Poverty Areas. In: Gervasi, O., Murgante, B., Laganà, A., Taniar, D., Mun, Y., Gavrilova, M.L. (eds.) ICCSA 2008, Part I. LNCS, vol. 5072, pp. 137–152. Springer, Heidelberg (2008)
5. Zadeh, L.A.: Fuzzy sets. Information and Control 8(3), 338–353 (1965)
6. Diamond, P.M.: Fuzzy Least Square. Information Sciences 46, 141–157 (1988)
7. Bilancia, M., Campobasso, F., Fanizzi, A.: The pricing of risky securities in a Fuzzy Least Square Regression model. In: Advances in Data Analysis and Classification 2010, pp. 639–646. Springer, Heidelberg (2010)
8. Montrone, S., Campobasso, F., Perchinunno, P., Fanizzi, A.: An Analysis of Poverty in Italy through a Fuzzy Regression Model. In: Murgante, B., Gervasi, O., Iglesias, A., Taniar, D., Apduhan, B.O. (eds.) ICCSA 2011, Part I. LNCS, vol. 6782, pp. 342–355. Springer, Heidelberg (2011)
9. Campobasso, F., Fanizzi, A., Tarantini, M.: Some results on a multivariate generalization of the Fuzzy Least Square Regression. In: Proceedings of the International Conference on Fuzzy Computation, Madeira, pp. 75–78 (2009)
10. Campobasso, F., Fanizzi, A.: A Fuzzy Approach To The Least Squares Regression Model With A Symmetric Fuzzy Intercept. In: Proceedings of the 14th Applied Stochastic Model and Data Analysis Conference, Roma, pp. 226–232 (2011)
11. Campobasso, F., Fanizzi, A.: A stepwise procedure to select variables in a fuzzy least square regression model. In: Proceedings of the International Conference on Fuzzy Computation Theory and Applications, pp. 417–426 (2011)
12. Chiandotto, B., Gola, M.: Rapporto finale del gruppo di ricerca per la valutazione della didattica da parte degli studenti. Murst - Osservatorio per la valutazione del sistema universitario (1999)
13. Lalla, M., Ferrari, D., Pirotti, T.: A fuzzy inference system for teaching evaluation. In: Proceedings of the International Conference on Statistical Modelling for University Evaluation: an International Overview, pp. 27–30. Foggia University (2008)
14. Takemura, K.: Fuzzy least squares regression analysis for social judgment study. Journal of Advanced Intelligent Computing and Intelligent Informatics 9(5), 461–466 (2005)

Cluster Analysis for Strategic Management: A Case Study of IKEA[*]

Paola Perchinunno[1] and Dario Antonio Schirone[2]

[1] Department of Statistical Science, University of Bari,
Via C. Rosalba 53, 70100 Bari, Italy
p.perchinunno@dss.uniba.it
[2] Department of Mediterranean Societies of Bari,
P.zza Umberto I, 70121 Bari, Italy
darioschirone@libero.it

Abstract. Business strategy, understood as the set of choices implemented in order to achieve long-term goals [1], or as identified through SWOT Analysis [2], to which reference is so frequently made during periods of economic boom, was surpassed during the 1970s and 80s by *strategic planning* [3] and *strategic management*. The current situation of the IKEA store in Bari (Apulia, Italy) may be understood within this framework. The objective of the present study is, therefore, to identify possible reasons for the likelihood of different consumption patterns and choices by particular groups of individuals in a Primary Market Area.

Keywords: strategic management, cluster analysis, primary market area.

1 Introduction

The continuous changes in the competitive landscape (the result of an absence of barriers) initiates, as a direct consequence, behavioural change in individuals yet also in companies, which, regardless of their size, may increase business as well as their overall turnover, should they possess the ability and flexibility to reinvent themselves.

Such changes, both in the relationship between business and environment as well as in time-space, have led to an evolution of company strategy which must be capable of metamorphosing in a highly complex and unstable environment [4], through decisions that allow, by responding effectively to highly unpredictable situations, for the maintaining of competitive advantage.

Business strategy, understood as the set of choices implemented in order to achieve long-term goals [1], or as identified through SWOT Analysis [2], to which reference is so frequently made during periods of economic boom, was surpassed during the 1970s and 80s by *strategic planning* [3] and *strategic management*. Consequently,

---

[*] The contribution is the result of joint reflections by the authors, with the following contributions attributed to P. Perchinunno (chapter 3.1, 3.2 and 4), and to D. Schirone (chapter 1, 2, 3.3, 3.4 and 5).

management can no longer apply strategies according to predetermined decisions, but must adapt to changing consumer needs in order to retain and increase market share.

The current situation of the IKEA store in Bari (Apulia, Italy) may be understood within this framework. Indeed, analysis conducted on the market share of the primary commercial activity (Figure 1) [1] of the IKEA store in Bari demonstrates that localities of the Bari North area record an average of 2.5 percentage points lower in comparison to other municipalities under examination.

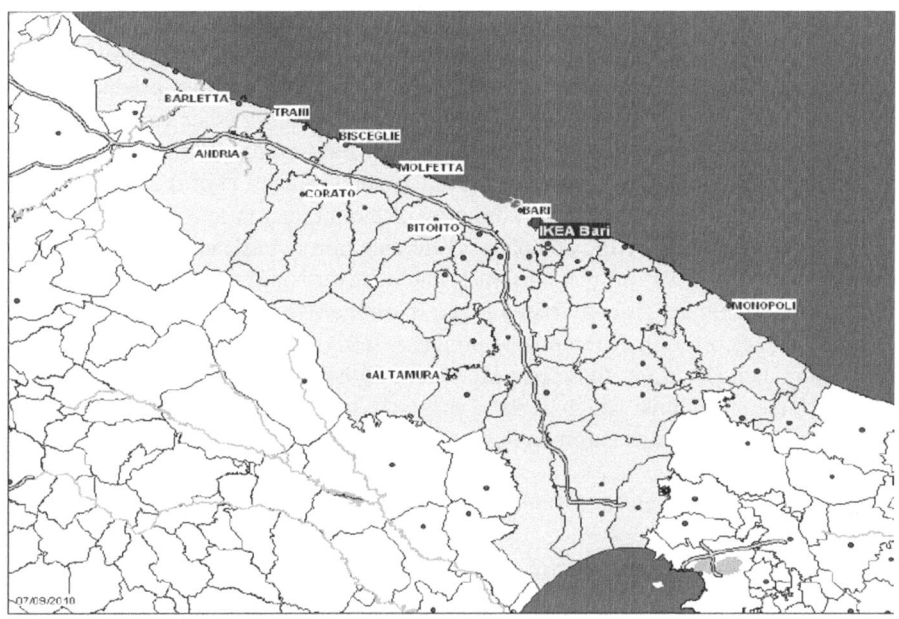

Fig. 1. Primary Market Area

The objective of the present study is, therefore, to identify possible rationales for different consumer choices made, with particular reference to furniture and furnishings, by the specific group of consumers identified as resident in the geographical area of the BAT province[2].

The methodology employed is that of face-to-face interviews with the top management and administrative employees of the store and through a questionnaire administered to visitors to a shopping mall located in the area of reference under investigation.

---

[1] It refers to the geographical area next to the point of sale with a flow of consumers between 50% and 70% of the total.

[2] The province of BAT (Barletta - Andria – Trani) is a new province in northern Apulia, which has 390,925 inhabitants.

2 Market Competition as a Possible Cause

Mechanisms of exchange have, for some time, taken place between similar and differentiated, rather than substantially differing, products and are generally more intra-industry than inter-industry in nature. Product differentiation is the basis for the existence of demand functions with negative gradients [5] and, above all, reveals decisions regarding production, pricing and marketing strategies of companies operating in markets with monopolistic competition [6].

Demand is a function, therefore, not only of the income of consumers, but also the style of a product, ancillary services and the unique marketing strategy pursued by the commercial offer. Demand may also vary with changes in pricing, production and sales policies of businesses belonging to the same sector and/or other sectors.

Differentiation is valid should products present their own peculiarities, i.e. when presenting differences in the inputs employed, in the localization of the company that determines the convenience for the consumer of the product in terms of accessibility and/or services offered by the company. Differentiation is misleading when products are essentially identical, yet the consumer, due to advertising, diversity in packaging, in design or trademarks, is convinced of their dissimilarity.

In companies offering differentiated products, selling costs, which relate to market penetration and, therefore, those of advertising, do not alter the identification of the quantity of equilibrium, which always responds to the marginal rule of equality between marginal revenue and marginal costs but, rather, increase average production costs and reduce profits.

Advertising investment is a sunk cost [7], as budgets are a substantial barrier to entry by potential competitors, discouraging even the intention to enter into competition and the formulation of positive expectations of profit derivable from its entrance.

3 The Field Survey

3.1 Socio-Demographic Characteristics of Respondents

In order to identify possible causes of the failure by IKEA to achieve its strategic objectives in a particular area of Apulia, a questionnaire was administered to a sample of potential customers. The survey was conducted within a complex comprising a hypermarket, multiplex cinema and shopping mall in the municipality of Molfetta during the period between 10/09/2011 and 18/09/2011.

The number of questionnaires administered totalled 303, of which only 252 were valid, namely those administered to residents of the municipalities of BAT or Bari North, as subjects of interest to the present investigation. The origin of respondents was, more specifically, distributed between the municipalities of Molfetta (16%), Bisceglie (13%), Trani (8%) and Barletta (7%).

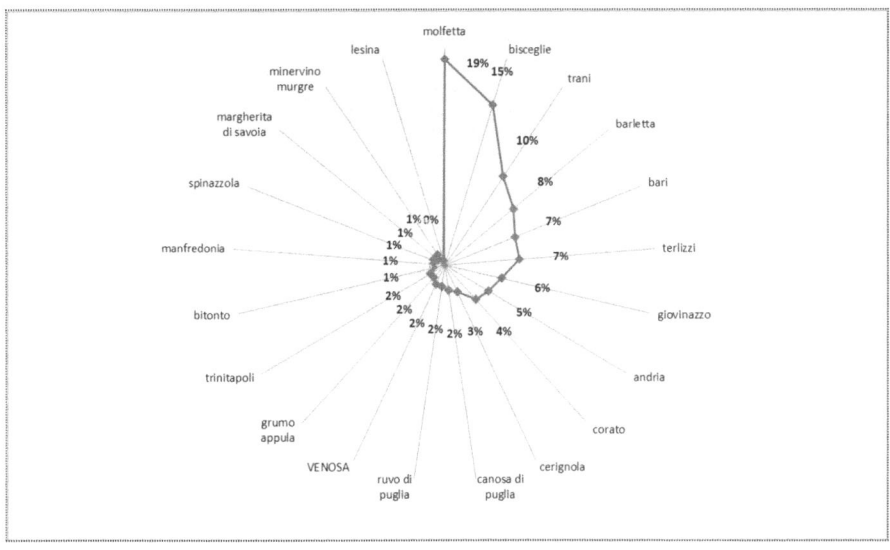

Fig. 2. Polar diagram of respondents according to the municipality of residence (Source: our elaborations on field survey)

The demographic and socio-economic characteristics of the sample (i.e. the 252 respondents) are as follows:

- *Gender*: 52% female and 48% male;
  Age: 60% aged between 25 and 44 years, 9% between 18 and 24 years and 31% over 45 years.
- *Level of education*: 13% graduates, 47% high school diploma graduates, and the remaining 40% educated to middle school level or below.
- *Occupation*: 61% employed, 28% unemployed, 6% students and 5% pensioners.

In addition, respondents were classified according to their relationship with IKEA. The majority of respondents (84% - 212 respondents) had shopped at IKEA at least once, while the remaining 16% (40 respondents) had never shopped at IKEA.
Existing IKEA customers can be further classified as:

- *Loyal IKEA customers*: 81 respondents (34% of total respondents), purchasing furniture and furnishings exclusively from IKEA;
- *Occasional IKEA customers*: 154 respondents (66% of total respondents) purchasing at IKEA, although not exclusively, and primarily furnishings rather than furniture.

3.2 Analysis of the "loyal" IKEA customer

Customers classified as "loyal" to the IKEA brand are those purchasing exclusively or principally from IKEA due to a good quality/price ratio (34% of respondents).

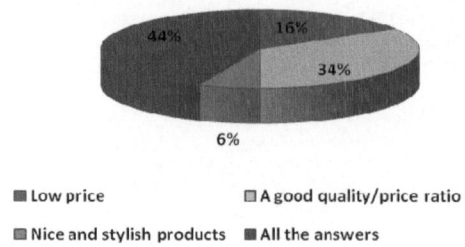

Fig. 3. Primary reasons for purchase by "loyal" IKEA customers (Source: our elaborations on field survey)

The "loyal" customer profile shows predominantly younger consumers of between 25 and 34 years (equal to 35%) with a high level of education (14% of respondents are graduates and 46% possess a high school diploma). The profile of the IKEA customer is, therefore, medium-high both in cultural and economic terms. This attribute is also confirmed by the Bari store that includes among its customer base a higher proportion of graduates than the Italian average with an average income, which, although lower than that of Italian client base in reference to the primary market segment, is almost equal to that of customers in Northern Italy [8].

The majority of such IKEA customers principally purchase both furniture and furnishings (62%) while 34% only purchase furnishings.

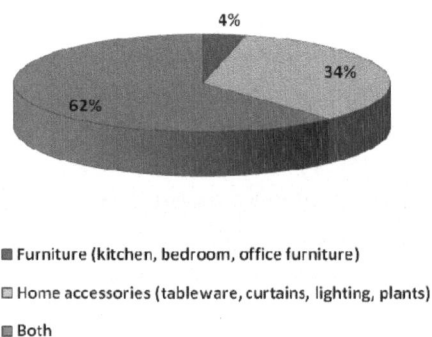

Fig. 4. Types of product purchased by "loyal" IKEA customers (Source: our elaborations on field survey)

The most popular styles of furniture are classified as "Popular Modern Ethnic" (25%) and "Popular Modern Contemporary" (21%).

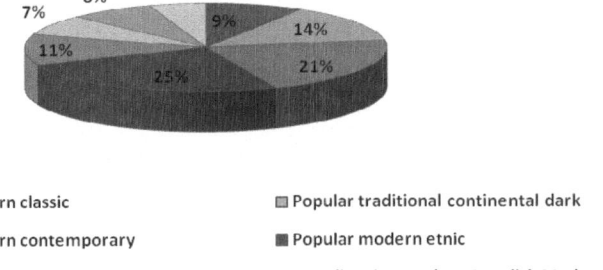

Fig. 5. Style preferred by "loyal" IKEA customers (Source: our elaborations on field survey)

3.3 Analysis of the "Occasional" IKEA Customer

"Occasional" IKEA customers do not buy exclusively at IKEA. Their profile, unlike that of "loyal" customers, is characterized by persons aged between 35 and 44 years (42%), with an educational level lower than the average IKEA customer inasmuch as it is characterized by 12% university graduates and 47% in possession of a high school diploma.

Compared to the "loyal" IKEA customer, "occasional" customers mainly purchase furnishings (65%), while 33% purchase both furniture and furnishings.

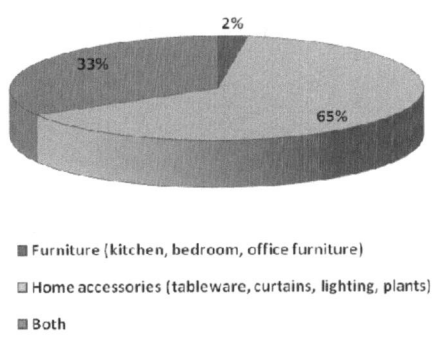

Fig. 6. Types of product purchased by "occasional" IKEA customers (Source: our elaborations on field survey)

The preferred styles of furniture are similar to those of "loyal" IKEA customers, "Popular Modern Ethnic" (26%) and "Popular Modern Contemporary" (17%).

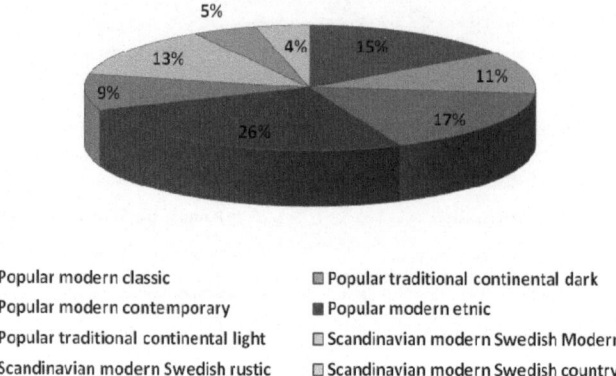

Fig. 7. Style preferred by "occasional" IKEA customers (Source: our elaborations on field survey)

Not buying exclusively from IKEA, "occasional" customers turn to competitors in the geographic area within proximity of their residency. Specifically, 40% indicated trusted stores in the local area while 29% turn to other competitors in the furniture sector. 17% of "occasional" IKEA customers, conversely, show no particular preference in purchasing as against 6% claiming to pay particular attention to value in purchases.

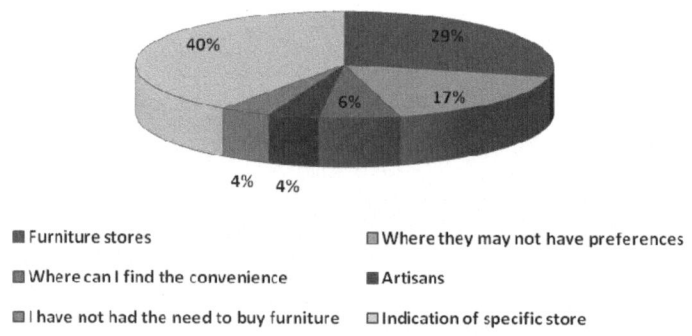

Fig. 8. Main competitors of IKEA in the "occasional" customers (Source: our elaborations on field survey)

3.4 Analysis of Non-IKEA Customers

The profile analysis of the "non-IKEA customers" is, therefore, of particular interest in this context and are shown as characterized by those of between 35 and 44 years (32%), with an educational level even less pronounced than the IKEA customers as profiled above; "Non-IKEA customers" are characterized by 12% of graduates and 51% in possession of a high-school diploma.

The main motivations driving this group of consumers not to purchase at an IKEA store are mainly due to the notable distance from home to the store (68% of respondents) and, to a much lesser degree, the unsuitability of the commercial products offered by IKEA for family needs (11% of respondents) or the presence of an alternative shopping centre (13% of respondents).

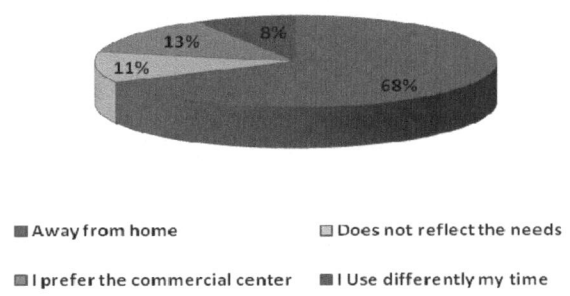

Fig. 9. Main reasons for not buying of "non-IKEA customers" (Source: our elaborations on field survey)

The main competitors of IKEA in the "Non-IKEA customer" group are either alternative stores, principally located in their area of residence (52%) or craftsmen (15%). By contrast, 26% of respondents expressed no particular preference in the choice of store for the purchase of furniture or furnishings.

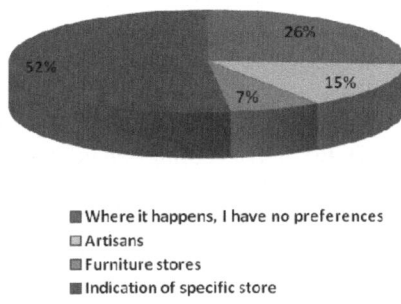

Fig. 10. Main competitors of IKEA in the "non-IKEA customer" (Source: our elaborations on field survey)

4 Identification of Profiles of Potential IKEA Customers

4.1 Cluster Analysis Using the Twostep Method

In order to define the profiles of consumers surveyed, a clustering procedure was employed. Cluster analysis is highly valuable as it provides clusters distinct from one another (or heterogeneous), each consisting of units with a high degree of natural association.

Differing approaches to cluster analysis share, nevertheless, the common need to define a matrix of dissimilarity or distance between the n pairs of observations, which represents the point from which each algorithm is generated. Most recent studies in the field of data mining are directed towards the identification of algorithms able to manage both very large data sets as well as data sets consisting of mixed variables.

A specific cluster analysis technique used for categorical data is often referred to as the *TwoStep*. This is an extension of the distance measures employed by Banfield and Raftery [9], introduced for data with continuous attributes. The *TwoStep* algorithm automatically determines the optimal number of clusters, although it allows for establishing the number of clusters required.

The *TwoStep* procedure, highly efficient with large data sets, is an algorithm of scalar cluster analysis and is able to simultaneously treat variables or categorical and continuous attributes. It is achieved through two steps:

1. in the first step, defined as **pre-cluster**, records are pre-classified into a number of small sub-clusters;
2. in the **second step** the sub-clusters (generated in the first step) are regrouped into a number of clusters that maximizes the BIC (Bayesion Information Criterion).

The *pre-cluster* is a segmentation process in which the results of the algorithm can result in an initial partition of the space in which the variables are defined (taking into account the order of their importance) or the distance between the cases. The representation of such a partition is a tree referred to as the *Cluster Features Tree* defined by levels of nodes [10].

All cases, commencing from the root node, are channelled through other nodes until becoming terminal nodes inasmuch as they consist of cases that are particularly close (within a distance threshold). In the *second step* the sub-clusters produced in the pre-clusters are further classified. In this second stage, given the modest dimension, traditional methods of clustering may prove effective.

The *TwoStep* considers the optimal partition through the use of the Bayesian Information Criterion (BIC) that for K cluster is defined as:

$$BIC_K = -2l_k + r_k \log n \qquad (1)$$

where r_k is the number of independent parameters, and:

$$l_k = \sum_{v=1}^{k} \xi_v \qquad (2)$$

is the function of *log-likelihood*, for the step with k clusters, which can be interpreted as the dispersion within the cluster. This, furthermore, represents the entropy within the k cluster in the case in which they are considered the only categorical variables (see analysis of ξ_v below).

The *TwoStep* joins the cluster at each iteration until all clusters are incorporated into one and, unlike hierarchical aggregative techniques, uses a statistical model. The model assumes that the continuous variables x_j (j=1,2,...,p) are within the i-th cluster distributed independently and normally with μ_{ij} mean and $\hat{\sigma}_{ij}^2$ variance, assuming that the categorical variables a_j are within the i-th independent clusters and multinomially distributed with probability π_{ijl}, where (jl) indicates the l-th category (l=1,2,...,) of the mutable a_j (j=1,2,..., p).

As in the hierarchical model, clusters with the shortest distance $d(i,s)$ are grouped at each step [11].

4.2 Variables and Factors Identified

The following variables were used in order to produce profiles of consumers surveyed: type of customer ("loyal" IKEA customer, "occasional" IKEA customer, non-IKEA customer), age groups, gender, educational level, occupation, place of residence, IKEA products purchased, motivation for purchase, motivation for non-purchase, preferred style.

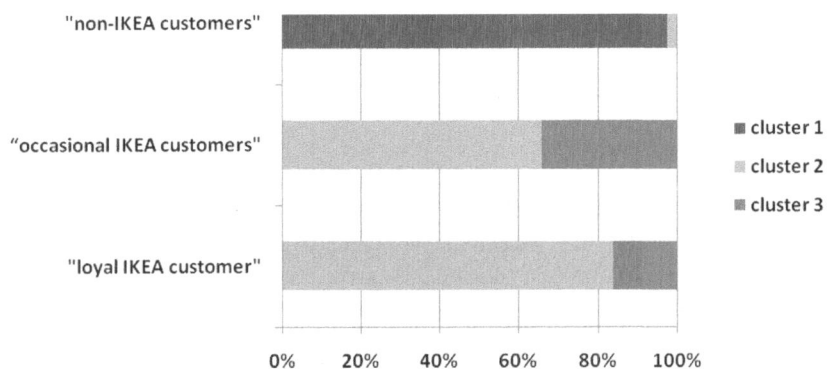

Fig. 11. Description of clusters based on consumer type (Source: our elaborations on field survey)

The *TwoStep* algorithm automatically determined an optimal number equal to 2 clusters related to IKEA customers and non-IKEA customers; at a successive stage the number of pre-defined clusters was determined as 3. Analysis was carried out, therefore, based on the profiles of the 3 clusters according to the different characteristics of each, illustrating those considered as most relevant to the present work.

In particular, with regards to the *type of consumer* it should be noted that (Fig.11):

- Cluster 1 is mainly characterized by "non-IKEA customers";
- Cluster 3 is characterized by "loyal IKEA customers";
- Cluster 2 is characterized by both "occasional" and "loyal" IKEA customers.

In terms of products purchased by IKEA customers, it is evident that those belonging to cluster 3 tend to purchase IKEA furnishings rather than furniture, as is the case for those belonging to cluster 2. Conversely, those belonging to cluster 1 were obviously unable to respond as "non-IKEA customers" (Fig.12).

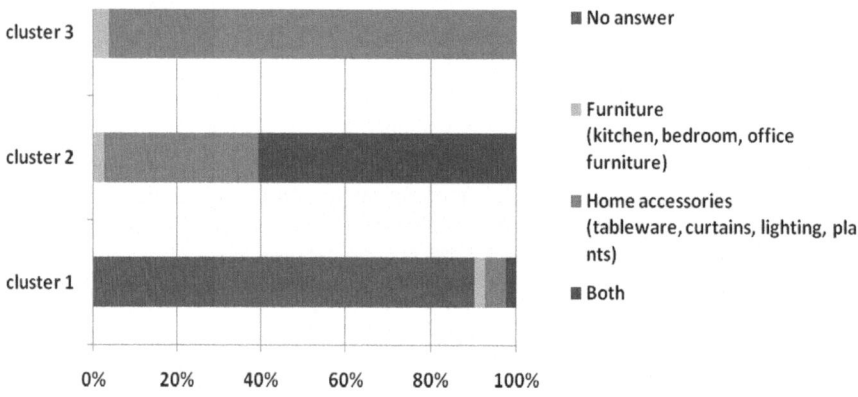

Fig. 12. Description of clusters based on type of products purchased at IKEA (Source: our elaborations on field survey)

As regards to the preferred *IKEA style*, it is evident that those belonging to clusters 2 and 3 tend to prefer the "Scandinavian modern style", "Swedish country" or "Rustic", while those of cluster 1 (non clients) prefer, based on products viewed in the catalogue, the "Popular modern classic" style (Fig.13).

The final aspect analysed regards the origin of respondents (place of residence) and their placement in different clusters. It is evident, and of particular interest, to highlight that those belonging to cluster 1 are also those residing furthest away from the IKEA store (Bari) or, specifically, within the municipalities of Lesina, Manfredonia, Margherita di Savoia, Trinitapoli and Ruvo di Puglia (Fig.14).

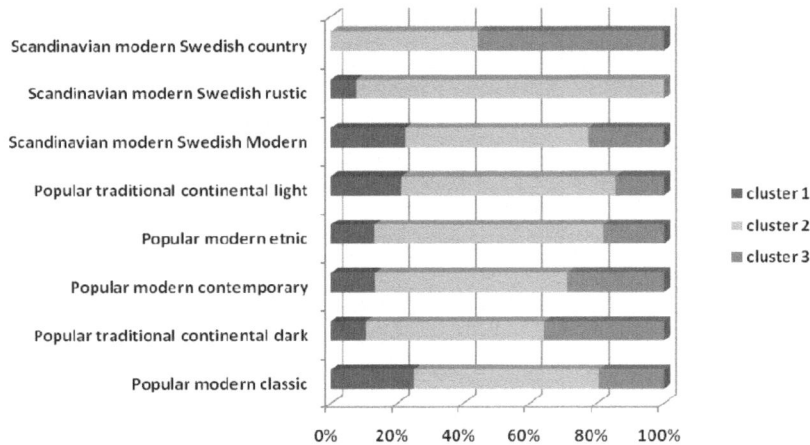

Fig. 13. Description of clusters based on preferred IKEA style (Source: our elaborations on field survey)

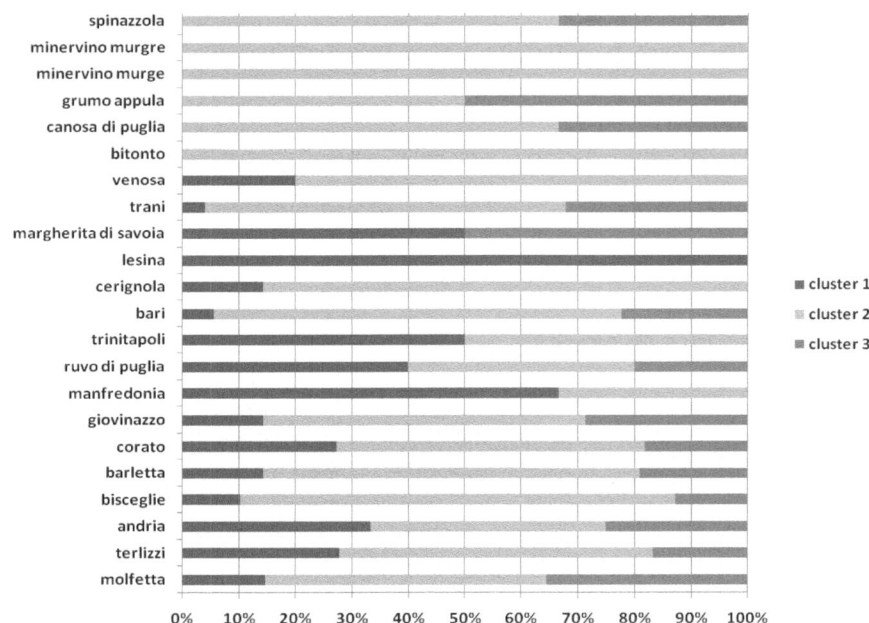

Fig. 14. Description of clusters on the basis of the municipality of residence of respondents (Source: our elaborations on field survey)

This finding is of particular interest as it supports the hypothesis that the main factor determining classification as non- customers is the "distance" of the store IKEA of Bari from places of residence (Fig. 15).

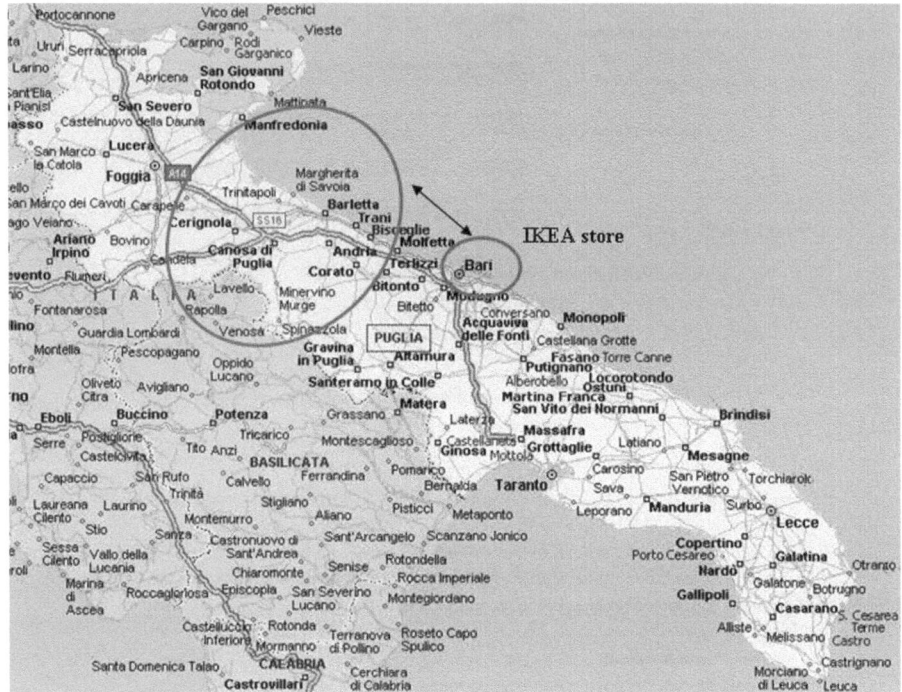

Fig. 15. Map of municipalities of Primary Market Area far from IKEA Store of Bari

5 Concluding Remarks

For differentiated products, consumers formulate a question based on a characteristic of the product, rather than the product itself. This may be defined based on a finite set of features, each of which meets a need of the consumer in an objective way. It therefore follows that indifference curves are not so much intended as an expression of preferences in the spacing of products but, rather, due to their characteristics and, therefore, each consumer is placed at a precise point in the spacing of the characteristics identified.

It is clear that should differentiation be vertical, the consumer prefers the product demonstrating higher quality, while in the horizontal, in which no one product is objectively superior to another, distance plays a crucial role in spacing. Indeed, it is possible that the consumer prefers to pay a higher price for a certain product in a store in the vicinity, rather than enjoying a lower cost in a more distant store [12].

Distance represents a disutility to consumers, which provides stores with market power in the case of purchases made in a spatially distant location, imposing higher transportation costs to the consumer than the price difference imposed by companies. Consumers also choose a store on the basis of "transportation costs" that they must sustain in order to reach the point of sale; indeed, such costs can counteract any positive effect that small reductions in price of the products on offer may bring. Products become effectively interchangeable in the absence of transport costs.

Indeed, *customer switching* is more easily realized should customers not be forced to bear the significant costs of transport; therefore strategy may be adapted according to the principle of "reducing geographical distance": transport cost promotions, the refunding of fuel costs incurred, or simply further promotion of value for money. The latter aspect could effectively justify travel costs. The final consideration could effectively justify transport costs. Indeed, searching for higher perceived product quality means increasing *value for money*, or increasing the level of utility derivable from the purchase and the amount of money spent. Indeed, a good quality/price ratio is based not only on the economy of expenditure but, rather, the maximum effectiveness and efficiency resulting from the purchase.

References

1. Chandler, A.D.: Strategia e struttura: storia della grande impresa Americana. F. Angeli, Milano (1976)
2. Andrew, K.R.: The Concept of Corporate Strategy. Irwin, Homewood (1971)
3. Ferrara, G.: Pianificazione strategica. In: Caselli, L. (ed.) Le parole dell'impresa. F. Angeli, Milano (1995)
4. Usai, G.: Le organizzazioni nella complessità. Lineamenti di teoria dell'organizzazione. CEDAM, Padova (2002)
5. Sraffa, P.: The Laws of Returns under Competitive Conditions. The Economic Journal 36(144), 535–550 (1926)
6. Chamberlin, E.H.: Theory of Monopolistic Competition. Harvard University Press, Cambridge (1933)
7. Sutton, J.: Sunk Costs and Market Structure. The MIT Press (1991)
8. Martucci, I., Schirone, D.A., Servodio, G.: Globalizzazione e strategia locale: l'esperienza IKEA in Puglia. In: Atti online del Convegno A.I.S.R.E., Torino. F. Angeli, Milano (2011)
9. Banfield, J.D., Raftery, A.E.: Model-based Gaussian and non-Gaussian clustering. Biometrics 49, 803–821 (1993)
10. Zhang, T., Ramakrishnon, R., Livny, M.: BIRCH: An Efficient Data Clustering Method for Very Large Databases. In: Proceedings of the ACM SIGMOD Conference on Management of Data, Montreal, Canada, pp. 103–114 (1996)
11. Montrone, S., Massari, A., Perchinunno, P., Altamura, L.: Un approccio integrato per l'analisi della povertà attraverso il matching redditi-consumi. In: Annali del Dipartimento di Scienze Statistiche "Carlo Cecchi", Facoltà di Economia, Università degli Studi di Bari. Tomo I, vol. 6, pp. 171–228. Cacucci Editore, Bari (2007)
12. Hotelling, H.: Stability in Competition. Economic Journal 39(153), 41–57 (1929)

Clustering for the Localization of Degraded Urban Areas[*]

Silvestro Montrone and Paola Perchinunno

Department of Statistical Science, University of Bari,
Via C. Rosalba 53, 70100 Bari, Italy
{s.montrone,p.perchinunno}@dss.uniba.it

Abstract. The presence of a varied range of definitions on the theme of poverty underlines the necessity of no longer relying on a single indicator yet rather on a group of indicators, useful in the definition of living conditions of various subjects. The starting point for this approach derives from the necessity of identifying, on the basis of statistical data, geographical zones of urban poverty, in the specific case of the area of the city of Bari. Rapid developments in the availability and access to spatially referenced information in a variety of areas have induced the need for better analysis techniques in order to understand the various phenomena. In particular, spatial clustering algorithms, which groups similar spatial objects into classes, can be used for the identification of areas sharing common characteristics. The aim of this paper is to present a density based algorithm for the discovery of clusters of units in large spatial data sets. These approaches have been improved using the SaTScan and DBSCAN or Seg-DBSCAN methodology. Further developments concern the application of a DENCLUE "weighed", obtained using the intensity of a phenomenon instead of the density kernel.

Keywords: urban poverty, SatScan, Seg-DBSCAN, fuzzy logic.

1 Introduction

Our work is prompted by the need to identify territorial areas and/or population subgroups characterized by situations of hardship or strong social exclusion through a fuzzy approach that allows the definition of a measure of the degree of belonging to the disadvantaged group.

Since the end of the 1970s, numerous studies have been based on a variety of approaches, each of which adopting an attentive definition and conceptualization of the phenomena. The concept of urban poverty is well expressed by Whelan [1], who hypothesized the existence of three factors of privation in the social field: primary (alimentation and clothing), secondary (holidays and leisure time) and tertiary (living conditions).

---

[*] The contribution is the result of joint reflections by the authors, with the following contributions attributed to S. Montrone (chapters 4 and 5) and to P. Perchinunno (chapters 1, 2 and 3).

The presence of a varied range of definitions on the theme of poverty underlines the necessity of no longer relying on a single indicator, but on a group of indicators which are useful in the definition of living conditions of various subjects (*multidimensional approach*) [2]. The different scientific research pathways are, as a consequence, directed towards the creation of multidimensional indicators, sometimes going beyond dichotomized logic in order to move towards a classification which is "fuzzy" in nature, in which every unit belongs to the category of poor with a degree in [0,1], where the value 1 means fully poor, 0 means not poor at the all, with the others values reflecting different levels of poverty. With the aim of going beyond the limits of the traditional approach it is necessary to enlarge the analysis to include a wide range of indicators on living conditions and, at the same time, adopt mathematical methods which adequately allow for the complexities and vague nature of poverty. A multidimensional index, which considers poverty as the overall condition of withdrawal and deprivation, would seem to be a more appropriate tool for a differential socio-economic analysis of demographic phenomena. The approach chosen is the so-called "Total Fuzzy and Relative"[3].

Grouping methods for territorial units are employed for areas with high (or low) intensity of the phenomenon by using clustering methods that permit the aggregation of spatial units that are both contiguous and homogeneous with respect to the phenomenon under study. The aim of this paper is to group territorial units in areas of high intensity, using SaTScan [4] and DBSCAN [5] clustering methods to aggregate adjacent spatial units that are homogeneous with respect to the phenomenon being studied. SaTScan scans the region of interest with a moving window and compares a smoothing of the intensity inside and outside it so that units belonging to contiguous windows with similar intensity are aggregated into a cluster. On the other hand, *Seg-DBSCAN*, a new version of DBSCAN, limits the arbitrariness of the choice of input parameters and identifies clusters as dense regions in space. As an application we analyze geo-referenced data concerning housing problems in the city of Bari (Puglia, Italy) and we propose a comparison between the two methods presented.

2 Methodological Aspects for Geographical Clustering

2.1 The SATScan Model

In order to examine the possibility of applying such methods to regeneration programs it is necessary to introduce a physical reference to urban spaces. In the field of epidemiological studies a range of research groups have developed different typologies of software; these are all based on the same approach, but usually differ from each other in terms of the shape of the window.

Among the various methods of zoning are SaTScan [4] that uses a circular window, FlexScan [6], the Upper Level Scan Statistics [7] and AMOEBA [8].

SaTScan scans the region of interest with a moving window and compares a smoothing of the intensity inside and outside it: units belonging to contiguous windows with similar intensity are aggregated into a cluster. Multiple different window sizes are used. The window with the maximum likelihood is the most likely cluster, that is, the cluster least likely to be due to chance. A *p*-value is assigned to this cluster.

In the following description of the SatScan method, it is assumed that the region under investigation can be divided into sub-areas with no common points, along with the existence of precisely one subset Z (constituted by the union of one or more areas) and two independent Poisson processes defined on Z and Z^c, respectively indicated with X_z and X_z^c whose intensity functions are:

$$\lambda_Z(x) = p\mu(x) \text{ and } \lambda_{Z^c}(x) = q\mu(x) \tag{1}$$

where *p* and *q* describe the individual probability of occurrence, respectively inside and outside the Z zone.

The intensity function of the "background" has a significant digit that varies depending on the particular application considered: for example, in epidemiological investigations, it models the spatial distribution of the population at risk.

The null hypothesis $H_0 : p = q$ that the probability of occurrence within the area considered is not higher than outside, is resolved by the use of the following ratio of likelihoods:

$$\Lambda_Z = \frac{\max_{p>q} L(Z, p, q)}{\max_{p=q} L(Z, p, q)} = \frac{\left(\frac{y_Z}{\mu(Z)}\right)^{y_Z} \left(\frac{y_G - y_Z}{\mu(G) - \mu(Z)}\right)^{y_G - y_Z}}{\left(\frac{y_G}{\mu(G)}\right)^{y_G}} \tag{2}$$

where y_Z and y_G respectively represent the number of events observed within the Z zone and the entire region under study, while $\mu(Z)$ and $\mu(G)$ are usually approximated by the consistency of the population "at risk" respectively within Z and the whole region under investigation.

The advantage of proceeding according to this method is that the most probable cluster is detected by the highest value of the likelihood ratio seen as a function of Z,

$$\Lambda = \max_Z \Lambda_Z \tag{3}$$

where Z is a suitable collection of subset of G, or at least a collection of putative spatial clusters [9].

Moreover, once the statistical significance of the cluster area defined on the Z zone that maximises the likelihood ratio has been evaluated, other secondary clusters can be significant: generally only clusters that do not overlap with the main cluster are considered [10].

In this case, SaTScan operates by locating a circular window of arbitrary radius, and calculating the probability of urban poverty, inside the circle, or the probability of urban poverty, outside the circle, and consequently by optimizing the dimension of the radius [11].

2.2 The DBSCAN and Seg- DBSCAN Models

DBSCAN (Density Based Spatial Clustering of Application with Noise) was the first density-based spatial clustering method proposed [5]. The key idea is that to define a new cluster or extend an existing cluster, a neighborhood around a point of a given radius ε must contain at least a minimum number of points *MinPts*, i.e. the density in the neighborhood is determined by the choice of a distance function for two points *p* and *q* denoted by *dist(p,q)*.

The greatest advantages of DBSCAN are that it can follow the shape of the clusters and that it requires only one distance function and two input parameters. Their choice is crucial because they determine whether a group is a cluster of points or a simple noise.

In order to limit the arbitrariness of the choice of a value to assign to ε, usually detected by a heuristic procedure, in this work we develop a new algorithm: *Segmented* DBSCAN (*Seg-DBSCAN*), a modified version of DBSCAN, in which the clusters are aggregated considering multiple levels of value of ε. Therefore, to define levels of ε, a value of *MinPts* is fixed and we analyze the distribution of the maximum radius of the cores that are groups formed by *MinPts* points. Then, we build a histogram of this distribution and we choose ε where there are the histogram peaks that indicate a proximity of the cores of a cluster.

As suggested in literature, we can fix the value of *MinPts* to 4, and a number of levels of ε equal to the number of the highest histogram peaks.

The final phase of the algorithm is to merge the clusters obtained. The merging of two clusters C_1 and C_2 characterized by different levels of density ε_1 and ε_2 is obtained if:

$$d(C_1, C_2) < \max(\varepsilon_1; \varepsilon_2). \qquad (4)$$

With this new algorithm, parameter ε is no longer established *a priori*.

The function that links in these terms two points A and B of coordinates $A(x_A, y_A, w_A)$ and $A(x_B, y_B, w_B)$ respectively, with $0 < \{w_A, w_B\} < 1$, is a weighted distance that is obtained by dividing the Euclidean distance by a mean of order integer $t > 0$:

$$d_{pesata}(A, B) = \frac{\sqrt{(x_A - x_B)^2 + (y_A - y_B)^2}}{\sqrt[t]{\left(\frac{w_A^{-t} + w_B^{-t}}{2}\right)^{-1}}} \qquad (5)$$

Observe that in this distance the triangle inequality does not hold, so it is a semi metric, but this restriction does not affect the definitions of density-reachability and density connectivity necessary for DBSCAN algorithm [5].

With this function the distance increases in matching pairs of points with low intensity value, so that they are penalized in the formation of clusters. Empirically it was verified that the most appropriate value of t is 5.

2.3 Future Developments: The DENCLUE Model

Other density based clustering methods are Optics [12] and another algorithm to clustering in large multimedia databases, called DENCLUE (DENsity based

CLUstEring). The basic idea of this approach is to model the overall point density analytically as the sum of influence functions of the data points.

The DENCLUE algorithm employs a cluster model based on kernel density estimation. The DENCLUE framework for clustering [13], [14] builds upon Schnells algorithm. There, clusters are defined by local maxima of the density estimate. Data points are assigned to local maxima by hill climbing. Those points which are assigned to the same local maximum are put into a single cluster. a simple equation based on the overall density function. The advantages of this approach are:

1. it has a strong statistical basis,
2. it has good clustering properties in data sets with large amounts of noise,
3. it allows a compact mathematical description of arbitrarily shaped clusters in high-dimensional data sets
4. it is significantly faster than existing algorithms.

A disadvantage of Denclue is, that the used hill climbing may make unnecessary small steps in the beginning and never converges exactly to the maximum, it just comes close. Nevertheless a comparison with DBSCAN shows the superiority of this new approach. Further developments concern the application of a DENCLUE "weighed", obtained using the intensity of a phenomenon instead of the density kernel.

3 Identifying Indicators of Housing Difficulty

3.1 Introduction

The subject of the case of study arises from the necessity to identify geographical areas characterized by situations of residential deprivation or urban poverty in the city of Bari (South of Italy). With the aim of analyzing the phenomena of residential poverty on a geographical basis, the work makes use of the data deriving from the most recent Population and Housing Census 2001 carried out by ISTAT; such information allows the geographical analysis in sections according to the census, albeit disadvantaged by the lack of the most recent data. The geographical units of the survey are the 1,421 census sections for the city of Bari (of which 1,312 sections represent all of the data relevant to housing, while the remaining sections are either uninhabitable areas or destined for other uses, for example parks or universities).

With regards to the choice of housing poverty indices there is, therefore, a consideration of various aspects associated to housing conditions along with the quality of housing. The indices were chosen with the aim of identifying the level of residential poverty and were calculated in order to align elevated levels on the indices with elevated levels of poverty.

Connected to the phenomena of residential poverty is the evaluation of the *classification of housing status* (in rented accommodation or homeownership)[1]. In particular, it is evident that homeownership is an indicator inversely correlated to poverty. A further measure is the *index of overcrowding*: the ratio between the total

---

[1] **Index 1** - Incidence of the number of dwellings occupied by rent-payers: ratio between the number of dwellings occupied by rent-paying residents and the total number of residents.

number of residents and size of dwellings occupied by residents[2]. Finally, aspects of residential poverty associated with the *availability of functional services* are included in the analysis, including goods of a certain durability destined for communal use such as the availability of a *landline telephone*; *heating systems* and *residential parking space*[3].

3.2 The Total Fuzzy and Relative Method

At the city level, policies can be addressed by investigating condition of unavailability of housing services and quality, to understand better some peculiar aspect of distribution of housing difficulty.

The approach chosen in order to arrive at the synthesis and measurement of the incidence of relative poverty in the population in question is the so-called *Total Fuzzy and Relative*, "which utilizes the techniques of the *Fuzzy Set* in order to obtain a measurement of the level of relative poverty within a population, beginning from statistical information gathered from a plurality of indicators" [15].

The TFR approach consists in the definition of the measurement of a degree of membership of an individual to the fuzzy totality of the poor, included in the interval between 0 (not poor) and 1 (poor). Mathematically, such a method consists of the construction of a function of membership to "the fuzzy totality of the poor" which is continuous in nature, and "able to provide a measurement of the degree of poverty present within each unit". Supposing the observation of k indicators of poverty for every family, the function of membership of i-th family to the fuzzy subset of the poor may be defined thus:

$$f(x_{i.}) = \frac{\sum_{j=1}^{k} g(x_{ij}).w_j}{\sum_{j=1}^{k} w_j} \quad i = 1,.....,n \quad (6)$$

For the definition of the function $g(x_{ij})$ please refer to other works [3], [15].

3.3 The TFR Application

The application of the TFR (*Total Fuzzy and Relative*) method begins from the presupposition of synthesizing the five indices elaborated in "*fuzzy*" values which are

---

[2] **Index 2** – Index of overcrowding: the ratio between the total number of residents and size of dwellings occupied by residents.
[3] **Index 3** - Incidence of the number of dwellings lacking a landline telephone: ratio between the number of dwellings occupied by residents without a landline telephone and the total number of dwellings occupied by residents.
Index 4 - Incidence of the number of dwellings lacking heating system: ratio between the number of dwellings occupied by residents without a heating system and the total number of dwellings occupied by residents.
Index 5 - Incidence of the number of dwellings residential lacking parking space: ratio between the number of dwellings occupied by residents without a parking space and the total number of dwellings occupied by residents.

able to measure the degree of membership of an individual to the totality of the poor, included in the interval between 0 (with an individual not demonstrating clear membership to the totality of the poor) and 1 (with an individual demonstrating clear membership to the totality of the poor). The data arising from various census sections are classified into 4 different typologies of poverty in accordance with the resulting fuzzy value: non-poor, slightly poor, almost poor and unquestionably poor [16], [17].

According to the set of indicators considered, a differing division of the total 1,322 census sections for conditions of housing poverty is produced. In relation to the set of indicators of housing difficulty, 6.6% of the census sections (86 sections) demonstrated present fuzzy values representative of *unquestionable poverty*, in comparison to 54.9% (722 sections) belonging to the fuzzy totality of *non-poor* (Fig.1).

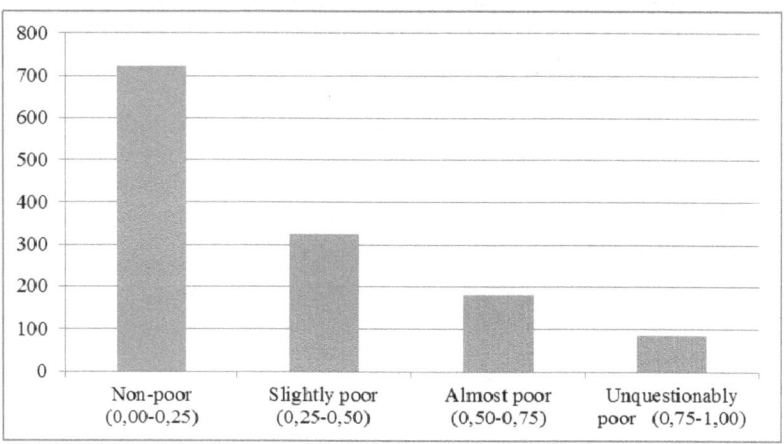

Fig. 1. Composition of absolute values of the census sections of Bari for conditions of housing poverty

Analysing the location of the census sections by "neighbourhood" in terms of percentage composition in the different categories under examination, it is possible to identify areas with lower levels of residential hardship, such as Murat (with 80% of the census section resulting as non-poor), Japigia (with 78%) and Picone (with 75%). The neighbourhoods in which a percentage of sections with situations of severe housing problems are detected are, however, San Nicola, the historical city centre (with 55% resulting disadvantaged) and Libertà, an urban area adjacent to the city centre (with 23% resulting as disadvantaged). The situation of neighbourhoods such as Maddonnella is also critical, presenting percentages of census sections resulting either *Unquestionably poor* or *Almost poor* of more than 35% of the total, while Carbonara and San Girolamo-Fesca present a percentage equal to approximately 25% (Fig.2).

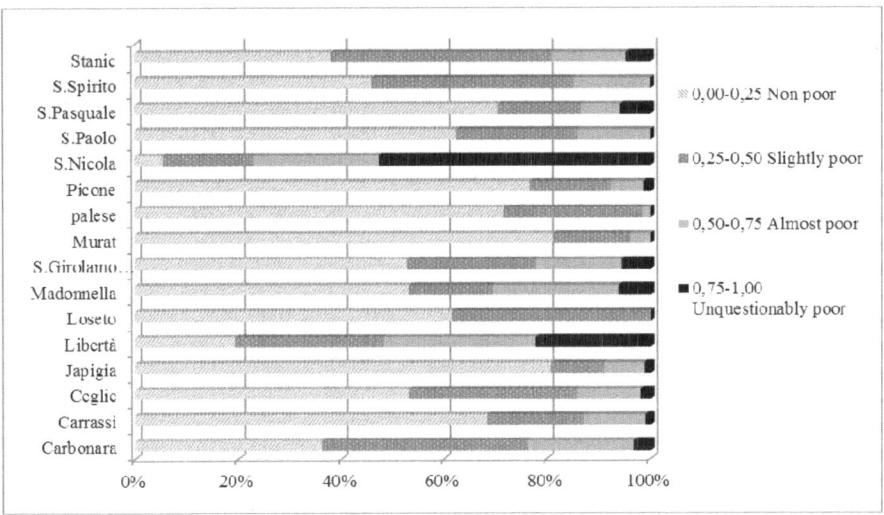

Fig. 2. Composition of percentage values of the census sections of Bari for housing poverty in 2001

4 Identification of Cluster of Degraded Areas in the City of Bari

As reported above, the core question of this paper derives from the need to identify hot spots presenting conditions of housing difficulty in the urban area of Bari. With the use of the SaTScan method [4] and DBSCAN method [5] a possible identification of hot spots of urban poverty has been obtained from the data generated by a fuzzy analysis starting from a set of indicators.

SaTScan operates locating a window of arbitrary radius, and calculating the probability of poverty (risk) p_1, mean inside, or the probability of poverty (risk) p_2, mean outside. The minimum *p-value* corresponds to the most important cluster.

Through the use of the *SaTScan model*, seven clusters of housing demonstrating hardship may be identified, consisting of a number of different sections of a census, totaling 311 sections (Fig. 3). The p-value demonstrates the presence of *four main clusters*. The level of housing hardship identified and defined by the internal average (Table 1).

Through the use of the *Seg*-DBSCAN model, twelve clusters of housing demonstrating hardship may be identified, consisting of a number of different sections of a census, totaling 388 sections. The *clusters* 1 (San Nicola), 2 (Libertà), 5 (Madonnella) and 10 (Carbonara, Ceglie and Loseto) identify the same areas found by SatScan. The major differences concern the number of sections involved (Fig. 4).

Table 1. Description of statistical measures of the SatScan clusters of housing difficulty

Cluster	Number of sections	Internal mean	p-value	Variance	C.V. index[4]
1	67	0.678	0.001	0.067	38.24
2	114	0.474	0.001	0.064	53.43
3	21	0.500	0.026	0.081	57.01
4	84	0.406	0.039	0.027	40.81
5	8	0.576	0.438	0.081	49.25
6	10	0.582	0.670	0.016	21.62
7	7	0.627	0.802	0.021	22.90

Source: Our elaboration on the data from the Population and Housing Census, 2001.

Table 2. Description of statistical measures of the *Seg*-DBSCAN clusters of housing difficulty

Cluster	Number of sections	Mean	Variance	C.V. index
1	72	0.709	0.052	32.07
2	94	0.567	0.048	38.66
3	9	0.523	0.052	43.79
4	15	0.501	0.025	31.35
5	24	0.575	0.050	38.75
6	22	0.514	0.060	47.80
7	5	0.406	0.020	34.40
8	27	0.479	0.032	37.55
9	91	0.435	0.025	36.10
10	14	0.525	0.032	34.29
11	7	0.421	0.030	41.09
12	8	0.419	0.020	33.67

Source: Our elaboration on the data from the Population and Housing Census, 2001.

We observe that SatScan identifies areas formed by contiguous spatial units in which a smoothing of the disadvantaged housing index is performed. This method is effective in identifying areas of high or low intensity and therefore may be a useful indication of areas "at risk" to be monitored. Like the SaTScan method, *Seg*-DBSCAN identifies areas in which the spatial units meet a criterion of adjacency, but *Seg*- DBSCAN can exactly identify sections of the city with housing problems, excluding those areas where the phenomenon is absent. For example, in the case of the central district of Bari, the SaTScan method identifies the whole districts while the

---

[4] Ratio between Standard Deviation and Mean *100.

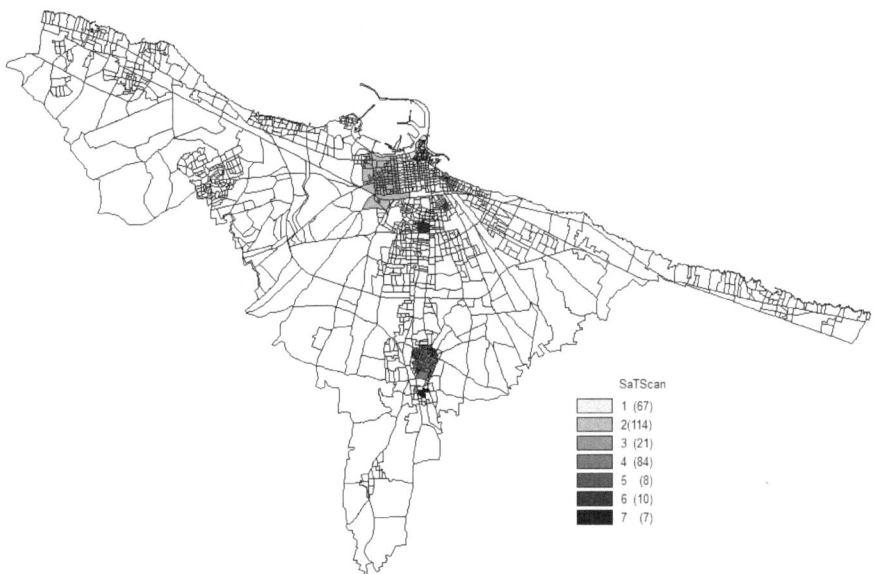

Fig. 3. SaTScan model for the identification of cluster of housing poverty in the cities of Bari

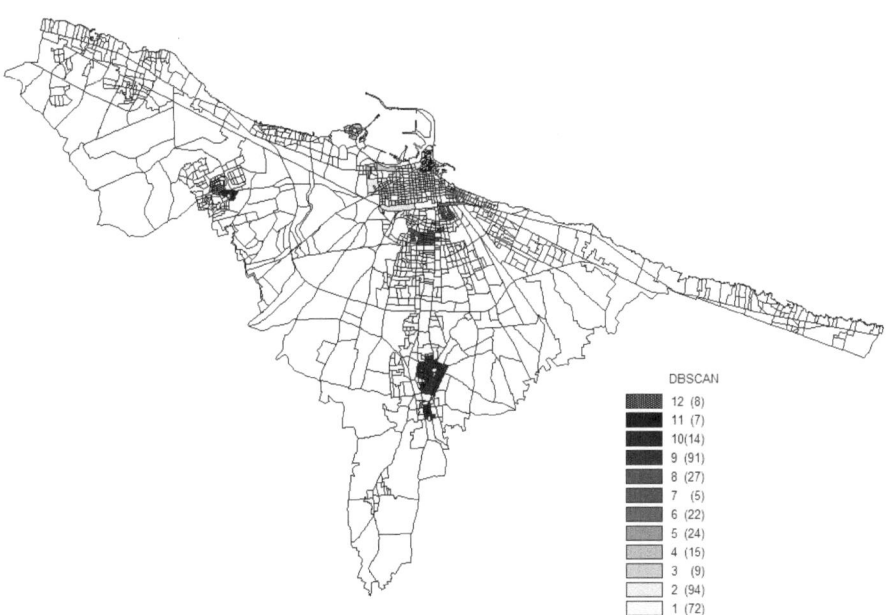

Fig. 4. Seg-DBSCAN model for the identification of cluster of housing poverty in the cities of Bari

Fig. 5. SaTScan model – central district in the city of Bari

Fig. 6. Seg-DBSCAN model – central district in the city of Bari

Seg- DBSCAN method identifies the same area of hardship but also analyzes the area in more detail (Fig. 5 and Fig. 6), excluding the sections in which the phenomenon is not present, since these areas are aimed at schools, gardens, public buildings not intended for habitation. Moreover, while in the model SatScan the identification of risk areas is essentially based on the Monte Carlo method, in the model DBSCAN the identification of cores is heuristically, identifying even small areas that are not indented in the Monte Carlo samples.

5 Concluding Remarks

The proposed methodologies identify areas where there is a high disadvantaged index. Starting from information obtained by the cluster intersection of housing hardship, it could be possible to obtain useful indications for planning urban regeneration policies, making decisional process more transparent and scientifically valid. The preliminary question leading towards the identification of town planning and architectonic solutions to the problem of urban regeneration, in a historical moment characterized by a lack of public resources for investment, focuses on the identification of areas characterized by the highest level of urban poverty in order to direct the choices of political decision-makers in a transparent, thought-out and objective manner. The model used in this study is able to provide, in the opinion of the authors, relevant data for the identification of such areas.

The authors believe that the model used in this study is able to provide relevant data for the identification of such areas.

The present study provides certain considerations for the future. The first stems from the importance of in-depth research based on methods which privilege groups of indicators of a limited number, as demonstrated above. The effectiveness of such a method is to some degree demonstrated by the specific case which can only lead to the wish to widen the investigation. In future studies these indicators could be integrated with further relational elements of, for example, the availability of social services provision by the city council to families in need or the number of requests for support in rent payments for those most at need. This type of data is, however, available at a different level of aggregation from that analyzed here, being available by street and house number of requesting families rather than from ISTAT census sections and, therefore would need to be made homogenous with ISTAT data prior to use. The second consideration regards the possibility of using this model as a form of evaluation "ex post" of the effectiveness of urban policy, to verify the consequences of urban regeneration on areas characterized by high levels of poverty, examining the nature of the variations measured according to the same statistical indicators used, between the Census period.

It is useful to underlining the ease of reading information offered by the SaTScan model compared to DBSCAN model. As we have noted above, a comparison of the two methods shows that the *Seg*- DBSCAN method is more accurate in identifying the spatial units in which there are housing problems. The future advancement of our work will be to seek a cluster validity index for spatial data, which takes into account the noise points, that is valid from a statistical point of view and that allows the accurate measurement of the *Seg-* DBSCAN method.

The applications with SaTScan or *Seg-DBSCAN* methodology in order to identify hot spots of housing hardship raises certain considerations for future research in the social field and for urban planning of regeneration areas, particularly relevant to the European Union policy agenda.

References

1. Whelan, C.T.: The role of income, life-style deprivation and financial strain in mediating the impact of unemployment on psychological distress. Working Paper 20. Economic and Social Research Institute, Dublin (1991)
2. Montrone, S., Campobasso, F., Perchinunno, P., Fanizzi, A.: An Analysis of Poverty in Italy through a Fuzzy Regression Model. In: Murgante, B., Gervasi, O., Iglesias, A., Taniar, D., Apduhan, B.O. (eds.) ICCSA 2011, Part I. LNCS, vol. 6782, pp. 342–355. Springer, Heidelberg (2011)
3. Cheli, B., Lemmi, A.: A "Totally" Fuzzy and Relative Approach to the Multidimensional Analysis of Poverty. Economic Notes 24(1), 115–134 (1995)
4. Kulldorff, M.: SaTScanTM User Guide (August 26, 2006), http://www.satscan.org/
5. Ester, M., Kriegel, H.P., Sander, J., Xiaowei, X.: A Density-Based Algorithm for Discovering Clusters in Large Spatial Databases with Noise. In: Proceeding of the 2nd International Conference on Knowledge Discovery and Data Mining, pp. 94–99 (1996)
6. Takahashi, K., Tango, T.: A flexibly shaped spatial scan statistic for detecting clusters. International Journal of Health Geographics 4, 11–13 (2005)
7. Patil, G.P., Taillie, C.: Upper level set scan statistic for detecting arbitrarily shaped hotspots. Environmental and Ecological Statistics 11, 183–197 (2004)
8. Aldstadt, J., Getis, A.: Using AMOEBA to create spatial weights matrix and identify spatial clusters. Geographical Analysis 38, 327–343 (2006)
9. Bilancia, M., Montrone, S., Perchinunno, P.: A Model-Based Scan Statistics for Detecting Geographical Clustering of Disease. In: Gervasi, O., Taniar, D., Murgante, B., Laganà, A., Mun, Y., Gavrilova, M.L. (eds.) ICCSA 2009, Part I. LNCS, vol. 5592, pp. 353–368. Springer, Heidelberg (2009)
10. Waller, L.A., Gotway, C.A.: Applied Spatial Statistics for Public Health Data. John Wiley & Sons, New York (2004)
11. Montrone, S., Perchinunno, P., Torre, C.M.: Analysis of Positional Aspects in the Variation of Real Estate Values in an Italian Southern Metropolitan Area. In: Taniar, D., Gervasi, O., Murgante, B., Pardede, E., Apduhan, B.O. (eds.) ICCSA 2010. LNCS, vol. 6016, pp. 17–31. Springer, Heidelberg (2010)
12. Ankerst, M., Breunig, M.M., Kriegel, H.P., Sander, J.: Optics: Ordering points to identify the clustering structure. In: Proceedings SIGMOD 1999, pp. 49–60. ACM Press (1999)
13. Hinneburg, A., Keim, D.: An efficient approach to clustering in large multimedia databases with noise. In: Proceedings KDD 1998, pp. 58–65. AAAI Press (1998)
14. Hinneburg, A., Keim, D.: A general approach to clustering in large databases with noise. Knowledge and Information Systems, KAIS 5(4), 387–415 (2003)
15. Cerioli, A., Zani, S.: A Fuzzy Approach to the Measurement of Poverty. In: Dugum, C., Zenga, M. (eds.) Income and Wealth Distribution, Inequality and Poverty, Springer, Berlin (1990)

16. Bilancia, M., Campobasso, F., Fanizzi, A.: The pricing of risky security in a Fuzzy Least Square Regression Model. In: Locarek-junge, Weishs (eds.) Studies in Classification, Data Analysis and Knowledge Organization – Classification as a Tool for Research, pp. 639–646. Springer, Heidelberg (2010)
17. Montrone, S., Bilancia, M., Perchinunno, P., Torre, C.M.: Economic Evaluation and Statistical Methods for Detecting Hot Spots of Social and Housing Difficulties in Urban Policies. In: Gervasi, O., Taniar, D., Murgante, B., Laganà, A., Mun, Y., Gavrilova, M.L. (eds.) ICCSA 2009, Part I. LNCS, vol. 5592, pp. 253–268. Springer, Heidelberg (2009)

A BEP Analysis of Energy Supply for Sustainable Urban Microgrids

Pasquale Balena, Giovanna Mangialardi, and Carmelo Maria Torre

Department of Civil Engineering and Architecture, Polytechnic of Bari,
(Via Orabona 4, 70125 Bari, Italy)
mangialardi.giovanna@libero.it,
{torre,p.balena}@poliba.it,

Abstract. The paper shows the use of GIS and other program to identify the technical and economic feasibility of a self-sufficient community in energy supply/demand, through the integration of ICT, intelligent infrastructure (smart grid) and the economic analysis of energy market constrains in housing property management, at the urban level. Here we report the development of a new methodology for assessing the potential capacity and benefits of installing rooftop photovoltaic systems in an urbanized area. It has been possible to analyze the climatic, morpho-typological and architectural characters of the place, in order to identify the optimal size and shape of rooftop area available for solar energy applications at different scales, looking for a minimum threshold to be discovered via an economic break-even point analysis. Computer simulation is included to predict the potential benefits of urban scale photovoltaic system implementation. G. Mangialardi wrote the first and the second paragraph, P. Balena wrote the third paragraph and C. Torre wrote the fourth and the last paragraph of the paper.

Keywords: Geographic Information Systems, BIPV, Urban Energy Consumption, Microgrids, Distributed Generation, BEP.

1 Introduction

The way to generate energy is moving from a "centralized", traditional supply model, (with a few main production nodes in a network aiming to transport energy to end users), to a management-based "distributed" model, where very often the end users (prosumer) become new energy self-suppliers. The new relationship between production and consumption represents the basis of the metamorphosis that looms in the field of self-generating distributive electricity supply. The transition to this new system seems to be well underway, but there are several barriers before of fully developing its potential. One of these constrain is that the infrastructure and network connections, which have become more and more capable of controlling bi-directional flows (from and to the user) of energy, control a plurality of inlet points, optimize the process of load management in a context of sources (eg renewables) with a sharp discontinuity and low predictability.

To meet these problems it is necessary that the network and the devices for electricity generation "distributed" become more "computerized" and able to talk to each other. A challenge that must be collected on two fronts: from ICT companies who must adjust to a very peculiar that the generation of energy and technology companies in the renewable sector, enabling true of so-called Smart Grid (SG) [1].

It goes from passive systems, where electricity is transported from the place of production to the consumer, to active systems, where the intelligent network manages an excellent two-way flows and discontinuous electrical energy generated on site. The design of the SG is to be an example of land management from the study of the energy of an isolated district to make self-sufficient through the use of renewable sources.

This level of detail defines the energy districts, according to broaden the study on an urban scale.

The rising on the fore of the Kyoto Protocol, (February 16, 2005), is one of the key elements that make local governments deal with problems related to energy saving and using renewable energy sources in order to reduce CO_2 emissions.

Specially interesting by such kind of installations, the use of solar systems in urban centers as a renewable source, that take advantage of solar energy low costs and consequently have more and more competitive without such problems as soil consumption (the technology for the generation of small wind still has high costs and problems of space and noise that make them still not widespread in urban scale) [2].

In the light of such issues, it is proposed an analytical procedure designed to evaluate the potential energy associated with the use of renewable sources within the city; speaking in detail, the procedure, is devoted to provide a tool able to estimate the usable surface for panels on the roofs of buildings and the urban environment with the purpose of capturing solar radiation for the production of electricity.

Local governments should have the tools to assess the actual propensity of the urban fabric to accommodate solar technologies.

To this end, the work will be stretched to the study of the interaction between an SG, in particular of a microgrid (MD), and the urban context, starting from the analysis of the latter by a climatic point of view, morpho-typological, energy and economic.

The ultimate goal is to pursue the principle of sustainability in order to promote low-carbon emissions [3] and to reduce the carbon footprint.

The aim is implemented by adopting an integrate approach that supports the process and moves toward a turning point in energy and land management, taking many resources and opportunities.

2 Estimate of Energy Performance of Settlements

The first aim of this work is to calculate the energetic performance of urban blocks, to refer their capacity to the needs of households, with the aim to proof if it is possible to assume energy systems based on new models of shared supply that uses the integration of renewable energy by a set MG, and gives a solution (as regards a balance between environment and economy) to the current energy problems [4]. In the first phase of the study we have identified the criteria used to define the "energy islands", and energy districts by focusing on simply morphological and territorial aspects. Then we passed to analyze the problem at an urban scale, trying to identify the relationships between microclimatic aspects (as the mean daily global solar radiation on inclined surface per month) and the conformation of the buildings (orientation and exposure of the shell). A general conceptual model of work has been developed in order to identify necessary elements, data and processes for the construction of knowledge bases and interpretative frameworks. Then buildings has been classified according to their real possibility of efficient use of the roofs for a good orientation of photovoltaic panels; finally manufacturability requirements have been considered.

The first classification of the work is at the municipal level, and enabled a overlay maps from which we deduced a second level of analysis, then focusing wich of the areas can be defined appropriate and responsive for good levels of energy self-sufficiency. These areas are better defined islands or district energy and form a MG community.

This process allows us to develop an usable model and at the same an available service for the community fed and continually updated by various inputs such as public and private research centers and the community itself.

A number of information layers was identified. The layers can be classified as follows:
- morphology,
- urban Fabrics,
- design Features,
- orientation and exposure,
- climate,
- socio-economic aspects

For the elaboration of a Spatial Model, GIS tools were used, (ESRI platform ArcGis 9.3), which helped to georeference spatial data base and to build frameworks of required knowledge.

More in detail, as regards the morphological aspects, (important to assess any shadow due to the conformation of the territory), both a DTM (Digital Terrain Model) and a DSM (Digital Surface Model) have been built, to analyze the morphology of the land in the surrounding of urban areas and the heights of the buildings.

The analysis and classification of the urban fabric is essential for the energy division of the island, as it allows to identify the type of tissue more compliant and suitable for inclusion of renewable sources (photovoltaic) and consequently the installation of a microgrid. The installation of a photovoltaic system integrated into the roofs of existing buildings (BIPV) is widely recommended as a sustainable solution that can address both energy consumption and environmental impact of that building. The fabrics that optimize these conditions are knitted regular building density that have extensive heights of buildings are not very different [2]. The extensive building density is also characterized by the presence of dense vegetation that must be properly evaluated for possible problems of shading surfaces capturing. Evergreen tall trees may pose significant obstacles if the heights of the trees are comparable to the heights of buildings.

Another key aspect of the requirements analysis of this work has focused on the structural characteristics of hedges. Have been identified to ensure existing building typologies and case studies generally applicable to any context (flat roof, lean-to roof, gabled, hipped and pitched roof with the head of the pavilion) [5].

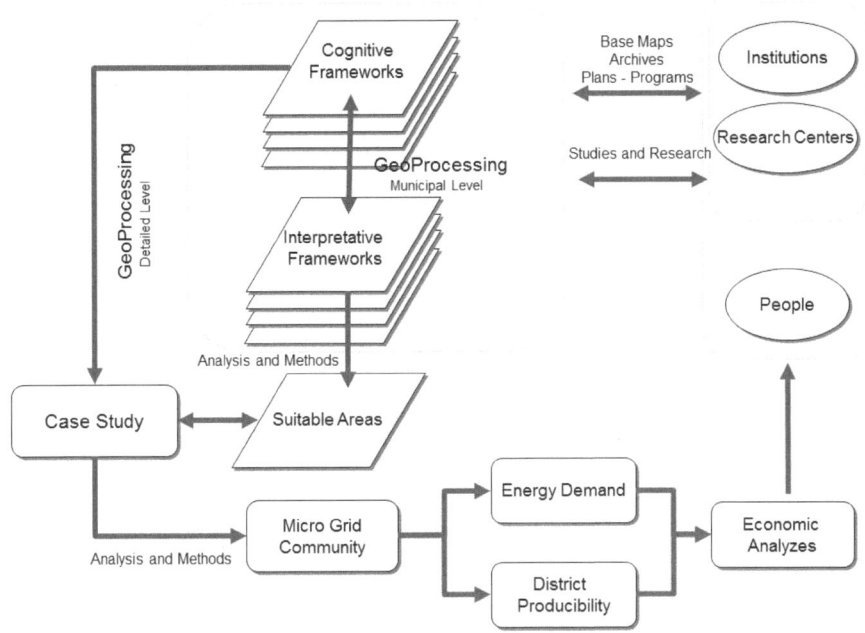

Fig. 1. Conceptual scheme of research

Where buildings present different typological characteristics, and therefore can not be classified in one category, were provided for the separation of the individual geometry of the building in a number equal to the types identified, while leaving an identification code that would allow you to associate the different geometries generated in the same building. The need to proceed with this classification articulated stems from the fact that, as many studies have shown [6], each cover gets a different amount of radiation as a function of location, its geometry, orientation, aspect and gradient of groundwater. A study of all the contextual factors can have a picture of the potential energy more or less accurate than the neighborhood scale, deduced from the model that is proposed, in which all the elements listed above are the inputs and the identification of the potential energy of each single island / district will be the output of the program. The analysis therefore lends itself to being expanded on an urban scale.

The following table, derived from simulations by the program Ecotect, you can see the percentage of radiation that have been allocated according orientation and slope of the pitch (angle of tilt) of fabrics, in the case of polycrystalline silicon photovoltaic coplanar panels at the foot of the building [7]. This study allows to understand how it is optimized uptake of solar energy (South orientation exposure angle of 30°) and what are the percentage reductions in exposure and orientation other than optimal [8]. For flat roofs the localization of the photovoltaic panels can follow the optimal conditions, in contrast to the pitched roofs to the installation of photovoltaic panels (hypothesized integrated to minimize the environmental impact) is bound to the morphology of the roof, and by this table have made the necessary reductions, reporting such data in a paper by a precise weighing of the various conditions.

Fig. 2. Solar showing energy yield relative to the inclination and orientation of the roof

The results of these analyzes have been organized in a geodatabase in which buildings were classified by construction type, height, orientation and exposure compared to the surface that will allow you to assign a score to each building energy.

The sum of the scores grouped by island will serve to establish a ranking of the eligibility of the islands with higher energy potential that makes them self-sufficient.

With Ecotech Analysis (Autodesk platform), instrument for the analysis and sustainable design, we build a 3D model of the block and will be simulated to generate shadows of the buildings on the winter solstice (21st December at 12: 00) when the shadows are in unfavorable conditions). These shadows have also been georeferenced, and entered into the database as an obstacle, and then weighted negatively: the conitione represents when a given fabric creates a shadow on the roof of another building and reduces the potential energy of manufacturability. Also you will calculate the annual energy production of the buildings on the island, while Arcgis has estimated the number of inhabitants and its energy needs in proportion of the volume of buildings.

The manufacturability of each surface will be calculated on the assumption of photovoltaic panels cover with a proportion of the surface which will vary depending on the type of coverage and its exposure and will be stored in the table associated to the buildings. Finally it will be purified of the shadow areas due to the presence of obstacles such as tall buildings or the presence of parapets, chimneys, or vegetation.

3 The Case of Study: Feltre

3.1 Analysis of Spatial Aggregation

The case study regards the application of the above in the previous paragraph in a real situation. For the simulation it was decided to try a small town that had a situation not altogether favorable microclimate (noticeable mainly in southern Italy), and had in perspective a sort of heterogeneous morpho-typological urbanization of municipalities.

The town of Feltre, was chosen in the north of the Italian Veneto region, nestled in a natural basin bounded on the north by the mountain peaks of the Dolomites and on the south by mountains and foothills of the Massif Tomatico Grappain. From the website of the Veneto Region were collected, the Regional Technical Map Numerical and orthophotos of the city, updated to 2005, on which it was possible to perform most of the processing required to the case study. Using ESRI ArcGIS 9.3 of companies has allowed the processing described in subsequent periods. In Figure 3 it is possible to observe processing urban fabric and the heights of the buildings, carried out elaborations a municipal scale

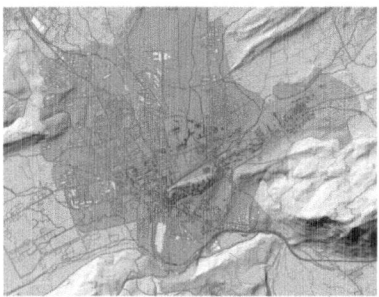
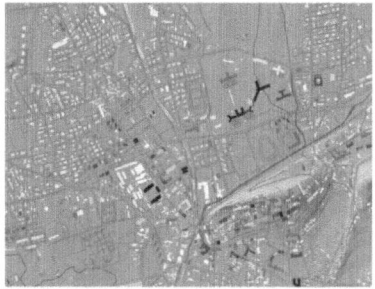

Fig. 3. In gray scales can be identified built-up areas by type of tissue (on the left). Heights: In gray scales can be identified built-up areas divided by height (on the right)

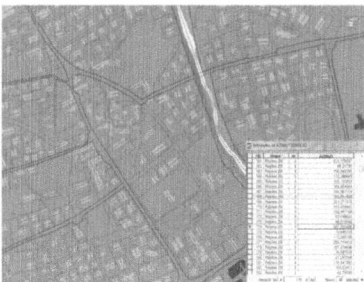

Fig. 4. Orientations: the lines drawn on buildings identify the main orientations (azimuth) of each building (on the left). In gray scales identifies a classification of buildings in relation to their orientation (on the right)

In particular, in the table of the tissues was detected first mesh labeled "historical compact city" which has been excluded from the potential energy islands for environmental reasons and regulatory standards. While they have been taken into account the other three types of mesh: "Intricate Fabric in Medium Grain", "Regular Knit Fabric", "Pieces of Cloth" and "Industrial and Public Buildings". Of such categories, the "Knit Fabric Set" is even more appropriate from an infrastructural point of view (topology of the grid) to organize a micro-grid. The table the heights confirms that the areas identified as being "Knit Fabric Set" have more homogeneous heights and consequently show less mutual shading problems for the installation of photovoltaic panels. Strong variations in height between the buildings produce many problems of obscuration of low buildings respect to the higher ones, going to significantly impact on relevant areas suitable for paneling. The result of this analysis has provided the opportunity to delimit the potential energy islands trying to meet the main roads and the consistency of the mesh and the heights of buildings. In Figure 4 one can observe the development of guidelines and their classification according to their best position in relation to solar radiation as well as in the previous section. of this kind five cards Have been produced one for each type of coverage.

The creation of a devoted software has enabled us to assign a score on the capacity of the same building to catch solar radiation and to what extent, by analyzing the relative azimuth for each building, exposure and construction type of roof (Fig. 5).

In order to rank among the isolates of eligibility has been defined an index:

$$Ind\_El = \frac{\sum_{i=1}^{n}(score)}{\sum_{i=1}^{n}(area\_)}$$

n= number of buildings for island
score = score
area= area of the building

dark gray areas more eligible

Fig. 5. Classification of energy districts and Index of eligibility of energy districts

Then it has gone to analyze at a more detailed scale a single enegetic island to quantify the energy capability of the residential buildings and to compare it with the demand of local residents. The tested Island (selected among the most suitable areas listed previously by the criteria) has a heterogeneity of situations, namely is composed of various types of coverage listed, presents both residential and public buildings (schools). First, the energy capability of residential areas was analyzed.

Reference frame in order to calculate the number of inhabitants became the standard for urban and residential Ministerial Decree 1444/68 that identify a building volume of 100 cubic meters per capita. It has been calculated the volume built up in the island equal to 95787.4 cubic meters (area and heights of the buildings were derived from it by using ArcGis 9.3) and consequently it is obtained the number of population equal to 958 by dividing the volume calculated per 100mc/ab. To estimate of the electricity requirements of residential buildings reference was made to the total annual consumption for a family dwellings on average 3.62 people (source: www.enea.it) which is equal to 2,996 kWh / year. The annual requirement of residential building energy analysis proceeds in the district amounted to 1,216,150.2 kWh / year.

Recourse is made to calculate the energy capability to the software Ecotect: a specific software function that allows to calculate with high accuracy the producibility of a surface in a specified period. In order to conduct the experiment a 3D digital model of the buildings on the island was made, drawing shapes and heights accurately and types of roofing construction, with special attention to the shape and steepness of slopes. A first draft has allowed to calculate the worst shadows and precisely calculated on the 21st of December at 12:00 and to ascertain with certainty that only one building was damaged by the shadow of the neighboring building. Figure (6) you can see the 3D model of the block or district energy in first processing by the program Ecotect that enabled to assess the daily and yearly path of the sun and the phenomena of mutual shading between buildings in unfavorable conditions.

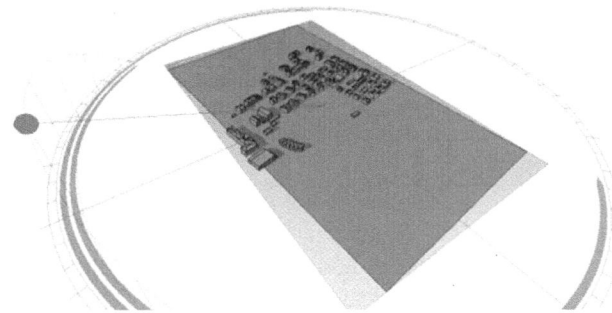

Fig. 6. Study of the shading of the buildings of energy district through Ecotect Software

It has been then compiled a field of the table referring to the buildings and indicating the piece of the roof that could be available for photovoltaic modules. Then we passed to estimate the yearly productivity of all surfaces. The input parameters required by the software were: the climate in the place (Venice - Italy, Lat: 45.4 ° and Long: 12.3 °), the performance of photovoltaic panels (electrical Efficiency: 12%) and material-type photovoltaic panels (poly-silicon), the refractive index of cover and visible transmittance.

We have selected the flat roofs of residential buildings and optimal slopes according to exposure and orientation (according to an adjusted percentage, due to the obstacles listed above). The total surface for paneling was approximately 8,487 square meters.

The software has consequently computed the annual energy supply amounting to 1,059,379kWh / year, that s to say the 87% of the earlier estimated demand of electricity. Below, Fig. 7 shows the evolution of energy production over the year that is optimal in the summer months (higher values per year in the months of June and July and August).

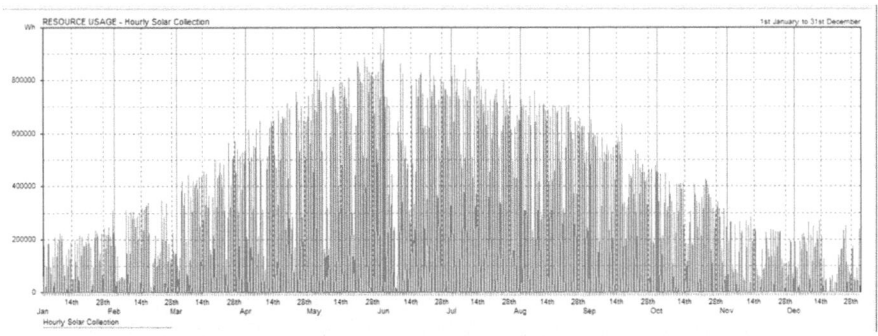

Fig. 7. Producibility of energy district - Residential Buildings

The same procedure was performed for public buildings inside the district: two schools and a public community building. In this case the manufacturability was estimated on very large surfaces, including a 100%-exploitable flat roof, for a total of paneled surface equal to 2.458 sq m. Manufacturability obtained from the software by entering the parameters listed above is equal to 306.630 kWh/year, that can both meet the needs of the structures that bear the capturing surfaces and put in the network the whole energy-surplus, meeting the needs of the "island" itself (public lighting, electric mobility possible, or to request more partly residential buildings) according to the logic core of smart grids.

4   Break even Point Analysis

The break even point analysis is aimed at identifying a break even between cost per unit of plant and surplus of energy production. The energy production has a rule of variation which takes into account the difference of price of the surplus created the increase of the plants belonging to the island of energy efficiency less the increase of the cost of mortgage (that grows according to the increase of the elements up the micro-grid). This difference must equalize in less than life-cycle cost of the plant the energy price standardized, which grows with increasing energy demand, and then with increasing mesh size.

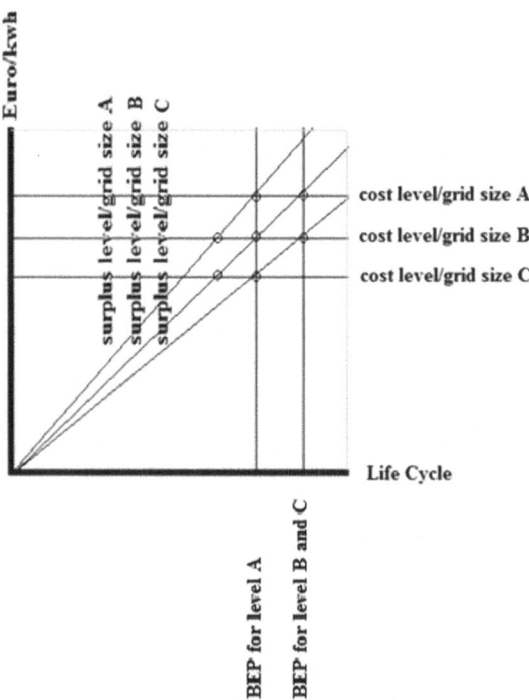

Fig. 8. Different BEP conditions according to different level of productivity (inclined lines) and different level of demand (horizontal line)

In Figure 8, the sloping lines show the increase in the surplus of production, for three levels of annual energy demand ABC, the lines represent the depreciation amount to be recovered annually by the end of the plant life cycle, the vertical lines represent the limits of break even point within which the resident household must recover the price paid to suppliers of energy as a function of varying levels of demand. The general assumption of the model are as follows:

1. the growth is linear, even if the cost function and the surplus function are not only depending on the element, but also on the characters of paneled surfaces (slope, orientation ecc)
2. prices are constant in the observed period
3. the comparison is made referring only to the most competitive offer by external suppliers.
4. the productivity and the costs depend both on the same variables
5. the increase of unit correspond on a single optimal increase of the dimension of surface of the grid. That is to say that the hypothesis of enlargement of the grid is unique and corresponding with the optimal one

In the case of study the referring market is the Italian one. According to new data published by the Energy and Gas Authority, since the liberalization of the electricity market, 1.6 million residential customers have switched suppliers and of these have gone in 1.2 million the open market thanks to former monopoly supplier, that is to say "Enel Energia", in the Italian Case. Table 1 shows the standard cost of one kWh without taxes and VAT, based on offers of "Enel Energia", for four different types of consumption with an annual installed capacity of 3kW.

Table 1. Price available in energy district - Residential Buildings -2011 (source Enel Energia)

Typology of consumption	yearly consumption 1500 kWh	yearly consumption 2700 kWh	yearly consumption 4500 kWh	yearly consumption 7500 kWh
Standard price	0,14 €/kWh	0,14 €/kWh	0,18 €/kWh	0,23 €/kWh
Number of grid's pieces for 3 kw of energy demand	500	900	1500	2500
Number of grid's pieces for 20 kw of energy demand	75	135	225	325

This means that the shift from the demand to energy suppliers towards the energetic self-sustainability should produce a lower yearly adjusted cost of mortgage than the offer of the external supplier in the efficient lifecycle period of the installation.

The supposed break even point is assumed in six year for a single fabric that expresses a demand of 1 up to 3 kw and in eight years for a single fabric that expresses a demand of 20 up to 100 kw.

The minimum size of the gird it will be therefore depending on the break even point and on the number of single unit inside the grid. The BEP should be reached by aggregating a number of units capable of producing enough energy not to exceed the ratio between the number of kilowatts consumed and the unit price of kilowatts.

5   Concluding Remarks

The proposed analysis procedure has allowed a rapid assessment of the economic vocation of an urban installation of solar systems. This research proposal also aims to strengthen, in cooperation with local authorities, research centers and the citizens themselves, a new methodology, through the use of modern technological means to achieve the best results in terms of creating the standard output second preliminary estimates of the potential of solar energy, producibility, and further analysis. Adequate exploitation of renewable resources such as the sun in cities highly urbanized represents a key objective to be achieved through the analysis of the territory to urban scale (for example, in case such as that indicated above, through the optimization of the positioning of the photovoltaic roofing on respecting the morphology of the same). The binding rules established by law and by means of 'non-traditional urban ensure "access to the sun" for owners of homes with solar systems, underestimates the right to use the resource as a natural right of ownership of solar energy [9]. This principle can be met in new buildings but should be well studied in the existing urban fabric through the means described above can be the base study for a possible adjustment of municipal building regulations, which take into account and provide for rules related to the promotion of renewable sources and energy savings.

The study carried out technically and economically allows to have an urban scale localization of micro grid community self-sufficient from an energy point of view that manage in an intelligent manner the intrinsic resources of territory [10]. The method, applicable to different levels of detail, can provide local governments a powerful tool of analysis, whose results can be decision support planning and management of economic resources from instruments such as incentives to local governments and the production shoud be increasing [11]. This will provide a further impetus to the spread of renewable energy sources in urban areas, with positive effects both on the consumption of energy from fossil fuels and in terms of improvement of the urban environment by reducing emissions of pollutants such technologies trigger.

6   Glossary

Smart Grid (SG): grid based on renewable energy generally, able to intelligently integrate the actions of all users connected to distribute energy efficiently, sustainable, cost effective and safe

BIPV: Building Integrated Photo-Voltaics

Prosumer: end users with systems that are both renewable electricity producers and consumers.

ICT: Information and communications technology is a more general term that stresses the role of unified communications and the integration of telecommunications, computers, middleware as well as necessary software, storage- and audio-visual systems, which enable users to create, access, store, transmit, and manipulate information

References

1. Zhao, H.: Energy Consumption Based Urban Texture Analysis. In: International Symposium on Urban Form. Concordia University, Montreal (2011)
2. Caputo, P., Manfren, M.: Strategie e modelli di efficienza energetica alla scala urbana. Modelli per la simulazione energetica. Report RSE 59 (2009)
3. City of Boulder. Solar Access Guide. Building Services Center, Boulder (2006)
4. Troglio, E., Doni, M., Haas, T.: Urban typologies and heat energy demand. A case-study in the Italian context. In: Proceedings of the 51st Congress of the European Regional Science Association, Barcelona, pp. 1–17 (2011)
5. Jo, J.H., Otanicar, T.P.: A hierarchical methodology for the mesoscale assessment of building integrated roof solar energy systems. Renewable Energy 36(11), 2992–3000 (2011)
6. Rifkin, J.: The Third Industrial Revolution: How Lateral Power is Transforming Energy, the Economy, and the World. Macmillan, London (2011)
7. Chiaroni, D., Frattini, F.: Solar Energy Report. Il sistema industriale italiano nel business dell'energia solare. Energy and Strategy Group, Milano (2011)
8. Ratti, C., Baker, N., Steemers, K.: Energy consumption and urban texture. Energy and Building 37, 762–776 (2005)
9. Lewis, G.: Optimum tilt of solar collector. Solar and Wind Energy 4, 407–410 (1987)
10. Fusco Girard, L., Nijkamp, P.: Energia, Bellezza, Partecipazione: le sfide della sostenibilità. Franco Angeli, Milan (2002)
11. Koch, B., Husain, B.: Smart Electricity. ABB Review 1, 6–10 (2010)

The Effect of Infrastructural Works on Urban Property Values: The *asse attrezzato*[*] in Pescara, Italy

Sebastiano Carbonara

Dipartimento di Architettura, Università G. d'Annunzio of Chieti-Pescara
Viale Pindaro, 42, 65127 Pescara, Italy
s.carbonara@unich.it

Abstract. The objective pursued in this text is that of verifying the effects on real estate values exercised by the *asse attrezzato*, the high-speed road artery connecting the city of Pescara with the A14 highway and its inland areas. More in detail, the investigation considered the final section of this infrastructure, which penetrates the consolidated urban fabric of the city by approximately two kilometres as an overpass running parallel to the Pescara River and the quays of the canal port, all the way to the coastline. The urban landscape is strongly characterised by the imposing and markedly functional presence of the viaduct, linked to its use as the privileged means of accessing the city, to which it is interconnected via three interchange ramps. The evaluation employed a technique of inferential statistics, a multiple linear regression analysis, with one of the independent variables represented by the distance of buildings from the interchanges of the *asse attrezzato*. Furthermore, the research experimented with the use of a complex variable: an expression of the grouping of various characteristics associated with the different urban contexts in which the buildings are located. The empirical results reveal the presence of a notable effect on real estate values related to accessibility, and suggest the practicability of using complex variables in regression models in relation to characteristics expressing urban quality.

Keywords: residential real estate market, hedonic pricing, accessibility, urban quality.

1 Introduction

Urban quality and building quality are multi-dimensional concepts that may be interpreted from different points of observation and perspectives. One possible declension refers to the analysis of real estate values, to some degree the complex synthesis, effect and measure of these phenomena.

A simplified and widely used schematisation consents the identification of two large categories of characteristics influencing these markets: the inherent qualities of each building and the external qualities of the urban context in which they are located.

---

[*] The Italian term *asse attrezzato* refers to a form of linear settlement serviced by and (generally) located along a high-speed link road – TN.

While these concepts have been implicitly expressed by others in the past, they referred exclusively to agricultural practices, given the historic period during which they were expressed. Traces can be found in *De re rustica* by Publio Catone, written between 164 and 154 BC and, more recently, in the *Trattato delle stime de' beni stabili*, published by Cosimo Trinci in 1755 [1]. The rich literature discussing the theme in recent years, which finds theoretical references and inspiration in the analyses of Lancaster [2] and the work of Rosen [3], proposes a series of studies that have gauged a multiplicity of characteristics that, to a more or less relevant degree, may orient consumer preferences in relation to real estate transactions.

The work presented here moves in this direction and proposes a model for the definition of urban real estate values for residential properties in multi-storey buildings situated in a central area of the Italian city of Pescara. This model can be utilised to measure the effects on the determination of real estate prices exercised on the one hand by external factors, and on the other by inherent factors.

A further element considered by the model is linked to the concept of accessibility by automobile, of particular relevance to the urban area analysed in light of the 'linear' development of this small Adriatic City.

As demonstrated in Figure 1, the built fabric of the city has developed parallel to the coastline, creating an uninterrupted extension that merges to the north with the town of Montesilvano and, to the south, with that of Francavilla, flowing into the towns of Spoltore and San Giovanni Teatino towards the interior.

Fig. 1. Principal routes of urban and extra-urban viability in the city of Pescara

The zone behind the coast is hilly (almost exclusively residential) and affected over the past two decades by intense phenomena of urbanisation that have not been accompanied by the realisation of new high-speed connections with the city centre, nor with the territory surrounding it.

The effects of this form of settlement are evident in the substantial difficulties in moving through the city (almost entirely concentrated along a line parallel to the coast), with movement to and from the exterior rendered even more problematic.

The *asse attrezzato* is thus the sole road infrastructure guaranteeing rapid movements between the city and the territories outside its borders. This corridor crosses into the consolidated urban fabric by approximately 3,500 meters and terminates in the form of a raised viaduct in its final 1,500 meters (Figures 2 & 3).

Fig. 2. Pescara. Aerial view of the *asse attrezzato*

Fig. 3. Pescara, with the profile of the *asse attrezzato* to the right

Cervero et al [4] have observed how the presence of elevated freeways crossing urban zones determine negative effects on residential real estate prices, which are significantly reduced when these infrastructures are transformed into surface boulevards. This would most likely also be true in the case examined. Nonetheless, this study intentionally avoids considering the impact on residential prices of the location of apartments in proximity to an infrastructural corridor, but rather the impact on prices resulting from their proximity to the interchanges of the *asse attrezzato*, which can be considered the 'terminus' of this freeway infrastructure (Figure 4).

Two preliminary considerations must be made with respect to the construction of the regression-based model elaborated here: a general discussion of the use of these models, and another circumscribed to the Italian situation.

In general terms, the majority of regression-based models utilise a relevant number of explicative variables, associating the formation of a price with a formally reassuring level of rationality that can be linked only with great difficulty to the decision-making process that characterises the acquisition-related choices of the real estate market. This is the case not only due to the imperfect nature of these markets, but precisely for the objective difficulties facing an individual attempting to manage a multiplicity of partially heterogeneous aspects, in order to proceed towards a rational synthesis that results in the decision to pay a particular price for an immobile asset.

It may thus prove more useful to consider a limited number of attributes.

This orientation is particularly useful to the Italian situation, where only with great difficulty is it possible to provide a quantitatively significant dataset of prices, in other words, capable of being related to a relevant number of explicative variables.

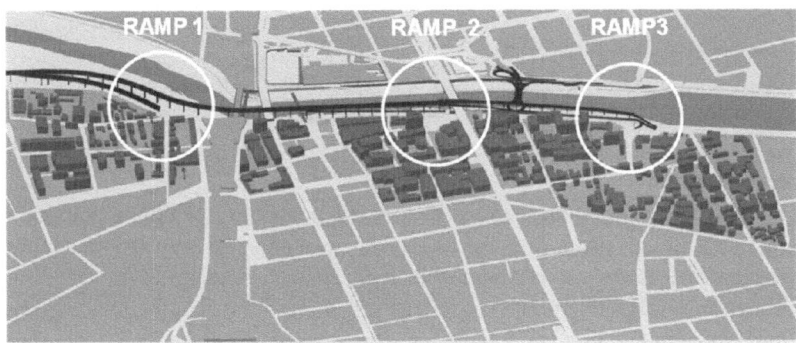

Fig. 4. The circles indicate the interchange ramps along the *asse attrezzato*

2 The Main Difficulty in Applying Hedonic Models in Italy

Studies utilising hedonic models to evaluate urban real estate assets are not particularly numerous in Italy - the first can be traced back to Simonotti in 1991 [5] - compared with an international literature rich with experiences.

There are many reasons why this is the case, though one factor more than any other can probably be considered the limiting factor: the structural lack of factual data related to real estate transactions. A recent and well-documented investigation [6] that analysed the use of these models in Italian and foreign studies, above all in the United States, clearly revealed the diverse consistency of data samples: the reduced number of Italian models, almost at the limit of the minimum statistically acceptable number, compared with the incredibly vast quantity of data provided by studies made in other countries.

This does not mean that there is a lack of official acts of sale indicating sale prices, regularly registered and consultable in dedicated public offices; the problem is that these prices, for the most part, cannot be considered truthful.

There exists a prevailing attitude in Italy that favours tax evasion among negotiating parties who, when drawing up the final act of sale, tend to declare lower prices than those effectively paid.

Anything but extraneous to the origin of this behaviour is national legislation governing the taxation of real estate transactions.

In fact, at the moment of sale, the purchaser is responsible for paying a tax (the so-called *imposta del registro*, or, registration tax), which varies between 5 and 10% of the price declared in the act of sale. Thus it is probable that the official price as declared is inferior, even significantly, to that effectively paid by the purchaser, in order to generate a reduced tax. The difference with respect to the real price is paid to the seller 'under the table'.

The difficulty faced by public entities responsible for verifying the correctness of the prices declared with respect to the actual market situation has led the Italian State to generate a mechanism of control that, while wholly insufficient and to certain degrees iniquitous, in any case impedes the buyer and seller from declaring prices that are exaggeratedly low with respect to those of the market.

Above and beyond the mechanisms generated, the reality is that for far too long those called upon to work with urban real estate appraisals were almost never able to utilise official prices because it is clear that they do not represent the reality of the market.

This has hindered, amongst other things, the possibility of utilising appraisal procedures capable of producing robust forms of understanding the mechanisms of developing real estate values, from which to derive orientations for urban policies, including those focused on mechanisms of surplus value recapture generated by interventions in the public sector.

An interesting novelty was brought about, in normative terms, during the summer of 2006, when Italian Law n. 248[1] declared that at the moment of stipulating the contract of sale, home buyers, so long as they are a natural person, may request that the base sum to which the registration, mortgage and land registry taxes are applied, be constituted of the so-called *valore catastale* (cadastral value)[2]. Furthermore, there also plans to reduce the fees paid to Notary Publics (obviously in relation to the price declared) by 30% [these fees are notoriously high in Italy – TN].

At present, the Italian Government, under the direction of Mario Monti, has introduced very rigid mechanisms for controlling the circulation of hard currency and has discussed the possibility of permitting The Italian Revenue Agency to verify the contents of bank accounts. While considered illiberal by many, this measure has nonetheless been generally welcomed by the majority of political forces and the population itself, who considering it a 'bitter pill' that must be swallowed at this moment in history.

All of this potentially represents an impulse towards a greater transparency of transactions and the increased trustworthiness of prices declared in acts of sale, whose effects may be perceived in the short term.

With regards to the specific conditions of the Pescara real estate market, in addition to the problems mentioned above, there is another issued tied to the drawn-out payment of the sums defined by the negotiating parties. Normal practice calls for the payment of approximately 20% of the final price as a down payment used by the purchaser to 'block' the property in question; the remaining 80% is to be paid upon stipulation of the act of sale. More frequently, the price declared in the act represents precisely the 80% of the price effectively paid. The point is that this behaviour, as widespread as it is, may not occur in all cases, or in others the down payment may be of a lesser amount (10-15%). This imposes a further verification of the prices ascertained from the examination of acts of sale, which must necessarily be carried out in the offices of the building contractors or real estate agents responsible for the sale of these properties. This is an extra and exceedingly time-consuming phase of work (it is

---

[1] Article 35, comma 21 of Law n.248 date 4 August 2006 n. 248, a conversion into law of Legislative Decree n. 223 from 4 July 2006.

[2] A conventional value, determined as per article 52, commas four and five of Presidential Decree 131/1986 (T.U. (Consolidating Act) govern registry taxes), for which law 248/2006 consents the payment of the tax 'independent of the sum agreed upon…indicated in the act'.

not always possible to identify and/or contact the subjects involved in a transaction), which may be conducted with only limited samples of data.

Faced with these difficulties, the research proceeded with the investigation of samples of market prices for residential properties in the city of Pescara over the course of eight months, between 2007-2008, using acts of sale stipulated during this period. The investigation was circumscribed to cadastral maps related to the location of the interchange ramps of the *asse attrezzato*. While not being able to define with absolute precision the number of sales made in the study area during the period considered, it was nonetheless possible to prudently estimate that approximately 65-70% of all transactions were registered (sales in the entire city of Pescara vary from 1,400 to 1,700/year, and thus an average of between 32 and 38 sales per sheet of the cadastral map). This was also a period of relative stability for real estate prices.

The prices listed in the model are those effectively declared in the acts of sale, without considering possible increases of 20%. This signifies that the marginal prices implicit of the characteristics considered, may be underestimated with respect to the effective market situation.

3 Predictors

The selection of the independent variables was made based on the results provided by a series of intentionally unstructured informant interviews (made in the city during the same period as the gathering of data), with approximately three hundred citizens, casually contacted in various points of social aggregation. The objective was that of clarifying the most important elements considered at the moment of purchasing the home they live in, or based on the hypothesis of a future purchase.

The research was not intent on realising a statistically rigorous investigation designed to define a hierarchy of consumer preferences with respect to the choice of a home, but rather to capture, though an informal approach, the elements that define the perception of the city of Pescara in relation to its inhabitation.

There was a full awareness of the risks that might accompany a similar approach, and the possible need to pursue other avenues was not excluded.

Instead, the results consented the definition of a sufficiently precise framework with respect to what proved most relevant in the selection of a dwelling. In general terms, it can be affirmed that specific attention is paid to a collection of the truly impossible to modify elements, that is, those characterised by an absolute rigidity, beginning with the general characteristics attributed to diverse urban zones. Less relevant are the inherent characteristics, defined in the strictest sense of the term (services, technological systems, quality of materials, etc.). In other words, there would appear to exist a series of 'non-negotiable' elements that serve as a limiting factor to the start-up of dialogue and the successive decision to purchase, set against the backdrop of a multiplicity of many other elements that undoubtedly influence the final price, however more related to its 'negotiation' than its 'determination'. To use a simple metaphor, what appears important is not the nest, but rather the tree and the branch that support it. The variables considered are as follows:

Amenities

The term *amenities* is used to indicate a complex variable, constructed by considering seven characteristics identified within each sheet of the cadastral map into which the city is subdivided (Figure 5).

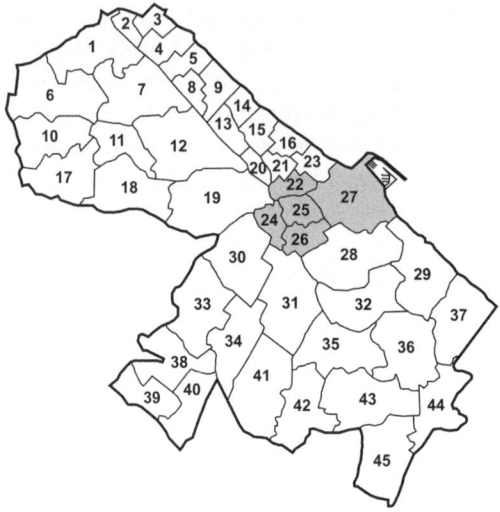

Fig. 5. Urban Zones of Pescara (cadastral map sheets) and Study Area

They are: morphological-environmental characteristics; the prevalent use of built areas; the socio-economic characteristics of built areas; public infrastructures and urban services; prevalent building typologies; building conditions.

Each of these categories was associated with an evaluation ranging between 1 and 7. The average of the seven points generated the measure of the variable.

In literature the use of complex variables in the study of the real estate market in unknown. Their determination is extremely difficult, and their elaboration may prove less than transparent. Nonetheless, this does not change that fact that, faced with complex phenomena, such as those related to the appreciation expressed by a buyer with respect to the comprehensive quality of an urban area, it may be useful to verify the possibility to determine and insert them within regression models. Statistic tests and, earlier still, the clarity and objectivity of the characteristics to be considered may validate their use.

In absolute terms, the element considered the object of the most attention by those interviewed is precisely that tied to the urban area in which the building is located.

In the case of the city of Pescara, which develops parallel to the coastline, dwelling along this strip is by far the most privileged option (7 points), compared with the immediate areas immediately behind it (plains, 5 points) and the hillside areas (3 points), characterised by more recent urban development, for the most part devoid of urban services and infrastructures, though guaranteeing various environmental amenities (Figure 6).

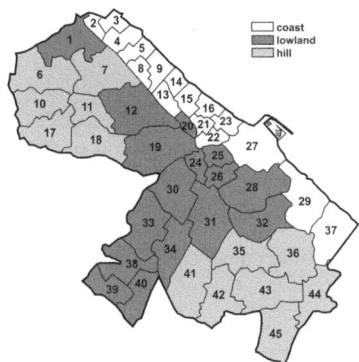

Fig. 6. Morphological-environmental characteristics

A second element with a strong impact is tied to the prevalent use of buildings in a given area. From this point of view there is an absolute preference for what we can refer to as a 'pure' residential destination (7 points), which excludes not only the presence of residential structures realised with partial or total public assistance, but also hotel structures (Pescara is also a tourist destination), large commercial settlements and industrial and/or artisanal areas (1 point).

A third aspect regards the quality of construction associated, more or less unconsciously, to the social structure of an area. From this point of view preferences move toward urban areas with '*signorili*' (upper class) buildings, an adjective that refers to the presence of historic buildings (in the city of Pescara prevalently from the first half of the twentieth century), in other words, buildings that express an evident architectural quality in formal and technological terms (this is equivalent to saying more recent constructions).

The fourth element is represented by the building typology. In the imagination of buyers, the prevalence of single-family *villas* (7 points) with annexed terrains is seen as a positive qualifier of neighbourhoods with respect to the presence of multi-storey buildings (3 points), or row houses (5 points).

The state of conservation and maintenance of buildings observable in diverse urban areas represents a further element of evaluation. The façades of buildings generate the most evident impact: the absence of water marks (thus in definitive terms their maintenance) and/or the presence of valuable cladding materials, constitute elements of significant attraction when they can be applied in general terms to all properties present in a given urban area.

Last but not least is a collection of services: transport networks, parkland, health and civic services (post office, banks, clinics, etc.), parking, sports facilities, and schools. These elements were treated as a whole: a value was attributed to each based on direct investigations and successive cartographic representation; this material was then used to determine an average value. The reason for this choice is to be found in the attitude manifested by those interviewed to consider these elements almost as if they were analysed within a specific 'common space of the mind' that results in their comprehensive evaluation, and not as single elements.

To better clarify the means of elaborating the value of each of these aspects, they have been presented in Table 1 below. Each was considered in relation to the single sheets of the cadastral map, attributing points according to the following system: optimum=7; good=5; sufficient=3; mediocre=1.

Table 1. Aspects that participate in the determination of the value assigned to the variable of *external factors*, by cadastral map sheet

morphological-environmental characteristics	assigned value
Coastal Area	7
Plains	5
Hilly Areas	3
prevalent use of built areas	
Private residential	7
Social Housing or Hotels	5
Public Housing or Commercial (shopping centres)	3
Artisanal, Industrial	1
socio-economic characteristics of built areas	
Upper Class	7
Average	5
Economic	3
Public / Social	1
public infrastructures and urban services	
Technological and Mobility Networks	1,3,5,7
Schools and Nurseries	1,3,5,7
Public Parks	1,3,5,7
Sport Facilities	1,3,5,7
Transport Services	1,3,5,7
Parking	1,3,5,7
Civic Centres & Social-Health Services	1,3,5,7
prevalent building typologies	
Single-Family and Multi-Family Buildings	7
Row Houses	5
Towers, Multi-Storey Buildings	3
Warehouses for Artisanal Activities, Industrial Facilities	1
building conditions	
Optimum	7
Good	5
Mediocre	3
Poor	1

Gross Floor Area
This variable was analytically verified using the cadastral maps for each building, expressed in square meters of *gross floor area*, and thus including the areas occupied by perimeter walls, to which was added the floor area of balconies, reduced by two thirds according to the indications expressed by the Pescara real estate market.

Accessibility
The analysis of the relationships between transport infrastructures and real estate values is amply present in literature, increasingly more often in relation to nodes of rail systems, railway stations and subway systems - of the many: Lin, Hwang [7], Nelson

[8], Voith [9] - or the relations created between real estate values and points of interchange with high-speed urban traffic arteries, i.e. Haughwout [10], Mikelbank [11], Frew [12].

Cervero [13] Kilpatrick et al. [14] have highlighted how networks of mobility influence real estate values both in the positive, due to the increased accessibility they offer, and in the negative, as a result of the dis-amenities tied to the same condition of proximity (acoustic, visual, atmospheric pollution, an increase in crime, etc.).

The study presented here was developed as part of a national research project[3] studying infrastructures [15] [16]. One of the themes of this research focused on verifying if and how the increased accessibility generated by new wheel-based transport infrastructures produces effects on the city and real estate values.

With the aim of analysing this latter aspect, and with reference to the *asse attrezzato* in Pescara, initial consultations were held with real estate agents working in the areas containing the interchanges with this infrastructure. Informant interviews were successively organised with the residents of buildings adjacent to the *asse attrezzato*. Both cases clearly revealed the existence of a specific preference, above all expressed by those living in the city but working elsewhere.

Accessibility was treated as a dichotomic variable, assigning the value of 1 to buildings located within a radius of 200 meters (Figure 7) of the focus represented by the interchanges with the *asse attrezzato*; the value of 0 was assigned to buildings located outside these zones. The value of 200 meters is that which, within the model, guaranteed the highest *t student* that, in any case, for all other distances verified, was statistically acceptable. This means of proceeding was made necessary by the fact that the market does not provide specific information regarding the space within which the benefits of proximity to the *asse attrezzato* are made manifest.

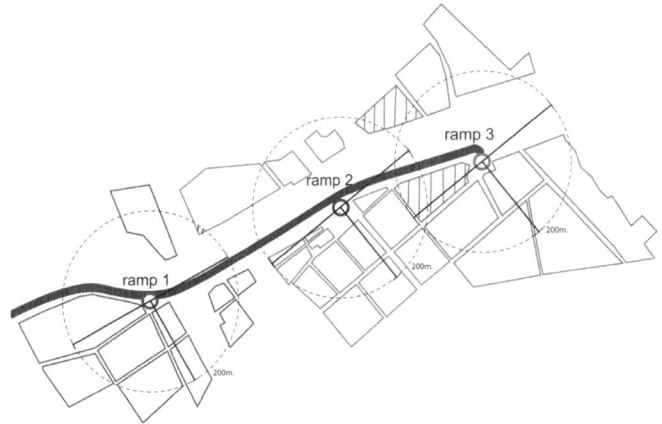

Fig. 7. Areas of influence of the interchanges along the *asse attrezzato*

---

[3] PRIN - *Progetti di Rilevante Interesse Nazionale*: *Infrastrutture per la mobilità e costruzione del territorio metropolitano*, national research coordinator Prof. Bernardo Secchi. Pescara Research Unit: *L'ultimo miglio: interconnessione tra porto e territorio*, research coordinator Prof. Rosario Pavia, Università degli Studi 'G. d'Annunzio' Chieti-Pescara.

Year (Building Age)
The year of construction of the building. This variable may be associated with a collection of attributes relative to technological materials, MEP systems, ceiling height, construction characteristics, etc.

Façades
The number of façades from which the unit receives natural light and ventilation, in virtue of the presence of windows and/or French doors.

Floor Level
The position/elevation of the residential unit within its building.

4 The Model

The model realised represents a typical hedonic price equation:

$$y = \beta_0 + \beta_1 x_1 + \beta_2 x_2 + \beta_n x_n + \varepsilon \tag{1}$$

where y represents the dependent variables; β_0 intercepts; β_n, marginal prices of the characteristics considered; x_n independent variables. The equation is presented in the determinist form, omitting the stochastic variable ε.

The observations of price were 114, expressed in Euro (€) per sq. m of conventional gross floor area, relative to residential real estate properties sold/bought during the period of analysis considered.

All properties are located in the area of study represented by sheets 22, 24, 25, 26 and 27 of the cadastral map (Figure 5).

4.1 Estimation Results

Tables 2 & 3 list a selection of descriptive statistics for each of the variables considered in the model and their index of correlation. As can be observed for the comparison of pairs, there are no relevant phenomena of co-linearity.

Table 2. Descriptive Statistics

	Mean	Std. Deviation	N
price_sq_m	1818,4309	379,93596	114
gross_floor_area	96,8888	31,97271	114
amenities	4,5352	,47391	114
accessibility	,4211	,49591	114
year	2001,4825	14,22837	114
floor_level	2,6228	1,49566	114
façades	2,0000	,70397	114

The variables were selected using the stepwise method, given the markedly explorative nature of the analysis; the relationship with the predictors is 1 to 19 observations of price.

The following Tables (4-9) present the results of the data elaboration.

With regards to the observations of price, there are neither outliers nor critical values of Cook's distance. Instead, there are four cases that manifest an excessive influence (Table 4).

Table 3. Correlations

		Gross _floor _area	amenities	accessibility	year	floor_ level	facades
Pearson Correlation	gross_floor_ area	1,000	0,034	0,171	-0,193	0,070	0,315
	amenities	0,034	1,000	0,218	-0,256	0,162	0,121
	accessibility	0,171	0,218	1,000	0,162	-0,058	-0,380
	year	-0,193	-0,256	0,162	1,000	0,311	-0,349
	floor_level	0,070	0,162	-0,058	0,311	1,000	-0,008
	façades	0,315	0,121	-0,380	-0,349	-0,008	1,000

Table 4. Outlier

Potential Outlier Cases abs(Standard Residual) > 3
Potential Influence Cases Cook's Distance > 1
Potential Leverage Cases Hat Matrix Diagonal > 0,1228 Case 71 - hat(i,i) = 0,182700 Case 72 - hat(i,i) = 0,199104 Case 109 - hat(i,i) = 0,125755 Case 111 - hat(i,i) = 0,141948

All six explicative variables were used in the model, providing an appreciable contribution to improving the capacities, as can be seen in Table 5. All the same, the predictors *amenities*, *gross floor area* and *floor level* on their own explain over 75% of the total variability. The contribution of the other three (*year*, *accessibility* and *façades*) are thus more limited, without exceeding 5%. The variable *accessibility* on its own has an effect of 1.1%. Its implicit marginal price is equal to approximately Euro 135.73/sq. m.

The standard error, with six variables, is equal to 9.4% of the average price observed.

Table 5. Variables Entered/Removed

Model	Variables Entered	Variables Removed	Method
1	amenities	.	Stepwise (Criteria: Probability-of-F-to-enter <= ,050, Probability-of-F-to-remove >= ,100).
2	gross_floor_area	.	Stepwise (Criteria: Probability-of-F-to-enter <= ,050, Probability-of-F-to-remove >= ,100).
3	floor_level	.	Stepwise (Criteria: Probability-of-F-to-enter <= ,050, Probability-of-F-to-remove >= ,100).
4	year	.	Stepwise (Criteria: Probability-of-F-to-enter <= ,050, Probability-of-F-to-remove >= ,100).
5	accessibility	.	Stepwise (Criteria: Probability-of-F-to-enter <= ,050, Probability-of-F-to-remove >= ,100).
6	façades	.	Stepwise (Criteria: Probability-of-F-to-enter <= ,050, Probability-of-F-to-remove >= ,100).

Table 6. Model Summary

Model	R	R Square	Adjusted R Square	Std. Error of the Estimate	Durbin-Watson
1	,684(a)	,467	,463	278,53998	
2	,809(b)	,655	,648	225,29196	
3	,870(c)	,756	,749	190,15988	
4	,885(d)	,784	,776	179,74629	
5	,891(e)	,795	,785	176,08238	
6	,898(f)	,806	,796	171,76486	1,015

Table 7. ANOVA

Model		Sum of Squares	df	Mean Square	F	Sig.
1	Regression	7.622.234,694	1	7.622.234,694	98,244	0,000
	Residual	8.689.466,220	112	77.584,520		
	Total	16.311.700,914	113			
2	Regression	10.677.733,111	2	5.338.866,555	105,186	0,000
	Residual	5.633.967,804	111	50.756,467		
	Total	16.311.700,914	113			
3	Regression		3	4.111.338,336	113,696	0,000
	Residual	3.977.685,907	110	36.160,781		
	Total	16.311.700,914	113			
4	Regression	12.790.049,349	4	3.197.512,337	98,967	0,000
	Residual	3.521.651,565	109	32.308,730		
	Total	16.311.700,914	113			
5	Regression	12.963.160,338	5	2.592.632,068	83,620	0,000
	Residual	3.348.540,576	108	31.005,005		
	Total	16.311.700,914	113			
6	Regression	13.154.861,903	6	2.192.476,984	74,313	0,000
	Residual	3.156.839,012	107	29.503,168		
	Total	16.311.700,914	113			

a. Predictors: (Constant), amenities
b. Predictors: (Constant), amenities, gross_floor_area
c. Predictors: (Constant), amenities, gross_floor_area, floor_level
d. Predictors: (Constant), amenities, gross_floor_area, floor_level, year
e. Predictors: (Constant), amenities, gross_floor_area, floor_level, year, accessibility
f. Predictors: (Constant), amenities, gross_floor_area, floor_level, year, accessibility, facades
g. Dependent Variable: price_sq_m

Table 8. Residuals Statistics

	Minimum	Maximum	Mean	Std. Deviation	N
Predicted Value	1149,7367	2575,4888	1818,4309	341,19600	114
Residual	-323,09802	334,70612	,00000	167,14254	114
Std. Predicted Value	-1,960	2,219	,000	1,000	114
Std. Residual	-1,881	1,949	,000	,973	114

a Dependent Variable: price_sq_m

Table 9. Coefficients

	Unstandardized Coefficients		t	95% Confidence Interval for B		VIF
	B	Std. Error		Lower Bound	Upper Bound	
(Constant)	-666,977	252,105	-2,646	-1.166,491	-167,463	
amenities	548,029	55,290	9,912	438,478	657,580	1,000
(Constant)	-221,851	211,828	-1,047	-641,601	197,900	
amenities	559,818	44,746	12,511	471,150	648,486	1,001
G.F.A.	-5,146	0,663	-7,759	-6,460	-3,832	1,001
(Constant)	-225,600	178,796	-1,262	-579,932	128,731	
amenities	518,401	38,261	13,549	442,576	594,225	1,027
G.F.A.	-5,394	0,561	-9,614	-6,506	-4,282	1,005
floor_level	82,203	12,146	6,768	58,132	106,273	1,031
(Constant)	-10.667,260	2.784,404	-3,831	-16.185,858	-5.148,663	
amenities	565,981	38,319	14,770	490,034	641,929	1,153
G.F.A	-4,919	0,545	-9,024	-6,000	-3,839	1,062
floor_level	63,938	12,468	5,128	39,227	88,649	1,216
year	5,110	1,360	3,757	2,414	7,806	1,310
(Constant)	-8.243,561	2.914,134	-2,829	-14.019,880	-2.467,243	
amenities	534,187	39,877	13,396	455,144	613,230	1,302
G.F.A	-5,259	0,553	-9,510	-6,355	-4,163	1,139
floor_level	71,180	12,592	5,653	46,220	96,141	1,293
year	3,960	1,419	2,791	1,148	6,772	1,485
accessibility	87,868	37,186	2,363	14,158	161,578	1,239
(Constant)	-9.320,983	2.873,931	-3,243	-15.018,217	-3.623,749	
amenities	514,756	39,639	12,986	436,176	593,335	1,352
G.F.A	-5,844	0,586	-9,969	-7,006	-4,682	1,346
floor_level	72,713	12,298	5,912	48,332	97,093	1,296
year	4,485	1,399	3,206	1,712	7,259	1,518
accessibility	135,730	40,846	3,323	54,757	216,703	1,571
facades	73,236	28,731	2,549	16,281	130,192	1,567

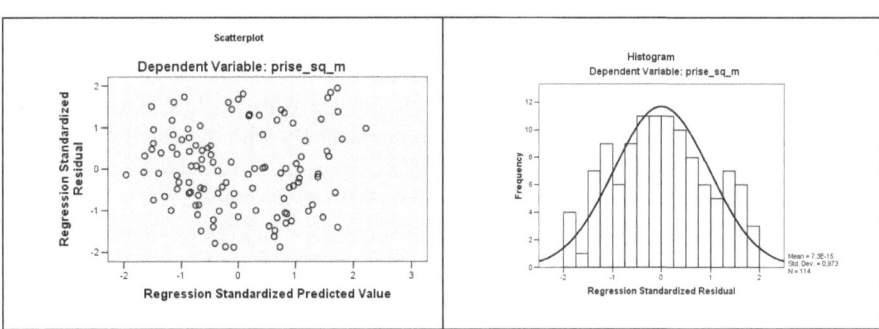

5 Concluding Remarks

The results of the analysis offer encouraging elements in relation to the initial assumptions.

There appears to be a foundation for the initial idea according to which the preferences of buyers of residential properties in multi-storey buildings are prevalently oriented by the position of the building within the city, and the residential unit within its building. These aspects have certainly not been ignored in the elaboration of other models of this nature, present in existing literature, however the novelty here is that of being able to significantly circumscribe the inherent characteristics.

All the same, it is not possible to exclude that this orientation may find some justification in the specific characteristics of the city of Pescara, which is not a historical city (like almost all other Italian cities) but rather the result of a law passed by the Fascist Government in 1927. The city grew from two small original nuclei of settlement, situated what is more on swampy lands. Thus the city contains no historical buildings expressing architectural and cultural values capable of influencing the real estate market, with the exception of a number of Liberty style buildings from the early twentieth century.

From this point of view, the positive nature of the coefficient relative to the variable *year of construction* proves entirely coherent, highlighting the preference for more recently constructed buildings offering formal qualities more aligned with current tastes, modern technologies and lesser maintenance costs.

What is more, given the linear confirmation of the urban fabric of Pescara, which renders wheel-based connections with the national viability network extremely slow, the model consents an understanding of the significant importance of the variable of *accessibility*, an expression of the characteristic of proximity to the three interchange ramps of the only high-speed traffic artery penetrating the urban fabric (the *asse attrezzato*): the coefficient of the variable *accessibility*, referred to the unit of floor area, as mentioned, provides a result of Euro 137.50/square meter.

This said, it is necessary in each case to underline that the experiment proposed, and its assumptions, require further verifications and investigations.

The variable *amenities* was determined for each sheet of the cadastral map of the city, though it was tested in only five of these. Its effective significance must be tested in a model that considers the entire urban territory.

There is also a need to understand to what degree the unexplained variability of the model depends on the greater or lesser truth of the prices declared in acts of sale. The model considered the prices effectively listed in the acts. If the problem were tied exclusively to the figure of 20% of the effective price that was hypothesised as not having been declared, the question would deal exclusively with the entity of marginal prices, without substantially modifying the statistic tests. Diversely, there could be more important alterations to the test that a new elaboration should dispel, above all in light of the normative measures mentioned above, designed to render the market more transparent.

Finally, a more in-depth reflection should be made on the intrinsic characteristics, in order to verify whether *floor level*, *year of construction* and *façades* are sufficient to satisfying the effect on price. Even in this case, it is not possible to exclude that the model may present important omissions responsible for a percentage of the unexplained variability. From this point of view, in perspective, one attribute that must surely be verified is related to the energy certification of buildings. This aspect begins to be considered in the Pescara real estate market, and a recent simulation has demonstrated that constructing new buildings capable of aspiring to a Class A energy rating[4], as opposed to a Class C, comports a cost increase of Euro 90.00 per square meter, compared to an increased sale price, estimated in the order of some Euro 140.00/sq. m.

References

1. Trinci, C.: Trattato delle stime de' beni stabili, Albizzini, Florence (1755)
2. Lancaster, K.: A new Approach to Consumer Theory. The Journal of Political Economy 74(2), 132–157 (1966)
3. Rosen, S.: Hedonic Prices and Implicit Markets: Product Differentiation in Pure Competition. The Journal of Political Economy 82(1), 34–55 (1974)
4. Cervero, R., Kang, J., Shively, K.: From Elevated Freeways to Surface Boulevards: Neighborhood, Traffic, and Housing Price Impacts in San Francisco. Department of City and Regional Planning University of California, Berkeley, Working Paper prepared for the University of California Transportation Center (2007)
5. Simonotti, M.: Un'applicazione dell'analisi di regressione multipla nella stima di appartamenti. Genio Rurale 2, 9–15 (1991)
6. Rosasco, P.: Modelli per il mass appraisal. Alinea, Florence (2010)
7. Lin, J.J., Hwang, C.H.: Analysis of property prices before and after the opening of the Taipei subway system. The Annals of Regional Science 38(4), 687–704 (2004)
8. Nelson, A.: Effects of elevated heavy-rail transit stations on house prices with respect to neighborhood income. Transportation Research Record 1359, 127–132 (1992)
9. Voith, R.P.: Changing Capitalization of CBD-Oriented Transportation Systems: Evidence from Philadelphia, 1970–1988. Journal of Urban Economics 33, 361–376 (1993)
10. Haughwout, A.F.: Central city infrastructure investment and suburban house values. Regional Science and Urban Economics 27, 199–215 (1997)
11. Mikelbank, B.A.: Spatial analysis of the relationship between housing values and investments in transportation infrastructure. The Annals of Regional Science 38, 705–726 (2004)
12. Frew, J., Wilson, B.: Estimating the Connection between Location and Property Value. Journal of Real Estate Practice and Education 5(1), 17–25 (2002)
13. Cervero, R.: Effects of Light and Commuter Rail Transit on Land Prices: Experiences in San Diego County. Department of City and Regional Planning, University of California–Berkeley. Working Paper (2003)
14. Kilpatrick, J.A., Throupe, R.L., Carruthers, J.I., Krause, A.: The Impact of Transit Corridors on Residential Property Values. Journal of Real Estate Research 29(3), 304–305 (2007)
15. Secchi, B. (ed.): On Mobility. Infrastrutture per la mobilità e costruzione del territorio metropolitano: linee guida per un progetto integrato. Marsilio, Venezia (2010)
16. Pavia, R., di Venosa, M. (eds.): Ultimo Miglio. Il progetto di interconnessione tra porto e città. Sala Editori, Pescara (2011)

---

[4] These ratings are based on the implementation of the Energy Performance of Buildings Directive (2002/91/EC, EPBD), which came into force on 4 January 2003, and had to be implmented by the EU Member States no later than 4 January 2006. The Directive was inspired by the Kyoto Protocol and commits the EU to reducing CO_2 emissions (http://www.epbd-ca.eu/) - TN.

Prospect of Integrate Monitoring:
A Multidimensional Approach

Marco Selicato[1], Carmelo Maria Torre[1], and Giovanni La Trofa[2]

[1] Department of Civil Engineering and Architecture, Polytechnic of Bari
Via Orabona 4, 70125 Bari, Italy
[2] GeoData SRL, via Tatarella 72, Noicattaro, Bari
maghi84@libero.it, torre@poliba.it, geodatasrl@gmail.com

Abstract. The paper shows as "Strategic Environmental Assessment (SEA)" can be considered as an essential "decision support system", supported by a systematic process for evaluating the environmental issues of plans and programs. It is shown how in coastal areas, with strong characters of mutability, the assessment procedure must be able to adapt environmental protection and local development and monitoring of feedbacks assumes critical importance. The issues of monitoring and modeling with spatial data infrastructure have been applied to the Coastal Plan of the Apulia Region: a possibility has been explored to implement the analysis of environmental sensitivity and propensity to Coastal erosion due to the level of human pressure on land. The system is based on assessing pressures due to different land uses; such assessment can be integrated without great difficulty with the analysis of criticality and sensitivity provided by the plan. Essential tools to aid the monitoring system are represented by an effective geographic information system for consulting and obtaining the necessary data and analysis from a methodological point of view by AHP. M. Selicato wrote the first paragraph, C. M. Torre wrote the second and the third paragraphs and G. La Trofa wrote the last paragraph.

Keywords: Monitoring, AHP, Coastal management, SEA, GIS.

1 Introduction

In history, the coastal areas have been an important pole of development of civilization. The ability to use the sea for transport and trade and the abundant availability of food derived from high-productivity coastal waters have encouraged and fostered the development of settlements.

The costs can be defined as the element of relationship and interaction between land and sea, and is considered a resource not only from an environmental perspective but also from social and economic.

Coastal areas and their natural resources play a strategic role in potentially meet the needs and aspirations of European citizens today and tomorrow.

The human pressures threaten to destroy habitats and coastal zone resources and consequently the ability of these same areas to carry out many of their essential functions.

Low-impact targets are often replaced by others that are in intensive and profitable in the short term but that the distance undermine the potential costs of reducing the "resilience".

It is unfortunately difficult to stem the spread of inappropriate uses of coastal areas and indeed, the growing number of residents and visitors, the pressure increases to unsustainable use.

It is clear that a sustainable development perspective, economically efficient and socially equitable use of coastal areas need to define strategies to correct these weaknesses. The definition of such strategies and their implementation in the "Strategic Environmental Assessment '(SEA) is an essential" tool of decision support "that is configured as a systematic process for evaluating the environmental consequences of plans and programs: SEA" permeates "the plan and becomes a constructive element, management and monitoring. Many authors [1] [2] [3] had recognized the need to follow the next steps for approval, but Fischer [4] find as many scholars treat it, define what it is, what must be done or how it is done in practice.

Partidario and Arts [5] argue that the implementation of the SEA can not be limited to what is prescribed, what should be done in the manner described in RA, accompanied by environmental monitoring carried out by means of appropriate indicators.

These aspects are still unclear, especially the transition from theory to practice, because there is still a theoretical debate about definitions, key concepts, approaches, tools, methods and techniques.

Morrison-Saunders, Marshall and Arts [6] introduced, to indicate the post-decision phases, the term follow up and provide a definition, identical to the EIA and SEA, that "the monitoring and evaluation of the impacts of a project or plan for management and communication of environmental performance of the project or plan."

The monitoring, evaluation of compliance, management and reporting of impacts are also elements of the follow-up according to Marshall [7].

In a direct follow-up can be defined as "what happens after" stages of approval, however the practice is not as simple as thinking about what might happen at the project level may be easier and easier to administer, at the project context, size, timing and predictable are well defined, while in the field of strategic decisions is very difficult to see the foreshadowing that are considered decisions based on the intentions or actions planned but provided long-term, you do not have much in reference to what will happen, what will be the embodiment and implementation, if there is a change in current policy and new policies, if implemented will be a project or program, and what will be your address.

As pointed out by Morrison-Saunders, Marshall and Arts [6] a strategic initiative can go in all directions, not necessarily in a linear fashion and not with the same amplitude, we add that the representation of planning as a linear or cycling is a simplification of reality.

Partidario and Arts [5] suggest that the follow up can be seen as an ex-post evaluation of the consequences of actions and can have four different approaches: compliance, performance, uncertainty and dissemination.

Indicate that you can follow five paths to follow in the later stages of the SEA:

1. monitor the actual changes;
2. assess achievement of stated objectives;
3. evaluate the performance of strategic initiatives;
4. test the compliance of the resulting decision making with strategic initiatives and the SEA;
5. monitor and evaluate the real impacts on the environment and sustainability strategic initiative.

These five approaches allow us to "confront and manage" the complexity of the later stages of an SEA. Each approach has different characteristics and require different resources have different objectives and techniques. You can use the approaches by using them individually and mixing them in different phases depending on the context, purpose.

Regarding the steps after making the Directive explicitly provides only monitoring and not provide information on evaluation activities, management and communication with regard to the impacts component part of the follow-up but they are implicit and connected to the first.

According Kornov [8] one can distinguish an assessment of environmental effects can be defined as normative evaluation, where the impact is assessed in relation to the objectives of sustainability, and a descriptive assessment in which the effects are described.

Although the Directive seems to emphasize the descriptive assessment, where the base-data plays an important role in defining the consequences, the correct approach to the SEA, according to the author, should be to describe the effects but then relate them to sustainability objectives; at each stage of planning the two evaluations have a specific function and must be made. Regarding the type of monitoring is required only monitoring of significant effects, but if you want to link the plan with its effects is necessary that they know the terms and timing of implementation, this means that monitoring must cover Also the plan.

The Directive 2001/42/EC is now the driving force to focus on later stages of the SEA, as with the provisions of Article 10 gives you a way to follow them, explicitly providing for monitoring of significant environmental effects of implementing plans and programs and the possibility of mitigation measures in the application but it is considered appropriate to "broaden the picture," not limited to environmental monitoring, explaining that it stems from it and what is required for the SEA has efficacy and is required when evaluation whose results are to be integrated in the post decision, yet also "limit" to monitor key indicators and environmental issues deemed most critical and sensitive that's a step for follow up.

Although the signs of the Directive on monitoring are limited and have limited the indications from the European guidelines. This applies even more if we refer to the Legislative Decrees 152/2006, 4/2008 and 128/2010: it is necessary to establish guidelines and criteria for monitoring so that the same is effective and VAS with it.

In reference to the construction and operation of the monitoring system are considered important indications of Mc Callun [9]:

- Plan in advance the necessary activities: what needs to be done, by whom and how, stakeholders and coordinate activities;
- Be clear about what you are doing;
- Manage information so that they are produced and made available;
- Provide adequate resources;
- Maintain the credibility of those involved in the process.

And of Partidario and Arts [5] that focus should be:

- First on the strategic nature of the initiative and its impacts on the direction, timing, scale and consequences of the initiative, the tangibility and concreteness and measurability so on;
- Secondly, objectives, implementation and controlling changes, learning, informing and communicating;
- Third on significant issues and approaches necessary: whatever the approach, the monitoring should follow the key indicators, identify areas sensitive to changes due to strategic initiative but first of all be aware of the information available.

To implement an effective monitoring system and adhering to the contents and meanings of the European Directive is necessary to verify the existence of a number of conditions, ie you must be methodological and contextual elements, as observed by Fischer and Gazzola [10] for those in Italy, a system based on rigid procedures and clear, with prescriptive rules, with an authority whose duties and responsibilities are clear with distinct roles evaluators / planners, policy makers, inspectors and must be primarily a definition of the thresholds of compatibility. All this implies effects also on the monitoring for which should be a clear procedure and at the same time flexible in order to change the controlled parameters where this has been a need, however, the same authors consider at the same time that in Italy it is very dangerous to give flexibility to the system.

2   A First Results of the Study: Criticality and Sensitivity

Having acquired the necessary data to define the state of the coastal territory and their interconnections were defined as "criticality" and "sensitivity" of the study area.

By the term criticality has been indicated the greater or lesser propensity to erosion of the coastal area, in addition to the causes that generated it; with the term sensitivity has been indicated a level of frailty associated with environmental features of the context.

The critical erosion of sandy coastline has been classified into high, medium and low. This was defined according to three indicators: the historical evolution trend of the coast, the evolutionary trend recently and the conservation status of dune systems.

The environmental sensitivity was defined as a complex multivariable function that represents the physical state of the coast, according to the system of legal protection standards that emphasize the environmental importance .

The sensitivity represents the state of the coastal environment from a historical perspective and for this reason have been identified a number of criteria, weighted appropriately, help to define it:

- The hydrography with a buffer zone of 300 meters on both sides;
- Sites of Community Importance (SCI), Special Protection Areas (SPAs);
- Protected Areas and the scope in the Putt;
- Other areas of the extended Heritage Plan;
- Distinguishable Areas of Regional Landscape Plan:
- The historic settlement patterns;
- Use of agricultural land.

The criteria are "weighted" by analyzing hierarchical AHP, proposed by T. L. Saaty [11] The acronym stands for AHP Analytic (decomposes the problem into its constituent elements) Hierarchy (structure of the constituent elements in a hierarchical manner to the main objective and the sub-goals) Process (processes the data and evaluations in order to achieve the result final).

Using the method AHP (by the use of the software Definite) and with the aid of "ratings experts", to each element of the hierarchy has been associated with a weight through the pairwise comparisons between the different alternatives.

The criteria were included in a matrix where each row contains the comparison of criterion present in the first cell in the row with the same criteria in the first row of the matrix: the comparison is knowing that you have respect for each of the 9 values of preference according to the scale of Saaty. Towards the end of the software calculates the weights attributed to each of the criteria by constructing a hierarchy between them. After each stretch of coast has been given a value given by

Value = i-th Σj (Score x Weight i-th criterion j-th):
Where i-th score is assigned based on the boolean method:

presence: Score 1 = i-th
absence: Score 0 = i-th

The result of this operation led to classify each part of the coast according to one of three values: high environmental sensitivity, environmental sensitivity medium, low environmental sensitivity.

The different levels of criticality and the erosion of environmental sensitivity were then crossed, giving rise to a classification with nine levels can provide reference information for the preparation of Municipal Coastal Plan (CCP).

In particular, the classification was as follows:

- C1S1. high criticality sensitivity and high;
- C1S2. high criticality and medium sensitivity;
- C1S3. high criticality and low sensitivity;
- C2S1. medium criticality and high sensitivity;
- C2S2. medium criticality and medium sensitivity;
- C2S3. medium criticality and low sensitivity;
- C3S1. low criticality and high sensitivity;
- C3S2. low criticality and medium sensitivity;
- C3S3. low criticality and low sensitivity.

Ultimately, the study has brought a significant contribution to the drafting of appropriate regulatory tools to ensure proper land management and the creation of a knowledge framework that must be continually updated.

For the purposes of the law classes have the critical task of conditioning the issuance of state concessions, while the classes of environmental sensitivity to influence the types of state concessions and how to contain its impacts.

3 Monitoring Values Change for Coastal Monitoring

3.1 General Data

The purpose of this second part of the study was to organize a monitoring system that can facilitate the control of the transformations on the coastal territory of Apulia Region: in particular, monitor and evaluate the real impact of the strategic initiative's plan on the environment and sustainability.

The methodology has been structured in relation to the objectives of the monitoring itself, so we opted for the structuring of an algorithm based on the feedback transmitter capable of communicating to the various phases and operate a continuous cycle.

It was considered to be appropriate for an assessment of "risk and vulnerability" for the most environmental, such as one arising from the plan, to ensure environmental aspects but also social and economic. The intersection between the classification of areas interested by the plan and the evaluation of the peculiarities and tendencies of development of the area at the base of the monitoring system so structured, allows a better understanding that facilitates the strategic assessment of the impacts of the initiative.

Briefly, the algorithm, starting from the evaluation of aspects such as to characterize the coastal areas (as classified by the plan based on criticality and sensitivity) from a point of view, socio-economic as well as natural, constitutes a "system of alerting", relatively to transformations land in contrast with its peculiarities.

3.2 The Evaluative Approach of Classification of Pressure Areas

To test the system structured as it is taken into account two coastal areas with different characteristics, namely the coastal territory of Monopoli, a medium sized city (about 50.000 inhabitants). The inland areas are bordered by a buffer variable that takes into account the physical characteristics of the terrain as defined by the Regional Coastal Plan.

The considered areas have a substantial variation in the morphology of the coastline, since Monopoli comes with a rocky coast north and south becomes quite sandy, are observed in the middle stretches of rocky coastline with sandy beaches to foot.

The coastal territory has been divided into three homogeneous areas: a first north of the (Monopoli 1) characterized by rocky shoreline and the presence of significant industrial areas, corresponding to a second 'urban area with the presence of the port (Monopoli 2); third that extends south from the end of the municipality (Monopoli 3), characterized by tourist sites of various kinds (holiday homes, villages, residences, beaches and entertainment venues) immersed in an agricultural and natural scenery of some significance given the presence of olive trees.

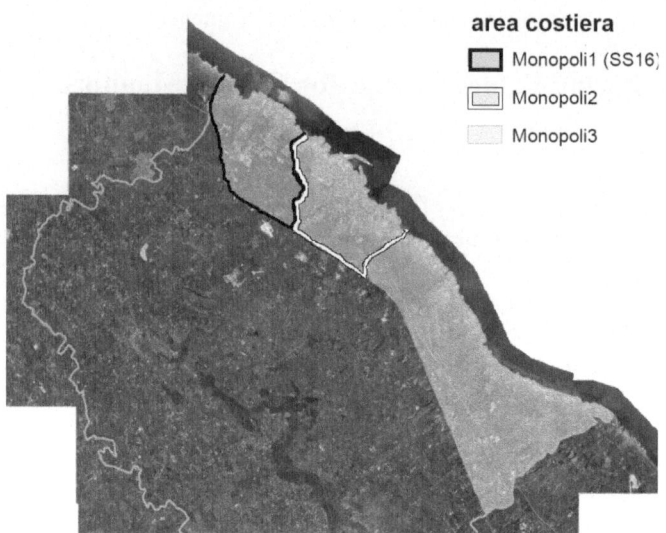

Fig. 1. The coastal areas of the case study (Monopoli 1, 2 and 3)

First step in the analysis of areas under consideration was the choice of data for the evaluation of naturalness of the characters, urban relevance, materiality port, agricultural relevance, importance of tourism, industrial relevance of the area.

To verify the effectiveness of the system are then assumed some plausible changes in the area, associated with the presence of e-planning dictated by the Municipality Plan. It is then evaluated the system's ability to read and grasp their greater or lesser compatibility with the regulatory classification of the plan.

Starting from classification based on the criticality and sensitivity, the following aspects are considered essential to characterize a coastal area:

naturalness (N),
relevance urban (U),
important agricultural (A),
industrial importance (I),
important tourist (T)
relevance port (P).

The choice of aspects to be monitored was made so that they are representing and explaining the action plan, simple and easy to interpret, based on readily available data and available, updated and upgraded at regular intervals, capable of showing the trend over time, sensitive and able to advise in relation to trends irreversible, measurable and have a space or geo-referenced "footprint". Based on the above shapefile from the land use have been created other documents describing aspects N, U, A, T, P and I, the uses for grouping categories as follows.

N: urban green areas, deciduous forests, coniferous forests, mixed coniferous and deciduous forests, meadows and pastures lined with trees, natural pasture, grassland, uncultivated, bushes and shrubs, areas in sclerophyll vegetation, tree-shrub areas evolving; areas to natural recolonization, recolonization areas at artificial surfaces to dense grass cover, beaches and sand dunes, bare rocks, cliffs, outcrops, areas with sparse vegetation, inland wetlands, salt marshes, salt marshes, intertidal marine areas, rivers, streams and ditches, canals and waterways, docks without overt productive uses, lagoons, coastal lakes and ponds, estuaries.

U: Continuous residential fabric, old and dense residential fabric continuous, dense, more recently, low; residential fabric continuous, dense, more recently, high, installation of large systems of public and private hospital settlements, settlements of technological systems; yards, spaces under construction and excavations, sports areas, cemeteries.

A: productive agricultural settlements; simple arable unirrigated areas; vegetable crops in open fields, greenhouses and under plastic in unirrigated areas; simple crops in irrigated areas, vegetable crops in open fields, greenhouses and under plastic in irrigated areas; vineyards, orchards and berry plantations, olive groves, other permanent crops, temporary crops associated with permanent crops, cropping systems and particle complexes, areas predominantly occupied by agricultural fields with significant areas of natural areas, forestry, soils and reworked artifacts.

T (receptive): campsites, tourist accommodation in bungalows or similar commercial establishment.

T (residential): residential fabric discontinuous residential fabric and rarely nucleiforme; scattered residential fabric.

P: port areas.

I: industrial or craft space with outbuildings, abandoned settlements, big plants concentration and sorting goods, networks and areas for distribution, production and transport of energy, mining areas, landfills, junkyards in the open, cemeteries of motor vehicles.

Note that Shapefile T (residential) was created by grouping all forms of residential fabric discontinuous that in most cases in coastal areas represent second homes, in T (receptive) were included in all commercial installations, which as classified in the paper of land uses include large hotels with attached bathing in these coastal areas are clearly prevalent forms of settlement on the other.

In the specific aspects have been ordered with respect to the relevance of the extension (covering of soil) and, as a function of potential changes, because of the danger of the transformation with respect to the criticality with respect to erosion and environmental sensitivity.

$$X = (\Sigma_i \alpha_i \gamma_i) \times \omega_C + (\Sigma_i \beta_i \gamma_i) \times \omega_S$$

$$\omega_C = 1 \ (C1) \ / \ 0.66 \ (C2) \ / \ 0.33 \ (C3); \ \omega_S = 1 \ (S1) \ / \ 0.66 \ (S2) \ / \ 0.33 \ (S3)$$

(1)

To facilitate the operation of pairwise comparison between the issues are first three classifications were made to facilitate the judgments of semantic Saaty: one concerning the importance of the extension. relative hazards of the transformation with respect to the critical coastal erosion, another relative hazards of the transformation with respect to environmental sensitivity. After the identification of Saaty's weights, the value have been transposed from the typical normalized eigenvalues of Saaty Matrix, to a standardized score (Table 1):

Table 1. Coefficient of extension γ

Land use	N	U	A	T (tour.)	T (res.)	P	I
Extension γ	0.16	0.08	0.03	0.38	0.12	1.00	0.27

The tables with the pressure values calculated as described for each weighted area with the values of ω_C and ω_S, are the following:

Table 2. Coefficient of criticality and sensitivity

Land use	N	U	A	T	P	I
criticality α	0.06	0.62	0.15	0.26	1.00	0.77
sensitivity β	0.06	0.73	0.15	0.28	0.59	1

Table 3. Adjusted pressure areas according to weighted coefficient of criticality and sensitivity for monitoring the change due to City plan implementation

Area	N (α_N,β_N)	U (α_U,β_U)	A (α_A,β_A)	I (α_I,β_I)	T (α_T,β_T)	P (α_P,β_P)
Weights	γ_N	γ_U	γ_A	γ_I	γ_T	γ_P
Monopoli1(c_x,s_y) (3.0, 2.6)	1.4800	0,3488	2.1258	0.5968	0.0000	3.5613
Monopoli2(c_x,s_y) (3.0, 2.6)	1.8705	2.5808	1.5036	0.5538	1.7000	0.6102
Monopoli3(c_x,s_y) (2.8, 2.2)	2.1152	0.1232	2.3502	1.4894	0.0000	0.0084

In the same way it has been possible to realize the matrices of the Saaty pairwise comparisons and determine the coefficients α_N, α_U, α_A, a_I, α_T, α_P, and the coefficients β_N, β_U, β_A, β_I, β_T, β_P, respectively for criticality and sensitivity (Table 3).

3.3 The Profiling of Coastal Municipalities

Based on this first trial, as part of a research project funded by the Region Apulia (identified by the acronym MOCA: Monitoring Of Coastal Areas), we proceeded to the realization of a software able to integrate the routine evaluation concerned with GIS technologies in collaboration with Polytechnic of Bari and the company Geodata SRL. The software is designed to manage a georeferenced database, which will facilitate the reading of the ongoing changes and potential, arising from plans, programs or interventions compared unplanned criticality and sensitivity to issues highlighted in the Coastal Plan.

The software can work on a database larger: in fact, the data relating to land use (aggregate indicators) are combined and joined together with other various useful to investigate situations of risk and danger as a local or ISTAT data relative to environmental components significant (disaggregated indicators).

The scales of analysis allowed by the software are different, the validation of the software was done working on a municipal scale, using the assessment of land use areas defined by administrative features.

A choice of this kind, however, involves the risk of evaluating the same manner similar transformations in the common characterized by a different "coastal character".

To remedy the problems highlighted above steps were taken in the testing phase to the implementation of data capable of characterizing simplicity with the characteristics of "coast-related" of each joint of the territory.

The means for this characterization is represented by a series of indicators, which are available and will be available for all common with part of the perimeter "wet". These are:

- Length of the coastline town;
- (Length of coast line city / municipal boundary) x 2;
- Sections of uniform classification legislation RCP / length of coast line city.

These indicators, suitably used in the routine evaluation help to relate the identified changes to the environment and coastal issues, "profiling" the territories of analysis.

If the first two indicators relate to the conformation of the analysis refers to the third floor as classified by the coast and serves to bind to each municipality impacts more or less probable because of the extent permitted by the standard resulting from the classification.

Very simply have been associated with each level of classification rules on the environmental sensitivity of potential impacts as follows. Levels of criticality has been connected the probability of occurrence of impacts identified.

The logical scheme implemented in the system is divided into the following phases:

- Identifying the scope of study;
- Definition of the coastal profile;
- Identification of potential impacts within the analysis (through classification of RCP);

- Combinations of land uses present on the CTR for N, U, A, T, P, I categories;
- Assessment of critical uses well defined with respect to coastal erosion and environmental sensitivity;
- Local and global analysis of variance;
- Local analysis of disaggregated indicators.

The software product, from both theoretical and practical gathered information, allows a uniform assessment of the environmental pressure caused by different land uses, with particular reference to critical coastal erosion and environmental sensitivity. The assessment may be conducted within the selected study, this according to some simple indicators is "profiled".

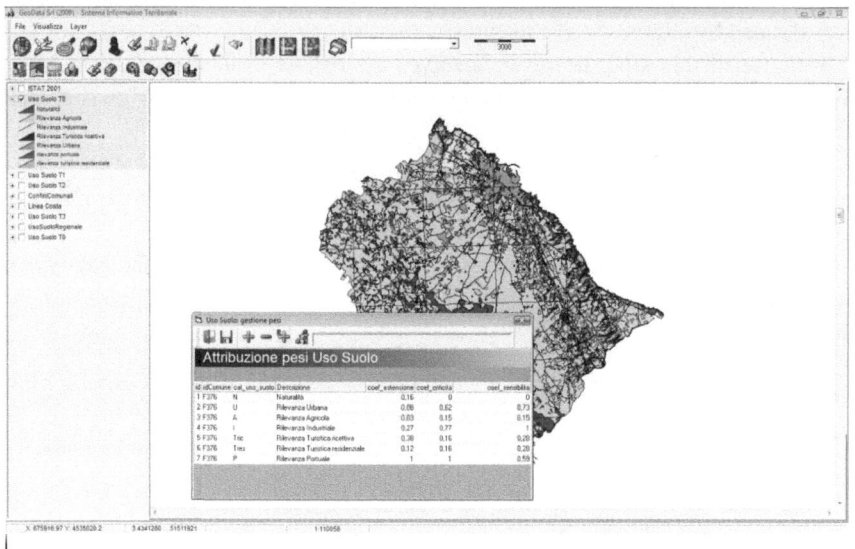

Fig. 2. The weighting of land use for the whole municipality area

The analyzes are thus relate field of study so as to be comparable between different areas. The assessment of the land use is a first information layer, follow this localized analysis of disaggregated indicators collected in databases that can be implemented continuously.

A significant aspect is related to adaptability to local contexts and coastal profiles of different sizes for analysis in different contexts and physical characteristics of size.

The possibility of identifying a field of study and the association of simple indicators for its characterization allows to opt for areas defined by administrative boundaries (as in the case of experimentation) but also through character definitions physico-morphological, sometimes more suited to 'analysis.

The following figure shows the computation of the coastal "shape coefficient" in the software.

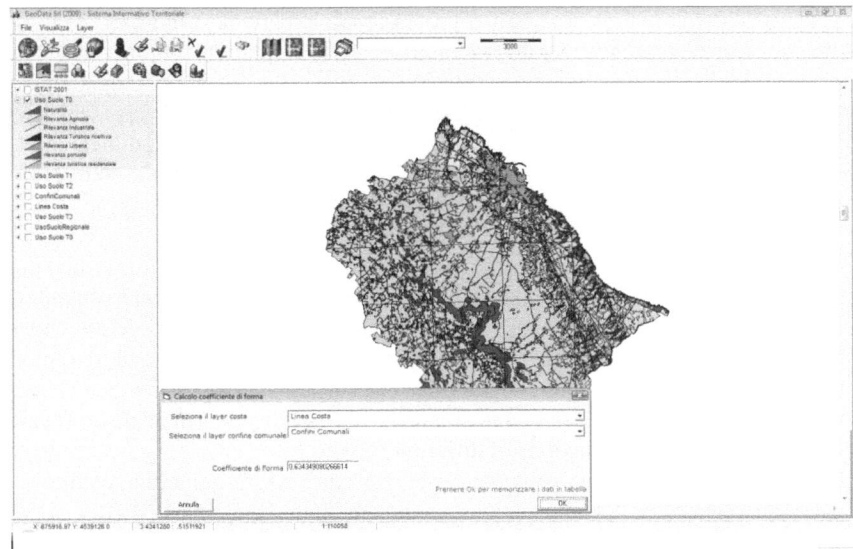

Fig. 3. The computation of the degree of coast influence on the municipality area, by the shape coefficient

The introduction of the "shape coefficient" allows, beyond the definition of the type of choice, of "weigh" the coastal character on the whole municipality area. The indicators chosen for profiling are valid for coastal areas of geometry and variable extension, this allows the possibility to perform the analyzes at any scale, relative to the needs identified.

The association to each area of a database allows disaggregated indicators in the areas of greatest interest, since they are subject to change or because exerting environmental pressure increased, more detailed analyzes. The indicators covered by this analysis may vary depending on the needs, because the databases are continuously updated and implemented. The cognitive maps produced by the software provides an excellent overview of the situations are monitored and very useful to guide the analysis of the disaggregated indicators.

4 Perspective and Remarks

The same theoretical and methodological steps taken to build the product are still replicable to other assessments, keeping fixed the basic knowledge on the classification of land uses.

However, it is not possible without an actual experimentation in other fields, to assess whether the routine structured as follows, although replicable, are the most appropriate for subjects of different nature. Either way, the product offers the possibility, through a simple user interface and at the same time flexible, to restructure the coefficients of impact in relation to different issues and to implement cognitive-different regulatory frameworks.

It seems clear, however, that only a professional, experienced in assessment methodologies, can consistently achieve a multi-criteria evaluation routines that can be imported into the system.

The evaluation system, fully implemented in software design, is sensitive to change in territory and allows an assessment with regard to global and local land use more or less compatible with coastal issues. It also allows you to render the results of analyzes using maps and cognitive evaluation.

Important results have shown the ability to monitor in addition to land use and classification of Coastal Plan (representing a fixed database and obviously updatable) any activity through indicators chosen appropriately according to local situations (in the trial were included but national statistic database nothing prevents you to widen or narrow the field of analysis as needed), the possibility of covering the entire region by comparing the analysis to settings with different coastal characteristics; the chance to work on different spatial scales, and finally by possibility to adapt the software to other developments in evaluations of different genres.

The analysis software can provide a useful framework to guide the synthesis and further analysis of the transformations, however, must be performed by competent figures.

Adaptability, flexibility, uniformity of analysis are the characteristics sought in the realization of the product, as tested meets these requirements.

References

1. Sadler, B., Verheem, R.: Strategic Environmental Assessment Status, Challenges and Future Directions. Ministry of Housing, Spatial Planning and the Environment WP 54, The Hague, Netherlands (1996)
2. Partidário, M.R.: Elements of an SEA framework. Improving the added-value of SEA. Environmental Impact Assessment Review 20(6), 647–663 (2000)
3. Sheate, W.R.: Tools, techniques & approaches for sustainability. World Scientific Publishing, Singapore (2010)
4. Fischer, T.B.: Theory and Practice of Strategic Environmental Assessment. Earthscan, London (2007)
5. Partidario, M.R., Arts, J.: Exploring the Concept of Strategic Environmental Assessment Follow-Up. Impact Assessment & Project Appraisal 23(3), 246–257 (2005)
6. Morrison-Saunders, A., Marshall, R., Arts, J.: Eia Follow up: International Best Practice Principles. International Association for Impact Assessment Special Publication Series n. 6, Fargo, USA (2007)
7. Marshall, R.: Environmental impact assessment follow-up and its benefits for industry. Impact Assessment and Project Appraisal 23(3), 191–198 (2005)
8. Kornov, L., Thissen, W.: Rationality in Decision and Policy-Making: Implications for Strategic Environmental Assessment. Impact Assessment & Project Appraisal 18(3), 191–200 (2000)
9. McCallum, D.: Planned follow-up, a basis for acting on EIAs. Working Paper, Annual Meeting of the International Association for Impact Assessment, Utrecht (1985)
10. Fischer, T.B., Gazzola, P.: SEA effectiveness criteria—equally valid in all countries? The case of Italy. Environmental Impact Assessment Review 26, 396–409 (2006)
11. Saaty, T.: The Analytic Hierarchy Process: Planning, Priority Setting, Resource Allocation. McGraw-Hill, New York (1980)

The Use of Ahp in a Multiactor Evaluation for Urban Development Programs: A Case Study

Luigi Fusco Girard[1] and Carmelo Maria Torre[2]

[1] Department of Conservation of Environmental and Architectural Heritage
University Federico II, Via Roma, 402 - 80134 Naples
girard@unina.it
[2] Department of Civil Engineering and Architecture
Polytechnic of Bari, Via Orabona 4, Bari Italy
torre@poliba.it

Abstract. The story of public policies and urban development plans is characterized by continuous deals, is affected by an dynamic institutional counterbalance among actors, a political re-formulation of actions that relevantly reduce the affordability of forecasts of success. The interaction among different actors affects the designed path of change, and the negotiation can change strategies, can accelerate or delay pieces of the predicted urban development and consequently modify the flow of positive/negative economic effect, according to the prevalence of one actor on the other inside the different step of the process. Since game theory to perspective theory economic literature describes each actor as afraid of losing the advantage obtained in a process, even in the case of urban transformation when another manages to get one more for himself. The relationship between major advantage of someone and less advantage of some other is relevant when two parties are in competition each other. In this paper a weighting system is applied to actors' view of reciprocal i8mportance, trying to understand who is going to affect more effectively the bargaining in a complex urban development process. L. Fusco Girard wrote the first and the second paragraph, C.M. Torre wrote the third and the fourth paragraph.

Keywords: Coevolution, AHP, Management Cash Flow, Scenarios.

1 Introduction

In the history of urban development policies, the conflict between actors has always been considered a crucial issue on which the field of urban and economic literature several authors have tried.

The bargaining is a fundamental element of any decision. The process is often lengthy, dynamic, constantly forming and burst coalitions, create alliances that change over the course of events. We attends a co-evolution of power relations, a political re-formulation of plans that relevantly reduce the affordability of forecasts of success.

The interaction among different actors affects the designed path of transformation, the negotiation can change strategies, can accelerate or delay pieces of the predicted urban development and consequently modify the flow of positive/negative economic effect, according to the prevalence of one actor on the other inside the different step of the process.

Each actor is afraid of losing the advantage obtained in a process of urban transformation when another manages to get one more for himself. He observes his/her potential enemy or allied, thinking who is going to weight mere at the end of the day, to gain the final prize. Economic rationality assumes a peculiar character in the multiactor process, as well explored in literature.

Someone look at the creativity, that can inspire the seek for new solution. In this process, as well it is possible to attribute a role to evaluation, as support to decision not only by assessing old solution, but also through the exploration of new perspectives.

Whereas a form of bias of the fact that each actor feels competing with all, we can refer to the perspective theory [1], which in some way does not consider the extent of the benefit as a whole, but in relation to a given threshold value; in this case the threshold can be related to the advantage obtained from the others. If a first subject obtains an advantage of equal to X, and a second subject obtains an advantage equal to Y, in an atmosphere of competition, the first subject perceive that he has obtained an advantage of X less Y.

This system of relationships between major advantage of some and less advantage of some other is true only when the parties are in competition each other. When a new solution appears, alliances can change, and the authoritativeness rises on the fore.

At least two aspects more have to be considered to explain why we face with a bias.

The first is that the interaction does not affect the actors involved in two by two, but all at once, so the change of balance in the process, considering all subjects, can not be interpreted as a sum of competition between pairs of antagonists. In a sense, every moment all parties can be potential antagonists or potential allies of one another;

The second aspect is that some processes are more beneficial overall than others. There are therefore some best coevolutionary processes and some worst coevolutionary processes from the point of view of mutual benefit.

Considering a crossroad between those theories (such as the perspective theory [1]) that explain some aspects of the relationship between future target and actor perception, and those theories (such as the Institutional Analysis [2]) that investigate the relationship between the evolution of contexts according to the changing importance of stakeholders in a process of urban transformation, the paper analyses the application of expected utility theory in a transformation managed by a number of actors, depending each other with imperfect sharing of objectives.

This is the case of the hypothesis of urban development carried on by the S.T.U. (society for urban development) of the Metropolitan City of Bari.

By the application of a original application of multi-group evaluation, the possibility of exploring the mutability of economic flows has been investigated and measured on different scenario, trying to join social complex dimension and financial values.

2   Public Private Partnership and Feasibility of Urban Policies

The decision making in urban processes is always more fragmented, related with multiple and conflicting interest, so that the uncertainty of pathways for developing the city increases, the results of town planning policies are pretty unpredictable, such as the future is unforeseeable.

Literature shows many seminal experiments especially involving actors and decision makers of strategic planning processes, where, quite often, the dimension of creativity in generation of idea is pretty emphasised.

In spite of literature pathways on ideas-making, the use of traditional indicators of economic feasibility is quite spread in the decisional arena; "feasibility" in practice still means a positive evidence coming from financial analyses, and the participation of public is reduced to meeting finalised at ratifying the evidence of analyses.

The speed of research and practices about collective interests in urban development therefore are quite inhomogeneous.

This paper investigates by referring to a case study, the possibility to create an hybridisation of economic and financial analysis used in support to the feasibility of great metropolitan transformation, starting from the experience of a feasibility study carried out to appraise the transformation in the Northern Area of the Metropolitan city of Bari, in the Italian "midday".

The work starts developing a drawing of the balance among the institutional and social actors representing counterparts in the decisional arena. Such approach starts from the assumption of the so-called prospect theory, to identify the way of observing each other used by actors in the multigroup decision-making process in the case study.

Utilising traditional qualitative weighting approaches the judgement about "who is relevant, where and when" a possible co-evolution of actor's relationship is commuted in factors affecting the future deviation of pathways of urban transformation from the predicted scenarios considered in traditional feasibility studies.

Finally, the different scenarios obtained in function of the role played by different counterparts is reduced to a limited universe of possible results and, consequently, the economic feasibility is pictured in a limited universe of possible revenues.

3   Methodological Aspects

In the concept of co-evolutionary institutional interaction, literature reminds some commonly accepted assumptions [3].

Actors try to pursue their future according to their objective; as a consequence, each actors do not simply accept the initial common agreement, but re-negotiate in any occasion the future development to make its own expectation real. Actors interact each other by negotiation; as a consequence, the future is a result of a continuous interaction where the most authoritative, or influent, or effective actor obtain a hypothesis of development close to its own vision of the World.

The weights of actors in the decision making process is dynamically variable. Each actor behaves reconsidering its actions according to the attribution of importance that she/he give to any other actors in every moment of negotiation. There is a perceived weight that affect the behaviour of actors, more than the real one.

A consistent part of this assumption create biases in rationality of actors [4], which tend to operate in function of the final objective, implementing an imperfect mediation, as traditionally reminded by most well known evidence of game theories [5].

Starting from the above assumption, we recognise that the investigation should make rising the perceived importance that each actor attributes in the real moment to the other. If we accept this way of developing the analysis, there is no relevant difference among private and public representative in the negotiate. All of they interact looking at objective, as individuals in a negotiate, thinking about which predictable event and which event due to the action of others can change the scenario in the future.

A simplest way to model this behaviour can be the attribution of a weight from each actor to all the others in order to obtain a general weight for each actor [6].

Fig. 1. The hierarchical process of weightings

A second level of weightings can be used to assess the power of events. In a similar way we can discuss bout the attribution of importance to dynamic events which can affect the revenue of a transformation. Starting from a perspective, we can consider that the attribution of importance and credibility to a given event is to be considered in function of the risk aversion more than to the general wellbeing due to a transformation. As a consequence, each couple of event can be by each actor weighted in a pairwise comparison other, to build a substitutive function of the well-known expected utility by von Neumann and Morgenstein model [7] . The incoherence admitted by the prospect theory, allows to use Saaty's eigenvalue for adjusting an obtaining a singular weight of events, in each transformation.

This model is applied in the case of study of this paper, where the model itself is applied to obtain a representation of the difference between the perceived revenue by the institutional actors, and the predicted net profit of the feasibility study included in the development plan carried out by a Company for Urban Transformation.

4 The Case Study of a Bargaining Game of Actors in the Society for Urban Development of "Bari Nord Area"

The experience of strategic planning in Italy regards metropolitan areas, and is still in fiery, since the deadline for proposals is established in 2008, September, 30.

Before of such events, some strategic dimension of urban development programs has been associated with the tray to start the constitution of Societies for Urban Development (named with the acronym 'STU'). The experimentation about STU had been carried out according with the several evolution of the proposed "Bassanini model", since the National Act 127/1997, until the National Act 166/2002.

That legislative document assumed that Companies for Urban Transformation can be constituted by municipalities and metropolitan cities. The main task of STU was to plan the urban transformation, to design and to implement the economic/physical realisation, by the involvement of the private sector, obtaining contemporary social and financial values [8]

The National Act 144/1999 disposed the realisation of a feasibility study for those STU that involve more than 50 million of euro, and, finally, the National Act 21/2001 provided by a competition to fund of the feasibility study of several Companies of urban transformation.

In this competition the Municipality of Bari obtained the sixth best result, and consequently received more or less thirty-nine thousand of euro, for producing a feasibility study.

At the beginning the City Council delimited two different great areas of interest ofr the feasibility study, which embraced more less the 30% of all the urbanised soil of the municipality.

Fig. 2. Evolution of areas of study: the first hypothesis (grey)and the final one (dark)

Finally, the study was reduced to a less consistent part of the urban area. The hypothesis of intervention defined two different context for two different STU: the object of this experiment was the "STU Nord Bari", regarding the most important Exhibition Area of Italy, after those ones of Milan and Naples, the so-called 'Fiera del Levante' (red area in Figure 3).

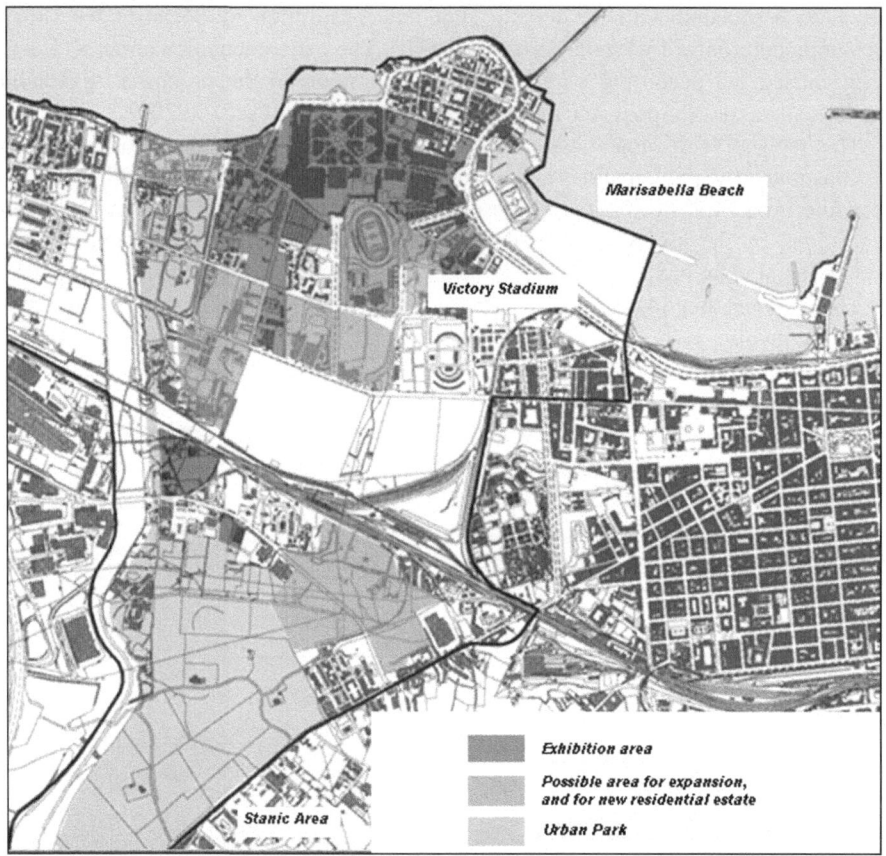

Fig. 3. The final delimitation of area of intervention, for STU North Area

Three different hypotheses of development were set as basis for future development:

the expansion on the seaside of the area, by the use of a artificial flow on the sea in the "Marisabella beach" (white coloured in Figure 3);

the expansion towards the internal side, near the Stadium of Victory (orange coloured in Figure 3)

the transfer in a partially dismissed industrial area, the so-called "Stanic" (green coloured in Figure 3)

Several different proposal had been developed in the past, and were object of public and political debate.

The success/failure of the intervention depend to some external factor and to the negotiation among at list the following social actors:

1. the Municipality of Bari, represented in a interview by the Major,
2. the Foundation of "Fiera del Levante", represented in a interview by the President
3. the Association of Entrepreneurs of Bari, represented in a interview by the President
4. the Authority of the Harbour of Bari, represented in a interview by the Head of the Port Authority
5. the local Association of residents, represented in a interview by some members
6. the political opposition, represented in a interview by the leader of the political minority in the City Council .

5   Evaluating Futures Development

All these futures has been seen differently connected with their own scopes by the different stakeholders.

The real state was that each future produced an improvement of condition for all of them, but as perspective theory affirm, they perceive as a loss any alternative future versus the most well expected.

Therefore a two level of interview was set: the first level identifies the perceived sthrengten of each actor, according to the interviewed stakeholder; the second level identifies the perception of possibility of happening of each external relevant event.

The interview were set to produce a variation of the traditional expected utility model. Remind that the traditional model of Von Neumann-Morgenstern is represented by an expected utility function:

$$U(c_1, c_2) = \pi_1 u(c_1) + \pi_2 u(c_2) \tag{1}$$

where $\pi 1$ and $\pi 2$ are known probability for the individual.

The state of the universe 1 is implementable with probability equal to $\pi 1$ and the individual has a payoff for c1, gaining the utility u(c1); the state of the universe 2 is implementable with probability equal to $\pi 2$ and the consumer has a payoff for c2, gaining the utility u(c2).

According to the prospect theory the individual does not perceive that both the state of the universe 1 and 2 generate utility, but he/she consider as a loss the smaller of the couple. Therefore if the payoff c1>c2 the payoff c2 is perceived as a loss.

The perceived counter position between gain and loss can be expressed by a pairwise comparison in a qualitative appraisal graduated by a Saaty's scale.

The loss can be considered extremely unacceptable, unacceptable, moderately unacceptable.

The Saaty eigenvalue will express different weightings that are complementary, supporting the realization of a function that in the same way of the Von Neumann-Morgenstern model generates a linear combination of the perceived utility of each event and of each different modality of occurring the event itself.

Steps are as follows:

1. reciprocal weighting of each actor, to evaluate the importance of actors and its ability to address the future towards some events; the weighting is based on the evidence given by the interviews;
2. weighting the relevance of events in function of their possibility to affect the transformation; the weighting is based on the expected value given by each one of the interviewed actors;
3. combining transformation according with their different perceived form by the actors in consequence of the different modality of occurring the events. The combination is based on the correction of the dimension of each project (for instance: the number of square meters of residential estate, and the consequent net present value varies according the possibility that the projects will be implemented more partially or completely).

The events are the following:

1. expansion of the "Fiera del Levante" exibition area,
2. new residential estate,
3. new offices and high tech buildings,
4. expansion of the harbor of Bari,
5. new buildings for cultural activities,
6. regeneration of seaside and tourism activity.

The modality of occurring the events are described below. The expansion of the exhibition area can occur according to three different addresses:

- expanding towards the sea, on the artificial platform of Marisabella Beach,
- expanding on the internal part, by a new urban renewal policy,
- transferring the Exhibition area to a dismissed brown field (named Stanic).

This expansion is counter-posed with the expansion of harbor activity, if it interests the sea side, and is counter-posed with residential estate if realized towards the internal part.

The new residential estate will be realized according to new demographic trends, and are counterposed with the office and high tech buildings, that can be located in the same area. The cultural activities will be realized in the hypothesis of total/partial shift of the exhibition centers in the Stanic area.

The external event that can affect the development of the area could address toward different pathways the use of soils.

For instance, a fast or slow demographic increase could activate the construction of a new area of residential estate near by the internal area south of "Fiera del Levante", and consequently could compromise the hypothesis of internal expansion of the exhibition area, especially considered as place for offices and high tech building.

In the same, the realisation of a road axis towards the sea line could compromise the use of the artificial flow of "Marisabella beach", in the north of the exhibition area.

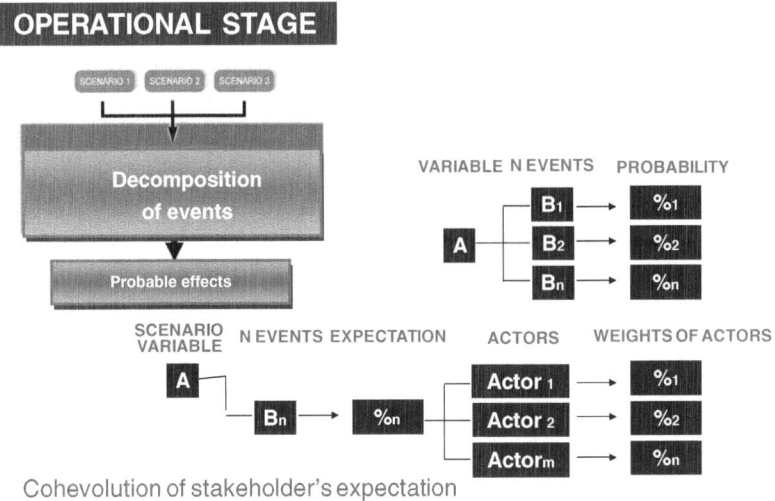

Fig. 4. The steps of reciprocal evaluation of actors' influence of events

Each actor has a different expectation regarding the listed events, and each actors consider more possible those events which are supported by the most influent stakeholder:

$$U(c_1,\ldots c_n) = w_1\, u(c_1) + w_2\, u(c_2) \ldots + w_n\, u(c_n) \qquad (2)$$

Fig. 5. An example of reciprocal assessment of weights in event by each actor

In the same, for each event a different revenue (neasured by its net present value) can be associated to its realizations.

The utility function u(ci) is therefore associated to the expression of the adjusted net present value NPV, for each event i:

$$NPV(c_1,...c_n) = w_1 \, NPV_1 + w_2 \, NPV_2 ... + w_n \, NPV_n \tag{3}$$

The final weights of each event is given by the following product:

$$w_i = w^{(1)}{}_i \cdot w^{(2)}{}_i \tag{4}$$

where

- the level (1) is affected by the combination of the weights representing the relevance of the actors, and
- the level (2) identifies the combination of the possibility of occurring a perceived modality of event.

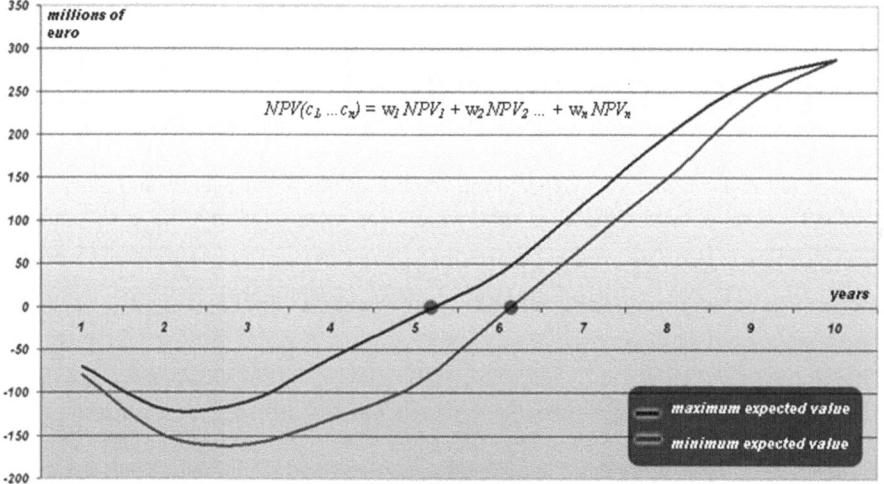

Fig. 6. The feasibility space of all the possible revenues deriving from the urban transformation of the STU Nord Area of Bari, is included inside the spindle-shaped area of the graph

The result can be expressed in a multidimensional form, in terms of each expected result (amount of new population, new square meters of area, new flows etc.) and in the same time can be expressed as an adjustment of the NPV already reported in the feasibility study produced for the STU Nord company.

The correction is shown in Figure 5, where we can see a spindle instead of a unique line of variation during time of the net present value.

6 Final Remarks

The prediction of future is always an interesting exercise, and in the same is a very heavy competence to carry out in the economic field. The future oriented behavior in the past was investigated by Khakee [9], and this experience seems to allow the evidence of his research.

Creativity comes from the comparison, however, and the search for better solutions to be shared in a humanistic perspective in which methods that favor the social budget plan recompose the conflicts between the actors [10].

In fact, part of the future is affected by the behavior of the same subject that are the most involved in predictions, that is to say in our case, the involved stakeholder in a urban transformation: they measure each other the potential in addressing transformation.

In the same time their acting is future oriented, and sometimes ethically, and not only rationally oriented, as supposed for instance by Akerlof and Kranton [11] The interviews that support the model seem to show such above aspects.

The use of quantitative parameters (as weightings), deriving from modelisation of interview's data, instead, can offer a useful simplification of future. The most utilized financial indicators can assume a value that can demonstrate the influence of imperfect rational future-oriented behavior.

References

1. Kahneman, D., Twersky, A.: Prospect Theory: An Analysis of Decision Making Under Risk. Econometrica 47(2), 263–291 (1979)
2. Funtowicz, S.O., Martinez-Alier, J., Munda, G., Ravetz, J.R.: Information tools for environmental policy under conditions of complexity. EC Environmental Issues Series 9, Bruxelles (1999)
3. Munda, G.: Social multi-criteria evaluation for a sustainable economy. Springer, New York (2007)
4. Fehr, E., Tyran, J.R.: Individual Irrationality and Aggregate Outcomes. Journal of Economic Perspectives 19(4), 43–65 (2005)
5. Nash, J.: The Bargaining Problem. Econometrica 18, 155–162 (1950)
6. Saaty, T.L., Vargas, L.G.: Decision Making with the Analytic Network Process. Springer, New York (2006)
7. Von Neumann, J., Morgenstern, O.: Theory of Games and Economic Behavior. Princeton, Univ. Press, Princeton (1944)
8. Fusco Girard, L.: Globalizzazione e città la sfida della diffusione delle opportunità. In: Gajo, P., Stanghellini, S. (eds.) La Valutazione Degli Investimenti Sul Territorio, pp. 11–28. Firenze University Press, Firenze (2003)
9. Khakee, A.: Relationship between futures studies and planning. European Journal of Operational Research 33(2), 200–211 (1988)
10. Lichfield, N.: Community Impact Evaluation. UCL Press, London (1996)
11. Akerlof, G.A., Kranton, R.: Economics and Identity. Journal of Economic Perspectives 115(3), 715–753 (2000)

Assessing Urban Transformations:
A SDSS for the Master Plan of Castel Capuano, Naples

Maria Cerreta and Pasquale De Toro

Department of Conservation of Architectural and Environmental Heritage, University of Naples
Federico II, via Roma 402, 80132 Naples, Italy
{cerreta,detoro}@unina.it

Abstract. The objective of this study is to present a spatial simulation modelling of real estate effects caused by urban transformations. The proposed approach extends the formalization of the "Monte Carlo" simulation methods in Geographical Information Systems (GIS), including spatial structure and temporal dynamics. The combined application can be useful in spatial decision making process for urban planning, supporting and modelling operations for urban land-use change. Analysing the new functions for the redevelopment of Castel Capuano, an historic building in Naples (Italy), the paper explores possible scenarios of transformations identifying the effects on the urban real estate market.

Keywords: Spatial Decision Support System, Monte Carlo simulation, Real estate market, Naples (Italy).

1 Introduction

In the recent years, spatially explicit simulation models of urban growth patterns have emerged. The economic versions of these models estimate land-use transition probabilities using discrete choice methods based on the behavior of agents making land-use decisions. Spatially explicit models use data from a Geographic Information System (GIS) to generate spatially disaggregated predictions of land-use change.

In recent works economic models of land-use change have been developed that are both spatially explicit and disaggregate, so that predicted outcomes may be to link ecological models of landscape changes. These modelling efforts require detailed parcel-level and GIS data that are often not widely available, limiting the possibility to apply the models to a broader region or transfer them to other areas altogether. Indeed, controlling for spatial effects in micro-level necessitates a range of analytical modifications varying from modest changes in data collection and the definition of variables to dramatic changes in the modelling of consumer and producer decision-making [1]. The main advantage of considering a micro-level is the opportunity to extend from using data at a scale that corresponds to the economic decision of interest; micro-level models can spatially aggregate up individual-level decisions to other relevant scales (e.g. city, labour market, agricultural market, real estate market, etc.) provide a unique means to assess the consequences of individual decisions and to analyse the impacts of policies directly. Then, because the unit of observation corresponds with the scale at which the underlying spatial process takes place, data measurement problems are minimised, which reduces a source of spatial error autocorrelation [1]. At the same time, the opportunity to overlay multiple layers of spatial data using GIS gives flexibility to describe the spatial aspects of economic, social and environmental problems, conceptualising spatial effects or patterns and showing the temporal and spatial distribution [2]. Visualising the results of policy analyses in map form may offer valuable information about the distributional impacts of specific policies or the potential cost savings from geographical targeting [3] [4]. The continuous changes in data collection, variable definition and communication of results have proved quite complementary to standard, empirical economic research methods. On the other side, relevant changes are evolving from the adoption of spatial econometric models and estimation approaches [5] [6].

In some cases, the changes coincide with modifications to traditional regression models: they reveal common approaches to the applied economics literature and empirical research, inspiring both different models and empirical methods (such as agent-based, Bayesian and geographically weighted regression models, etc.). Spatial and spatiotemporal econometric methods modify the representation of *consumer* and *producer* decision-making by bringing attention to spatial interactions among these decision-makers [1] [7].

In order to assess the potential effectiveness of urban policies, in this paper we examine correlations between real estate market and urban transformation, underlying the need to develop a Spatial Decision Support System (SDSS) in order to analyse the different impacts and represent their spatial implications. Through the empirical investigation in an operative case study related to the effects of the Master Plan of Castel Capuano, in the historic center of Naples, in the South of Italy, it has been possible to elaborate a dynamic map of impacts on the real estate market, result of the integration of Monte Carlo simulation and GIS. The SDSS emphasises the spatial distribution of the urban use/cover units, the real estate values and the spatio-temporal patterns, which were modelled by urban use/cover change trajectories over a series of observation years. Using the integrated GIS, several spatial variables were derived, including the proximity to historic monuments, major roads and public areas/piazzas, but also to economic and training activities. A simulation model was implemented to establish relationships between urban transformations and above spatial variables, capable of estimating the spatial probability of the Castel Capuano transformation effects on local real estate market.

2 Castel Capuano Master Plan: Analysis of Urban Transformations

Castel Capuano is an ancient castle, located in the historic center of Naples and, in particular, in the area that, in 1995, was proclaimed by UNESCO "World Heritage". In fact, the historical center of Naples has been listed as a UNESCO World Heritage

Site and the inscription refers to the extension of Old Town introduced with the approval of the General Plan of the city in 1972, and part of the historic city located in the new General Plan approved in 2004. The identification of new functions for Castel Capuano, not more a court since 2005, can be seen as part of a wider strategy for the redevelopment of the historic center, which could have a significant impact on the entire city, especially if interventions were to be activated with a high index of employment, to improve the socio-economic context (fig. 1).

In this perspective, the Management Plan of UNESCO sites, according to the Plan proposed by the Ministry of Heritage and Culture, explicitly provides for a phase of "integrated development", both cultural and economic, aims to promote the cultural value sites and emergencies affecting them, and to foster growth opportunities for enjoyment of all economic sectors and industries (tourism, hospitality, commercial, cultural, etc.).

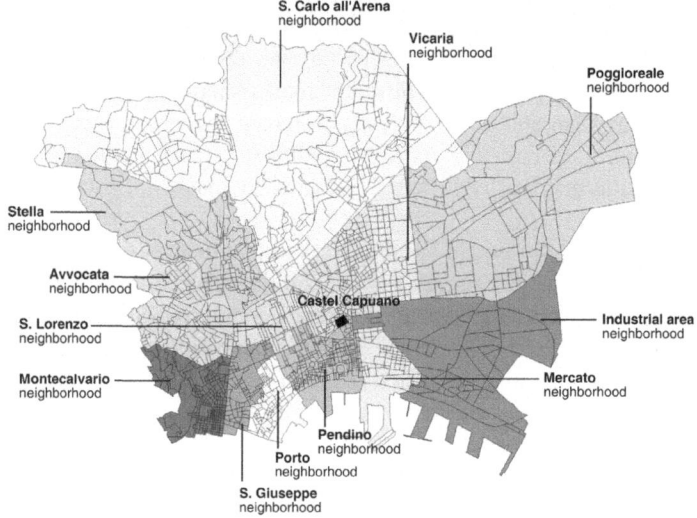

Fig. 1. Castel Capuano: localization in the historic centre of Naples

Protection and enhancement of historic and architectural heritage means, therefore, providing appropriate interventions to establish new functions that not only support the system of cultural heritage, but also the potential related to the existing socio-economic functions and/or future ones.

Taking into account the broader context of the city of Naples, the socio-economic framework and the proposed transformations, it was decided to evaluate the real estate impacts and the overall improvement of urban environment as a result of the settlement in the architectural complex of new functions.

In particular, the methodological process was articulated into the following phases:

1. *first phase*, in which it has been selected a preliminary set of socio-economic indicators, able to describe the characteristics of the urban environment in question, with reference to data already available from official sources. We

analyzed two main categories of indicators: the first category for the economic activities in the area of study, analyzing the dynamics starting from 2001 (year of the last Census Istat), the second category relating to real estate values recorded and available at Borsa Immobiliare of Naples, which define and characterize the study area, considering the trend over time too. The study area has been defined as a function of the characteristics and trends found in the course of the analysis of the data;

2. *second phase*, in which it has been constructed an adequate Geographic Information System (GIS), with the different types of data collected, useful to look at trends and possible relationships between the different variables considered, with reference to an area of influence within the perimeter of the selected study and significant for the determination of the impacts related to different activities. The GIS contains processed data available regarding the characteristics of the population and buildings, the presence of businesses and activities, with the number of employees, and the dynamics of the housing market, with reference to market values of transactions that took place. Specifically, data relating to real estate market consider the period between 2007 and 2010, divided into eight semesters, and refers both to houses (Market Value and Rental Value) and the shops (Market Value and Rental Value). The data on population, housing stock, businesses and activities refer to Istat Census 2001. All data are reported in detail for particles and are related to the census districts (or parts of neighborhoods) are considered significant for the study (fig. 2);

3. *third phase*, in which it has been examined the functions to be set up in terms of induced effects on the urban context, using tools of multidimensional assessments. By building simulation models of the impacts produced by the installation of the new features, we analyzed the likely consequences of urban change and tested the relevance of their implications. In particular, at this stage we have identified, described and assessed the significant effects of the intervention analyzing the context of reference through: a. the construction of suitable indicators, in order to describe the effects of the proposed transformations; b. the assessment of potential environmental, social, and economic functions of the proposals through the use of appropriate simulation models applied to the dynamics of real estate values.

In order to understand the dynamics of socio-economic context relating to Castel Capuano, has been defined the perimeter of the study area (fig. 2), selected from the available data.

The phases of the methodological path are summarized in the diagram of fig. 3, showing the relationships between the GIS and simulation models of the impacts, which are a prerequisite for the analysis of the two main scenarios: *Scenario 1*, which identifies the features of the environment after 2005, the year from discontinued original functions present in Castel Capuano; *Scenario 2*, which identifies the potential impacts on the environment analyzed in real estate dynamics, determined by the inclusion of new functions in Castel Capuano in synergy with some territorial changes already planned by the municipality.

Fig. 2. Castel Capuano: perimeter of the study area

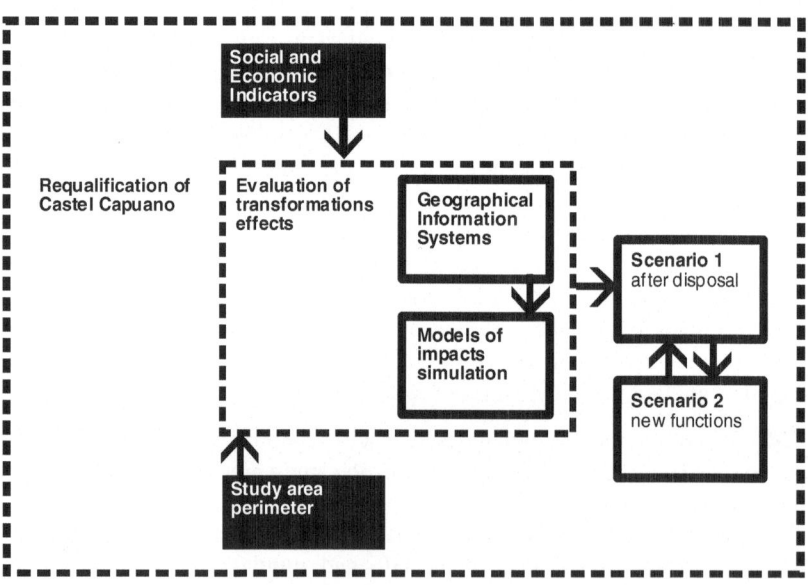

Fig. 3. The methodological phases

In particular, *Scenario 2* was projected to 2020, distinguishing between three possible developments: *2020 minimum scenario*, *2020 middle scenario* and *2020 maximum scenario*.

3 Evaluation of Transformations Effects: A Perspective to 2020

The cognitive framework, taking into account the socio-economic characteristics of the study area and the dynamics of the housing market, has identified the potential and the critical context affected by the changes envisaged for Castel Capuano, outlining the main features characterizing the Scenario 1 (see § 2), and refers to the period after 2005, the year of the original functions disposal.

In order to understand the impacts following the redevelopment, it has been analyzed the Scenario 2, which represents a projection in 2020 of three possible developments: a *low scenario*, an *intermediate scenario* and a *maximum scenario*. Therefore, Scenario 2 analyzes the features identified, considering the wider context of urban transformation.

Thus, the proposal to create a "Legal Culture Centre" evoking the previous functions, but hosting new activities full fitting the current needs of the community, together with the idea of turning part of the building into a museum and a training centre, has been examined referring to the possible impacts that the project could have on the surrounding urban context. In particular, the intervention strategy was focused on design alternatives that can balance the demands of functions for the Judiciary and Lawyers with those of a cultural tourism linked to a route developing along the Decumanus major, that recognizes Castel Capuano as a landmark for the eastern approach to the ancient centre of the city.

The functions identified by the Master Plan are related to the operations of transformation/valorisation provided at the municipal level. In particular, the "Plan of the 100 stations" aims to improve and redevelop the area served by the network of public transport, with interventions to get better the accessibility and quality of architecture and urban stations (such as the metro station of Porta Capuana), public areas and squares. Besides, places, car parks and bus stations are tourist "nodes" that form the modal interchange between the different transport services. Furthermore, the enhancement process suggested by the Master Plan fits consistently into the broader Integrated Program for the Historic Urban Center UNESCO Heritage that aims to enable development processes and significantly develop the environment and quality of life of residents.

Taking into account the different redevelopment strategies planned and in progress, it was possible to develop a scenario of future transformation of Castel Capuano and its context. The impacts assessment focused on the dynamics of real estate market projected to 2020, assuming the consequences that may occur following the implementation of the planned changes. In particular, using an appropriate simulation model of the impacts, which falls into the family of models of Monte Carlo simulation [8] [9] [10], integrated into the GIS platform, it was possible to analyze the dynamics of the rental value and market value, both of houses and shops, for three scenarios [11] [12] [13] [14] [15] [16] [17]:

1. *2020 minimum scenario*, where it is assumed that the transformations can have effects that do not significantly interact with the context;
2. *2020 middle scenario*, where it is assumed that the changes can get together with the context and result in significant effects;
3. *2020 maximum scenario*, where it is assumed that the transformations can be integrated with the social and economic context, resulting in a substantial improvement in quality of life and welfare conditions of the local community.

The three scenarios have been simulated referring to the average of the values found in 2010 and defining, within the study area, a radius of influence of 500 meters, useful to identify differences between the two main types of impact: *direct impacts* (within the radius of influence) and *indirect impacts* (outside the range of influence) (fig. 4). Indeed, it is believed that the planned changes can be more significant in the areas overlooking Castel Capuano less so in more distant areas. Clearly, the distance cannot be the only parameter to be taken into account, but it can help changing some dynamics closely related to economic and social uses.

Finding the area of influence has led to be take two different coefficients of simulation, two "risk coefficients" [18], one for the area inside the perimeter and another for the outdoor area, selected according to the socio-economic context and in analogy with other studies in structured contexts considered similar for both types of functions established and planned changes.

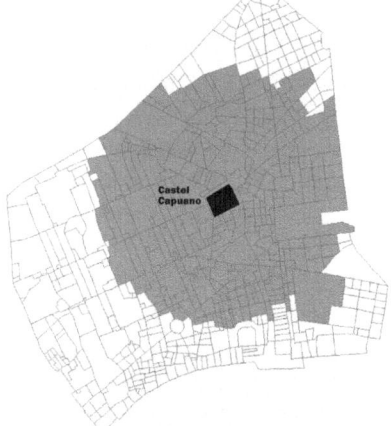

Fig. 4. The area of influence for the evaluation of direct impacts

Referring to the rental value of dwellings, for the three scenarios analyzed (fig. 5), it is possible to show how values have changed:
1. for the *2020 minimum scenario*, the rental values in the areas close to Castel Capuano are included within two main ranges, 4.98 to 5.67 euros/sqm and 5.68 to 6, 34 euros/sqm. The values do not differ greatly from those experienced in 2008 and 2009, but are lower than those recorded in 2010;

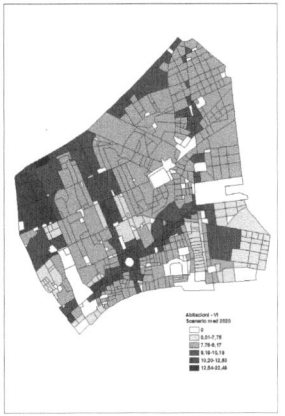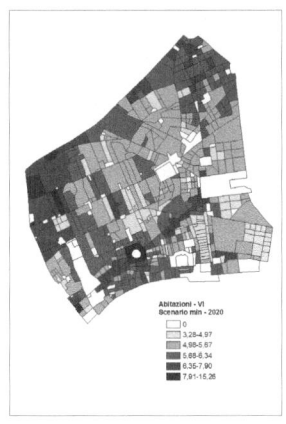

Fig. 5. Rental values of dwelling: three scenarios

2. for the *2020 middle scenario*, the rental values in the areas close to Castel Capuano are in the range 7.76 to 9.17 euros/sqm, with values slightly higher than those recorded in 2010. This result allows to emphasize that the ongoing changes have resulted in a steady improvement in economic and social environment and a slight recovery from the general state of crisis;
3. for the *2020 maximum scenario*, the rental values in the areas close to Castel Capuano vary more, identifying three possible ranges of variation (12.78 to 14.30 euros/sqm, 14.31 to 16, 53 euro/sqm, 16.54 to 31.99), the highest of which for a single particle census corresponds to the highest value expected in Piazza Nicola Amore. The maximum scenario, therefore, outlines a more dynamic perspective with effects that can make the areas near Castel Capuano more attractive.

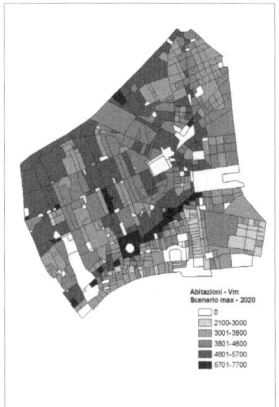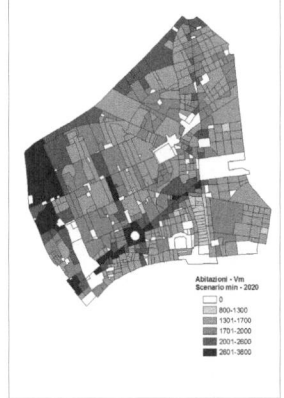

Fig. 6. Market values of dwelling: three scenarios

Referring to the market value of houses, analyzed for the three scenarios (fig. 6), it is possible to show that the values have changed as follows:

1. for the *2020 minimum scenario*, the market values in areas close to Castel Capuano are characterized by three main ranges of variation (1300-1700 euros/sqm, 1700-2000 euros/sqm, 2000-2600 euros/sqm). The estimated values are consistent with surveys conducted from 2007 to 2009, even if they present the lowest average;
2. for the *2020 middle scenario*, the market values in areas close to Castel Capuano are included in two main ranges of variation (2500-3000 euros/sqm and 3000-3750 euros/sqm), with values that are below the maximum values observed in 2010 (4000 euros/sqm);
3. for the 2020 maximum scenario, the market values in areas close to Castel Capuano appear divided into two main ranges of variation (3800-4600 euros/sqm and 4600-5700 euros/sqm). In this case, the values significantly increase in consistency with the implementation of strategies of transformation able to activate positive dynamics of development for the context.

Referring to the rental value of the shops, analyzed for the three scenarios (fig. 7), it is possible to show how values have changed:

1. for the *2020 minimum scenario*, the rental values in the areas close to Castel Capuano vary in two major ranges (11.29 to 14.7 euros/sqm and from 14.76 to 19.46 euros/sqm). The estimated values are consistent with the average values in the period 2007-2010 and that they are not affected by the possible positive effects of the planned changes;
2. for the *2020 middle scenario*, the rental values in the areas close to Castel Capuano are characterized by a single range of variation between the values from 17.82 to 27.08 euros/sqm; these values are very similar to those of 2010, which outline a pattern that does not significantly increase;
3. for the *2020 maximum scenario*, the rental values in the areas close to Castel Capuano are characterized by two main ranges of variation (26.60 to 35.65 euros/sqm and from 35.66 to 46.03 euros/sqm) which are definitely higher than the maximum values (25.30 euros/sqm) reported in the same area in 2010. In this scenario, the dynamics of the redevelopment can make the area more attractive for retailers.

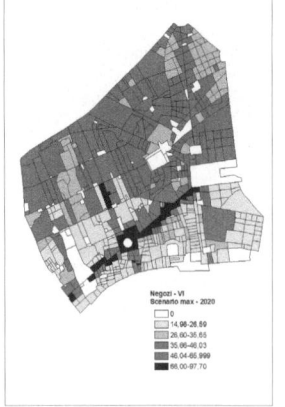  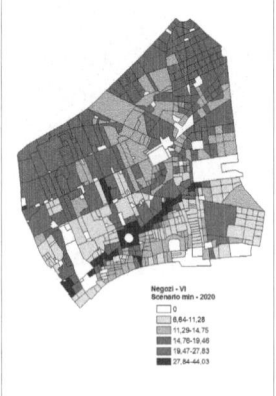

Fig. 7. Rental values of stores: three scenarios

Referring to the market value of the shops, analyzed for the three scenarios (fig. 8), it is possible to show that the values have changed as follows:

4. for the *2020 minimum scenario*, the market values in areas close to Castel Capuano vary between 2250-3130 euros/sqm, with a few examples of the lower interval (1100-2250 euros/sqm) and the higher (3130-4450 euros/sqm);
5. for the *2020 middle scenario*, the market values in areas close to Castel Capuano are characterized mainly by two intervals of significant changes to the values reported 3251-4100 euros/sqm and 4100-5450 euros/sqm, higher than values observed in 2010 and able to detect a slight recovery of the dynamic properties in the area in question;
6. for the *2020 maximum scenario*, the market values in areas close to Castel Capuano are characterized by a predominance of the value range 5400-7600 euros/sqm, with a few appearances in the range of lower values (3300-5400 euros/sqm) and the one with the highest values (7600 to 10900 euros/sqm). The highest values are, however, below the maximum ones in the same area in 2010, and indicate a significant difficulty in investing in businesses, as it emerges from the comparison with the maximum values found in Piazza Nicola Amore and that are in the range 16200-20900 euros/sqm.

Fig. 8. Market values of stores: three scenarios

The *2020 maximum scenario* identify a perspective of development of the area cautiously optimistic, recognizing the opportunity that Castel Capuano becomes a catalyst in an area that currently suffers from various negative influences, due to the disposal of property, and a more general state of crisis and decay that characterizes the environment and the city. It is likely that the *2020 middle scenario* will outline a more realistic perspective of exploitation, that sees economic activities as leverages able to structure a system of synergies and mutual support.

In order to explore the potential of the intermediate scenario, two other simulations have been developed to compare the results of the *2020 middle scenario* with the average scenario of 2010, for both the rental value and for the market value of the stores (fig. 9).

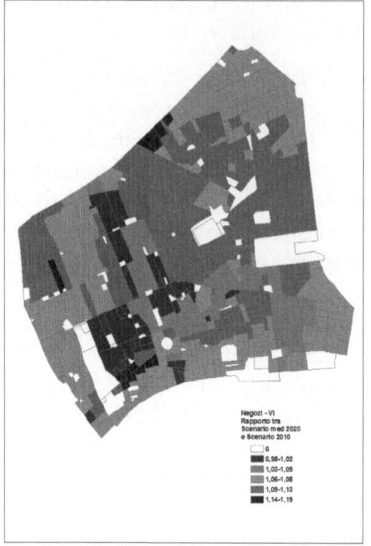 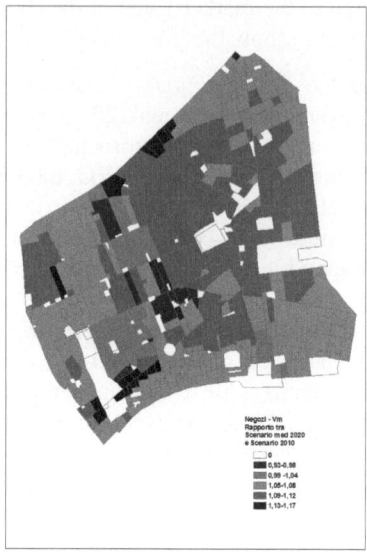

Fig. 9. Rental value of stores (€/mq) and market value of stores (€/mq): comparison between *2020 middle scenario* and average scenario of 2010

It was considered appropriate to carry out the analysis only on the values for the shops, because they are able to better reflect the impacts of urban change determined by the new functions located in Castel Capuano. The analysis of the maps shows that the change in value of location is in the range 1.09 to 1.13, and the change in market value considered interval 1.09 to 1.12. Basically the values are similar, allowing to define a common trend in increasing values, expression of the consequences envisaged after the implementation of planned interventions.

4 Conclusions

The results of the study highlight that the changes envisaged in the Master Plan for Castel Capuano are able to promote a process of active exploitation of the surrounding context, but need to relate to a broader and complex urban renewal strategy, capable to influence the social and economic fabric.

The application of Monte Carlo simulations of the variables, which are based on statistical measures and probability distributions, make it possible to incorporate the uncertainty of valuation parameters, in particular of future real estate values, using empirical data to extract information about the probability distributions of the related parameters and suggest a simple model to analyse the Castel Capuano transformation effects. Our empirical results suggest that simulations are expression of the distribution of values able to improve long-term decisions in real estate, but above all in urban transformation [18] [19] [20] [21].

From the analysis of impacts on the dynamics of real estate market, it seems clear that they can consolidate and increase according to the relevant values that characterize urban areas already considered as landmarks, such as Piazza Nicola Amore and Corso Umberto I. Therefore, it may be significant to establish relationships, tangible and intangible ones, including areas already known to lead the local economy and that are object of requalification, in order to define the common synergistic relationships aiming to get a total redevelopment of the city.

The results of elaborations emphasize the need to promote strategies and interventions that can be implemented at different scales, by acting on the monumental building, on minor building, on urban spaces and the system of connections, thereby establishing a relationship of constant interaction between the various interventions in order to implement a concrete process of development and enhancement of a relevant area of the city of Naples. Indeed, future studies will further investigate the impacts and interplay of different spatial variables on urban transformation patterns in order to analyse the interaction among use values, no-use values and intrinsic values that may also play important roles in Castel Capuano valorisation process.

References

1. Bell, K.P., Timothy, J.D.: Spatial Economic Analysis in Data-Rich Environments. Journal of Agricultural Economics 58(3), 487–501 (2007)
2. Zhang, Z., Fan, S., Cai, X.: The path of technology diffusion: Which neighbors to learn from. Contemporary Economic Policy 20, 470–478 (2002)
3. Yang, W., Khanna, M., Farnsworth, R., Onal, H.: Is geographical targeting cost-effective? The case of the conservation reserve enhancement program in Illinois. Review of Agricultural Economics 27, 70–88 (2005)
4. Newburn, D.A., Berck, P., Merenlender, A.M.: Habitat and open space at risk of land use conversion: Targeting strategies for land conservation. American Journal of Agricultural Economics 88, 28–42 (2006)
5. Anselin, L., Florax, R.J.G.M., Rey, S.J.: Econometrics for spatial models: Recent advances. In: Anselin, L., Florax, R.J.G.M., Rey, S.J. (eds.) Advances in Spatial Econometrics. Methodology, Tools and Application, pp. 1–25. Springer, Berlin (2004)
6. LeSage, J.P., Pace, R.K. (eds.): Spatial and Spatiotemporal Econometrics, Advances in Econometrics, vol. 18. Elsevier Publishing, Oxford (2004)
7. Liu, H., Zhou, Q.: Developing urban growth predictions from spatial indicators based on multi-temporal images. Computers, Environment and Urban Systems 29, 580–594 (2005)
8. Kalos, M.H., Whitlock, P.A.: Monte Carlo Methods. John Wiley and Sons, New York (1986)
9. Kelliher, C.F., Mahoney, L.S.: Using Monte Carlo simulation to improve long term investment decisions. The Appraisal Journal 68(1), 44–56 (2000)
10. Kroese, D.P., Taimre, T., Botev, Z.I.: Handbook of Monte Carlo Methods. John Wiley and Sons, New York (2011)
11. Webster, C., Wu, F.: Coase, spatial pricing and self-organising cities. Urban Studies 38, 2037–2054 (2001)

12. White, R., Engelen, G.: High resolution integrated modelling of the spatial dynamics of urban and regional systems. Computers, Environment and Urban Systems 24, 383–400 (2000)
13. Appleton, K., Lovett, A.: GIS-based visualisation of development proposals: reactions from planning and related professionals. Computers, Environment and Urban Systems 29, 321–339 (2005)
14. El Yacoubi, S., El Jai, A., Jacewicz, P., Pausas, J.G.: LUCAS: an original tool for landscape modelling. Environmental Modelling and Software 18, 429–437 (2003)
15. Moreno, N., Ablan, M., Tonella, G.: SpaSim: a software to simulate cellular automata models. In: Proceedings of Integrated Assessment and Decision Support, First Biennial Meeting of the International Environmental Modelling and Software Society, Lugano, Switzerland, June 24-27 (2002)
16. Wu, F.: Simulating temporal fluctuations of real estate development in a cellular automata city. Transactions in GIS 7, 193–210 (2003)
17. Pang, M.Y.C., Shi, W.Z.: Development of a process-based model for dynamic interaction in spatio-temporal GIS. Geoinformatica 6, 323–344 (2002)
18. Hoesli, M., Elion Jani, E., Bender, A.: Monte Carlo Simulations for Real Estate Valuation. Research Paper 148. FAME, International Center for Financial Asset Management and Engineering, Université de Genève (2005)
19. Kelliher, C.F., Mahoney, L.S.: Using Monte Carlo simulation to improve long term investment decisions. The Appraisal Journal 68(1), 44–56 (2000)
20. Tucker, W.F.: A real estate portfolio optimizer utilizing spreadsheet modelling with Markowitz mean-variance optimization and Monte-Carlo simulation. Working paper. John Hopkins University Maryland (2001)
21. French, N., Gabrielli, L.: The uncertainty of valuation. Journal of Property Investment and Finance 22(6), 484–500 (2004)

Computational Context to Promote Geographic Information Systems toward Human-Centric Perspectives

Luis Paulo da Silva Carvalho[1,2] and Paulo Caetano da Silva[2]

[1] IRT – Recôncavo Institute of Technology, Av. Tancredo Neves, 805, Terceiro andar,
Caminho da árvores, Salvador, Bahia, Brazil
luisp@reconcavo.org.br
[2] UNIFACS – University of Salvador, Rua Ponciano de Oliveira, 126, Rio Vermelho,
Salvador, BA, Brasil
paulocaetano.dasilva@gmail.com

Abstract. Spatial information is of vital importance to Geographic Information Systems, but it does not suffice to its application to solve problems that require a human-centric perspective. For this to be accomplished, it must be considered other context dimensions, such as the personal data of users. With this purpose, GIS can be combined with Computational Context in order to produce software parameterized by information other than user's location. Moreover, considering the shift of GIS toward web/distributed approaches, in this paper, it is also surveyed the placement of the proposed solution in composite Web Services to enable the coupling of geographic systems with worldwide client software.

Keywords: GIS, Context-awareness, CCMF, Web Services.

1 Introduction

The Brazilian government released, in 2011, November, a social program by which patients of the public healthcare system would be visited, at their houses, by healthcare professionals [1]. The intention of the program is to reduce the crowding and the costs related to the maintenance of public hospitals and clinics and, as well, to provide to patients the opportunity to be treated in an environment where they feel comfortable, relaxed and accommodated: their homes. Considering this scenario, it is important to investigate how software can benefit people who must carry outdoor missions with the necessary tools and mechanisms to run their tasks, for instance, geographic maps showing the location of patients who must be treated by the healthcare professionals of the aforementioned Brazilian program.

According to [2], conventional spatial information is important to Geographic Information Systems (GIS), but it is not enough. Information related to time, activities and personal data must also be taken in account in order to enhance this type of software toward a human-centric perspective. For instance, healthcare professionals might want to know about a certain patient's medical profile in order to prioritize visits during a mission. In this case, geographical software could show maps on which the location of patients who suffer from chronic health conditions would be highlighted.

Computational context is defined by [3] as any information used to characterize the situation of an entity, in which an entity can be a place, a person or an object that is considered relevant to the interaction between users and software. As stated by [4], context is used when it is necessary to parameterize software in order to provide automated adaptations. For instance, geographic software embedded into handheld smart devices could be parameterized by information about patients, e.g. their location and medical profiles, to provide maps showing routes to favor patients who suffer from critical health conditions. In this example, "location" is the information that characterizes the situation of an entity, "patient" and it is used to provide a software-based adaptation: provision of contextualized maps.

As maintained by [5], geographic software has undertaken a shift from merely providing desktop solutions to distributed web/internet architectures, e.g the ones oriented to the utilization of Web Services. Web Services are platform-independent autonomous entities which can be used to execute tasks ranging from providing trivial answers to users request to the processing of complex algorithms. According to [6] Web Services have also become of vital importance to context-aware software since it promotes dynamic behavior, content adaptation and simplicity of use to end users. Thus, provided that service orientation serves to both context-aware and geographic software, in this paper, it is covered the combination of GIS and Computational Context around SOA, Service Oriented Architecture, in order to produce contextualized human-centric geographic systems. Table 1 summarizes the contributions from each area in the production of this assortment of software.

Table 1. Contributions of considered areas

Computer Science Area	Contribution
Computational Context	Representation of context information within human-centric perspective, e.g. time-related (temporal) data, preferences, location.
GIS	Provision of geographic-related resources and functionalities, e.g. maps, geographic location of objects and people, calculation of routes.
Service Oriented Architecture	Placement of the generated software in the internet so that remote computational clients might be benefited by contextual adaptation related to geographic systems.

This paper is organized as follows: in Section 2, related works are reviewed. In Section 3, it is proposed the application of the framework presented in [7] to develop contextualized geographic software. The framework is instantiated in Section 4 to develop a domiciliary health care geographic system in order to illustrate its application. Section 5 encloses conclusions and future work.

2   Related Works

In this section, related works are discussed. In specific, it is surveyed proposals which are related to the binding of Computational Context and Web Services to GIS with the purpose of enabling human-centric approaches in the making of geographic systems.

In [2], it is proposed a middleware infrastructure which enables the representation and utilization of context dimensions by geographic systems. Context dimensions have the purpose of categorizing information related to situations of use of context-aware software [8]. The dimensions, commonly referred as the four W's, are: (i) "Who", which identifies people who run a task (e.g. healthcare professionals who visit patients); (ii) "What", that represents the activity being performed (e.g. visiting patients at their houses); (iii) "Where", which represents information related to people and objects location (e.g. patient's house) and (iv) "When", that has the purpose of symbolizing time-related information about context (e.g. the scheduled date and time of a visit). The infrastructure is integrated with web technologies, e.g. Google Earth and Google Maps, in order to interface with the user. This approach, however, does not take full advantages of the integration with remote services. Other than providing only interface to end users, the calculation and retrieval of routes, for instance, could also be trusted to Web Services.

[9] present an approach, CM-GIS, Context-Mobile-GIS, intended to integrate Computational Context and mobile geographic software in order to add value to the information displayed to users. CM-GIS relies on ontologies to describe dimensions of context information used by Web Services. The utilization of Web Services enables the transference of processing from mobile devices to robust remote machines. Software generated by CM-GIS, nevertheless, has to be parameterized by information provided by the user, e.g. name of places and people whose locations must be shown on maps, which opposes to a key feature that Computational Context is intended to grant to adaptable software [4]: automatic parameterization of its functionalities to avoid manual entries from users. For instance, healthcare professionals should not be required to enter the patient's name in order to visualize his location.

[10] proposes a framework to combine Computational Context, Web Services and GIS in the making of tourist guide applications. The applications are intended to show maps and information concerning to nearest restaurants, commercial and cultural centers, etc. The software must be able to interface with Web Services in order to indicate destinations in consideration to user's preferences. The context information used to parameterize the software is, however, specific to a domain, tourist guiding, which prevents the framework from serving to other areas of interest, such as domiciliary healthcare.

An example of human-centric approach in the use of computational context, web services and geographic location-aware systems, with the intention of enhancing community life, can be found in [11]. Resulting applications would, for instance, alert users, via mobile devices, that a group of volunteers are planting flowers at a park in the neighborhood. Thus, considering the location of the users, blogs and maps are filled with significant information about nearby events. However, [11] focus on the evaluation of such software, but not on the provision of either a generic-purpose framework or architecture to facilitate the development of contextualized geographic systems.

In [12], it is investigated how context makes possible the creation of adaptive geographic systems with the purpose of improving the usability of relevant information to users. They propose a framework that relates services, geographic systems and context in which context is used to provide adaptation rules to select geographic content without requiring many actions from users. The rules are categorized according to the context information accessed by different groups of users. In this case, new users can have their context profiles derived from a certain group's profile in order to inherit rules automatically. The framework automates the support for collaborative use of context, but it does not provide for the determination of the context information to be used by different geographic systems.

In this section, related works were surveyed in order to determine issues that arise from the development of human-centric geographic systems: (i) lack of adequate human-centric view on the combination of Computational Context and Geographic Information Systems; (ii) Web Services are not used in their fullest capacities as to provide automated contextualized adaptation regarding to GIS; (iii) it is not taken full advantage of context information to parameterize geographic software; (iv) the application of context information toward specific domains does not promote its use under different areas of interest concerning geographic systems; (v) it is not usually determined what are the context information that affects the utilization of GIS. Considering the aforementioned inadequacies, in Section 3 it is proposed the integration of the framework proposed by [7] and GIS in order to automate the creation of contextualized human-centric geographic systems.

3 A Framework to Integrate Computational Context, Web Services and Geographic Systems

The CCMF [7], Computational Context Modeling Framework, contains a set of activities dedicated to the development of context-aware web-oriented applications. It enables the integration of such activities with external third-part technologies, tools and other frameworks in order to cover the modeling of context structures and the definition of adaptation mechanisms. Its main purpose is that of lessening the effort applied in the integration between Computational Context and Web Services, i.e. it is intended to support the generation of context-aware composite Web Services to serve functionalities that automate adjustments to situations of use to remote client applications. In this section, it is proposed the adaptation of CCMF to support the creation of contextualized human-centric geographic software.

Figure 1 shows the adapted framework. The development activities, enumerated from 1 to 7, are divided into two layers: (a) *Context Structure Modeling*, which defines sets of information that influences the use of software; (b) *Context Adaptation Modeling*, that contains the activities related to the creation of composite Web Services that automate context-awareness mechanisms. The first activity defines the structures of context information. Considering that XSD [13], XML Schema

Definition, is indicated to interoperate software [14], context schemas are used by the second activity of the framework the make structures of context information available to Web Services. To fulfill this purpose, the schemas are used as input by the third activity of the framework to produce serializable language-specific classes (e.g. JAVA classes). The generated classes are capable of serializing context information from one Web Service to another using, for instance, XML documents. Once the context structures are modeled and the corresponding serializable classes are made available, the adaptation mechanisms can be generated. For this to be accomplished, CCMF relies on the WSBPEL [15], Web Service Business Process Execution Language, to interoperate Web Services around context information. This step is performed by the fourth activity of the framework. WSBPEL is also used by the fifth activity to generate composite Web Services that populate sources of context information. Later one, the sources are queried by the Web Services produced by the fourth activity in order to parameterize adaptations. The execution of the sixth activity produces WSDL [16], Web Service Description Language, documents, describing the Web Services generated by the fourth and fifth activities. The WSDL documents are then used as input of the seventh activity to generate source codes of client software.

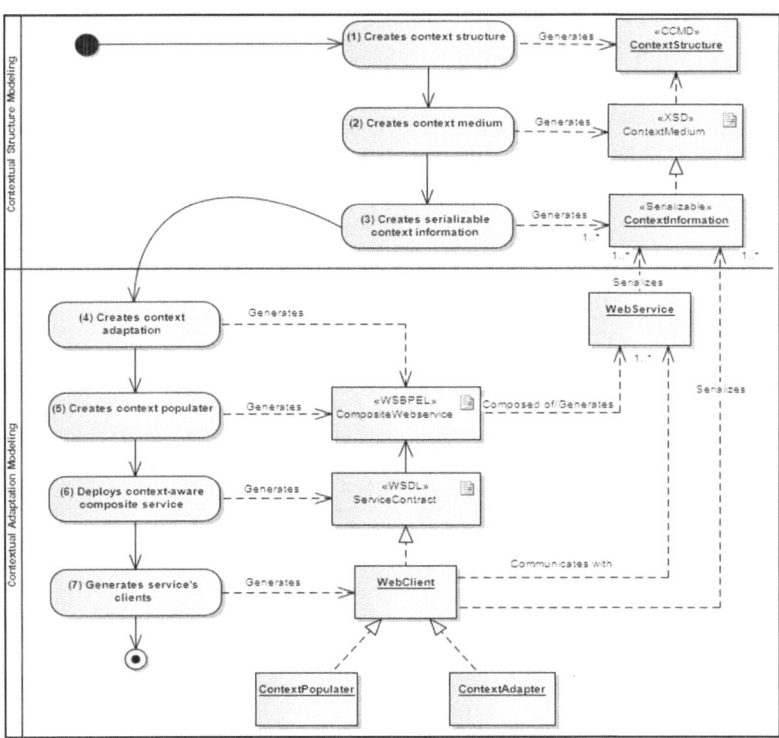

Fig. 1. Adapted architecture of CCMF [7]

Considering the original architecture of CCMF, the adapted one presented by Figure 1 introduces the apparatus of context populating, i.e. it provides functionalities for the insertion of structures of context information which parameterizes the adaptation. For instance, the Web Services created to populate the context structures can be used to save medical profiles into on-line databases. The saved profiles will then feed other Web Services to parameterize the actions performed to automate the context adaptation mechanisms, such as the calculation of routes toward chronic patients.

CCMF relies on CCMD [17], Computational Context Modeling Diagram, to enable the modeling of context structures (performed by the first activity of the framework). CCMD, illustrated in Figure 2, contains stereotypes that represent: (i) next to item 1, the task influenced by the context of use (e.g. healthcare professionals go to patient's house to provide medical assistance); (ii) item 2 represents the focus originated from the task (e.g. provision of geographic maps to help healthcare professionals to find patients during a mission); (iii) item 3 symbolizes a class that models characteristics related to the preferences that influence context adaptations (e.g. chronic conditions of certain patients may indicate that, preferably, they must be visited first); (iv) item 4 defines a class to describe information about participants (e.g. the patients that must be visited); (v) item 5 corresponds to a class that represents location (e.g. the geographic location of patients); (vi) a class to define the activities that influence the performance of the task is represented by item 6 (e.g. certain medical procedures may take more time to be completed in comparison with others); (vii) item 7 is the class that represents temporal information (e.g. intervals of time during which patients are found at their houses).

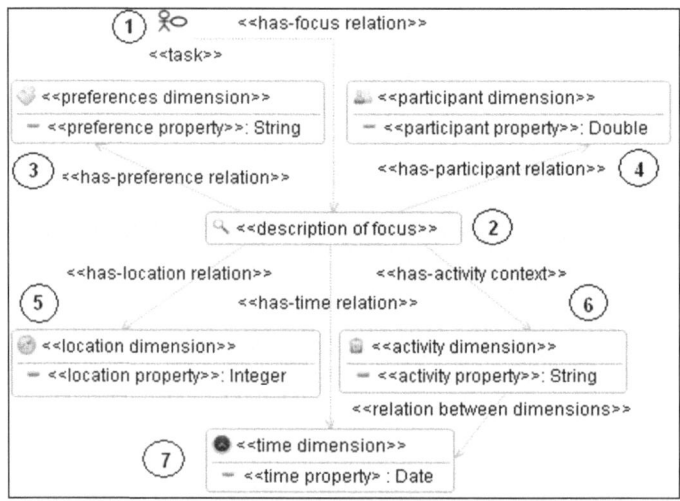

Fig. 2. Computational Context Modeling Diagram [17]

As a proof of concept, in section 4, the framework is instantiated to develop context-aware composite Web Services with the purpose of providing adaptations to domiciliary healthcare geographic systems.

4 Instantiating the Adapted Framework to Develop Context-Aware Web Oriented Geographic System

The first activity of the framework utilizes CCMD to model the structures of context information that parameterizes adaptations. Figure 3 shows an instance of the diagram containing context information related to the domiciliary healthcare geographic system. Item 1 represents the centre task performed by a healthcare professional during a mission. A focus originated from the task is symbolized by the item 2. The context information is categorized by the structures numbered from 3 to 8: (i) item 3, which represents the profile of the patient, provides information about health conditions; (ii) item 4 represents the patient (its name); (iii) item 5 represents health care professionals; (iv) the time of the day during which the patient can be found at his house is represented by item 6; (vi) item 7 corresponds to the geographic location of the patient and healthcare professionals.

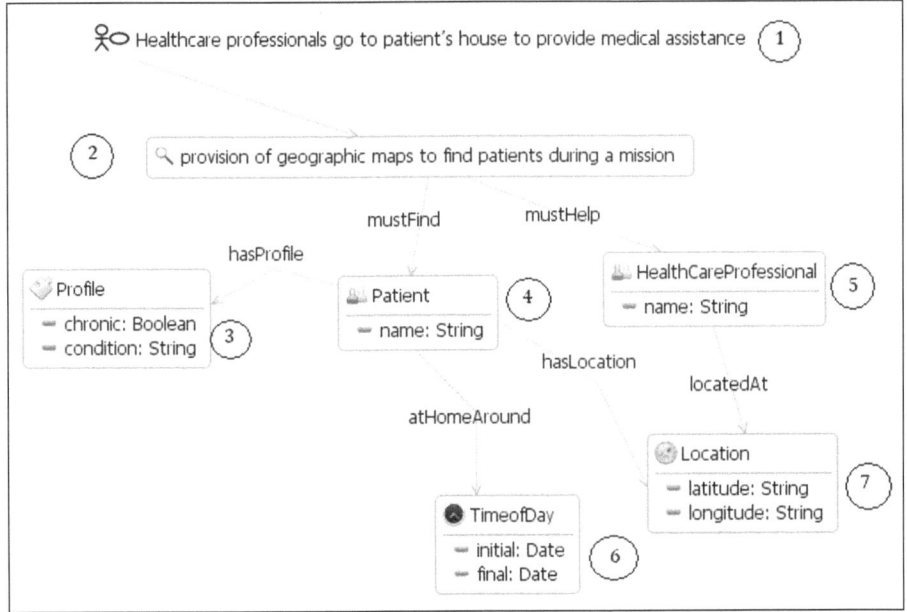

Fig. 3. Context information that influences domiciliary healthcare geographic system

CCMD is capable of transforming its stereotypes into XSD documents in order to textually represent context information (performing the second activity of the framework). An excerpt from the generated document is shown in Figure 4. Item 1 corresponds to the name of the patient. Item 2 identifies the relation of the patient with its profile, location and the time of the day during which he is at home.

```
<xs:complexType name="Patient">
  <xs:sequence>
    <xs:element name="name" type="xs:string" minOccurs="1" maxOccurs="1"/>   ①

    <xs:element name="Profile" type="Profile" minOccurs="1" maxOccurs="unbounded"/>
    <xs:element name="TimeOfDay" type="TimeOfDay" minOccurs="1" maxOccurs="1"/>   ②
    <xs:element name="Location" type="Location" minOccurs="1" maxOccurs="1"/>
  </xs:sequence>
</xs:complexType>
```

Fig. 4. Context medium for the domiciliary healthcare system

The XSD document that represents the context medium can be transformed by the XMLBeans API [18] into serializable JAVA classes (the output of the third activity of the framework). The generated classes are capable of encapsulating the structures of context information within structured XML documents. An example of such documents is illustrated in Figure 5. Next to item 1 the name of the patient, "Luis Paulo", is shown. The item 2 represents an entry of patient's profile describing a chronic disease, "High blood pressure". Item 3 shows the interval of time during which the patient can be found at home. Item 4 corresponds to the location of the patient.

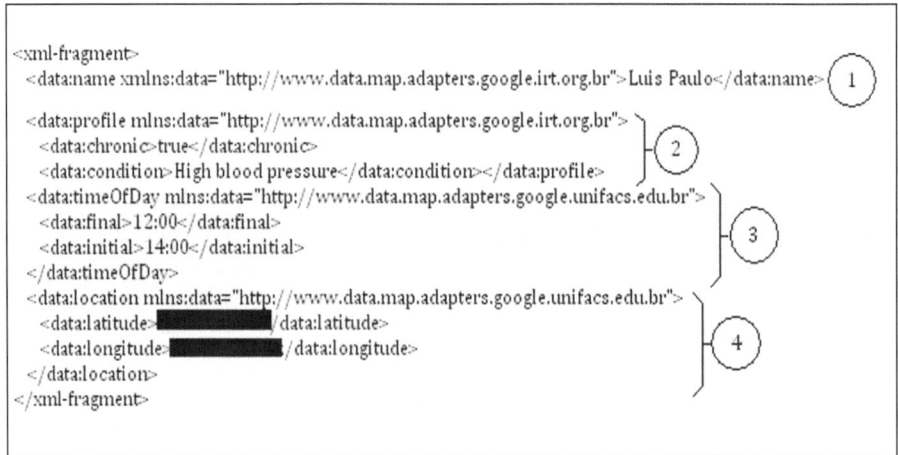

Fig. 5. XML document representing patient's structures of context information

After defining the contextual structures, the framework can be used to create composite Web Services to base context-aware adaptations on the information exemplified by Figure 5. The fourth activity of the framework includes finding adequate Web Services to take part in the composition. Table 2 enumerates the Web Services necessary to automate adaptations of the domiciliary healthcare system.

Table 2. Services APIs for the domiciliary healthcare system

Service API	Usage
Google Maps Data	It allows the storage and retrieval of information related to maps. It is used to keep contextualized information about patient's profile.
Google Geocoding	It converts the address of patients to latitude-longitude pairs.
Google Directions	It calculates the trajectory toward nearby patients.

The fifth activity of the framework is automated by the integration of the framework with WSBEL language. Its utilization is intended to create composite Web Services to allow the insertion of the context information used by domiciliary healthcare system. Figure 5 shows an excerpt of a WSBPEL workflow that receives context information related to patient's profiles and stores it in the Google Maps Data service. Item 1 represents the action by which context information is collected from remote client software. The sequences of processes next to items 2 and 3 are responsible for converting the context information to the format shown in Figure 5. Patient's name, health problems and interval of availability are processed by the item 2. Item 3 executes Google Geocoding to convert the address of the patient into geographic coordinates. Item 4 utilizes Google Maps Data to insert the context information into maps.

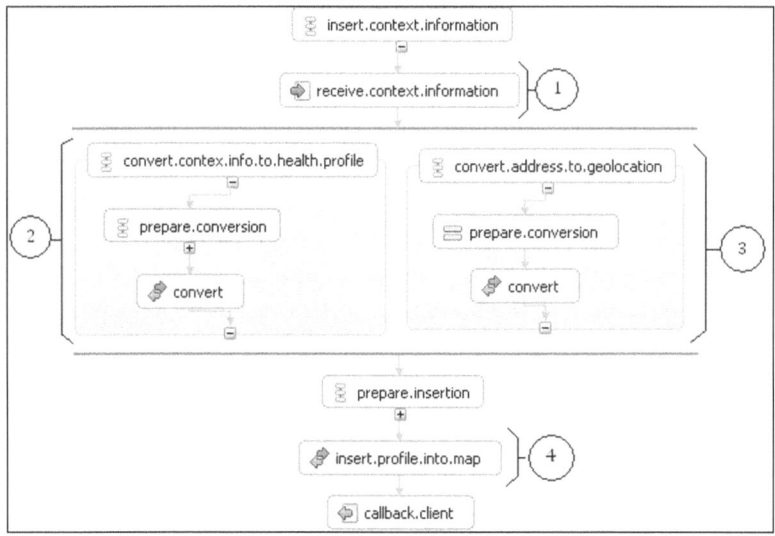

Fig. 6. WSBPEL workflow to insert context information

After the insertion of medical profiles in the databases of Google Maps Data, the information is made available to be retrieved by other Web Services. Figure 7 shows an example of the stored profile. Item 1 identifies a fictional healthcare professional, "Dr Who", who treats the patient represented by item 2, "Luis Paulo". The description of the placemark is utilized to store the document illustrated in Figure 5.

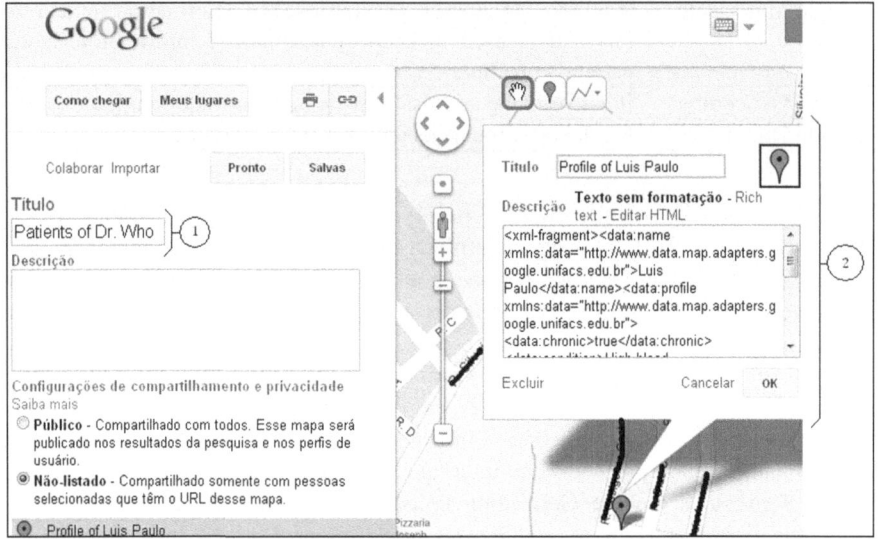

Fig. 7. Google Maps Data storing medical profile

The execution of the sixth activity of the framework generates another composite Web Service to access information about medical profiles of patients. Figure 8 shows a WSBPEL workflow which uses Web Services to automate context-aware adaptation of the domiciliary healthcare system. Item 1 represents a call to a Web Service based on the Google Maps Data API to retrieve medical profiles (exemplified in Figure 7). After the retrieval, each profile is processed in order to determine the route that the healthcare professional must follow to visit the patient. The first test (item 3) considers the availability of the patient (the interval of time he is likely to be found at his house) and the proximity of the professional (a fixed radius of 3 kilometers). If the patient is available and the healthcare professional is near to him, his health condition is evaluated by item 2. In case it is a chronic problem, the visit to the patient is prioritized over others. As the execution of the workflow ends, a map generated by Google Directions is returned to client software. The map must show a route to patients houses considering the adaptations related to the context.

Figure 9 shows a map drawn by Google Maps indicating the route from a healthcare professional to three nearby patients. In spite of the fact that the patient identified by the item D is at a close distance to the professional, patients B and C must be visited first since they have chronic health problems. Thus, D is visited last, since his condition does not demand urgent treatment.

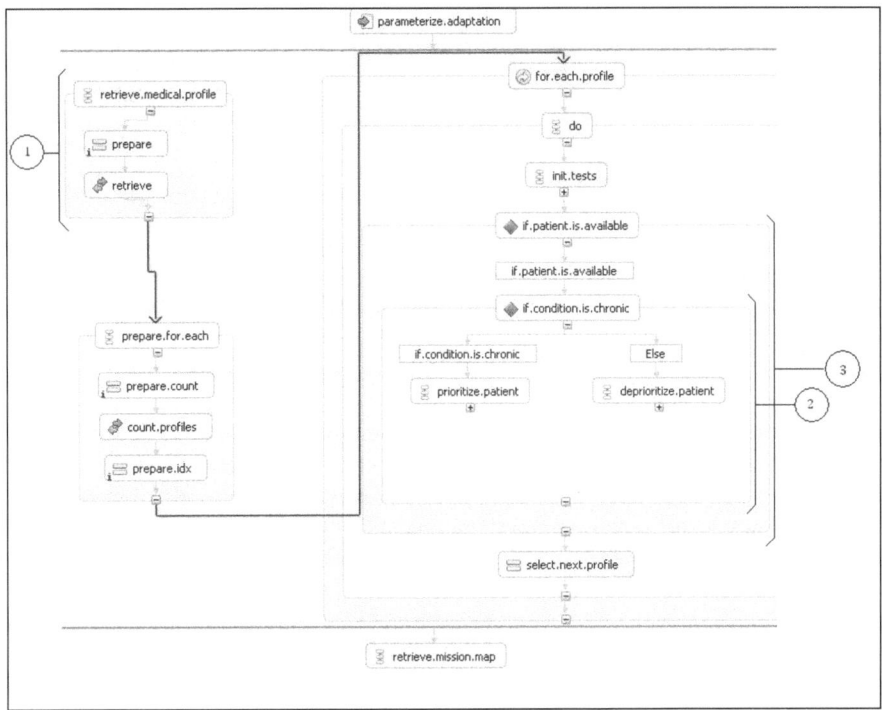

Fig. 8. WSBPEL workflow to provide adaptation for the domiciliary healthcare system

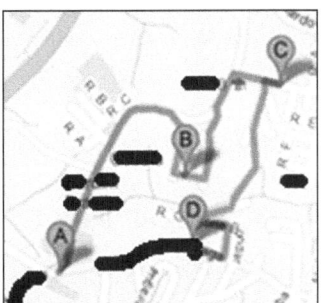

Fig. 9. Contextualized route to patients

By executing the last activity of the framework, the WSDL documents generated by activities 5 and 6 are used to generate parts of the source codes of client software. WSDL2JAVA [19] is capable of automating this development step. The resulting source code is a pack of JAVA classes that is able to invoke the Web Services. Clients must provide locally sensed context information, e.g. the current location of the healthcare professional who is executing the software on a handheld GPS-enabled device.

In this section, it was described how to combine Geographic software, Computational Context and Web Services, via the CCMF framework, in order to produce human-centric geographic systems.

5   Conclusion and Future Work

This paper proposes the integration of Computational Context and Web Services in order to provide adaptable service-oriented Geographic Information System. In comparison to the related works discussed in Section 2, the approach herein presented provides the following advantages: (i) automatic parameterization of geographic software by context information in replacement to the conventional utilization of locations only (ii) application of the generated software to problems that require a consideration of human-centric perspectives; (iii) promotion of reuse via the internet, since the adaptation logics are placed in composite Web Services. This approach enables the coupling of the generated solutions with worldwide distributed clients; (iv) lessening of the effort required by the development of contextualized GIS, because the framework promotes its integration with available third-part tools and sub-sequential reuse of development artifacts as input to development activities.

In order to complement this work, the proof of concept described in Section 4 must be evaluated against real case scenarios, e.g. healthcare missions. To fulfill this purpose, it has to be covered the complete set of context information that influences adaptations of such software. For instance, determining that a health problem is chronic might not be sufficient to indicate which, out of two patients, should be visited first if it is considered that both patients have chronic medical problems. In this case, the age of patients could be used to enhance the adaptation mechanism illustrated in Figure 8.

References

1. Portal do Ministério da Saúde do Brazil (Portal of the Brazilian Health Office), http://portalsaude.saude.gov.br/portalsaude/?portal=pagina.visualizarArea&codArea=364
2. Castelli, G., Rosi, A., Mamei, M., Zambonelli, F.: A Simple Model and Infrastructure for Context-Aware Browsing of the World. In: Fifth IEEE International Conference on Pervasive Computing and Communications (PERCOM 2007), pp. 229–238. IEEE Computer Society, Washington, DC (2007)
3. Dey, A.K.: Understanding and Using Context. Personal Ubiquitous Computing 5, 1 (2001)
4. Dey, A., Abowd, G.: Towards a Better Understanding of Context and Context-Awareness. In: Workshop on the What, Who, Where, When and How of Context-Awareness (2000)
5. Predic, B., Stojanovic, D., Djordjevic-Kajan, S.: Developing Context-Aware Support in Mobile GIS Framework. In: 9th AGILE Conference (2006)
6. Kapitsaki, G.M., Prezerakos, G.N., Tselikas, N.D., Venieris, I.S.: Context-Aware Service Engineering: A Survey. Journal of Systems and Software, 1285–1297 (2009)
7. Carvalho, L.P.S., Silva, P.C.: A Transformation-Driven Framework to Model Context-Aware Web Applications. In: IADIS Applied Computing (2011)

8. Abowd, G.D., Mynatt, E.D.: Charting Past, Present and Future Research in Ubiquitous Computing. ACM Transactions on Computer-Human Interaction, Special Issue on HCI in the New Millenium, 29–58 (2000)
9. Lamas, A.R., Filho, J.L., Oliveira, A.P., Júnior, R.M.A.B.: A Mobile Geographic Information System Managing Context-aware Information Based On Ontologies. In: First International Workshop on Managing Data with Mobile Devices (2009)
10. Abbaspour, R.A., SamadZadegam, F.: Building a Context-aware Mobile Tourist Guide System based on Service Oriented Architecture. In: The International Archives of the Photogrammetry, Remote Sensing and Spatial Information Sciences, Beijing, vol. XXXVII, Part B4, pp. 871–874 (2008)
11. Ganoe, C.H., Robinson, H.R., Horning, M.A., Xie, X., Carroll, J.M.: Mobile awareness and participation in community-oriented activities. In: First International Conference and Exhibition on Computing for Geospatial Research. ACM, New York (2010)
12. Petit, M., Ray, C., Claramunt, C.: A user context approach for adaptive and distributed GIS. In: Proceedings of the 10th International Conference on Geographic Information Science: AGILE 2007. Lecture Notes in Geo-information and Cartography, pp. 121–133 (2007)
13. XML Schema Definition (XSD) (2007), http://www.w3.org/XML/Schema
14. Alboaie, S., Buraga, S., Alboaie, L.: An XML based Serialization of Information Exchanged by Software Agents. International Informatic Journal (2003)
15. Web Service Business Process Execution Language, WSBPEL (2007), http://docs.oasis-open.org/wsbpel/2.0/wsbpel-v2.0.pdf
16. W3C Web Service Description Language, WSDL (2007), http://www.w3.org/TR/wsdl
17. Carvalho, L.P.S., Silva, P.C.: CCMD – Computational Context Modeling Diagram and WSBPEL Integration. In: IADIS Applied Computing (2011)
18. XMLBeans JAVA API (2009), http://xmlbeans.apache.org/
19. Apache WSDL2JAVA (2010), http://sourceforge.net/projects/wsdl2javawizard

Voronoi-Based Curve Reconstruction: Issues and Solutions

Mehran Ghandehari and Farid Karimipour

Department of Surveying and Geomatics Engineering, College of Engineering,
University of Tehran, Iran
{ghandehary,fkarimipr}@ut.ac.ir

Abstract. Continuous curves are approximated by sampling. If sampling is sufficiently dense, the sample points carry the shape information of the curve and so can be used to reconstruct the original curve. There have been lots of efforts to reconstruct curves from sample points. This paper reviews the curve reconstruction methods that use Voronoi diagram in their approach. We, then, describe the main issues of these methods and suggest solutions to deal with them. Especially, we improve one of the Voronoi-based curve reconstruction algorithms (called one-step crust algorithm) by labeling the sample points as a preprocessing. The highlights of our proposed approach are (1) It is simple and easy to implement; (2) It is robust to boundary perturbations and noises; (3) Special cases in sampling like sharp corners can be handled; and (4) It can be used for reconstructing open curves.

Keywords: Sample points, Curve reconstruction, Voronoi diagram, Delaunay triangulation.

1 Introduction

Continuous curves are approximated by sampling. Sample points carry the shape information of the curve and are used for reconstructing the original curve: The input is a set of sample points in R^2, without any structure or order, and the output is a curve (Fig. 1). The problem can be extended to 3D [1-3] where a set of sample points in R^3 are used to reconstruct a surface (Fig. 2). The focus of this paper is on 2D space, but the results can properly be extended to 3D.

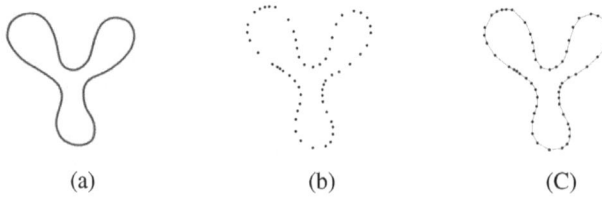

Fig. 1. (a) A 2D continuous curve; (b) Sampling; (c) Curve reconstruction from samples [4]

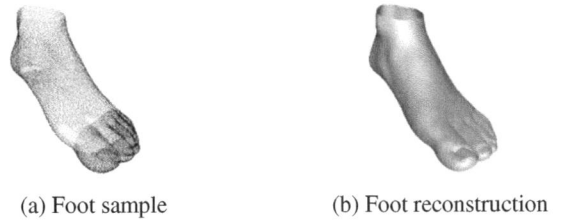

(a) Foot sample (b) Foot reconstruction

Fig. 2. (a) An example of a 3D sampling and (b) its surface reconstruction [4]

The accuracy of reconstruction depends, among other parameters, on the density of sample points. The proper density of sample points vary for different parts of the curve: simple parts can be reconstructed from fewer samples, while other parts may need to be approximated by more sample points.

Amenta *et al.* [5] represented some theories for a proper sampling. They determined a lower bound for sampling that guarantees a proper reconstruction, but no upper bound was defined. Furthermore, such theoretical criteria may not be useful in practice. For instance, from a theoretical point of view, for sharp corners and noisy samples, infinite dense sampling is needed to guarantee the proper reconstruction, which is not practically possible [6-8].

There have been lots of efforts to reconstruct curves from sample points. Figueiredo and Gomes [9] introduced a method based on minimum spanning tree. Bernardini and Bajaj [10] reconstructed curves in the plane using α–shapes. Attali [1] proposed an algorithm for uniformly sampled curves, which theoretically guarantees a proper reconstruction. Amenta *et al.* [11] introduced an algorithm for shape approximation using Voronoi balls. A well-known algorithms that uses the crust structure for curve reconstruction was introduced by Amenta *et al.* [5], which handles non-uniform samples. This algorithm was improved by Gold and Snoeyink [12] (called one-step crust algorithm) for curve reconstruction and medial axis approximation, which is very fast and easy to implement. The focus of this paper is on the methods that uses Voronoi diagrams for curve reconstruction, i.e., Voronoi balls, curst and one-step crust algorithms.

The rest of the paper is structured as follows: Section 2 represents some geometric preliminaries, including Delaunay triangulation, Voronoi diagram, medial axis and two definitions related to sampling. In section 3, the Voronoi-based curve reconstruction methods are reviewed. Section 4 represents the problems we encountered in using the one-step crust algorithm and addresses the main issues that may occur. It led us to an improvement to this algorithm by labeling the sample points as a pre-processing, which is introduced in section 5. We, then, compare our results with the crust and the one-step crust algorithms in this section. Finally, section 6 concludes the paper and represents ideas for future work.

2   Geometric Preliminaries

This section represents some geometric preliminaries, including Delaunay triangulation, Voronoi diagram, medial axis and two definitions related to sampling. In this section, O is a 2D object, ∂O is its boundary and $S \subset \partial O$ is a dense sampling of ∂O.

2.1 Delaunay Triangulation

Definition 1. Given a point set S in the plane, the *Delaunay triangulation* (DT) is a unique triangulation (if the points are in general position) of the points in S that satisfies the circum-circle property: the circum-circle of each triangle does not contain any other point $s \in S$ [13]. Fig. 3.a illustrates a 2D example.

2.2 Voronoi Diagram

Definition 2. Let S be a set of points in R^2. The *Voronoi cell* of a point $p \in S$, denoted as $V_p(S)$, is the set of points $x \in R^2$ that are closer to p than to any other point in S:

$$V_p(S) = \{ x \in R^2 \mid \|x - p\| \leq \|x - q\|, q \in S, q \neq p \} \tag{1}$$

The union of the Voronoi cells of all points $s \in S$ forms the *Voronoi diagram* of S, denoted as $VD(S)$:

$$VD(S) = \cup V_p(S), p \in S \tag{2}$$

Fig. 3.b shows the Voronoi diagrams of a set of 2D points. Delaunay triangulation and Voronoi diagram are dual structures: the centers of circum-circles of Delaunay triangulation are the Voronoi vertices; and joining the adjacent generator points in a Voronoi diagram yields their Delaunay triangulation (Fig. 3.c) [14].

For Voronoi diagram of sample points S, the Voronoi vertices are classified into *inner* and *outer vertices*, which lie inside and outside O, respectively. Then, the Voronoi edges are classified into three groups: edges between two inner vertices (*inner Voronoi edges*), edges between two outer vertices (*outer Voronoi edges*), and edges between an inner and an outer vertices (*mixed Voronoi edges*).

A *Voronoi ball* is centered at a Voronoi vertex and its radius is its distance to the closest sample point. Again, Voronoi balls are classified into *inner* and *outer balls* depending on type of their center points [15].

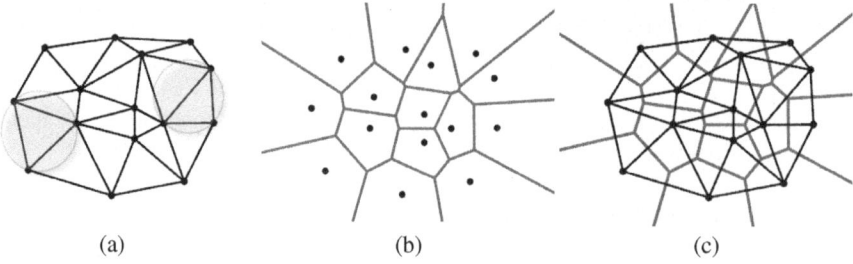

(a)   (b)   (c)

Fig. 3. (a) Delaunay triangulation and (b) Voronoi diagram of a set of points in the plane; and (c) their duality

2.3 Medial Axis

The Medial Axis (MA) was first introduced by Blum [16] to describe biological shape and as a tool in image analysis. Grassfire model is an intuitive concept that simply describes MA: consider starting a fire on the boundary of a shape in the plane. The fire starts at the same moment everywhere on the boundary and it propagates with homogeneous velocity in all directions. The medial axis is the set of points where the front of the fire collides with itself, or other fire front. MA is used in sampling criteria for curve reconstruction.

Definition 3. The *medial axis* is (the closure of) the set of points in $\mathcal{O}$ that have at least two closest points on the object's boundary $\partial\mathcal{O}$ [5]. In other words, the medial axis of a plane curve $\mathcal{O}$ is the set of points in $\mathcal{O}$ that are equidistant from at least two points on the boundary of the shape (Fig. 4).

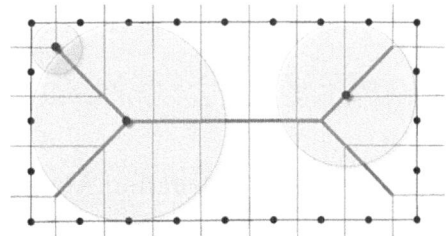

Fig. 4. Medial axis of a 2D curve (rectangle)

2.4 Local Feature Size and *r*-sampling

As stated before, quality of sample points S has a direct effect on curve reconstruction. *Local feature size* is a quantitative measure to determine the level of details at a point on a curve, and the sampling density needed for curve reconstruction.

Definition 4. The *local feature size* of a point $p \in \partial\mathcal{O}$, denoted as $LFS(p)$, is the distance from p to the nearest point m on the medial axis [5].

Note that $LFS(p)$ is different from radius of medial circle, which is tangent to curve in p (Fig. 5).

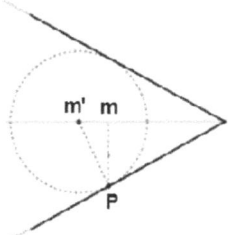

Fig. 5. The local feature size of a point p (line segment pm) is not necessarily the same as the smallest radius of the medial circle touching p (line segment pm') [4]

Definition 5. The object O is *r-sampled* by a set of sample points S if for each point $p \in \partial O$, there is at least one sample point $s \in S$ that $\|p - s\| \leq r * \text{LFS}(p)$ [5].

The value of r is less than 1; and usually $r=0.4$ is considered a reasonably dense sampling [5]. Fig. 6 shows an example where sample points around the center are denser to provide a proper sampling.

Fig. 6. (a) A curve with its medial axis (green curves); (b) An *r*-sampling of the curve [4]

3 Voronoi-Based Curve Reconstruction Algorithms

This section reviews the Voronoi-based algorithms for curve reconstruction.

3.1 Voronoi Ball Algorithm

This algorithm was proposed by Amenta *et al.* [11] for shape approximation. They showed that any shape O with smooth boundary ∂O can be approximated by the union of Voronoi balls. The steps of this algorithm are:

1. Compute the Voronoi diagram of the sample points S (Fig. 7.b)
2. Identify the inner Voronoi vertices (Fig. 7.c)
3. Compute the inner Voronoi balls (Fig. 7.d)
4. Union of the inner Voronoi balls approximates the shape (Fig. 7.e and 7.f)

3.2 Crust Algorithm

Amenta *et al.* [5] proposed a Voronoi-based algorithm (called crust algorithm) to reconstruct the boundary from a set of sample points forming the boundary of a shape. In this algorithm, the crust is a subset of the edges of the Delaunay triangulation of the sample points.

To compute the crust, let S be the sample points and V be the vertices of the Voronoi diagram of the sample points. Then:

1. Compute the Voronoi diagram of the sample points S (Fig. 8.a).
2. Compute the Delaunay triangulation of $S \cup V$ (Fig. 8.b).
3. The edges of the above Delaunay triangulation whose endpoints belong to S form the crust, which is an approximation of the shape (Fig. 8.b).

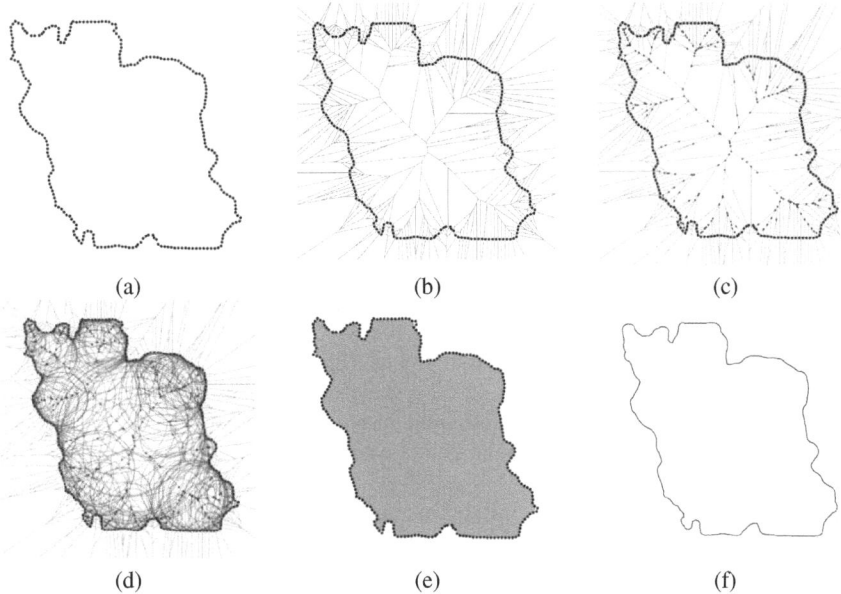

Fig. 7. Shape approximation using Voronoi balls: (a) Sample points on the boundary of the shape; (b) Voronoi diagram of the sample points; (c) Inner Voronoi vertices (red points); (d) Inner Voronoi balls; (e) and (f) Union of inner Voronoi balls approximates the shape

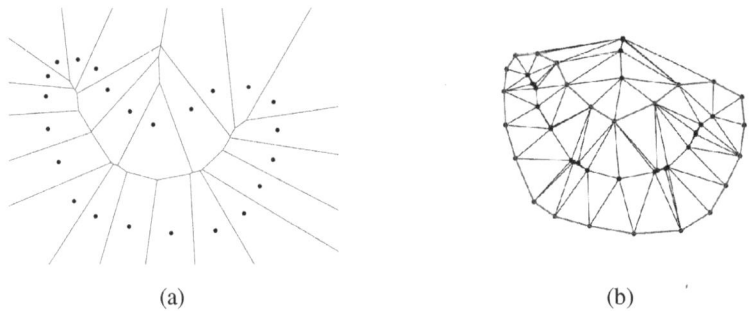

Fig. 8. Curve reconstruction using the crust algorithm: (a) Voronoi diagram of the sample points; (b) Delaunay triangulation of the sample points and Voronoi vertices; and selecting the Delaunay edges whose endpoints belong to S (red lines), which approximate the curve

The crust algorithm is based on the fact that an edge e of the DT belongs to the crust if e has a circum-circle that contains neither sample points nor Voronoi vertices of S. It means that a global test is needed to check the position of every sample points and Voronoi vertices respect to this circle. There are also three theorems related to this algorithm as follows (see [5] for complete proofs and theories):

Theorem 1: "*Let F be an r-sampled smooth curve in the plane, $r < 1$. The Delaunay triangulation of the set S of samples contains an edge between every adjacent pair of samples*" [5].

Theorem 2: *"The crust of an r-sampled smooth curve, r <0.40, contains an edge between every pair of adjacent samples"* [5].

Theorem 3: *"The crust of an r-sampled smooth curve does not contain any edge between nonadjacent vertices, for r <0.252"* [5].

3.3 One-Step Crust Algorithm

The above crust algorithm was improved by Gold and Snoeyink [12]. They coined the name "one-step crust and skeleton" for this algorithm, because it extracts both crust and skeleton at the same time (in the literature, medial axis and skeleton are considered equivalent [17]).

This algorithm is fast and easy to implement. Here, every Voronoi/Delaunay edge is either part of the crust (Delaunay) or the skeleton (Voronoi), which can be determined by a simple *inCircle* test. Each Delaunay edge (D_1D_2 in Fig. 9.a) belongs to two triangles ($D_1D_2D_3$ and $D_1D_2D_4$ in Fig. 9.a). For each Delaunay edge, there is a dual Voronoi edge (V_1V_2 in Fig. 9.a).

In the crust algorithm (section 3.2), a Delaunay edge belongs to the crust if there is a circle that contains the edge, but does not contain any Voronoi vertices. However, in the one-step crust algorithm, this global test is replaced with a local test that uses only the two endpoints of the dual Voronoi edge.

Suppose two triangles $D_1D_2D_3$ and $D_1D_2D_4$ have a common edge D_1D_2 whose dual Voronoi edge is V_1V_2. The *InCircle*(D_1, D_2, V_1, V_2) determines the position of V_2 respect to the circle passes through D_1, D_2 and V_1. If V_2 is outside the circle, D_1D_2 belongs to the crust (Fig. 9.b). If V_2 is inside, however, V_1V_2 belongs to the skeleton (Fig. 9.c).

The value of *InCircle*(D_1, D_2, V_1, V_2) test is calculated using the following determinant:

$$InCircle\ (D_1, D_2, V_1, V_2) = \begin{bmatrix} x_{D1} & y_{D1} & x_{D1}^2 + y_{D1}^2 & 1 \\ x_{D2} & y_{D2} & x_{D2}^2 + y_{D2}^2 & 1 \\ x_{V1} & y_{V1} & x_{V1}^2 + y_{V1}^2 & 1 \\ x_{V2} & y_{V2} & x_{V2}^2 + y_{V2}^2 & 1 \end{bmatrix} \qquad (3)$$

D_1D_2 belongs to the crust if this determinant is negative, otherwise V_1V_2 belongs to the skeleton [12, 18].

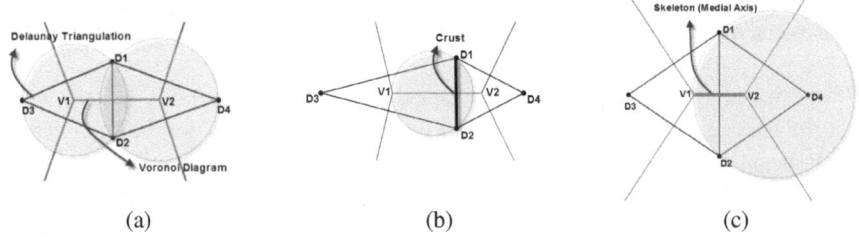

Fig. 9. One-step crust and skeleton extraction algorithm: (a) Delaunay triangulation and Voronoi diagram of four sample points D_1 to D_4; (b) V_2 is outside the circle passes through D_1, D_2 and V_1, so D_1D_2 belongs to the crust; (c) V_2 is inside the circle passes through D_1, D_2 and V_1, so V_1V_2 belongs to the skeleton

The pseudo-code for this algorithm is as follows:

One-step crust and skeleton extraction
Input : Sample point S
Output: Crust and skeleton of the shape approximated by S

1. DT ← Delaunay Triangulation of S
2. E ← Edges of DT
3. For every $e \in E$ do
4. S_1, S_2 ← triangles that contain e
5. D_1, D_2 ← end points of e
6. V_1, V_2 ← centers of the circum-circles of S_1 and S_2
7. H ← $InCircle(D_1, D_2, V_1, V_2)$
8.   If $H < 0$ then $D_1 D_2 \in$ Crust
9.   else $V_1 V_2 \in$ Skeleton

4 Issues and Solutions

This section presents the problems we encountered in using the one-step crust algorithm and addresses the main issues that may occur. It led us to an improvement by labeling the sample points as a pre-processing, which is introduced afterwards. Our results are compared with the crust and the one-step crust algorithms at the end.

As mentioned before, the global circle test used in the crust algorithm is replaced with a local test in the one-step crust algorithm to assign the Delaunay/Voronoi edges to the crust and skeleton. Although it is simpler and faster, it may lead to assigning wrong edges to the crust. For example, in Fig. 10 the edge e is in the locally-defined crust because the circle passes through e does not contain the other Voronoi vertices of its dual Voronoi edge. However, e is not in the globally-defined crust because the circle passes through e includes some Voronoi vertices [12]. This problem is solved by satisfying the sampling conditions.

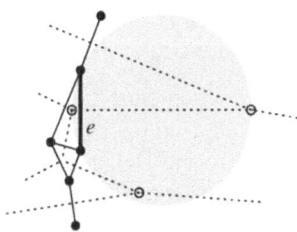

Fig. 10. The edge e (bold line) is in the locally-defined crust but it is not in the globally-defined crust [12]

Another issue of the one-step curst algorithm is dealing with the boundary Delaunay edges, which from the convex hull of the sampling points. These edges are adjacent to only one triangle, so the local test cannot be performed (Fig. 11). Fig. 12 illustrates that the boundary Delaunay edges could be a crust edge or not.

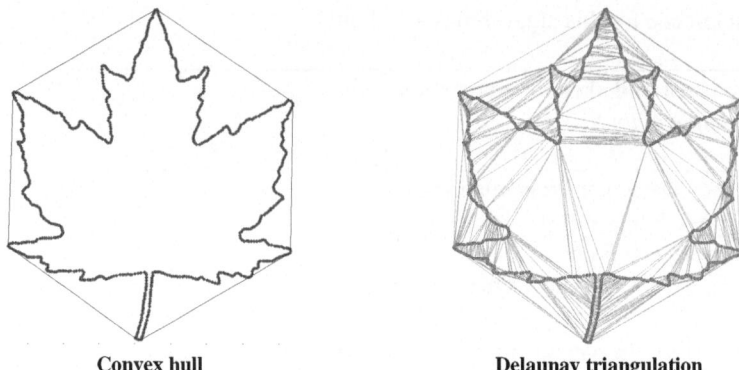

Fig. 11. Convex hull edges are a subset of the Delaunay edges but, they are adjacent to only one triangle and local test cannot be performed

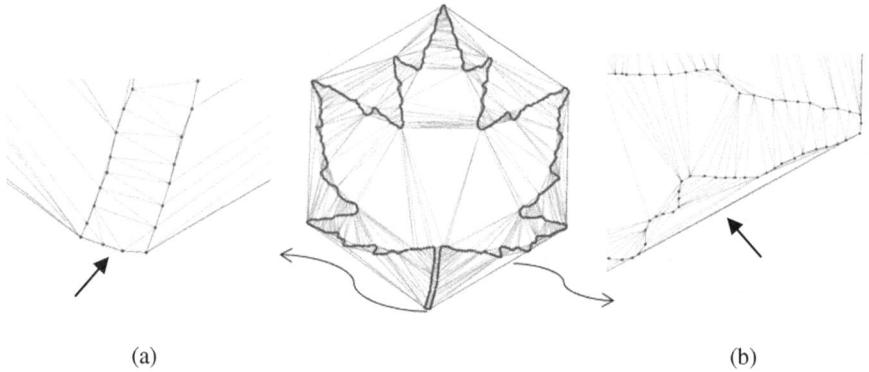

Fig. 12. A convex hull edge could be a crust edge (a) or not (b)

To solve the later problem, we propose the following lemma:

Lemma 1. A convex hull edge (which is a boundary Delaunay edge) is a crust edge if and only if the center of the circum-circle of the corresponding triangle is an inner Voronoi vertex.

Proof. Inner and outer Voronoi edges do not intersect with $\partial \mathcal{O}$, but mixed Voronoi edges do [15]. The same applied to the Delaunay edges: Delaunay edges of sample points S are classified into three classes: *Mixed Delaunay edges* that join two consecutive points and belong to the crust; And *inner/outer Delaunay edges* that join two non-consecutive points and are completely inside/outside $\mathcal{O}$ (note that all Delaunay vertices lie on the $\partial \mathcal{O}$).

We use the fact that the inner/outer/mixed Voronoi edges are dual to the inner/outer/mixed Delaunay edges. A convex hull edge that is not a crust edge is an outer Delaunay edge, which means that its dual Voronoi edge is outer (the vertices of an outer Voronoi edge are out of the boundary). The dual of a convex hull edge that is a crust edge is a mixed Voronoi edge, which means that one of its Voronoi vertices is inner and the other is at infinite (i.e., outer).

5 Labeling Sample Points

In this section we propose an improvement to the one-step crust algorithm for curve reconstruction using labeling the sample points as a pre-processing; and show how our proposed approach improves the results.

Figure 13 illustrates the medial axes of a shape extracted using the one-step crust algorithm. As this figure shows, this algorithm detects some extraneous edges as parts of the medial axis. A so called *pruning* post-processing step is used to detect and remove such edges [12]. However, we observed that such extraneous edges are the Voronoi edges created between the sample points that lie on the same segment of the curve (Fig 13). It led us to the idea of labeling the sample points in order to automatically avoid such edges in the media axis.

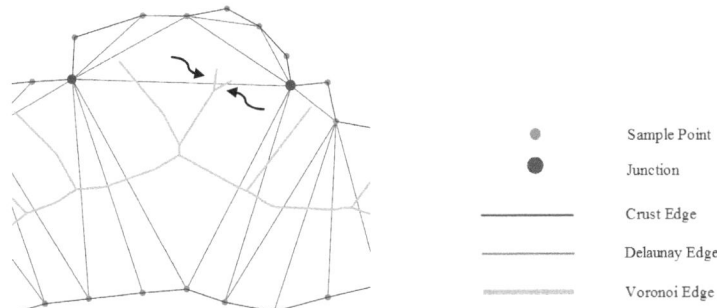

Fig. 13. One-step crust algorithm detects some extraneous edges (two of them are indicated by arrows) as parts of the medial axis. They are the Voronoi edges created between the sample points that lie on the same segment of the curve.

We start with labeling the sample points: Each segment of the shape is assigned a unique label; and all of its sample points are assigned the same label. The points that are common between two curve segments are called *junctions*. We assign a unique negative label to junctions to distinguish them from other sample points.

To extract the crust and skeleton, each Delaunay edge passes the *InCircle* test: If the determinant is negative and the corresponding Delaunay vertices have the same labels or one of them is a junction, that Delaunay edge is added to the crust. Otherwise, if the determinant is positive and the corresponding Delaunay vertices have different labels, its dual is added to the skeleton.

To apply our proposed approach in the one-step crust algorithm, the lines 8 and 9 of the pseudo-code presented in section 3.3 are modified as follows:

8.   If $H < 0$ and label(D_1)=label(D_2) or label(D_1)*label(D_2)<0 then $D_1D_2 \in$ Crust
9.   else if label(D_1) ~= label(D_2) then $V_1V_2 \in$ Skeleton

The highlights of our proposed approach are:

- It is simple and easy to implement.
- It is robust to boundary perturbations and noises.
- Special cases in sampling like sharp corners are handled.
- It can be used for reconstructing open curves.

These issues are described in more details in the following:

Robustness to Boundary Perturbations and Noises: Medial axis is very sensitive to small changes of the boundary. Such small perturbations may produce many irrelevant branches in the medial axis so as two similar shapes may have significantly different medial axes. Filtering irrelevant branches is a common solution [19-21]. However, this issue is automatically solved in our labeling approach: The dual of proper edges in the medial axis are inner Delaunay edges whose end points lie on two different curve segments. Thus, if the end points of an inner Delaunay edge lie on the same curve segment, its dual inner Voronoi edge will be an irrelevant edge, which does not appear in the skeleton. Fig. 14 compares the results of the one-step algorithm and our proposed approach.

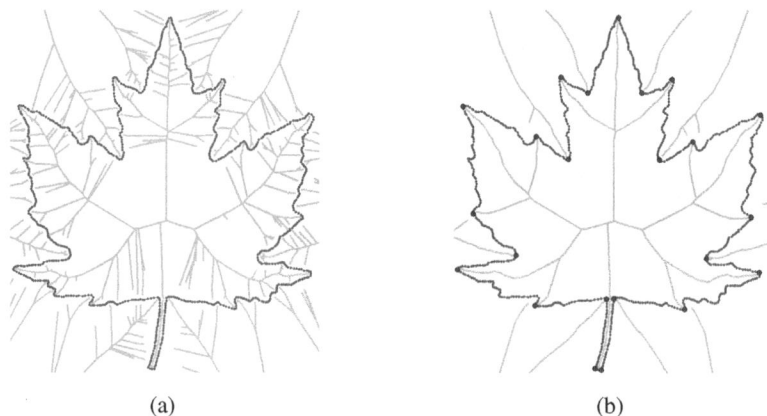

Fig. 14. Curve reconstruction and medial axis extraction: (a) One-step algorithm; (b) Our proposed algorithm

Handling Sharp Corners. Crust algorithm sometimes have problems in reconstructing curves at sharp corners, where the medial axis is very close to the boundary (Fig. 15). Based on sampling criteria, it requires infinite density sampling to guarantee the reconstruction process, which is not practically possible (high density of sample points leads to increasing the data volume and decreasing the speed of the algorithm). Another solution is arranging the sample points around all corners in an appropriate way, which is time-consuming for high volume data.

In our proposed approach, we detect the problematic shape corners through a post-processing step and only the sample points around these problematic corners needs rearrangement: After computing the crust, the number of crust lines joined at each junction are counted. If this number is less than a predefined threshold (usually 2 or 3), a rearrangement of sampling points is needed around this corner.

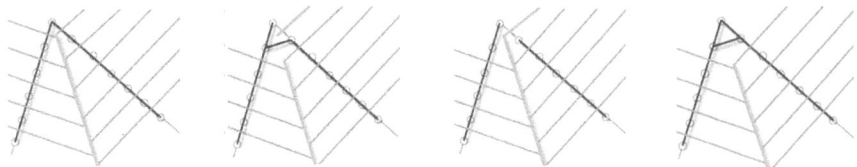

Fig. 15. Different states that may occur at sharp angles

Open Curves. The existing Voronoi-based algorithms for curve reconstruction and medial axis extraction are suitable only for closed curves, whereas our proposed approach can be properly used for open curves as well. Fig. 16 illustrates the results of curve reconstruction using the crust, the one-step crust and our proposed algorithms.

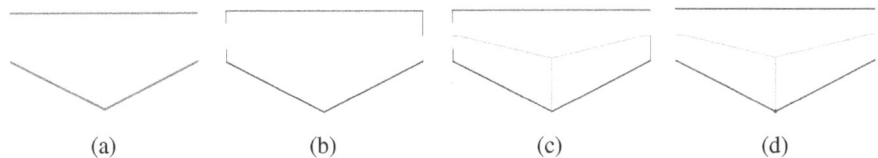

Fig. 16. Curve reconstruction for open curves: (a) Original curve; (b) Crust algorithm (which only extracts the crust); (c) One-step crust algrithm; (d) Our proposed algorithm. In (b) and (c) the extracted crusts have two extraneous vertical edges, which do not exist in the original curve.

Special Reconstruction Issues. Conjunctions are usually a special case in curve reconstruction (Fig. 17). Changing the sampling density may solve the problem. We used different densities (r=1, 0.5, 0.42, 0.24) in this example and the result is only correct for r=0.24. It shows that increasing the density for the whole shape may not necessarily lead to a more accurate result. Two solutions for this issue are suggested:

- Increasing the density of sampling around the conjunctions (Fig 18.a). However, in practice, this solution is very time consuming for large data sets.
- Removing the conjunction points (Fig. 18.b).

In our proposed approach, such problems are automatically avoided, because a crust edge can be created between two points with the same label or two points with different labels if one of them is a junction.

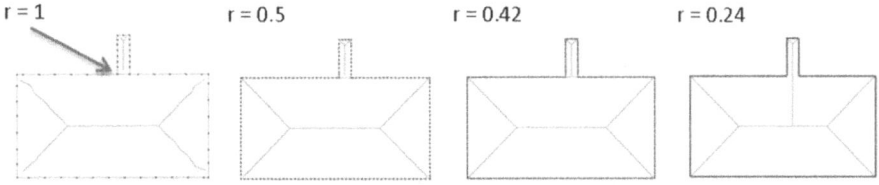

Fig. 17. Curve reconstruction and medial axes extraxtion using the one-step algorithm for different r-samplings (the result is only correct for r=0.24)

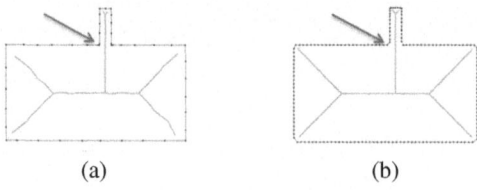

Fig. 18. Two solutions for a proper reconstruction at conjunctions: (a) Dense sampling at conjunctions; (b) Removing the conjunction points

The above problem may happen for open curves, too, especially when two line segments have the same direction. Our proposed approach properly works for such cases, as well (Fig. 19).

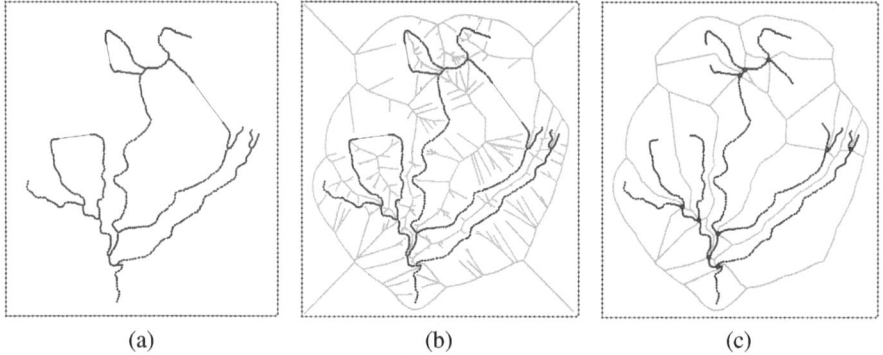

Fig. 19. The crust for open curves when two line segments have the same direction: (a) Crust algorithm; (b) One-step crust algorithm; (c) Our proposed algorithm

6 Conclusion and Future Works

In this paper, we reviewed the curve reconstruction methods that use Voronoi diagram in their approach and improve one of the Voronoi-based curve reconstruction algorithms (i.e., one-step crust algorithm) by labeling the sample points as a pre-processing. It leads to a solution that is simple and easy to implement, robust to boundary perturbations, able to handle sharp corners and open curves. The results show that our proposed approach deals elegantly with different cases of sample points and solves the problems that may occur in other algorithms.

In the future, we will extend the approach for surface reconstruction and 3D medial axis extraction. We will also study in more details the relationship between curve reconstruction and medial axis extraction and its application in some new fields.

References

1. Attali, D.: r-Regular shape reconstruction from unorganized points. Computational Geometry 10, 239–247 (1998)

2. Amenta, N., Bern, M.: Surface reconstruction by Voronoi filtering. Discrete and Computational Geometry 22, 481–504 (1999)
3. Hoppe, H., DeRose, T., Duchamp, T., McDonald, J., Stuetzle, W.: Surface reconstruction from unorganized points. Computer Graphics - NewYork Association for Computing Machinery 26, 71 (1992)
4. Wenger, R.: Shape and Medial Axis Approximation from Samples. PhD Thesis. The Ohio State University (2003)
5. Amenta, N., Bern, M.W., Eppstein, D.: The crust and the beta-skeleton: combinatorial curve reconstruction. Graphical Models and Image Processing 60, 125–135 (1998)
6. Cheng, S.W., Funke, S., Golin, M., Kumar, P., Poon, S.H., Ramos, E.: Curve reconstruction from noisy samples. Computational Geometry 31, 63–100 (2005)
7. Dey, T.K., Wenger, R.: Fast reconstruction of curves with sharp corners. International Journal of Computational Geometry and Applications 12, 353–400 (2002)
8. Dey, T.K., Wenger, R.: Reconstruction curves with sharp corners. In: Proceedings of the Sixteenth Annual Symposium on Computational Geometry, pp. 233–241 (2000)
9. De Figueiredo, L.H., de Miranda Gomes, J.: Computational morphology of curves. The Visual Computer 11, 105–112 (1994)
10. Bernardini, F., Bajaj, C.L.: Sampling and reconstructing manifolds using alpha-shapes. In: Proc. 9th Canad. Conf. Comput. Geom. (1997)
11. Amenta, N., Choi, S., Kolluri, R.K.: The power crust. In: Proceedings of the Sixth ACM Symposium on Solid Modeling and Applications, pp. 249–266 (2001)
12. Gold, C., Snoeyink, J.: A one-step crust and skeleton extraction algorithm. Algorithmica 30, 144–163 (2001)
13. Ledoux, H.: Modelling Three-dimensional Fields in Geo-Science with the Voronoi Diagram and its Dual. Ph.D. School of Computing, University of Glamorgan, Pontypridd, Wales, UK (2006)
14. Karimipour, F., Delavar, M.R., Frank, A.U.: A Simplex-Based Approach to Implement Dimension Independent Spatial Analyses. Journal of Computer and Geosciences 36, 1123–1134 (2010)
15. Giesen, J., Miklos, B., Pauly, M.: Medial axis approximation of planar shapes from union of balls: A simpler and more robust algorithm. In: Proc. Canad. Conf. Comput. Geom., pp. 105–108 (2007)
16. Blum, H., et al.: A transformation for extracting new descriptors of shape. Models for the Perception of Speech and Visual Form 19, 362–380 (1967)
17. Gonzalez, R.C., Woods, R.E.: Digital image processing. Prentice Hall, Upper Saddle River (2002)
18. Karimipour, F., Delavar, M.R., Frank, A.U.: A Mathematical Tool to Extend 2D Spatial Operations to Higher Dimensions. In: Gervasi, O., Murgante, B., Laganà, A., Taniar, D., Mun, Y., Gavrilova, M.L. (eds.) ICCSA 2008, Part I. LNCS, vol. 5072, pp. 153–164. Springer, Heidelberg (2008)
19. Chazal, F., Lieutier, A.: The "λ-medial axis". Graphical Models 67, 304–331 (2005)
20. Attali, D., Montanvert, A.: Computing and Simplifying 2D and 3D Continuous Skeletons. Computer Vision and Image Understanding 67, 261–273 (1997)
21. Tam, R., Heidrich, W.: Feature-Preserving Medial Axis Noise Removal. In: Heyden, A., Sparr, G., Nielsen, M., Johansen, P. (eds.) ECCV 2002. LNCS, vol. 2351, pp. 672–686. Springer, Heidelberg (2002)

Geovisualization and Geostatistics: A Concept for the Numerical and Visual Analysis of Geographic Mass Data

Julia Gonschorek and Lucia Tyrallová

Geoinformation Research Group, Department of Geography, University of Potsdam, Germany
{julia.gonschorek,lucia.tyrallova}@uni-potsdam.de

Abstract. Scientific visualization as an interdisciplinary research field offers a wide range of methods and techniques to efficiently analyze and visualize the spatial and temporal data and information. In this article a discipline of scientific visualization is needed to explore information and construct knowledge from geodata: geovisualization. This research field offers tools and techniques to discover relationships, clusters and trends in geodata. Adapting methods from computer science and geographic knowledge discovery spatial and spatio-temporal data-patterns and information can be visualized in different types of charts, plots and combined with elements of cartography.

This article presents selected methods and techniques for two different scientific tasks: (1) To detect spatial attribute relationships of a landform linked to map. Specifically: spatial relationships of size, shape and location of thermokarst lakes, formed through permafrost degradation. We used star charts to visualize multivariable data and to make visible how landforms attributes are relate to each other. (2) A concept is developed to visually explore the specific distribution of emergency services and service clusters due to space and time. It depends on the example of emergency services of the City of Cologne. This geodata set contains more than 500,000 emergency services beginning 01/2007 and ending in 08/2011. The concept's core is a new hierarchical data representation based on tree diagrams linked with a cartographic visualization of the results received from explorative statistics (e.g., kernel density estimation).

Keywords: Scientific Visualization, Geovisualization, GIS-based Data Analysis.

1 Adapting the Geovisualization Pipeline

Nowadays, systems focused on target groups and specific user-oriented applications become ever more important. It seems that theories, concepts and visualization processes are getting put aside. Same is true for basic ideas of generic methods, such as explorative analysis, visual analysis, spatiotemporal analytics etc. These, based on numerous publications and scientific papers presented at international conferences, are considered as a trend to be followed. The empiric visualization is regularly used as a research activity and there exists its methodological concept. Data and information visualization is the primary aim of numerous scientific studies. We shall, however, focus also on discussion and improvement of the analytical methods as well as on informative contents and quality of the resultant data and its user-oriented communicability.

In the beginning we consider the visualization pipeline as a three-step process (see Fig. 1) which makes it obvious that data and information visualization are related. They should, however, be understood as individual units. Initial formulation of visualization process is KREITEL´s conception of data, information and knowledge. Data can be defined as a flow of characters carrying semantic meaning from common source between the sender and receiver. Receiver defines which characters are to be sent within the given character-region. Information originates always when the data reach certain meaning. Explicit knowledge is documented knowledge in any form and it becomes knowledge when it reaches its receiver. Implicit knowledge is understood as knowledge bound to a person, unspoken and embedded, based on the subjective perception, personal experience, educational and learning process (KREITEL, Ressource Wissen, p. 14). Thus knowledge originates by interconnection of data and information.

This process is outlined in Fig.1 where the data are initial material. The possible and necessary first analyses can be visualized in form of data displays. These data displays can be diagrams or the overview maps. The results gained from the analyses and visualizations are the basis for explorative analyses, i.e. more complex statistics as regression, time-series analyses and tests on spatial correlations, as well as for cluster analyses regarding the central problem of space and time. Following the analyses results, the techniques of visualization that are used can be those that are known, or adapted or restored so that information visualizations are achieved.

They can be separately analyzed or connected to each other via interactive linkages. That offers access to multidimensional information and their usage as well as their communication. The information's complexity is increasing from simple maps and charts to multiple, brushed views. Hence the need to proof the consumer's cognitive skills and the visualization's usability in practice.

So the mental combination of analytic and visual results broadens the individual knowledge and may lead to a deep understanding and decision making knowledge of the specific task and its possible relations for specific and related tasks.

A more detailed version of this three-step model represents a more complex process. Geodata contain spatial, temporal and factual information and have to be preprocessed for analysis purposes. Therefore the dataset's quality and completeness has to be proven and prepared for analysis. An import into an analytics system, e.g. a GIS, R etc., is following. First analyses, e.g. descriptive statistics and simple spatial analysis, are performed. Their results can be numeric or a diagram or map output. This very first data visualization – the so called data displays – may offer an insight into the dataset in a non-numerical way. Based on assumptions concerning a hypothetic spatial and/or temporal distribution of the incidents and their scale measurements, suitable methods for explorative data analysis can be used and where appropriate customized. Regression analysis, density estimation, time series analysis, cluster (and hotspot) analysis and much more have to be named. Numeric information is transformed into (carto-) graphic information (rendering pipeline).

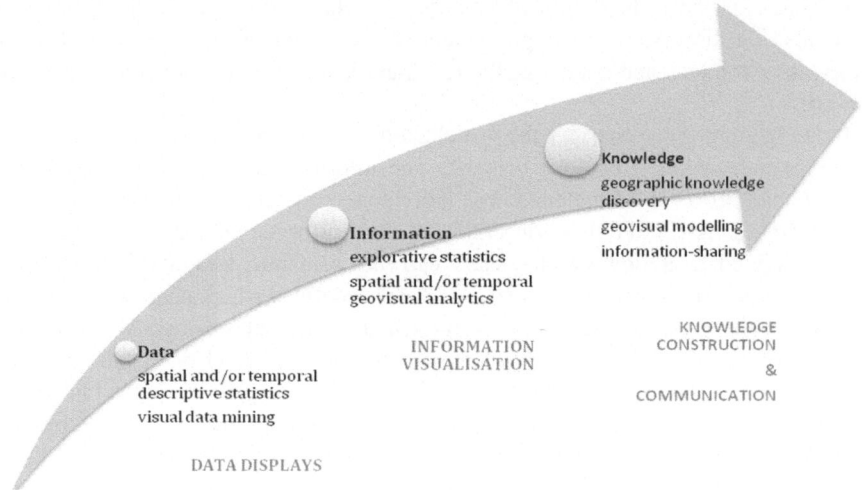

Fig. 1. Communication of Geoinformation (own work)

These graphic results give way to performance of the geovisual analyses and further validation and extension of the user's knowledge. The knowledge will be built and further communicable as based on the forms of information visualization or on the possible combination of various data displays and information visualization.

In our work, the partial processes of the pipeline will be dealt with and discussed within two complementary examples, as follows.

Thousands of thermokarst lakes as multivariate natural features in different regions of Canada, Alaska and Siberia over large areas are investigated to detect spatial attribute relationships and spatial relationships of size, shape and location of the landforms, formed through permafrost degradation. We try to make visible how landforms attributes are relate to each other and what are their specifics.

Also, more than half a million of firefighter interventions in the area of Cologne (Germany) are currently being investigated, taking into account their spatiotemporal distribution. The aim is to identify hotspots, detect the possible movement patterns to set the prognoses and help to induce the focused prevention measures. Various types of interventions will be differentiated among (i.e. surgical emergencies, fires, traffic accidents) and efficient data and information visualizations as well as suitable geovisualizations will be presented.

2 Geovisualization of Multivariate Natural Features for a Better Understanding of Their Spatial Relations

Geovisualization of multivariate natural features data and information serves not only as a documentation of a research (visual support), but also as a tool for discovering spatial and spatiotemporal relations. This article deals with selected methods for detecting spatial attribute relationships of landforms linked to a map. Specifically:

spatial relationships of size, shape and location of thermokarst lakes, formed through permafrost degradation. Thermokarst lakes are aftermath of thawing of the subsurface permafrost in form of a depression filled with water (FRENCH, 2007). One of the reasons, why we have chosen thermokarst as an example for a multivariate natural feature was that it is a prime indicator of changes in the subsurface permafrost. Also, this transient isolated landform can be automated detected in GIS (TYRALLOVÁ, 2011; TYRALLOVÁ, et.al., 2011).

The innovation is to merge well-known visualization methods with techniques typical for other fields (design, informatics, medicine, art). A map is valuable and useful, but contains only very limited amount and type of information about the features. This can be improved by combining maps with outputs of statistical analyses and different visualizations of desired or needed information. Visual links can help to avoid the occlusion of relevant information (STEINBERGER, at. al.).

Star charts are used to visualize multivariable data and information to make visible how landforms attributes are related to each other. Additionally, the general need of geovisualization is expressed by using the star charts. Variables, such as perimeter, radius and area, shape index, asymmetry, elliptical fit help to describe thermokarst lake as a feature.

There are a number of visualization methods for multivariable data. However, some methods, such as parallel plot, are not suitable for larger amounts of data. Although it is possible to optimize the parallel plot settings, these should not be the first choice when visualizing data and information which are in big volumes. However, only the tendency and deviations are then visible, while respective visualizations of individual information are difficult to read. Therefore, for larger amounts of multivariate data, a star chart is more appropriate. The attributes of a feature are depicted in a star chart counter-clockwise beginning with axis x (west to east) in the following order: area, perimeter, maximal distance from the centroid of the feature (major axis length), minimal distance from the centroid of the feature (minor axis length), shape index and asymmetry (Fig. 2; Fig. 3 a,c).

The shape index describes the smoothness of the object border, the smaller the index value, the less complex the object is. To calculate shape index (see Eq. 1), the following formula, as defined by Hinkel (Hinkel et. al. 2005), is used.

$$S_i = \frac{perimeter}{4 \cdot \sqrt{area}} \quad (1)$$

Calculation of asymmetry, assuming the shape of the evaluated feature is near elliptic, has been calculated according to Hinkel (Hinkel et. al. 2005) as well (See Eq. 2).

$$A = 1 - \frac{l_b}{l_a} \quad (2)$$

With length (L), minor axis (b) and major axis (a).

The value of A is between 0 and 1. The bigger the value, the more asymmetrical is the landform. The centre of the star chart is identical to the centroid of the detected feature.

The star chart depicts information about the feature and its spatial relation to other features and about attributes of the feature. The element is characterized by attributes, location and by relation to other elements. A standardization of the variables has been necessary to enable comparisons of the respective attribute values. Data and information were generated in R and ArcGIS.

This representation is a good example of the supportive role of visualization in a decision-making process (Figure 2). It brings a compact picture of all the landform's attributes and relations between each other. This form of representation shows clearly how the added value can be gained. The work on linking the maps with star charts is currently in process. Identified objects depicted on a map have a unique ID, which corresponds with a unique key for every star in the diagram. A centroid of a thermokarst lake has the same coordinates as a centroid of its corresponding star chart. The goal is to explore spatial relations of the landforms, their distribution and whether they form some sort of pattern.

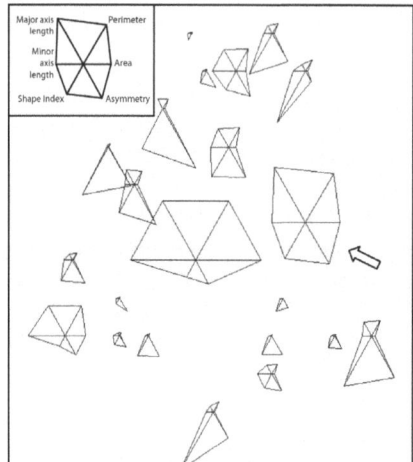 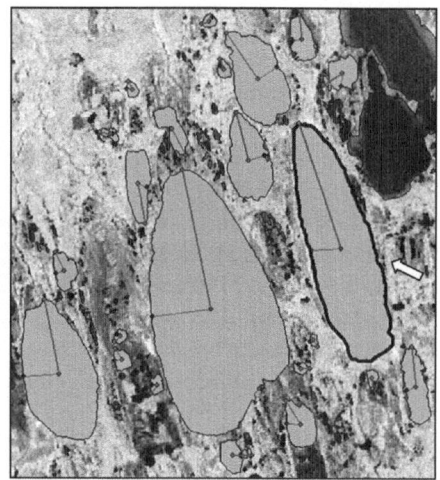

Fig. 2. Spatial attribute relationships of a landform linked to map represented by star charts

Our intention is to help experts in the field of geology and geomorphology to be able to identify objects and gather extra information about these objects even over large areas, in order to allow them to select the results relevant to their respective goals and, if needed, verify them in situ.

We have chosen a section typical for the large areas in different regions. It contains various shapes of the same landform type. Figure 3 depicts thermokarst lakes (b) and rivers (d) respectively. These are typical water bodies that are usually very difficult to automatically identify and to distinguish one from the other. However, linking the advantages of the star charts to the map-like presentation, as in Figure 3, one can easily distinguish these two landform types purely on the basis of geovisualization.

Thanks to the data display (perimeter, area, length of minor and major axis) and information display (shape index, asymmetry) and the connection between them, it will be easier to differentiate between these types of water bodies.

We can see that all the subtypes of thermokarst lakes (Fig. 3, a) are characterized by pronounced shape index and asymmetry in comparison to the other parameters. In the case of thermokarst lakes with bigger area (50,000 m^2 or more) are the parameters more balanced. That points to the tendency that bigger thermokarst lakes are usually more symmetrical, i.e. the major and minor axis are similarly long and the value of the asymmetry nears "0". This is caused by the fact that the bigger thermokarst lakes are generally older and have developed by merging of more smaller lakes, and it is less likely that the merging would develop predominantly only in one direction. Of course this also depends on the area's relief.

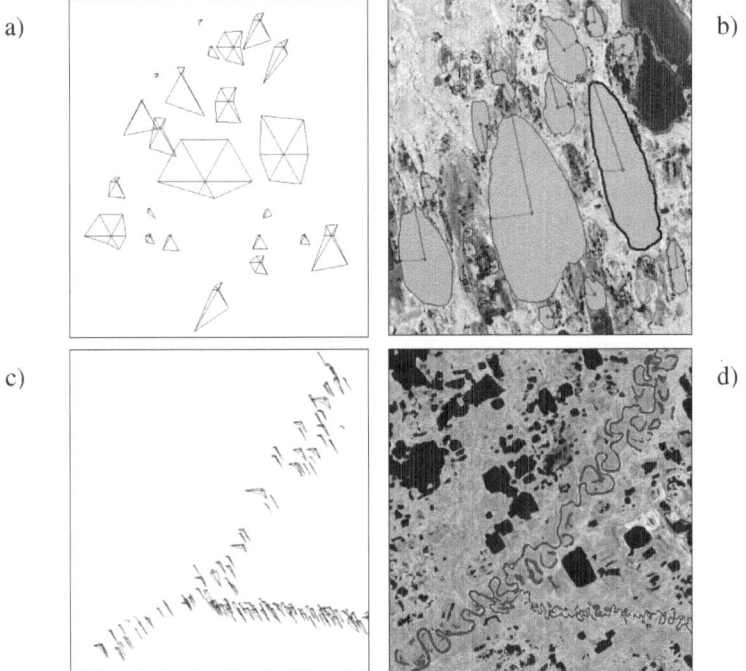

Fig. 3. Parallel representation of the water bodies (b, d) and geovisualization of their parameters (a, c)

Meandering rivers are difficult to identify based only on optical remote sensing data. Multispectral data allow for identification of river segments under the condition that the riverbed (river-basin) hadn't gone completely dry. Hence, every river segment is represented by a star chart. Of the river segments (Figure 3, c) is true that the parameter of asymmetry is dominant. This is only natural since rivers are obviously far more asymmetrical than thermokarst lakes. The major axis is significantly longer than the minor axis, i.e. the value of the asymmetry nears "1".

3 Geovisual Analysis of Mass Data for Civil Security Purposes

The purpose of this work is a geovisual analysis of emergency services mass data. Firefighters from the fire department of the City of Cologne in western Germany answered round about half a million emergency calls from January 2007 to August 2011. These emergency services are characterized by a large bandwidth, e.g. extinct fire, rescue casualty person as well as animal and offer medical assistance.

This mass dataset contains all geocoded emergency locations, temporal information (date, day of the week, daytime) and type of emergency case. Our goal is to detect spatiotemporal relations, distributions and pattern for identifying emergency agglomerations in general and hotspots in particular as well as to explore possibly existing (ir-) regularities in frequency and moving patterns. Thus a basis for strategic and requirements-planning is developed: explorative statistics and information visualization allow for user- and benefit-oriented views of the past and prognoses. Hence suitable prevention measurements like fire security or mobile heart-defibrillators at public places can be realized and evaluated.

Our purposes are strongly related to crime analysis and crime mapping, focusing especially on (carto-) graphic methods and techniques of strategic, operational and tactical crime maps. The geovisualizaion process described in chapter 1 is realized by adopting, broadening and combination of these crime analysis methods and techniques. One central question is, whether an increasing complexity of analysis means complex and sophisticated visualizations. This has to be discussed for the target group "fire department and administration".

A sample-dataset that contains all surgical emergencies from July 2007 till June 2008 is generated. In ESRI ArcGIS and R simple map-like representations like point maps, grid maps and also bar plots can be produced – the so called data displays. Based on the enormous data (ca. 17,000 surgical emergencies) the data display's clarity is in dispute. Point signatures for example do overlay each other. That depends on the map's scale and the neighborhood of emergency locations. There is no need to say, that an optic quantification isn't practicable. Grid maps on the other hand may hide information or special local features because of the chosen cell size. In consequence of this, knowledge about potential emergency-agglomerations is speculative. Simple diagrams for the data visualization (e.g. bar or line plots) present data in a well-arranged form, because of the multidimensionality of the dataset these representation methods are inadequate. An adaptation of a tree-diagram, as a method which combines hierarchical structures with quantity-information, is a sunburst diagram. Sunbursts do have their origin in interactive computer design. Tree-diagrams in general are a graphic output of hierarchical data. That means the interlaced structures (data levels) are the basis of this diagram type. Hence a sunburst diagram constructs a radial structure, with minimum two circles, one circle for each hierarchy-layer that starts with the inner circle and ends with the last outer one. The circle's segments represent quantitative information (e.g. frequency of emergencies). At a glance, mass data can be visualized in a non-cartographic way but containing all (grouped and weighted) information. OpenProcessing, Excel and other proprietary and open source software or developing environments offer the opportunity to create sunbursts.

Figure 4 shows such a diagram of surgical emergencies during July 2007 and June 2008. The inner ring represents the year with the belonging months (in the second circle) and the day of the week in the outer circle. A circle's segment size results from the absolute emergency frequency at the rate of the total number of emergencies. One can get a complex insight into the data by this type of data display. There is a direct correlation between the independent time frames year, month and day of the week in reference to the emergency type. However, one can easily imagine a more complex sunburst model where more (or all) emergency cases and locations (e.g. administrative groups) are implied.

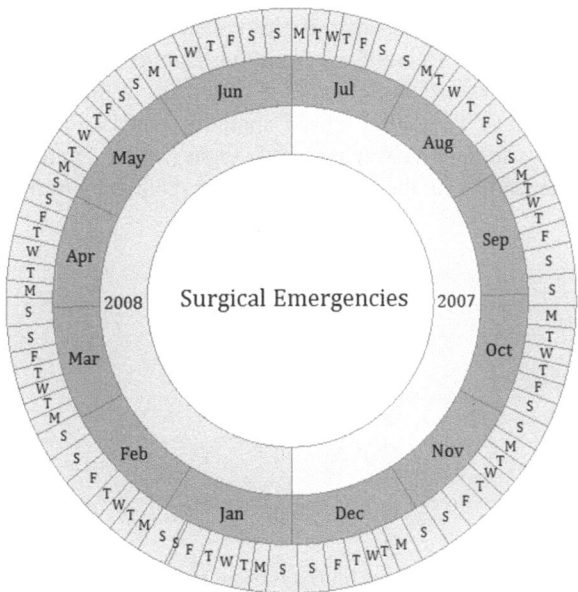

Fig. 4. Sketch of a sunburst diagram of all incoming emergency calls related to surgical emergencies

Obviously the temporal distribution of surgical emergencies is quite homogeneous during the months, but there are big differences to the day of the week distribution: The weekdays show a kind of temporal pattern. There are an increasing number of emergency services especially on Fridays, Saturdays and Sundays. This is quite important and necessary for further spatial and spatiotemporal analysis. To get more detailed information and to generate knowledge an explorative analysis and information visualization has to be realized. The kernel density estimation (kde) is suitable for the analysis and visualization of spatial distribution of the geocoded emergency services (points). CHAINEY and RATCLIFFE (2005) use this nonparametric method generates continuous variables from the existing point-data that allows for areal display due to the estimation of the probability distribution of variables. We apply the Epanechnikov kernel (K) in ESRI ArcGIS with the form (see Eq. 3):

$$\hat{f}_\lambda = \frac{1}{n\lambda}\sum_{i=1}^n K_0\left(\frac{y-y_i}{\lambda}\right) \text{ with } K_0(y) = \begin{cases} \frac{3}{4}(1-y^2) \text{ for } |y| \leq 1 \\ 0 \text{ else} \end{cases} \quad (3)$$

Highly important for estimation and density-visualization is the selected bandwidth (λ) and cell size of the output-grid. Both parameters are chosen depending on Cologne's areal size and are fix during the whole analysis, comparisons aren't possible else. To set the bandwidth different settings were tested and compared with other techniques of hotspot analysis like K-means. This trial-and-error method isn't quite scientific. The development of rules for bandwidth settings with respect to cartographic design and usability for the fire department of the City of Cologne is one issue of our current work.

We apply the kde on surgical emergencies partitioned into day of the week. One can easily see that the first impressions from our sunburst are approved (see Fig. 5). First of all there are two small agglomerations at the left of the Rhine and are increasing up to the weekends. They also expand spatially. One can identify optically five hotspots on Sundays in the city's center of Cologne, where innumerous shops, bars, night clubs and, depending on the historic structures, confusing traffic areas as well as nursing homes.

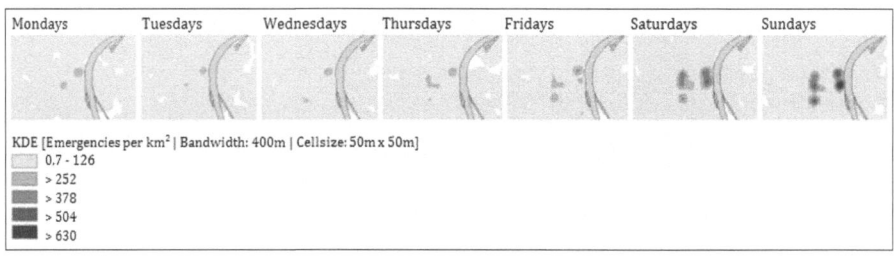

Fig. 5. Surgical emergencies partitioned into days of the week located at Colognes city center at the left of the Rhine

We also have detected spatiotemporal patterns and relations of other emergency types. As a matter of fact the knowledge about such distributions (analytical and visual) and their long-term experience lead the fire department of the City of Cologne to a modified strategic planning. At ordinary weekends (not including public holidays), for instance, more dispatcher and fire fighters as well as emergency doctors are on stand-by for emergency services.

4 Conclusion and Outlook

HAMILTON estimates that 95% of all data have a spatial reference (PERKINS, 2010). Following the purpose and the target group of users it is important not only to have the data prepared in a database, but they have to be preprocessed and used to gain information and gather as much knowledge as possible about these data.

It was necessary to adapt the geovisualization pipeline and the partial processes of the pipeline were discussed within two complementary examples.

Our case studies show that is meaningful to differentiate between data-displays and information visualization. We have introduced into a concept which offers a general framework for spatial and spatiotemporal analyses and visualization techniques, as well as visualization methods. Consequently we have shown on diverse examination subjects, how this concept can work. The goal, to extract information from data and then process them visually, so they are user-fitting and eligible, is partial accomplished. The next step is to test and expand other techniques, comparisons and prognoses for spatiotemporal analyses. Our plan is to implement the concept of multiple, more interactive and linked information. To be able to achieve this, observation and evaluation of target groups are necessary.

Our methods and results are important for future explorative analyses and geovisual analytics. We will analyze all emergency services and their distributions in space and time automatically. Also time-series-analysis is to apply to detect trends. We'll realize this mathematical analysis in R and ArcGIS. A very important aspect of the analyses is the comparative analysis of different types of emergency services. It is necessary. On the one hand such analysis doesn't exist. On the other hand, as a matter of fact, specific emergency types do have specific distribution patterns. E.g. fuel leaks within a very small probability in nighttime, forests, parks and playgrounds. However, internistic emergencies take place on highways and in/nearby nursing homes in the early morning and at the weekends.

Information and a deep understanding of specific distributions and patterns as well as (ir-) regularities of intensity are absolutely necessary for prevention measurements to be well-directed and needs based. Time-series-analysis and prognoses are suitable for the personnel, strategic and tactical planning.

Another goal of our visualization-concept is the implementation of interactive multiple linked views (see Fig. 6). Thus data-displays and information visualization could be combined and may lead to a deeper understanding of spatiotemporal relations and characteristics of emergency services. But we have to consider, that visualization must be user- and benefit-orientated. Complex and sophisticated views are inoperative for civil security purposes, where most of planning action is done analogue on the "black board". Also the mathematical formula must be communicated in a user-centered way. Visualization has to speak for itself: simplicity is the key to knowledge and to a redefined strategic and tactical planning for civil security purposes.

On the example of a thermokarst lake it has been shown that geovisualization of multivariate natural features data and information should not serve only as a documentation of any research, but also as a tool for discovering spatial and spatiotemporal relations. The method of linking map to suitable types of charts can be used basically for any landform type. One must only select the right parameters to describe the landform the best and the right type of their graphic visualization.

The focus of the work to come is to explore spatiotemporal relations of the landforms, their distribution and whether they form some sort of pattern. The results of geovisualization of multivariable features will be used for further geospatial analyses and automated identification.

Our concept for the numerical and visual analysis used different examples and combined methods and techniques of statistics and visualization with elements of cartography. We have shown the differences between data displays and information visualization and its necessity and usage for decision support – e.g. in civil security or permafrost research, ecological and carbon balance.

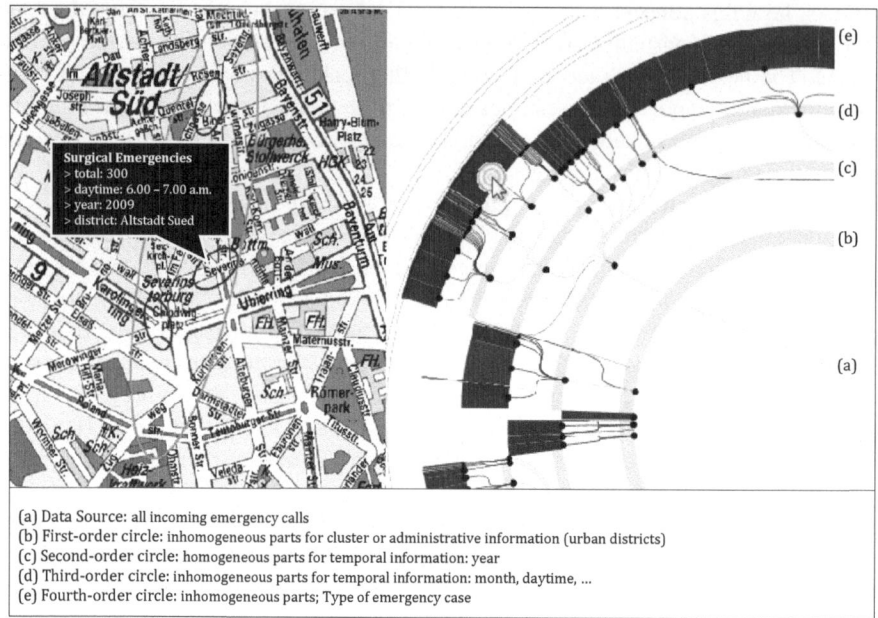

(a) Data Source: all incoming emergency calls
(b) First-order circle: inhomogeneous parts for cluster or administrative information (urban districts)
(c) Second-order circle: homogeneous parts for temporal information: year
(d) Third-order circle: inhomogeneous parts for temporal information: month, daytime, ...
(e) Fourth-order circle: inhomogeneous parts; Type of emergency case

Fig. 6. Sketch of linked views: a conceptual design

Acknowledgements. Special thanks go to Prof. Dr. HartmutAsche (Geoinformation Research Group at the University of Potsdam) and Prof. Dr. Stephan van Gasselt (Research Field of Planetology and Remote Sensing at the Freie Universität Berlin) who support our research activities in the context of our dissertations.

References

Bodendorf, F.: Daten- und Wissensmanagement, 2nd edn. Springer, Heidelberg (2006)
Bohnacker, H., Groß, B., Laub, J.: Generative Gestaltung – Entwerfen, Programmieren, Visualisieren, 2nd edn. Verlag Hermann Schmidt, Mainz (2010)
Chainey, S., Ratcliffe, J.: GIS and Crime Mapping. John Wiley & Sons Ltd., Chichester (2005)
Dodge, M., McDerby, M., Turner, M. (eds.): Geographic Visualization: Concepts, Tools and Applications. John Wiley & Sons Ltd., Chichester (2008)
Dykes, J., MacEachren, A.M., Kraak, M.-J. (eds.): Exploring Geovisualization. International Cartographic Association/Elsevier, Amsterdam (2005)
Ebert, A., Dix, A., Gershon, N.D., Pohl, M. (eds.): HCIV (INTERACT) 2009. LNCS, vol. 6431. Springer, Heidelberg (2011)

French, H.M.: The Periglacial Environment, 3rd edn. John Wiley & Sons Ltd., Chichester (2007)

Kreitel, W.A.: Ressource Wissen: Wissensbasiertes Projektmanagement erfolgreich im Unternehmen einführen und nutzen. GablerVerlag Springer Fachmedien GmbH, Wiesbaden (2008)

Perkins, B.: Have You mapped your data today? In: Computerworld Online (2010), http://www.computerworld.com/s/article/350588/Have_You_Mapped_Your_Data_Today_ (access: January 6, 2012)

Schumann, H., Müller, W.: Visualisierung: Grundlagen und allgemeine Methoden. Springer, Berlin (2000)

Spies, T.: Generische Architektursichten: Erzeugung und Visualisierung kontextspezifischer Sichten am Beispiel serviceorientierter Architekturen. Gabler Verlag Springer Fachmedien GmbH, Wiesbaden (2011)

Steele, J., Iliinsky, N. (eds.): Beautiful Visualization: Looking at Data Through the Eyes of Experts. O'Reilly Media Inc., Sebastopol (2010)

Steinberger, M., Waldner, M., Streit, M., Lex, A., Schmalstieg, D.: Beautiful Context-Preserving Visual Links. Visualization and Computer Graphics 17, 2249–2258 (2011)

Taylor, R., Boulanger, P., Krüger, A., Olovier, P. (eds.): Proceedings, 10th International Symposium: Smart Graphics: Banff/Canada, June, 24-26. Springer, Berlin (2010)

Tyrallová, L.: Automated object detection of climate tracers in remote-sensing data. In: Proc. SPIE 8174, Remote Sensing for Agriculture, Ecosystems, and Hydrology XIII, Praque, September 19-22, vol. 8174 (2011), doi:10.1117/12.898111

Tyrallová, L., Schernthanner, H., van Gasselt, S.: Identifikation von Klimaprozessen durch automatisierte Objektidentifikation isolierter Oberflächenmorphologien. In: Strobl, J., Blaschke, T., Griesebner, G. (eds.) Angewandte Geoinformatik 2011. Beiträge zum 23. AGIT-Symposium Salzburg, July 6-8, Universität Salzburg, Heidelberg (2011)

Yau, N.: Visualize This – The FlowingData Guide to Design, Visualization and Statistics. Wiley Publishing Inc., Indianapolis (2011)

Spatio-Explorative Analysis and Its Benefits for a GIS-integrated Automated Feature Identification

Lucia Tyrallová and Julia Gonschorek

Geoinformation Research Group, Department of Geography, University of Potsdam, Germany
{lucia.tyrallova,julia.gonschorek}@uni-potsdam.de

Abstract. This paper deals with an automated feature identification process, specifically identification of landforms and their attributes. The feature identification process is GIS-integrated and is carried out on the commercial platform ArcGIS. Spatio-explorative analysis offers a wide range of methods and techniques for the feature identification. An automated process is helpful for experts to identify many feature types over large areas. Our study case are thermokarst lakes (as prime climate indicators) in two different areas: North Canada and North Siberia. For the analysis of variance and correlation we have established a required significance level up to five percent. We have found correlations between the existing feature parameters and use regression analysis to optimize the identification process as well as to be able to better distinguish individual landforms from each other. Our goal is to provide GIS-integrated object-identification tools to identify and characterize landforms indicative of climate change, to allow extracting parameters required to assess climatic boundary conditions.

Keywords: GIS, Automated Feature Identification, Spatio-Explorative Analysis.

1 Background and Objectives

Automated feature identification is, in geomorphology, mainly used to support mapping (Dragut and Blaschke 2006); (Hese, et.al. 2010); (Hinkel, et.al. 2005); (Taramelli and Melelli 2008). Landforms and their changes, related to anthropogenic and climatic influences are best detected and analysed using multitemporal data sets and elevation models. Geomorphic landforms are, to disadvantage for automated process, complex features that are neither geometrically nor spatially unambiguously constrained and they usually represent a snapshot stage in landscape development that has been continually modified by external factors.

A thermokarst is an example of landforms responding directly to climatic conditions (Büdel 1984). Within this, the recent or past climate changes can be traced. Extracting characteristic geomorpholometric parameters, we can derive quantitative and spatial statistics over large areas and many landform types.

The larger scientific group concentrates on the shape and surface expression to achieve classification and mapping of the single landform types. These are mostly glacial or fluvial landforms. This approach is followed by Schneevoigt et.al. (2010), Eisank et.al. (2010), Hese et. al. (2010) und Taramelli and Melelli (2008) and includes topics such as detection of landforms in alpine environments based on digital terrain models (DTMs) and ASTER data (Schneevoigt et.al. 2010), or the object-based classification of glacial cirques through terrain-model properties (Eisank et.al. 2010). The authors conducted the classification on the basis of the curvature parameters, segmentation and neighborhood relationships. Vertical curvature has been applied as well, to track sand dune dynamics (Mitasova et.al. 2005).

Significant examples of semi-automated classifications of landforms can be found in the recent literature (Dragut and Blaschke 2006); (Dragut et.al. 2010); (Hese et.al. 2010); (Moore et.al. 2003). As a part of the detection and mapping of debris cones, methods using fuzzy logic and quantitative roughness analysis were discussed by Taramelli and Melelli (2008). The database consists of SRTM (Shuttle Radar Topography Mission), SAR (Synthetic Aperture Radar) and C-band radar data. Dragut and Blaschke (2006) have dealt with the object-oriented classification of landform elements in several works. They analyzed, segmented and classified elevation models. Derivatives of the relief height, except the slope and slope exposure, entered this analysis.

Frohn et.al. (2005) have dealt with satellite image data classification of thaw lakes and basins in Alaska. Consequently, Hinkel et.al. (2005) used object-oriented classification and spatial analysis to sort thaw lakes and drained thaw lake basins in the western arctic coastal plain, in Alaska. They chose three subregions in Alaska, where they employed the spatial analysis on all the mentioned forms mutually. To define these varied forms, they have used, besides other approaches, also the shape index.

2 The Automated Feature Identification Process

The automated feature identification process input depicted in the flowchart (Figure 1) are the remote-sensing data, specifically multispectral data (I1) such as satellite-image data Landsat-based camera suits (e.g. Enhanced Thematic Mapper (ETM+) or the Rapid Eye-based data; airborne image data from local airborne-mapping campaigns. Additionaly, multitemporal image data can also be used in change detection. Characteristics of the surface features in the third dimension are presented by terrain-model data that we consider as the other important input element (I2). These can be derived from a diversity of sensors either stereophotogrammetrically or by introducing RADAR (RAdio Detection And Ranging) or LIDAR (LIght Detection And Ranging) techniques. Finally, environmental factors such as wind, temperature, moisture and precipitation as well as their derivatives form the (I3) input group. The feature-identification process is GIS-integrated.

After atmospheric correction and radiance conversion, an important element follows (I4): an object catalog and a well-defined decision tree, which is the basis for the spatial analysis. The core of the automated feature, which represents landforms, consists of the extraction of geomorphologic and morphometric parameters from terrain-model data (if available) and of image data resulting from object-based classification and other calculations. Filtering the negative features we get an output of positive features (O1). They are identified and defined through vector-based delineations. The landforms identified within the automated feature identification process can be further characterized for the spatial distribution assessment and type classification. Geostatistical analyses, specifically regression analysis, analysis of variance and non-parametric tests provide better understanding of these landforms (O2). Output data of the analysis of variance (O2) are iterated by means of the geostatistical evaluation operations from positively identified features (O1), see iteration loop in the workflow (Figure 1).

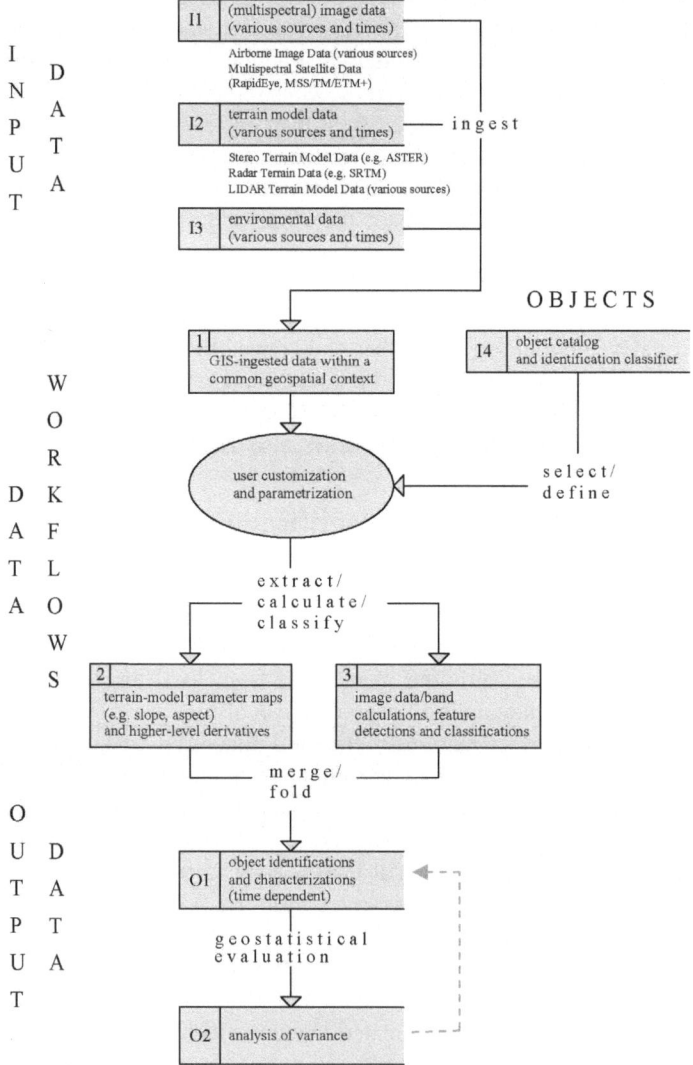

Fig. 1. Simplified workflow of automated feature identification and geostatistical evaluation

This work makes use of multispectral satellite-image data Landsat ETM+ in the resolution-range of 15 to 30 m/px. Medium spatial resolution, i.e. scales of 2-30 m, do suffice for the identification of thermokarst. The automated identification is suitable for isolated landforms.

3 Thermokarsts – Isolated Landforms

Geomorphologic landforms are complex features, with a number of modifications caused by external factors. When monitoring the effects of climate variability, prime indicators are very important. There are more types of prime indicators. The type that we focus on is isolated landforms. One of such landforms which can be defined as a prime indicator is a thermokarst because it responses sensitively on the changes of sub-soil temperature as well as surface temperature. A thermokarst is an aftermath of thawing of the sub-surface permafrost in form of a depression often filled with water, which dries up in time as the permafrost degrades. Thermokarsts can be described qualitatively in image data as elliptic or circular shapes, whose main axes have the same orientation. They can be found in groups or clusters in polar permafrost regions.

We have chosen thermokarst depressions as an initial landform type because it is known as to be variable on a scale of years and it is a prime indicator of changes in the subsurface permafrost that represent actual climate changes. Thermokarst depressions are associated with frozen ground and periglacial environments and are formed and developed through permafrost degradation (Washburn 2003). They are common and widely spread in both Northern Siberia and Canada which makes for a good basis for identification and comparison work as the features are here in abundance and the geomorphologic settings are quite different.

Thermokarsts usually have sizes and shapes that are identifiable in a wide range of imaging sensors and terrain model data and can also be mapped on scales of common Landsat-based sensors. This allows us to detect the changes in the span of up to three decades. Surface age, distance from the coast, ground ice content, sediments, and local relief have influence on thermokarsts morphometry (French 2007), (Hinkel, et.al. 2005). The focus is laid on the identification of characteristics of a landform. This identification is based on the landform's morphometric parameters and on the statistical compilation of its characteristics.

4 Exploratory Analysis for the Automated Feature Identification

To obtain the knowledge concerning specific landforms exploratory spatial data analysis methods are applied. Techniques from applied mathematics are used to detect relationships, variances and possible similarities between different landform-features.

The prime data-set is about Canadian thermokarst lakes. A second one (Siberian thermokarst lakes) is used to compare test results and functional correlations.

4.1 Detecting Relationships via Regression Analysis

Regression analysis is a very important and in various (geo-) sciences an often used statistical method to answer diverse questions and to efficiently explore mass-data. To explore functional dependencies in more or less complex data structure is an important issue of statistics. A regression model should describe the reality within a well-defined relationship. Also the mathematical modelling results in a better understanding of relations between two or more variables in a dataset. The points of interest are, on the one hand, whether a relation actually exists, and on the other hand, how strong the calculated bonding is. (Sachs 2009)

Perimeter, surface area and a radius (difference between a centroid and farthest point of the polygon) are three given metric parameters of the 250 Canadian thermokarst depressions. They vary between 85 and 18,329 metres (perimeter) and 307 and 11,361,109 square metres (surface area). The length of the radius varies from 20 to 3,261 metres.

The main goal is to answer the question whether or not any causal correlation between the parameters exists. Such a characterisation of this landform could help to detect it automatically in larger datasets, not only in Canada. Also spatial-temporal change detection could be improved.

It is to be mentioned that for further analyses, the thermokarst lakes are grouped into two categories: Group A means a maximum area of less than 500,000 square metres; Group B includes all features equal to or larger than the limit of Group A.

The variables are neither normally distributed nor approximately normally distributed. Therefore Normal Quantile-Quantile-Plots (QQ-Plots) are generated. This is a quite popular graphical method to prove the regression's necessary precondition of normal distributed data. Figure 2 shows the QQ-Plots of the standardised residuals. One can easily see large deviations from the transformed normal distribution (in Figure 2 the black line) at the left and right corner. Considering the "area" parameter of the thermokarsts in the average QQ-Plots we observe extreme aberrations also in the medial values. Various visual methods clearly verified that the observed landforms and their parameters are not linear dependent that is why further tests were not considered as requisite.

Figure 2 request us to apply a non-linear model. To detect the relations between any two of the three parameters – the dependent and one or more response variables - a (simple) non-linear regression analysis has been applied. Using the free open source statistics-software R (The R Project for Statistical Computing 2011) or other analytic software the mathematical function can be modelled within a few steps. In practice, an exponential regression-function is calculated as well as quality indicators.

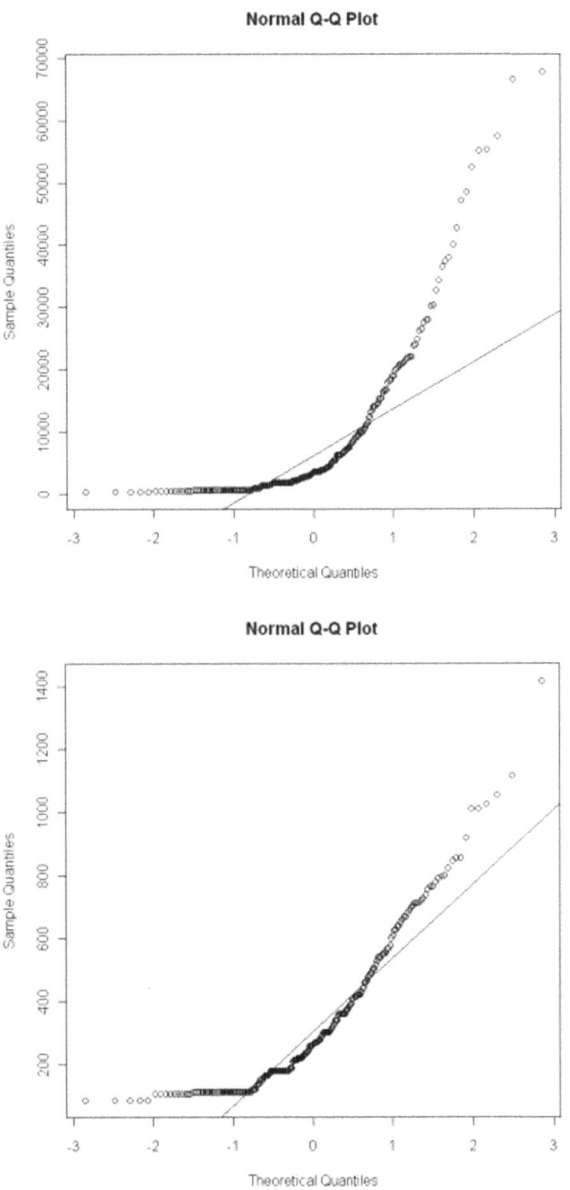

Fig. 2. The Normal QQ-Plots for the variables "area" (upper) and "perimeter" (lower) of the Group-A thermokarst lakes

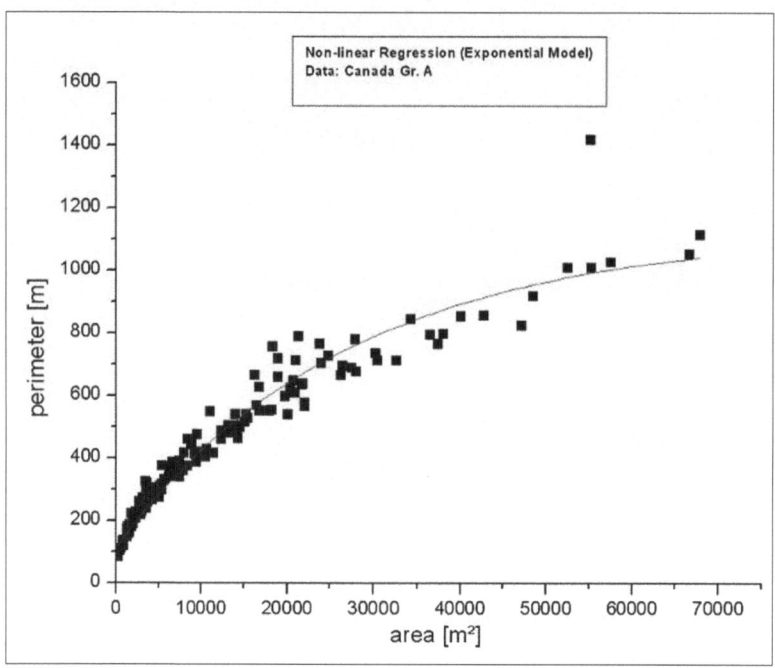

Fig. 3. Non-linear regression using perimeter and surface area of Canadian climate tracers

Figure 3 shows the point plot and an exponential regression function (black line) which has a general form (see Eq. 1) of (Originlab 2011):

$$f(x) = y = y_0 + A\left(e^{-(x-x_0/t)}\right) \qquad (1)$$

with the following parameters (see Eq. 1):

- y_0 Minimum value of the perimeter-dataset
- x_0 Minimum value of the area-dataset
- x Values of the area-dataset
- A Amplitude
- t Decay constant. (1)

The data-points at the left corner are close to the regression function (the black line) while there is an increasing variation of a few data-points in the right corner. In general the distribution of the Canadian Group A seems to be cornet-like. Getting an idea of the regression's quality, the adjusted R-squared should be calculated. It is based on the overall variance and the error variance (Crawley 2007). This exponential regression model works within an Adjusted R-squared of $r^2 = 0.91$. So the Canadian Group-A-Model matches the data's reality quite well. The next task will be to test differences and to compare this regression model with thermokarst lakes in a different geographic region.

4.2 Data-Comparison due to Further Variance Analyses

More regression-based data should be integrated into this complex statistical analysis. The relatively good adaptation of the Canadian landform-data within the exponential regression leads to the question whether thermokarst depressions with different geographic position, e.g. Siberian landforms, show similar functional correlations. It needs to be mentioned that should a second example prove positive, it would not be the general evidence for this geomorphological structure and its characterisation. But it actually could work as an indicator and may help the automated detection of thermokarst depressions and similar landforms.

Siberian thermokarst lake's "perimeter" and "area" values show a quite similar distribution (compare Figure Canada-A and Siberia-A). The same exponential regression model is used with different coefficients. While the Canada Group-A-Model showed a slim cornet-like distribution (Figure 4), the data-points show a higher variation in the right corner. As a consequence, the Adjusted R-squared is good but not equal to the Canadian Group-A-Model ($r^2=0.83$).

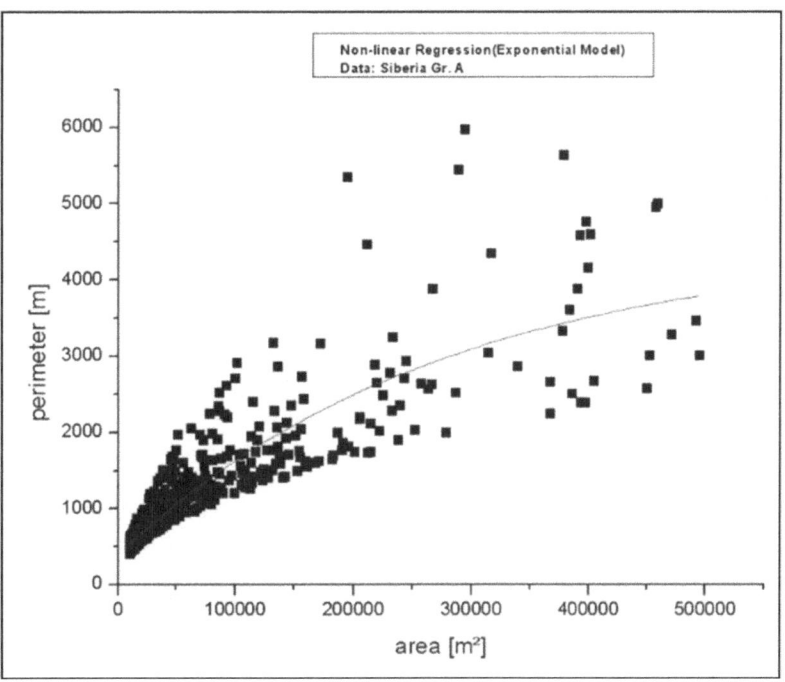

Fig. 4. Non-linear regression using perimeter and surface area of Siberian climate tracers

Box-and-Whisker-Plots support an optical comparison between Canadian and Siberian landform-data. Therefore the so called "shape index (Si)" (see Eq. 2) is calculated and plotted (Hinkel et.al. 2005). To define these forms, the shape index has been used:

$$S_i = \frac{perimeter}{4 \cdot \sqrt{area}} \qquad (2)$$

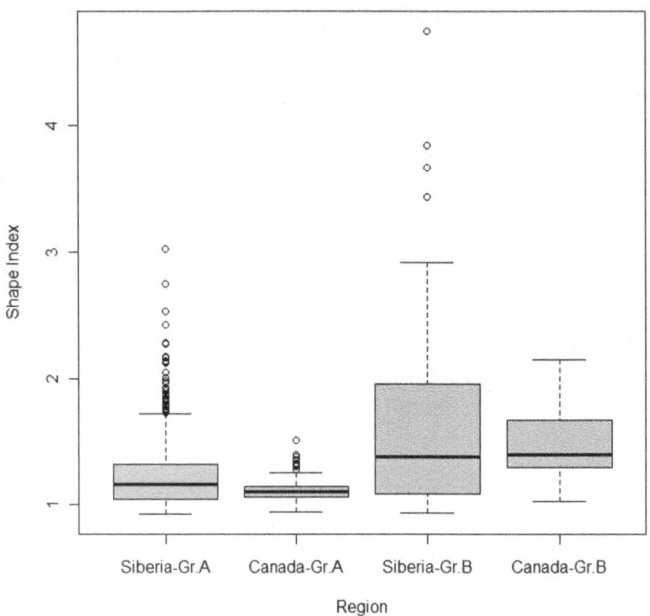

Fig. 5. Box-and-Whisker-Plots of the shape index values of the Canadian and Siberian thermokarst lakes

Figure 5 shows the Box-and-Whisker-Plots based on shape index of the region's grouping. One can visually identify following differences:

A huge variation of the median's position and the size of the boxes in the Canadian as well as in the Siberian dataset between Group A (area<500,000 sqm) and Group B (area >=500,000 sqm);

Variations between the median's position of the Siberian and Canadian Group A and almost equal median's position of the Siberian and Canadian Group B.

Using non-parametric tests, one can gain information about how similar or different the sample-sets are. The Wilcoxon-Rank-Sum-Test helps to identify, within a significance level of less than five percent, if there are significant differences between the perimeter and surface area of Canadian and Siberian thermokarst lakes. Therefore the whole sample-set has to be well-ordered and ranked. Table 1 lists the results of this non-parametric test.

Observed differences showed in the Box-and-Whisker-Plots are numerical verified by the Wilcoxon-Rank-Sum-Test. One can easily see, that the variation between two groups is significant. There is only one exception: The Siberian and Canadian Group B do not vary significantly. Except shape index we have calculated also some other parameters, for example asymmetry of a landform.

Table 1. Variation between Canadian and Siberian thermokarst lakes

Wilcoxon-Rank-Sum-Test	Siberia Gr. A – Canada Gr. A	Siberia Gr. A – Siberia Gr. B	Canada Gr. A – Canada Gr. B	Siberia Gr. B – Canada Gr. B
Test Score	W = 99813	W = 18294	W = 522	W = 679
Probability	p = 7.03e-08	p = 8.17e-06	p = 3.23e-08	p = 0.6925
Significance Level		$\alpha = 0.05$ (5%)		

5   Results and Discussion

Thermokarst lakes cover an area of 53 036 ha, which represents almost 27% of the study area (197 493 ha). The automated feature identification process can be a significant asset for identification of isolated landforms over large areas.

Figure 6 depicts the feature identification process (from left to right): image data after preprocessing (atmospheric correction and radiance conversion). NDVI (Normalized Difference Vegetation Index), one of the indicators used in the analysis. Water bodies contain biomass with low or no activity, so the value of NDVI ranges from -1 to 0. Output data of the spatial analysis were polyline-based delineation of features. Features smaller than 10,000 sqm were excluded. Next important step was filtering the negative features. River segments were excluded. Finally, on the right end of the figure 6, we see positive feature identification.

Figure 7 depicts vector-based delineations of the identified thermokarst lakes, calculated centroid and also maximum distance (points connected by lines on the figure 7) and minimum distance from centroid to delineation of the feature.

We have created a script that automatically computes a centroid of a feature and points with maximal and minimal distance from the centroid. Other steps of the automated feature identification process have been carried out in the model builder and are a part of ArcGIS toolbox. The script creates two files at user defined locations on the filesystem, a polyline-shapefile and a polygon-shapefile. The polyline-shapefile holds data of the geometry of the semi-axes, the axis-types (semi-minor or semi-major-axis), the lengths of the semi-axes and the identifiers of the original polygon. The semi-axes are determined in the following way; for each polygon of the original polygon-shapefile the distance from its centroid to each vertex is calculated iteratively. If the distance is smaller or greater than the recently found smallest or greatest distance, the coordinates of the vertex in question are saved. Finally, after the last vertex was processed, the semi-axes will be constructed with the centroid as the first vertex of the line and the semi-axes will be saved in the polyline-shapefile. The polygon-shapefile contains copies of the polyline data to be processed. The calculations of the attribute-values of both polylines and polygons are then being performed by the ArcToolbox-tool "Calculate Field" which is accessed by a geoprocessing-object.

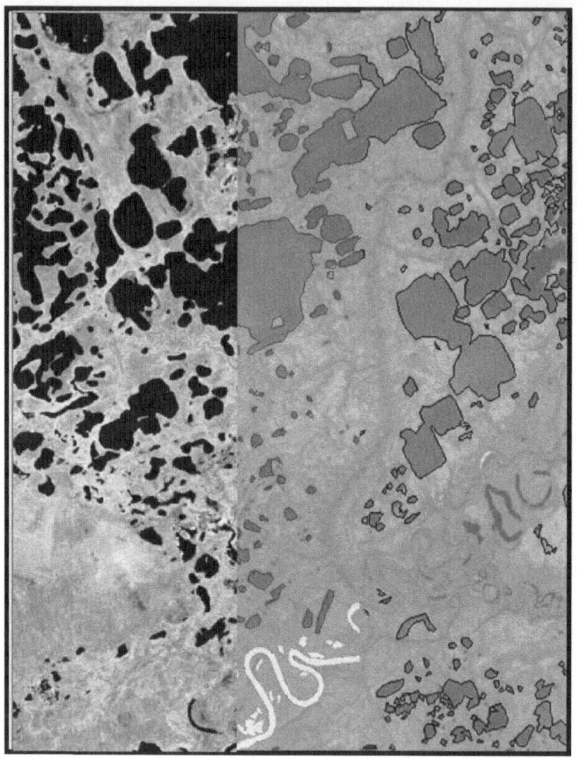

Fig. 6. Image data after preprocessing; output data from NDVI; polyline-based delineation of features; positive feature identification (from left to right)

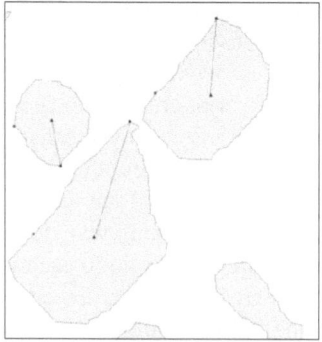

Fig. 7. Polyline-based delineations of the identified thermokarst lakes

In the course of this work, it has been demonstrated on 2 examples, that particular dependencies between thermokarst lake variables do exist (areas up to 500,000 sqm) but they can vary slightly. On the other hand, functional dependencies between variables of greater thermokarst lakes (areas bigger than 500,000 sqm) were computed and it has been established that variations in shape index value are not significant. Those are interesting results which call for further and more detailed scientific research. Obviously, structures of exponential distribution exist between thermokarst lake variables, such as perimeter, radius and area.

Thermokarst lake variables, such as perimeter, radius and area (shape index) or asymmetry are essential for correct feature identification. Water bodies are detected with high precision in remote sensing image data, but without knowing the aforementioned parameters and their comparisons it is not possible to differentiate natural water bodies from artificial, or rivers from lakes. Frozen water surfaces, however, are still problematic, because their reflections are quite different to those of liquid surfaces. As already mentioned, additional data would enable more advanced analyses. We have tried to show, which conclusions can be made based on satellite image data without a DEM or any additional data.

6 Conclusions and Outlook

The workflow of automated feature identification (Figure 1) has been created on a conceptual level for available and suitable data. The whole process has been effectively transposed to multispectral satellite-image data Landsat ETM+. In order for geomorphologists and geologists to be able to take advantage of the automated feature identification process, we plan to apply this conceptual process on terrain model data. The advantage of this workflow is that its use not limited to particular data. Accuracy assessment of the automated feature identification process via visual comparison and interpreter knowledge is 86 percent. 14 % are represented by small negative features. This process can be optimized. The Accuracy of the automated feature identification process will be verified statistically.

We are communicating and exchanging information with scientists from ESA Data User Element (DUE) Permafrost Project. This project is a result of a cooperation of several scientific organizations and studies permafrost and related climatic changes. As a result of this communication we should be able to obtain the needed terrain model data with sufficient accuracy and optical data with higher spatial resolution than Landsat.

Automated feature identification process and retrogressive analysis will be applied on other landforms and on the data collected in other and different regions. Possible use of this process will be proved on aeolian forms-sand dunes and on some other regions.

Through the further research it shall be important to examine whether only this exponential form leads to good results or could they be achieved using other forms, with high adjusted R-squared (greater than 0,75) as an indicator of goodness.

It has to be mentioned that the Siberian and Canadian Group B thermokarst lakes do not show any correlation. The data shows big variances. Hence, the regression analysis is not a useful analysis technique for this data group. Other explorative methods have to be proved.

Inferring further information from an image data and terrain model data is to be considered as an important outlook. Specifically (for example), the depth of thermokarst lakes should be deducted from terrain model data. The lake depth is an important parameter for thermokarst (unlike for other types of lakes). However, the DEM in these latitudes are not sufficient (missing information in the DEM). Furthermore, environmental data would be very applicable to infer the wind data and the water temperature.

References

1. Büdel, J.: Climatic Geomorphology. Princeton University Press, Princeton (1984)
2. Crawley, M.J.: The R Book. John Wiley & Sonst Ltd., Chichester/West Sussex (2007)
3. Dragut, L., Blaschke, T.: Automated classification of landform elements using object-based image analysis. SID Int. Geomorphology 81, 330–344 (2006)
4. Dragut, L., Tiede, D., Levick, S.R.: ESP: a tool to estimate scale parameter for multiresolution image segmentation of remotely sensed data. SID Int. International Journal of Geographical Information Science 24, 859–871 (2010)
5. Eisanak, C., Dragut, L., Götz, J., Blaschke, T.: Developing a semantic model of glacial landforms for object-based terrain classification – the example of glacial cirques. In: Proc. the International Archives of the Photogrammetry, Remote Sensing and Spatial Information Sciences, vol. XXXVIII-4/C7 (2010)
6. French, H.M.: The Periglacial Envionment. John Wiley and Sons, Chichester (2007)
7. Frohn, R.C., Hinkel, K.M., Eisner, W.R.: Satellite remote sensing classification of thaw lakes and drained thaw lake basins on the North Slope of Alaska. SID Int. Remote Sensing and Environment 97, 116–126 (2005)
8. Hese, S., Grosse, G., Pöcking, S.: Object based thermokarst lake change mapping as part of the ESA Data User Element (DUE) permafrost. In: Proc. OBIA Conference 2010, Genth, Belgien (2010)
9. Hinkel, K.M., Frohn, R.C., Nelson, F.E., Eisner, W.R., Beck, R.A.: Morphometric and spatial analysis of thaw lakes and drained thaw lake basins in the western arctic coastal plain, pp. 327–341. Wiley InterScience, Alaska (2005)
10. Mitasova, H., Mitas, L., Harmon, R.S.: Simultaneous spline approximation and topographic analysis for lidar elevation data in open source GIS. Proc. IEEE Geoscience and Remote Sensing Letters 2(4), 375–379 (2005)
11. Moore, A.B., Morris, K.P., Blackwell, G.K., Jones, A.R., Sims, P.C.: Using geomorphological rules to classify photogrammetrically - derived digital elevation models. SID Int. International Journal of Remote Sensing 24, 2613–2626 (2003)
12. Originlab, Exponential Model, User´s guide, Originlab, USA (2011), http://www.originlab.com/www/helponline/Origin/en/UserGuide/ExpDec1.html (last date accessed: July 2011)
13. Sachs, L., Hedderich, J.: Angewandte Statistik – Methodensammlung mit R, 13th edn., p. 122. Springer, Heidelberg (2009)

14. Schneevoigt, N.J., van der Linden, S., Thamm, H.: Detecting Alpine landforms from remotely sensed imagery. A pilot study in the Bavarian Alps. SID Int. Geomorphology 93, 104–119 (2010)
15. Taramelli, A., Melelli, L.: Detecting Alluvial Fans Using Quantitative Roughness Characterization and Fuzzy Logic Analysis. In: Gervasi, O., Murgante, B., Laganà, A., Taniar, D., Mun, Y., Gavrilova, M.L. (eds.) ICCSA 2008, Part I. LNCS, vol. 5072, pp. 1–15. Springer, Heidelberg (2008)
16. The R Project for Statistical Computing, http://www.r-project.org (last date accessed: September 2011)
17. Washburn, A.L.: Geocryology, a survey of periglacial processes and environments, p. 406. Edward Arnold, London (2003)

Peer Selection in P2P Service Overlays Using Geographical Location Criteria*

Adriano Fiorese[1,2], Paulo Simões[1], and Fernando Boavida[1]

[1] Centre for Informatics and Systems of the University of Coimbra - CISUC
Department of Informatics Engineering - DEI
University of Coimbra - UC
{fiorese,psimoes,boavida}@dei.uc.pt
[2] Department of Computer Science - DCC
Santa Catarina State University - UDESC
890233-100 Joinville, SC, Brazil
fiorese@joinville.udesc.br

Abstract. Peer-to-peer service overlay networks (P2P SON) are increasingly being infrastructure providers for networking services, allowing service providers to cooperatively offer and run a flexible set of services. Regarding this condition, the selection of peers is a key issue for improving resource usage; service performance, and ultimately end users Quality of Experience (QoE). This paper presents an approach to best peer selection in a three-tier P2P SON architecture, allowing the splitting of service business functions and peer selection functions. The proposed best peer selection approach is evaluated by simulation, using a literature available geographic positioning metrich that takes into account real delay and jitter made available by the CAIDA project and MaxMind's free database. The simulation results show the consistency and good performance of the proposed peer selection approach.

Keywords: Peer Selection, P2P, Services Management.

1 Introduction

Services are becoming one of the primary sources of revenue on the Internet. Currently, data transport is being sold as a commodity for a broad audience. This business model allows rising a wide service consumer market. On the other hand, it allows the appearing of new service providers at the stage. This also leads to a competition for the service consumers among the whole set of service providers, eventually leading to a global competition depending on the quality and originality of the offered service.

In this context, service providers can enhance their ability to make their service or service components available to a broader set of customers, through the utilization of a Service Overlay Network (SON) [17, 19]. In this case, a SON acts

---
* This work was partially funded by Portuguese Foundation for Science and Technology (grant SFRH/BD/45683/2008).

as an infrastructure where services are published/offered and to which the users access in order to select and use these services. Moreover, a SON can be created by a consortium of service providers using the Internet to make their services available to the user community at large. The peer-to-peer (P2P) technology is a well suited way for constructing that kind of SON. This technology leads to a self-organizing overlay and, additionally, to sharing maintenance costs among service providers.

Although P2P eases the construction and maintenance of SONs, it does not guarantee adequate service performance. In order to maximize performance, the best peer must be found in the P2P SON, among all the potential partners that provide the desired service. Naturally, the choice of best peer should take into account one or more of a set of Quality of Service (QoS) parameters, such as delay, jitter, available bandwidth, etc.. Nevertheless, to minimize inter-provider traffic and, thus, reduce costs for the user and for the service provider, the choice of peers belonging to a different, remote domain should be avoided as much as possible. Thus, locality should also be taken into consideration when choosing a SON peer as the best peer to serve a requested service.

Previous work from the same authors [7–9] proposed an architecture for services management in P2P Service Overlay Networks (SON). The architecture, named OMAN, takes into account several aspects of services management, particularly the use of a second overlay - called Aggregation Service (AgS) - to provide efficient service search in the context of multi-domain P2P SONs. In its turn, this paper deals with the proposal and assessment of a third-tier component of OMAN, whose purpose includes the searching and selection of the best peer with which a service-requesting peer should interact in the context of the P2P SON.

Having in mind the stated goal and approach, this paper is organized as follows. Section 2 discusses related work and also provides the context of the current work, by briefly presenting the OMAN architecture, which the BPSS component belongs to. Section 3 describes the proposed BPSS service. Subsequently, Section 4 presents the evaluation and discusses the simulation results, after describing the simulated scenarios. Finally, Section 5 summarizes the contributions and presents further work.

2 Related Work

2.1 Service Overlay Networks

According to Tran and Dziong [17], a Service Overlay Network (SON) is a network composed of interconnected nodes, whose generic purpose is to provide the required Quality of Service (QoS) to applications that execute on those nodes. The same authors establish a difference between SON and P2P overlay network claiming that the purpose of the latter is related with providing efficient searching and retrieval.

The bandwidth provisioning in a SON composed of nodes that lease links from several link providers is studied in [6].

We advocate a P2P overlay network can also provide QoS services when the participants are in a consortium of service providers that establish well-defined SLAs to regulate the contribution of each participant to the network. In this sense, a platform called ALASA is presented in [19]. It uses a structured P2P overlay network over the Internet to describe, discover, compose, and reputation of services.

P2P is also used in [15] as support for the SON architecture. In that piece of work the authors address discovery of services considering QoS aspects in their approach.

2.2 Peer Selection

Haase et. al [10] explores neighboring peers relationships and shared peers expertise in order to select peers. The use of artificial intelligence techniques, like machine learning, is another approach to peer selection, which also takes advantage of the peers' expertise [4]. This latter work aims at adapting the selection process to the peers' requirements.

In file sharing, the free-riding problem encourages the adoption of incentive mechanisms as part of the selection scheme. Thus, the fairness between uploads and downloads is used as a metric to the best peer selection. Bittorrent [12] is an emblematic example for this. However, our scheme does not target file(data)-sharing environments. Rather, the problem is to select the best peer that satisfies the requirements of the intended service. Therefore, in our case, performance instead fairness is used. In addition, unlike file sharing applications, our approach considers the best peer selection process for long lasting sessions, as opposed to relatively short burst chunk downloads/uploads.

In the P2P multimedia stream services field, several pieces of work proposing the use of P2P as the delivery mechanism face peer selection issues [11, 18]. Similarly to our approach they aim at optimizing peer choice regarding the performance of the service.

Furthermore, P2P traffic is an issue faced by ISPs and over the Internet in general. In order to assist ISPs in avoiding costs with the choice of best peers out of their own domains, several proposals have been put forward [3, 5, 20]. These advocate the collaboration between providers and P2P applications. On the other hand, our BPSS scheme selects best peers for service interactions inside the geographical domain of the service requester, without the need for explicit collaboration between the service providers.

2.3 OMAN

In order to contextualize the Best Peer Selection Service (BPSS), this section briefly presents the underlying OMAN architecture, previously proposed by the authors in [7–9].

OMAN is a P2P SON architecture that handles aspects ranging from the composition of the SON until the interaction aspects between the services and the SON, including how to take advantage of the information at the P2P overlay level to leverage the services and applications. Fig. 1 provides an overview of the OMAN architecture.

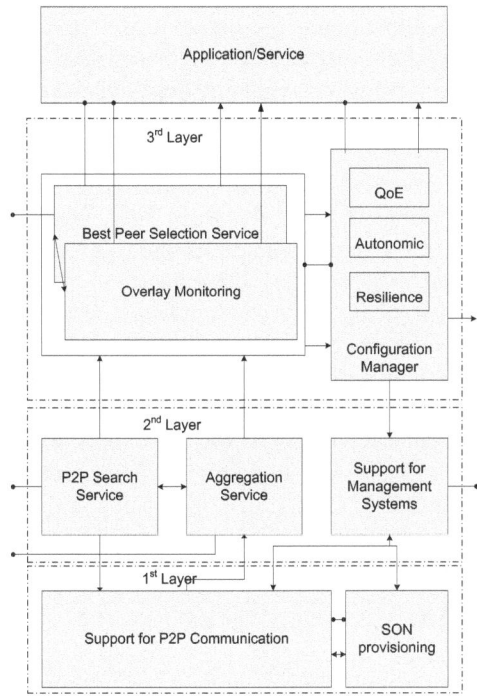

Fig. 1. The OMAN architecture [8]

The lower layer of the architecture is the basic P2P SON layer. A service provider can make available more than one peer to be represented at the P2P SON layer.

The central module of the second tier of OMAN is the Aggregation Service (AgS) [7, 9]. AgS is an unstructured P2P overlay-tier, meaning there is no tight coupling between overlay topology and information location/placement. AgS executes on top of the P2P SON and it consists of peers that belong to potential providers interested in advertising/offering their services. The purpose of AgS is to aggregate the offerings of services and service components. This is accomplished by concentrating the service offerings in its peers (nodes), in order to facilitate and optimize the services searching process.

The peers that form the AgS P2P overlay are called *aggregation peers (AgS peers)*. They are chosen among the peers that form the underlying P2P SON tier, which makes them specialized *SON peers*.

Each SON peer plays a double role: to execute the services (as any other P2P SON peer) and to maintain and publish references to the available services. Note that service references are made available (i.e., published) to AgS peers in order to optimize service searching. Each SON peer involved in the AgS can take care of several service offerings. A single service offering can be spread over multiple AgS peers in order to allow some redundancy and to overcome churn. SON peers make the services indirectly available (through interfaces encapsulated in a service profile or template) to external entities (such as service composers and aggregators, users, and other services) likely located in other network domains. The service template also contains reference to the SON peer that is publishing it. SON peers make these interfaces available by publishing the template of a service offer at several aggregation peers in the P2P AgS overlay. These aggregation peers may be located at the same domain as the SON peers or, in some cases, at different domains.

Searching for a service, using the AgS framework, therefore results in a set of references to SON peers that offer an interface to services matching the search criteria. This preserves the internal details of the service, since the external entity, i.e. a user, a third party service provider, or other SON peer in the P2P SON, is only granted with a mediated access (by means of the SON peer), which may hide sensitive information and filter undesired operations.

AgS is built on a number of elementary operations. Table 1 presents key AgS operations and the corresponding messages exchanged among peers.

Table 1. AgS Table of Operations and Messages

Operation	Goal	Executor	Message sent
Join	Form the Aggregation Service.	aggregation peer	JoinMessage sent by the requesting peer to its successor and predecessor in the overlay.
Leave	Leave the Aggregation Service (in a normal way).	aggregation peer	LeaveMessage sent by the requesting peer to its successor and predecessor in the overlay.
Query	Look for a peer that provides a particular service/service component.	aggregation peer	QueryMessage sent by the requesting peer to its successor in the overlay ring in a clockwise manner. The message is forwarded clockwise until it arrives at its goal or until the message reaches the requesting peer. When the desired information is found, a QueryReply message containing it, is then created. This latter message is directly transmitted to the requesting peer of the Query´s operation.
Publish	Make the services to be searched available.	SON peer	PublishMessage sent by the SON peer to its aggregation peer(s), which makes the service(s) public.

The Best Peer Selection Service (BPSS) is one of the modules of the third tier of the OMAN architecture. The third layer is the most specialized layer on the OMAN architecture and it interfaces the application. Therefore, BPSS is cornerstone for the architecture and for services provided using the envisaged business model made available by P2P SON.

3 Best Peer Selection Service

The objective of the Best Peer Selection Service (BPSS) is to provide SON peers with the identification of the best peer for a particular service, in the context of services offering in a P2P SON. Naturally, the best peer depends on several aspects, including the service objectives and characteristics. Key to the choice of a suitable peer is the intended service performance, which is tied not only to measurable networking characteristics, such as bandwidth, delay, loss, etc., but also to the peers underlay location.

3.1 BPSS Architecture

Regarding the use on OMAN, service developers can implement an interface with the BPSS module in order to request and receive best peer information, allowing the splitting of service business functions and best peer determination. This decoupling enhances modularity and best peer selection metric independence, thus leading to high flexibility when choosing the particular metric to use for a particular service type. In order to take advantage of this aspect, the BPSS builds on the AgS [7] service toward select the best peer on request of a P2P SON peer.

Fig. 2 illustrates the use of BPSS. SON peers can request best peer information (select_BP), regarding a particular service, from the BPSS module. On the reception of a best peer request, the BPSS module asks the AgS service the list of all SON peers that have published a service profile for the intended service. After receiving the requested list, the BPSS module calculates the best peer and returns its reference to the requesting SON peer. The selection of the best peer is done using one of the supported metrics. In Section 3.3 one of such metrics is presented.

It is worth mentioning that with this decoupled approach it is also possible for an external entity (e.g. a user or a particular application/service or service component from outside the P2P SON) to request a best peer selection, as long as the request is compliant with the BPSS interface and the requested metric is supported.

3.2 BPSS Model

Based on its architecture, BPSS comprises the model depicted in Fig. 3. The BPSS model is represented as a UMLlike diagram. The BPSS model comprises the roles of the involving parts; their interactions; and the information exchanged among them.

The service or service component object and the best peer are the central pieces in the BPSS model. They make part of the interactions with every subject

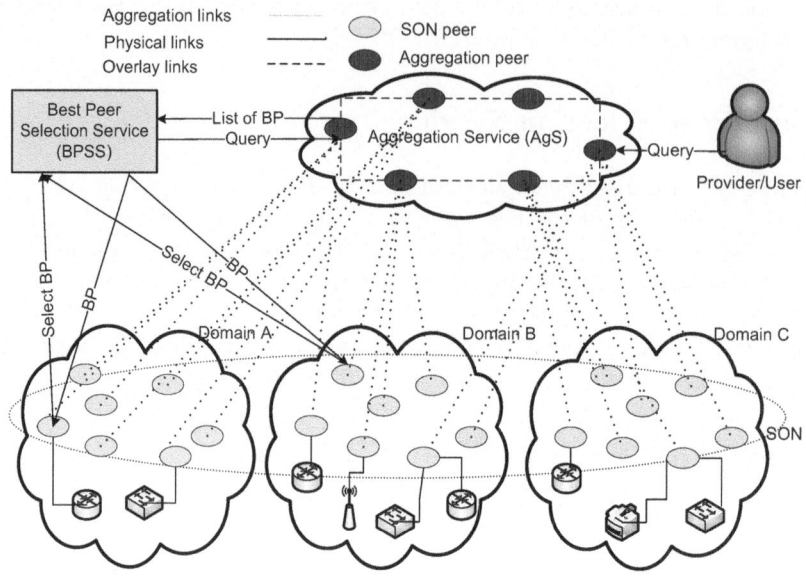

Fig. 2. BPSS Architecture

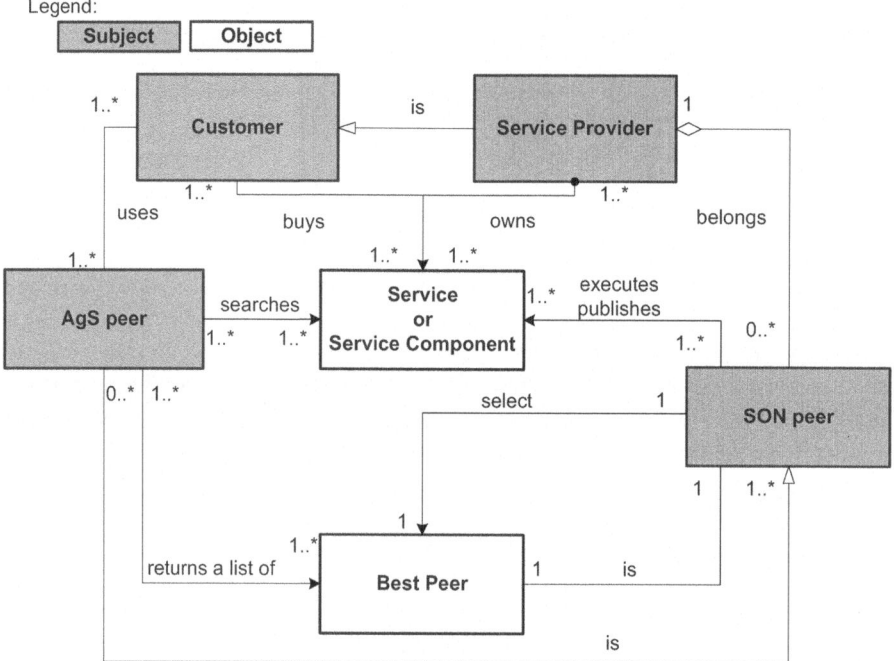

Fig. 3. BPSS Model

in the model. The service object represents the Service or Service Component that actually executes on the SON peer and that is published in the aggregation peers. Aggregation peers are the peers that form the AgS P2P overlay. The Best Peer object is a SON peer resulting of a best peer selection started by a particular SON peer.

The Service Provider is another cornerstone piece in the BPSS model. It is the owner of the Service or Service Component as well as the SON peers and the AgS peers (since AgS peers are specialized SON peers). Service Providers are Customers as well, so they can act as third party consumers of service components of other service providers. In this case, this transaction can involve service composition and provisioning, which is out of the scope of this paper. They also can use third party services in a service chain, in order to offer to a home user (customer) a complete set of services. Nonetheless, to accomplish this step, the service provider needs to find the best peer to which its SON peer will interact in order to offer that set of services. Following the BPSS model, the Customer, which can be a third Service Provider, uses the AgS peer to search for services in the AgS service. The AgS peer by its turn returns a list of all SON peers that executes the searched services. Ergo, the SON peer then selects the best one to be used. The interaction among the services involved is not covered by AgS either BPSS.

The threefold role of the SON peer is emphasized at the BPSS model by the actions: execute, publishes, and select. These actions together the searching services action executed by the AgS peers define the whole behavior of the services cycle offering at the foreseen P2P SON.

3.3 BPSS Metric

The proposed BPSS service is based on one or more service performance metrics to determine best peers. In other words, the peer selected as the best peer should offer optimized interaction with the requesting service/peer, according to some performance criteria. Although BPSS provides independence from the used performance metrics, for the purpose of implementation and evaluation a specific metric was used.

Several performance criteria can be used when determining the best peer. Performance of P2P systems is often very sensitive to the underlying delay characteristics. These are influenced, among other factors, by bandwidth, load and also geographic location. In fact, according to [13], the geographical location of nodes heavily influences jitter and packet loss. This observation points to the need for the node's geographic location when developing a delay prediction model. Having this in mind, the authors of [13] developed a predictive model of the Internet delay space that takes into account the geographical location of the nodes and the delay between them.

Using a rich set of real data, namely measured end-to-end round trip time (rtt) [1] and measured end-to-end link jitter [2], those authors mapped the measured end-to-end nodes into a 5-dimensions Euclidean space model of the Internet, by combining this information with global network positioning information [16]. Using the coordinates of each peer in this 5-dimensions model, it is then possible to calculate the Euclidean distance between peers, which takes into account not only network conditions but also peer location.

For the purpose of the work presented in this paper, BPSS used the mentioned distance metric in order to select the best peer for a particular SON peer.

4 Evaluation Framework

In order to assess the BPSS behavior, we conducted a simulation study to observe the best peer and the second-best peer selection distributions over well-constrained geographical domains.

4.1 Used Simulator

The PeerFactSim.KOM [14] discrete event simulator was used in all simulations. Unlike mainstream network simulators, the PeerFactSim.KOM simulator was specifically designed to simulate P2P networks. Among its advantages are the use of modular abstractions for the network and transport layers, and the utilization of well-known and useful software engineering design-patterns. These characteristics ease the experimentation and facilitate code reutilization and maintenance.

4.2 Simulations Setup

The simulation environment was constituted of SON peers whose network identifiers (IP addresses) were taken from the CAIDA project and MaxMind GeoIP database. Therefore, the simulated peers belong to real geographical domains. The Internet delay space scheme presented in Section 3.3 was used.

The scenarios were modeled based on real nodes available in the compiled data set. To accomplish that, peers belonging to specific geographical country domains were split between SON and aggregation peers. The number of AgS peers was 10% of the total number of SON peers used in each domain. Thus, if a country had 50 SON peers at the P2P SON - which, for instance, can mean that it comprised 50 service providers - then 5 AgS peers belonging to that country would be part of the Aggregation Service. The chosen geographical domains were the following European countries: Portugal, Spain, France, Italy and Germany.

The simulation comprised 11 sets of individual simulations. Each set simulated a particular number of SON and AgS peers. The initial set simulated a scenario with the total of 50 SON peers, corresponding to 10 randomly chosen SON peers from each of the aforementioned countries. The second simulated set comprised the total of 75 SON peers (15 for each country). Thus, with steps of 25 SON peers between each simulated scenario, the last simulated set comprised 300 SON peers.

4.3 Simulations Strategy

For the sake of simplicity, without loss of generality, a particular SON peer could only offer, at most, seven randomly chosen services or service components (using a uniform distribution) from the service set S={S1,S2,S3,S4,S5,S6,S7}. In addition, each SON peer could only publish its service subset on, at most, 10 distinct randomly-chosen aggregation peers (also following a uniform distribution). Nevertheless, it was possible that more than one SON peer could offer the same services and publish them on the same or different aggregation peer.

Each experiment execution had simulated 50 hours of work and it had been repeated 10 times in order to get averaged values. Every operation (e.g. joining, leaving, publishing, searching, and select) was specified in time. Each simulation executed 100 searching operations, i.e., 100 best peer selection requests, since a searching operation is triggered by a best peer selection request.

The SON peers that execute this selection process are chosen randomly using a uniform distribution. This happens when the number of SON peers composing the P2P SON is greater than the 100 best peer select operations, which composes each experiment. Otherwise, each SON peer executes that operation at least once.

The experiments also had chosen second-best peers. They are the peers whose performance regarding the used metric puts them in the second place in an hypothetical ordered list of best peers. Actually, the second-best peer is selected from the same list of SON peers provided by the AgS service, by removing the best peer from that list and repeating the measurement process.

The determination of the second-best peer can aid in the validation of the used metric in two ways: 1) by checking if the service requesting the best peer behaves better in an interaction with the selected best peer, then this provides a measure of the metrics consistency; 2) by measuring the average improvement of the best peer over the second-best peer, an indication of the metrics effectiveness can be obtained.

The wide range of simulated P2P SON sizes intends to cover scenarios with few service providers (e.g. small P2P SON for very specialized services) until scenarios composed of many service providers (e.g. a more competitive scenario).

4.4 Results

The results presented in this section show the distribution of best and second-best peers regarding requests made by SON peers belonging to the Portuguese geographical domain only, though it is obvious that the used methodology can be applied to any other geographical domain.

Fig. 4 depicts the geographical location of the SON peers selected as best peers. There are eleven 5-bar clusters, each one corresponding to one of the eleven simulated scenarios. In each cluster, each of the five bars represents the number of best peer occurrences in each of the five geographical domains, namely Portugal, Spain, France, Italy, and Germany, respectively.

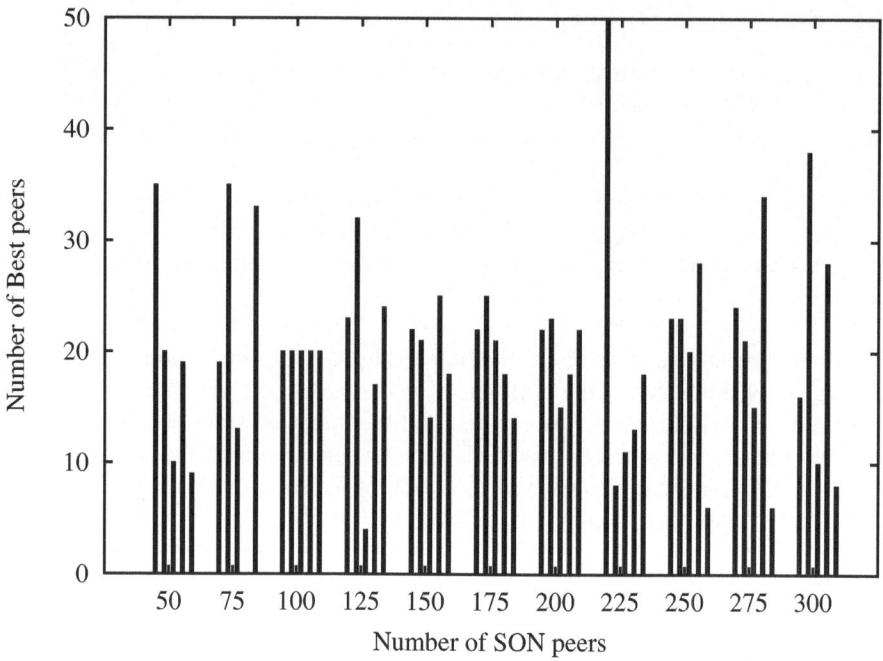

Fig. 4. Clustering Best Peer Ocurrences by Domain by Number of SON peers

One can expect the highest number of selected best peers is in the domain of the requester (Portugal, in the case of these simulations), due to geographical distance considerations. Nevertheless, the obtained results clearly show the effect of two key aspects of the OMAN architecture and of its AgS and BPSS services: on one side, in some cases, service searching performed by the AgS service determined that the desired services were not available at any of the SON peers of the requester's domain; on the other hand, the metric used by the BPSS service - based on the Internet model proposed in [13], which takes into account not only the geographical position but also delay and jitter - led to the fact that the closest peer, in terms of the 5-dimension Euclidean distance, resided in a different geographical domain.

Similar results building on the same explanations are obtained regarding the experiments with the second-best peers. Fig. 5 depicts them.

Nevertheless, even with the mentioned constraints regarding the statistical availability (or unavailability) of the desired services in the requester's domain, averaging the results of all simulations shows that the highest number of best peers was selected in the same domain of the requester. This can be seen in Fig. 6 and Fig. 7. It is worth mentioning the results rely on and are presented based on a confidence interval of 95% for the mean number of best peer selections on each geographical domain.

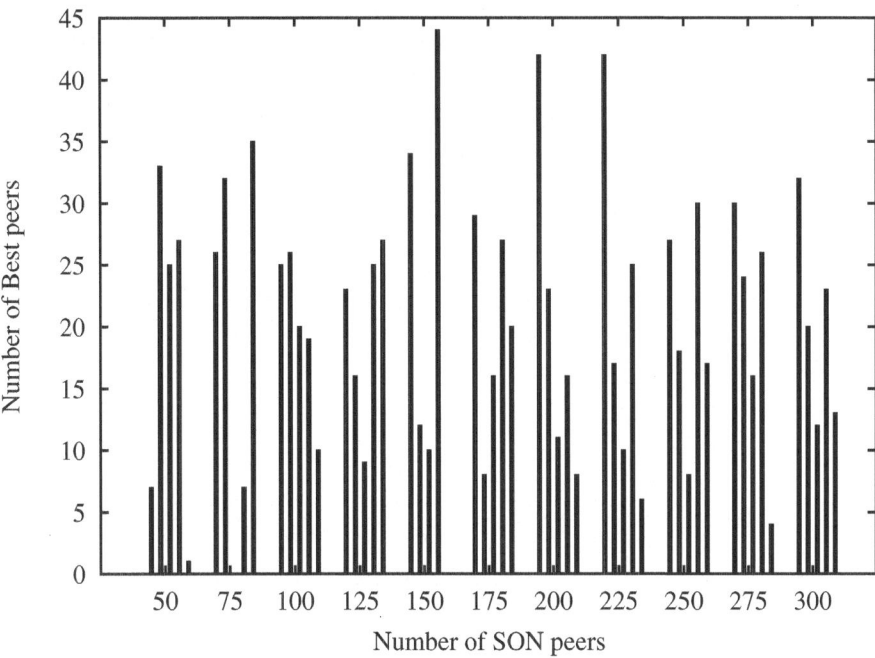

Fig. 5. Clustering Second-Best Peer Ocurrences by Domain by Number of SON peers

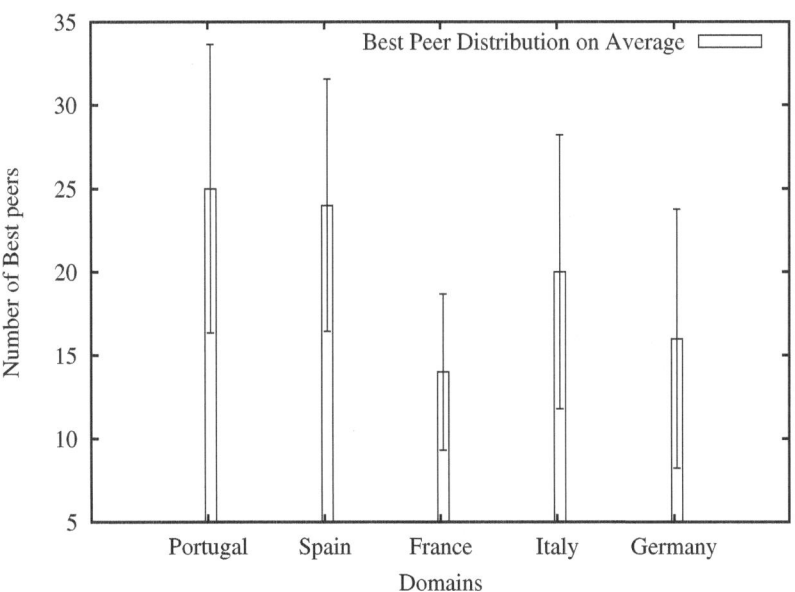

Fig. 6. Best Peer Distribution by Geographical Domain

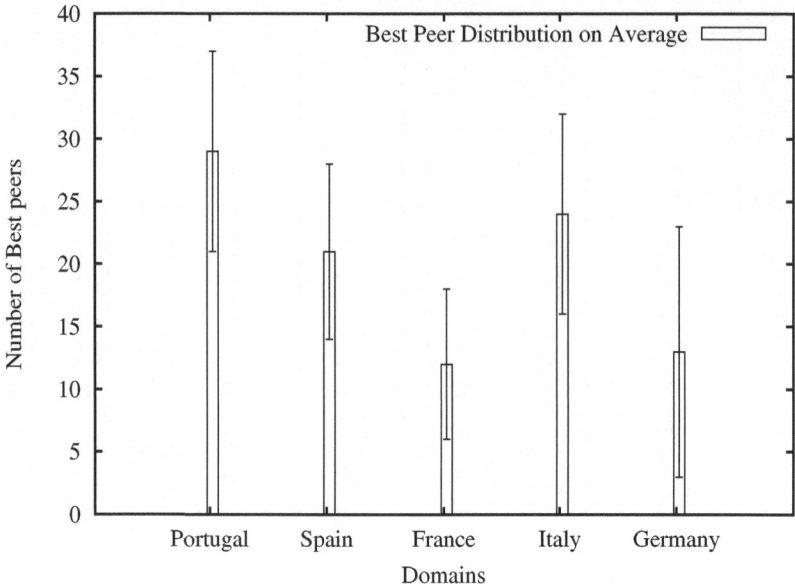

Fig. 7. Second-Best Peer Distribution by Geographical Domain

Therefore, taking the sum of best peers and second-best peers by domain, SON peers in the requester's domain (Portugal) were selected as best peers in 27% of the time, followed by Spain (22.5%), Italy (22%), Germany (14.5%) and France (13%). This means that almost half the best peer selections resulted in peers belonging to the same geographical domain or to the neighboring geographical domain. This suggests the consistency of the used metric and the good operation of OMAN's AgS and BPSS services.

It is also worth mentioning that the BPSS overhead is negligible. The time taken by the selection process for both the best and second best peer is around 2ms. On the other hand, the overhead associated with the maintenance of the P2P SON and with the AgS service is around 164ms. These values strongly suggest the efficiency, feasibility, and practicality of the proposed BPSS service, in conjunction with AgS and the other modules of the OMAN architecture.

5 Conclusion

In this paper, an approach to the problem of best peer selection in peer-to-peer service overlay networks has been proposed and studied. The proposed Best Peer Selection Service (BPSS) is one of the components of a three-tier service management overlay network architecture (OMAN).

After identifying key issues in related work and providing an overview of the underlying OMAN architecture for the sake of contextualization, the BPSS service was presented. The main features of the presented BPSS are the use of a very efficient aggregation service (AgS) for service searching, and its independence from the best peer selection metric. The latter feature provides flexibility, adaptability and modularity to the overall process of best peer selection.

In order to test and assess the BPSS behavior, a specific distance metric was used. The used metric combines measured delay and jitter real data with geographical location data. The performed simulations involved the selection of best and second-best peers in a universe of five distinct geographical domains.

Obtained results have shown that BPSS performs well and that the overall OMAN architecture - of which the AgS service is a key component - is very effective. Simulations have also shown that the BPSS overhead is negligible and the AgS overhead, which derives from its operations, is very low.

Further work can compare different best peer selection metrics, based not only on different performance parameters, but also on factors such as inter-provider link cost. Still another line of research can be the use of the OMAN approach by service providers in order to identify ways of maximizing user quality of experience or ways of reducing inter-provider traffic, e.g., by deploying specific peers inside their own domain.

References

1. Macroscopic topology measurements, http://www.caida.org/projects/macroscopic/
2. The PingER project, http://www-iepm.slac.stanford.edu/pinger/
3. Aggarwal, V., Akonjang, O., Feldmann, A.: Improving user and ISP experience through ISP-aided P2P locality. In: INFOCOM Workshops 2008, pp. 1–6. IEEE (2008)
4. Bernstein, D.S., Feng, Z., Levine, B.N., Zilberstein, S.: Adaptive Peer Selection. In: Kaashoek, M.F., Stoica, I. (eds.) IPTPS 2003. LNCS, vol. 2735, pp. 237–246. Springer, Heidelberg (2003)
5. Choffnes, D.R., Bustamante, F.E.: Taming the torrent: a practical approach to reducing cross-isp traffic in peer-to-peer systems. In: ACM SIGCOMM Computer Communication Review, vol. 38, pp. 363–374. ACM, New York (2008)
6. Duan, Z., Zhang, Z., Hou, Y.T.: Service overlay networks: SLAs, QoS, and bandwidth provisioning. IEEE/ACM Transactions on Networking 11(6), 870–883 (2003)
7. Fiorese, A., Simões, P., Boavida, F.: An aggregation scheme for the optimisation of service search in Peer-to-Peer overlays. In: 2010 International Conference on Network and Service Management (CNSM), Niagara Falls, Canada, pp. 481–486 (October 2010)
8. Fiorese, A., Simões, P., Boavida, F.: OMAN – A Management Architecture for P2P Service Overlay Networks. In: Stiller, B., De Turck, F. (eds.) AIMS 2010. LNCS, vol. 6155, pp. 14–25. Springer, Heidelberg (2010)
9. Fiorese, A., Simões, P., Boavida, F.: Service searching based on P2P aggregation. In: Proceedings of International Conference on Information Networking 2010 (ICOIN 2010), Busan, South Korea (2010)
10. Haase, P., Siebes, R., Harmelen, F.: Expertise-based peer selection in Peer-to-Peer networks. Knowledge and Information Systems 15(1), 75–107 (2007)

11. Habib, A., Chuang, J.: Service differentiated peer selection: an incentive mechanism for peer-to-peer media streaming. IEEE Transactions on Multimedia 8(3), 610–621 (2006)
12. Huang, K., Wang, L., Zhang, D., Liu, Y.: Optimizing the BitTorrent performance using an adaptive peer selection strategy. Future Generation Computer Systems 24, 621–630 (2008)
13. Kaune, S., Pussep, K., Leng, C., Kovacevic, A., Tyson, G., Steinmetz, R.: Modelling the internet delay space based on geographical locations. In: 2009 17th Euromicro International Conference on Parallel, Distributed and Network-based Processing, pp. 301–310 (2009)
14. Kovacevic, A., Kaune, S., Liebau, N., Steinmetz, R., Mukherjee, P.: Benchmarking platform for peer-to-peer systems. IT - Information Technology 49(5), 312–319 (2007)
15. Lavinal, E., Simoni, N., Song, M., Mathieu, B.: A next-generation service overlay architecture. Annals of Telecommunications 64(3), 175–185 (2009)
16. Lee, S., Zhang, Z., Sahu, S., Saha, D.: On suitability of euclidean embedding of internet hosts. In: ACM SIGMETRICS Performance Evaluation Review, pp. 157–168. ACM, New York (2006)
17. Tran, C., Dziong, Z.: Service overlay network capacity adaptation for profit maximization. IEEE Transactions on Network and Service Management 7(2), 72–82 (2010)
18. Wu, C., Li, B.: Optimal peer selection for minimum-delay peer-to-peer streaming with rateless codes. In: Proceedings of the ACM Workshop on Advances in Peer-to-Peer Multimedia Streaming, pp. 69–78. ACM, New York (2005)
19. Zhou, S., Hogan, M., Ardon, S., Portman, M., Hu, T., Wongrujira, K., Seneviratne, A.: ALASA: when service overlay networks meet peer-to-peer networks. In: 2005 Asia-Pacific Conference on Communications, Perth, WA, USA, pp. 1053–1057 (2005)
20. Zhuang, H., Xu, Y., Hu, Y., Lin, X.: A peer selection mechanism in P2P networks based on the collaboration of ISPs and P2P systems. In: International Conference on Information Science and Engineering, pp. 1573–1576. IEEE Computer Society Press, Los Alamitos (2009)

Models for Spatial Interaction Data: Computation and Interpretation of Accessibility

Morton E. O'Kelly

Department of Geography, The Ohio State University
1036 Derby Hall, 154 N Oval Mall, Columbus, OH 43210

Abstract. The paper is in the area of spatial optimization, for simulating and understanding spatial interaction. The model foundation is from Wilson's "doubly constrained" model of spatial interaction. The idea is to perform a sensitivity analysis on the parameters of this model and to interpret the results in terms of accessibility. The main purpose of this type of analysis is to use data from interaction systems to uncover structural effects that help to understand the role of origin and destination location, and accessibility. US air passenger traffic is used as the starting point for the model. The model reproduces many of the features of the data with a parsimonious set of parameters, leaving some aspects of the analysis open to interpretation. An innovative idea in this paper is to compute averages and other measures directly from the data, fit a model to these data, and then to use the fitted (and observed) matrices to evaluate numerous theoretically inspired measurements. This paper (in a modular way) develops the introduction and context, and then moves to theory, spatial disaggregation, and empirical applications.

Keywords: origin- and destination-specific parameters, interpreting sensitivity analysis, computational properties, dual variables.

1 Introduction

A series of papers has been developed in the recent literature on the theme of spatial interaction, efficiency, and the relationship between trip length and selected sensitivity measures in the style of Alan Wilson's pioneering work on entropy models (Wilson 1974; see O'Kelly 2010 for a retrospective and references to a wide range of related literature). The foundation is from the Wilson "doubly constrained" model of spatial interaction. Basically the model is a well-known platform for computing expected flows in the presence of constraints derived from data, and is therefore amenable to interpretation (Fotheringham and O'Kelly, 1989; Rabino and Occelli, 1997; O'Kelly, 2009). This is also an area that continues to receive active interest from the applied computing community (Okuhara, et al. 2010; Fang et al., 1997; Kapur, 1982). Among the common interpretations are the forces that origin and destinations exert (push and pull respectively), and these are significant effects in understanding the role of distance as an impedance to interaction. It is expected that a carefully calibrated model can provide quantitative support for measuring accessibility and the relative advantages enjoyed by core *vs* peripheral locations, with the important caveat that the measures are adjusted to reflect actual spatial structure. It is also important to base such a model on detailed data, and that the data reflect the spatial arrangement of origins *vs* destinations and vice versa. Because traffic from peripheral nodes is often routed through intermediate hub stops, it is also important to measure the distance to destinations by the actual average travel distance.

The idea in this paper is to perform a sensitivity analysis on the parameters of a doubly constrained spatial interaction model (with origin- and destination-specific effects) and to interpret the results in terms that help to understand and explain accessibility. These measures can be considered as dual variables, and show the rate of change of the structural balancing factors with respect to changes in trip length. These measures have interesting interpretations, and tie together models from optimization, computation, and spatial interaction viewpoints. The literature has also connected these models to a theme referred to as excess commuting models (O'Kelly and Niedzielski, 2009). However, to date, these ideas have not been transferred to other forms of spatial interaction. For example the extension of these ideas to models in the area of air passenger movement seems to be a useful way to calibrate and define issues of spatial accessibility in an air passenger system where it can be difficult to determine structure from the massive amounts of data. This paper continues such efforts in the spatial interaction tradition. In addition, the theory, disaggregation and empirical work have been closely intermeshed in a series of papers by O'Kelly and his colleagues (O'Kelly, Niedzielski and Gleeson, 2012). An extension to this idea, in the context of visualizing key features of a large air passenger data base is the subject of this paper. The data is the well-known DB1B data set available from the US Bureau of Transport Statistics web site. The particular application in this paper is designed to provide an index of the place accessibility – by measuring both emissivity (the push from the origin) and the experience of that origin in reaching its active destination in comparison to others who also travel to those same destinations (in a technical way that is defined in the paper). The analysis therefore goes some way to providing a logical measure for complex accessibility in a summary index. In this regard the paper provides a quantitative index for the quality of service, and this topic has received considerable attention in the literature from Grubesic, Matisziw and Zook (see references).

The basic spatial situation can be illustrated with respect to an origin, looking towards a number of destinations. The conditional probability of interacting with destinations j1, j2, ... from i, is shown in the following diagram (Figure 1). The measurements that can be made at i include the origin-specific average trip length and the comparative weighted average of the incoming trip lengths to the places i interacts with (the averages of the experiences in the four surrounding places for example).

Similarly from a destination point of view, we can compute the set of places that contribute to j, namely origins i1, i2, ... as shown in Figure 2. The destination-specific average in-bound trip length can be compared to the experiences of the trip makers on average, in the surrounding places that contribute to j (the four surrounding areas in the following example).

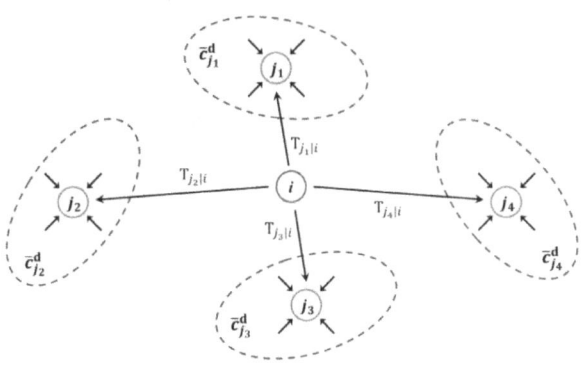

$\bar{c}_i^o$ = Origin specific average trip length

Fig. 1. Origin i relative to its active destinations

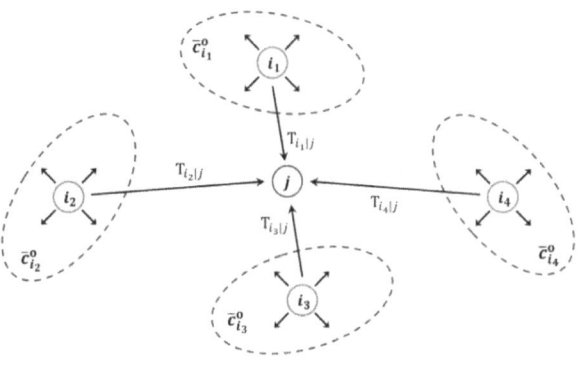

$\bar{c}_j^d$ = Destination specific average trip length

Fig. 2. Destination j relative to its active origins

These figures explain some notation that is used in the following section. The model is designed to reproduce the origin and destination specific trip lengths, as well as the trips produced from and attracted to each place.

2   Models

The maximum entropy derivation of the doubly constrained trip distribution model is found from:

MAX $H = -\sum_i \sum_j T_{ij} \ln T_{ij}$

Subject to:

$\sum_j T_{ij} = O_i$ [λ_i is the associated Lagrangean multiplier to ensure the model reproduces the observed trip origins and O_i is the number of air passengers departing from city i],

$\sum_i T_{ij} = D_j$ [μ_j is the associated Lagrangean multiplier to ensure the model reproduces the observed trip destinations and D_j is the number of air passengers arriving to city j],

$\sum_j T_{ij} C_{ij} = c_i^o$ [β_i is the associated (origin specific) Lagrangean multiplier to ensure the model reproduces the average trip length as observed from the origin],

$\sum_i T_{ij} C_{ij} = c_j^d$ [γ_j is the associated (destination specific) Lagrangean multiplier to ensure the model reproduces the average trip length as observed at the destination].

It can be shown in many different derivations that

$$T_{ij} = \exp(\lambda_i + \mu_j - (\beta_i + \gamma_j) C_{ij}) , \qquad (1)$$

is the first order condition for the objective function to reach a maximum, where the beta and gamma parameters are generally expected to be positive numbers, and are written with a negative sign, per widely accepted convention. This convention holds that larger beta reduces trip length, other things being equal.

This model is often re-written as a version of the standard doubly constrained spatial interaction model (Wilson, 1974) which is given by:

$$T_{ij} = A_i O_i B_j D_j \exp(-(\beta_i + \gamma_j) C_{ij}) , \qquad (2)$$

where, A_i = origin balancing factor
O_i = number of air passengers departing from city i
B_j = destination balancing factor
D_j = number of air passengers arriving to city j

$f(C_{ij}) = \exp(-(\beta_i + \gamma_j) C_{ij})$ contains an origin and destination specific cost function between zones i and j; in brief beta ensures that the origin observed trip length is reproduced by the model, and gamma performs a similar role for destinations. The cost function, $f(C_{ij})$, is taken to be the exponential function and uses the parameters to model empirically defined distance decay effects (Fotheringham and O'Kelly, 1989). In this paper distance is measured as the average actual route distance between i and j (not the direct distance). For example, if 50% of the passengers from i to j travel via a hub and the other 50% travel direct, the distance will be a combination of the two types of path. (In reality, city pairs interact with each other by a variety of paths, not just two. In general hub cities tend to have more of their destinations reachable in a direct way.)

3 The New Problem: Sensitivity Analysis

The initial empirical fitting of the model derived a common set of beta and gamma parameters (one for each origin and one for each destination). Assume now that all these parameters have been estimated and we are now prepared to consider a gradual shift of all the distance decay parameters for each zone by a constant ϕ. We are willing to hold the distance parameters constant reflecting the local effects needed to control the origin and destination specific trip lengths.

3.1 From the Origin End Constraint

Connecting the model notation and introducing a new shift parameter ϕ that is initially set to zero:

$$T_{ij} = \exp(\lambda_i + \mu_j - (\phi + \beta_i + \gamma_j)C_{ij})$$
$$= \exp(\lambda_i)\exp(\mu_j)\exp(-(\phi + \beta_i + \gamma_j)C_{ij})$$
$$= A_i O_i B_j D_j \exp(-(\phi + \beta_i + \gamma_j)C_{ij}) . \quad (3)$$

Since $\sum_j T_{ij} = O_i$,

$$\sum_j \exp(\lambda_i + \mu_j - (\phi + \beta_i + \gamma_j)C_{ij}) = O_i , \quad (4)$$

$$O_i \exp(-\lambda_i) = \sum_j \exp(\mu_j - (\phi + \beta_i + \gamma_j)C_{ij}) , \quad (5)$$

where ϕ is a shift parameter designed to move all the local distance decay parameters to the right, decreasing trip lengths a little compared to the empirical fitted $\bar{c}_i$. While beta is held constant, after the introduction of ϕ > 0, we assume that λ and μ need to be adjusted.

Following a technique introduced in O'Kelly et al. (2012) and differentiating the above terms with respect to ϕ, and assuming β_i and γ_j are calibrated from initial conditions,

$$-O_i \exp(-\lambda_i)\frac{d\lambda_i}{d\phi} = \sum_j \exp(\mu_j - (\phi + \beta_i + \gamma_j)C_{ij})(\frac{d\mu_j}{d\phi} - C_{ij}) . \quad (6)$$

Recognizing that $T_{ij} = \exp(\lambda_i + \mu_j - (\phi + \beta_i + \gamma_j)C_{ij})$ and after rearranging:

$$-O_i \frac{d\lambda_i}{d\phi} = \sum_j \exp(\lambda_i + \mu_j - (\phi + \beta_i + \gamma_j)C_{ij})(\frac{d\mu_j}{d\phi} - C_{ij}) , \quad (7)$$

$$\frac{d\lambda_i}{d\phi} = \frac{1}{O_i}\sum_j T_{ij}(C_{ij} - \frac{d\mu_j}{d\phi}) . \quad (8)$$

Given that $O_i = \sum_j T_{ij}$, we can restate as:

$$\frac{d\lambda_i}{d\phi} + \frac{\sum_j T_{ij}\frac{d\mu_j}{d\phi}}{\sum_j T_{ij}} = \bar{c}_i^o . \quad (9)$$

In the case where the ith origin is the numeraire, the $d\lambda_i/d\phi = 0$ so the average trip length and average dμ are equal, i.e. $\overline{P}_i = \bar{c}_i^o$ for the numeraire zone, which gives the key result for the interpretation of the sign of the origin effects shown in the maps:

$\frac{d\lambda_i}{d\phi} > 0$ if $\bar{c}_i^o > \sum_j T_{j|i} \frac{d\mu_j}{d\phi}$, and $\frac{d\lambda_i}{d\phi} < 0$ otherwise, where $T_{j|i}$ is defined as $T_{ij}/\sum_j T_{ij}$, the conditional probability of interacting with j from i, since $T_{ij} = T_{j|i} O_i$

3.2 From the Destination End Constraint

Similarly, in the case of destinations, connecting the model notation:

Since $\sum_i T_{ij} = D_j$,

$$\sum_i \exp(\lambda_i + \mu_j - (\phi + \beta_i + \gamma_j)C_{ij}) = D_j, \tag{10}$$

$$D_j \exp(-\mu_j) = \sum_i \exp(\lambda_i - (\phi + \beta_i + \gamma_j)C_{ij}). \tag{11}$$

Differentiating these expressions with respect to ϕ, and assuming $\beta_i + \gamma_j$ is a pre-defined constant

$$-D_j \exp(-\mu_j) \frac{d\mu_j}{d\phi} = \sum_i \exp(\lambda_i - (\phi + \beta_i + \gamma_j)C_{ij})(\frac{d\lambda_i}{d\phi} - C_{ij}). \tag{12}$$

Rearranging as before:

$$-D_j \frac{d\mu_j}{d\phi} = \sum_i \exp(\lambda_i + \mu_j - (\phi + \beta_i + \gamma_j)C_{ij})(\frac{d\lambda_i}{d\phi} - C_{ij}), \tag{13}$$

$$\frac{d\mu_j}{d\phi} = \frac{1}{D_j} \sum_i T_{ij}(C_{ij} - \frac{d\lambda_i}{d\phi}). \tag{14}$$

Given that $D_j = \sum_i T_{ij} D$, we can restate as:

$$\frac{d\mu_j}{d\phi} + \frac{\sum_i T_{ij} \frac{d\lambda_i}{d\phi}}{\sum_i T_{ij}} = \bar{c}_j^d, \tag{15}$$

which gives the key result for the interpretation of the sign of the destination effects: $\frac{d\mu_j}{d\phi} > 0$ if $\bar{c}_j^d > \sum_i T_{i|j} \frac{d\lambda_i}{d\phi}$, and $\frac{d\mu_j}{d\phi} < 0$ otherwise, where $T_{i|j}$ is defined as $T_{ij}/\sum_i T_{ij}$, the conditional probability of interacting with i from j, since $T_{i|j} = T_{i|j} D_j$,

4 Matrix Solution for dλ and dμ

Assume now that the model has been fitted and that we have a converged set of β_i and γ_j parameters as well as all the associated A_i and B_j balancing factors. Note that the development so far has established that there is a relationship between dλ and the average of the surrounding dμ values and the associated origin specific trip length:

$$\lambda' + T^d \mu' = c^o. \tag{16}$$

Similarly for the destinations:

$$\mu' + T^o\lambda' = c^d . \tag{17}$$

Where for now, c^o is the origin based average trip length, and c^d is the destination based trip length. It is important to note that since (β_i, γ_j) were used to calibrate the models, the right hand sides of (3) and (4) are equal to the empirically observed data.

We are using T^d and T^o to mean the $T_{j|i}$ and $(T_{i|j})^T$ matrices which are respectively the conditional probability of interacting with destination j given origin i, and the conditional probability of an interaction coming to j from i. We use λ' to mean the vector of partial derivatives and the same for μ'. Note that it assumed that the vectors and matrices are typed in bold face and are arranged to be conformable for multiplication, taking transpose as needed.

Arrange these conditions as a set of linear equations in matrix form; from (16):

$$\lambda' = c^o - T^d\mu' . \tag{18}$$

Therefore from (17)

$$\mu' + T^o(c^o - T^d\mu') = c^d . \tag{19}$$

Therefore

$$\mu' + T^oc^o - T^oT^d\mu' = c^d , \tag{20}$$

$$(I - T^oT^d)\mu' = c^d - T^oc^o , \tag{21}$$

$$\mu' = (I - T^oT^d)^{-1}(c^d - T^oc^o) . \tag{22}$$

Now turning to the destination side, from (17)

$$\mu' = c^d - T^o\lambda' . \tag{23}$$

Therefore substituting into (16)

$$\lambda' + T^d(c^d - T^o\lambda') = c^o , \tag{24}$$

$$\lambda' = (I - T^dT^o)^{-1}(c^o - T^dc^d) . \tag{25}$$

Notice that these derivatives can be computed explicitly from the data and the model. Special care to deal with the numeraire – set one row of $T_{j|i} = 0$ (it is going to set $d\lambda = 0$, and set the corresponding r.h.s. = 0).

5 Numerical Example

The data are from 344 Continental US Airports with the passenger flows from Q1 2011, DB1B data set (this is the successor data base to the original CAB data used a lot in spatial models). This is a 344 x 344 US Air interaction matrix. Several areas are excluded because of their special features: Alaska, Hawaii, Virgin Islands, Puerto Rico, and the Territories. An important way in which the data are used is to measure the inter-nodal distance as the average of the actual market distance traveled and NOT the straight line distance. This will help to explain how cities relatively close to each other (Denver and Boulder, or Los Angeles and Santa Barbara) can have different experiences based on the number of direct services).

The fitted model is chosen to balance the origin and destination outflows and inflows respectively, and to reproduce the average origin and destination specific trip length. Three particular properties are now illustrated and explained in the context of these data.

6 Properties

Once the origin and destination λ and μ values are set, the beta values ensure that the c_i^{mod} column matches the observed trip lengths from each origin, and the gamma values ensure that the destination specific trip lengths can be computed, and they also match the observations for these zones.

Think of this as the solution with $\phi = 0$. Now, if we make a small change in ϕ, we can numerically evaluate the new values of λ and μ, and compute to a good degree of accuracy the derivative. It remains to see then that the expected properties hold: (a) that the origin and destination trips add up both before and after to a high degree of accuracy; (b) that the relationships between the values obey the expected aggregation properties; and, (c) that the expected origin/destination specific trip lengths stay close to the original empirical observations, to be consistent with assumptions that we can hold beta and gamma parameters constant. We also need to confirm that there is still a very close correspondence between the new origin specific trip length and the original observed data.

6.1 Property 1

A number of highly significant properties emerge from the data: The relationship between $d\lambda$ and the average of the surrounding $d\mu$ values and the associated origin specific trip length: from (16) and (17) and since $\mathbf{c^o}$ **and** $\mathbf{c^d}$ are very similar (not exact, but expected to be quite close because there is symmetry in the data due to round trip air patterns):

$$\lambda' + \mathbf{T^d}\mu' \approx \mu' + \mathbf{T^o}\lambda' . \qquad (26)$$

Therefore

$$\lambda' - \mathbf{T^o}\lambda' \approx \mu' - \mathbf{T^d}\mu' . \qquad (27)$$

Or the deviation between $d\lambda$ and its average surrounding values is the same as the deviation between $d\mu$ and its average surrounding values.

6.2 Property 2. λ vs O and μ vs D

We know from the origin constraint that:

$$\sum_j \exp(\lambda_i + \mu_j - (\phi + \beta_i + \gamma_j)C_{ij}) = O_i . \tag{28}$$

Now note as an approximation for analysis, that C_{ij} is replaced on average by $\bar{c}_i^o$ so that

$$\exp(\lambda_i - \beta_i \bar{c}_i^o) \sum_j \exp(\mu_j - (\phi + \gamma_j)C_{ij}) \approx O_i , \tag{29}$$

$$\exp(\lambda_i - \beta_i \bar{c}_i^o) \approx O_i / \sum_j \exp(\mu_j - (\phi + \gamma_j)C_{ij}) . \tag{30}$$

So

$$\lambda_i - \beta_i \bar{c}_i^o \sim \ln(O_i) . \tag{31}$$

Very similar analysis shows

$$\mu_j - \gamma_j \bar{c}_j^d \sim \ln(D_j) . \tag{32}$$

Specifically, from numerical results:

$$\lambda_i - \beta_i \bar{c}_i^o \sim \ln(O_i) = a_1 + b_1 \ln(O_i) , \tag{33}$$

$$\mu_j - \gamma_j \bar{c}_j^d \sim \ln(D_j) = a_2 + b_2 \ln(D_j) . \tag{34}$$

where the last two equations are empirically observed from fitting l.h.s terms against $\ln(O_i)$ and $\ln(D_j)$ respectively. These results show that λ and μ are increasing functions of the size of origins and destinations respectively, so that emissivity is correlated with the size (as well as location) of the place.

6.3 Property 3. $T^d c^d$ and $T^o c^o$

To measure the share of traffic from i that goes to places that themselves have particular trip length characteristics, consider $T^d c^d$. This returns a measure of each origin that tells us on average how big or small the trip length is in the places to which this origin sends its passengers. In that sense it is an indicator of the types of places that i interacts with, in terms of their position (see Fig. 1 for a graphical explanation).

It has been discovered through a combination of analysis (see r.h.s. of equation 25), and data examination, that the sign of $d\lambda$ largely depends on the sign of $(c^o - T^d c^d)$ from equation (25) and the sign/magnitude of $d\mu$ largely depends on the sign of $(c^d - T^o c^o)$ from equation (22).

Interpretation: the $d\lambda$ is positive where the corresponding element $(c^o > T^d c^d)$, typically occurs at places where the average trip length from the origin is larger than the average of the inbound trips to the destinations to which it connects (weighted by the strength of the interaction T^d). Thus, this origin is traveling further on average than the typical interaction to the places where it goes. It is somewhat more isolated than average and would therefore have to become more emissive (more pushy) if distance impedance increased. It is suggested that such places do not command a locational premium, and actually have a lower rent (or even a subsidy).

Conversely, a place with $d\lambda < 0$ has a shorter origin based average trip length than the typically blend of interactions at those destinations, and so is probably more favorably located than average, and would do well under a restriction in trip length. Figure 3 shows the result of plotting the model values of $d\lambda$ vs the fitted model deviation between origin average trip length and surrounding community experience (as explained in Figure 1). The result shows that $d\lambda$ is almost perfectly explained by this deviation.

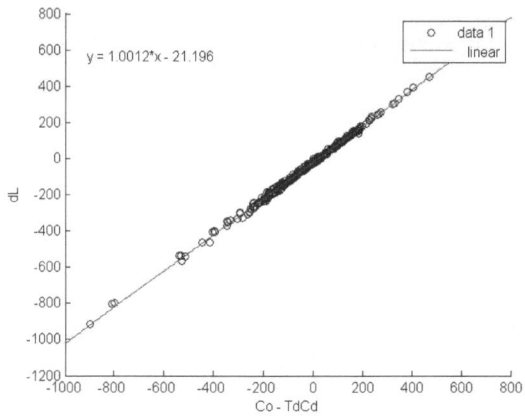

Fig. 3. $d\lambda$ vs $(c^o - T^d c^d)$ from the fitted model

Similarly on the destination side: $T^o c^o$ measures the share of the incoming travel to node j that comes from each surrounding i (as illustrated in theory in Figure 2). As mentioned above, T_{ilj} returns a measure for each destination defining the average trip lengths of the sources that contribute to this destination. If a place for example receives a lot of its traffic from places with particular trip length characteristics, they will show up in this measure. In this sense it is an indicator of the place that contributes to j, in terms of their average tip length. These results can be used, as above to explain the pattern and locations of the size of $d\mu$.

As was the case for Figure 3, Figure 4 shows that there is a very strong empirical fit between $d\mu$ and the fitted values of $(c^d - T^o c^o)$. The above graphics are only an approximation, however, because they plot the local $d\lambda$ against $(c^o - T^d c^d)$ and local $d\mu$ against $(c^d - T^o c^o)$. Of course $d\lambda$ is not simply related to the local value, as equations (22) and (25) show it to be the sum of many weighted values.

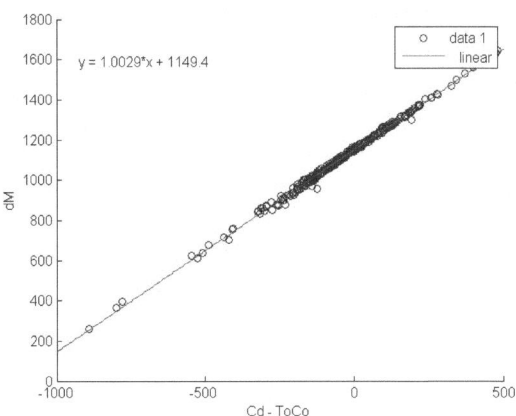

Fig. 4. $d\mu$ vs $(c^d - T^o c^o)$ from the fitted model

This result points the way to a more fundamental identity which is fairly easy to prove: $\overline{\lambda'} + \overline{\mu'} = \overline{c}$.

As expected from the derivations and algebra, this result may be confirmed from either the origin or destination identities (16) or (17). The result is also consistent with the idea that the sum of $d\lambda$ and the sum of $d\mu$ add up to the average trip length and in turn, in the limit these values would match the dual values of the LP for minimum feasible trips for the associated OD pattern (see again O'Kelly, 2010).

Figure 5 shows a map of the signs of λ and $d\lambda$ as well as size of origin. The spatial pattern is that red and blue are places with $\lambda > 0$ and are generally larger (no green or yellow node is large enough at this scale to register as a significant origin source, though of course all have positive originating flows). On the other hand, all yellow and green nodes are smaller places. Green and blue represent two sizes of places that have negative $d\lambda$. These are places that in general have a favorable degree of centrality and so have $c^o < T^d c^d$. In terms of location and accessibility, these green and blue nodes might be thought of as paying a premium 'rent' for their accessibility advantages. Yellow and red nodes (small and large respectively) are the opposite case in terms of sign of $d\lambda$, where there is a positive $d\lambda$, and so the value of $c^o > T^d c^d$, or in other words the place has a slightly disadvantageous location which earns a small rent, or even perhaps what might be seen as a locational subsidy. These geographical contrasts for the most part make sense, though there are still of course some anomalies, and the generalization is not completely valid as the sign of $d\lambda$ is only approximated by $c^o - T^d c^d$ as explained above.

Nevertheless, there are some reasons to believe that this is a solid and meaningful classification: note Chicago, Denver, and Columbus all enjoy good access scores. South Florida (peripheral and large) is red, and the upper plains are yellow all as expected and consistent with the verbal discussion above.

Airport Accessibility from Passenger Flows

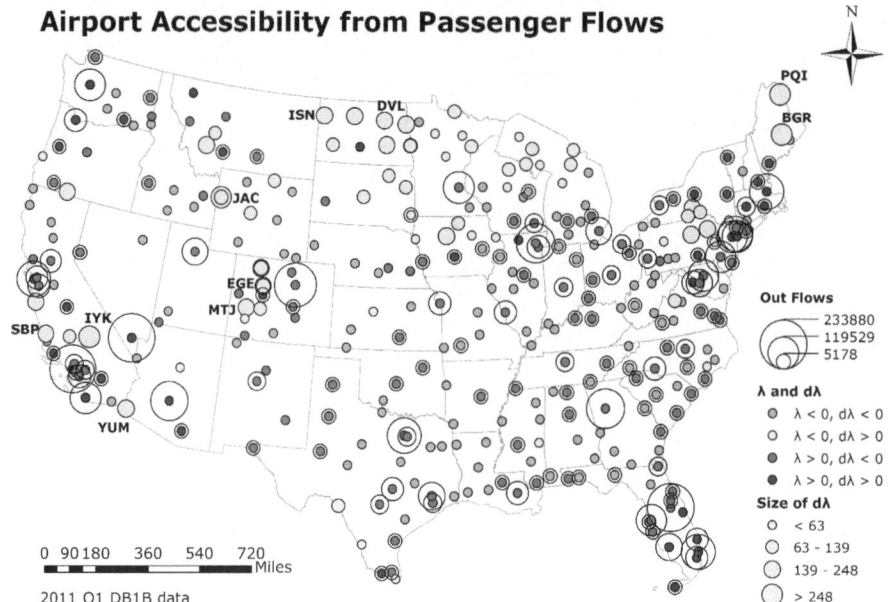

Fig. 5. Map of signs of λ and dλ as well as size of origin

The broad spatial results are as follows: the yellow nodes are not strong sources of emission and have a positive value of dλ, signifying a locational subsidy. The larger yellow circles are potentially the most underserved. The top 10 such places (DVL in North Dakota, etc) are labeled and are listed in Table 1 below.

Table 1. The 10 places with the largest values of dλ and negative λ (and therefore theoretically eligible for locational subsidy)

Airport Name	IATA Code	State
Inyokern Airport	IYK	CA
Jackson Hole Airport	JAC	WY
Bangor International Airport	BGR	ME
Northern Maine Regional Airport at Presque Isle	PQI	ME
Eagle County Airport	EGE	CO
Devils Lake Regional Airport	DVL	ND
San Luis Obispo County Regional Airport	SBP	CA
Montrose Regional Airport	MTJ	CO
Yuma International Airport	YUM	AZ
Sloulin Field International Airport	ISN	ND

One might well ask why the red nodes (which also have positive dλ) are not discussed in terms of subsidy, we note that these are peripheral but large cities. It is not expected that places with a strong viable commercial market will need to receive a subsidy for their location, even though the results are clearly indicating their locational disadvantages. They are typically on the edge of the region.

The blue nodes are large places and comparatively well served in terms of average trip accessibility, and the green nodes are locationally similar but are smaller places (i.e. often on the edge of larger blue centers).

7 Conclusions

This paper has begun the task of dissecting the role of location in a detailed set of intercity (inter airport) data where it is important to note that the distance used to separate the places is the average of the actual market flight for the OD pair. To bring these ideas home to a practical level, and bearing in mind the interpretation of dλ as a location rent or subsidy, some small local airports have especially wide divergence between origin trip length and the trip lengths into the corresponding destinations. As a result they would either be eligible for a low rent term in this model's terminology, or a subsidy. This idea may be relevant to quantifying subsidies along the lines of those provided by the essential air service. Several places with low emissivity and very strong locational disadvantages are listed in the following table. Further work will determine if there is a good logical relevance of these results with the places that are eligible for the essential air service subsidy, (Matisziw, Lee and Grubesic, 2011). In addition, as well as places with low emissivity/accessibility, the technique of course also identifies places with exceptionally good levels of service.

Acknowledgements. The author is very grateful to Meng Guo, Kejing Peng, and Yongha Park for excellent research assistance. Support from NSF (BCS-1125840) and from Ohio State University's Center for Aviation Studies is also acknowledged with gratitude.

References

Fang, S.-C., Rajasekera, J.R., Tsao, H.S.J.: Entropy Optimization and Mathematical Programming. Kluwer, Dordrecht (1977)

Fotheringham, A.S., O'Kelly, M.E.: Spatial Interaction Models: Formulations and Applications. Kluwer, Dordrecht (1989)

Grubesic, T.H., Matisziw, T.C.: A spatial analysis of air transport access and the essential air service program in the United States. Journal of Transport Geography 19(1), 93–105 (2011)

Grubesic, T.H., Matisziw, T.C., Zook, M.A.: Global airline networks and nodal regions. GeoJournal 71, 53–66 (2008)

Grubesic, T.H., Matisziw, T.C., Zook, M.A.: Spatio-temporal fluctuations in the global airport hierarchies. Journal of Transport Geography 17(4), 264–275 (2009)

Grubesic, T.H., Zook, M.: A ticket to ride: evolving landscapes of air travel accessibility in the United States. Journal of Transport Geography 15(6), 417–430 (2007)

Kapur, J.N.: Entropy maximization models in regional and urban planning. Int. J. Math. Educ. Sci. Technol. 13(6), 693–714 (1982)

Matisziw, T., Lee, C.-L., Grubesic, T.: An analysis of essential air service structure and performance. Journal of Air Transport Management 18, 5–11 (2011)

O'Kelly, M.E.: Spatial Interaction Models. In: Kitchin, R., Thrift, N. (eds.) International Encyclopedia of Human Geography, vol. 10, pp. 365–368. Elsevier, Oxford (2009)

O'Kelly, M.E.: Entropy Based Spatial Interaction Models for Trip Distribution. Geographical Analysis 42(4), 472–487 (2010)

O'Kelly, M.E., Niedzielski, M.: Are long commute distances inefficient and disorderly? Environment and Planning A 41(11), 2741–2759 (2009)

O'Kelly, M.E., Niedzielski, M., Gleeson, J.: Spatial Interaction Models from Irish Commuting Data: Variations in Trip Length by Occupation and Gender. Forthcoming Journal of Geographical Systems (2012)

Okuhara, K., Yeh, K.-Y., Hsia, H.-C., Ishii, H.: Geographical advantage from accessibility based on spatial interaction model. International Journal of Innovative Computing, Information and Control 6(9), 1–09-0653 (2010)

Rabino, G., Occelli, S.: Understanding spatial structure from network data: theoretical considerations and applications. Cybergeo: European Journal of Geography [Online], Systems, Modelling, Geostatistics, article 29 (Online since June 26, 1997), http://cybergeo.revues.org/2199, doi:10.4000/cybergeo.2199

Wilson, A.G.: Urban and Regional Models in Geography and Planning. Wiley, London (1974)

Am I Safe in My Home? Fear of Crime Analyzed with Spatial Statistics Methods in a Central European City

Daniel Lederer

KFV (Austrian Road Safety Board), Research and Knowledge Management,
Schleiergasse 18, 1100 Vienna, Austria
daniel.lederer@kfv.at

Abstract. This article presents the results of a quantitative survey in a central European city, where more than 1,500 citizens were asked about their fear of becoming a victim of burglary. Additionally, vulnerabilities to crimes were measured. A large set of spatial data was analyzed with different spatial-statistic methods and visualized in maps intended to serve as a summarized overview of the citizens' fear of crime. First results show that there are specific hot spots in fear of burglary, majorly in the core of the city, and statistically significant differences in the pattern of fear of burglary between the districts. Furthermore, areas with a lack of technical safety standards were identified. This information shall help to start local crime prevention programs to reduce fear of crime and increase the quantity of protected homes.

Keywords: Environmental criminology, Fear of crime, Crime mapping, GIS, Spatial statistical analyses, Urban area studies.

1 Introduction

Most of the crimes committed in a country occur in the cities. Being that a majority of the population resides in these cities, personal security is an important factor for the quality of life and the well-being of the citizens. This is the background as to why the Austrian Road Safety Board (KFV) started the project, "Urban Crime Analysis and Mapping" (UCAM). Specifically, the intended goal was to connect different kinds of crime data and provide a comprehensive report, which would include the most important information on the crime situation and problems in nine major European cities. This monitoring includes the personal perception of crime from the citizen's view and the reported crime from police. The main purpose of the project is to find hot spots of crime and fear of crime by using different methods of spatial statistics and visualizing the results in maps.

In a quantitative survey, 1,505 citizens were asked about (1) fear of crime and the location of 'fear-of-crime'-places in the city, (2) victimizations and (3) contentedness with and perception of the police. In the present study, there existed some issues of fear of crime selected, like fear of burglary or anti-victimizations strategies to protect personal property. These items are very complex when it comes to measuring and may have a high degree of spatial association. As a consequence, the main research questions of this study are as follows: Are there differences in the level of fear of becoming a victim of a residential burglary between the districts in the city? Within the city, are there certain areas with a lack of technical safety measures, which may lead to an increased vulnerability to burglary? The aim of this study is to answer these questions by using GIS and specific analysis methods. Furthermore, the local resident's perception of crime shall help to better understand the reasons why some districts are more vulnerable to crime than others.

2 Methods and Techniques Used

The linkage of crime and place has a very long history and traces back to the early work of French social ecologists in the middle of the 19th century. First, in the 20th century, representatives of the Chicago School, within their socio-ecological studies, began analyzing the relationship of crime and place. A renaissance of these approaches was reached with the "new" Chicago School in the 1980's, where modern computer systems were used to create maps with ecological aspects of crime [1]. The placed-based theories of crime (environmental criminology) are based on this knowledge and the theoretical perspectives derived from the routine activities theory [2], the rational choice theory [3] and the studies of crime patterns and hot spots [4]. The use of Geographic Information Systems in analyzing and interpreting patterns of crime became, in the last two decades, very important in research and practice [5].

Initially, publications on the research of fear of crime started in the 1960's (e.g. McIntyre, 1967 [6]) and many quantitative and qualitative studies in this topic have been published over the last decades. It is considered cutting edge to analyze the fear of crime with spatial statistics and to put the results with GIS in a map. In the last years, various studies have been published. For example, Doran & Burgess (2012) studied the crime perception in the city of Wollongong, Australia. In their study, they combined a fear-of-crime survey (standardized interviews) among 260 working people in the CBD area with techniques from behavioral geography, such as cognitive mapping and an activity diary analysis. This study shows the importance of analyzing place-based information of fear of crime with GIS. The results helped the city to find problem-areas and showed ways to revitalize these areas, for instance, with land-use planning changes [7].

The present study tries to extend on the previous research in fear of crime and GIS by analyzing the perception of crime in a representative value of citizens spread all around the city. There are two major complications in this quantitative study: Firstly, the measure of fear of crime is especially difficult, hence, the reason the definition is of high importance. Secondly, the dataset is very large and the spatial information must be standardized to have a clear result of the personal fear-of-crime-area. Due to these complications, the following research setting was created.

2.1 Research Setting

The study site is a city in central Europe with a population of approximately 190,000. Due to protection of the personal safety of the respondents the name of the city is not mentioned. The city is subdivided into 43 statistical districts, which are used as the basic spatial study-site-level. For these areas, socio-demographic data is available.

2.2 Measuring Fear of Crime

Fear of Crime is described as '...*a diffuse psychological construct affected by a number of aspects of urban life*' [8, p. 11]. It is a subjective phenomenon that cannot be measured in the same way as crime, for instance, by reporting the quantity of victims and offenders from the police data. In the present study, the fear of crime is defined as the personal fear of becoming a victim of a certain crime (e.g. residential burglary, robbery, sexual assault). By describing the frequency of anxiety in relation to crime on a five-point scale (very often, often, sometimes, rarely, and never), the level of fear of crime can be estimated. In the present study the citizen's information about the personal level of fear will be used as a subjective description of the urban crime situation.

Fear of crime is a serious social problem, which has different negative impacts on affected people. The impacts ranges from physiological changes (e.g. increased heart rate) to psychological reactions (e.g. distrust of others, frustration, violation). Furthermore fear of crime disrupts neighborhood cohesion and reduces the personal quality of life [7].

In a quantitative survey with computer assisted telephone interviews (CATI), conducted in September 2011, 1,505 randomly selected citizens were asked about their fear of crime and other relevant topics related to personal security. The dataset includes two important characteristics for spatial analysis, (1) selection from every district and (2) the use of personal addresses. Specifically, 35 inhabitants were selected in a disproportional stratified random sampling from every district, so that local differences could be measured in fear of crime. Additionally, permission was asked of the participants to allow the use of personal addresses. The assignment to a specific district and the localization of home addresses make it possible to have spatial-based information in detail.

Addresses are considered highly sensitive personal information and, therefore, protected by national data safety laws. It is for this reason that the addresses were directly geocoded and coded in a random method by the survey institute. This method consists of a latitude and longitude transformation with a deviation of 30 to 60 m from the origin position. This measure leads to a loss of data exactness, but allows working with highly sensitive personal data. Furthermore, the results of the study are available on two levels: Firstly, on the district level, as polygon data and secondly, on the level of coordinates, as point data. This brings the advantage of analyzing the data with local spatial statistic methods and reduces certain sources of errors. Such as the Modifiable Areal Unit Problem (MAUP), which means that spatial analysis with areal units (polygon data) are influenced by the level of aggregating [9] [10].

2.3 Analyzing and Visualizing Fear of Crime

There are many spatial data analysis methods, which are useful in identifying patterns. In the present study, the three descriptive and exploratory spatial data analysis (ESDA) methods, Spatial Autocorrelation, Kernel Density Estimation (KDE) and Nearest Neighbor Hierarchical Clustering (NNHC), were used to identify and analyze the patterns of fear of crime in the city [1] [11]. These cluster analysis methods can be divided into two groups: (1) areal and (2) point pattern analysis.

The first group uses areal (polygon) georeferenced data to describe the distribution of certain events and includes spatial association and autocorrelation. These techniques test the spatial relation to each other. In the present study, the global indicator, Moran's I, and the Local Indicator of Spatial Association (LISA), were applied. The global Moran's I can be used to understand general spatial patterns and measures the deviation from spatial randomness by comparing the value at any one location with the value at all other locations. The result of Moran's I statistic varies from -1 to +1. A positive value indicates a positive spatial autocorrelation. The higher the value (near +1), the higher the positive spatial autocorrelation, which means that more polygons with similar high, middle or low values, are neighbored. On the contrary, a negative Moran's I shows a negative spatial autocorrelation. In this case, more polygons with different values are neighbored (spatial outliers). A Moran's I approximately zero indicates a random spatial autocorrelation and the attribute values are randomly spread over space. LISA helps to identify statistically significant local spatial clusters (e.g. hot or cold spots) and is based on the local Moran statistics. This statistical indicator compares local averages to global averages and assesses the local association of certain events. In comparison to the global indicators, the local Moran statistic provides much more than a summarized single statistic for the whole study site, as it identifies, in a set of polygons, clusters of high or low values [1] [12] [13] [14] [15].

In the second group, point data was analyzed with a density (KDE) and a hierarchical (NNHC) type of cluster analysis method. The Kernel Density Estimation (KDE) is a interpolation method, which creates a smooth surface of the point data with a variation in the density of enclosed points. Areas with a high quantity of points result in a high density. The method is based on two parameters, the grid cell size and the bandwidth (search radius), and works in a 3-step process. Within the first step, a fine grid with a user defined cell size is placed all over the point events of the study area. In the next step, a '...*moving three-dimensional function (the kernel)...*' [16, p. 259] is laid on the cells of the grid and calculates weights for each point within the kernel's radius. Finally, the values of all circle surfaces for each location of the study area will be summed and calculated for each cell of the fine grid. Concentrations of point events are visualized in circular clusters, which depend on the Euclidean distance of the kernel function. Extensions of the KDE, such as the Network Density Estimation (NDE) are based on network distances, which results in linear clusters along networks (e.g. streets) [1] [13] [16] [17]. The NNHC method is used for grouping spatially close points into hierarchical clusters. It depends on the Nearest Neighbor Index (NNI) test, which compares the distances between the points of the actual distribution against a random distributed data set of the same sample size. The NNHC identifies clusters of points with two fixed parameters: (1) the threshold distance, which defines the search radius of enclosing or not enclosing points into the cluster, and (2) the minimum number of points that are required for each cluster [13] [14].

All three mentioned spatial analysis methods are based on traditional statistical research from the early part of the last century. Spatial Autocorrelation is based on the studies from Moran, who studied the interpretation of statistical maps in 1948 [18] and the relations between continuous and discontinuous processes in 1950 [19]. Kernel Density Estimation reaches back to the studies of Fix and Hodges (1951) [20]. Additionally, Clark and Evans developed the NNI in 1954 [21] and King described the step-wise clustering procedures in 1967 [22], both of which are necessary basics for the hierarchical clustering techniques [14] [23].

The developments in spatial data analysis and the technological advances, such as computers with a high efficiency or spatial statistical applications in combination with geographic information systems, help to better understand place-based data. In the present study the combination of the mentioned three spatial data analysis techniques helps in analyzing clusters of certain spatial events, like crime or fear of crime. This information was visualized in different types of thematic maps. In common, there are four types of used maps: (1) choropleth maps, (2) LISA-maps, (3) density maps, and (4) cluster maps. The three following applications were used: ArcGIS 10 was applied for creating choropleth maps and visualizing spatial distributions (http://www.esri.com/software/arcgis/arcgis10/index.html). For exploratory spatial data analysis, the programs CrimeStat® III (http://www.icpsr.umich.edu/CrimeStat/) and GeoDa (http://geodacenter.asu.edu/) were applied.

3 Analysis and Results

In the first step, the distribution of persons who fear becoming a victim of burglary were aggregated on the level of statistical districts and shown in a choropleth map (see figure 1). The darker the district is colored, the higher the rate of people in fear. The pattern of fear of crime varies all over the city. There are four districts with values higher than 13.7 %. District 14 has the highest rate (22.4 %) of anxious citizens.

The spatial autocorrelation is visualized on the right side of figure 2 (LISA-map). The global Moran's I for fear-of-burglary rate is 0.14 and was calculated with a rook-based contiguity matrix. The value indicates a low positive autocorrelation across the districts. The local spatial autocorrelation results in two types of spatial clustering: On the one hand, there is a fear of crime hot spot located on the Westside, with a significantly ($p = 0.05$) high quantity of people fearing a burglary. This area contains the three districts 10, 11, 12 and the directly neighboring districts (IDs: 9, 13, 14, 15, 20, 22). On the other hand, on the north-western part of the study site, there is a cold spot of two significant districts ($p = 0.05$) and four neighboring districts, where the fear of residential burglary is very low.

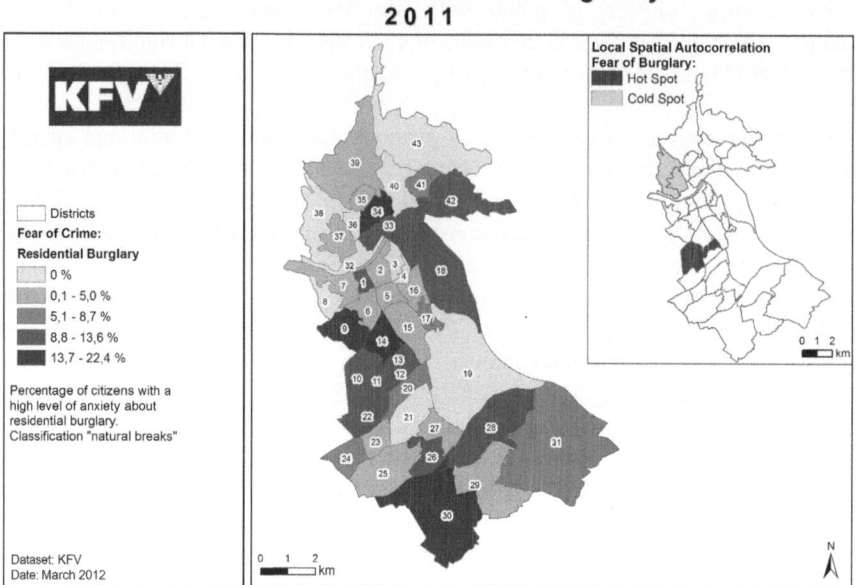

Fig. 1. Fear of residential burglary in the districts of a central European city analyzed with spatial autocorrelation

In the next step, the dataset of people who are afraid of becoming a residential burglary-victim was analyzed with point pattern analysis methods (e.g. kernel density estimation). In comparison to the last method, this dataset is not aggregated on the level of districts. As it was mentioned in chapter 2.2 (Measuring Fear of Crime), the results of the survey (addresses of the asked persons) are geocoded on the level of coordinates and available as point data. The dataset consists of 104 addresses of people experiencing fear throughout the city. The kernel density estimation was used to identify concentrations of fear of crime and calculated with CrimeStat® III. The following parameters were chosen:

- single kernel density estimation
- "normal" interpolation
- bandwidth as a "fixed interval" of 200 m

Figure two shows the density of people who are afraid of becoming a victim of residential burglary. The darker the area is red colored, the higher the quantity of people experiencing fear. There are approximately eight areas with the highest density (up to 0.17 points per square kilometers). These fear-of-crime hot spots are widening in the center, western and south of the city. The hot spot with the largest area is located in the center of the city and reaches from district 14 in the north, to district 20 in the south. This hot spot contains approximately eleven people experiencing fear.

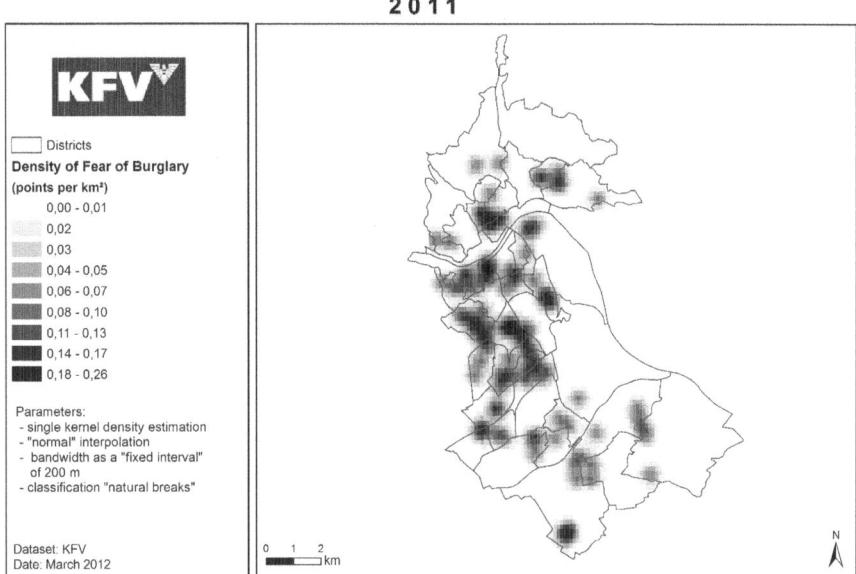

Fig. 2. Fear of residential burglary in a central European city analyzed with the kernel density estimation

The Nearest Neighbor Hierarchical Clustering (NNHC) is another method to identify spatial clusters of people experiencing fear in the study site. The clusters were generated in the first order with a fixed distance of 500 m and five minimum points per cluster. The results are visualized in figure three. There are three clusters which fulfill the parameters: The first cluster (Hull 1) is located in the northern part of the city and contains seven persons. The area is 0.31 km² and has a density of 22.29 points per km². On the Westside of the city, there are two clusters. Additionally, Hull two includes seven persons and has the largest area (0.58 km²) of all clusters. The last cluster (Hull 3) has the highest density (32.1 points per km²) and includes six people who fear becoming a burglary-victim.

The three used clustering methods (spatial autocorrelation, kernel density estimation, and nearest neighbor hierarchical clustering) explore in different ways the topic of fear-of-crime hot spots. One of the most important differences is the used dataset (polygon vs. point data) and the problems behind the aggregation level (e.g. MAUP). A comparison of the results show one very important match in the location of hot spots: This area is located in the Westside of the city and includes the entire area of districts 9, 12, 13, 14 and parts of districts 10, 11, 20, and 22. A second spatial cluster is located in the northern part of the study site (IDs 34, 36), which was calculated with the NNHC (Hull 1) and the KDE. The results of the spatial autocorrelation do not show a statistical significant cluster in this area. This is a very good example to demonstrate the importance of using different analysis methods and combining the results.

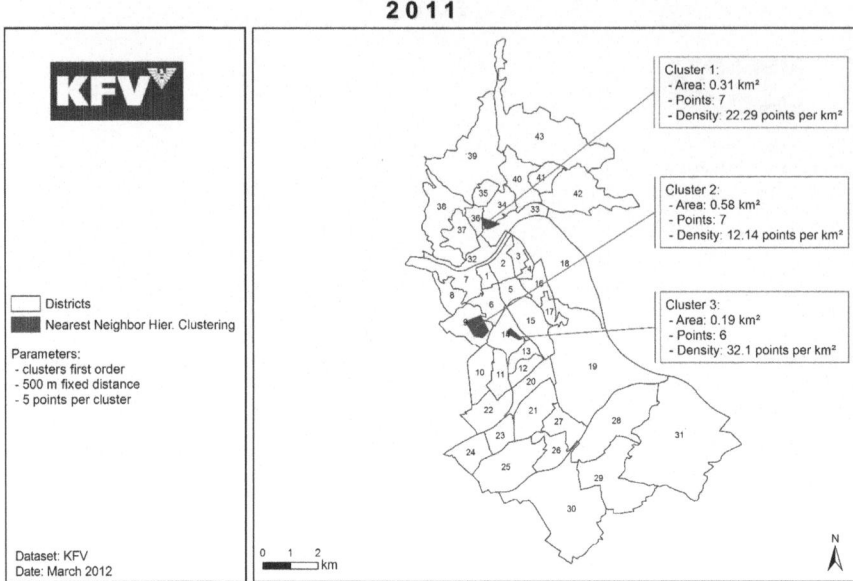

Fig. 3. Fear of residential burglary in a central European city analyzed with nearest neighbor hierarchical clustering

For estimating the vulnerability to residential burglary of an area, the distribution of technical measures to protect the citizen's apartments and houses were analyzed. Hence, the reason why the citizens were asked about personal safety measures, such as safety locks, safety doors, death bolts, automatic lights with timers, or alarm systems. In the first map, the distribution of people who do not protect their homes on a high safety level (less than one technical safety measure) is visualized on the level of districts (see figure 4). The darker the district is colored, the higher the quantity of those people and the lower the distribution of safety measures. Most of the districts with high values of low protected homes are located in the inner city (e.g. ID 1). The results of the spatial autocorrelation are shown on the right side of figure 4. The concentration in downtown is a statistically significant ($p = 0.05$) hot spot and includes five districts with a significant level of clustering and nine directly neighboring districts. In comparison to the other districts of the city, the quantity of high protected homes in downtown is relatively low. A cold spot was identified in the southern part of the study site, which includes four districts (ID 28-31).

Visualized in the kernel density map (figure 5) is the pattern of people with a high vulnerability to burglary. The highest densities are located in downtown and in the districts 34 and 36. These areas contain more than 0.79 points per square kilometers. The majority of all low-equipped homes are in the inner city region. Further high densities are located in the south of downtown (ID 11 to 14) and in the southwestern part of the city, which reaches from district 24 in the West, to district 27 in the East.

Am I Safe in My Home? Fear of Crime Analyzed with Spatial Statistics Methods 271

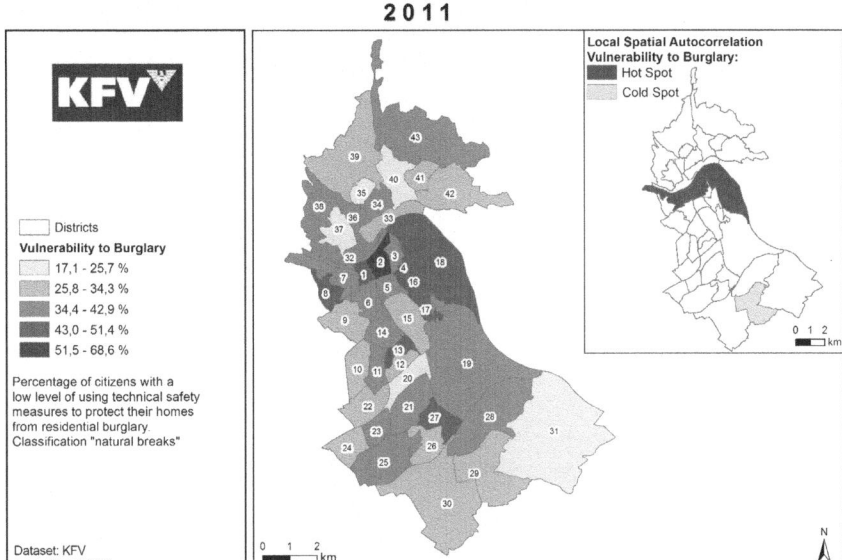

Fig. 4. Vulnerability to residential burglary in the districts of a central European city analyzed with spatial autocorrelation

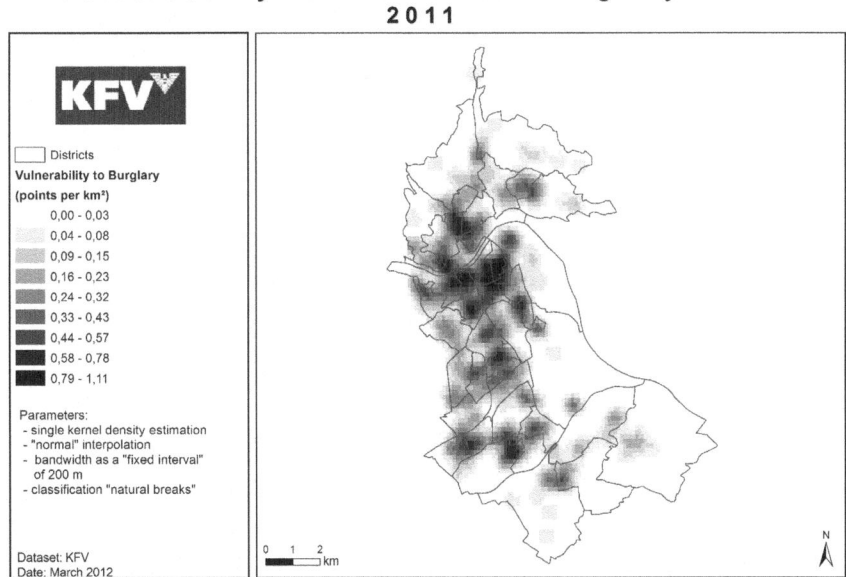

Fig. 5. Vulnerability to residential burglary in a central European city analyzed with kernel density estimation

The results of the Nearest Neighbor Hierarchical Clustering method are visualized in figure 6. There are eight clusters in the first order, which are calculated with a fixed distance of 500 m and a minimum of 20 points per cluster. The clusters are mostly located in the inner city, similar to the hot spots of the Kernel Density Estimation. The highest value of enclosed points was reached in the clusters 1, 5, 7, and 4. Between 40 and 23 people, who do not protect their homes on a high safety level, were identified in these areas.

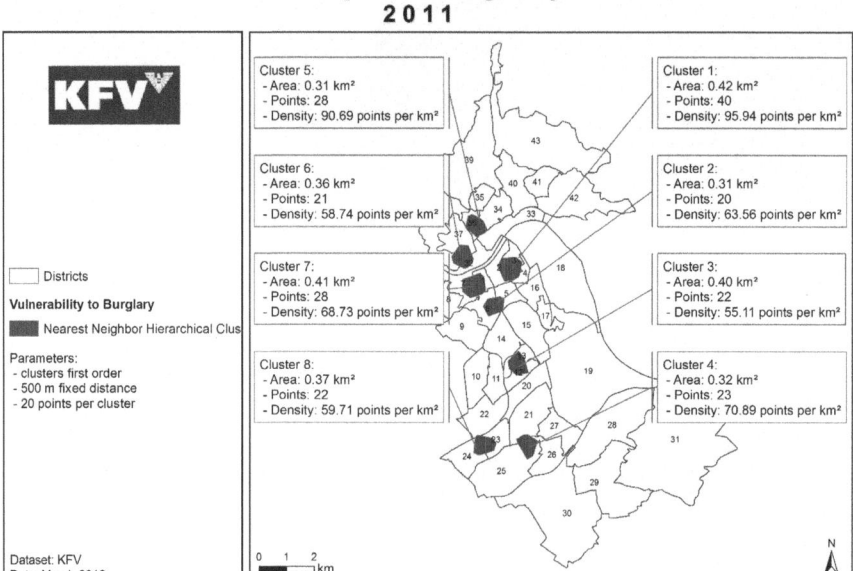

Fig. 6. Vulnerability to residential burglary in a central European city analyzed with nearest neighbor hierarchical clustering

The combination of all three used methods shows one very important area of high vulnerability to burglary, which is located in downtown. This area encloses the districts 1, 2, 3, 4, 5, 6, and 7. In comparison to the other districts, the people of this region show more disregard for personal technical safety measures.

4 Conclusion and Future Research

The research demonstrates that spatial analysis methods help to better understand special topics in fear of crime in the selected European city. By using different techniques in clustering spatial data, a large amount of complex information could be compressed in thematic maps. One of the main finding is the identifying of an important fear-of-crime hot spot, which is located in the Westside. In this area, the quantity of people who fear a burglary is statistically significantly higher than in other districts of the town.

Another subject of research was the analyzing of the technical safety standards of the citizen's homes. The investigations show that there are several differences in the distribution between the 43 districts. Furthermore, there is one statistically significant hot spot of low technical safety measures. Especially, the homes in downtown show a high level of vulnerability to burglary for the reason that many citizens disregard using safety locks or other safety techniques. The combination of hot spots in fear of burglary with the high vulnerability to burglary areas matches one area with an overlapping result: ID 11 to 15. In this region, it would be very interesting to start a local crime prevention program to support people in protecting their homes. This may lead to a lower level of fear of crime and, in addition, to a better quality of life in this neighborhood.

In the future, the results of the exploratory analysis can be used to find out why the mentioned hot spots are located in these areas. The spatial analysis will be enlarged by using confirmatory spatial statistical methods, like ordinary least squares (OLS) regressions or predictive models. Furthermore, the clustering methods will be used for analyzing other topics in fear-of-crime research. One of the future aims of the project is to investigate links between fear of crime and the actual occurrence of crime. In addition, a very important step of the project will be to analyze the victimizations of the citizens from the survey and combine these results with police data. On the one hand, this will be a very interesting dark field study of the unreported crimes in the cities and, on the other hand, the combination of these different kinds of data may identify new hot spots in crime. Finally, the analyses will be continued in eight other central European cities with the aim to protect the citizens from crime.

References

1. Anselin, L., Cohen, J., Cook, D., Gorr, W., Tita, G.: Spatial Analyses of Crime. Criminal Justice 4, 213–262 (2000)
2. Cohen, L.E., Felson, M.: Social Change and Crime Rate Trends: A Routine Activity Approach. American Sociological Review 44, 588–608 (1979)
3. Cornish, D.B., Clarke, R.V. (eds.): The Reasoning Criminal: Rational Choice Perspectives on Offending. Springer, New York (1986)
4. Brantingham, P.L., Brantingham, P.J.: Mobility, Notoriety, and Crime: A Study of Crime Patterns in Urban Nodal Points. Journal of Environmental Systems 11, 89–99 (1982)
5. Harries, K.: Mapping Crime: Principle and Practice. U.S. Department of Justice, National Institute of Justice, Washington, DC (1999)
6. McIntyre, J.: Public Attitudes toward Crime and Law Enforcement. The Annals of the American Academy of Political and Social Science 374(1), 34–46 (1967)
7. Doran, B.J., Burgess, M.B.: Putting Fear of Crime on the Map. Investigating Perceptions of Crime Using Geographic Information Systems. Springer, New York (2012)
8. Skogan, W.G.: Public Policy and the Fear of Crime in Large American Cities. In: Gardiner, J.A. (ed.) Public Law and Public Policy, pp. 1–18. Praeger, New York (1977)

9. Holt, J.B., Lo, C.P., Hodler, T.W.: Dasymetric Estimation of Population Density and Areal Interpolation of Census Data. Cartography and Geoinformation Science 31(2), 103–121 (2004)
10. Openshaw, S.: The Modifiable Areal Unit Problem. Concepts and Techniques in Modern Geography 38 (1984)
11. Bailey, T.C., Gatrell, A.C.: Interactive Spatial Data Analysis. Longman (1995)
12. Anselin, L.: Local Indicators of Spatial Association-LISA. Geographical Analysis 27(2), 93–115 (1995)
13. Eck, J., Chainey, S.P., Cameron, J., Leitner, M., Wilson, R. (eds.): Mapping Crime: Understanding Hotspots. National Institute of Justice, Washington DC (2005)
14. Levine, N.: CrimeStat 3.0. A Spatial Statistics Program for the Analysis of Crime Incident Locations. Ned Levine & Associates, Houston and U.S. Department of Justice, Washington DC (2004),
 http://www.icpsr.umich.edu/CrimeStat/download.html
15. Getis, A., Ord, J.K.: Local Spatial Statistics: An Overview. In: Longley, P., Batty, M. (eds.) Spatial Analysis: Modelling in a GIS Environment. John Wiley & Sons (1996)
16. Gatrell, A.C., Bailey, T.C., Diggle, P.J., Rowlingson, B.S.: Spatial Point Pattern Analysis and Its Application in Geographical Epidemiology. Transactions of the Institute of British Geographers 21, 256–274 (1996)
17. Borruso, G.: Network Density Estimation: Analysis of Point Patterns over a Network. In: Gervasi, O., Gavrilova, M.L., Kumar, V., Laganá, A., Lee, H.P., Mun, Y., Taniar, D., Tan, C.J.K. (eds.) ICCSA 2005. LNCS, vol. 3482, pp. 126–132. Springer, Heidelberg (2005)
18. Moran, P.A.P.: The Interpretation of Statistical Maps. Journal of the Royal Statistical Society. Series B. Methodological. 10(2), 243–251 (1948)
19. Moran, P.A.P.: Notes on Continuous Stochastic Phenomena. Biometrika 37(1/2), 17–23 (1950)
20. Silverman, B.W., Jones, M.C., Fix, E., Hodges, J.L.: An Important Contribution to Nonparametric Discriminant Analysis and Density Estimation. Commentary on Fix and Hodges (1951). International Statistical Review 57(3), 233–238 (1951)
21. Clark, P.J., Evans, F.C.: Distance to Nearest Neighbor as a Measure of Spatial Relationships in Populations. Ecology 35, 445–453 (1954)
22. King, B.: Step-Wise Clustering Procedures. Journal of the American Statistical Association 62(317), 86–101 (1967)
23. Danese, M., Lazzari, M., Murgante, B.: Kernel Density Estimation Methods for a Geostatistical Approach in Seismic Risk Analysis: The Case Study of Potenza Hilltop Town (Southern Italy). In: Gervasi, O., Murgante, B., Laganà, A., Taniar, D., Mun, Y., Gavrilova, M.L. (eds.) ICCSA 2008, Part I. LNCS, vol. 5072, pp. 415–429. Springer, Heidelberg (2008)

Developing a GIS Based Decision Support System for Resource Allocation in Earthquake Search and Rescue Operation

Abolfazl Rasekh and Ali Reza Vafaeinezhad

Department of GIS/RS, Faculty of Environment and Energy, Science and Research Branch,
Islamic Azad University, Tehran, Iran
{A.Rasekh,A_Vafaei}@srbiau.ac.ir

Abstract. After an extreme earthquake strikes an urban area, the main objective of a Search and rescue operation is to minimize the number of fatalities. Therefore the assignment of rescue resources (Rescue Groups) to operational areas is of great importance to disaster managers. This paper presents a Decision support system based on Geographic Information System (GIS) and queuing theory in order to determine the resource allocation of an operational area. First, a simulation of an earthquake is implemented in GIS environment. Second, a queuing model simulation is executed and the resulting performance measures are evaluated by considering the survival time for entrapped occupants in the area and the desired Disaster management parameters. Additionally, the best assignment of rescue groups to an operational area is determined. Further work will be dedicated to applying the survival rate into the task allocation model.

Keywords: Disaster Management, GIS, Queuing Theory, Resource Allocation, Search and Rescue.

1 Introduction

Major earthquakes result in disasters that cause large buildings to collapse, in consequence of this there will be a great number of casualties, deaths and entrapped occupants. The survival of entrapped occupants relies on the aid of trained rescue teams such as life-detectors and rubble-removers [1].

The main goal of a primary search and rescue period is to minimize the disaster's death toll. Therefore, the performance of search and rescue (SAR) operations during the survival span can be an essential factor [2]. One of the challenges in this period is to determine the number and combination of rescue groups for each operational district. GIS can be used as a fundamental tool for location mapping, positioning resources, damage assessment and decision support systems in disaster management [3], [4]. A framework that allows the examination of GIS utilization in the context of emergency management is known as "comprehensive emergency management", encompassing the temporal factor the comprehensive emergency management is divided into a four phase cycle (Mitigation, Preparedness, Response, Recovery) [5]. In this paper our attention is towards the response phase and Resource assignment to different urban districts with the employment of GIS.

Within the response phase of disaster management after an earthquake, the management of spatially distributed resources is one of the most significant but challenging tasks in disaster response [6]. Such challenges are damage assessment, resource (SAR groups) allocation and task allocation to rescue groups.

Many researchers have dedicated their work to the challenge of task allocation by applying auction heuristics [7], [8]. Vafaeinezhad et al. [9] have gone further by presenting a GIS-based auction heuristic model for task allocation of rescue teams after an earthquake.

Fiedrich et al. [2] presented a dynamic optimization model by using simulated annealing heuristics for task allocation scheduling.

The utilization of GIS as a tool for routing and resource distribution after an earthquake was developed by Chen et al. [6], the objective of the framework is to facilitate the deployment of equipment as a decision support system for disaster response operations.

It's a well known fact that allocation of resources for earthquake response operations is an essential aspect of disaster management; therefore, the focus of this paper will be bounded to defining the number of resources (search and rescue groups) for each district in an urban area. For the first time the mentioned objective will be achieved with the aid of GIS, auction heuristics and queuing theory methods.

2 SAR Queuing System

Operations research (OR) has had a long history of applications in emergency response management [10], one of the problem solving techniques and methods presented by OR is queuing theory. Queuing theory is the mathematical study of congestion and waiting lines. In every search and rescue operation we are dealing with waiting queues and congestions, hence, such a system can be modeled and studied by computer simulation, this is an appropriate option once certain parameters are unknown [11]. But before simulation, the first thing that must be considered is the queuing model. Hence, the server type and its service time should be defined. The SAR system can be viewed as a queuing system in the following sense:

- The search and rescue groups are assumed as the servers and their service rate will be estimated by using vafaeinezhad et al.'s [9] auction based simulation model in a GIS environment.
- The entrapped occupants are customers of the SAR queuing system and their arrival rate is assumed as a constant value which will be presented in detail.

2.1 Assumptions

- It has been assumed that SAR groups have been distributed in the operational districts in the city, shown with points in a GIS environment.

- Having estimated the number of injured occupants in each parcel the total number of injured occupants in a street is assigned to a damage point which is placed in the middle of each street adjacent to the parcels.
- Each Group is composed of three people, whom skill priorities are similar to each other. Together, the three group members are facilitated with a vehicle.
- The three different tasks of securing, searching and rescuing have been considered for task allocation.
- For simplicity we have only considered the major earthquake casualties without a secondary disaster such as the aftershocks or resulting fires.

2.2 Service Rate and Arrival Rate

As it was mentioned before, having executed a simulation of the task allocation model presented by vafaeinezhad et al. [9], the resulting output contains the service time data of the operational area. It should be pointed that the service time is calculated using the sum of two parameters, the time spent by search and rescue groups to reach a damage point and the time spent at the damage point for searching and rescuing the entrapped occupants.

Carrying out the task allocation auctions in a simulation for the three assumed tasks the total time spent by SAR groups at each damage point is evaluated. By applying the statistical Kolmogorov-smirnov test, the nearest statistical distribution for the service time is NORM(1.22e + 003, 167) which is a Normal distribution with the mean of 20.28 hours and a standard deviation of 2.78 hours (Fig. 1).

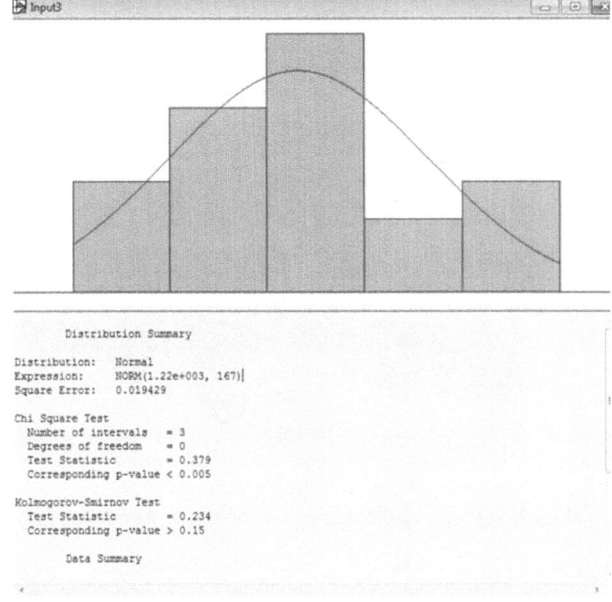

Fig. 1. Histogram of service time distributions and the matching Normal distribution curve

On the other hand if λ is the arrival rate of entrapped occupants (Customers), bearing in mind that we have only considered the major earthquake casualties without a secondary disaster such as the aftershocks or resulting fires, using equation 1, the arrival rate at the time t = 0 is equal to 20.

$$\lambda = f(t) = \begin{cases} 20, & t = 0 \\ 0, & t > 0 \end{cases} \tag{1}$$

2.3 Area of Study and Loss Estimation

The chosen study area is located in the capital city of Iran, more specific the central section of Tehran's municipality district 17, with its center's geographical coordination of 35° 39' 10" and 51° 21' 33" latitudinal and longitudinal respectively. The area of interest has a population of 32,239 people, whom are the occupants of 4843 building parcels. The selected site is located on the ray fault which makes the area one of the most earthquake prone districts. Encompassing a vast number of vulnerable unreinforced masonry buildings, the area has been chosen as our case study. Figure 2, is a 3D view presentation of the study area and the damage points.

The loss estimation of the area has been carried out by Mansouri et al. [12] hence, their proposed model is applied to this study for the simulation of task allocation and furthermore, for the resource allocation.

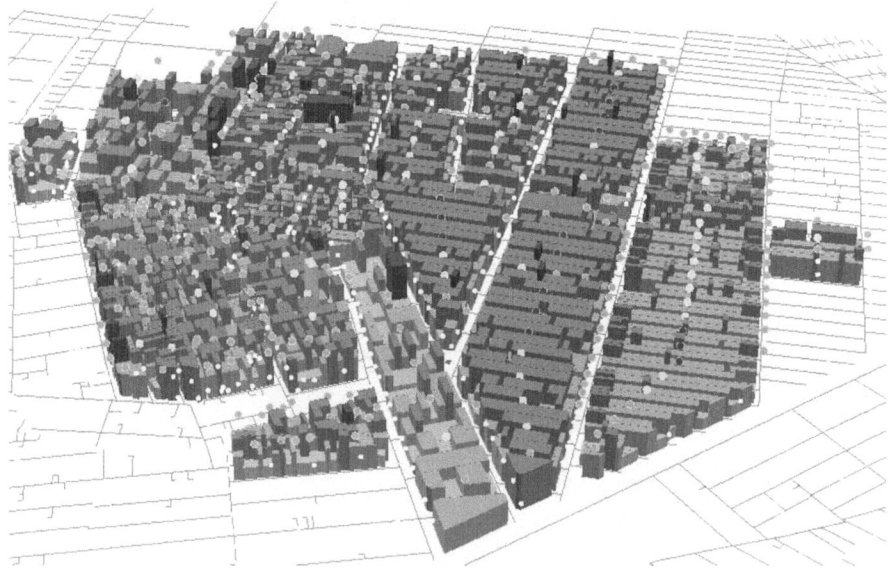

Fig. 2. 3D view of the study area and the damage points

3 Design and Simulation of the SAR Queuing Model $\mu/\lambda/c$

Once the topic has been fully studied and the data collection and analyzing has been fulfilled we can proceed to design and implement a model with a computer simulation [13]. For this purpose we have used the Arena simulation software, the resulting model is presented in figure 3.

- The Casualty Arrival module contains the information regarding to the arrival rate of the casualties (Damage points) into the SAR queuing system. As mentioned earlier in this paper, the arrival rate is equal to the constant number of 20 ($\lambda = 20$).
- The Search and Rescue module is the heart of the SAR queuing system. This module is initialized with the number of SAR groups (c = 18) and simulates the service rate of the system by using the service time distribution of NORM(1.22e + 003, 167) and therefore $\mu = {1}/{\text{NORM}(1.22e + 003, 167)}$.
- The End of Rescue process module records the time that a service to a specific damage point has been completed and leaves the system (damage points exit time).

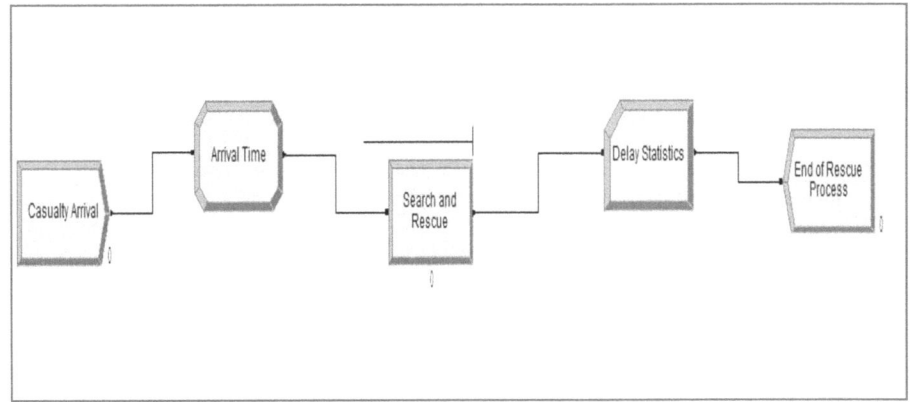

Fig. 3. Simulated SAR process queuing model

3.1 Execution of the Model

Having designed and defined the module parameters for the model, now it's time to run a replication of our simulation. For this purpose the replication's duration must be determined. Since the final task exits the queuing system after approximately 26 hours, therefore, the replication should be equal to this number.

With the replication's duration in place, the simulation is ready to be executed. The resulting performance measures are presented as reports, but different parameters must be examined to reach the desired performance measures.

3.2 Parametric Process Analysis

The term parametric analysis refers to the activities of executing a simulation model multiple times with a different set of input parameters for each run, and then comparing the resultant performance measures; a collection of input parameters and performance measures are called a scenario [13]. To evaluate the performance measures there needs to be a standard value to use as a comparison or as a determination factor. The influential parameters for the SAR group allocations to different operational areas are as follows: Time, Distance, magnitude, population, number of rescuers, building structures and ... [1].

The parameters of distance, magnitude, population and building structures have been considered within vafaeinezhad et al.'s [9] task allocation model. The time parameter must also be taken into account as it is undoubtedly the most important factor in a SAR activity and allocation [1], [2], therefore the focus of this paper will be further towards this factor.

3.3 Survival Rate

After an earthquake occurs the probability of survival fades away in time. In general by referring to the gathered earthquake data from around the world, one can say that if an entrapped person is rescued in the first 24 hours, then they have a survival rate of 80% [1].

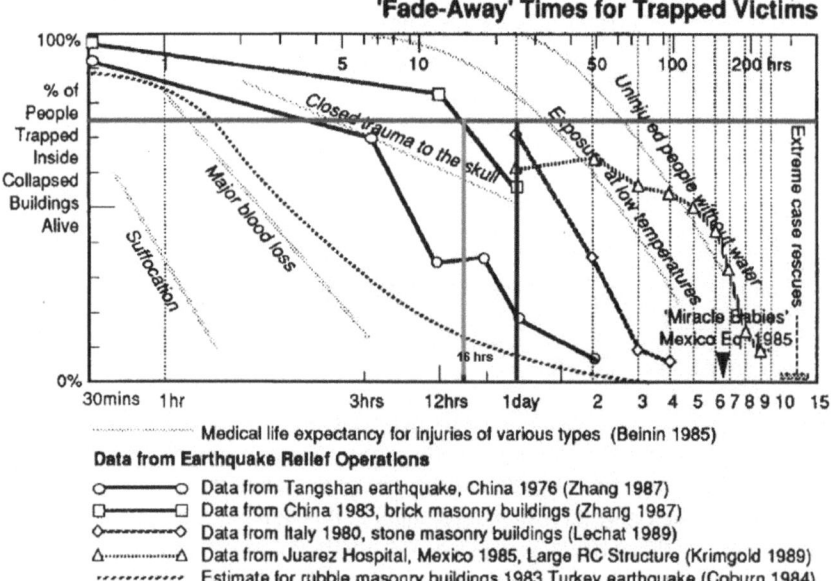

Fig. 4. Survival rate of entrapped victims in collapsed dwellings (taken from [14])

The total life span of a person trapped under a collapsed building depends on many factors such as the amount of air, physical conditions, type of injury and the degree of injury [2], [14], [15]. Since the injury factor is not known before rescuing the casualties, we need to use a probability model that deals with a factor other than injury conditions.

Different construction types such as reinforced concrete, masonry or wooden-frame, have different collapse mechanisms, therefore each structure has its own specific void to volume ratio [14] and As a result each construction type has its own survival rate. Coburn et al. [14] and Ohta et al. [15] have presented a model based on this factor (structure type) (Fig. 4 & 5).

Allocating a defined number of rescue groups to different operational zones, in advance we must determine an estimation of the life span for the entrapped people inside the damaged dwellings. Presented in figures 4 and 5 are the survival rates depending on the building structures in six different earthquakes around the world.

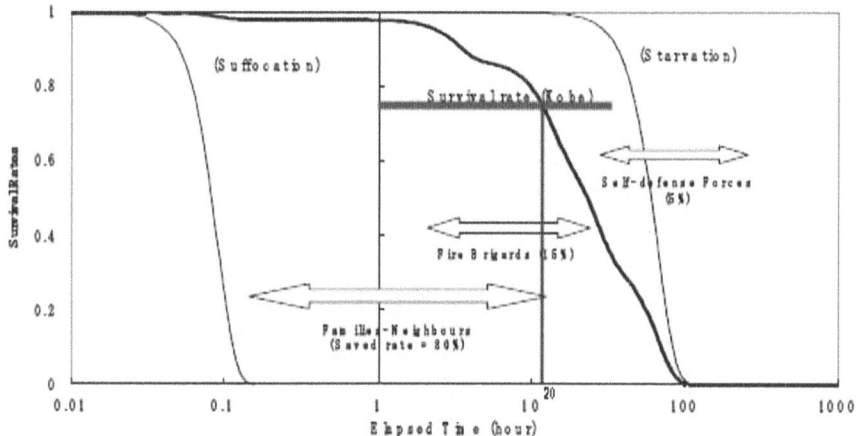

Fig. 5. Survival rate of entrapped victims in collapsed wooden structures (taken from [15])

3.3.1 Implementing the Survival Rate to the Area of Study

The simulated earthquake loss estimation that we have utilized has the following characteristics:

- The time of the earthquake is around 5:30 am.
- Its magnitude is near 6.5 on the Richter scale.
- There are 4 different classes of building structures in the Area of study (Tehran's municipality district 17):

1. Reinforced concrete
2. Brick masonry
3. Wooden structures
4. Stone masonry

These classes must be matched to their corresponding classes in the survival rate model (Fig. 4 & 5). Before the matching procedure a number of matters must be taken into account, these are the structure types, time of occurrence, magnitude and the number of storeys for the majority of the buildings. As a result of this procedure the four corresponding classes are as presented in table 1.

For determining the life span of each structure we have selected the third quartile in order to cover 75% of the survival probability, this value has been indicated with a red line in figures 5 & 6.

Table 1. Corresponding construction types and their survival time

Class number	Construction Types in the Study Area	Matching construction Types in the survival Model	Survival time (Hours)
1	Reinforced concrete	Reinforced Concrete (Mexico 1985)	24
2	Brick masonry	Brick masonry (China 1983)	16
3	Wooden structures	Wooden-Frame structures (Japan 1995)	20
4	Stone masonry	Stone masonry (Italy 1980)	24

Having determined the entrapped occupants survival time for each structure type we must now calculate the survival rate for each street or in other words for each damage point in the GIS environment. For this purpose the following assumptions are presented.

If there are 4 different structure types in each street (damage point), then:

- Let n_1 be the number of buildings with the structure type of **A**, then the survival time within the third quartile for the entrapped occupants of this structure is T_1.
- Let n_2 be the number of buildings with the structure type of **B**, then the survival time within the third quartile for the entrapped occupants of this structure is T_2.
- Let n_3 be the number of buildings with the structure type of **C**, then the survival time within the third quartile for the entrapped occupants of this structure is T_3.
- Let n_4 be the number of buildings with the structure type of **D**, then the survival time within the third quartile for the entrapped occupants of this structure is T_4.
- Let **N** be the total number of buildings in a street (damage point).
- Let T_L be the total life span for each street (damage point).

$$T_L = \sum_{i=1}^{4} \left(\frac{n_i}{N} \times T_i \right) \tag{2}$$

Developing a GIS Based Decision Support System for Resource Allocation   283

By applying equation 2 Within the GIS environment and with the help of VBA script, the life span for each damage point is calculated and added into the damage point attribute table under the Life Span data field (Fig. 6). Prioritizing the task allocation model by using the life span data for each damage point is presented as one of the future works in this area.

For simplicity, we have only considered the damage points with high destruction and within our area of study there are 20 points of this type, therefore the average life span for these points is 21.5 hours. This value is going to be used as a performance evaluation measure of SAR queuing system.

UIDnet	ParcelCount	InjuredCount	DestructionArea	Total_LifeSpan
278	44	34	3	21
279	43	30	3	23
315	36	28	3	24
316	44	33	3	24
402	51	31	3	24
441	46	29	3	24
442	47	32	3	24
445	41	30	3	18
446	42	33	3	16
449	44	35	3	20
789	51	40	3	24
821	17	29	3	17
838	35	28	3	23
932	42	30	3	24
1157	40	32	3	18
1159	40	33	3	20
1161	39	28	3	24
1162	40	30	3	23
1163	40	29	3	22
1252	38	29	3	16

Fig. 6. The Attribute table of Damage points and the Life Span for each point in GIS

4   Scenario Analysis

As it has been mentioned in section 4.2, a number of different scenarios must be created and each scenario has its own input parameter. The two input parameters that we have used for the creation of scenarios are the number of rescue groups and duration of the replication (Fig. 7).

The utilization performance measure represents the percent of time that the rescuers have been busy. In a disaster response operation, this parameter is of high importance and therefore, we have assumed the acceptable value to be more than 80%. On the other hand the SAR operations must be accomplished within the calculated average life span which was 21.5 hours to insure the maximum number of people rescued from the rubbles. Hence, the replication time must be less than 21.5 hours.

By taking a close look at the resulting performance measures for each scenario it can be seen that there are two scenarios within the acceptable utilization factor (scenario 3 & 6). But only one of these two completes within the average survival time, therefore scenario 6 is the most suitable option with the total number of 72 rescuers (18 groups of 4).

S	Scenario Properties		Controls			Responses						
	Name	Program File	Rescue Groups	Num Reps	Rep Length	Casualties Number In	Casualties Number Out	Search and Rescue.Queue.Num	Search and Rescue.Queue.WaitingTim	Rescue Groups.Number	Rescue Groups.Utilization	Rescue Delay
1	Scenario 1	22 : Search and Rescue.p	18.0000	10	24.0000	20.000	16.100	1.228	1.474	15.769	0.876	19.733
2	Scenario 2	22 : Search and Rescue.p	19.0000	10	24.0000	20.000	16.900	0.583	0.700	16.322	0.859	19.755
3	Scenario 3	22 : Search and Rescue.p	20.0000	10	24.0000	20.000	17.800	0.000	0.000	16.758	0.838	19.721
4	Scenario 4	22 : Search and Rescue.p	21.0000	20	24.0000	20.000	17.350	0.000	0.000	16.837	0.802	19.723
5	Scenario 5	1 : Search and Rescue 2.p	18.0000	10	20.0000	20.000	17.700	1.084	1.084	14.455	0.803	15.100
6	Scenario 6	1 : Search and Rescue 2.p	18.0000	20	20.2000	20.000	17.750	1.107	1.118	14.380	0.799	15.190
7	Scenario 7	1 : Search and Rescue 3.p	18.0000	10	18.0000	20.000	18.000	0.952	0.856	13.070	0.726	12.147
8	Scenario 8	1 : Search and Rescue 3.p	17.0000	10	18.0000	20.000	17.000	1.482	1.334	12.794	0.753	12.132

Fig. 7. The Table of Scenarios for evaluation of performance measures

5 Conclusion

The applications of GIS in disaster management have been successful and developed in the past recent years. GIS allows the joint spatial and attribute analysis of data and because of this, they can aid responsible organizations for various stages of (earthquake) disaster management. One of these aspects is the resource management of the rescue groups to different operational districts. Since the number of available resources for the SAR operations is limited the need for an allocation model that considers the road networks together with the information on population in each area, loss and injured estimates is vital.

The presented research has developed a decision support system for earthquake disaster management in a way that first a loss and injury estimation is simulated for the desired operation area in GIS environment, and then a task allocation model was implemented within the GIS framework to estimate the service times needed for SAR operations and finally by applying queuing theory concepts, the appropriate queuing model was designed and simulated. The results of the simulation were carried out by a number of scenarios to determine the best combination and number of rescue groups. Therefore, a survival model was used and implemented into the GIS as an evaluation measure of the scenarios.

6 Future Works

Queuing theory has been widely used for emergency service vehicle system evaluations [11], [16]. In order to develop the presented work in this field the following items are recommended as future works:

- Adapting the presented model to other resource allocation challenges for earthquake disaster management.
- Developing the survival rate model by applying effective stochastic parameters.
- Implementing the survival rate into the task allocation algorithm.

Acknowledgements. The authors wish to thank Professor Abdolrahman Rasekh and Dr. Mehdi Ghatee for their helpful suggestions in the required areas of statistics and queuing theory respectively.

References

1. Metzger, M.D.: Formulating EarthQuake Response Models in Iran. Masters Thesis, MIT (2004)
2. Fiedrich, F., Gehbauer, F., Rickers, U.: Optimized Resource Allocation for Emergency Response after Earthquake Disasters. Safety Science 35, 41–57 (2000)
3. Cutter, S.L.: GI Science, Disasters, and Emergency Management. Transactions in GIS 7(4), 439–445 (2003)
4. Pradhan, A., Laefer, D.F., Rasdorf, W.J.: Infrastructure management information system framework. Journal of Computing in Civil Engineering 21(2), 90–101 (2007)
5. Cova, T.J.: GIS in Emergency Management. In: Longley, P.A., Goodchild, M.F., Maguire, D.J., Rhind, D.W. (eds.) Geographical Information Systems: Principles, Techniques, Applications, and Management 2, pp. 845–858. John Wiley & Sons, New York (1999)
6. Chen, A.Y., Peña-Mora, F., Ouyang, Y.: A Collaborative GIS framework to support equipment distribution for civil engineering disaster response operations. Automation in Construction 20, 637–648 (2011)
7. Gerkey, B.P., Mataric, M.J.: Sold!: Auction Methods for Multirobot Coordination. IEEE Transactions on Robotics and Automation 18, 758–768 (2002)
8. Dias, M.B., Zlot, R., Kalra, N., Stentz, A.: Market-Based Multirobot Coordination: A Survey and Analysis. Proceedings of the IEEE 94, 1257–1270 (2006)
9. Vafaeinezhad, A.R., Alesheikh, A.A., Hamrah, M., Nourjou, R., Shad, R.: Using GIS to Develop an Efficient Spatio-temporal Task Allocation Algorithm to Human Groups in an Entirely Dynamic Environment Case Study: Earthquake Rescue Teams. In: Gervasi, O., Taniar, D., Murgante, B., Laganà, A., Mun, Y., Gavrilova, M.L. (eds.) ICCSA 2009, Part I. LNCS, vol. 5592, pp. 66–78. Springer, Heidelberg (2009)
10. Wright, P.D., Liberatore, M.J., Nydick, R.L.: A Survey of Operations Research Models and Applications in Homeland Security. Interfaces 36(6), 514–529 (2006)
11. Goldberg, J.B.: Operations research models for the Deployment of emergency services vehicles. EMS Management Journal 1(1), 20–39 (2004)
12. Mansouri, B., Hosseini, K.A., Nourjou, R.: Seismic Human Loss Estimation In Tehran Using GIS. In: 14th World Conference on Earthquake Engineering, Beijing (2008)
13. Altiok, T., Melamed, B.: Simulation Modeling & Analysis with Arena. Elsevier, USA (2007)
14. Coburn, A., Spence, R.: Earthquake Protection. John Wiley, London (2002)
15. Ohta, Y., Murakami, H., Watoh, Y., Koyama, M.: A model for Evaluating Life Span Characteristics of Entrapped Occupants By An EarthQuake. In: 13th World Conference on Earthquake Engineering, Vancouver, Canada (2004)
16. Singer, M., Donoso, P.: Assessing an ambulance service with queuing theory. Computers & Operations Research 35, 2549–2560 (2008)

Concepts, Compass and Computation: Models for Directional Part-Whole Relationships

Gaurav Singh, Rolf A. de By, and Ivana Ivánová

University of Twente
Faculty of Geo-Information Science and Earth Observation (ITC)
Department of Geo-Information Processing (GIP)
Hengelosestraat 99, 7500 AE Enschede, The Netherlands
{singh09721,deby,ivanova}@itc.nl
http://www.itc.nl

Abstract. We present a conceptual framework and computational mechanism to allow the interpretation of text phrases such as "in southeastern Bahia," and "in central Goiás" as spatial features in GIS context. The framework recognises different notions of centre, central sector and outer sector, from which a variety of interpretation models is derived.

We subsequently evaluate these models for the performance characteristics of precision and recall, against a digitised, natural history gazetteer for Brazil, and draw conclusions on the cognition of, and computation with directional part-whole relationships.

Keywords: GIS, part-whole relationships, cognition, direction, geographic information retrieval, VGI.

1 Problem Statement and Outlook

1.1 Problem Statement

In the field of geographic information retrieval, users often pose queries wanting to know and retrieve information 'about' or 'within' some portion of a region, described with reference to a compass direction, such as *'northern'*, *'southeastern'* or *'central'*. Although, 'central' does not belong to the category of compass directions, it is often found along with other direction notions in messages from SMS, twitter, blogs and Facebook. Phrases such as *'last month I camped in eastern Paraná'*, or *'hiking in central Mato Grosso do Sul'* well capture the cognitive conceptualisation of direction notions used by humans.

In the past as well, expeditioners used to record in their travel diaries a variety of spatial descriptions involving distance and direction, to describe places with reference to one another. Lesser known places were described with reference to well-known places. For example, in Figure 1 the location of ALAGOINHAS is identified with reference to the well-known place Salvador, indicating distance and direction relations, besides being located in northeastern Bahia.

```
ALAGOINHAS;  Bahia                                    1207/3826 (USBGN)
   158 m (MHA); 93 km N of Salvador [1259/3831 (USBGN)], north-
   eastern Bahia (ICWB); Reiser, "Alagoinhas ?", 28 Feb., Reiser, 1 Mar.
   1903 (Reiser, 1910:57, 80; 1925:107).
```

Fig. 1. Example entry from a natural history gazetteer for Brazil [10]

In this paper, we address a small yet important interpretation problem of text-based geographic message content. Our main objective is to identify interpretation models for English phrases such as 'northeastern Bahia' or 'central Bahia', with the aim to derive for such part-whole relations a best-as-possible, spatial representation. The problem of interpreting location with reference to distance and direction has been addressed in [8,7,15] while identification of cardinal direction relations between geographic entities has been studied in [5,6,12] and identification of internal cardinal directions parts by [9].

We are aware that there has been research in computational geometry to cut up polygons based on different hypotheses [3]. The important body of work comes from [13], who developed algorithms for partitioning polygons into cardinal direction-based sectors. Their work addressed mainly the four cardinal directions for partitioning whereas our work puts forward approaches for cardinal as well as ordinal directions, including the central regions. With this work, we hope to improve substantially on many of the cases, and revive and improve on the fundamental work by Paynter and Traylor [10,11].

Upfront, we need to make the observation that this interpretation problem is heavily parameterised by context: the messenger, the intended audience, and the content field often play a role in interpretation. In addition, when the messenger has topographic knowledge of the region, such as prominent rivers, coastlines, or mountain ranges, this often affects her/his phrasing: 'southern Amazonia', for instance, in Brazil is typically associated with that part of state that is south of the Rio Amazon. This paper does not address the use of such specific topographic knowledge.

By directional part-whole relation we mean a text phrase that combines a cardinal or ordinal compass direction with the name of a region, for which we know the spatial extent, as in the examples given above.

1.2 Results and Outlook

We look at *regions* (such as Bahia) with known extent, and want to derive sensibly an extent for a *region sector* that is textually described (such as "se Bahia"). We develop a conceptual framework based on the independent notions of *centre*, *central sector*, and *border sector*. By centre, we mean a single point in the region that is, by some definition, centric to that region. We will use multiple definitions of 'centre'. Likewise, we will postulate a number of definitions for 'central sector', as well as for 'border sector'. Different (sensible) combinations of ⟨centre/central sector/border sectors⟩ subsequently give us different models of interpretation for directional part-whole phrases. One example model is illustrated on the right, showing the case of Bahia state, as well as actual gazetteer entries from our study data. Every of our models covers all sectors of all states.

Next, we evaluate our models as information retrieval machines against a corpus of data that was derived from the two-book series of ornithological gazetteers for Brazil [10,11]. That evaluation, performed with the information retrieval tools of precision and recall [14], gives rise to conceptual improvements on our models, on issues of sizes of sectors and angles used, leading to a second generation of interpretation models that we also evaluate. The outcome of those evaluations is both surprising and useful.

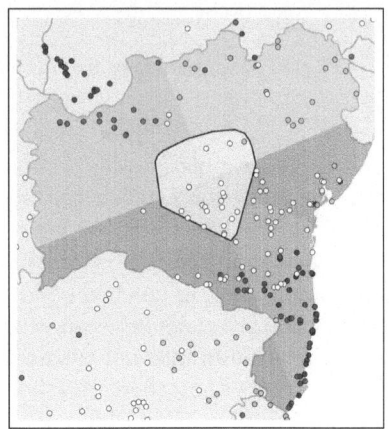

In Section 2, we provide an overview of work on the subject matter before ours, and we discuss the relationships with our paper. The data that we use for validation of these models is presented and discussed in Section 3. Section 4 carves out the conceptual framework, and defines eventually the spatial sectoral models that we use in the paper. The results of our validation work are presented in Section 5. In that same section, we review our conceptual points of departure, and present also an improved conceptual framework, leading to more advanced spatial sectoral models, and that we discuss against a cognitive background as well. Finally, Section 6 identifies open issues that we are currently pursuing.

2 Related Work

A substantial body of work has laid the foundation on direction relationships between spatial features, and much work has been carried out on qualitative reasoning over such relationships. Characteristically, that work started with determining directional relations between point objects, and reasoning over these. Location descriptions are not a novelty of the era of neogeography, but have been with us for a long time. Large amounts of geographic information have been collected historically in the form of textual content, instead of maps or spatial data. An important case is that of specimen labels in collections of biological specimens in museums, and that of travel logs of expeditioners of the past. Descriptions in these collections exhibit a variety of spatial relations between features [15]. In natural language, identification and understanding of spatial relations is important, and is fundamental to building geospatial semantics [2]. Typically, three types of spatial relations are recognised: topological, direction and distance relations.

Early work by [5] deserves reference because it suggested two methods for sector recognition: angle-based and grid-based sectors, used to interpret cardinal directions between two spatial objects. These models were applied mostly for static applications of point object referencing, in which the grid-based approach is cognitively more acceptable, whereas angle-based directions proved to be a better model for movement applications. Subsequently, [6] proposed an envelope approach for representing cardinal direction relations between two non-point objects. In this approach, the partition lines of the envelope of the reference object are extended until they intersect the envelope lines of the other object. This approach clearly gives unequal cardinal direction zones that depend on the shape of the reference objects. Some other challenges of that model were discussed in [12]. The latter approach does not approximate a region to a point.

Various methods and models have been proposed to compute direction from a reference object to a target object. In [15], a point-and-radius method is proposed to georeference locality descriptions. Many factors — like distance, direction, map scale — are taken into consideration while computing the georeference of a locality. The method determines not only the georeference, but also the uncertainty associated with the respective inputs. We plan to publish separately on these associated uncertainties.

The above approaches identify cardinal direction relations between two objects externally, not between a region and its subregions. Recent work has addressed the latter to some extent. In [9], external techniques are used to determine region/subregion compass directions. Three approaches are offered, one of which is similar to the model of [6]. Amongst others, in a cognitive experiment, subjects were asked to assign a direction and level of accuracy to a number of presented points on a map [9], in an attempt to determine the ideal value of ρ, being the scaling factor from region to central sector. Varying over values for ρ from $\frac{1}{12}$ to $\frac{2}{3}$, an value of $\rho = \frac{1}{3}$ was determined as best fitting with the experiment subjects' cognition. Using this value, the other eight sectors were determined.

The problem of determining direction-based sectors was also addressed in [13]. That body of work addressed only the four cardinal directions, and presented both criteria for splitting the original, as well as efficient algorithms for determining the extents of sectors, meeting those criteria. Our work is not on the complexity of the general case algorithmics as that of [13], but is an attempt to address more pragmatically the same problems for a wider range of directions, including the central sector, while evaluating also different models against a body of data. We believe this presents a complimentary view on the problem.

3 Data and Tools Used

Our gazetteer data derives from the two volumes of the Ornithological Gazetteer of Brazil [10,11], which provides over 8,000 entry descriptions in natural language text. These descriptions are about over 3,200 sites visited by expeditioners in the past. The book is a publication in a longer series by the same authors, covering most other Latin American countries, and were published between 1975 and 1991.

The entry descriptions use spatial relations such as distance, topological and direction, relating with other places or features such as rivers, mountains, and forests. In our gazetteer, place name descriptions from 25 states of Brazil are mentioned with a variety of spatial relations as well. The most important spatial relation in our gazetteer descriptions is regional containment. Here, we focus on directional containment cases as described in the earlier sections. In our gazetteer, as many as 25 compass directions are found in containment relation patterns. In these, ordinal directions (38%) occur more commonly than cardinal directions (32%) and various half-directions (30%), like 'central northern' and so forth. We have discarded the third group for the present study, giving a total number of 1304 entries with known location and known regional containment. Of these, 190 entries are tagged as "in central part of state," while the remainder is tagged as being in a border sector of indicated compass direction.

The entries are not evenly spatially distributed, as can be seen in Figure 2(a). Unsurprisingly, there is a rather strong coastal and southeastern bias. This has much to do with historic accessibility and expeditioners' region preferences, with the Brazilian Atlantic Forest being a main target for their early explorations. Notwithstanding this history, it is also apparent that entries tagged with ordinal compass directions (ne, se, sw, ne) outnumber rather sharply those with cardinal directions (n, e, s, w); see Figure 2(b). Not a single cardinal direction outnumbers any of its next neighbour ordinals.

State boundary polygons were obtained from Esri's ArcWorld Supplement Map data, dated 1998. We merged the states of Goiás, Tocantins and the Distrito Federal as a single state 'Pre-1988 Goiás' to reflect the gazetteer's notion of Goiás state.

We stored the gazetteer corpus as well as support spatial data sources in a PostGIS spatial database (postgresql 8.4; postgis 1.4.4). Most of our spatial computations were straightforward applications of spatial functions, with a small number of exceptions. A handful of special plpgsql functions was developed a.o. to construct border sectors with variable angle, ovals with variable size and angle, central sectors of appropriate size even after back-intersection with polygon of origin, and finally a function that determines the *average* distance of all points in a polygon to a fixed, given point [4]. The code of these functions will be made available online.

4 Interpretation Models for Directional Part-Whole Relations

4.1 Models of Interpretation

A *model of interpretation* for directional part-whole (sector-region) relations works on the basis of a known spatial extent for the whole region, and associates with every combination of compass direction and the region's name a representative spatial extent for the indicated sector. The sector's extent is naturally restricted to fall inside the region's extent. For this paper, we recognise two levels of compass direction:

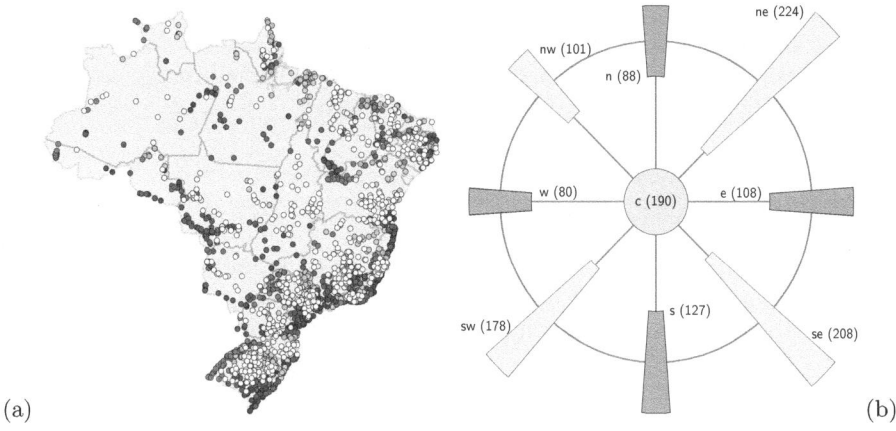

Fig. 2. (**a**): Spatial distribution of entries over the country ($n = 1304$); (**b**): Histogram of distribution of non-central entries over the compass directions ($n = 1114$), and central entries ($n = 190$)

cardinal directions namely: north, east, south and west;
ordinal directions namely: northeast, southeast, southwest and northwest;

In addition to those listed, we add 'central' as a special case of cardinal direction. In part, this has pragmatic reasons, as the term is common in our corpus, but also because models that include 'central' can be generalised to ones that do not. In the sequel, we will abbreviate all compass directions with the well-known 1- or 2-character acronym, written in lowercase, such as in 'se Bahia'.

4.2 Organisational Principles for Models of Interpretation

Different principles apply to models for directional part-whole relations. The distinctions that one can draw directly affect the spatial assignments to the sectors, as well as the reasoning that one can perform over those assignments. We have organised our work using the following principles:

1. every sector falls within its region,
2. different sectors of the same region do not overlap,
3. angle computations take place in a geographic reference system, while length and area computations take place in a (metric) projected reference system,
4. interpretation models consist of choices for centre, central, and border sectors,
5. border sectors in the same direction category have potentially same size,
6. the central sector has preferably same size as border sectors,
7. all sectors together preferably cover the region,

This list partially coincides with that of [13]. In Sections 4.3 to 4.5, we discuss choices of centre, central sector, and border sectors, respectively. We construct our interpretation models by a single choice each for these three variables. The choice for centre determines placement of central and border sectors.

4.3 Models of Centre

By centre of mass, we mean the geometric centre of a region's polygon. For any arbitrary region, we recognise four notions of centre:

region mass centre is the centre of mass of the region,
envelope mass centre is the centre of mass of the region's envelope,
circle centre is the centre of mass of the region's minimum bounding circle,
mass box centre is obtained by first determining the horizontal and vertical $\frac{1}{3}$ and $\frac{2}{3}$ mass cut lines of the region, which give rise to a roughly one-ninth-of-region central sector, of which we determine the centre of mass.

Observe that the four notions, as illustrated in Figure 3(a), are different from each other, and that none of them is guaranteed to fall within the original region.

4.4 Models of Central Sector

Models for the notion of 'central', as in "central Bahia," determine a sector in the middle of the region. There are three important characteristics of central sectors: their *shape*, *placement*, and *size*.

On *shape*, we allow boxes, circles and ovals, pure clones and convex hulls. Other options are certainly possible but they are not in the toolbox for this paper. On *placement*, in principle, we centre the central sector over a choice of centre taken from Section 4.3. On *size*, at first we apply the principle that the central sector occupies $\kappa = \frac{1}{9}$ of total mass of the region, for the simple reason that the central sector is one of nine mutually exclusive sectors, with the eight border sectors being the others. Observe that our κ behaves as $\kappa \approx \rho^2$, with ρ as in [9].

Boxes are simple shapes based on region envelope or region mass distribution. We have applied the κ-principle rudimentarily to them by defining the 'cut lines on thirds'. For region mass, this means we obtain horizontal and vertical cut lines that cut the region into $\frac{1}{3}$ and $\frac{2}{3}$ chunks. The mass box of the region is the central part obtained from those four cut lines. For envelopes, we do the same but using the envelope mass distribution, not the region's. Observe that these two notions are not scaled out of principle of design. The envelope and mass grids are illustrated in Figure 3(a).

The following six models for central sectors are proposed:

Envelope (i.e., bounding box) of the original region is subdivided into nine equally sized and shaped rectangles, by using cut lines at $\frac{1}{3}$ and $\frac{2}{3}$ of envelope height and width. The central rectangle thus obtained is used for intersection with the original region, giving the *envelope central*.
Mass like above, uses cut lines for latitudinal and longitudinal values that define $\frac{1}{3}$ and $\frac{2}{3}$ of the original region's mass, giving like the envelope model, nine rectangles, of which the central one is intersected with the original region, as the *mass central*.

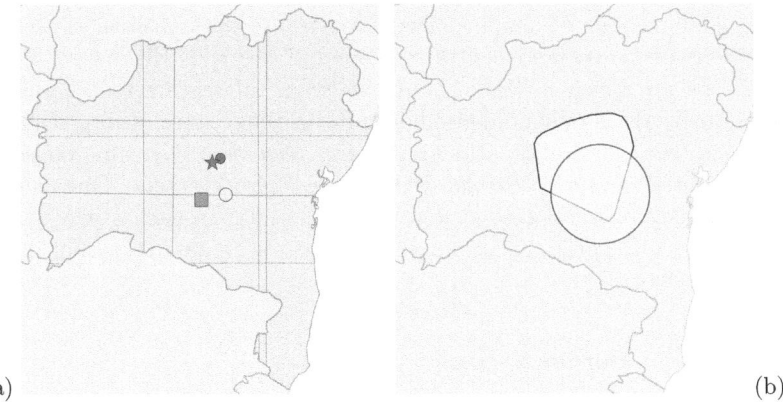

Fig. 3. **(a):** The envelope (blue) and mass grids (brown) for Bahia, with region mass centre (star), envelope mass centre (square), circle centre (big circle), mass box centre (small circle); **(b):** Two notions of central sectors: circle and hull

Circle The circle model determines the minimum bounding circle around the original region, to obtain the *circle central*.
Oval The oval model constructs an ellipse rotated such that it aligns with the dominant direction of the region through the region mass centre. It determines the longest diagonal distance chord (*lddc*) through it, as well as its perpendicular distance chord (*apdc*). The ratio of their lengths determine the ratio of semi-major / semi-minor axes of an ellipse: $\frac{|lddc|}{|apdc|} = \frac{a}{b}$. The constructed ellipse is used as outer boundary of an oval to give us *oval central*.
Clone The clone model clones the original region, to obtain the *clone central*. We include it here because it underlies the hull central model.
Hull The hull model obtains the clone central (see above) and determines its convex hull, giving us the *hull central*.

Envelope and mass models have fixed placement and size. The circle, oval, clone and hull models still require placement and scaling. We scale circle/oval/clone/hull in such a way that after placement their area is exactly nine times smaller than that of the region. Two of these notions are illustrated in Figure 3(b).

Table 1. Placement options for centre/central sector combinations

centre	central acronym	envelope bbox	mass mbox	circle circ	oval oval	clone clon	hull hull
region mass centre	*mass*		☑	☑	☑	☑	☑
envelope mass centre	*envbox*	☑					
circle centre	*circle*			☑			
mass box centre	*massbox*		☑				

Some combinations of shape/placement/size are not sensible, especially on combinations of centre and central sector, and are thus dismissed. Plausible combinations are given in Table 1. As we want all sectors to fall inside the region, the central sector should also be, and with the constraints imposed, not all shape/placement/size combinations meet this condition. Certain combinations cause the central sector to extend beyond the region's extent. Our algorithms ensure that they position and scale the central sector in such a way that after intersection with the original region, a central sector area remains that is exactly κ of original size.

4.5 Models of Border Sectors

Models for the notion of 'northeastern', such as used in the phrase "in northeastern Bahia," determine a region part in the border sectors of the region. As with central sectors, there are three important characteristics of these sectors: their *shape*, *placement*, and *size*.

Fig. 4. Two notions of border sectors for Santa Catarina state. (**a**): mass grid; (**b**): angular sectors placed at region mass centre.

On *shape*, we allow grid-based border sector boundaries that naturally combine with the central boxes described above. We also allow angular sector boundaries, which come naturally with the cardinal and ordinal compass directions. *Placement* is again associated with a choice for centre, where this seems natural. A choice of shape and placement essentially fixates the size of each border sector, being obtained as the intersection of the proposed border sector with the original region. In some cases, this approach leads to sectors without extent, caused by shape of original region. We retain the κ-principle up to where possible. This leads to the following three models for border sectors:

Envelope grid This model applies the same principles as the envelope central model above, and applies cut lines at $\frac{1}{3}$ and $\frac{2}{3}$ of envelope height and width. This gives us the eight *envelope grid border sectors*.

Mass grid As above but applying the principles as for the mass central model, leading to eight sectors, under the name *mass grid border sectors*.

Angle The angle model forms cone-shaped sectors originating from some choice of centre. In our model, all eight directions have identical fan-out angle of 45°, giving us the *angular border sectors*.

The last two are illustrated in Figure 4. Various combination possibilities for centre and border sector options are listed in Table 2.

Table 2. Placement options for centre/border sector combinations

centre	border acronym	envelope grid egrid	mass grid mgrid	angle angle
region mass centre	*mass*		✓	✓
envelope mass centre	*envbox*	✓		✓
circle centre	*circle*			✓
mass box centre	*massbox*		✓	✓

4.6 Complete Models of Interpretation

We construct complete models out of the central sectors and border sectors by simple overlays and intersections, favouring central sector areas over border sectors. By this, we mean that regions of overlap between central and border sectors are assigned to the central sector. In this way, 34 different models were created. These models were applied to all Brazilian states. Two examples are illustrated in Figure 5. We analyse the outcomes in Section 5.

Fig. 5. Two complete models for Santa Catarina state. **(a)**: circle centre/hull/angle; **(b)**: mass centre/oval/angle.

5 Results of Model Comparisons

For all of our original 34 models of interpretation, we computed precision/recall scores [14] for the 1304 entries tagged to be in a regional sector. This resulted in recall scores R ranging between 0.50 and 0.60, and precision scores P between 0.50 and 0.64. The P, R and (balanced mean) F scores found are linearly and similarly distributed (see Figure 6). Below, we say that a model performs poorly if it is found in the lower third of the P/R score range, and that it performs well if found in the upper third of P/R scores.

5.1 Standard Models Compared

In Figure 6, we compare the results of our models on three parameters: choice of centre, of central sector, and of border sectors. Analysing these results renders a few immediate finds.

Following Figure 6, we observe that models using *circle centre* have poor performance, and those with *mass box centres* are at best mediocre. The four model with *envelope box centre* are amongst the best and the worst, while models using *region mass centre* are on average the best, and outperform the others.

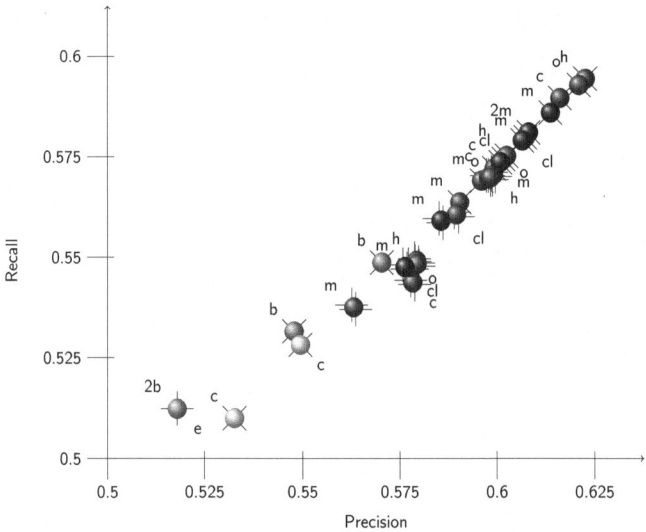

Fig. 6. Comparison of performance as precision/recall of 34 models against gazetteer entries ($n = 1304$). Indicated are choice of centre (red: region mass, green: envelope mass, yellow: circle, blue: mass box), choice of central sector (*e*nvelope, *m*ass, circle, *o*val, *cl*one, *h*ull), and choice of border sectors (+: envelope grid, #: mass grid, ×: angle).

Classifying models by their central sector (Figure 6), we find that *envelope central* performs badly, and *oval central* gives rather mediocre results. In addition, *mass central* displays the whole gamut of scores: one is bad, another is really good, most are somewhere in between. *Hull* but especially also *circle central* outperforms the others. We must remark that by construction, envelope and mass centrals have a size substantially below κ of total region size, and this is a disadvantage against other models of central, all sized at κ.

As choice of border sector, using an *envelope grid* is simply not a good idea, and *mass grids* lead to mediocre results at best. All the well-performing models have *angular border sectors*.

In summary, models making use of region mass centre and having angular sectors appear to be the best choice. A variety of central sectors constitute a viable choice, with hull and circle central sectors as important, primary candidates. In the following three sections, we make a more in-depth study of three parameters that are at the basis of our model formation. These are: position of centre (see Section 5.2), size of central sector (see Section 5.3), and angle sizes for border sectors (see Section 5.4).

5.2 Centres of Variable Position

To study impact of choice of centre, we specifically looked at how that choice affected reading correctly the azimuth of entries from centre. To this end, for all eight directions, and all four notions of centre we developed circular boxplots [1] with 10/25/50/75/90-percentiles for deviation from the main direction. For two of the cases, these are provided in Figure 7.

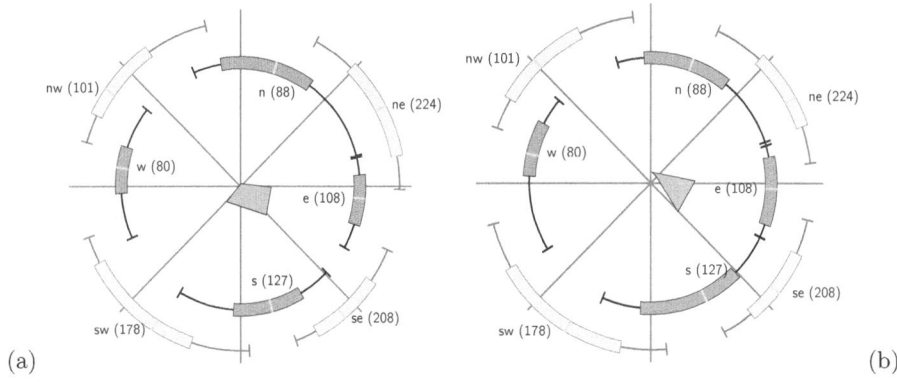

Fig. 7. Azimuth spread for the eight compass directions, indicated as 10/25/50/75/90% boxplots over the compass rose for two notions of centre: **(a)**: region mass centre, **(b)**: envelope mass centre. Cardinal (blue) and ordinal directions (green). In red: area of predicted optimal centre.

We found that the two mass-based centres (region mass centre (a), but also massbox centre) perform very similarly, and so does the pair of bounding box-based centres (envelope mass centre (b), and circle centre). The latter two display areas of overlap between cardinal and ordinal directions, which is an important explanation for their lower model performance. All models have a tendency of 'hanging towards east,' which no doubt is caused by the eastern skew of the data that we found in discussion around Figure 2(b). This is further highlighted in the preferred central position for all four displays of azimuth spread, illustrated as an area in red in the two compass roses of Figure 7.

A final important observation is that, at least for our data set, ordinal (direction) entries are more common than cardinal entries. This has prompted us to consider wider angles around ordinals, at the cost of angles around cardinal directions. We come back to this in Section 6.

5.3 Central Sectors of Variable Size

On average, our eight compass directions have close to 140 entries each; compared to this total, the 190 central entries are an important separate group. We have always believed that our axiomatic $\kappa = \frac{1}{9}$ rule for the central sector was both natural, yet also arbitrary [9]. As discussed above, [9] found $\rho = \frac{1}{3}$, thus $\kappa \approx \frac{1}{9}$, to be optimal. We have therefore studied the impact of varying the size of the central sector on model performance, and have generally found that, with limits, bigger central size improves model performance. The high frequency of central entries, however, complicates a direct study into central sector size optimality.

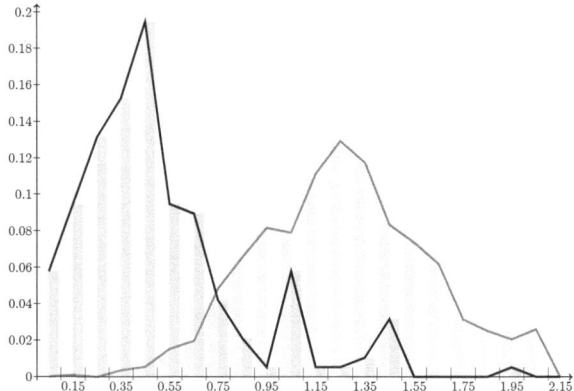

Fig. 8. Frequency of normalised distance-to-centroid for central (blue) and non-central entries (green) in the study

Our well-performing model under name M16 (mass centre/hull/angle), was optimised against the data by varying the value of factor κ for size of central sector. We found that M16 performance gradually increased to optimality at $\kappa = 0.177$, meaning $\rho \approx 0.434$. With that factor the model showed a 5% better precision/recall score, see Figure 9(a). There is obviously no guarantee that against other data sets the same improvement would show.

Seeking for an alternative corroboration, we devised a metric named average_dist2centroid, per region P, defined as

$$\text{average\_dist2centroid}(P) = \frac{1}{\text{area}(P)} \int_{p \in P} \text{dist}(\text{centroid}(P), p) \, \delta p.$$

The metric determines the average distance to centroid over all points within a region P. We use this metric to normalise distances to centroid within different regions, so as to make them comparable [4]. We analysed the distribution of all normalised distances-to-centroid, for both central and non-central entries. From the two histograms (Figure 8), probability density functions were derived by fitting a standard normal distribution with the sample of non-central entries, giving $N[\mu = 1.2628; \sigma = 0.3558; n = 1114]$, and a Rayleigh distribution with the central entries, giving $R[\hat{\sigma} = 0.4234; n = 190]$.

The intersection point of distributions N and R lies at the normalised distance of 0.8470, and marks the distance at which entries can be assigned with equally strong evidence to either the central sector or a non-central sector. It defines an appropriate average distance of the central sector boundary to the region's centroid. Taking the argument a little further, if it would be a circle central sector, the normalised distance value found above translates to κ values around 0.17 for larger states, and towards 0.30 for smaller states in our study. In other words, an optimal choice for κ depends on size of the region.

Independently of these finds, we ran three of our best performing standard models with a range of κ values, varying between 0.10 and 0.40. That model ran against our current data set displayed optimal performance at $\kappa = 0.19$ (giving $P = 0.64$), well above our start position of $\kappa = 0.11$.

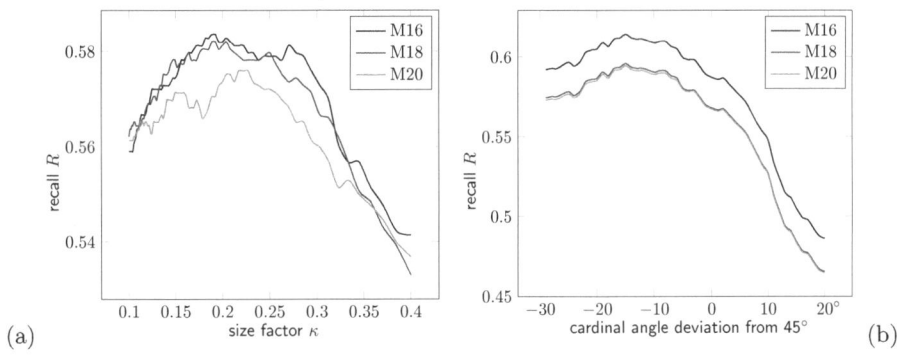

Fig. 9. Optimisation of three models against size of central and angle of cardinals. **(a):** recall for three models with varying values for size factor κ; **(b):** recall with varying values for deviation from the 45° angle for cardinal directions. M16=mass centre/hull/angle, M18=mass centre/oval/angle, M20=mass centre/clone/angle.

5.4 Border Sectors of Variable Angle Size

Given that angle border sectors very clearly outperform other border sectors, we also studied variation in the angle sizes associated with different compass directions. Somewhat naïvely, we expected models to do better with larger angles for cardinal directions, but we actually found the opposite. Given the frequencies of cardinal and ordinal directions provided in Figure 2(b), as well as the measures of angle deviation around the eight compass directions of Figure 7, we found ample empirical reason for considering variable angle size per direction.

The variation we studied is that of making angles for ordinal directions larger at the expense of those for cardinal directions. Angular spread around the mean per direction turned out essentially the same for both groups of directions, and if there was a tendency the data appeared to indicate that cardinal directions display a somewhat bigger angular spread ($\sigma_{\text{card}} = 34°; \sigma_{\text{ord}} = 27°$ when outliers included, and $\sigma_{\text{card}} = 25°; \sigma_{\text{ord}} = 25°$ when outliers are excluded).

Still, our models performed better when angles associated with ordinal border sectors were larger, even to the extent that 60° for ordinals versus 30° for cardinals showed to be the best choice. These finds are illustrated in Figure 9, for three models on our current data set. Our interpretation is that this can be solely attributed to the higher frequencies with which entries were assigned to ordinal sectors.

6 Paper Contribution and Our Current Work

We presented a computational framework to derive for a directional part-whole phrase, and whole with known geometric extent an extent for the part. Thus, we developed a number of models, and validated these against a body of gazetteer entries. Best performing models used a region mass centre (or the closely related mass box centre), and angular model for border sectors. Best choice of central sector was hull, oval, or circle, in that order. Further analysis against the data indicated that our original choice of $\frac{1}{9}$ for size of central sector, matching Liu's [9] $\rho \approx \frac{1}{3}$, compared to overall region is non-optimal, and that choices of $\frac{3}{10}$ to $\frac{4}{10}$ are probably wiser, where the lower range is for the larger regions.

There is certainly a cognition side to this story, even though our corpus originates from just two collaborating authors. Perceptions of 'central sector' are typically larger than just the natural one ninth, and come close to central sectors that stretch out midway to the region border.

Our current work is addressing subordinal directions like 'nnw', as well as cases of 'central nw', 'western central', and so forth. We have no models for these at present. We also are working on extending our findings on optimal angle choices for border sectors. In all of this work, we hope to grow our understanding also on the role of region compactness and region size, and do state-by-state comparisons in our corpus data.

In an upcoming paper, we hope to extend our framework to other than directional part-whole relations, allowing more textual interpretation to geometry, even including measures of spatial uncertainty in this context. This aims to allow for synthetic interpretation models that assign linguistically steered spatial uncertainty to features in this context.

Acknowledgements. Gaurav Singh's research is supported by the EU Erasmus program External Cooperation Window 2009-LOT 15, and by the ITC research fund. Rolf de By's work on this publication was in part supported by the Dutch national program COMMIT. Ivana Ivánová's work was supported by the 'UT stimuleringsfonds'.

References

1. Abuzaid, A.H., Mohamed, I.B., Hussin, A.G.: Boxplot for circular variables. Comp. Stat., 1–12 (2011)
2. Arpinar, I.B., Sheth, A., Ramakrishnan, Usery, E.L., Azami, M., Kwan, M.-P.: Geospatial ontology development and semantic analytics. Trans. in GIS 10, 551–575 (2006)
3. Bose, P., Czyzowicz, J., Kranakis, E., Krizanc, D., Lessard, D.: Near optimal-partitioning of rectangles and prisms. In: Canadian Conf. on Comput. Geom., pp. 162–165 (1999)
4. de By, R.A.: A funny distance computation and... the power of spatial SQL. Technical report, University of Twente, ITC (2012)
5. Frank, A.U.: Qualitative spatial reasoning about distances and directions in geographic space. J. of Visual Lang. & Comput. 3(4), 343–371 (1992)
6. Goyal, R.K., Egenhofer, M.J.: Consistent queries over cardinal directions across different levels of detail. In: Proc. 11th Int. Workshop on Database and Expert Systems Applications, Greenwich, U.K., pp. 876–880 (2000)
7. Guo, Q., Liu, Y., Wieczorek, J.: Georeferencing locality descriptions and computing associated uncertainty using a probabilistic approach. Int. J. of Geog. Info. Sci. 22, 1067–1090 (2008)
8. Liu, Y., Guo, Q., Kelly, M.: A framework of region-based spatial relations for non-overlapping features and its application in object-based image analysis. ISPRS J. of Photogr. and Rem. Sens. 63(4), 461–475 (2008)
9. Liu, Y., Wang, X., Jin, X., Wu, L.: On Internal Cardinal Direction Relations. In: Cohn, A.G., Mark, D.M. (eds.) COSIT 2005. LNCS, vol. 3693, pp. 283–299. Springer, Heidelberg (2005)
10. Paynter Jr., R.A., Traylor Jr., M.A.: Ornithological Gazetteer of Brazil, vol. A–M, 1. Mus. of Comp. Zoology, Cambridge (1991)
11. Paynter Jr., R.A., Traylor Jr., M.A.: Ornithological Gazetteer of Brazil, vol. N–Z, 2. Harvard University, Mus. of Comp. Zoology, Cambridge, Ma (1991)
12. Skiadopoulos, S., Koubarakis, M.: Composing cardinal direction relations. Artif. Intell. 152(2), 143–171 (2004)
13. van Kreveld, M.J., Reinbacher, I.: Good NEWS: Partitioning a simple polygon by compass directions. Int. J. of Comput. Geom. & Appl., (14), 233–259 (2004)
14. van Rijsbergen, C.J.: Information Retrieval, 2nd edn. Butterworth, London (1979)
15. Wieczorek, J., Guo, Q., Hijmans, R.J.: The point-radius method for georeferencing locality descriptions and calculating associated uncertainty. Int. J. of Geog. Info. Sci. 18(8), 745–767 (2004)

SIGHabitar – Business Intelligence Based Approach for the Development of Land Information Systems: The Multipurpose Technical Cadastre of Ouro Preto, Brazil

João Tácio C. Silva, José Francisco V. Rezende, Érika Fidêncio, Tarick Melo, Brayan Neves, and Joubert C. Lima e Tiago G.S. Carneiro

Computer Science Department (DECOM)
Federal University of Ouro Preto (UFOP) – Ouro Preto, MG - Brazil
{joaotacio,erika.fidencio,brayan.ufop,chico10.civil,tarick.melo,
jouberlima,tiagogsc}@gmail.com

Abstract. This paper evaluates the use of Business Intelligence(BI)technologies as a viable approach to the efficient engineering, update and evolution of Multipurpose Technical Cadastres (MTC) in the context of Land Information Systems (LIS). For this, the MTC of a small Brazilian city named Ouro Preto, located in the Minas Gerais state, has been built as a spatiotemporal data warehouse and BI tools have been used to developing the LIS. The system architecture, conceptual data model and tools are presented. Evaluation experiments have shown that, in two months, 92% of 3,037 realties registered in the study area have been correctly georeferenced and associated to legated information systems. We estimate that fifty percent of project investments will be recovered in the first year in which the LIS will be used to collect taxes and tributes.

Resumo. Este artigo avalia o uso de tecnologias de Inteligência de Negócios (BI) como uma abordagem viável para a construção, atualização e evolução eficiente de um Cadastro Técnico Multifnalitário (CTM) no contexto de Sistemas de Informação Territorial (SIT). Para isso, o CTM do município Ouro Preto, MG, foi construído como um armazém de dados espaço-temporal e ferramentas de BI foram utilizadas para o desenvolvimento do SIT. A arquitetura, modelos conceitual de dados e ferramentas desenvolvidas são apresentadas. Análises mostram que, em dois meses, 92% dos 3.037 imóveis registrados na área de estudo foram corretamente georreferenciados e associados a informações de sistemas legados. Estima-se que os investimentos no projeto sejam recuperados após o primeiro ano de uso para cobranças de taxas e impostos.

Keywords: GIS, Land Information Systems, Multipurpose Technical Cadastres, spatiotemporal data warehouse, Business Intelligence.

1 Introduction

There is a long time that free technologies for the development of information systems for land administration are available. However, the study by Getulio Vargas

Foundation in 2007 pointed out that, in Brazil, only 95 of the 3.359 municipalities evaluated (2,82%) could be considered efficient on the tax management. According to Pereira (2002), the investments in technology are quite restricted in most municipalities with small and medium-sized. The absence of a standardized and recognized methodology for the construction and maintenance of a Land Information System is a limitation that discourages a lot of municipalities to invest resources on the cadastral reform, because they have no guarantee of success.

Land Information Systems (LIS) are information systems that promote the orderly growth of a city and the effective and fair distribution of its resources [Fourie and Nino-Fluck, 2000] [Wyatt and Ralphs, 2003] [Loch and Erba, 2007]. Despite the development of LIS be a theme much researched in recent decades, the initiatives reported on the literature that explore *Business Intelligence (BI)* methods, techniques and tools in this niche are few [Ting and Williamson 2000] [Abdel-Rahman et. al 2001] [Ahmed, et. al 2004]. The early works sought to answer issues related with the requirements and with the design of a unified data repository called Multipurpose Technical Cadastre (MTC) [Kaufmann and Steudler 1998] [Williamson 2000]. From this repository, completed and updated analysis would support the process of decision-making. However, due to the city complex dynamics, in which changes are common, the update and evolution of this repository is still the main challenge.

On the other hand, BI researches led to the development of methods and tools to collect, store, analyze and serve data that help organizations, such which municipalities, to make decisions from completed and updated information [Turban and Aronson 1998], [Shmueli et. al 2007]. BI systems support decisions on the levels tactical (a technician who monitors the resources availability), operational (a manager tracking results to planning future actions) and strategic (an executive analyzing performance indicators) of the organizations. Spatial Temporal Data Warehouses *(STDW)* hierarchically organize a massive amount of data. The Extract, Transform and Load (ETL) tools allow the efficient update of STDW from heterogeneous spatial and alphanumeric data sources, spread in several information systems. Changes in these sources or in the STDW data model can be absorbed by changes on ETL flows, making systems flexible. The use of Data Mining (DM) and Online Analytical Processing (OLAP) tools results on the effective customization of reports.

In this scenario, this work evaluates the use of BI methods, techniques and tools as a viable approach for the construction and efficient maintenance of MTCs in the context of LISs. For this, the MTC of the Ouro Preto city, a small town located in the Brazilian Minas Gerais state, was implemented as a STDW and BI technologies were used to develop a LIS focused on geographical, economical and legal aspects. Along two years, this LIS was matured in incremental development cycles. Improved versions of the STDW, of ETL flows and of customized reports were produced at each cycle. They aim to answer administrative issues that evolved in time. Object oriented requirement analysis subsidized the design of the STDW data model. ETL flows were developed to integrate data from several geographic databases with data from the real estate cadastre. Field surveys and tools for data quality control subsidized the continuous improvement of the STDW and ETL flows. Tools integrated with Geographic Information Systems (GIS) were developed to generate custom reports in the form of dynamic and interactive maps. Evaluation experiments were conducted on the region, where the process of land occupation is most intense. This experiments aim to measure the ability of the proposed approach to evolve and keep updated a MTC.

The use of BI for the development of a LIS sounds like an effective approach and also an efficient solution to the continuous update and evolution of a MTC. The integration of the legacy real estate cadastre with the LIS resulted in the cadastre sanitation. Six month experiments showed that 81% of the 3,037 land properties, registered in the study area, were correctly georeferenced and associated to data from other information systems. Moreover, other 1.415 new land properties were identified on field and registered on the LIS. During this project, a team basically formed by students and trainees could maintain the system working and updated. The team quickly absorbed several changes suffered by the data sources, by the MTC data models and by the analysis demanded by the city administrative staff. It is estimated that fifty percent of the investment on this project will be recovered in the first year after its use for charging of fees and taxes. This payback can be explained by the expansion of the real estate cadastre and due to more effective taxation.

The remainder of this paper is organized as follows: Section 2 introduces fundamental concepts and discusses related works. The basic principles that guided the approach proposed in this work are presented in section 3. Section 4 discusses how a LIS can be architected as a BI platform. It also discusses the STDW data models and the update, maintenance and analysis tools developed in this work. In section 5, is described the case study that allowed the evaluation of this approach. The main lessons learned during this work and proposals of future work are presented in section 6.

2 Related Work

The first advances on LISs sought to identify the requirements and to design the data model of a MTC, i. e., to conceive a unified data repository used to storage, recovering, and analysis of spatial and alphanumeric data. In 1998, the International Federation of Surveyors (FIG) published the document *Cadastre 2014* that anticipates trends to the cadastral system of 2014 [Kaufmann and Steudler 1998]. This document highlights the need of (i) precision improvement in the spatial information; (ii) change on the focus of analysis from land parcels for territorial objects; (iii) the use of three-dimensional (3D) and time (4D) to deal with the dynamics and the verticalization of the municipalities; and (iv) services to update and query cadastral information in real-time. The work of Williamson (2000) puts together the best practices used on developing countries. It also highlights that each country should develop specific LIS, due to differences in social, legal, cultural, economical and institutional factors and due to available infrastructure. Considering these initiatives, Van Oosterom et al. (2006) proposed a data model called *Core Cadastral Domain Model* (CCDM), able to meet the common needs of many countries in land administration. To deal with the dynamics of a city, Heo et al. (2006) proposes the use of Geographic Information

Systems (GIS) to analyze changing data sources. According to Bennett et al. (2010) during the last 30 years, the cadastre won multidisciplinary functions, mainly driven by the diffusion of the spatial information technologies and theory of sustainability. However, the authors point out that we still need to concentrate efforts to develop 3D/4D, real-time and global cadastres.

In Brazil, the works of Loch and Erba (2007) and Gonçalves (2008) identify the requirements of a MTC. They also provide a conceptual data model and guidelines to its implementation. Despite of national and international efforts, technical difficulties involved in the use of these technologies have prevented they to become a reality in Brazilian municipalities.

To manage the complexity of rights, restrictions and responsibilities about a territory, Ting and Williamson (2000) proposed the use of GIS integrated with decision support tools organized as BI systems. However, they did not evaluate this approach. According to Šimonová and Komárková (2004), regional data warehouses are able to make planning and urban development more accurate.

3 Basic Principles of This Approach

Despite all the problem domains administered by a LIS, including those related with economical, legal and social issues, a LIS should mainly be able to deal with the dynamics of a city. It should **be prepared for changes and evolve with them,** besides being able to update a **massive volume of data**. Therefore, **flexibility** to absorb changes with a few effort and minimal rework is an indispensable feature. This way, designing an extensible MTC is more important than designing one that accurately represents each problem domain. Data integration and data analysis tools should not depend on the MTC data model. These tools should allow updated analysis based on data summarized into spatial (country, state, city, land property) and temporal (year, month, day) hierarchies.

The development of a LIS is expensive. However, the **sustainability** of the system should be guaranteed. The development involves (i) the hiring and training of a multidisciplinary and specialized team; (ii) the continued implementation of field survey; and (iii) the acquisition of costly spatial data, such as high resolution satellite images, topographic maps, parcel maps and street maps. Decisions about the need, the amount and the level of detail of data should take into account the benefits that will result from its use. The expansion of the real estate cadastre and efficient charging of fees and taxes should pay the LIS development costs in the first years after its implantation. In later years, the earnings will be smaller, but should be enough to finance the system maintenance and evolution.

In Brazil, to be comply with the **national guidelines for E-Government,** projects of development of public Information Systems should (i) prioritize solutions, programs and services based on **free software**; (ii) should adopt open standards for the development of information technologies; (iii) should guarantee the full auditability and security of systems; (iv) should restrict the growth of legacy systems based on proprietary technologies; (v) should guarantee the free distribution of

306   J.T.C. Silva et al.

systems; (vi) should promote the voluntary and collaborative development of systems; and (vii) should improve the capacity of public staff for the use of free software. For these reasons, this approach emphasizes the use of free and open source technologies, which maturity ensures staff productivity and the long-term **continuity** of the project.

In 2009, the Brazilian Ministry of Cities published the 511 ordinance that presents the **national guidelines to create, institute and update a MTC.** These guidelines are adopted in this work. They define the procedures that should be followed by the Brazilian municipalities to evolve the real estate cadastre into a MTC. This ordinance discusses issues related to spatial information needs, e. g., what spatial features should be represented, the way they should be modeled, and the cartographic projections that should be used. The ordinance evidences the need of a multidisciplinary team dedicated to cadastre maintenance. It points out the possibility of the income increments generated by MTC to finance the cadastral reform. It establishes the ideal period between two real estate evaluations. It also presents some practices that aim to improve the distribution of territorial resources, the real estate evaluation process, and the tax collection process.

4   LISs Structured as Business Intelligence Systems

This section details the approach proposed in this work. It describes the LIS development methodology, the LIS software architecture, the MTC data model, and the MTC construction and update processes.

4.1   Evolutionary Development Methodology

The development methodology adopted considers cycles in which conception, elaboration, construction and evaluation activities are performed to produce improved versions of the LIS and of the MTC. In addition to ensuring the quality of these components, the evaluations sought to mature the approach proposed, making it incrementally more reliable and efficient. For this, during six cycles of development, team productivity and data quality measures were collected. In the first year and six months, cycles had duration of six months because of the volume of integration and analysis tools to be developed. In these cycles, only a few samples of filed survey data were collect. After that, three cycles of two months focused on the development of the MTC. In this period, the field survey and database sanitation activities were intensified.

4.2   Software Architecture

Figure 1 illustrates a LIS structured as a BI system. The MTC logical data model is organized as a STDW in which update tools, implemented as ETL flows, integrate data from heterogeneous sources. Legacy Data sources are heterogeneous data sources located different places, e. g., spreadsheets, shapefiles, text files and relational databases. Data collection tools are mobile application to support collecting data during field surveys. ETL flows are scripts to extract data from legacy sources, transform them to a desired format and load them into the MTC. Tools for spatiotemporal data analysis combine techniques of DM, OLAP and GIS. Mobile applications support field surveys.

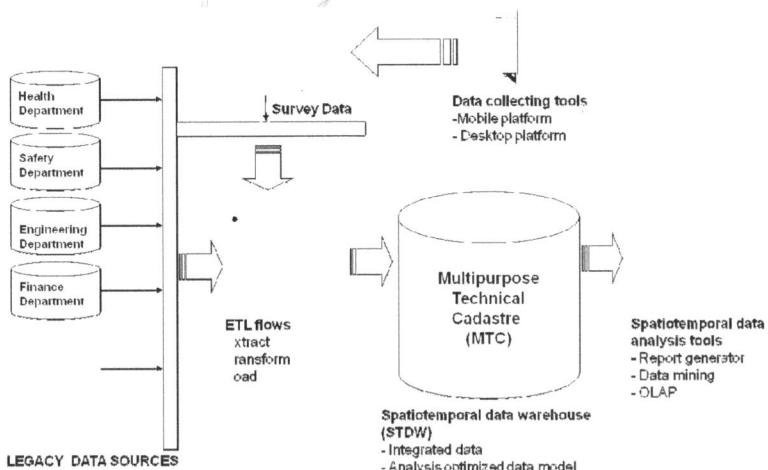

Fig. 1. Software architecture of a Land Information System structured as a Business Intelligence system [Inmon, W. 1996]

4.3 Data Model

Figure 2 shows the MTC conceptual data model. The object oriented paradigm use, which abstractions are close to the application domains, allows the collaborative design of the conceptual model, involving LIS developers and municipal administration experts. Moreover, the inheritance mechanism allows easy model specialization in each domain.

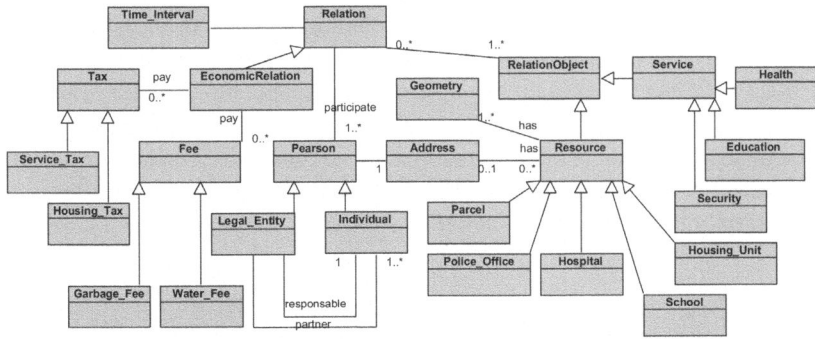

Fig. 2. MTC object oriented conceptual data model

The structure of the conceptual data model determines that one natural or legal *person* can get involved in different temporal *relations*. Each relation receives a label that identifies its validity period. They can have one or more *relation objects*. Relation objects can be resources or services. They can be associated to a unique *address* and can have one or more associated *geometries* to determine their location and shape. Taxes and fees act over *economic relations*. This way, concepts used on land administration, e. g., parcels, land property, schools, and police office can be modeled as resource types.

The STDW development strategy is bottom-up, in other words, from the construction of thematic *data marts* to the construction of complete data warehouse. Figure 3 shows a *data mart* to represent economic relations.

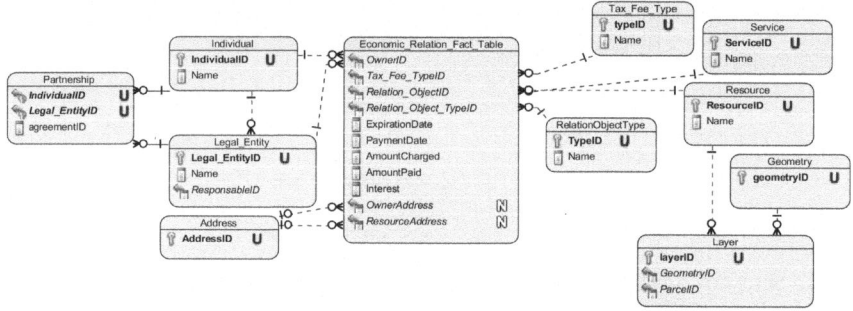

Fig. 3. Entity relationship model of MTC organized as a STDW

Transactional models are designed to meet operational (cadastre) issues of information systems. Multidimensional models are designed to obtain high performance in process queries in which measures summarizes data aggregated by hierarchically organized dimensions. In time dimension, measures can summarize data per day, months and years. In the space, they can summarize data per lot, block, neighborhoods, and districts. It is fundamental to guarantee that the model design allows changes to be performed in the analytical dimensions or in the hierarchies, in which they are organized. Changes should not bring big impacts for the data stored or for the analysis tools. For these reasons, the logical model of the Figure 3 is designed as a STDW extensible and structured according to the *star* scheme. This scheme prioritizes efficient data indexing, in opposite to the greater model normalization offered by the *snowflakes* scheme. In this example, economic relations are modeled as fact table, from which, relations can be analysed by measures (*count, sum, average, max*) about the attributes *AmoutPaid, AmoutCharged,* and *Interest*. The attributes *ExpirationDate, PaymentDate, OwnerAdress,* and *ResourceAdress* are useful for navigation on the measures hierarchically organized into time and space dimensions.

4.4 Development Process

Each development cycle, the process described in Figure 4 can be total or partly executed for the construction/update of the spatiotemporal MTC. Control points acquired by high precision GPS equipments (*Global Position System*) are used to register high resolution images of remote sensors in a determined geographic reference system. Topographic and drainage lines are used to generate the Triangular Irregular Network (TIN) to model the relief. So, the TIN is discretized into a grid of regular cells resulting on the Digital Elevation Model (DEM) that, on the other hand, is used for the orthorectification of images. This process gives accuracy to distance and area calculations performed on images. After that, the radiometric correction of each image is performed to equalize differences in lighting, color and saturation. So, resulting images are putted together to form a unique mosaic that covers the whole city. Then, this mosaic is used as a background on which geometric features of several geographic objects are manually drawn: street axis, lot limits, block limits, etc. These geometric features and the survey data containing the location of resources are incorporated into the MTC by automated ETL flows. The location of a resource is, in general, determined by a geographic coordinate (latitude, longitude) and an address (street, neighborhood, number and complement). Finally, the ETL flows are used to incorporate the data of legacy information systems. They use the resource location as a key to link data from legacy systems to resources registered in the MTC. When the ETL flows fail, the problem is corrected surveying the field. This way, the locations of the resources become known and the data sources are improved.

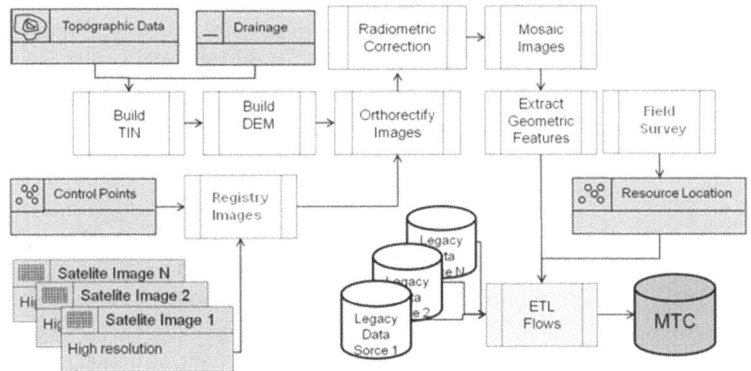

Fig. 4. OMT-G notation of the construction/update process for the spatiotemporal MTC

5   A Case Study – The City of Ouro Preto, MG

For the evaluation of the proposed approach in a real study case, it was adopted to develop the MTC of Ouro Preto, Mina Gerais, Brasil.

5.1 Study Area

Since the MTC update is considered a critical task in a LIS, the region of Ouro Preto, where the occupation process is recent and more intense, was selected as study area.

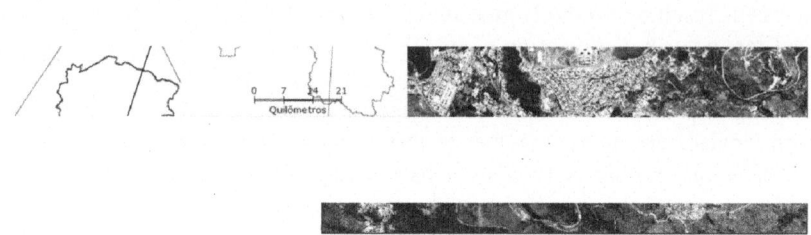

Fig. 5. Study area in the city of Ouro Preto, Minas Gerais, Brazil

5.2 The Land Information System SIGHabitar

The Minister of Cities recommends the use of free tools to MTC development and provides the TerraSIG application for this purpose. Both TerraSIG and TerraView are GISs developed by the C++ TerraLib library use. They are developed by the National Institute for Space Research (INPE). TerraLib allows the development of geographic database over different SGBDs. But, the project TerraView has evolved faster than TerraSIG. The importance of the GIS TerraView in projects of national interests, like the monitoring and modeling of deforestation in the Brazilian Amazon region (DETER[1] e TerraME[2]), is warranty of TerraView continuity and evolution. In this work, it is adopted as the platform for the implementation of the MTC. Tools for data collecting, update, quality control and analysis were developed as TerraView's *plugins*.

5.2.1 The Multipurpose Technical Cadastre of Ouro Preto

The logical data model in Figure 3 was implemented as a TerraLib database over the SGBD Postgre with the PostGIS extension PostGis[3]. Panchromatic images of the *Quickbird* satellite, 60x60 cm, were registered using 77 control points. In the Ouro Preto main district, terrain was modeled from 5 meters equidistant topographic curves. In other regions, it was used 20 meters curves. A large effort was needed in filed surveys and to analyze the cadastral documents provided by the municipality, in order to extract the geometry of the street network, block map, and lot map.

---

[1] http://www.obt.inpe.br/deter/
[2] http://www.terrame.org
[3] http://www.postgis.org

5.2.2 Tools to Collect, Update and Quality Control Data

The ETL flows responsible for data collecting and updating were implemented in the free framework GeoKettle[4]. ETL flows can be edited graphically as data flow diagrams. This fact facilitates their development and maintenance by people with a few experiences in programming. One of the TerraView plug-ins trigger the ETL flows registered in system in programmed periods, or over demand. An *Android*[5] mobile application was developed to support field surveys. Other TerraView plug-in was developed to the improve the quality of the CTM data. These plug-ins are free and open tools. In Figure 6, it is possible to see the screens of the quality control tool (left), which presents resources correctly associated to the data from legacy systems as circles. Rectangles indicate the location of resources that exists in the MTC, but that could not be associated to legacy data. The cross sign indicates the location suggested to a resource that it was only founded in the legacy data source. The collecting tool (right) and of the update tool (center) show the location of the resources found in a field survey and incorporated in the MTC.

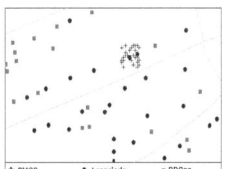  

Fig. 6. Screens of tools to data quality control (left), data update (center) and data collect (right)

5.2.3 Data Analysis Tools

Two TerraView *plugins* were developed to analyze the MTC data. The former is a customizable report generator that presents SQL query results in a form of dynamic maps. The later is a spatiotemporal OLAP tool that is able to generate aggregated measures from spatial operations like union and intersection. Row-up and drill-down operations are available over temporal and spatial hierarchies. Unfortunately, the detailed description of these tools is out of this paper scope.

5.3 Experiments to Evaluate the Proposed Approach

In order to evaluate the proposed approach, three MTC versions were development following the process depicted in Figure 4. Data quality and team productivity measures were collected on each development cycle. The team was composed by a programmer responsible for the construction of ETL flows and for the customization of reports, a building technician responsible for field surveys, and a geoprocessing

---

[4] http://www.spatialytics.org/projects/geokettle/
[5] http://android.com/

technician responsible for spatial data treatment and for the integrated evolution of the MTC. They work twenty hours per week.

During the six months of experiments, the team managed to build and to maintain the MTC and ETL flows updated, quickly absorbing several changes in data sources, by the MTC data model and in the analysis demanded by administrative technicians. Between 3,037 land properties previously registered in study area, 92% were correctly georeferenced and associated to information from legacy systems. In the end of the first development cycle, only 25.6% and, after the second cycle, only 68.5%. Different causes can justify the fail on incorporation of 8% of the real estates. Most of the time, the addresses were not find in the field: (i) The citizens change the number of the address by free will; (ii) The realty does not exist anymore; and (iii) The real estate are closed, there are no numbers posted on the real estate, and the information provided by the neighbors are not enough to determine the location of an address.

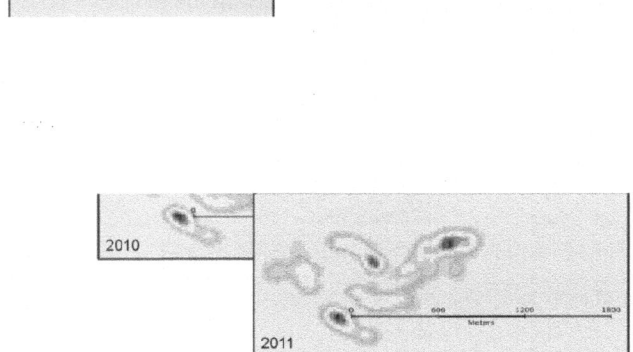

Fig. 7. Evolution of the land tax value (IPTU) spatial pattern during the years of 2009, 2010 and 2011

However, 1.415 new real estate were identified in field surveys, incorporated to the MTC, and after that, associated to legacy system information. Figure 7 shows the spatial evolution of the Brazilian land tax (IPTU) value, during the years of 2009, 2010 and 2011. Maps have been produced using kernel estimators method applied to point data.

6 Lessons Learned and Future Works

The following lessons were learned during this work:

- **Lesson 1 – It is needed a multidisciplinary team:** Having a multidisciplinary team is the main factor for the success of an LIS. In addition to the knowledge of city administration experts, it is necessary specialists in the following areas: building techniques and civil engineering, computation and geoprocessing. During the field works, experts on civil engineering had better performance on the use of maps, on the understanding of the urbanization process from satellite images and on register the geometry of land properties and streets. The development of a geographic database requires experts on geoprocessing. Computation experts should be responsible for the MTC management and for the customization of tools for data collecting, update, and analysis.

- **Lesson 2 – It is necessary the direct involvement of stakeholders:** The effort of decision makers, administrative technicians and legacy systems experts is vital to the evaluation of versions data repositories and tools. They have to ensure the incremental evolution of the LIS. The knowledge they have about the city territory and history is essential to the data sanitation.
- **Lesson 3 – It is possible to develop spatiotemporal BI systems with free tools. However, staff productivity is affected:** There are excellent GIS integrated frameworks for the development of BI systems. They provide ETL, DM and OLAP tools for dealing with spatiotemporal data GeoKettle[6], Talend [7] and Cloveretl[8]. However, the deficiency or absence of training and technical support services limits the performance of developers. The lack of exemplified documentation and success cases to be copied may also prejudice the team performance.
- **Lesson 4 – Must exist a balance between the CTM temporal and spatial accuracy:** Keeping updated a MTC, based on the parcel concept, becomes expensive and exhausting the task due to the accelerated city dynamics. Parcels are contiguous parts of land, each one with single legal regime. They are represented by a set of disjoint polygons that covers the whole territory [Ministério das Cidades, 2009]. However, several types of resources can be represented by 2D points. This is a geometrical representation that can be easily updated and that allows the mapping of vertical constructions. It also allows the use of space analysis methods, e. g., kernel estimators, kriging, spatial clustering detection, etc. The LIS managers have to consider the effort involved on data update and the financial return taken by the chosen spatial representation. The spatial and temporal accuracy have to be balanced.
- **Lesson 5 – Mobile tools to collect the data bring productivity gain:** During the field surveys, the use of mobile platforms with GPS and broadband data connection allows navigation assisted by digital maps. It also reduces the mistakes induced by manual data collecting process. Manual processes require much more efforts to achieve same quality results.
- **Lesson 6: Approaching LIS as BI systems is a viable strategy:** The use of technologies of Business Intelligence satisfies the expectations of the research. It allows the analysis of a massive volume of data. The developed LIS is in accordance with the national policies for e-government, with the

---

[6] http://www.spatialytics.org/projects/geokettle/
[7] http://www.talend.com
[8] http://www.cloveretl.com

ordinance 511 of the Brazilian Minister of Cities, and with guidelines of the project *Cadastre 2014*. The decision makers reported a significant advantage on know indicators summarized by different temporal and spatial hierarchies. The LIS continuity is guaranteed by the maturity of the BI frameworks that provide services for customizing tools through the use of diagrams and graphical interfaces. The use of ETL flows showed to be a flexible and efficient strategy for updating the MTC. It is estimated that due to the real estate cadastre expansion and due to a more effective taxation, fifty per cent of the investments on this project will be recovered on the first year after its use for collecting fees and taxes. This demonstrates the sustainability of the proposed approach. However, from a certain point, the effort on field surveys will not result on the MTC evolution until the reorganization of the geographical space (set land property limits, official naming, standardize numbering, etc) and the registration of real estate take place. A residual number of land properties will not be correctly registered and the effort to register them will not be profitable.

The future works will aim to expand the use the proposed approach and to evaluate it on city administration domains beyond of the cadastral and taxation management. We will keep the development of a TerraLib based framework for STDW implementation and maintenance. Methods and tools for the computation of multidimensional cubes and for the mining of spatiotemporal data will be the focus of future researches.

Acknowledgments. Authors thank the Pos-Graduate Program in Computer Science and the TerraLAB modeling and simulation laboratory of the Federal University of Ouro Preto (UFOP), Brazil. This work was supported by Ouro Preto, Minas Gerais, municipality.

References

Vargas, F.G.: Estudo Comprova Eficiência Fiscal dos Municípios do PNAFM (2007), http://www.ucp.fazenda.gov.br/PNAFM/ (accessed November 2011)

Pereira, N.E.C.: Repensando o valor do cadastro técnico urbano. In: Congresso Brasileiro de Cadastro Técnico Multifinalitário, Florianópolis, Anais, vol. 5 (2002)

Abdel-Rahman, M.S., El Bahy, M.M., Malone, J.B., Thompson, R.A., El Bahy, N.M.: Geographic information systems as a tool for control program management for schistosomiasis in Egypt. Acta Tropica 79, 49–57 (2001)

Ahmed, I., Azhar, S., Lukauskis, P.: Development of a decision support system using data warehousing to assist builders/developers in site selection. Automation in Construction 13(4), 525–542 (2004)

Fourie, C., Nino-Fluck, O.: Cadastre and Land Information Systems for Decision Makers in the Developing World. Geomatica 54(3), 335–342 (2000)

Wyatt, P., Ralphs, M.: GIS in Land and Property Management. Great Britain, p. 240. Spoon Press (2003) ISBN 0-415-24065-4

Loch, C., Erba, D.A.: Cadastro Técnico Multifinalitário - Rutal e Urbano, p. 142. Lincoln Institute of Land Policy, Cambridge (2007)

Ting, L., Williamson, I.P.: Spatial Data Infrastructures and Good Governance: Frameworks for Land Administration Reform to Support Sustainable Development. In: Proceedings of the 4th Global Spatial Data Infrastructure Conference, Cape Town, South Africa, pp. 13–15 (2000)

Kaufmann, J., Steudler, D.: Cadastre 2014, A vision for a future cadastral system, FIG (1998), http://www.swisstopo.ch/Wg-wg71/cad2014.html (accessed October 2001)

Williamson, I.P.: Best Practices for Land Administration Systems in Developing Countries. In: International Conference on Land Policy Reform, Jakarta, July 25-27 (2000), http://www.landpolicy.org/conference/index.html (accessed July 2011)

Van Oosterom, P.J.M., Lemmen, C.H.J., Ingvarsson, T., van der Molen, P., Ploeger, H., Quak, W., Stoter, J., Zevenbergen, J.: The core cadastral domain model. Computers, Environment and Urban Systems 30, 627–660 (2006)

das Cidades, M.: Portaria 511: Diretrizes para a Criação, Instituição e Atualização do Cadastro Territorial Multifinalitário (CTM) nos Municípios Brasileiros (2009), http://www.cidades.gov.br (accessed August 2011) ISSN 1676-2339

Heo, J., Hyun Kim, J., Kang, S.: Temporal Land Information System (TLIS) for Dynamically Changing Cadastral Data. In: Gavrilova, M.L., Gervasi, O., Kumar, V., Tan, C.J.K., Taniar, D., Laganá, A., Mun, Y., Choo, H. (eds.) ICCSA 2006. LNCS, vol. 3981, pp. 1066–1073. Springer, Heidelberg (2006)

Bennett, R., Rajabifard, A., Kalantari, M., Wallace, J., Williamson, I.: Cadastral Futures: Building a New Vision for the Nature and Role of Cadastres. In: FIG Congress 2010: Facing the Challenges – Building the Capacity, Sydney, Australia, April 11-16 (2010)

Turban, E., Aronson, J.E.: Decision Support Systems and Intelligent Systems. Prentice Hall, Upper Saddle River (1998)

Shmueli, G., Patel, N.R., Bruce, P.: Data Mining for Business Intelligence: Concepts, Techniques and Applications in Microsoft Office Ecel with XLMiner. Wiley-Interscience (2007)

Gonçalves, R.: Modelagem Conceitual de Bancos de Dados Geográficos para Cadastro Técnico Multifinalitário. Tese de Mestrado. Universidade Federal de Viçosa (2008)

Šimonová, S., Komárková, J.: Data and Data Warehousing for Management of Small Region– from the Czech republic Point of View. WSEAS Transactions on Information Science and Applications 1(1), s.184–s.188 (2004) ISSN 1790-0832

Inmon, W.: Building the Data Warehouse, 2nd edn. J. Wiley & Sons, New York (1996)

Rehabilitation and Reconstruction of Asphalts Pavement Decision Making Based on Rough Set Theory

Shaaban M. Shaaban and Hossam A. Nabwey

Department of Engineering Basic Science, Faculty of
Menofia University, Shebin Elkom, Egypt

Abstract. Every year a great amount of money is expended for the rehabilitation and reconstruction of roads and pavements in most countries. One of the most usual methods in evaluating pavements distresses is to determine the PCI of the Pavement Condition Index. As a result of large number of variables and complicated decision - making algorithm using the information obtained in this method, may have some difficulties. Presenting an analytic - theoretical method mixed with the PCI method may be the bases for the development of a theoretical empirical method in evaluation of concrete pavements distresses & can remove the difficulties. This paper presents a new approach to the rough-set theory in a Pavement Management System PMS database that enables pavement engineers to discover the shortest subsets of condition attributes having quality equal to the general quality of defined characteristics in the information system, to assess and describe pavement conditions, and to derive decision rules for rehabilitation and reconstruction of the pavements. To evaluate the results, the best algorithm of defined attributes in the information system is determined by making use of Artificial neural network (ANN) method and the result is compared with rough-set ones. The results of the research indicate that the rough-set theory has a better and stronger operational capability in identifying the effective parameters for the severity evaluation of typical distresses in pavements and in decision-making for selecting the type of repair.

Keywords: Pavement Management System (PMS), Rough-set theory, Pavement distress, Decision making, Rough sets, Pavement Condition Index (PCI), Artificial neural network (ANN).

1 Introduction

Pavement distresses are classified into two different categories. The first is known as the functional failure. In this case, the pavement does not carry out its Intended function without either causing discomfort to Passengers, or high stresses to vehicles. The second Known as the structural failure includes a collapse of Pavement structure or the breakdown of one or more Components of the pavement with such magnitude that the pavement becomes incapable of sustaining the loads Imposed upon its surface [1]. In some cases one type of failure may be accompanied by the other type but mostly there is only one type of failure.

One of the most appropriate methods for evaluating pavement distresses is the Pavement Condition Index (PCI) procedure. This procedure has been recommended by the U. S. Corps of Engineers and is considered as a standard method in many organizations all over the world [5]. The PCI is in fact a number between 0 and 100 that shows pavement conditions from poor to excellent. The PCI number is calculated by evaluating several segments of a pavement to determine the severity of distress. The information gained in this procedure can provide a complete reorganization of the main causes of failure and their relation to traffic load, climatic conditions or other effective factors [6].

After evaluating the distresses by the PCI method, it is important to make an economical decision about the rehabilitation or repair of the pavement. In most cases, it is difficult to select which type of repair is suitable for the pavement, because the type and severity of distress may be different in each segment of the road and cannot be repaired using one procedure. On the other hand, a general PCI number cannot provide sufficient information to make the ideal decision for different parts of a pavement [7]. Thus, development of an adequate algorithm to reduce the number of variables effective on decision making with the PCI evaluation results can be very helpful.

Many approaches have been proposed for extracting hidden relationships holding among pieces of information stored in a given database [2]. Rough set (RS) theory is first proposed by Z.Pawlak [3] in 1982. It is a kind of very useful mathematical tool to deal with vagueness and uncertainty information. Recently, this theory attracts more attentions in the fields of data mining, knowledge discovery in database (KDD), pattern recognition, decision support systems (DSS) etc. The main idea of this theory is that it provides us with a kind of mechanism of extracting the classification rules by knowledge reduction, while keeping the satisfactory capacity of classification. There are many successful applications by using RS theory in the following areas such as machine learning, data mining, knowledge discovery, decision analysis, and knowledge acquisition discover knowledge and rules from database.

In this paper, a new simplified method for applying rough set theory, to determine the severity levels of distresses in high way pavements with experimental knowledge the. This method deals with the application of the rough set theory in decision making and takes into account all relevant factors for the judgment procedure.

The proposed method based on the symbolic value partition technique which divides each attribute domain of a data table into a family of disjoint subsets, and constructs a new data table with less attributes and smaller attribute domains.

2 Preliminaries

2.1 Decision Tables

Data are often presented as a table, columns of which are labelled by attributes, rows by objects of interest and entries of the table are attribute values. This paper only concerns decision tables with only one decision attribute. Formally, a decision table is a triple $S = (U, C, \{d\})$ where $d \notin C$ is the decision attribute and elements of C are called conditional attributes or conditions for briefness. Table 1 lists a decision table where U = {S_1, S_2.. S_{27}}, C = {a, b, d, e, f, g, h, i, j, k, L} and d = severity level. It will be employed throughout the paper.

Table 1. Decision Table

Segments	Conditional Attributes											Decision Levels
	a	b	c	d	e	f	g	h	i	j	k	L	
S_1	2	1	4	2	3	2	1	1	3	2	1	4	M
S_2	3	3	3	3	2	1	2	2	4	3	2	1	M
S_3	2	2	2	4	4	4	4	2	3	2	2	4	L
S_4	4	4	1	4	3	3	2	2	3	3	3	4	L
S_5	1	4	3	4	3	2	2	1	1	1	2	3	M
S_6	3	2	2	1	1	1	2	1	2	3	2	3	H
S_7	1	4	4	3	2	4	3	1	2	4	3	4	L
S_8	4	4	4	3	3	4	3	2	4	4	2	3	L
S_9	3	1	3	2	1	2	1	2	1	2	1	3	H
S_{10}	4	3	2	3	2	1	2	1	1	3	3	2	M
S_{11}	3	3	1	2	2	1	1	1	1	2	3	1	H
S_{12}	2	3	4	2	3	2	1	1	2	2	2	2	M
S_{13}	3	2	3	4	4	4	3	1	4	3	2	2	L
S_{14}	4	3	2	2	3	4	3	2	2	1	3	3	M
S_{15}	4	3	1	2	1	1	3	1	2	1	2	2	H
S_{16}	4	4	4	3	2	4	3	2	3	3	2	4	L
S_{17}	3	3	4	2	3	1	3	1	3	3	1	4	M
S_{18}	3	1	3	1	1	2	3	2	1	1	3	2	H
S_{19}	2	1	2	3	2	4	2	1	1	1	2	1	H
S_{20}	1	2	1	3	4	3	3	2	4	3	2	2	M
S_{21}	1	2	1	2	3	4	2	2	3	3	1	4	M
S_{22}	2	2	1	2	3	1	2	2	1	1	1	1	H
S_{23}	3	3	4	2	4	4	3	1	4	4	1	3	L
S_{24}	4	3	3	2	2	2	4	1	2	3	2	3	M
S_{25}	3	1	1	2	2	4	1	1	1	2	3	1	H
S_{26}	4	3	2	2	3	4	3	2	2	1	3	3	M
S_{27}	2	3	4	3	2	3	3	2	1	4	3	4	M

2.2 M-Relative Reduct

The number of relative reducts of a decision table may be large [4-5], and there may exist many minimal reducts of a decision table. Bazan presented the concept of dynamic reduct [6] to obtain stable reducts.

Definition 1.
Given a decision table $S = (U, C, \{d\})$ and a set of specified attributes $M \subseteq C$, any $B \subseteq C$ is called an M-relative reduct of S iff:

(1) $M \subseteq B$;
(2) $POS_B(\{d\}) = POS_C(\{d\})$;
(3) $\forall a \in (B - M)$, $POS_{B-\{a\}}(\{d\}) \subset POS_C(\{d\})$.

The set of all M-relative reducts of S is denoted by Red(S,M).

2.3 Partitions

Let $S = (U, C, \{d\})$ be a decision table where $C = \{a_i : U \rightarrow V_{ai}\}$ for $i \in \{1 \ldots |C|\}$. Any function $P_i = P_{ai} : V_{ai} \rightarrow W_{ai} \cup V_{ai}$ where $W_{ai} \cap V_{ai} = \phi$, $P_i(v) \in W_{ai}$ or $P_i(v) = v$ is called a partition of V_{ai}. The function P_i defines a new partition attribute $a_i^{P_i} = P_i \circ a_i$, i.e., $a_i^{P_i}(u) = P_i(a_i(u))$ for any object $u \in U$. The range of $a_i^{P_i}$ is $V_{ai}^{P_i} = \bigcup_{u \in U} \{a_i^{P_i}(u)\}$. Often P_i is also expressed by a set of value pairs, i.e., $P_i(v_1) = v_2 \Leftrightarrow (v_1, v_2) \in P_i$. Any array of partition $P = \left[P_1, \ldots, P_{|C|}\right]$ is called a partition scheme of S. P defines from S a new decision table $SP = (U, CP, \{d\})$ where $C^P = \left[a_1^{P_1}, \ldots, a_{|C|}^{P_{|C|}}\right]$. The rank of S is the value $\sum_{i=1}^{|C|}|V_{ai}|$. The rank of P_{ai} is the value $rank(P_{ai}) = \left|V_{a_i}^{P_i}\right|$.

Similar to the definition of a reduct [7], we propose two concepts as follows. A partition scheme P is consistent iff: $POS^P(\{d\}) = POS_C(\{d\})$. A decision table S is unpartitionable iff there does not exist a partition scheme P such that P is consistent and rank $(S^P) <$ rank(S).

2.4 Scaling

In some theories, especially Formal Context Analysis [5], there is a need to transform a many-valued attribute into a number of binary valued attributes. This process is called scaling. The scaling process has no influence on the indiscernibility relations or positive regions of attribute sets. Here we require that the decision attribute not to be changed in the scaling process.

Definition 2.

Given a decision table S = (U, C, {d}), the set of scaled attributes of Q $\subseteq$ C is: QB = {(a, v)|a $\in$ Q, v $\in$ Va} (1)

Where (a, v):U $\rightarrow$ {0, 1} and $(a,v)(u) = \begin{cases} 1 & \text{if } a(u) = v \\ 0 & \text{otherwise} \end{cases}$ (2)

Table 2. The class numbers of conditional attributes

Conditional Attributes	Classification of Individual Situations
(a) faulting	1- the difference in evaluation in two sides of the joint is more than 19mm 2- the difference in evaluation in two sides of the joint between 10mm to 19mm 3- the difference in evaluation in two sides of the joint between 3mm to 10mm 4- the difference in evaluation in two sides of the joint is less than 3mm or no difference
(b) corner spalling	1- spalling depth is more than 51mm 2- spalling depth is between 25mm to51mm 3- spalling depth is less than 25mm 4- none
(c) joint steal damage of tramsverse joints	1- joint sealant is in poor condition over the entire surveyed selection with one or more types of damage occurring to a sever degree 2- joint sealant is in fair condition over the entire surveyed selection with one or more types a1 damage to a moderate degree 3- joint sealant is in good condition throughout the selection with only a minor amount of any of the above damage 4- none
(d) Lane/ Shoulder drop-off or Heave	1- the difference in evaluation between the traffic lane and the shoulder is more than 102mm 2- the difference in evaluation between 51mm to 102mm 3- the difference in evaluation between 25mm to 51mm 4- the difference in evaluation is less than 25mm
(e) popouts 1	1- more than 15 percent of pavement area 2- less than 15 percent of pavement area 3- very few area 4- none

Table 2. (*Continued*)

(f) Blow up	1- blow up causes excessive bounce of the vehicles which creates substantial discomfort 2- blow up causes a significant bounce of the vehicles which creates some discomfort 3- blow up has occurred but only causes some discomfort 4- none
(g) Corner Break	1- Crack is spalled at high severity 2- Crack is working and spalled at medium severity 3- Crack is tight (hair line) and is not spalled 4- none
(h) Polishing of aggregates	1- Noticeable 2- very few or no polishing
(i) Edge punchout	1- high severity 2-medium severity 3- low severity 4- none
(j) Linear cracking	1- high severity 2-medium severity 3- low severity 4- none
(k) Durability("D") cracking	1- in more than 15 percent of slab area, high severity level of joints / cracks exists, so that most pieces can be separated from the slab 2- in more than 15 percent of slab area, sever level of joints / cracks exists, but pieces is not separable 3- in less than 15 percent of slab area, most of the cracking are sealed and can not separated 4- none
(l)Slab dividation	1- the number of pieces is high 2- the number of pieces is medium 3- the number of pieces is low 4- none

3 Determination of Minimal Decision Algorithm

Firstly, for evaluating of pavements distresses, using rough set theory, 27 different cases of information obtained from parts of a reinforced concrete highway pavement, were used in the study. The concrete pavement section belonging to National Road 7, situated in the Andes Range and close to the international frontier between Argentina and Chile. This paper deals with preliminary results obtained after a pavement distress survey conducted during year 2001 for that section (World Bank and PIARC, 2002).]. These information was performed by experts in accordance with the diagnostic method prepared by American Army Corps of Engineering in 1984.

The collected data categorized as a table called decision table in which each row reflects the specifications of a particular segment. Each column of the mentioned table indicates one of the characteristics considered (conditional attributes) for the segment as (a) faulting; (b) corner spalling and (c) joint steal damage of tramsverse joints. The last column shows the suitability of that segment for severity level determination (decision attribute).Table 1 shows the decision table.

Secondly, to utilize the rough-set theory and analyze the information placed in the mentioned table, it is required to classify the information. So, each conditional attribute is provided with 4 or 2 classes, which show high, medium, low and no severity and the decision parameter (attribute) is classified by three levels, which describe high, medium and low suitability conditions, H; M and L. Consequently, by defining the specified levels and allocating a digit code to each defined attribute in the rows of the table, the classification of all attributes has been undertaken. Table 2 shows the class numbers of conditional attributes and danger levels for 27 segments diagnosed by experts.

Table 3. Simplified Decision table S

Segments	Conditional Attributes			Decision Levels
	a	c	e	
S_1	2	4	3	M
S_2	3	3	2	M
S_3	2	2	4	L
S_4	4	1	3	L
S_5	1	3	3	M
S_6	3	2	1	H
S_7	1	4	2	L
S_8	4	4	3	L
S_9	3	3	1	H
S_{10}	4	2	2	M
S_{11}	3	1	2	H
S_{13}	3	3	4	L
S_{14}	4	2	3	M
S_{15}	4	1	1	H
S_{16}	4	4	2	L
S_{17}	3	4	3	M
S_{18}	3	3	1	H
S_{19}	2	2	2	H
S_{20}	1	1	4	M
S_{21}	1	1	3	M
S_{22}	2	1	3	H
S_{23}	3	4	4	L
S_{24}	4	3	2	M
S_{27}	2	4	2	M

Alternatively, the locations, S_{14} and S_{26} were governed by one and the same rule. In such a case, it suffices to remove one segment and consider only the other. Accordingly, the segment S_{26} was removed from Table 2. In order to simplify the decision table it is required to find the insignificant conditional attributes in the diagnoses, i.e. it is required to find a minimum number of conditional attributes, but, are still able to diagnose the problem. This is done by number of conditional attributes should be removed each time and the decision table should be checked to make sure any contradiction has not occurred. In this paper this computing the reduct is done by using software called ROSETTA.

In case of Table 1, the conditional attributes, other than those of (a), (c) and (e), were removed from Table 1, minus the segment S26, so a new table obtained. In the new table, a pair of segments S1 and S12 and S9 and S18 and so on, has been classified by one and the same rule used for each. The segments of each pair were removed except one and the result is given in Table 3.

Finally, the proposed algorithm is applied to the simplified decision table shown in table 3. In order to make the algorithm easier to comprehend, an example will be analyzed the prior to any theoretic analysis. A sample decision table shown in Table 3 is used to illustrate the idea then the OSGP-problem is proposed as the basis of the algorithm and finally the algorithm is listed.

Step 1. Scaling

Table 4 lists the scaled decision table SB = (U, CB, {d}) of the decision table S listed in Table 3, where a1, a2.., b1 stand for (a, 1), (a, 2)......., (b, 1), respectively.

Step 2. Reduction of the scaled decision table

One can apply attribute reduction to the scaled decision table. For example, R1 = {a1 ,a2 ,a3 ,c1, c4, e1, e4} is a minimal reduct of S^B listed in Table 4 and a reduced decision table (U, R1, {d}) can be obtained.

Step 3. Converting Back to a "normal " decision table

However, it is more interesting to convert (U, R1, {d}) back to a "normal" decision table as listed in Table 5, where duplicated objects are removed.

Because a3 $\notin$ R1, we do not distinguish S_6 from S_9. for the extending purpose a is used instead.

Table 4. S_B, the scaled decision table of S

Segments	Conditional Attributes											Decision Levels
	a1	a2	a3	a4	c1	c2	c3	c4	e1	e2	e3	e4	
S_1	0	1	0	0	0	0	0	1	0	0	1	0	M
S_2	0	0	1	0	0	0	1	0	0	1	0	0	M
S_3	0	1	0	0	0	1	0	0	0	0	0	1	L
S_4	0	0	0	1	1	0	0	0	0	0	1	0	L
S_5	1	0	0	0	0	0	1	0	0	0	1	0	M
S_6	0	0	1	0	0	1	0	0	1	0	0	0	H

Table 4. (*Continued*)

S_7	1	0	0	0	0	0	0	1	0	1	0	0	L
S_8	0	0	0	1	0	0	0	1	0	0	1	0	L
S_9	0	0	1	0	0	0	1	0	1	0	0	0	H
S_{10}	0	0	0	1	0	1	0	0	0	1	0	0	M
S_{11}	0	0	1	0	1	0	0	0	0	1	0	0	H
S_{13}	0	0	1	0	0	0	1	0	0	0	0	1	L
S_{14}	0	0	0	1	0	1	0	0	0	0	1	0	M
S_{15}	0	0	0	1	1	0	0	0	1	0	0	0	H
S_{16}	0	0	0	1	0	0	0	1	0	1	0	0	L
S_{17}	0	0	1	0	0	0	0	1	0	0	1	0	M
S_{18}	0	0	1	0	0	0	1	0	1	0	0	0	H
S_{19}	0	1	0	0	0	1	0	0	0	1	0	0	H
S_{20}	1	0	0	0	1	0	0	0	0	0	0	1	M
S_{21}	1	0	0	0	1	0	0	0	0	0	1	0	M
S_{22}	0	1	0	0	1	0	0	0	0	0	1	0	H
S_{23}	0	0	1	0	0	0	0	1	0	0	0	1	L
S_{24}	0	0	0	1	0	0	1	0	0	1	0	0	M
S_{27}	0	1	0	0	0	0	0	1	0	1	0	0	M

Step 4. Repeating steps from step 1 to step 4

The scaling, reduction and converting back process (where new values such as a2, a3 . . . instead of 1 should be used) could be repeated until the cardinality of any attribute cannot be reduced further. $S_B^{P_1}$ is obtained as listed in Table 6.

R2 = {a1 ,a2 ,a3 ,c1, cc, e1, ee} is a reduct of $S_B^{P_1}$ and SP2 is obtained as listed in Table 7. Since S^{P2} is unpartitionable, the whole process terminates..

Step 5. Rule Selection

The minimal decision algorithm can be described by rules such as "If conditional part (conditional attribute), THEN, conclusive part (decision level of distress or decision attribute)". For example, one can describe the minimal decision algorithm of Table 7 by 11 rules. The decision rules of locations S_1 and S_2 are represented by rules 1 and 2, respectively. The decision rule of S_6 in the tables indicates that if a segment were classified into class number 1 of conditional attributes (e), the decision level would be "H", or: if (e) = 1 THEN decision level = H.

Table 8 shows the rules generated from the present research data set.

Table 5. S^{P1}

Segments	Conditional Attributes			Decision Levels
	a	C	e	
S_1	2	4	E	M
S_2	3	C	E	M
S_3	2	C	4	L

Table 5. (*Continued*)

S_4	a	1	E	L
S_5	1	C	E	M
S_6	3	C	1	H
S_7	1	4	E	L
S_8	a	4	E	L
S_{10}	a	C	E	M
S_{11}	3	1	E	H
S_{13}	3	C	4	L
S_{15}	a	1	1	H
S_{17}	3	4	E	M
S_{19}	2	C	E	H
S_{20}	1	1	4	M
S_{21}	1	1	E	M
S_{22}	2	1	E	H
S_{23}	3	4	4	L

3.1 Algorithm Structure

The main structure of the algorithm is repeating the three steps, namely, scaling, reduction and converting back. Until the new decision table is unpartitionable. In other words, a single group partition reduct (SGPR) of the new decision table should be computed recursively and new values such as 2, 3 ... should be assigned to k in Equation (3). By doing so an optimal partition reduct can be obtained [9], where the optimal metric is defined by the cardinality sum of new attribute domains.

Table 6. The scaled decision table of S^{P1}

Segments	Conditional Attributes							Decision Levels
	a1	a2	a3	c1	cc	e1	ee	
S_1	0	1	0	0	0	0	1	M
S_2	0	0	1	0	1	0	1	M
S_3	0	1	0	0	1	0	0	L
S_4	0	0	0	1	0	0	1	L
S_5	1	0	0	0	1	0	1	M
S_6	0	0	1	0	1	1	0	H
S_7	1	0	0	0	0	0	1	L
S_8	0	0	0	0	0	0	1	L
S_{10}	0	0	0	0	1	0	1	M
S_{11}	0	0	1	1	0	0	1	H
S_{13}	0	0	1	0	1	0	0	L
S_{15}	0	0	0	1	0	1	0	H
S_{17}	0	0	1	0	0	0	1	M
S_{19}	0	1	0	0	1	0	1	H
S_{20}	1	0	0	1	0	0	0	M
S_{21}	1	0	0	1	0	0	1	M
S_{22}	0	1	0	1	0	0	1	H
S_{23}	0	0	1	0	0	0	0	L

Definition 3.
A partition scheme P = [P1...... P|C|] of S is called a single group partition scheme (SGPS) if for any $i \in \{1, \ldots, |C|\}$, |Wai | = 1. Given an SGPS P = [P1, ..., P|C|], for any $i \in \{1, \ldots, |C|\}$, Pi essentially divides V_{a_i} into two disjoint subsets $V_{a_i}^F$ and $V_{a_i}^G$, and

$$P_i(v) = \begin{cases} v & \text{if } v \in V_{a_i}^F \\ k & \text{if } v \in V_{a_i}^G \end{cases} \quad \text{where } k \in W_{ai} \tag{3}$$

3.2 Algorithm Description

The pseudo code of the introduced algorithm is listed below:

Algorithm
{Input: A decision table S.}
{Output: A set of decision rules}
 //Initialize. Mi is used for M-relative reduct.
Step 1. M1 = ϕ
 //The initial partition scheme P0. In fact SP0 = S.
Step 2. $P^0 = \left[P_1^0, \ldots, P_{|C|}^0 \right]$ Where $P_i^0(v_i) = v_i$ for any $i \in \{1, \ldots, |C|\}$ and vi ∈ Vai ;
 //Initialize unprocessed attribute-values pairs for each attribute
 //Now all attribute-values pairs are unprocessed.
Step 3. For (i = 1; i ≤ |C|; i++) $H_i^0 = \{a_i\}_B$;
 //Attack the OSVP-problem through attacking the OSGP-problem recursively.
Step 4. For (i = 1; ; i++) begin
 //**scaling.**
Step 4.1 compute $S_B^{P^{i-1}}$;
 //**Reduction.**
Step 4.2 Ri = an optimal M-relative reduct of $S_B^{P^{i-1}}$ where M = Mi;
Step 4.3 Mi+1 = Mi;//Initialize Mi+1.
Step 4.4 for (j = 1; j ≤ |C|; j ++) begin
 //Compute Pi.
Step 4.4.1 $\forall (a_j, v) \notin H_j^{i-1} - R^i, P_j^i(v) = P_j^{i-1}(v)$;
Step 4.4.2 $\forall (a_j, v) \notin H_j^{i-1} - R^i, P_j^i(v) = i$;
Step 4.4.3 $H_i^j = H_j^{i-1} \cap R^i$ //Remove processed attribute-values pairs
 //Compute Mi+1
Step 4.4.4 if $\left(H_i^j \neq \phi \right)$ Mi+1 = Mi+1 ∪ {(aj, i)};
end//of for j;

//**Converting back to a "normal" decision table **

Step 4.5 compute SPi where $P^i = \left[P_1^i, ..., P_{|C|}^i \right]$;

//See if all attribute-values pairs have been processed

Step 4.6 if $H^i = \bigcup_{j=1}^{|C|} H_j^i = \phi$ break; end //of for i

Table 7. S^{P2}

Segments	Conditional Attributes			Decision Levels
	a	c	e	
S_1	2	CC	E	M
S_2	3	C	E	M
S_3	2	C	E	L
S_4	a	1	E	L
S_5	1	C	E	M
S_6	3	C	1	H
S_7	1	CC	E	L
S_8	a	CC	E	L
S_{10}	a	C	E	M
S_{11}	3	1	E	H
S_{13}	3	C	EE	L
S_{15}	a	1	1	H
S_{17}	3	CC	E	M
S_{19}	2	C	E	H
S_{20}	1	1	EE	M
S_{21}	1	1	E	M
S_{22}	2	1	E	H
S_{23}	3	CC	EE	L

Table 8. Rules generated by rough set analysi

Deterministic Rules	
Rule 1	(a=2)&(c=4) $\Rightarrow$ (severity= M)
Rule 2	(c=3)&(e=2) $\Rightarrow$ (severity= M)
Rule 3	(c=2)&(e=4) $\Rightarrow$ (severity= L)
Rule 4	(a=4)&(c=1)&(e=3) $\Rightarrow$ (severity= L)
Rule 5	((c=3)&(e=3) $\Rightarrow$ (severity= M)
Rule 6	(e=1) $\Rightarrow$ (severity= H)
Rule 7	(a=1)&(e=2) $\Rightarrow$ (severity= L)
Rule 8	(a=4)&(c=4) $\Rightarrow$ (severity= L)
Rule 9	(a=4)&(c=2) $\Rightarrow$ (severity= M)
Rule 10	(c=1)&(e=2) $\Rightarrow$ (severity= H)
Rule 11	(c=3)&(e=4) $\Rightarrow$ (severity= L)
Rule 12	(c=2)&(e=3) $\Rightarrow$ (severity= M)
Rule 13	(a=3)&(e=3) $\Rightarrow$ (severity= M)

Table 8. (*Continued*)

Rule 14	(a=2)&(c=2)&(e=2) ⟹ (severity= H)
Rule 15	(c=1)&(e=4) ⟹ (severity= M)
Rule 16	(a=1)&(e=3) ⟹ (severity= M)
Rule 17	(a=2)&(c=1) ⟹ (severity= H)
Rule 18	(C=4)&(e=4) ⟹ (severity= L)

4 Comparison of the Results Obtained from Rough Set Theory and Artificial Neural Network

To compare the results of the assessment of distress severity level in concrete pavements using two different methods; i.e. rough set theory and artificial neural network, five more segments of the selected road were investigated which their distress characteristics of other 27 segments are tabulated in Table1. These distresses are evaluated in the same manner by experts and the results are shown in Table 6. As it can be seen from the table, segments X1 , X5 and X2 , X4 and X3 have the severity level high (H), medium (M) and low (L), respectively.

Table 9. Observation information regarding assessment of distress severity level for five segments of a concrete pavement

Segments	Conditional Attributes											Decision Levels
	a	b	c	d	e	f	g	h	i	j	k	L	
S_1	1	3	2	2	1	2	1	1	2	1	3	1	H
S_2	2	4	2	1	1	2	2	1	2	2	1	2	M
S_3	3	4	4	3	2	4	2	2	3	4	3	3	L
S_4	4	2	2	2	1	2	4	2	3	2	2	1	M
S_5	1	2	3	1	2	2	1	1	2	1	2	3	H

To assess the distress severity level of these segments using artificial neural network, observation data in Table 1 were used as training input using Matlab 2010 software. This software is one of the most powerful and easy-to-use programs in this field.. It could also use different criteria to finish the operation of net training like reaching a specified value for cycles, specified value for time elapsed, specified error value, etc. it can present the effect of input and output data. Applying this important ability, the most effective data can be recognized through lots of data variables. The neural network used here has one hidden layer including 20 neurons which have been determined by trial and error method. The criteria for stop – based training were based on target error less than 0.00001. This is a very low target error and shows that the training of model has high accuracy. After the network training procedure and confirming the authenticity of the results, the observation data of Table 6 regarding segments X1, X2, X3, X4 and X5 were processed using this method. Consequently, these data were assessed using rough set theory. The results of these investigations are tabulated in Table 10.

Table 10. Comparison of the results of distress analysis fo5 segments of a concrete pavement for two different methods; i.e. artificial neural network and rough set theory

segments	Evaluation of experts	Assessment using RST	Assessment using ANN
X1	H	H	M
X2	M	M	M
X3	L	L	L
X4	M	M	L
X5	H	H	H

As it can be seen in Table 10, the results of rough set theory are compatible with assessment of experts in all 5 cases, where as in artificial neural network method only the results of segments X1 and X3 are in accordance with the assessment of experts and rough set theory. Therefore, it can be concluded that the rough set theory has more reliability level in assessment of distress severity level of concrete pavements than the artificial neural network method.

5 Conclusion

In this paper, a rough set theory-based approach as a new method is presented and evaluated to determine severity levels of distress in highway pavements using a complete practical example. Also the results of this method and artificial neural network method have been compared. The most important results of the present paper are as follows:

a) The decision table determines the most convenient decision for a certain condition. Formulating such decision table may help for any decision making problem.
b) By constructing an Expert System using the rules derived from the minimal decision algorithm, it is possible to renew the knowledge speedy reasoning, efficiently.
c) The minimal decision algorithm is extracted by identification of the significant conditional attributes determined from the diagnostic results of pavements distress by experts, and is used to make the best decision..
d) The rough set theory has more reliability level in assessment of distress severity level of concrete pavements than the artificial neural network method.
e) Rough set theory is used for expressing the set of decision attributes by the set of paired conditional attributes, thus we can use this theory as a method to acquire informations from diagnostic cases which exist in civil engineering problems.

References

[1] Ujorth, P., et al.: Operation monitoring and decommissioning of dams final version: November 2000. In: Secretariat of the World Commission on Dams, Vlaeberg, Cape Town 8018, South Africa (2000)
[2] Fayyad, U., Piatetsky-Shapiro, G., Smyth, P., Uthurusamy, R.: Advances in Knowledge Discovery and Data Mining. AAAI Press/MIT Press (1996)
[3] Pawlak, Z., Skowron, A.: Rough sets and Boolean reasoning. Information Sciences 177, 41–73 (2007a)
[4] Skowron, A., Rauszer, C.: The discernibility matrices and functions in information systems. In: Slowiński [4], pp. 331–362
[5] Ganter, B., Wille, R. (eds.): Formal Concept Analysis: Mathematical Foundations. Springer, New York (1999)
[6] Xu, C., Min, F.: Weighted Reduction for Decision Tables. In: Wang, L., Jiao, L., Shi, G., Li, X., Liu, J. (eds.) FSKD 2006. LNCS (LNAI), vol. 4223, pp. 246–255. Springer, Heidelberg (2006)
[7] Min, F., Liu, Q., Fang, C.: Rough sets approach to symbolic value partition. Internat. J. Approx. Reason (2008)
[8] Mahab Ghods (MG) Consulting Engineers, Karkhe dam, the feasibility study report. Ministry of Power (2000)
[9] Min, F., Liu, Q., Tan, H., Chen, L.: The M-relative reduct problem. In: Wang, et al. (eds.) [27], pp. 170–175
[10] Arabani, M.: Evaluation of rough set theory for decision. Scientia Iranica 13(2), 152–158 (2006)

Cartographic Circuits Inside GIS Environment for the Construction of the Landscape Sensitivity Map in the Case of Cremona

Pier Luigi Paolillo, Umberto Baresi, and Roberto Bisceglie

Politecnico di Milano, Dipartimento Architettura e Pianificazione
via Bonardi 3, 20133 Milano, Italy
`pierluigi.paolillo@polimi.it`

Abstract. The centrality of the landscape, inside territorial planning, has been influencing, for years, the testing of innovative analytical techniques aimed to gather the peculiarities of urban and suburban context. The experience upon construction of landscape sensitivity maps, written for the Cremona's Urban Variant, brings out the wide variety of cartographic outputs resulting from the large amount of investigations conducted on the various aspects of the local landscape. The fundamental operations of preliminary information recognition are mainly aimed to obtain the specific territorial units of inquiry, the next step concerns the development of detailed evaluation using multidimensional analysis applications, which allow to lead to the different portions of the Cremona area, specific landscape sensitivity classes, by combining synthetic indicators: i) insularisation of non-built spaces ii) morphological / structural values iii) perceptual aspects of the landscape; iv) permanence of the urban system; v) the degree of imperativeness of the environmental constraints; vi) the integrity of land use.

Keywords: Data Mining, Spatial Data Mining, Geostatistics and Spatial Simulation, Urban Modeling, Spatial Statistical Models, Spatial Data Analysis.

1 The knowledge Organization for the Analysis Conduction Inside Geographical Information System Environment

1.1 The Methodology and Operating Basis

Rejecting the notion that flattens the concept of landscape in the simple natural / environmental way, the same overnational, national and regional laws of Italy, have identified, over the years, a number of complex factors to be considered for the identification of the structural elements of real value.

In reference to these ideas, the preparatory operations for the realization of the landscape sensitivity map in Cremona's study, were made of detailed discussion focusing on: i) the understanding of the specifics of the settlement and environmental context going through the already in force local and overlocal plan documents, ii a) the collection and preparation of the corresponding GIS map layers.

After the formalization of the knowledge kept, a methodological/procedural diagram was set, which illustrates the symbiosis between the GIS software used for the preparation and the information processing to produce thematic indicators, and their subsequent reconduction to the homogeneous areas for thematic characterization and events entity.

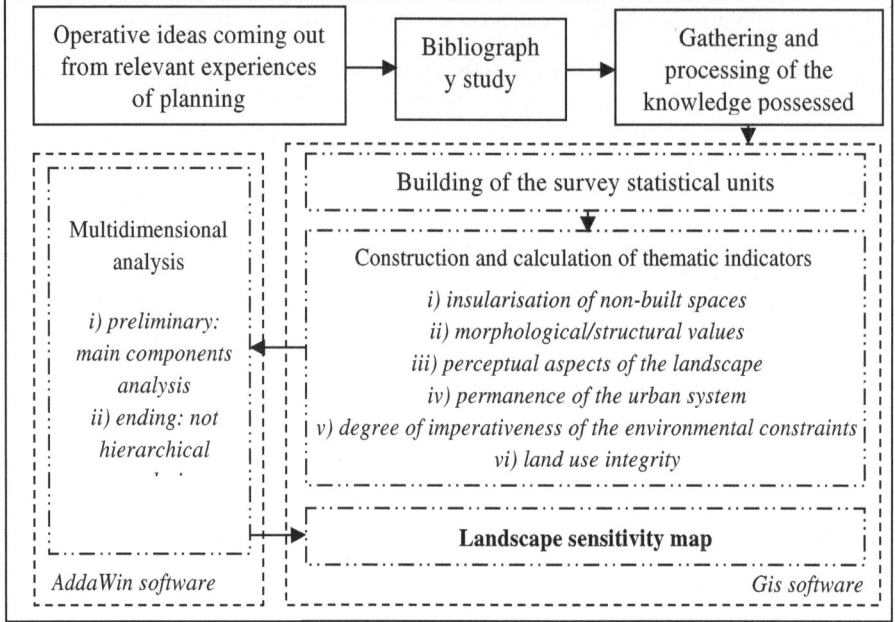

Fig. 1. Procedural method diagram, as applied to the case of Cremona

1.2 The Construction of the Survey Statistics Units

Keeping within the preliminary operations of the thematic indicators estimation, it turned out to be crucial the construction of the survey units to whom conneting the information collected, depending on the subsequent treatment with dedicated GIS software and, later, with geostatistical applications inside AddaWin environment. Concerning different analysis, it was considered appropriate to operate at two different scales: to investigate the exclusive events of suburban areas, continues dimension survey units were considered, identifing, by processes resulting from landscape ecology, landscape units (UDP) generated from the conformation of natural and anthropic elements, setting barriers of different degree of intensity; vice versa, for all other events, in the study area, it was discretized by constructing a grid of cells of 25 m side, assigning to each one the information derived from both complex analysis made in continuous geometries, and other analysis.

2 The Calculation of the Thematic Indicators to Recognize the Specific Characteristics of the Cremona Area

2.1 The Insularization Degree of Unbuilt Areas

A starting operation aims to investigate the conformation of these unbuilt areas, making possible to understand which ones are most intact and have the most probability to remain intact with no anthropic use, separating them from those compromised by human transformation, for which specific guidelines need to be made, to prevent further degradation, by an index expressed as:

$$Ins = f^{I}(A_{UdP}) \cdot \left(1 + f^{II}(P_{UdP}, F_{UdP})\right)$$

dove:

Ins = insularization index (significant measure of integrity of the *Udp*);
A_{Udp} = extension factor of *Udp*;
P_{Udp} = permeability degree of the *Udp*'s perimeter;
F_{Udp} = shape factor of the *Udp*
(Udp:landscape unit)

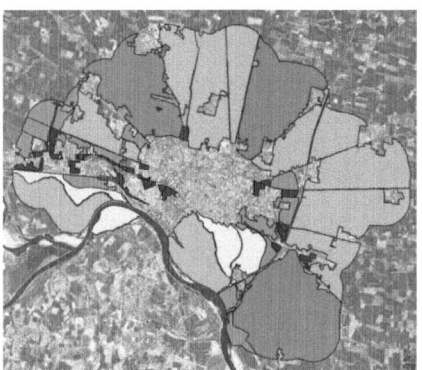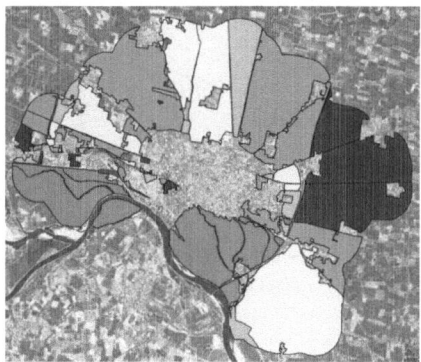

Fig. 2. Representation of the area factor (1) and of permeability (2), for each landscape unit

The *area factor* is involved in the analysis of the insularisation index for his factorial influence over the extension of UDP: in fact, it takes maximum values at the Udp> 350 ha and below where the surface is affected by the insularisation consequences. The *permeability factor* expresses the impact of anthropogenic pressures on the relationship between the UDP and its context, returning an index of propensity to go through, outside to inside, the perimeter of UDP, and its area represents a significant element since the UDP extension> 350 has reached the maximum value of the index, also considering that these cases have conditioned permeability values by the ratio of the area and perimeter of each field; beyond 150 m from the perimeter edge it does not happen to suffer any negative effect.

The index of insularisation identifies the state of structural *integrity of landscape systems* in which the UDP that have a low or low-medium value have to be considered integrated (or integrable) with the anthropic areas because they have lost their landscape identity.

Contrariwise, the landscape units, characterized by high values of structural integrity, correspond to portions of territory that, over time, were not involved by disturbing factors, presenting good chances of developing relations between the existing natural systems/structures, that suggests for these UDP to gather a greater protection and preservation.

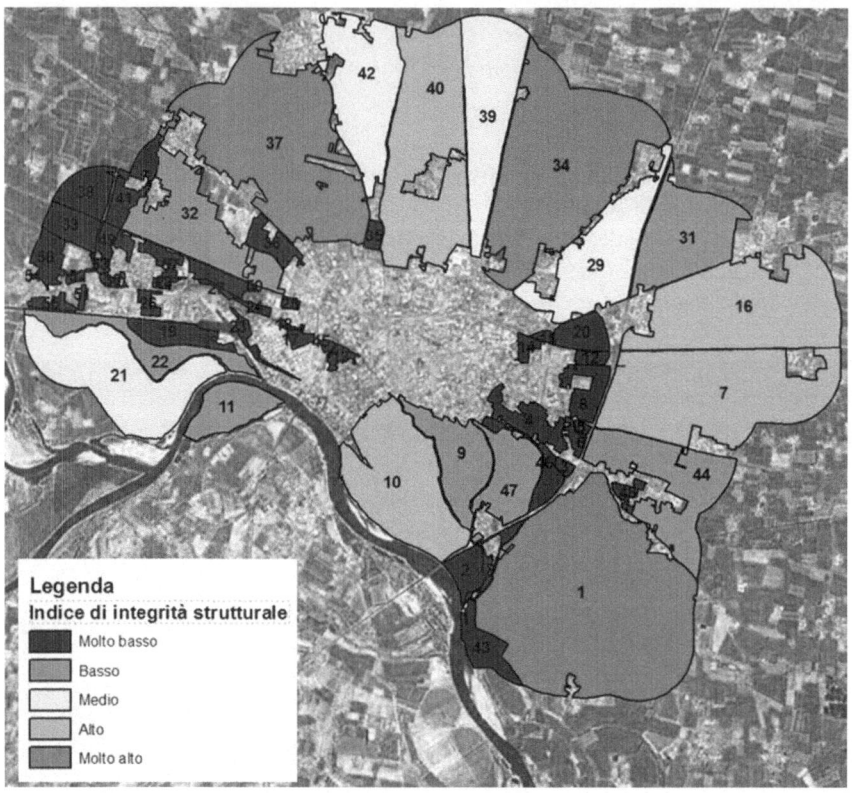

Fig. 3. Representation of the Insularisation index, or structural integrity

Particular attention should be given to the UDP that have medium and low values of the indicator, because they reveal higher risk areas of degradation due to human contingencies; here you find unstable situations that require aggressive actions to prevent the progressive decline from to the original integrity condition, compared to the global scenario represented by the structural integrity index, particular attention should be given to the landscape units in medium and low range, that will be discussed below: while preserving natural character, they are in some parts surrounded by urban areas and infrastructure of various impact, and are also interested in impacts made by spread and spot settlement.

2.2 The Disposition of the Morphological/Structural Values

Primary importance must be given, also, to the topic of morphological / structural evaluation of the landscape in the awareness that any area of analysis is affected by multiple characters of this type, and it is their special integration that helps to determine the landscape quality of the sites; very important is therefore the research of the distinguishing characteristics of an area, identified as follows in Cremona; **over local scale:** i) the morphological structures of particular importance in the configuration of the landscape contexts (ridges, edges of terraces, river banks), ii) the areas or elements of environmental significance that have relationships with other relational elements in the composition of systems of greater wideness (components of natural water, green corridors, protected areas, wooded areas and scrub), iii) the key elements of the historical settlement pattern (path, channels, artifacts and art works, historical centers, important buildings such as villas, abbeys, castles, fortifications), iv) the evidence of formal and material culture of an historical / geographical area (eg "that" terrace morphology, or " that portion "of terrace morphology), with stylistic solutions typical use of specific materials and techniques (building stone or wood, dry-wall, etc..), together with the treatment of public spaces; **local scale:** a) signs of the space morphology (gradients of altitude, slope morphology, minor elements of the surface hydrography); b) the natural/environmental significant elements (trees, natural monuments, springs or wetlands not connected to larger systems, green spaces with nodal role in the local frame of the green areas), c) the components of the historical-agricultural landscape (rows, elements of the irrigation network and artifacts such as sluices, bridges, etc.., paths tracks, historical centers and rural buildings), d) the elements of historical / artistic interest (centers and historical centers, monuments, churches and chapels, historic walls), e) the basic elements of relational importance at local level (urban parks, linear elements of green or water, "doors" of the center or town center, railway station); f) the proximity or belonging to a distinct place with high levels of linguistic, typological and iconographic consistency.

These factors were considered to derive two synthetic analysis blocks, led to synthesis by pairwise comparison, in this way the values given to each element were georeferenced in descret shape (step 25 m), giving to each cell, affected by the presence of a specific element, the value corresponding to the weight obtained; once the sum of all values associated with the individual elements present in the cell was calculated, the morphological/structural value of the sites was derived, properly unbundled into five classes of intensity.

The functions of normalized value, expressed by the coefficients (weighted factors) attributable to each element of investigation for measuring the synthetic index, are derived going through the matrix A (nx • n), formed by pairwise comparison of the referred to 100 items (with $\alpha_{ij} \alpha_{ji} + = 100$) to the matrix B (always nx • n) whose single element is obtained from the relationship with its complementary to 100 (i.e. $\beta_{ij} = \alpha_{ij} / \alpha_{ji}$), and whose processing is performed by calculating many column vectors (V_j) equal in number to the n elements corresponding to the sum of the values of the analyzed column, with:

$$V_j = \sum_{j=1}^{n} \beta_{ij},$$

and then normalizing the elements of matrix B with the corresponding column vectors Vj obtaining the normalized matrix C.
The coefficients of the value function are obtained as standardized average values on the maximum value (*Best Positioned One*) of the elements sum of each row of the matrix C normalized as follows:

$$E_j = \left(\sum C_{ij}\right)/n \quad \text{con} \quad \sum_{j=1}^{n} E_j$$

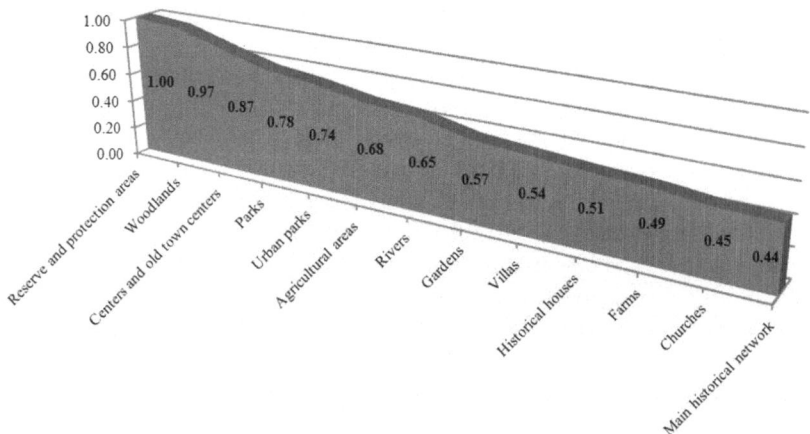

Matrix A of the pairwise comparison, where $\alpha_{ij} + \alpha_{ji} = 100$

$$A = \begin{bmatrix} \alpha_{1,1} & \alpha_{1,2} & \alpha_{1,3} & \dots & \dots & \alpha_{1,n} \\ \alpha_{2,1} & \alpha_{2,2} & \alpha_{2,3} & \dots & \dots & \alpha_{2,n} \\ \alpha_{3,1} & \alpha_{3,2} & \alpha_{3,3} & \dots & \dots & \alpha_{3,n} \\ \dots & \dots & \dots & \dots & \dots & \dots \\ \dots & \dots & \dots & \dots & \dots & \dots \\ \alpha_{n,1} & \alpha_{n,2} & \alpha_{n,3} & \dots & \dots & \alpha_{n,n} \end{bmatrix}$$

Transposed Matrix B, where $\beta_{ij} = \alpha_{ij}/\alpha_{ji}$

$$B = \begin{bmatrix} \alpha_{1,1}/\alpha_{1,1} & \alpha_{1,2}/\alpha_{2,1} & \alpha_{1,3}/\alpha_{3,1} & \dots & \dots & \alpha_{1,n}/\alpha_{n,} \\ \alpha_{2,1}/\alpha_{1,2} & \alpha_{2,2}/\alpha_{2,2} & \alpha_{2,3}/\alpha_{3,2} & \dots & \dots & \alpha_{2,n}/\alpha_{n,} \\ \alpha_{3,1}/\alpha_{1,3} & \alpha_{3,2}/\alpha_{2,3} & \alpha_{3,3}/\alpha_{3,3} & \dots & \dots & \alpha_{3,n}/\alpha_{n,} \\ \dots & \dots & \dots & \dots & \dots & \dots \\ \dots & \dots & \dots & \dots & \dots & \dots \\ \alpha_{n,1}/\alpha_{1,n} & \alpha_{n,2}/\alpha_{2,n} & \alpha_{n,3}/\alpha_{3,n} & \dots & \dots & \alpha_{n,n}/\alpha_{n,} \end{bmatrix}$$

Fig. 4. Coefficients of the function value chart resulting from the pairwise comparison

In summary, evaluating the morphological characters sensitivity, the physical/geological parts were examined with the hydrographic elements, parks and gardens, the vegetation cover, the ecological network, the agricultural landscape, the infrastructural and the historical dimension of mobility, attributing to each layer the value resulting from the pairs comparison starting from the consideration that the more the morphological/structural elements match and interact, the more a site is unique and sensitive; therefore, it was determined, for each cell, the Σ of the values assigned to the different elements that were inside it, to take into account both the greater importance of some of them, and the added-value of the co-presence of the elements.

Fig. 5. Representation of the morphological/structural index of the local landscape

2.3 The Investigation of the Landscape Perceptual Aspects

Starting point for the analysis of the landscape aspects is the creation of a DTM (Digital Terrain Model) raster, able to take account of the morphology of the Cremona area; on this basis it was consequently entered the presence of coincident perceptual disturbance elements, in the Cremona reality, with the buildings and wood areas, by using the command "Mosaic to New Raster" of the program Esri Arcgis, obtaining in this way, a DSM (Digital Shape Model).

The subsequent landscape analysis, conducted with the *Viewshed* application identifying the observation points and target points, allowed to identify, for all items of each category, the analytical cell of 25 m visually stressed and, on the other hand, those without relationship with these visual elements.

These categories shall be constructed as follows: i) historical and scenic trails; ii) viewpoints iii) architectural and religious sites of historical-landscape interest iv) defensive artifacts and ruins of historical-landscape interest, v) buildings and rural artifacts of historical-landscape interest vi) architectures and historical-industrial artifacts vii) places, buildings and civil artifacts, residential and military of historical-landscape interest viii) elements of the natural landscape of scenic interest; ix) water surface items and connected artifacts of scenic interest.

The nine rasters for each category, obtained from the *viewshed* analysis, were subsequently converted into points and combined into a single information layer, deriving cumulated values with:

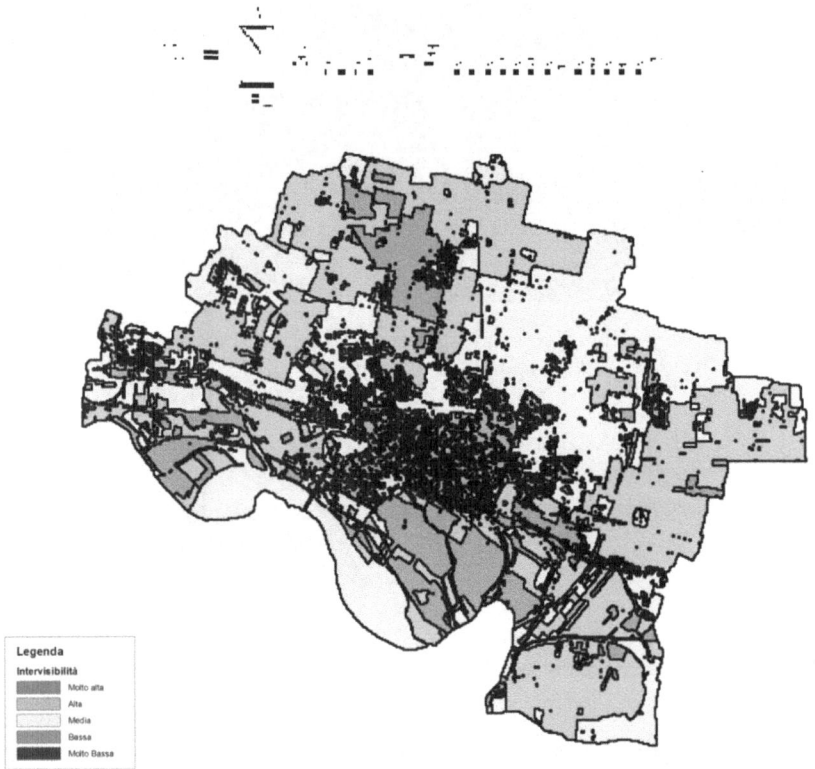

Fig. 6. Representation of the landscape intensity value

2.4 The Permanence Differentials of the Urban Settlement

Another indicator of primary importance is the degree of transformation or, reciprocally, of the landscape integrity among the natural and anthropogenic historical forms; in this way, we examined the historical evolution from 1723 to nowadays defining, the urban constructions, for each time threshold, identifying the core settlements and the corresponding expansions to various thresholds and calculating, thus, the degree of permanence of urban areas over the years: it has been classified in this way, the permanence at eight historical thresholds (discretized with 25 m step), obtaining the total frequency per cell, adding horizontally all the column vectors, and then estimating the persistence value, then, with the *Natural Breaks* classification

(Arcgis software internal algorithm) it has been allocated the degree of permanence in five classes, excluding the value 0, corresponding to the absence of urbanized. From these values we showed that two thirds of the municipality are not settled and that a considerable part of the built areas has medium/high permanence values , indicating a good stability of the historical settlements against a lasting progressive expansion until the latest thresholds; among the problematic areas we have registered the past Tamoil refinery area, already disused, and the isolated villages of recent growth; it's well located the city inside the medieval walls, as well as the historical town centers outside the walls.

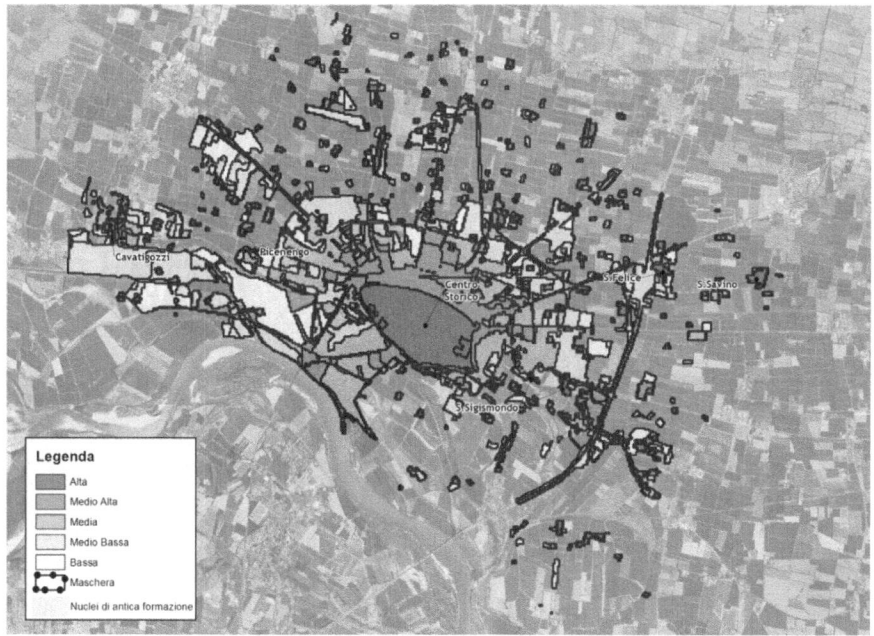

Fig. 7. Representation of the historical permanence of the urban settlement

2.5 The Limint Coming out from the Intensity of the Constraints Application

To highlight the relevances of the constraints system, all the contributing factors were considered for the construction of a constraints map, made by the collection of the invariants identified in the territory among the instruments of government at different spatial scales, that reports:

i) the areas where restrictions for rail and road setback operate, cemetery buffer zone, absolute protection and human use wells buffer zone, the buffer zone of power lines;
ii) the areas with limitations for historical-artistic and landscape-environmental reasons: spaces with environmental constraints, the buildings of historical- art interest, the water system;

iii) the areas where urban planning previsions operate: the over-local Park of Po and Morbasco, the Special Protection Area of Spinadesco - Cremona, the oasis of forest protection.

With these factors we also considered, in this experience, those risk factors resulting from human activities or natural events that determine geography limits; the reading of disaggregated groups of constraints allowed us to derive synthetic values stratifying the information described above, and the estimate of the Total Synthetic Index of the constraints application has been obtained from a calculation that takes into account the different peculiarities expressed by orienting the subsequent assumptions of residual territorial government, with En = [(+ AIVA Aibc Aili +) / Aiqta] / Aimax.

To enhance the visibility of the municipality constraints application intensity, we operated with discrete mapping, based on 25 m steps grid. Within each cell we put the number of overlays of each layer examined before, so the cumulative effects of overlaps was possible to return the degree of application on the constraints no more following the continuous areas but following the statistical units not predetermined.

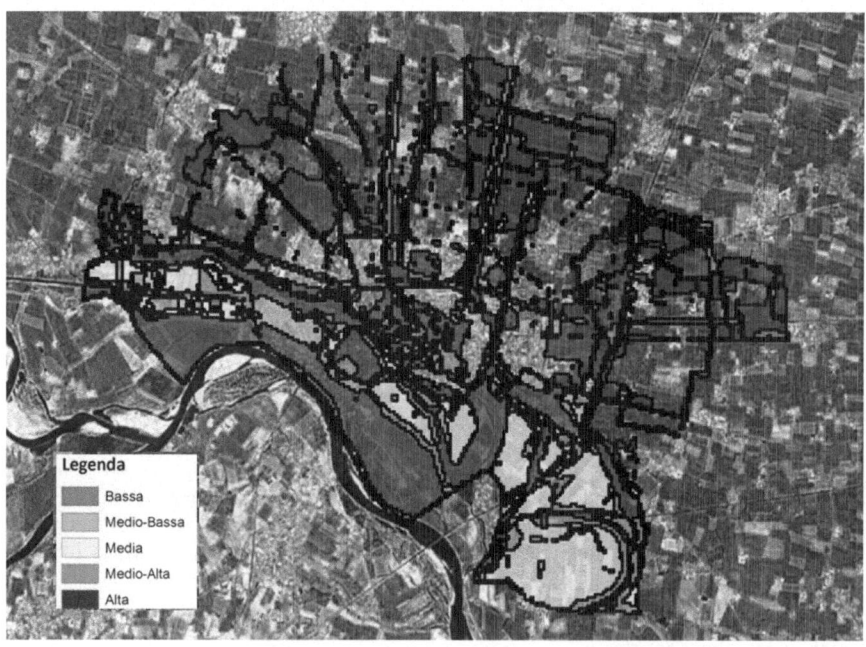

Fig. 8. Representation of the landscape constraints relevances

2.6 The Geography of the Land Use Integrity

The formation of the land use multitemporal archive found the analytical basis with the *Dusaf* database, updated to 2009 and compared with the seven other time thresholds 1994, 1981, 1967, 1925, 1890, 1805, 1723; the data sources that needed to be examined were selected from those available based primarily on three criteria: i) we preferred the one with the most detailed scale, ii) we wanted to create a temporal scan being able to represent the speed of the changes succession, identifying thresholds, little by little getting closer until they reach the current one, iii) the historic thresholds were preferred with sufficiently detailed available material to face a depth analysis.

To the different types of destination use it was overlayed a grid of discrete 25 m steps, assigning the identifiers of each cell uses, calculating the area under each use (and choosing not to identify the prevalent use of the unit, but to consider the possible coexistence of different land uses for each of the six historical thresholds considered), deriving a data matrix of 64 variables: through the software AddaWin analysis the variance was recognized (the degree of mutation of each cell compared to the variation of the land use in the time range considered), thus resulting in the municipality space the following results.

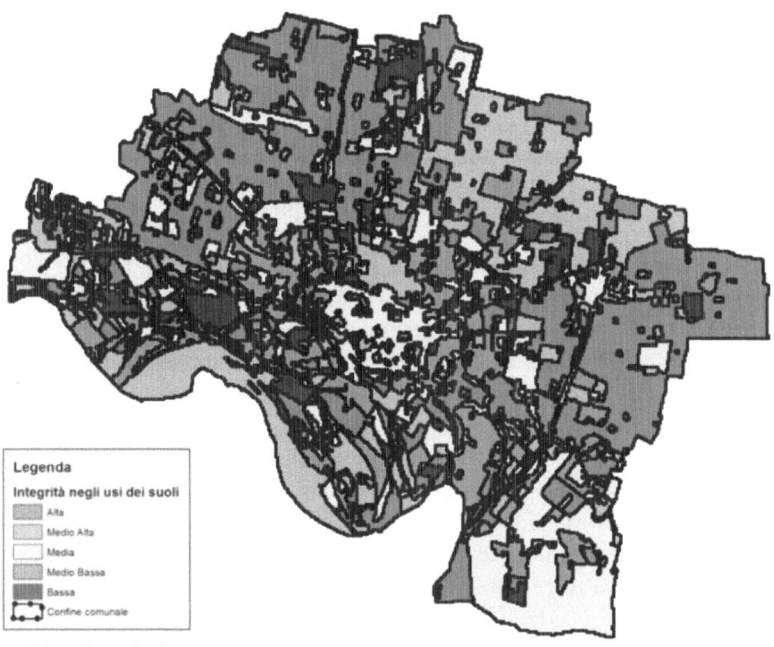

Fig. 9. Representation of the land use degree of integrity

Large portion of territory is characterized by high and medium/high land use integrity classes, confirming the maintenance of traditional agricultural use; it should be underlined, how a good quantity of cells, with medium and medium/low integrity, come out from progressive substitutions in agricultural assets; meanwhile we detect low-and medium-low degree of integrity in the south/west portions of the area and along the main routes against the main alteration given by the strategic role of these areas.

3 The Multidimensional Analysis, the Final Summary of the Investigated Vectors to Identify the Landscape Sensitivity Areas

The landscape sensitivity classes were built with the intensity of landscape values, estimated and taken to synthesis through a multivariate geostatistical method, whose use has permitted to interpret the events that determine the complex municipality landscape structure through observation of the set of k variables on n statistical units (cells).

In this case the exploration of the matrices produced was made from the classification of data into ordinal categorical variables, none (N), low (B), medium (M), high (A) and matrices treated in AddaWin environment, are related to the outcome of the investigation for:

a) the insularisation degree of unbuilt spaces (INS);
b) morphological /structural values (IMS);
c) landscape perception aspects (landscape survey) (VIS);
d) the degree of permanence of the urban settlement (PER);
e) the constraints degree of application (COG);
f) the integrity factors of the land use (IUS).

After identifying a group of twenty variables, we proceeded with the main components analysis; for each one, the absolute magnitude of the eigen values was provided (Eigen Value, inertia explained for each component), and the proportion (ie the portion of variance explained by each component relative to the total), accumulated from previous reports (sum of eigen values), to assess from how many main components results a specific portion of variance was explained.

In the usual applications rarely all the n principal components are considered (CP) and, therefore, you must select the number of components to be considered in the analysis according to criteria of optimality, consistent: i) in the parsimony (minimum possible number of principal components) ii) in the minimum loss of information; iii) in the minimum deformation of the quality of representation.

Aiming to what written above we need to set a % of total variance explained (which satisfies the three criteria mentioned above) and, under the assumed type of matrix (cells of 25 m side), it was not necessary to assume greater cumulative inertia higher than the range of the 65 to 70% since the first 7 principal components explain about 69% of the model, entity (for this analysis) more than sufficient for the subsequent identification of the iso-events characterization clusters.

In order to obtain the basins of propensity to urban transformation and/or environmental conservation, we considered important to use a non-hierarchical cluster analysis to be performed on the results of the analysis for main components, carried out earlier: the application processes the variables chosen deriving a target, able to describe in synthetic words the whole area considered, and shows how the value of the target function decreases following the classes number lowering through mergers and optimizations.

Consequently, after the non-hierarchical classification, the portions of the territory (cells 25 m) have been grouped in "equal areas" with similar behavior, and in this way, a further reduction of complexity has taken place able to generate the following situation: i) number of classes identified = 14; ii) degree of inertia explained = 76% (the target curve, in fact, tends to the maximum value of inertia for the classes from 88 to 15, where it expresses a more linear trend).

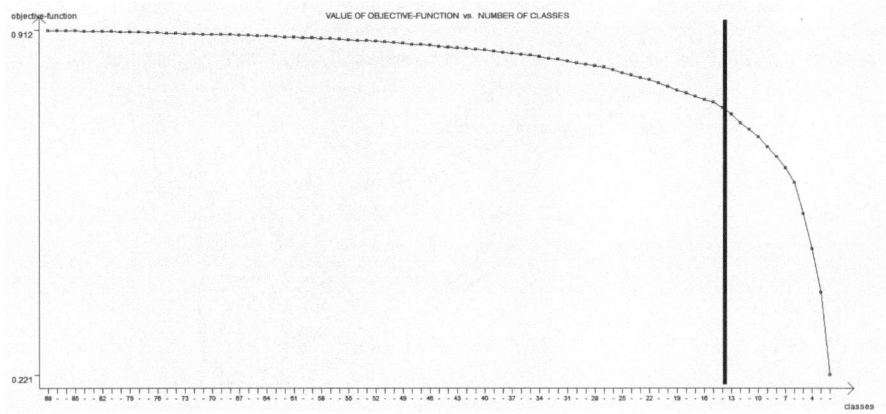

Fig. 10. Target function, with the intersection point identification (XIV class)

Knowing the number of classes to describe, that equals 14, we obtain a process that, for each class identified, shows from which of the 20 active variables (above choices) is more characterized; the subsequent description of the 14 stable classes identified, through the interpretation of the reports provided by the geostatistical AddaWin package, is made of a matrix with classes on the ordinate and variables in abscissa; for each junction, through the symbols (+) and (-) the software realizes the degree of the single variable characterization for each class.

From this analysis the summary table of each class is derived; from the table interpretation classes with low and very low characterization came out; so the forst step of the cartography elaboration started relatively to the 14 cluster iso-events performing operations of table joining in GIS environment between the mother matrix of 25 m step and the text file produced by AddaWin as a result of non hierarchical analysis, properly treated and made compatible with the format required by the GIS software.

The matrix, in this way produced, was then georeferred to create an orthophoto map, formulating afterwards a quality and similarity judgment of the "equal-areas" obtained, and so arising an appropriate classification by the adjustment of the data output, their dismemberments and the resulting combinations of classes: in this way, from 14 original, 10 synthetic classes of multidimensional landscape value characterization were aggregated, whose further synthesis made us able to derive five landscape sensitivity classes: i) *very high* (areas that have boundary conditions simultaneously at a high protection for landscape safeguard and mitigation), ii) *high*

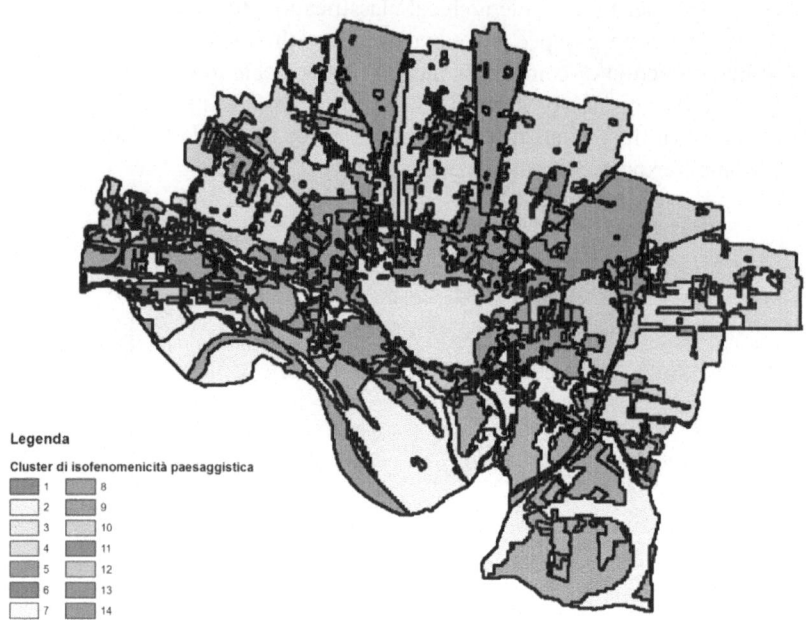

Fig. 11. Representation of the iso-events landscape cluster recombined in 10 classes

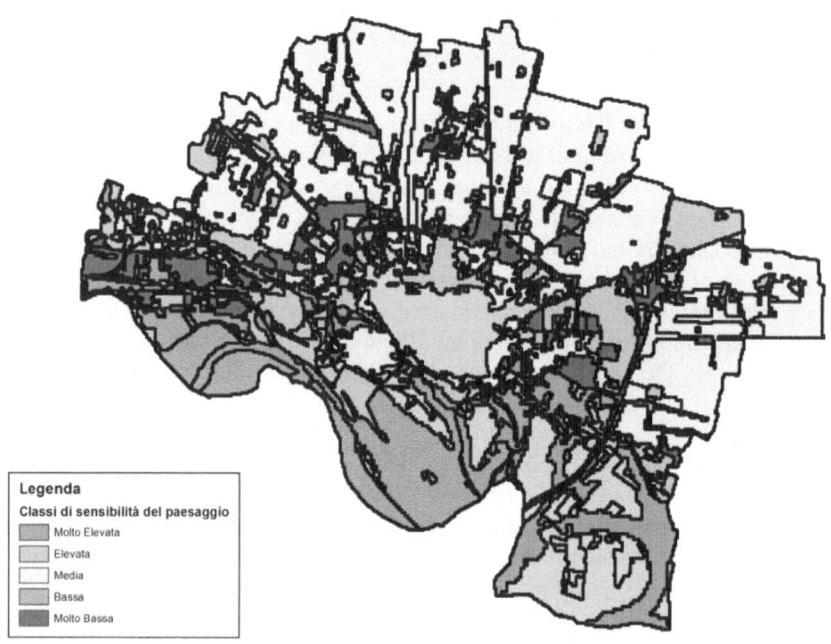

Fig. 12. Representation of the degree of landscape sensitivity

(areas that have more than one boundary condition, not to be exceeded, with high landscape protection and maintenance of existing original features), iii) *medium* (areas with landscape protection conditions, with a limit determined by the landscape relevance), iv) *low* (areas that have landscape "edge conditions", with protections related to the residual values of permanence and importance of the landascape agricultural matrix); v) *very low* (areas that have only "edge conditions", with no landscape qualification).

References

1. Bartel, A.: Analysis of landscape pattern: towards a 'top down' indicator for evaluation of landuse. Ecological Modelling 130, 87–94 (2000)
2. Bertinelli Spotti, C., Mantovani, M.T.: Cremona momenti di storia cittadina, Turris (1996)
3. Borachia, V., Paolillo, P.L.: Territorio sistema complesso. Franco Angeli Editore, Milano (1993)
4. Brenna, S., Fasolini, D., Sale, V.M.: Suoli e paesaggi della provincia di Cremona, Milano (2004)
5. Cataldo, A.: Trasformazioni del paesaggio: categorie territoriali e indicatori per il riconoscimento e il governo di nuovi paesaggi. In: XXVII Conferenza Italiana di Scienze Regionali (2006)
6. Cosgrove, D.: Realtà sociali e paesaggio simbolico, Unicopli, Milano (1990)
7. Debinski, D.M., Holt, R.D.: A survey and overview of habitat fragmentation experiments. Conservation Biology 14, 342–355 (2000)
8. Duchateau, K.: PAIS: proposal on indicators for landscape, agriculture practices and rural development at EU level. Organization for Economic Co – operation and Development, Paris (2002)
9. Duelli, P.: Biodiversity evaluation in agricultural landscapes: an approach at two different scales. Agriculture, Ecosystems and Environment 62, 81–91 (1997)
10. Duhme, F., Pauleit, F.: Some examples of different landscape systems and their biodiversity potential. Landscape and Urban Planning 41, 249–261 (1998)
11. Fabbris, L.: Statistica multivariata. Analisi esplorativa dei dati. McGraw Hill, Milano (1997)
12. Griguolo, S.: Addati. Un pacchetto per l'analisi esplorativa dei dati – Guida all'uso. Istituto Universitario di Architettura di Venezia, Venezia (2008)
13. Haila, Y.: A conceptual genealogy of fragmentation research: from island biogeography to landscape ecology. Ecological Applications 12, 321–334 (2002)
14. Hoffmann, J., Greef, J.M.: Mosaic indicators theoretical approach for the development of indicators for species diversity in agricultural landscapes. Agriculture, Ecosystems and Environment 98, 387–394 (2003)
15. Ingegnoli, V.: Fragmentation and Connectivity Processes. In: Landscape Ecology: a Widening Foundation, Springer, Berlin (2002)
16. Loreau, M., Mouquet, N., Holdt, R.D.: Meta – ecosystems: a theoretical framework for a spatial ecosystem ecology. Ecology Letters 6, 673–679 (2003)
17. Malcevschi, S., Poli, G.: Indicatori per il paesaggio in Italia. Raccolta di esperienze. Catap, Trieste (2008)
18. McAlpine, C.A., Eyre, T.J.: Testing landscape metrics as indicators of habitat loss and fragmentation in continuous eucalypt forests (Queensland, Australia). Landscape Ecology 17, 711–728 (2002)

19. Paolillo, P.L., Mariani, L., Rasio, R.: Climi e suoli lombardi.. Il contributo dell'Ersal alla conoscenza, conservazione e uso delle risorse fisiche. Rubbettino, Catanzaro (2001)
20. Paolillo, P.L.: Rendiconti cremonesi, Clup, Milano (2005)
21. Paolillo, P.L.: New survey instruments: studies for the environmental assessment report of the general plan in a case in Lombardy. In: Conferenza nazionale in Informatica e pianificazione urbana e territoriale, Lecco, Politecnico di Milano, Facoltà di Ingegneria, Polo regionale di Lecco, INPUT 2008, Marzo 4-6, pp. 1–10 (2009a)
22. Paolillo, P.L., Benedetti, A., Terlizzi, L.: New survey instruments: studies for the environmental assessment report of the general plan in a case in Lombardy. In: Rabino, G., Caglioni, M. (eds.) Planning, complexity and New Ict, Alinea, Firenze, pp. 215–224 (2009b)
23. Paolillo, P.L., Benedetti, A., Baresi, U., Terlizzi, L., Graj, G.: An Assessment-Based Process for Modifying the Built Fabric of Historic Centres: The Case of Como in Lombardy. In: Murgante, B., Gervasi, O., Iglesias, A., Taniar, D., Apduhan, B.O. (eds.) ICCSA 2011, Part I. LNCS, vol. 6782, pp. 162–176. Springer, Heidelberg (2011)
24. Paolillo, P.L., Benedetti, A., Graj, G., Terlizzi, L., Bisceglie, R.: The decisions support scenarios in the first phases of the strategic environmental evaluation: the Barzio territory government plan experience. In: Conferenza Internazionale in Informatica e Pianificazione Urbana e Territoriale, INPUT10, Cagliari, Maggio 10-12 (2012)
25. Simmel, G.: Saggi sul paesaggio, Armando, Roma (2006)
26. Socco, C.: Città, ambiente, paesaggio. Lineamenti di progettazione urbanistica, Utet, Torino (2000)

Cloud Classification in JPEG-compressed Remote Sensing Data (LANDSAT 7/ETM+)

Erik Borg[1], Bernd Fichtelmann[1], and Hartmut Asche[2]

[1] German Aerospace Center, German Remote Sensing Data Center, 17235 Neustrelitz
[2] Potsdam University, Department of Geography, 14476 Potsdam, Germany
Erik.Borg@dlr.de

Abstract. Environmental parameters required for geo-information modelling are subject to spatial and temporal dynamics. Remote sensing data can contribute to measure those parameters. For that purpose high-accuracy classifications of remote sensing data are required which can be very time-consuming due to the large data volumes involved. In many applications, however, the rapid provision of classified mass data is of higher priority than classification accuracy. One important focus on research and development efforts in the past years has been to optimise the automated interpretation of remote sensing data. Different investigators have shown that this interpretation can both be effective and efficient in JPEG compressed data with acceptable accuracy. This paper presents an operational processing chain for cloud detection in JPEG-compressed quick-look products of LANDSAT 7/ETM+-scenes (compression ratio is 10:1). Two well-developed conventional algorithms are applied to these datasets for cloud detection. Results show that the processing chain developed is stable and produces quality results with substantially compressed mass data.

Keywords: operational cloud detection, classification, JPEG, LANDSAT 7.

1 Introduction

Permanently increasing volumes of remote sensing data and the resulting necessity to constantly adapt storage facilities have been dealt with as early as the late 1990s. One solution to the data storage problem has been the development of lossy and lossless compression methods. Their primary objective was minimisation and handling of those data essential for thematic information mining. In this context, relevant issues, such as thematic interpretation (e.g. geo-correction, classification) based on compressed remote sensing data or achievable accuracy in relation to compression rate and quality losses, have been discussed intensively [1], [2], [3], [4].

Studies on classification accuracy of compressed remote sensing imagery have used multispectral data which were subjected to a step-wise compression of 5% in an ascending order [1], [2]. Compressed and un-compressed reference data sets were subsequently classified by supervised as well as un-supervised classifications. The investigations revealed that classification accuracy decreases in the interval [1:1; 35:1] from 95 % to 60%. At compression ratios higher than 35:1 the decrease is even more marked compared against the original image data. Contrasting investigations, such as [3] and [4], have shown that JPEG compression effects some kind of data homogenisation. As a result the so-called salt-and-pepper-effect in noisy data is reduced significantly so that better classification results can be achieved.

This work deals with the development of a cloud detection algorithm for data-reduced JPEG-quick-look-data of LANDSAT 7/ETM+-scenes at a compression ratio of 10:1. Considering the results of [1], [2] a classification accuracy of some 70% is feasible.

2 Material and Methods

2.1 Quick-Look and Metadata of the LANDSAT 7 / ETM+

Investigations presented here are based on 2,480 quick-look[1] and metadata[2] from the period of 2000 to 2003. In a first step the quick-look data are pre-processed. The radiometric correction of the data is based on a simple linear stretch of raw data based on predefined look-up tables. Geometric pre-processing includes, i.a., a data reduction [5]:
- *Bands 1-5, 7:* 6 x 6 pixels of the original data are reduced to 1 selected pixel,
- *Thermal band 6:* 3 x 3 pixels are reduced to 1 selected pixel.
- *Panchromatic band:* 12 x 12 pixels are reduced to 1 selected pixel.

As a consequence, all bands have a ground resolution of 180 m. Following data reduction the quick-look data are stored in the JPEG format. As has been mentioned the JPEG compression method is lossy. Hence quality and information losses as well as artefacts are possible at a compression too strong. A detailed description of the procedure can be found in [6], [7]. The data available for this study were compressed with a ratio of 10:1 [5]. Although the level of compression depends on the image content of a remote sensing scene, this represents a JPEG quality metric Q-factor of 35 [2]. Consequentially, quality loss is of significance for thematic post-processing of these data.

2.2 Classification Algorithms

Classification generally aims at labeling of feature properties to corresponding pre-defined property scale. For that purpose supervised and unsupervised procedures are used. Both classification strategies require interactive operation of an interpreter to, e.g., define training targets or interpretation schemes. In contrast, automatic classification procedures operate on pre-defined schemes based on extensive analysis to derive a representative attribution list.

---

[1] Quick-look data are preview images derived from original remote sensing data.
[2] Metadata describes remote sensing data (e.g. satellite mission, orbit, track, frame).

With NASA-ACCA (NASA - National Aeronautics and Space Administration) [8], [9] and ACRES-ACCA-procedure (National Earth Observation Group - previously known as ACRES) [10] two established Automatic Cloud Cover Assessment (ACCA) procedure are used for cloud detection within LANDSAT-data. The NASA-procedure is integrated in the operative processing chain of the LANDSAT Ground station at EROS Data Center in Sioux Falls [8], [9]. The ACRES-procedure was developed to minimise personal and time consuming data quality analysis as well as to minimise the subjectivity of operator evaluations in the operative LANDSAT-processing chain of ACRES [10].

NASA-ACCA-Procedure: The cloud detection algorithm uses radiometric corrected data of LANDSAT bands 2 to 6 [9]. Bands 2 to 5 are calibrated and converted to spectral reflectance. The data of band 6 are transformed into temperature values. The automatic NASA-procedure is a two-level procedure. Level 1 is based on an eight conservative filter-cascade for pixel-wise cloud detection to derive a temporary cloud mask. Level 2 uses the classification results of level 1 processing as input. Subsequently, a confident thermal cloud signature is derived for verified cloud pixels by means of the cloud-pixels detected. For this reason data of band 6 (thermal band) are exclusively statistically evaluated. To differentiate snow and clouds the snow area within the image matrix will be calculated. For that purpose a corresponding filter is used. A quality control procedure selecting misclassified pixels is integrated in this processing level. A detailed description of the NASA-procedure is given in [9].

ACRES-ACCA-Procedure: Pre-processing of the ACRES procedure includes different processing levels. First, the effective spatial resolution of the data is reduced to 240 m. The result is a reduction of the required processing duration and the oppression of isolated clouds at low height (< 200 m). Subsequent processing of the data is carried out in segments of 3 km x 3 km. Once clouds are identified in one of the segments, this segment is marked as cloudy. Additionally all seven bands are calibrated and converted to reflectance. The data of the thermal band are converted to radiation temperatures. The following description of the processor is based on [10]. The ACRES procedure uses 12 filters to derive cloud masks. An additional filter is required to distinguish cold clouds from warm surface features [10]. The method is only based on a processing cascade unlike the NASA-procedure. To distinguish cloud and non-cloud features in the data special band thresholds or ratios of different spectral band combinations are used.

3 Method

Data quality can be defined as ratio of cloud-pixel-number to the total number of pixels in an assessment unit (e.g. quadrant, scene). If different cloud detection algorithms are exchanged for one another, the algorithms can be subjected to comparative analysis under standardised conditions.

At implementation of additional algorithms the possibility for derivation of value added data products insists on basis of quick-look-data. Mass data processing under realistic conditions requires a solution of the organisational problems of demand driven data supply as well as of the functional problems of data processing as a prerequisite. Moreover, the processing chain includes 7 modules (Fig. 1) for provision of data for processing JPEG-compressed remote sensing data and/or for coordination of information processes.

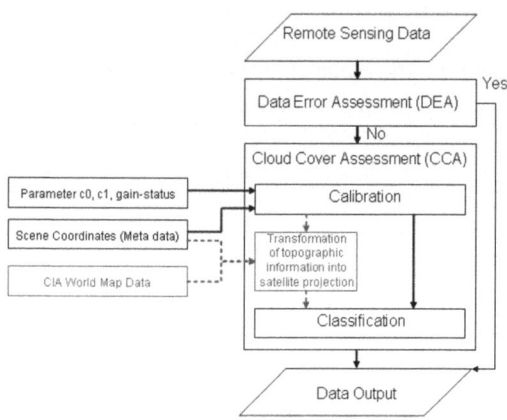

Fig. 1. Scheme of the processing chain for the determination of the data quality of LANDSAT 7/ETM+ data

3.1 Pre-processing Step

Conversation to Top-of-Atmosphere reflectance: With sensor-specific calibration coefficients (offset c_0, gain c_1) digital numbers (DN) recorded can be transformed into measured at sensor radiance $L(\lambda_k)$ [11]:

$$L(\lambda_k) = c_0(\lambda_k) + c_1(\lambda_k) DN(\lambda_k) . \tag{1}$$

$c_0(\lambda_k)$ and $c_1(\lambda_k)$ are the calibration coefficients of band k of LANDSAT 7/ETM+. The relation between planetary top-of-atmosphere reflectance ρ_p and the digital number of a pixel (DN) in a specified band k can be expressed by equation 2 [13], [14]:

$$\rho_p = \left(\pi L(\lambda_k) d^2\right) / \left(E_s(\lambda_k) \cos\theta_s\right) . \tag{2}$$

$L(\lambda_k)$ is the spectral radiance of band k at sensor's aperture $\left[W/(m^2 sr \mu m)\right]$, $E_s(\lambda_k)$ is the mean exoatmospheric solar irradiance of band k $\left[W/(m^2 \mu m)\right]$, d is the Earth-Sun distance [Astronomical Units], and θ_s is the solar zenith angle [degrees].

Sensors like LANDSAT 7/ETM+ can switch the gain of bands to control the sensitivity of the sensor. Parameters for deriving the planetary top-of-atmosphere reflectance ρ_p are represented in Table 1.

Table 1. Solar spectral irradiances and sensor-specific calibration coefficients (c0, c1) according to LANDSAT-bands [12] [3,4]

Band	1	2	3	4	5	7
Sol^5 [W/(m² µm)]	1969	1840	1551	1044	225,7	82,07
Low Gain						
c_0^6 [W/(m²sr µm)]	-6.2	-6.4	-5.0	-5.1	-1.0	-0.35
c_1^7 [W/(m²sr µm)]	1.1760	1.2051	0.9388	0.9654	0.1905	0.0668
High Gain						
c_0^8 [W/(m²sr µm)]	-6.2	-6.4	-5.0	-5.1	-1.0	-0.35
c_1^9 [W/(m²sr µm)]	0.7757	0.7956	0.6192	0.6372	0.1257	0.0437

The Earth-Sun-distance can be calculated approximately using equation 3 [15].

$$d = 1 + 0.0167 \sin[(\pi(Doy - 93.5))/Loy] \ . \qquad (3)$$

Doy is the Day of year, and Loy the Length of year (365 or 366 in the leap year).

Conversation to at sensor brightness temperature: Band 6 of LANDSAT 7/ETM+ covers the wavelength from 10.4 µm to 12.5 µm recording emitted thermal radiation of the earth. According to equation 4 digital numbers recorded of band 6 (DN) can be transformed into effective measured temperatures T [12].

$$T = K_2 / \ln((K_1 / L(\lambda)) + 1) \ . \qquad (4)$$

T is the effective at-sensor brightness temperature [K], $L(\lambda)$ the spectral radiance at the sensor's aperture $[W/(m^2 sr \mu m)]$, K_1 the calibration constant 1 $[666{,}09 \cdot W/(m^2 sr \cdot \mu m)]$ [10], and K_2 the calibration constant 2 [1282,71 K] [11].

---

[3] http://landsathandbook.gsfc.nasa.gov/cpf/prog_sect9_2.html (last access 04.01.2012).
[4] http://landsathandbook.gsfc.nasa.gov/data_prod/prog_sect11_3.html (last access 04.01.2012).
[5] LANDSAT-Handbook: chapter 9.2.4 Table 9.1 Solar Spectral Irradiances.
[6] LANDSAT-Handbook: chapter 11.3.1 Conversation to Radiance.
[7] LANDSAT-Handbook: chapter 11.3.1 Table 11.2, $c_1 = (L_{max} - L_{min})/(Q_{calmax} - Q_{calmin})$.
[8] LANDSAT-Handbook: chapter 11.3.1 Conversation to Radiance.
[9] LANDSAT-Handbook: chapter 11.3.1 Table 11.2, $c_1 = (L_{max} - L_{min})/(Q_{calmax} - Q_{calmin})$.

4 Results and Discussion

4.1 Processor Adaption (Europe)

The cloud classification algorithms (NASA- and ACRES-procedure) were tested for their applicability to JPEG-compressed quick-look-data and acquisition conditions prevalent in Europe.

Table 2. Basic ACRES-filter-configuration and modified configuration of ACRES-procedure (design, threshold values, relations and results)

Filter	Filter definition	Re-lation	ACRESS threshold	Modified threshold	Result	Input for filter
1	B6	<	253 K	253 K	Cold cloud	2
2	B4	<	55	50	Dark objects	3
3	B4/B7	>	2.0	5.5	Snow e.g.	4
4	B4/B5	>	3.1	4.5	Snow	5
5	B1/B7	>	4,5	5.2	Snow	6
6	B1	</>	80 / 240	60	Land/Sand	7
7	B5/B7	<	1.3	0.9	Salt	8
8	B2/B3	<	0.9	0.6	Salt	9
9	B4	</>	140 / 210	30 / 250	Sand	10
10	B1/B2	<	1.03	0.73	Sand	11
11	B7	<	80	20	Snow / Salt	12
12	B1/B3	<	0.9	0.6	Land	13
13	B6	<	170-180	310	No	Cloud mask

While the NASA-procedure could directly be applied to the quick-look-data without quality restriction, the ACRES-procedure required adaptation to recording conditions of Central Europe. That is why different data sets were used. The results were checked visually for the iterative optimisation of the filter threshold values. The resulting new threshold values are listed in Table 2. The absolute error ε_i was calculated for each quadrant. The respective interpreter vote was accepted as true value x and the results of the original ACRES and modified ACRES-procedures were defined as measurements $x_1, x_2,...x_j$. The absolute error ε_i can be written as follows:

$$\varepsilon_i = x_i - x . \qquad (5)$$

The absolute error ε_i of classification results is shown before (Fig. 2a) and after modification (Fig. 2b). The interpretation of negative skewness of frequency distribution (Fig. 2a) suggests that clouds are underestimated when using the original

---

[10] LANDSAT-Handbook: chapter 9.2.4, Table 9.2 ETM+ Thermal Constants.
[11] LANDSAT-Handbook: chapter 9.2.4, Table 9.2 ETM+ Thermal Constants.

ACRES-procedure. Modification of the algorithm thresholds yields skewness and frequency distribution curves (Fig. 2b) that indicates a more stable and more precise result when employing the modified ACRES-procedure. Differences that occur between assessment results by the automatic procedure and by interpreter can partly be explained by the subjectivity of interpreters [16].

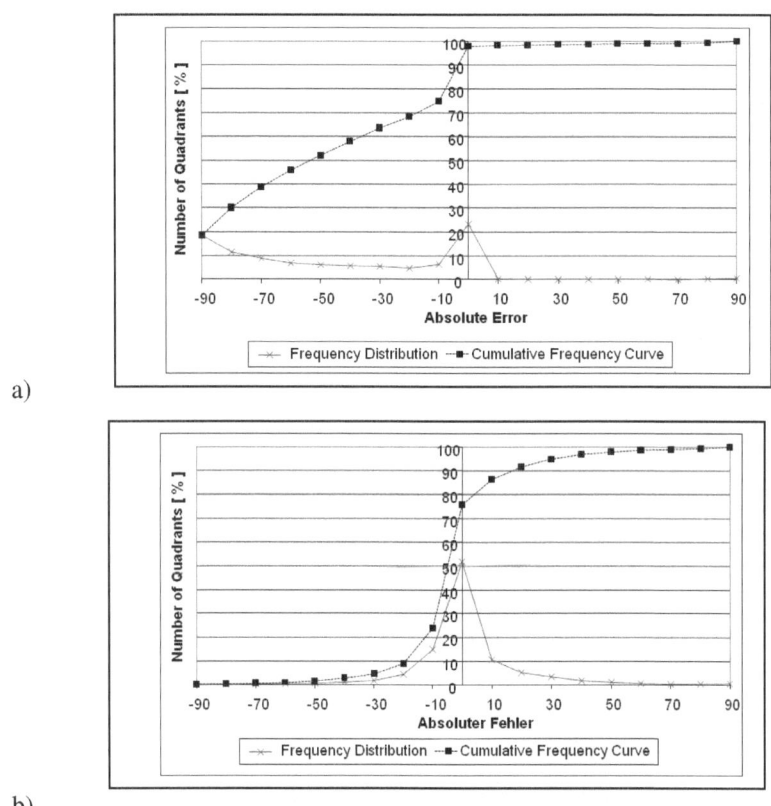

Fig. 2. Comparison of a) original ACRES- and b) modified ACRES-procedure

4.2 Validation of Classification Power

Both NASA-ACCA and modified ACRES-ACCA classification procedures assign each pixel either as "cloud" or "non-cloud". In this binary classification unclassified pixels are impossible.

The classification accuracy of both procedures is investigated on the basis of a pixel by pixel cloud mask comparison (Tab. 3). For that purpose two classes are defined. Pixels classified identically by both procedures as "cloud" or "cloud free" are labeled as "identically classified". The remaining pixels are labeled as "differently classified". Fig. 3 shows exemplarily a quick-look image (above; left site) with its auxiliary data (above; right site) in comparison to the overlaid classification results of NASA- and

ACRES-algorithms (below; left site) and the interpretation key of the mask (below; on right site). The visual control of the classification allows assessing the classification accuracy of both classification procedures.

Table 3. Confusion matrix for the entire classified data set (NASA-procedure (Reference) and ACRES-procedure (Classification))

		NASA-ACCA (Reference)		
		Cloud free	Cloud	Total
ACRES-ACCA	Cloud free	1260119847	163828367	1423948214
	Cloud	99051006	957000780	1056051786
	Total	1359170853	1120829147	2480000000

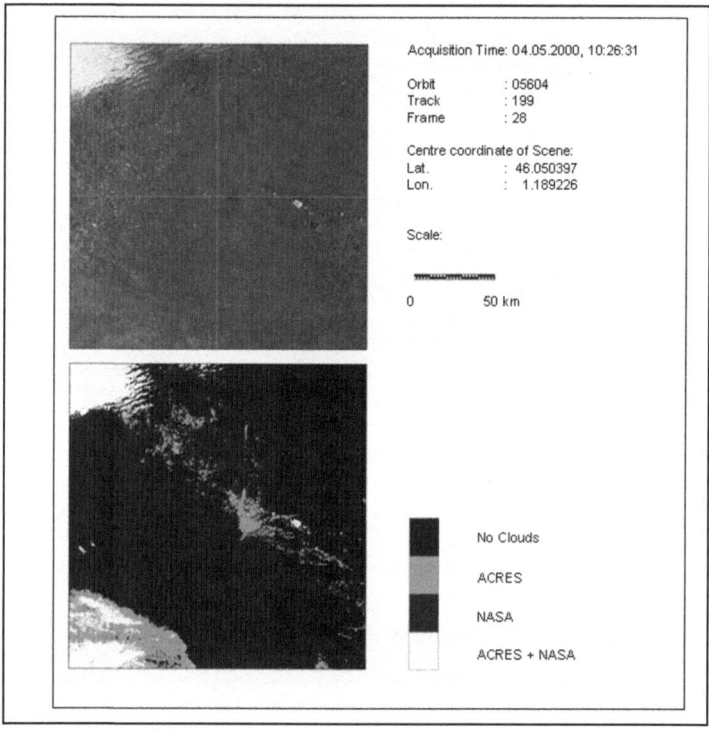

Fig. 3. Comparative analysis of classification results of the modified ACRES procedure overlaid by the classification result of NASA procedure

The qualitative assessment by visual control is complemented by a quantitative analysis. Accuracy of a classification can be assessed by a confusion matrix [17]. Hence a confusion-matrix is calculated pixel-wise for all cloud masks to obtain a classification accuracy statement of both procedures which allows further analysis. Table 3 shows the confusion matrix; and Table 4 shows the quality indices obtained from the confusion matrix. Additionally the Kappa coefficient KC (equation 6) and the overall accuracy OA (equation 7) are calculated [18].

$$KC = \left(n\sum_{i=1}^{k} n_{ii} - \sum_{i=1}^{k} n_{i+}n_{+i} \right) \bigg/ \left(n^2 - \sum_{i=1}^{k} n_{i+}n_{+i} \right) . \qquad (6)$$

$$OA = \sum_{i=1}^{K} n_{ii} \bigg/ n . \qquad (7)$$

k is the number of valuation classes, n is the number of reference pixels. n_{ii} are the elements of the principle diagonal, $n_{i+}n_{+i}$ are the elements of the matrix above and below the principle diagonal. The mean overall accuracy of method comparison for the NASA and ACRES procedures is calculated from the overall accuracies of the classified single quick-look data. In addition the corresponding standard deviation is calculated. The mean overall accuracy is 89.40 percent and standard deviation is 12.60 percent.

Table 4. Kappa -Coefficient, Overall Accuracy, Users Accuracy, Producer Accuracy of overall comparison

Kappa-Coefficient	Overall Accuracy [%]	Users Accuracy [%] ACRES-ACCA		Producer Accuracy [%] NASA-ACCA	
		Cloud free	Cloud	Cloud free	Cloud
0,78	89,40	92,71	85,38	88,49	90,62

The Kappa coefficient (scaled at interval [-1.0, 1.0]) of the combined NASA and ACRES procedures over all quick-look-data is 0.78. A validation of the classification accuracy can be carried out by using the valuation scale by [19]. Kappa coefficients in interval [0.81 – 1.00] are interpreted as "almost perfect" and in interval [0.61 -0.80] as "substantial". The overall accuracy values show that both algorithms can be applied to data-reduced quick-look products and to produce acceptable accurate and stable classification results.

The presented investigations demonstrate that a primarily physical evaluation of remote sensing data produces stable and accurate results for data-reduced JPEG-compressed quick-look data. Further examinations will show whether more complex evaluations are possible on the basis of reduced remote sensing datasets.

5   Conclusions and Discussion

In this paper a summary of some of the most used indices for measuring segregation or diversity in the distribution of migrant groups at urban level have been proposed, considering in particular the spatial aspects of such indices and the need to examine more in depth the articulated structure and characteristics of the population. Some problems still need to be addressed. On one side limitations can be noticed in the availability of disaggregated data, as individual nationals from different countries are often aggregated according to some zoning system that can affect the results from further analyses. However, if the zoning system produces sufficiently small areas some analytical methods reduce such problem. With reference to the indices used, an issue is still concerned with the choice of the bandwidth or distances of observation, although efforts in this direction are under exam [11] [25]. Other issues concern the multi-group analysis, therefore not limiting this to two subgroups but to the overall variety of countries represented in a given study region. Furthermore, qualitative, multivariate attribute of population data should be considered. There is also the need to explore the opportunity to develop and implement entropy-based diversity indices, as well as to examining the relations between economic activities, residential locations and segregation [25] [31] [32] as emerging migration issues to analyse [33] [34]. As a partial conclusion concerning the use of quantitative methods, and particularly those based on density or those based on a choice of distances, it is worth noting that these must be refined and that not 'easy' solutions of the problem at stake can be found, nevertheless they provide a good starting point for more in depth and local analysis by the researchers, that can focus their attention over a micro scale of analysis, going further than the administrative divisions of space and reducing the minimum distance of observation to examine locally the dynamics at urban scale.

Acknowledgements. This study was made possible by substantial data sponsoring from ESA. In particular, the authors would like to thank Drs. Berutti, Biasutti, and Pitella for their interest in the investigations presented. The authors would also like to thank the ground station operation team at DLR Neustrelitz.

References

1. Lam, K.W.-K., Lau, W.-L., Li, Z.-L.: Effects of JPEG Compression on Accuracy of Image Classification. In: Proceedings of GIS Developments, ACRS, p. 3 (1999), http://www.gisdevelopment.net/aars/acrs/1999/ts11/ts11009.shtml (last access: August 13, 2005)
2. Lau, W.-L., Li, Z.-L., Lam, K.W.-K.: Effects of JPEG compression on image classification. Int. J. Remote Sensing 24(7), 1535–1544 (2003)
3. Zabala, A., Pons, X., Masó, J., García, F., Aulí, F., Serra, J.: Evaluation of JPEG and JPEG2000 effects on remote sensing image classification for mapping natural areas. WSEAS Transactions on Information Science and Applications 6(2), 717–725 (2005)
4. Zabala, A., Pons, X., Diaz-Delgado, R., García-Vílchez, F., Aulí, F., Serra-Sagristà, J.: Effects of JPEG and JPEG2000 lossy compression on remote sensing image classification for mapping crops and forest areas. In: 26th International Geoscience and Remote Sensing Symposium. IEEE Press, Denver (2006)
5. Spaventa, V.D.: Personal communication (2004)
6. Lammi, J., Sarjakoski, T.: Image Compression by the JPEG algorithm. Photogrammetric Engineering and Remote Sensing 61, 1261–1266 (1995)
7. Lane, T.: JPEG image compression FAQ, part 1 and part 2 (1999), http://www.faqs.org/faqs/jpeg-faq/ (last access: February 20, 2012)

8. Irish, R.: Automatic Cloud Cover Assessment (ACCA) LANDSAT 7 ACCA. Goddard Space Flight Center. LANDSAT -7 Science Team Meeting (December 1-3, 1998), http://landsathandbook.gsfc.nasa.gov/pdfs/ACCA_slides.pdf (last access: February 20, 2012)
9. Irish, R.: LANDSAT 7 Automatic Cloud Cover Assessment. In: Sylvia, S.S., Descour, M.R. (eds.) Algorithms for Multispectral, Hyperspectral, and Ultraspectral Imagery VI. Proceedings of SPIE, vol. 4049, pp. 348–355 (2000)
10. Xu, Q., Wu, W.: ACRES Automatic Cloud Cover Assessment of LANDSAT 7 Images. In: Spatial Sciences Conference 2003 – Spatial Knowledge without Boundaries, Canberra, September 23-26, p. 10 (2003)
11. Slater, P.N., Biggar, S.F., Holm, R.G., Jackson, R.D., Mao, Y., Moran, M.S., Palmer, J.M., Yuan, B.: Reflectance and Radiance-Based Methods for the In-Flight Absolute Calibration of Multispectral Sensors. Remote Sens. Environ. 22(1), 11–37 (1987)
12. NASA (2011), http://landsathandbook.gsfc.nasa.gov/cpf/prog_sect9_2.html (last access: January 04, 2012)
13. Markham, B.L., Barker, J.L.: Spectral characterization of the LANDSAT Thematic Mapper Sensors. Int. J. Remote Sensing 6(5), 697–716 (1985)
14. Chander, G., Markham, B.L., Helder, D.L.: Summary of Current Radiometric Calibration Coefficients for Landsat MSS, TM, ETM+, and EO-1 ALI Sensors. Remote Sens. Environ. 113, 893–903 (2009), http://landportal.gsfc.nasa.gov/Documents/Landsat_Calibration_Summary.pdf (last access: February 20, 2012)
15. Gurney, R.J., Hall, D.K.: Satellite-derived surface energy balance estimates in the Alaskan Sub-Arctic. J. Clim. Appl. Meteor. 22(1), 115–125 (1983)

A Probabilistic Rough Set Approach for Water Reservoirs Site Location Decision Making

Shaaban M. Shaaban and Hossam A. Nabwey

Department of Engineering Basic science, Faculty of Engineering,
Menofia University, Menofia, Egypt

Abstract. Recently, advanced methods have been developed for selection of suitable sites for water reservoirs. Although these methods being developed and some new approaches, like GIS techniques, are being used currently, all these methods are mostly dependent on engineering decision making and a need for high cost. In this study, a new approach in the determination of water reservoirs location is proposed. The core of the proposed approach is a soft hybrid induction system called the Generalized Distribution Table and Rough Set System (GDT-RS) to discover classification rules. The system is based on a combination of Generalized Distribution Table (GDT) and the Rough Set methodologies the proposed approach is applied for water reservoirs site location and the results indicate that it is very effective and accurate.

Keywords: water reservoirs, Rough set theory, Decision making, Generalized Distribution Table (GDT).

1 Introduction

Water reservoirs are commonly used in nearly all water supply systems. Water storage is provided to ensure the reliability of supply, maintain pressure, balance pumping and treatment rates, reduce the size of transmission mains, and improve operational flexibility and efficiency. Site location investigations should be performed to select the appropriate site for reservoirs. In this paper, rough set theory which is introduce by pawlak (1982) [1] is applied as a machine learning method to select the most appropriate site for constructing a reservoirs. The process is rewired in details and the results are compared with those of a case study of a 15000 m^3 semi-buried concrete water reservoirs under construction in the North of Iran.

Many approaches have been proposed for extracting hidden relationships holding among pieces of information stored in a given database [2]. Rough set (RS) theory is first proposed by Z.Pawlak [3] in 1982. It is a kind of very useful mathematical tool to deal with vagueness and uncertainty information. Recently, this theory attracts more attentions in the fields of data mining, knowledge discovery in database (KDD) [4], pattern recognition, decision support systems (DSS) etc. The main idea of this theory is that it provides us with a kind of mechanism of extracting the classification rules by knowledge reduction, while keeping the satisfactory capacity of classification. There are many successful applications by using RS theory in the following areas such as machine learning, data mining, knowledge discovery, decision analysis, and knowledge acquisition discover knowledge and rules from database [5].

In this paper, a new simplified method for applying rough set theory, to determine the severity levels of distresses in high way pavements with experimental knowledge the. This method deals with the application of the rough set theory in decision making and takes into account all relevant factors for the judgment procedure.

The core of the approach is a soft hybrid induction system called the Generalized Distribution Table and Rough Set System (GDT-RS) for discovering classification rules. The system is based on a combination of Generalized Distribution Table (GDT) and the Rough Set methodologies.

2 Review of Rough Set Theory

The Rough Set Theory [4] is a new mathematical tool presented to dispose incomplete and uncertainty problem by Pawlak in 1982. He defined the knowledge according to new point of view, and regarded it as partition of universe. The concept of a rough set can be defined quite generally by means of topological operation, interior and closure, called approximations. Rough set theory can discover implicit knowledge and open out potentially useful rule by efficiently analyzing and dealing with all kinds of imprecise, incomplete and disaccord information [11].

2.1 Decision System

Definition 1): In rough set theory, an information system can be considered as system $S = (U, A, V, f)$, where U is the universe; $A = C \cup D$ is the sets of conditions attribute, the subset C and D are disjoint sets of fault symptoms attribute and decision attributes respectively; $V = \bigcup_{r \in R} V_a$, where Va is the value set of symptoms attribute a, is named the domain of attribute a. Each attribute $a \in A$; f is an information function $f : U \times A \to V$, and $f(x,a) \in V_a$, in which $x \in U$. Table 1 lists a decision table where U = {S1, S2.. S22}, C = {a, b, d, e, f, g, h, i, j, k, L, m} and d = decision levels. It will be employed throughout the paper.

Table 1. Decision Table

Sites	Conditional Attributes													Decision Levels
	a	b	c	d	E	f	g	h	i	j	k	L	m	
S_1	2	3	1	3	3	1	2	1	1	2	1	1	2	L
S_2	2	2	1	1	2	2	2	2	3	2	1	2	1	M
S_3	1	2	1	3	3	1	2	2	2	1	1	1	3	L

Table 1. (*Continued*)

S_4	1	3	2	2	1	1	2	1	1	1	1	1	H
S_5	1	3	1	3	3	2	2	2	1	2	1	2	3	L
S_6	2	2	2	2	2	1	2	1	1	1	2	1	2	M
S_7	1	2	2	2	2	1	2	1	1	1	2	1	2	M
S_8	1	2	1	1	2	1	2	1	1	1	2	1	2	H
S_9	2	2	1	3	3	1	2	2	1	2	3	1	3	L
S_{10}	1	2	1	1	2	1	2	1	1	1	1	1	1	H
S_{11}	2	3	2	3	3	1	2	2	2	3	3	1	3	L
S_{12}	1	2	1	1	2	1	2	1	1	2	1	1	1	H
S_{13}	2	2	1	2	3	3	2	3	2	3	2	2	3	L
S_{14}	1	2	1	1	1	1	2	1	2	1	1	1	1	H
S_{15}	2	2	1	2	3	1	2	1	2	1	1	1	2	L
S_{16}	1	3	2	1	2	1	2	2	3	3	2	1	2	M
S_{17}	1	1	2	2	2	2	2	1	2	1	1	2	1	M
S_{18}	1	3	1	3	3	2	2	2	1	2	3	2	2	L
S_{19}	1	1	2	1	1	1	2	1	1	1	2	1	2	H
S_{20}	1	1	2	2	1	1	2	1	1	1	1	1	1	H
S_{21}	1	2	1	2	1	1	2	1	1	1	2	1	2	H
S_{22}	2	2	2	2	1	2	2	3	2	2	2	2	1	L

2.2 Equivalent Relations

Definition 2): In decision system S = (U, A, V, f), every attributes subset, an indiscernible relation (or equivalence relation) IND(B) defined in the following way:

$$IND(B) = \{(x,y) \in U \times U \mid \forall a \in B, f(x,a) = f(y,a)\} \quad (1)$$

The family of all equivalence relation of IND(B), a partition determined by B, denoted by U/IND(B),$[x]_B$ can be considered as equivalence classes, and defined as follows:

$$[x]_B = \{y \in U \mid \forall a \in B, f(x,a) = f(y,a)\} \quad (2)$$

And

$$[x]_{IND(B)} = \cap [x]_B \quad (3)$$

2.3 Reduction and Core

Definition 3): In decision system S=(U, A,V ,f),Let $b \in B$ and $B \subseteq A$, if $pos_B(D) = pos_{B-\{b\}}(D)$, attribute b is redundant to B, which relatives to D, otherwise the attribute b is indispensable.

If IND(B) = IND(A) and POSB (D) ≠ POSB −{b} (D) , then B is called a reduction for information system S , are denoted as RED(A) ; the intersection of these reduction sets is called core, denoted as CORE = $\cap$ RED(A) .

3 Generalized Distribution Table

Generalized Distribution Table (GDT) is a table in which the probabilistic relationships between concepts and instances over discrete domains are represented [6], [7]. Any GDT consists of three components: possible instances, possible generalizations of instances, and probabilistic relationships between possible instances and possible generalizations. The possible instances, which are represented at the top row of GDT, are defined by all possible combinations of attribute values from a database, and the number of the possible instances is:

$$\prod_{i=1}^{m} n_i \qquad (4)$$

Where m is the number of attributes, n is the number of different data values in each attribute.

The possible generalizations for instances, which are represented by the left column of a GDT, are all possible cases of generalization for all possible instances, and the number of the possible generalizations is:

$$\left(\prod_{i=1}^{m}(n_i+1)\right) - \left(\prod_{i=1}^{m} n_i\right) - 1 \qquad (5)$$

The probabilistic relationships between possible instances and possible generalizations, represented by entries Gij of a given GDT, are defined by means of a probabilistic distribution describing the strength of the relationship between every possible instance and every possible generalization. The prior distribution is assumed to be uniform if background knowledge is not available. Thus, it is defined by:

$$G_{ij} = p(PI_j \setminus PG_i)$$
$$= \begin{cases} \dfrac{1}{N_{PG_i}} & \text{if } PG_i \text{ is a generalization of } PI_j \\ 0 & \text{otherwise} \end{cases} \qquad (6)$$

where

PI_j is the j th possible instance,

PG_i is the ith possible generalization,

and N_{PG_i} is the number of the possible instances satisfying the ith possible generalization, that is :

$$N_{PG_i} = \prod_j^m n_j \qquad (7)$$

where j = 1,. . . , m, and j # the attribute that is contained by the ith possible generalization (i.e., j just contains the attributes expressed by the wild card) .

Table 2. The class numbers of conditional attributes

Conditional Attributes	Classification of Individual Situations
(a)Topography	1- flat area 2- low hills 3- fairly high hills 4- mountains or relatively high hills
(b) Geology and tectonics	1- very fine compacted soil 2- sedimentary layers 3- possibility of layers movement or landslide 4- low quality soils with harmful minerals, fault zone area
(c) Accessibility of electric power	1- availability near the site 2- relatively near the site 3- far from the site
(d) Distance from water resource	1- water resource near the site 2- relatively near the site 3- relatively long distance 4- very far from the site
(e) Ground flooding	1- without any structure 2- existence of roads 3- existence of farmland houses 4- existence of farms, houses and other facilites
(f) Flood risk	1- low severity 2- low risk 3- high risk
(g) Seismicity background	1- low severity 2- medium severity 3- high severity 4- very high severity

Table 2. (*Continued*)

(h) Bearing capacity of soil	1- high bearing capacity 2- quite good bearing capacity 3- relatively low bearing capacity 4- very low bearing capacity
(i) Water distribution system	1- gravity 2-gravity and pumping 3- pumping 4- very difficult to transit
(j) Future development consideration	1- can be developed with no problem 2- need to destroy some roads 3- need to destroy farmland houses 4- need to destroy buildings or other facilities
(k) Access to project	1- existence of access road 2- quite easy access to site by miner roads 3- need to construct access road to site
(l) Environmental impacts	1- no Environmental impacts 2- creating some solvable Environmental problems 3- sever Environmental impacts
(l) Economical consideration	1- very economic 2- partly economic 3- none economical

4   Determination of Minimal Decision Algorithm

Firstly, to be able to make use of rough set theory to detect effective parameters in relation to water reservoirs location, the germane data and information have to be gathered and provided. In this paper, the mentioned information has been prepared and collected from a practical case, water reservoirs Maklavan (Guilan Province, North of Iran).

The collected data categorized as a table called decision table in which each row reflects the specifications of a particular segment. Each column of the mentioned table indicates one of the characteristics considered (conditional attributes). Data tables are usually difficult to evaluate. They may store an enormous quantity of data, which is hard to manage for decision making. Due to geotechnical and civil uncertainties, some data may not be reliable for the project [8]. One of the main objectives of Rough Sets data analysis is to reduce data size. In order to show the applicability of the Rough Sets theory, the information table was constructed, according to standards that are prepared by experts in water industry [9]. All the knowledge available about the site is the corresponding row in the table as (a) Topography; (b) Geology and tectonics and (c) Accessibility of electric power. The last column shows the suitability of that site for the project (decision attribute).Table 1 shows the decision table.

Secondly, to utilize the rough-set theory and analyze the information placed in the mentioned table, it is required to classify the information. So, each conditional attribute is provided with 4 classes, which show high, medium, low and no severity and the decision parameter (attribute) is classified by three levels, which describe high, medium and low suitability conditions, H; M and L. Consequently, by defining the specified levels and allocating a digit code to each defined attribute in the rows of the table, the classification of all attributes has been undertaken. Table 2 shows the class numbers of conditional attributes and danger levels for 22 sites diagnosed by experts.

Alternatively, in order to simplify the decision table it is required to find the insignificant conditional attributes in the diagnoses, i.e. it is required to find a minimum number of conditional attributes, but, are still able to diagnose the problem. This is done by number of conditional attributes should be removed each time and the decision table should be checked to make sure any contradiction has not occurred. In this paper this computing the reduct is done by using software called ROSETTA.

In case of Table 1, the conditional attributes, other than those of (a), (c) and (e), were removed from Table 1, so a new table obtained. In the new table, a pair of segments S1 and S9 and S14 and S21 and so on, has been classified by one and the same rule used for each. The segments of each pair were removed except one and the result is given in Table 3.

Finally, the proposed algorithm is applied to the simplified decision table shown in table 3. In order to make the algorithm easier to comprehend, an example will be analyzed the prior to any theoretic analysis. A sample decision table shown in Table 3 is used to illustrate the idea then the GDT is proposed as the basis of the algorithm and finally the algorithm is listed.

Table 3. Simplified Decision table S

Sites	Conditional Attributes			Decision Levels
	a	c	e	
S_1	2	1	3	L
S_2	2	1	2	M
S_3	1	1	3	L
S_4	1	2	1	H
S_6	2	2	2	M
S_7	1	2	2	M
S_8	1	1	2	H
S_{11}	2	2	3	L
S_{14}	1	1	1	H
S_{22}	2	2	1	L

Step 1. Create the GDT

Since:

Topography $\in \{1, 2\}$ $\Rightarrow n_1 = 2$

Accessibility of electric power $\in \{1, 2\}$ $\Rightarrow n_2 = 2$

Ground flooding $\in \{1, 2, 3\}$ $\Rightarrow n_3 = 3$

Hence :
the number of attributes (m) = 3 ,
from Eq.(2) number of the possible instances is 12 ,
from Eq.(3) number of the possible generalizations is 23 ,

Table 4. GDT table

(Note the elements that are not displayed are all zero)

	111	112	113	121	122	123	211	212	213	221	222	223
*11	½						½					
*12		½						½				
*13			½						½			
*21				½						½		
*22					½						½	
*23						½						½
1*1	½			½								
1*2		½			½							
1*3			½			½						
2*1							½			½		
2*2								½			½	
2*3									½			½
11*	1/3	1/3	1/3									
12*				1/3	1/3	1/3						
21*							1/3	1/3	1/3			
22*										1/3	1/3	1/3
**1	1/4			1/4			1/4			1/4		
**2		1/4			1/4			1/4			1/4	
**3			1/4			1/4			1/4			1/4
1**	1/6	1/6	1/6	1/6	1/6	1/6						
2**							1/6	1/6	1/6	½	1/6	1/6
1	1/6	1/6	1/6				1/6	1/6	1/6			
2				1/6	1/6	1/6				1/6	1/6	1/6

Step 2. Simplify the GDT.

By deleting all of the instances and generalizations un-appeared in the example database shown in Table 4. So the simplified GDT is shown in Table 5.

Table 5. The simplified GDT

(Note the elements that are not displayed are all zero)

	111	112	113	121	122	211	212	213	222
*11	½					½			
*12		½					½		
*13			½					½	
*21				½					
*22					½				½
1*1	½			½					
1*2		½			½				
1*3			½						
2*1						½			
2*2							½		½
2*3								½	
11*	1/3	1/3	1/3						
12*				1/3	1/3				
21*						1/3	1/3	1/3	
22*									1/3
**1	1/4			1/4		1/4			
**2		1/4			1/4		1/4		1/4
**3			1/4					1/4	
1**	1/6	1/6	1/6	1/6	1/6				
2**						1/6	1/6	1/6	1/6
1	1/6	1/6	1/6			1/6	1/6	1/6	
2				1/6	1/6				1/6

Step 3. Group the generalizations.

Generalizations can be divided into four groups contradictory, belonging to class High, belonging to class Medium, and belonging to class Low. The contradictory generalizations, containing the instances belonging to different decision classes, cannot be used as the rules. Hence they are ignored. In other words, we are just interested in the generalizations belonging to class High or Medium or Low, which will be selected as the rules.

Table 6 lists the generalizations belonging to class High, Table 7 lists the generalizations belonging to class Medium and Table 9 lists the generalizations belonging to class Low.

Table 6. The generalizations belonging to class High

	111	112	121
*11	½		
*12		½	
*21			½

Table 6. (*Continued*)

1*1	½		½
1*2		½	
11*	1/3	1/3	
12*			1/3
**1	1/4		1/4
**2		1/4	
1**	1/6	1/6	1/6
1	1/6	1/6	
2			1/6

Table 7. The generalizations belonging to class Medium

	122	212	222
*12		½	
*22	½		½
1*2	½		
2*2		½	½
12*	1/3		
21*		1/3	
22*			1/3
**2	1/4	1/4	1/4
1**	1/6		
2**		1/6	1/6
1		1/6	
2	1/6		1/6

Table 8. The generalizations belonging to class Low

	113	211	213
*11		½	
*13	½		½
1*3	½		
2*1		½	
2*3			½
11*	1/3		
21*		1/3	1/3
**1		1/4	
**3	1/4		1/4
1**	1/6		
2**		1/6	1/6
1	1/6	1/6	1/6

Step 4. Rule Selection.
The algorithm has an advantage that it allows to the user to select the decision rules based on the previous experience or any required demands. There are several possible ways for rule selection. For example:

- Selecting the rules that contain as many instances as possible.
- Selecting the rules in the levels of generalization as high as possible according to the number of " * " in a generalization.
- Selecting the rules with larger strengths.

Here All the decision rules that can be induced:

Rule 1: If topography is flat area Then decision level is High, with strength S=1/2

Rule 2: If accessibility of electric power is relatively near the site and ground flooding is existence of roads Then decision level is Medium, with strength S=1.

Rule 3: If topography is low hills and ground flooding is existence of roads Then decision level is Medium, with strength S=1.

Rule 4: ground flooding is existence of farmland houses Then decision level is Low, with strength S=1/2.

Rule 5: If topography is low hills and ground flooding is without any structure Then decision level is Low, with strength S=1/2.

5 Analyzing the Information and Detecting Parameters, Using Linear Regression Method

The linear regression method is a mathematical tool reflecting the linear correlation between independent and dependent parameters [10]. Through the said method, the changes existed in the dependent parameters can be evaluated. In this method, mathematical correlative equations are used. The selection of the best form of the said equations and the determination of their parameters require a great deal of study, concerning the aforementioned issues. The current and common form of the correlation model used in the linear regression model is as follows:

$$Y = a_0 + a_1 X_1 + a_2 X_2 + \ldots + a_n X_n \tag{8}$$

Here, Y is the dependent variable, $X_{1\ldots n}$ are independent variables and $a_{0\ldots n}$ are coefficients of the equation, which have already been determined. Since the purpose of this project is the recognition of the most important and effective attributes, concerning the site location evaluation of water reservoirs, the dependent parameters ought to be the indicator of the selective level of water reservoirs site location. Independent parameters must be the indicator of water reservoirs Characteristics.

However, there is a great problem in the construction and formation of regression equations which are used to select the best site, because using all the parameters and factors to form the regression equation is practically difficult and even at times impossible. Consequently, it is very hard and even infeasible to gather a lot of input information, required for the said issue. Thus, the stepwise method can be used to determine the shortest and the most suitable and possible combinations of attributes or to detect the most important defined attributes in the information system which are considered to define the observed changes concerning the values of the dependent parameter. In this method, different parameters are used to gain the best linear correlation in the model in question to obtain the highest value of R^2 by dependent parameters. In this process, at first, the value of the correlation coefficient between each independent parameter and dependent variable is evaluated. This is accomplished to determine which independent parameter enjoys the highest degree of correlation coefficient with the dependent parameter. In the next step, the said process is continued by adding each independent parameter to the primary one, under the framework of a linear regression equation with two independent variables. So, at each step, the value of R^2 is assessed. This trend continues until the best secondary parameter of the independent attributes is gained (and they do not have a high value of correlation with each other). This process continues until, with the addition of another independent parameter to the model, some changes happen in the value of R^2, which are trivial and also negligible. Consequently, the existed parameters in the linear regression equation obtained from this method, are treated as the most significant defined parameters in the information system. According to the linear regression method, these parameters can analyze the observed changes at the levels of dependent parameters in the best way and can be used in the site location evaluation of water reservoirs. In this investigation, the analysis of the information has been accomplished via a stepwise method and it is concluded that a Ground flooding parameter is the most significant attribute, enjoying the highest degree of correlation coefficient with a dependent parameter. By the same taken, the parameters regarding Topography and Distance from water resource are treated as the second and third most important parameters in the information system respectively. The said parameters form a four variable regression equation, which has the highest value of R^2 between the other four-variable equations. Table 9 indicates the gained linear

Table 9. Stepwise Regression Equation

step	parameters	Equations	R^2
1	Ground flooding	Decision= 0.243+0.862e	60.5%
2	Topography	Decision= -0.485+0.755e+0.675a	79.3%
3	Distance from water resource	Decision= -0.714+0.412e+0.525d+0.609a	87.3%
4	Water distribution system	Decision= 0.1.14+0.434e+.0468a+0.562d+0.347i	92.4%

regression equations. At the end of this step-by-step method, the result indicates that the Water distribution system, after the three above-mentioned parameters, is the most significant factor among independent parameters, for estimating the site location of water reservoirs. As it is, the set of these four parameters resulted from the stepwise regression analysis is considered as the best algorithm for the site location of water reservoirs.

6 Comparing the Results of Rough Set Theory with The Stepwise Regression Method

In order to investigate and evaluate the results of rough-set analysis, the linear regression equations have been formed. Therefore, as the first criterion, the R^2 is obtained from the multiple linear regression equations by various decision-making reducts. These are accomplished by statistical software, the results of which are reflected in Table 10. As mentioned earlier, the algorithms presented in this investigation are decision-making algorithms resulted from the rough-set analysis and stepwise regression method. In the second part, to evaluate the results of the stepwise method, the values of the accuracy and quality of approximation have been studied. These values represent how precisely the independent defined parameters, regarding the reducts, can predict the dependent parameter. In other words, they represent the accuracy and quality of the algorithms studied in the analysis of water reservoirs site location. In essence, although the stepwise algorithm studied confirms the accuracy and quality of approximation resulted from the rough-set analysis, the differences between these values with the algorithms concerning the rough-set analysis are 34% and 20%, respectively. Therefore, the three important indicators (Coefficient of Determination, Accuracy and Quality of Approximation) studied in this research are true for the reducts resulted from the rough-set analysis. First of all, the multiple

Table 10. Coefficient of determination, Quality and Accuracy of Approximation of different Reducts

Reducts	R^2	Accuracy of Approx.	Quality of Approx.
{a, c, e} [a]	81.6%	1	1
{c, e, f} [a]	72.4%	1	1
{c, e, h} [a]	81.4%	1	1
{d, e, h} [a]	86.9%	1	1
{d, e, i} [a]	84.8%	1	1
{c, e, i} [a]	68.8%	1	1
{a, d, e, i, k} [a]	92.5%	1	1
{a, d, e, i} [b]	92.4%	0.66	0.8

[a] The shortest decision-making Reducts resulted from Rough Sets analysis.
[b] The Reduct resulted from stepwise regression analysis.

linear regression equations resulted from these decision-making algorithms confirm the desirable values of R^2, to estimate the site location of water reservoirs and to explain the reasons for the site location. Second, the above-mentioned algorithms confirm the highest values of accuracy and quality of approximation in the processing of information and the evaluation of water reservoirs site location. Consequently, reducts resulted from the presented rough-set analysis can better explain the differences in water reservoirs locations. So, these algorithms can be considered as the most confident in the information process.

7   Conclusion

In this paper, a probabilistic rough set theory-based approach as a new method is presented and evaluated to determine site locations of water reservoirs using a complete practical example. Also the results of this method and linear regression method have been compared. . Rough set theory is a powerful mathematical tool that enables the users to decide and judge the minimal algorithms required for a particular problem, so that the accuracy of the selected algorithm is kept while the number of attributes is remarkably reduced. The major advantages of this method are: The simplicity of input data construction, the high speed and precision of the method during recognition of the simplified algorithm and the avoidance of the formation of weighted matrixes for comparison of alternatives. Therefore, in this method, there is no need to compare parameters at the beginning and all relevant parameters are equally entered in the calculation procedure, which results in a remarkable saving in the time and cost of the calculation. The most important results of the present paper are as follows:

a) The advantages of the rough set theory in water reservoirs location present the high reliability of this method in the investigated example and can be introduced as a well-built useful mathematical based method.
b) By using the presented method, the cost of the process of water reservoirs location, in this particular case, can be reduced by up to 77% and can lessen the overall cost of the feasibility study.
c) The rough set theory is used for expressing a set of decision attributes by a set of paired conditional attributes. Thus, one can use this theory as a method of acquiring information from the diagnostic cases that exist in civil engineering.

References

[1] Pawlak, Z.: Rough sets: Theoretical Aspects of Reasoning about Data. Kluwer Academic Publishers (1991)
[2] Hung, C.T., Chang, J.R., Lin, J.D., Tzeng, G.H.: Rough Set Theory in Pavement Maintenance Decision. Springer (2009)
[3] Chang, J.R., Hung, C.T., Tzeng, G.H., Hsiao, W.H.: Pavement maintenance and rehabilitation decisions derived by Rough Set theory. In: Joint International Conference on Computing and Decision Making in Civil and Building Engineering, Montreal, Canada, June 14-16 (2006)

[4] Miradi, M., Molenaar, A.A.A., van de Ven, M.F.C.: Knowledge Discovery and Data Mining Using Artificial Intelligence to Unravel Porous Asphalt Concrete in the Netherlands. In: Gopalakrishnan, K., Ceylan, H., Attoh-Okine, N.O. (eds.) Intelligent and Soft Computing in Infrastructure Systems Engineering. SCI, vol. 259, pp. 107–176. Springer, Heidelberg (2009)
[5] Zhou, Y., Fried, S., Stiemer, F.: Engineering analysis with uncertainties and complexities, Using reasoning approaches. In: Joint International Conference on Computing and Decision Making in Civil and Building Engineering, Montreal, Canada, June 14-16 (2006)
[6] Zhu, W., Zhang, W., Fu, Y.: An Incomplete Data Analysis Approach using Rough Set Theory. In: Proceedings of the 2009 International Conference on Intelligent Mechatronics and Automation, Chengdu, China (August 2009)
[7] Pawlak, Z., Busse, J.G., Slowinski, R., Ziarko, W.: Rough Sets. Communication of the ACM 38(11), 89–95 (1995)
[8] Arabani, M., Lashteh Neshaei, M.A.: Application of Rough Set theory as a new approach to simplify dams location. Scientia Iranica 13(2), 152–158 (2006)
[9] Iran bureau for water and wastewater engineering system and standards, code 436, Guideline for hydraulics of water treatment plant (2008)
[10] Chattefuee, S., Hadi, A.S.: Regression Analysis by Example, 4th edn. John Wiley & Sons (2006)

Definition and Analysis of New Agricultural Farm Energetic Indicators Using Spatial OLAP

Sandro Bimonte, Kamal Boulil, Jean-Pierre Chanet, and Marilys Pradel

Irstea, UR TSCF, 24 Avenue des Landais, 63172 Aubière, France
name.surname@irstea.fr

Abstract. Agricultural energy consumption is an important environmental and social issue. Several diagnoses have been proposed to define indicators for analyzing energy consumption at large scale of agricultural farm activities (year, farm, family of production, etc.). However, to define ad-hoc environmental energetic policies to better monitor and control energy consumption, new indicators at a most detailed scale are needed. Moreover, by defining detailed scale indicators, large quantities of geo-referenced data need to be collected to feed these energetic diagnoses. This huge volume of data represents another important limitation of systems that implement these diagnoses because they are usually based on classical data storage systems (such as spreadsheet tools and Database Management Systems). These systems do not allow for interactive analysis at different granularities/scales of huge volumes of data and do not provide any cartographic representation. By contrast, Spatial OLAP (SOLAP) and spatial data warehouse (SDW) systems allow for the analysis of huge volumes of geo-referenced data by providing aggregated numerical values visualized by means of interactive tabular, graphical and cartographic displays. Thus, in this paper, we (i) propose new appropriate indicators to analyze agricultural farm energy performance at a detailed scale and (ii) show how SDW and SOLAP technologies can be used to represent, store and analyze these indicators by simultaneously producing expressive reports.

Keywords: Spatial Data Warehouses, Spatial OLAP, Energetic indicators.

1 Introduction

Agriculture energy consumption depends on the method of production used. Direct agricultural energy consumption was estimated to be 28 Mtpe of a total consumption of 1142 Mtpe, which was 2.5% of the energy directly consumed by the EU25 in 2004 [17]; 55% of this energy was the result of fuel consumption. With the planned reduction in oil and rising oil prices, agriculture must reduce energy consumption to improve its economic development and decrease its environmental impact. Awareness of the importance of preserving non-renewable energy resources is a certainty, as evidenced by the energy development policies adopted in recent years by different governments (Energy Policy of 2005, the Grenelle Environment, etc.). Applied to the agricultural context, this reality requires a better assessment of the energy balance of farms in terms of energy performance. At present, many diagnoses that define a set of indicators exist to assess the energy performance of agricultural farms [13]. These diagnoses, which are specially adapted to a comprehensive assessment at the farm scale, are not necessarily relevant to evaluate the energy performance at a detailed scale (plot, technical operation, etc.). Moreover, by defining detailed scale indicators, large quantities of geo-referenced data have been collected to feed these energy diagnoses. This huge volume of data represents another important limitation of systems that implement these diagnoses because they are usually based on classical data storage systems, such as spreadsheet tools, which do not allow interactive analysis at different granularities/scales of huge volumes of geographic data. Moreover, diagnosis systems do not provide any cartographic visualization of energy indicators by limiting important analysis capabilities associated with geographic data [9].

By contrast, Geo-Business Intelligence technologies such as Spatial Data Warehouses (SDW) and Spatial OLAP (SOLAP) systems are widely recognized as efficient tools for the on-line analysis of huge volumes of geo-referenced datasets. SOLAP systems have been defined by Y. Bédard as "*Visual platforms built especially to support rapid and easy spatiotemporal analysis and exploration of data following a multidimensional approach comprised of aggregation levels available in cartographic displays as well as in tabular and diagram displays*" [3]. In the last years, SOLAP technology has been successfully used in several application domains such as geo-marketing, health monitoring, and agriculture [11].

Thus, in this paper we propose (i) *some new appropriate indicators to analyze agricultural farm energy performance at a most detailed scale* and (ii) *show how SDW and SOLAP technologies can be used to represent, store and analyze these indicators by simultaneously producing cartographic and tabular reports*.

The case study of this work is a result of the EnergeTIC project. This project aims to use a scientific and technical solution to assess the energetic performance of farms through the use of ICT (Information and Communication Technologies) on the finest scale. Installed on agricultural equipment, the identified technological solutions (low-cost sensors, RFID, etc. provide reliable and continuous data to calculate energetic performance indicators at the finest scale (field, technical operation...).

The paper is organized as follows. Section 2 presents related work concerning agricultural energy diagnoses, SDW and SOLAP. New indicators for farm energy consumption at a detailed scale are presented in Section 3. Section 4 describes how the SDW system is used to represent these new indicators and their implementation in the SOLAP tool JMap-SOLAP. Conclusions and future work are presented in Section 5.

2 Related Work

In this section, we introduce the main concepts of Spatial OLAP and the Spatial data warehouse (Sec. 2.2), and provide a survey of agricultural farm energy consumption diagnoses (Sec. 2.1).

2.1 Agricultural Farm Energy Consumption Diagnosis and Related Indicators

In the last few years, several studies dealing with direct and indirect energy consumption at the farm scale have been proposed [2][12][19]. These works aim to create complete diagnosis tools and/or energy performance assessment methods. The term "diagnosis" will be used in this paper to refer to both "tool" and "method". These diagnoses, based on the energy consumption of farms and the energy value of agricultural products, allow for the calculation of energy balance and the assessment of the energy efficiency of farms.

In particular, the energy diagnosis aims to:

— quantify energy consumption per processes to identify possible improvements by acting either on the production system, practices or equipment
— compare energy performance of livestock and crop farming, and
— establish reference values for the above production.

These diagnoses are based on the calculation of energy indicators, which are variables that provide information on less accessible data. These indicators are references used to make decisions [8], as they allow for the understanding of a complex system to assist in the realization of objectives [10].

We performed a survey of farm energy performance diagnoses used in France (Table 1). These diagnoses can be grouped into 6 categories (some examples of indicators are also provided):

— *Global agro-environmental diagnoses*: these consider the farm as a whole and assess the global environmental impact of the farm (nitrogen excess, energy consumption, etc.) (*Sum of the quantities of direct and indirect energy used by the farm, expressed in MJ per year*)
— *Field agro-environmental diagnoses*: these assess the environmental performance of the farm at the field scale (*Sum of the quantities of direct and indirect energy used by the farm, expressed in MJ per hectare per year*)
— *Sustainability diagnoses*: these assess farm sustainability performance through environmental, social and economic sustainability issues
— *Energetic diagnoses:* these allow an energy balance at the farm scale by quantifying energy inputs and outputs (*Energy efficiency: $\sum$(energy produced) / $\sum$(direct and indirect energy consumed)*)
— *Prediagnoses or autodiagnoses*: these assess the energy consumption of farm equipment and give an idea of the main possible improvements (based on the farmer's qualitative assumptions)
— *Life cycle assessment*: these methods assess the environmental impact on a system by inventory pollutant emissions, raw materials and energy during its whole life cycle (*Energy consumption based on the functional unit used, either the hectare or milk liter*)

The main inputs of these diagnoses are direct and indirect energy flows and energy consuming operations (energy used for irrigation, for example). Direct energy flows are electricity, fuels and gases (propane, butane and city gas). Indirect energies are energies used to produce outputs such as fertilizers, crop protection products, etc.

Table 1. Farm energy consumption diagnoses

Diagnosis	Scale	Category
PLANÈTE	Farm	Energy diagnoses
IDEA	Farm	Sustainability diagnoses
DIALECTE	Farm	Global agro-environmental diagnoses
DIALOGUE	Field Farm	Global/field agro-environmental diagnoses
INDIGO	Field Farm	Field agro-environmental diagnoses
BILAN CARBONE	Economic activity	Life cycle assessment
AUTO DIAGNOSTIC ÉNERGÉTIQUE DES BATIMENTS D'ÉLEVAGE	Farm	Auto diagnosis
GESTIM	Farm Operation	Energy diagnoses
AUDIT ÉNERGÉTIQUE EN PRODUCTION LAITIÈRE	Farm	Auto diagnosis
KUL	Farm	Global agro-environmental diagnoses
SALCA	Farm	Life cycle assessment

The most common indicators are simple or aggregated indicators. For example, a simple indicator is the fuel quantity and two aggregated indicators are the energy balance and energy efficiency of farms. The calculation of these quantities is based on direct and indirect energy data collected from accountancy, paper documents or direct communication between farmers (equipment and building characteristics, produced quantities, etc.). Indeed, to the best of our knowledge, there is actually no technical solution implementation to collect farm energy data at this fine scale

Information obtained from these diagnoses allows for a good analysis of the global farm energy performance by identifying the most energy-consuming activities. As shown in Table 1, indicator calculation is most often limited to the global scale (i.e., farm scale) due to the lack of reliable data. An analysis at a most detailed scale (field, production activity or operation) will require more precise data acquisition systems.

This means that a set of new indicators at a most detailed scale is needed. Moreover, because collecting data at a fine scale produces a huge volume of data, classical systems used to analyze these indicators (e.g., spreadsheet tools and database management systems) are not sufficient in this context because they do not allow visual/cartographic interactive analysis at different granularities/scales of huge data volumes.

2.2 Spatial OLAP Main Concepts

A data warehouse is defined as "*a collection of historical, integrated uniformed collection of data to support decision making*" [15]. Warehoused data are organized according to the multidimensional model, which defines analysis axes, named dimensions, and analysis subjects (facts). Facts are described by numerical indicators (measures) and are analyzed along dimensions at different granularities or scales defined by hierarchical structures that compose the dimensions. A measure, when "observed" at coarser hierarchical levels, is aggregated using classical SQL aggregation functions such as SUM, MIN, MAX, etc. Multidimensional data are analyzed using OLAP tools, which provide operators to interactively analyze and explore data. These tools include Slice, which selects a part of the data warehouse; Dice, which projects one dimension; Roll-Up, which allows climbing into hierarchy-aggregating measures; and Drill-Down, which is the inverse of Roll-Up. OLAP tools, contrary to classical DBMS systems, are effective decision support systems, as they allow decision-makers to explore large quantities of data online by triggering OLAP operators with simple interaction through visual interfaces to produce graphical and tabular reports.

Introduction of spatial data in dimensions and facts leads to the concepts Spatial Data Warehouses (SDWs) and Spatial OLAP (SOLAP). SOLAP tools integrate OLAP and Geographic Information Systems (GIS) advanced functionalities to explore, visualize and analyze multidimensional geo-referenced data by means of tabular, graphic and interactive cartographic visualization [3]. In this way, spatial decision makers can visually detect unknown patterns and spatial phenomena and verify and/or formulate hypotheses. An example of the use of SDW to analyze pollution by French departments is presented in [4]. Here dimensions include (i) the temporal dimension organized into a classical calendar hierarchy (day, month, year), (ii) a dimension representing pollutants and (iii) a spatial dimension representing the French administrative organization into departments and regions. The measure is the pollution value that is aggregated using the average. Using this SDW, users can answer questions like: "*What was the average pollution value per pollutant and department in 2000?*" or "*What is the average pollution value per month per region for inorganic pollutants?*".

A typical SOLAP architecture is composed of a Spatial DBMS to store warehoused (spatial) data; a SOLAP server, which implements OLAP operators; and a SOLAP client, which combines and synchronizes tabular, graphical and interactive maps to visualize and trigger SOLAP queries. An example of SOLAP visualization is shown in Figure 1 using the environmental SOLAP application previously described.

There are various SOLAP application domains, such as health, urban monitoring and marketing. Recently, SOLAP has also been applied to the agricultural context [18]. The monitoring of pollutants has been investigated in [1][4]. Economic analysis of agriculture productions is presented in [6][7][11][14][16]. These works highlight the relevance of using spatial decision support tools such as SDW and SOLAP in the agricultural context, and include evidence for particular issues concerning design of dimensions (complex hierarchies), facts (measures at different granularities, measure types, etc.) and architectural solutions (integration of GIS and Spatial Data Mining, etc.). However, to the best of our knowledge no works have studied the use of SDW and SOLAP systems to produce, aggregate and visualize energy consumption indicators of agricultural farms.

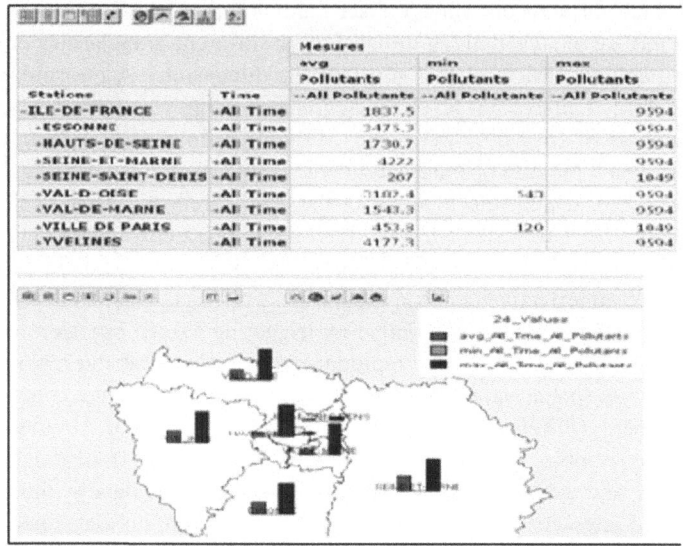

Fig. 1. SOLAP application concerning the monitoring of pollutants [4]

3 New Energy Indicators to Assess Farm Performance at a Detailed Scale

The assessment of energy performance of farms at a detailed scale aims to establish what types of indicators are needed and the related assessment scale. We choose to design two types of indicators (Table 2):

— Indicators based on invoices, direct communication between farmers or administrative documents to assess indirect energy consumption (foodstuffs, fertilizers, pesticides, etc.).
— Indicators based on direct measurements collected by means of technological solutions installed on mobile equipment to assess direct energy consumption.

The indicators were designed according to the most relevant analysis scale:

— Spatial scale: the hectare is the spatial reference unit used to express the technical-economic results on a farm. We will use it to express all the direct energy flows, such as fuel or gas for mobile equipment, or to express indirect energy flows such as inputs in crop management (e.g., fertilizers, pesticides).

— Temporal scale: the hour is the temporal reference unit for farmers. This unit is frequently used to express fuel consumption for mobile equipment and electric/gas consumption for electric equipment.
— Production scale: all of the energy flows involved in farm activities (crop or cattle management) can also be expressed at the production scale. The results will be expressed with the common production unit used by the farmer depending on the type of farm production (e.g., tons of dry matter, liters of milk, etc.).

These indicators were calculated for the three main farm activities:

— Crop management activities, which include all the technical operations on crops (sowing, plowing, fertilization, harvest, etc.);
— Cattle management activities, which include all the technical operations on cattle, such as care, feeding and milk/meat production (milking, slaughtering, etc.);
— General activities, which cannot be allocated to cattle or crop management (cleaning, logistics, transport, etc.).

These indicators can be aggregated at the global scale using the sum.

An example indicator is the fuel consumption per plot, technical operation, year and production. An example of the value of this indicator is "140 liters of fuel used for the parcel '13 pal' of the farm of Montoldre in 2010 during the plowing operation for the production of wheat". An example of the aggregated indicator obtained from the previous indicator by aggregation on the spatial scale is the fuel consumption per department, technical operation, year and production. The aggregates are calculated by summing all the plots' values belonging to the same department (i.e., Allier).

4 Spatial OLAP Analysis of New Indicators

In this section, we propose a system to implement (represent, store and analyze) the indicators defined in Section 3. The main idea is to represent detailed indicators defined in Section 3 as measures of a spatial data warehouse and to represent their inputs (energy, time, spatial scale, technical operation and production) as dimensions. Thus, aggregated indicators correspond to aggregated measures.

In this way, the spatial data warehouse can provide answers to the questions of two types of users: farm managers and life-cycle assessment (LCA) practitioners. From the perspective of farm managers, useful SOLAP (i.e. (aggregated) indicators) queries are the "*number of interventions by culture*", etc. For LCA practitioners, questions arise in terms of life cycle assessment inventories. For example, it may be interesting to know the "*fuel to weed a plot of wheat*" and the "*average consumption of fuel to weed the entire farm per year*".

Using a SOLAP system to represent indicators overcomes the limits of existing diagnosis systems with respect to two aspects: it allows the interactive analysis of huge volumes of (aggregated) indicator values and it provides a cartographic representation.

Table 2. New indicators to assess farm performance at a detailed scale

Energy flow			Indicator	Example	Indicator objective
Direct energy	Spatial scale	Fuels	FU[1]/crop ha/plot/technical operation	Energy consumption to weed 1 ha of wheat for plot X	Assess the most energy consuming operations for each technical operation for each crop. Compare the energy consumption of the same operation for different crops
		Fuels, gas, renewable energies	FU/m²/type of building	Energy consumption to heat 1 m² of the milking parlor using electricity	Assess the most energy consuming buildings
	Temporal scale	Fuels, butane, propane gas	FU/hour/technical operation/equipment/crop	Energy consumption for 1 hour of weeding wheat with the XY equipment	Identify the most energy consuming equipment for each technical operation for each crop system
Indirect energy	Spatial scale	All inputs	FU/crop ha/technical operation	Energy consumption to weed 1 ha of wheat	Assess the most energy consuming operations for each technical operation for each crop
		Cleaning products	FU/m²/building	Energy consumption to clean 1 m² of the milking parlor	Assess the most energy consuming buildings
	Temporal scale	All inputs	FU/year/production cycle/technical operation/crop	Annual (or production cycle) consumption to weed wheat	Identify the most energy consuming operations for each technical operation

Figure 2 presents the conceptual schema of the spatial data warehouse we propose, using the multidimensional conceptual model based on the UML presented in [5]. The conceptual model presents stereotypes for each spatio-multidimensional element, such as "Fact" for the facts, "SpatialAggLevel" for spatial dimension levels (i.e., levels having a geometric attribute "LevelGeometry" that represents the locations of their members), "AggRelationship" for hierarchical associations between dimension levels, "DimRelationship "for associations between levels and facts, etc.

The SDW presents several measures that represent the previous indicators defined in Section 3. In particular, the measures are: the area worked (*surface_w*), the number of animals (*animaux_nb*), the amount of product (input represented with "*intrant*" and

---

[1] FU = Evaluated Flow unit (litre, kWh, kg…).

output denoted with "*extrant*") used during work or no work (denoted with "w" and "nw", respectively), the duration in hours (*duree_w* and *dure_nw*) and the distance traveled (*distance_parcourue_w* and *distance_parcourue_nw*). The measures are aggregated using the sum on all dimensions.

The difference between work and no work (*w* and *nw*) is used to quantify energy consumption both directly related to and not attributable to work. For example, the amount of fuel used to plow a field is associated with work, while the amount consumed during the turn of the machine, with the plow raised, is not considered directly related to work.

The different dimensions are:

— Campaigns (*Campagne*): production cycles expressed in years (e.g., wheat produced in 2009)
— Time (*Temps*): classical temporal dimension, in which days are grouped by month and year.
— Products (*Produits*): the input and output products (intrant and extrant). Products are grouped recursively into larger classes of products
— Operators (Opérateurs): people who perform the operation
— Equipment (Attelage): machines and tools used
— Location: the spatial dimension that groups plots (parcelle) by farm (exploitation), department (département) and region (région)
— Productions: type of production (e.g. wheat)
— Technical Operations (Opérations Técniques): the technical operations performed, which are grouped by functions.

Using this spatio-multidimensional model, it is possible to represent and aggregate all indicators presented in section 3 such as the fuel consumption per plot, technical operation, year and production.

For the implementation of our application we have chosen the SOLAP tool JMap-SOLAP because it is one of the best business solutions that incorporates all existing OLAP and GIS functionalities. JMap-SOLAP is based on a three-tiers of Relational SOLAP architecture as described in Section 2.2. The SOLAP server allows the definition of the elements of the spatial data warehouse and the various data access policies for different users through a simple visual interface. The SOLAP web client integrates intelligent mapping concepts that support the automatic creation of thematic maps, while ensuring compliance with the semiotics rules (colors, symbols, frames, etc.). The client provides simple and multi-maps with synchronized diagrams and tabular displays to visualize SOLAP queries results. The client also implements the SOLAP operators described in Section 2.2.

In the following figure we show how to formulate and visualize spatio-multidimensional queries to obtain, for example, the aggregated indicator "the amount of fuel used for cultivating one hectare of wheat on the Montoldre farm". Once the user has selected interesting items (e.g., dimension elements) for the indicator calculation, the table in Figure 3 is displayed showing the quantities of fuel consumed per hectare over the entire Montoldre farm. The result can also be displayed on a map (Figure 3).

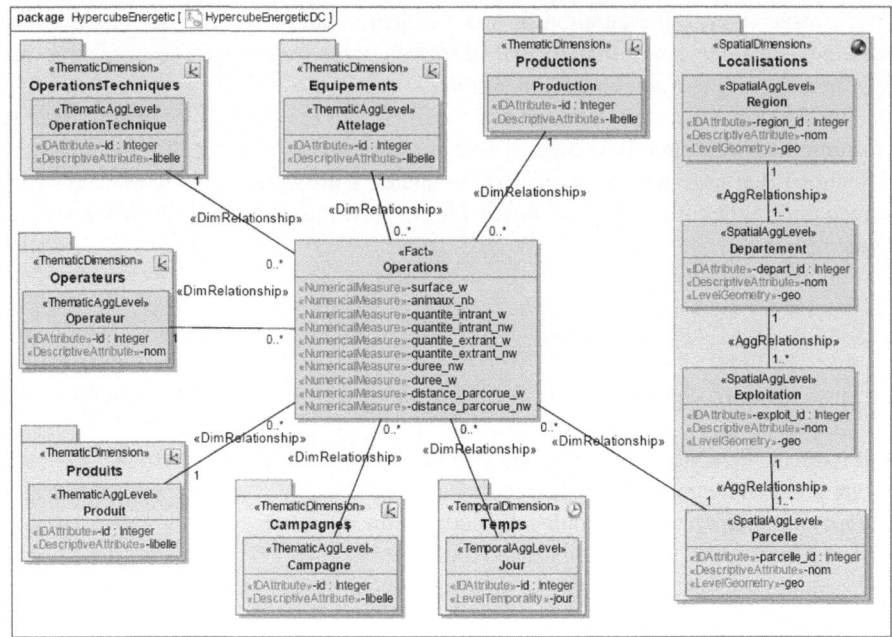

Fig. 2. SDW for indicators of Table 2

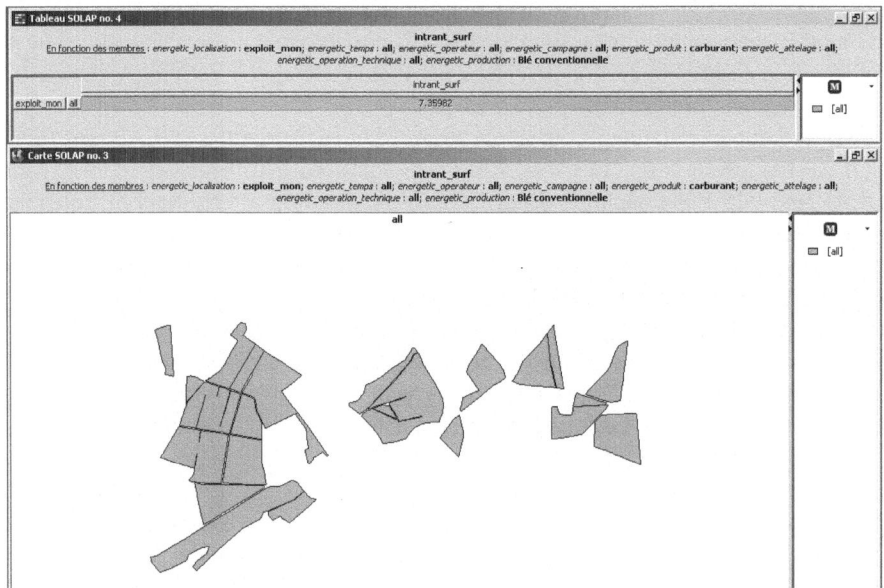

Fig. 3. Visualization of the aggregated indicator "the amount of fuel used for cultivating one hectare of wheat on the Montoldre farm"

The decision maker can apply a spatial drill-down operation directly on the map to obtain the energy consumption per plot. The result, shown in Figure 4, indicates that "PIQ 1" and "2PIQ" are the parcels that have consumed the most energy. It is also possible to determine the most expensive technical operations in terms of fuel. Using the operator to drill-down on the technical operations dimension of the previous table, a new table is shown (Figure 5), which gives more details on energy consumption by plot.

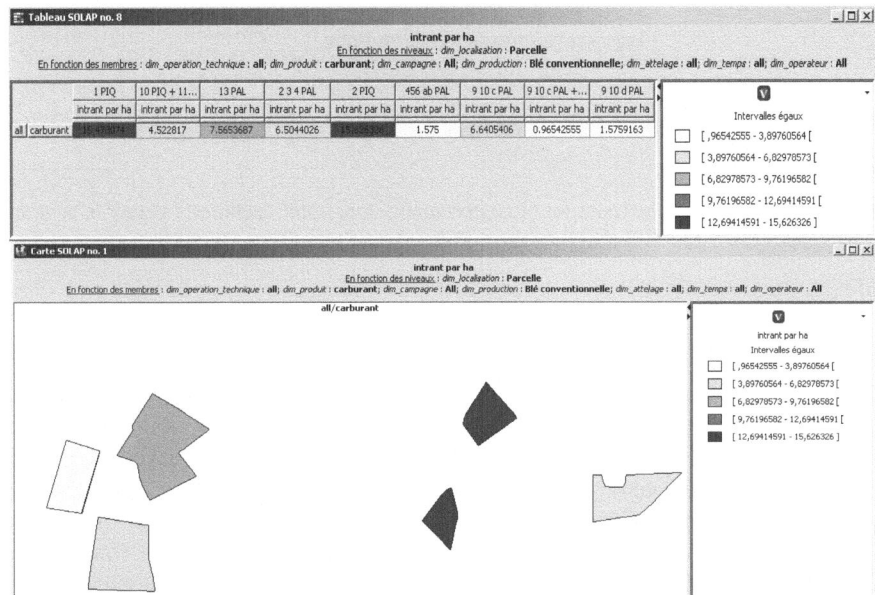

Fig. 4. Visualization of the indicator "the amount of fuel used for cultivating one hectare of wheat for each plot of the Montoldre farm"

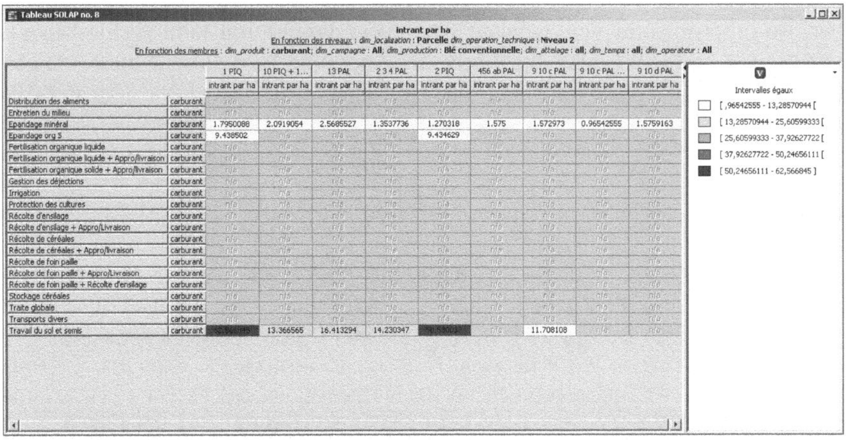

Fig. 5. Visualization of the indicator "the amount of fuel used for cultivating one hectare of wheat for each technical operation for each plot of the Montoldre farm"

In the same way, the example of the indicator described in Section 3, "fuel consumption per plot, technical operation, year and production", can be easily visualized in our system by using the SOLAP query, the result of which is shown in Figure 6.

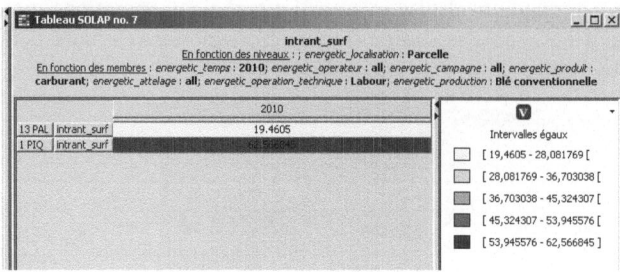

Fig. 6. Visualization of the indicator "fuel consumption per plot, technical operation, year and production"

5 Conclusions

Quantifying the impacts of human activities has now become an important issue for society, including agriculture. This quantification may eventually reduce these impacts by changing the most polluting practices, which is the reason several diagnoses defining a set of indicators exist to assess the energy performance of agricultural farms. However, these diagnoses present two important limitations: 1) they define indicators only at a global scale, avoiding the finest analysis of energy consumption related to technical operations, and 2) they are based on classical data storage systems that do not allow interactive cartographic analysis of huge volumes of data.

By contrast, SDW and SOLAP technologies provide tools to interactively analyze massive geo-referenced data sets. Recently, SOLAP has been successfully applied in the agriculture domain. Thus, to overcome the previously described limits, in this paper, we propose some new indicators to assess agricultural farm performance at a most detailed scale. We also show how it is possible to represent these indicators using a spatial data warehouse and analyze them by means of a classical SOLAP system.

References

1. Abdullah, A., Brobst, S., Umer, M., Khan, M.: The Case for an Agri Data Warehouse: Enabling Analytical Exploration of Integrated Agricultural Data. In: International Conference on Databases and Applications, pp. 139–144. IASTED/ACTA Press (2004)
2. Bailey, A.P., Basford, W., Penlington, N., Park, J., Keatinge, J., Rehmana, T., Tranter, R.B., Yates, C.: A comparison of energy use in conventional and integrated arable farming systems in the UK Agriculture. Ecosystems and Environment 97, 241–253 (2003)

3. Bédard, Y., Han, J.: Fundamentals of Spatial Data Warehousing for Geographic Knowledge Discovery. In: Geographic Data Mining and Knowledge Discovery. Taylor & Francis (2001)
4. Bimonte, S., Tchounikine, A., Miquel, M., Pinet, F.: When Spatial Analysis Meets OLAP: Multidimensional Model and Operators. International Journal of Data Warehousing and Mining 6(4), 33–60 (2010)
5. Boulil, K., Bimonte, S., Pinet, F.: Un modèle UML et des contraintes OCL pour les entrepôts de données spatiales. De la représentation conceptuelle à l'implémentation. Ingénierie des Systèmes d'Information 16(6), 11–39 (2011)
6. Chaudhary, S., Sorathia, V., Laliwala, Z.: Architecture of sensor based agricultural information system for effective planning of farm activities. In: IEEE International Conference on Services Computing, pp. 93–100. IEEE (2004)
7. Clements, A., Pfeiffera, D., Otteb, M., Morteoc, K., Chenb, L.: A global livestock production and health atlas (GLiPHA) for interactive presentation, integration and analysis of livestock data. Preventive Veterinary Medicine 56(1), 19–32 (2004)
8. Gras, R., Benoît, M., Deffontaines, J.P., Duru, M., Lafarge, M., Langlet, A., Osty, P.L.: Le Fait Technique en Agronomic: Activité Agricole, Concepts et Méthodes d'Étude. L'Harmattan, Paris (1989)
9. Maceachren, A.M., Gahegan, M., Pike, W., Brewer, I., Cai, G., Lengerich, E., Hardistyn, F.: Geovisualization for Knowledge Construction and Decision Support. Computer Graphics and Application 24(1), 13–17 (2004)
10. Mitchell, G., May, A., McDonald, A.: PICABUE: A methodological framework for the development of indicators of sustainable development. International Journal of Sustainable Development and World Ecology 2(2), 104–123 (1995)
11. Nilakanta, S., Scheibe, K., Rai, A.: Dimensional issues in agricultural data warehouse designs. Journal Computers and Electronics in Agriculture Archive 60(2), 263–278 (2008)
12. Pervanchon, F., Bockstaller, C., Girardin, P.: Assessment of energy use in arable farming systems by means of an agro-ecological indicator: The energy indicator. Agricultural Systems 72(2), 149–172 (2002)
13. Pradel, M., Boffety, D.: Quels indicateurs et solutions technologiques adaptés pour évaluer finement les performances énergétiques des exploitations agricoles? Sciences, Eaux et Territoires 7, 16–29 (2012)
14. Rai, A., Dubeya, V., Chaturvedia, V., Malhotraa, K.: Design and development of data mart for animal resources. Computers and Electronics in Agriculture 64(2), 111–119 (2008)
15. Kimball, R.: The Data Warehouse Toolkit: Practical Techniques for Building Dimensional Data Warehouses. John Wiley & Sons, New York (1996)
16. Schulze, C., Spilke, J., Lehner, W.: Data modeling for Precision Dairy Farming within the competitive field of operational and analytical tasks. Computers and Electronics in Agriculture 59(1), 39–55 (2007)
17. SOLAGRO: Energie dans les exploitations agricoles: état des lieux en Europe et éléments de réflexion pour la France. Etude ADEME/ MAP 0671C0036 (2007)
18. Thornsbury, S., Davis, K., Minton, T.: Adding Value to Agricultural Data: A Golden Opportunity. Review of Agricultural Economics 25(2), 550–568 (2003)
19. Vilain, L.: La méthode IDEA, Indicateurs de durabilité des exploitations, Guide d'utilisation, Educagri Edition (2003)

Validating a Smartphone-Based Pedestrian Navigation System Prototype

An Informal Eye-Tracking Pilot Test

Mario Kluge and Hartmut Asche

University of Potsdam, Department of Geography,
Karl-Liebknecht-Strasse 24/25, 14476 Potsdam, Germany
{mario.kluge,hartmut.asche}@uni-potsdam.de
http://www.geographie.uni-potsdam.de

Abstract. Pedestrian navigation is an area which has been researched in a variety of projects and prototypical developments in recent years. This work describes the concept of a pedestrian navigation system prototype, Reality View, which is based on real images in real-time navigation. The navigation instruction is depicted by a virtual route object presented in an augmented reality (AR) environment. This paper discusses a pilot test of the pedestrian navigation system prototype developed on eye-tracking. An additional questionnaire shows the test subject's mental reaction towards the system and the usability of the system itself. The evaluation of the eye-tracking pilot test allows the analysis of prominent objects in the environment and describes the relationship between reality and navigation instructions.

Keywords: Augmented Reality, Eye-Tracking, Pedestrian Navigation System, 3D-Augmented Geographical Environment.

1 Introduction

Pedestrian traffic is an integrated part of traffic. Pedestrians are both: road users and passengers in public transport. Everyone moves every day by foot because only foot traffic guarantees a seamless connection from door to door. Pedestrian traffic is often underestimated and is regarded as *"self-evident and natural"* and is not associated with the general term of traffic (Thomas & Schweizer 2003). Pedestrian traffic is characterized by a free and unbounded mobility and thus is not subject to restrictions such as road traffic regulations for vehicles. Pedestrians make certain demands on a system which must be satisfied to allow a user-friendly navigation.

- The basis of the navigation is a precise instruction with current information about location, position and orientation.
- The use of the system should be intuitive and easy to understand.
- The system should be robust against external environmental factors such as weather or lighting conditions.

The hypothesis of this paper is that pedestrians relate AR instructions on prominent objects in order to implement this in the real environment. This paper describes a prototypical application including the main characteristics of AR and a detailed discussion about the relationship between navigation instruction and real environment.

2 Concept

The concept of the prototype implemented in this paper is based on an AR navigation application. A virtual route will be generated and positioned with respect to the real location (Fig. 1). The structure of the concept consists of the processes registration, tracking and presentation (Zlatanova 2002).

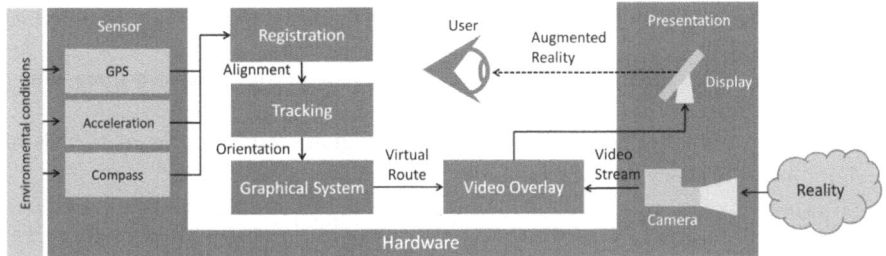

Fig. 1. The concept of the prototype

The registration process captures, at defined intervals, position, direction and alignment information from various sensors. The tracking process specifies the trace of virtual objects aligned on the basis of measured sensor information. The presentation process describes the output of the AR image display on the screen of the device and serves as an interface between user and system.

3 Augmentation

"An augmented reality system is a system that creates a view of a real scene that visually incorporates into the scene computer-generated images of three-dimensional (3D) virtual objects." (Vallino & Kutulakos 2001) AR is part of the virtual continuum of mixed reality and combines different forms of representation with the realistic image of the environment. AR allows the user to see the real world, with virtual objects superimposed with the image of the real world. In contrast virtual reality technologies completely immerse a user inside a synthetic environment so that the user cannot see the real world around him (Azuma 1997). The classification according to Milgram and Kishino (1994) describes the relationship of the different forms of reality based on their graphical expression (Fig. 2).

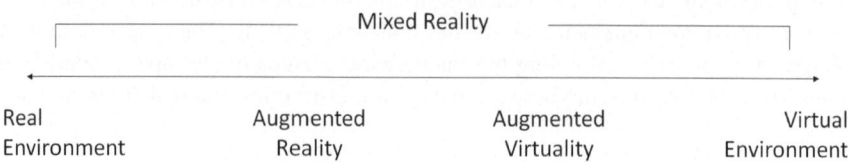

Fig. 2. Mixed Reality, Source: (Milgram & Kishino 1994)

The fundamental difference between Virtual Reality and AR is the fact that AR supplements the environment with extensive information instead of recreating or replacing it. According to the definition of Azuma, there are three requirements that a system must comply with in order to characterize this as AR (Azuma 1997).

(1) The presentation must consist of real and virtual elements to an expanded view.
(2) The information between real and virtual object must have a three dimensional reference.
(3) Any form of interaction with the system and the resulting presentation should be done in real time.

AR allows a level of immersion that no virtual equipment can provide and has already been used in many applications such as medical, engineering, advertising and, more recently, navigation (Azuma 1997; Zlatanova 2002).

4 Instruction

Real-life orientation of pedestrians is based on signs and landmarks such as prominent buildings or infrastructure objects like bridges, crossroads or different road types. Individuals recognize objects in the environment with different perceptions and memorize the result in the form of a mental map (Lynch 1965).

4.1 Navigation Instructions

Current navigation systems are based on instruction forms which are presented sequentially. The formulation is based on absolute distance *"In 50 meters turn right"* or by relative text descriptions *"Soon, turn right"* (Sester 2004). Pedestrians have a need for individual navigation adapted to the personal preferences of the user. For this reason, many people find it harder to estimate distances and times whilst walking as opposed to travelling in vehicles. „*Incremental guidance is a service that is frequently associated with the concept of a mobile assistant.*" (Kray et al. 2003). The idea to visualize the route instruction in the prototype is based on a concept taken from vehicle navigation and is called Virtual Cable (VC). The VC navigates a user along a route without a detailed description and is *"safe, simple and intuitive"*[1]. The VC

---

[1] http://www.mvs.net/

adapts to the surrounding conditions to obtain the reference to the environment. There has been some evidence (May et al. 2003; Walther-Franks 2007) that people recognize prominent objects and landmarks in the real environment and find their route in the cities easier using a realistic AR-mode, rather than using a 2D map. In order to produce AR-based route instructions, several processes have to be completed (Kray et al. 2003).

(1) determination of the origin and target location
(2) computation of a suitable route
(3) presentation of the route to the user

4.2 RealityView-Prototype

The development and implementation of the RealityView prototype show that the concept of an AR-based pedestrian navigation system is achievable. However, the results also show the weaknesses of this novel approach. In particular, the accuracy and tolerance of the sensors are one of the main reasons for errors in the placement of virtual objects (Holloway 2001; Malik 2002). The screen display of the RealityView prototype can be operated in 2 presentation modes: AR and map display (Fig. 3). The activation or the change of the presentation mode is done by changing the alignment of the unit in a horizontal or vertical direction (Kluge & Asche 2008).

Fig. 3. Screen Display Reality View Prototype: (a) AR, (b) map display

„*The most common way to present route instructions graphically is a geographic 2D map that is annotated with route information.*" (Kray et al. 2003). The map provides a general overview of the environment and the route. Without information about distance and time to the next decision point, the execution of the navigation instruction is difficult.

Augmented reality provides precise instructions and relates the operation to the real environment in a more natural way. *"The three-dimensional visualization has a model character, i.e. the shown objects of the real world shall be represented in a geometrically correct way and in the right position."* (Kray et al. 2003). The disadvantage is the low visual range and the fact that only the next statement is visible. The user has the freedom to decide which of the two view forms are used. This makes it possible to implement the navigation appropriately and is therefore adaptable.

5 Eye-Tracking

5.1 Experimental Setup

No formal testing of users under controlled conditions has been carried out to date in order to evaluate the navigation prototype's design, functionality and interactive map use. The prototype was, however, informally validated and discussed with potential users in addition to a critical analysis by all people involved in its development. The eye tracking equipment (SMI[2] iView X Hed) is specifically designed for mobile use and could be easily adapted to different test candidates. The following properties would be measured:

(1) the pattern of various fixations (scan paths)
(2) the time spent looking at display elements
(3) the deployment of prominent objects in the real environment

The experimental setup consists of a user-centered eye camera to capture a person's eye motion or the point of gaze (Fig. 4a). In addition, a scene camera is used, which makes it possible to record the environment. Both camera modules are mounted on headgear such as a helmet or a cap so that a test subject is able to wear it on the head (Fig. 4b). The storage of the recorded information is provided by a mobile computer in the backpack of the test subject (Heidmann 1999).

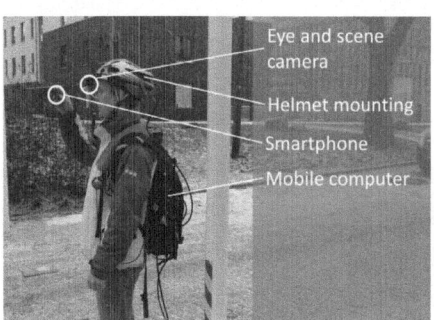

Fig. 4. Experimental Setup: (a) eye-tracking equipment, (b) headgear

5.2 Technology

Compared to eye-tracking studies in the laboratory the use of a lightweight headset and a laptop allow the recording in outdoor areas. The technical process for recording the eye movements of humans is described as a non-invasive, video-based eye-

---

[2] http://www.smivision.com

tracking technology (Fig. 5a). The tracking process consists of the following steps (Sareika 2005; Smi-Vision 2012):

(1) Calibration procedure to measure the properties of the eye
(2) Measure the locations of the pupil centre
(3) Locate the relative position of the corneal reflection
(4) Calculate the direction of gaze

The mobile head-mounted eye and gaze tracker must be calibrated to the properties of the test subject (Fig. 5b). Using multiple fixation points, the orientation of the pupil will be saved. The next step is to measure the position of the pupil center and to locate the relative position of the corneal reflection. The end result is to calculate the direction of the gaze. The data obtained with the help of the eye-tracker is being recorded into a database. The evaluation is done in special software which allows the eye movement and the gaze to be presented graphically.

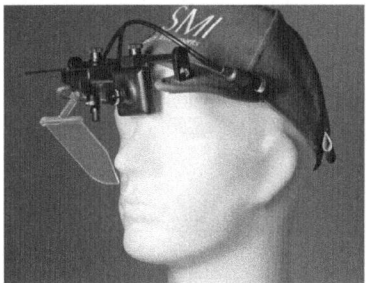
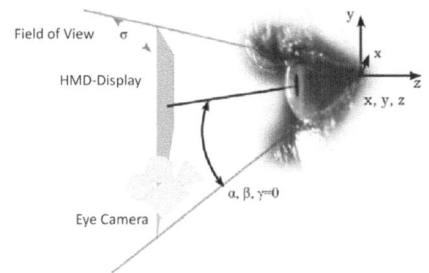

Fig. 5. Eye-tracking technology: (a) mobile tracker, (b) calibration procedure, Source: Sareika 2005, SMI-Vision 2012

5.3 Experimental Procedure

Due to the cumbersome preparation and calibration of eye tracker only a pilot test with a total of three users was performed. Each test session consisted of three parts: introduction, test run and questionnaire. In the first part the project and the application were introduced to the user. In the second part the test run was implemented. Finally the user filled out a questionnaire, in which his personal experience and evaluation were incorporated. The test runs were separately evaluated and compared at various landmark objects in terms of duration and form of presentation. The advantage for carrying out the test run on the campus area was the low traffic volume and the fact that routes could be followed in the middle of the road so that a difference between street and footpath were avoided.

The total length of the test run was 500m, with an average walking time of about 5 minutes[3]. All three test subjects were familiar with the terrain, but they had no prior information on the course route. To evaluation the data from the test run, the manu-

---

[3] http://maps.google.de/

facturer-supplied software program BeGaze version 2.4 was used. The software allows scientific analysis of eye movement and displays the data as meaningful graphs in front of the video sequence (Fig. 6).

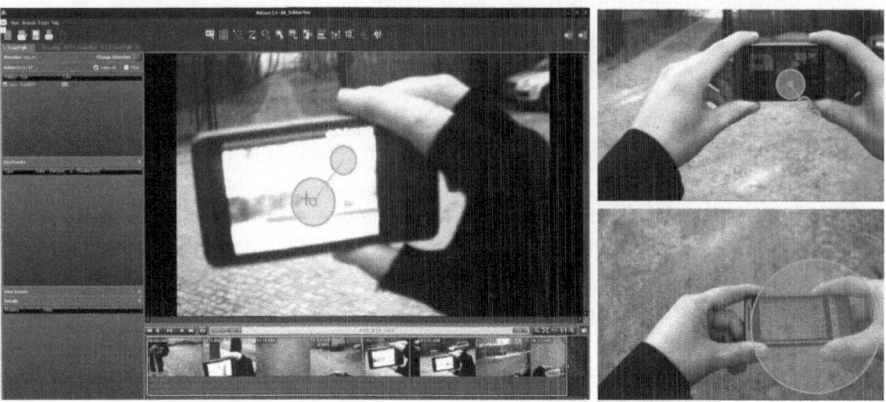

Fig. 6. Evaluation of the eye-tracking results

6 Results

6.1 Eye-Tracking

Everything a person does in the environment is distinguished by the fact that information is perceived through the human senses, is cognitively processed and reissued (Streitz 1987). Navigation problems arise because the user is not able to refer the information from the instruction of the real environment. The reason is often the formulation of the navigation instruction, which has no clear relation to the environment. The user creates a visual relationship between what appears on the display and what exists in the real environment. Based on this experience it can be concluded that the inclusion of real images helps to improve the orientation and increases the understanding of the navigation instructions. The trial was divided into four sections which are composed as follows (Fig. 7).

1. *Distinctive buildings* like shops or restaurants can easily be found in the real environment and helps the user to refer to the instruction.
2. *Parked vehicles* are at first view not a good reference point for orientation but can be easily distinguished by specifying the manufacturer, model and color.
3. *Signage* can be distinguished primarily by their semantic content.
4. *Target point* with a great visual attention for the user.

The aim of the eye-tracking pilot test was to determine how often a user relies on the assistance of the system and what presentation form, AR or 2D map, is chosen. Another focus of the research was to identify what kind of relationship between AR and real environment a user creates. Objects with high recognition such as prominent buildings, signs or parked vehicles, which are visible in both the AR-representation as well as in reality, can serve as reference points in order to implement the navigation instructions (Sareika 2005).

With one exception, all test subjects had an identical favorite at the individual stations (Table 1). Most of the users tried to use the AR-mode at decision points with more than one possibility in order to go further. In contrast, the map display was used more often along two decision points. All of them used prominent objects successfully to recognize the instruction in the real environment. Apparently, a connection exists between the more frequent uses of AR at intersections as opposed to the map display. Apart from matching prominent objects the most common navigation strategy of a user was to follow the virtual cable in the AR-mode, and to display the target and current location in the 2D map. In conclusion the AR display provides the user a more detailed navigation at decision points whereas the map display allows a better overview.

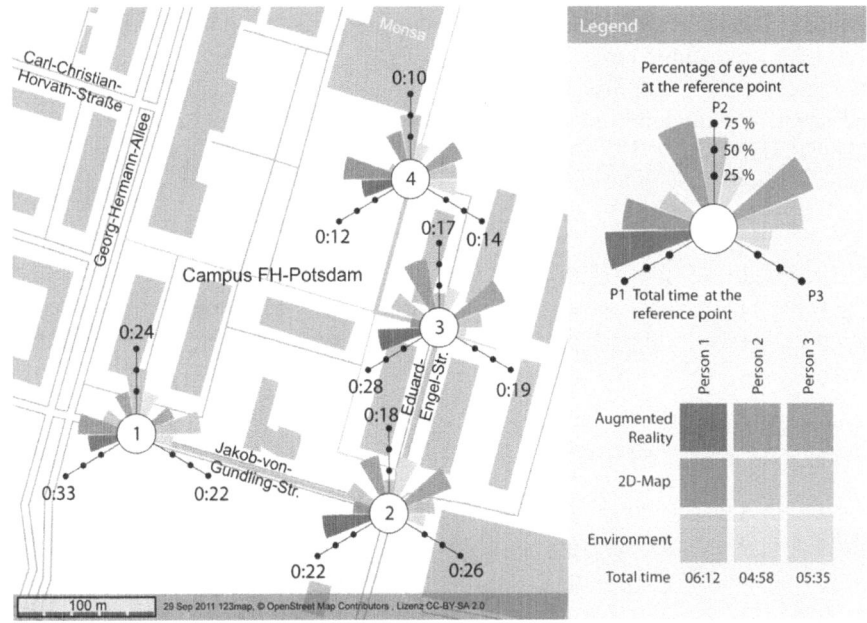

Fig. 7. Eye-Tracking Test run and results

Table 1. Total and percentage of time at the reference point (AR, Map, Environment)

	Person 1	Person 2	Person 3
Total time	0:06:12	0:04:58	0:05:35
Point 1	0:00:33 (33, **45**, 21)	0:00:24 (25, **50**, 25)	0:00:22 (23, **59**, 18)
Point 2	0:00:22 (**55**, 18, 27)	0:00:18 (**44**, 17, 39)	0:00:26 (**54**, 27, 19)
Point 3	0:00:28 (**50**, 14, 36)	0:00:17 (**59**, 18, 24)	0:00:19 (**63**, 26, 11)
Point 4	0:00:12 (33, **58**, 8)	0:00:10 (30, **50**, 20)	0:00:14 (**43**, 29, 29)

6.2 Questionnaire

In addition to the eye-tracking pilot test the test subjects had to complete a questionnaire to evaluate the usability of the prototype. The questionnaire had an extent of ten questions and was answered on the basis of the Likert-scale (Fig. 8).

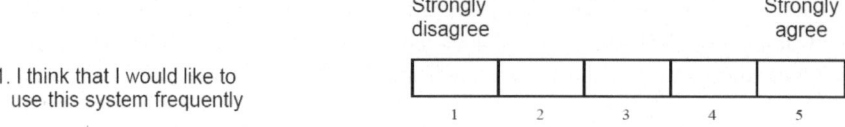

Fig. 8. Excerpt from the SUS-questionnaire, (Brooke 1996)

The questionnaire used for this purpose is based on the system Usability Scale (SUS). *"Usability is an attribute of every product-just like functionality."* (Dumas & Redish 1999). Testing usability makes sure that the product works according to specifications. The structure of the SUS questionnaire is divided into ten different questions with a maximum range of values between 0 - 100. The rating scale was divided into agreement (5) and disagreement (1). The statements switches between positive and negative formulation to prevent the survey results by arbitrary responses (Kirchhoff et al. 2006).

Table 2. Questionnaire results

Nr.	1	2	3
1	4	3	4
2	1	2	3
3	5	5	5
4	1	1	2
5	3	3	4
6	2	2	1
7	5	4	3
8	3	2	2
9	4	3	4
10	2	2	2
Sum	32	29	30
Score	80	72,5	75

The validation results in terms of usability provided a mean score of 75 out of 100 points which corresponds to a percentile rank of 70%[4]. A SUS score of 75 can be interpreted as a school grade of a B and therefore a positive validation of the proto-

---

[4] http://www.measuringusability.com/sus.php

type (Table 2). One of the reasons why the result has not been even better would be that 2 of 3 test subjects found the various functions in the prototype not well integrated although 2 out of 3 felt very confident using the system. It was striking that all test subjects answered the third statement, *"I thought the system was easy to use"* rating with the value 5 (agree). At the same time two of three test subjects disagreed with the fourth statement, *"I would need the support of a technical person to be able to use this system."* It can be concluded that one of the main requirements (intuitive use) of a PNS has been fulfilled.

7 Summary

This paper described the conceptual structure of an AR-based PNS and the validation based on eye-tracking technology. The use of prominent and distinctive landmark objects for route instructions simplifies the orientation for pedestrians and allows the implementation of the navigation instructions in relation to the environment. This paper discussed only technical and cognitive recourses. However, the age and the gender of the test subject are very important factors in the context of selecting a presentation mode and to solve spatial problems. The main purpose of this pilot study was to collect feedback about using augmented reality in geographical environments. Whether or not and to what extent the navigation based on AR is a success, depends on various criteria. A key role in the success of future developments is located in the precision and quality of the hardware devices and especially to the accuracy of the built-in sensor components.

References

Azuma, R.: A Survey of Augmented Reality. Presence. Teleoperators and Virtual Environments 6(4), 355–385 (1997)

Brooke, J.: SUS: a "quick and dirty" usability scale. Jordan, P.W., Thomas, B., Weerdmeester, B.A., McClelland, A.L. (eds.) Usability Evaluation in Industry. Taylor and Francis, londen (1996)

Dumas, J.S., Redish, J.C.: A practicle Guide to Usability Testing. Intellect Books, Exeter (1999)

Heidmann, F.: Aufgaben- und nutzerorientierte Unterstützung kartographischer Kommunikationsprozesse durch Arbeitsgraphik. Fachbereichs Geographie/Geowissenschaften. Universität Trier, Trier (1999)

Holloway, R.: Registration Error Analysis for Augmented Reality Systems. In: Barfield, W., Caudell, T. (eds.) Fundamentals of Wearable Computers and Augumented Reality. Lawrence Erlbaum, Mahwah (2001)

Kirchhoff, S., Kuhnt, S., et al.: Der Fragebogen. VS Verlag für Sozialwissenschaften, Wiesbaden (2006)

Kluge, M., Asche, H.: 3D Pedestrian Navigation - RealityView. In: 5th International Symposium on LBS & TeleCartography, Salzburg (2008)

Kray, C., Laakso, K., et al.: Presenting Route Instructions on Mobile Devices. In: Johnson, W.L., Andre, E., Domingue, J. (eds.) Proceedings of IUI 2003, pp. 117–124. ACM Press, Miami (2003)

Lynch, K.: Das Bild der Stadt. Birkhäuser, Basel (1965)

Malik, M.: Robust Registration of Virtual Objects for Real-Time Augmented Reality. Institute for Computer Science. Carleton University, Ottawa (2002)

May, A., Ross, T., et al.: Pedestrian Navigation Aids: Information Requirements and Design Implications. Personal and Ubiquitous Computing, 331–338 (2003)

Milgram, P., Kishino, F.: A Taxonomy Of Mixed Reality Visual Displays. IECE Transactions on Information and Systems, 1321–1329 (1994)

Sareika, M.: Einsatz von Eye-Tracking zur Interaktion in Mixed Reality Umgebungen. Medien- und Informationswesen. Fachhochschule Offenburg, Offenburg (2005)

Sester, M.: Präsentation von Geodaten für ortsbezogene Anwendungen. Akademie der Geowissenschaften zu Hannover 24, 60–65 (2004)

SMI-Vision, SMI Gaze & EYE Tracking Systems (2012), http://www.smivision.com/

Streitz, N.: Die Rolle der Psychologie. In: Fähnrich, K.-P. (ed.) Software-Ergonomie, pp. 43–53. Oldenburg Verlag, Oldenburg (1987)

Thomas, C., Schweizer, T.: Zugang zum öffentlichen Verkehr: Der Fussverkehr als «First and Last Mile». Strasse und Verkehr 10, 16–19 (2003)

Vallino, J.R., Kutulakos, K.N.: Augmenting Reality Using Affine Object Representations. In: Barfield, W., Caudell, T. (eds.) Fundamentals of Wearable Computers and Augmented Reality, pp. 157–182. Lawrence Erlbaum, Mahwah (2001)

Walther-Franks, B.: Augmented Reality on Handhelds for Pedestrian Navigation. In: Informatik. Universität Bremen, Bremen (2007)

Zlatanova, S.: Augmented Reality Technology. GIS Technology Report. Delft, Universität Delft 17, 1–76 (2002)

Open Access to Historical Atlas: Sources of Information and Services for Landscape Analysis in an SDI Framework

Raffaella Brumana[1], Daniela Oreni[1], Branka Cuca[1], Anna Rampini[2], and Monica Pepe[2]

[1] Politecnico di Milano, p.za Leonardo da Vinci 32,
20133 Milano, Italy
[2] National Research Council, IREA-CNR, via Bassini 15,
20133 Milano, Italy
{daniela.oreni,raffaella.brumana}@polimi.it,
branka.cuca@mail.polimi.it,
{rampini.a,pepe.m}@irea.cnr.it

Abstract. The paper illustrates the potentials of geospatial data and services to access historical digital atlas for landscape analysis and territorial government. The experience of a historical geo-portal, the 'Atl@nte dei Catasti Storici', in the management of geo-referenced and non-geo-referenced maps - ancient cadastral and topographic maps of Lombardy Region - can be considered a case study with common aspects to many European regions having an extensive cartographic heritage. The development of downstream web based services to integrate other data sources (current maps, satellite and UAV airborne photogrammetry, multi-spectral images and derived products) provides new scenarios for retrieving geospatial knowledge of territory, bridging the gap in supporting a sustainable management of the territory.

Keywords: Historic cartography, landscape, SDI, Geospatial Web Services, satellite data, on site data collection, multispectral images, downstream services.

1 Introduction

The European Landscape Convention established the necessity for European countries 'to integrate landscape into its regional and town planning policies and in its cultural, environmental, agricultural, social and economic policies' [1]. Guidance on the assessment of the impacts of the projects are progressively enhancing the use of integrated instrument to support the stages of the projects [2]. Analysis of the territorial transformation are increasingly relying upon services based on geospatial and space-based information, creating a progressive demand of data sets and service providers and resulting in an extensive availability of these information. The main objective of the geoportal Atl@nte is to fully integrate historical data within geospatial information into decision making processes, addressing the government of territory by Public Authorities (PA) and supporting professional related activities.

The open source platform (http://www.atlantestoricolombardia.it) 'Atl@nte dei Catasti Storici e delle Carte Topografiche della Lombardia' (Atlas of Historical Cadastres and Topographic maps of Lombardy), is an on-going project, funded by Fondazione Cariplo that aims to implement a geo-portal for management of historical geo-referenced maps at territorial-regional and cadastral-local scale (section 2). The advanced Spatial Data Infrastructure (SDI) of Lombardy Region (one of the most developed among European regions) gives lot of perspectives to the evolution of this kind of services [3], under various domains of multi-scale territorial dimension and technological progress. However, many experiences on spatial management of historical map among PAs is growing without coordinated structured action [4].

From the technological point of view, the geo-portal Web Service Architecture has been designed to support the request for data and services, progressively distinguished in many aspects such as time-scale or frequency, resolution or territorial extension, query of geospatial functionalities, such as Web Map Services (WMS) and Web Feature Services (WFS), in order to interact with the end user exact requests and applications, guarantying context-aware downstream services (section 3). An approach based on information harmonization has been considered acting upon global synthesis level of spatial and satellite data, and in-situ data collection, for local detailed analysis. Integration of historic maps with other data sources like satellite data and multi-spectral images, airborne photogrammetry, mobile-mapping vehicles (UAV) and laser scans, in order to provide new scenarios for retrieving geospatial knowledge of territory, are here experimented. A sustainable management of built environment and landscape, in the presence of 'vulnerable and sensible areas', can benefit by open access to historical atlas development, as explained through some extracted samples, to identify agricultural biodiversity, water resources, secondary channel networks, and their riparian areas, allowing to enhance and protect them (section 4).

The potentials of Atl@nte, as a planning instrument developed to share multi-temporal and multi-scale spatial data, are here discussed together with critical aspect and barriers, responding to multiple demands of end users for landscape analysis and in relation with other INSPIRE themes. For example water courses and coastal zones are areas of a great strategic importance for the EU. Hence, in this sense, the INSPIRE Directive [5], an instrument that, together with Shared Environment Information System (SEIS), has been identified as a main tool to facilitate the information flow in these areas. The progressive implementation of different spatial data source has underlined the necessity of improving WPS to support adaptive process context-aware processing by the users (conclusions).

2 The Atl@nte Geoportal (web portal): Spatial Data Management of Temporal Map Series

In order to support Spatial Data Management of Temporal Map Series, the structure of the Atl@as geo-portal has been based on two main geo-referenced level of analysis, strictly integrated among each other through the services developed (Fig.1): the level of territorial scale (regional domain), and the one of local scale (municipality domain).

2.1 The Level of Territorial Scale

Methodological indications for landscape management of territorial transformation are progressively involving the importance of historical maps as an important source for landscape analysis and management of territorial transformation [6]. The availability across many European regions and countries of historical maps representing the territory at a small scale but with huge level of detail and precious qualitative information. These data can act in support of landscape interpretation and re-discovering, if strictly related to the current spatial data using innovative instruments. Also, current SDIs can represent a common geographic framework, in which it is possible to correlate the cadastral maps (old and current) of municipalities, as for example in Lombardy the ones from Austrian, as well as the Napoleonic period.

Furthermore, regional level is recognized by the European institutions as a driver for sustainable growth and administrative level capable of properly managing space and geospatial information policies. EU Digital Libraries initiative and protocols represent a powerful instrument towards a progressive digitalization and availability of map sources [7]. Territorial maps used by the historians and experts of landscape analysis are substantially unexploited within planning process, due to the lack of experience in treating this kind of spatial data. The generation of geo-referenced layers of this particular map series, treated like a photogrammetric block [8], has been experimented within this project. In this way double tolerance values were obtained respect to the current scale used for the spatial correlation for each reconstructed mosaics of the region Lombardy at different thresholds, using current maps such as IGM 100,000 via WMS [9]. In order to accomplish this task, many different problems have been faced due to the geodetic survey context carried out in the past centuries (XVIII-XVIII) to generate maps at the scale of 1:100,000 without GPS, without the current estimation of geoid, and other instruments nowadays available. The Astronomic Observatory of Paris and the Astronomic Observatory of Brera (Milan) at that time were considered avant-guard in the territorial surveys addressed to obtain a modern representation of the territory though a rigorous innovative methods [10].

Georeferenced small-scale territorial maps of Lombardy Region published:
- 'Carta del Territorio del Milanese e del Mantovano' (Map of Milan and Mantova Territories, Astronomers of Brera, 18th century), 1788-1796 (1:86,400) and 'Carta degli Astronomi della Lombardia austriaca (con confini tracciati)', (Astronomers' map of Austrian Lombardy, with traced borders), 1791 (1:86,400).
- 'Corografie delle Provincie del Regno Lombardo-Veneto', (Corographies of Provinces of Lombardy-Venetia Kingdom), from 1836 (1:115,000).
- 'Carta Topografica ITM del Regno Lombardo-Veneto', (Topographic ITM Map of Lombardy and Venetian Kingdom), 1833 (1:86,400).

2.2 The Level of the Local Scale

The cartographic heritage conserved by the National Archive of Milan (ASMi) is one of the biggest collections in Europe that counts about 200,000 maps. The high numbers of maps is represented principally as cadastral historical series at the large scale 1:2,000 with lot of synthesis sheets at the scale 1:5,000 and 1:8,000. Experts have estimated that in the only in Lombardy Region each historical cadastral level contains ~60,000÷70,000 map sheets (periods from the 18th to the 19th centuries), roughly approximated to a total of more than 250,000 sheet units within the decentred Lombardy National Archives. The map patrimony made available to the Atl@nte by ASMi amounts to the 28,000 map units digitalized till now (new sets of scans are progressively on-going). The granularity level of the cadastre local scale required to identify a gradual and sustainable methodology to reconstruct the continuity of territory through the geo-referencing process, and at the same time to develop geospatial services to support landscape analysis in the areas where the geo-referencing has not been yet carried out.

Sample of historical cadastral maps series georeferenced and published:
- Cadaster of Maria Theresa of Austria, 1718-1757.
- 'Catasto Lombardo-Veneto', 1865-1887.
- 'Cessato Catasto', 1897-1902.
- "Cessato Catasto Terreni in conservazione" maps (1886-1962) (Agenzia del Territorio -AdT- the cadastral administration that conserve Current Cadastre).
- 'Catasto Fabbricati' (Cadastre Maps of Buildings), 1939-1962 (AdT).
- "'mpianto in Conservazione' maps, 1962-1994 (AdT).

Fig. 1. Atl@nte Geoportal: open access to the territorial and large scale level. A mosaic of Chorographic territorial maps of Lombardy Region (left). A detail of Milan and Mantova territorial map (right).

3 System Architecture: Historical Map Spatial Services

Digital approach to cartographic heritage is defined as a meeting area of two spheres dealing with cartography: the humanistic component of historians and the scientists and engineers of cartography [11]. Atl@nte is based on the concept of user-defined geospatial services, to allow to access and extract data of interest to compare multi-scale objects with different data sources. Open source tools are being developed following Open Geospatial Consortium (OGC) standards for Geographic Information System (GIS) interoperability [12], which supports the application developers in the integration of many geo-processing and location services, mainly WMS, WFS [13].

The Atl@nte web-platform application relies on a multi-tiered client-server system (Fig. 2). Core components are located on the server side. At the inner tier (tier 2) it was found a standard postgresql server, with geographic capabilities (postgis): the DB stores metadata of georeferenced maps, vector features, and a file storage device for data persistence (raster data and others). The middle tier (tier 1) is made up of a standard web server to host the CMS application (Typo3) and the map viewer engine (PHP). The open source GIS server providing WMS, WFS and WPS to the map viewer is Geoserver 2.0.2, runs as a web application within a servlet container (Apache Tomcat 6). At the client-side (tier 0), a standard web browser allows the end user to navigate . The map viewer includes the entire OpenLayers bundle (supporting proprietary imagery, i.e. Google Maps), combined with MapFish 1.2 (for controls and widgets) and ExtJS (for page layout).

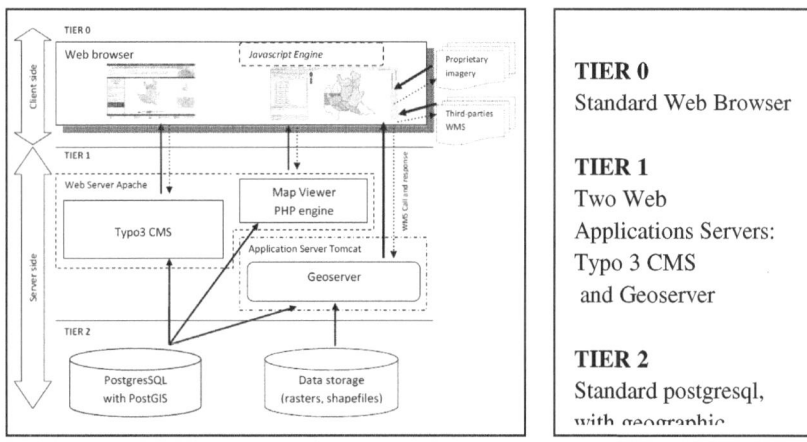

Fig. 2. A logical schema of the System Architecture of Atl@nte Geo-database application

3.1 WMS and Map Viewer Functionalities to Navigate Temporal Map Series

The map viewer provides several functions (Fig.3) developed with OpenLayers [14]:

- seamless browsing between historical geo-referenced map series (raster) and thematic layers (vector),
- a complex set of controls for opacity changing and real-time layer reordering to support user analysis of chronological sequence, including zooming and panning,
- custom choice of the background layer, selecting among an empty base or proprietary imagery (e.g. Google Imagery ©, i.e. Satellite, street guide), and of the coordinates reference system (native reference system of map georeferentation (UTM/WGS84 or Gauss-Boaga) versus Spherical Mercator, required to display proprietary imagery, i.e. Google Imagery ©,
- integration with OGC-compliant external WMS resources.

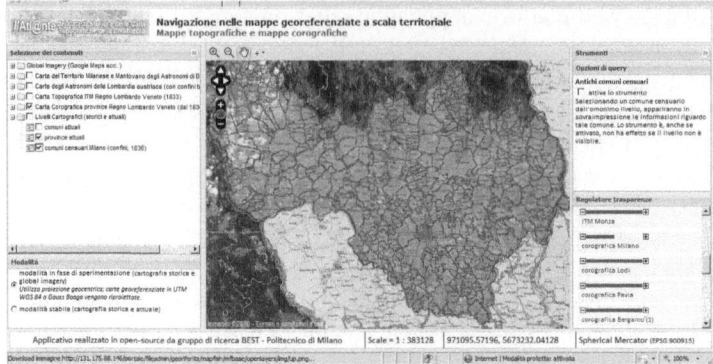

Fig. 3. Map viewer functionalities developed to navigate the temporal map series

3.2 Bridging External Services to Overcome Access Map Catalogue Gap

A relevant barrier to an extensive map catalogue access by the users is represented by the difficult language of archivists, keywords and hierarchical map sequences, depending on the multiple historic maps updating carried out by the coeval official surveyor, on the boundary transformation of municipalities during the centuries. To this aim, together with ASMi, a spatial functionality has been developed to facilitate the map catalogue retrieving to the web user, which have difficulties to extract the needed map within the catalogue of 28,000 maps (Fig.4).

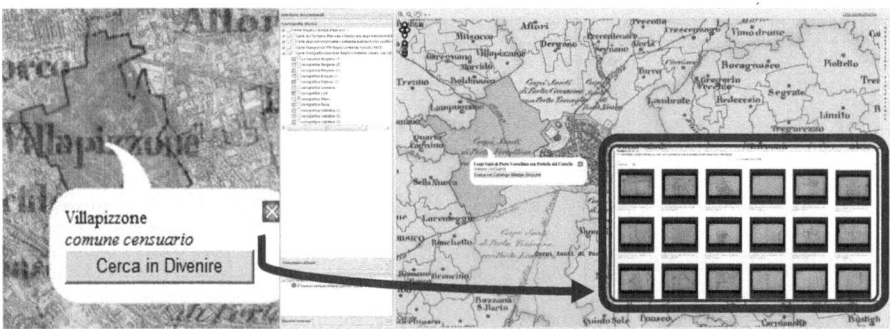

Fig. 4. An example of a query to map catalogue external service using WFS

A WFS functionality has been implemented within the map viewer: it enables user to select a single feature from vector layers and retrieve attributes that may be used to query an external service. Starting from the geo-referenced territorial map, the 'Chorography of the Province of Milan' - made during XIX century with the ancient municipalities (census units) - all the boundaries have been digitized, implementing a shape file (.shp) containing all the ancient census boundaries. Selecting a polygon from the map, the current municipalities or the old units, the user sends an HTTP query to the digital catalogue of ASMi (Divenire©) and obtains the complete gallery, among all the cadastral maps digitalized (~28,000), thus extracting the maps belonging to the selected municipality through the census name, the ISTAT code or the new Historical code introduced for each census parcel automatically recognized in function of the position.

4 Multi-temporal SD Integration for Landscape Analysis

The main aim of this project is to promote a progressive involvement of historic map within planning processes, through the systematic availability of multi-temporal and multi-scale data and services, in order to support analysis operation oriented towards governmental plan of territory, sustainability projects, Environmental Impact Assessment (EIA), mitigation actions, indications and intervention. Data management based on an SDI framework through geo-portal functionalities allows to enhance recognition of environment transformation along the centuries, helping the interpretation of permanencies and mutations by natural and anthropogenic actions. The result of Atl@nte is foreseen as methodological instrument for a systematic comparison of temporal maps series, that can contribute to a better knowledge and comprehension of landscape, in-land water basins, historical settlement, offering support to hydro-morphologic and ecological modelling, hydro-geological risk analysis, planned conservation, re-development and protection. It has been addressed to several end-user typologies. On one side it can support public bodies such as the regional agencies for environmental protection (ex. ARPA in Italy), Civil Protection and PAs. This planning instrument can be used at general regional level in an SDI framework for enhancing multi-temporal data acquisition, to address territorial government policies and European directives implementation (ex. Water Framework Directive, others); also, it could be of help to local government, like provinces and local municipalities, for the development of the territory government plan (PGT in Italy), strictly connected to the previous level as requested by the regional laws. On the other side, the certainty of ready to use structured data and their availability, can make the difference, moving from 'advanced isolated case studies' towards a daily practice of multi-temporal-spatial data use among professionals. In fact, Atl@nte requires only basic browsing skills, overcoming thus the barrier of the specific competences needed for operations like geo-referencing, statistic interpolation and geographic system transformation requested to consult digitalised historic layers.

4.1 Landscape Analysis at Different Map Scale. Potentials and Barriers

The integrated use of geo-referenced ancient maps and spatial data generated at similar comparable scale of representation, is an important auxiliary tool in reading of transformations that occurred in a territory over the centuries. Land planning is currently based on as-built situation analysis (GeoDB), social and economic analysis, integration with the surrounding region and historical documents. Added-value of integrating historical maps with new source data within planning instruments could be: respect the traditional character of a given area (agricultural, residential, touristic, religious), identification of the ancient water courses, canal courses and networks, recognition of old settlements and archaeological sites to be preserved.

The possibility to overlay historical maps, current technical maps, aerial or satellite photography using WMS, makes it possible to identify many changes occurred in the landscape over the time and to combine data in an easy and open manner. The chance to compare and overlay cartographic data, different even for generative origins, first of all requires a consideration on the scales of representation of the maps and images, according to the phenomenon to be evaluated and monitored. For example, if for the identification of a large river course, whose profile can be considerably varied in the past, will be sufficient to use maps and images on a territorial scale, ex. 1: 100,000 and 1: 50,000, it will be necessary to scale down to recognize the course of the minor canals. There are still a number of difficulties relating to the use of different map layers via WMS, not only for the limited availability of maps and images over the network, for the incomplete geographic data on the maps, but also for the operations that this service does not allow. For example, the impossibility to perform common operations, such as the export of geo-referenced images available via WMS, (ex. geo-tiff, regional orthoimages), obtained by overlaying map layers, makes it difficult to exchange data between different operators and to support processing services and advanced classification between different spatial sources, introducing a great barrier.

Here are few examples of research themes that could be improved by this instruments from various ancient and current maps at different scales of representation.

4.2.1 Agricultural Species Recognition and Biodiversity Preservation

An innovative ready to use instrument offers various types of analysis on agricultural landscapes, considering their cultural value [15], studies on ancient wetlands areas, and on changes of river beds/in land waters over time [16], permanencies and mutations of parcels properties directly facing the water courses, changes of road paths, and many more.

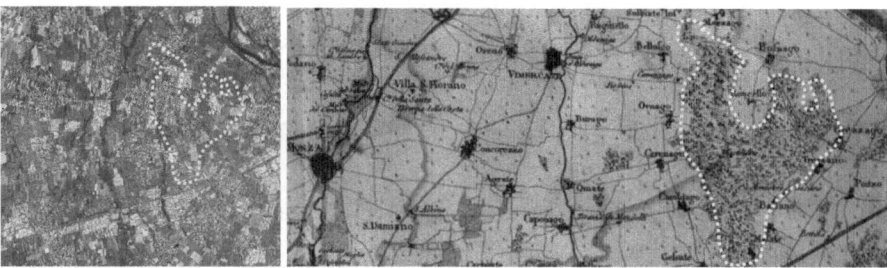

Fig. 5. Agricultural species detection and safeguarding: the wooded area in Castellazzo (Roncello) at northeast of Milan. The current ortophoto 1: 50,000 (left) and 'Map of Milan and Mantova Territories' 1: 86,400, Astronomers of Brera, 1788-1796 (right).

Topics such as agricultural vocation, biodiversity safeguarding, extension of agricultural species and forests detection in the past, primary and secondary network of irrigation, may find interesting applications, allowing to quickly identify and recognize the old tracks and sediments. The recognition of a wooded area today greatly reduced in its size in Castellazzo (Roncello) at northeast of Milan is shown using the regional historical maps (Fig.5).

Such information can be fundamental for the planning of territorial transformations. Environmental sustainability, cultivation with compatible existing ecosystem, biodiversity, are issues that require instruments of analysis of the territory.

4.2.2 Ancient Riverbed Transformation and Water Courses Secondary Canals: Built Anthropogenic Environment Rediscovery

Great interest, for many applications, lies in the possibility to verify the presence of ancient riverbed segments, related to the current state of the art: archaeological inspections, planning, flooding monitoring and prevention, as well as the valorisation of ancient natural and artificial traces of the human settlements along the water courses, and touristic divulgation of these traces. Transformations of prehistoric riverbeds (paleoalvei), related paleo-environmental adjacencies, of the primary and secondary water paths, can be progressively analysed to develop hydrological and morphological model based on the systematic use of diachronic temporal map series.

In order to observe the variation of the course of Ticino river, the area at south of the town of Boffalora (Milan) was chosen, where comparisons between different ancient maps clearly showed the significant changes occurred to ancient river path. This event caused significant transformations in the morphology of the surrounding territories.

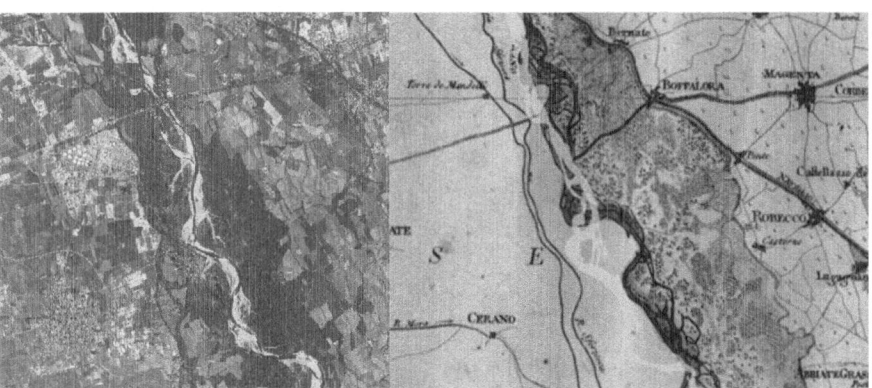

Fig. 6. The variation of the Ticino river flow at Boffalora, (south of Milan): Current ortophoto 1:10,000 from the Portale Nazionale© (left) and the 'Map of Milan and Mantova Territories' 1: 86,400 (right) with shape file of the current flow, obtained from regional WMS (in blue)

An important issue is tied to the secondary system of canals detection. In the past they assured a proper supply of water to cultivated land and guaranteed an efficient system of outlet in case of flooding of the main courses. The dense network of irrigation ditches and drains canals, allowed to control and manage water in times of great rainfall. Most of these canals, though now closed or covered, are actually still present in the tracings of the lines of division of lands, and easily recognizable overlapping old maps with actual orthoimages and maps.

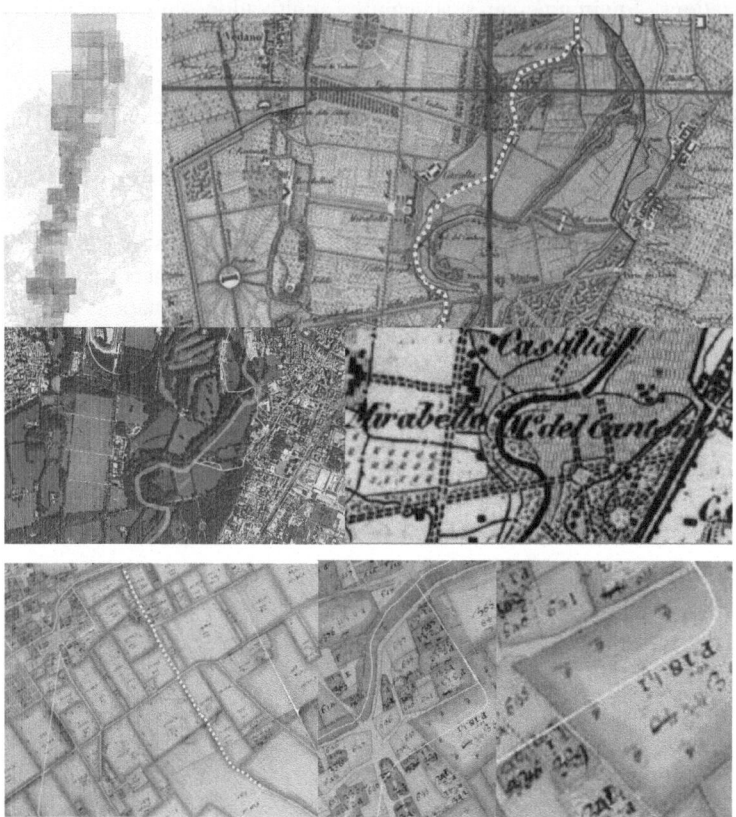

Fig. 7. Secondary canals: 'Roggia della Pelucca' in Park of Royal Villa (Monza), map of Giovanni Brenna (above); Ortophoto 1:50,000 and detail of Lombardy-Venetia ITM Topographic Map (middle). Canal 'Pescapera' deriving from the Martesana canal at Gorgonzola (Milan) situation in XVIII cen., 1: 2,000 (below), today dry and partially buried.

Particular attention is given to sample of Lambro, Adda rivers and Martesana Canal, which maps are georeferenced and published in Atl@nte, in the action of rediscovering and restoration of side channels as issues of interest for Water Framework Directive implementation. In Lombardy, infact, the articulated network of rivers, lakes, natural elements, artificial canals and other manmade artefacts, like ditches and fountains, deserve to be rediscovered and protected. The variations of Lambro river can be observed as a river 'forced' to pass within dense urban fabric and in an natural-agricultural context characterized by wetlands area in the south of Milan, for example the water meadow Chiaravalle Abbey, or the detection of agricultural species in the area of the Royal Villa in Monza (Milan province), as well as many agricultural species in Lodi area. The large scale historical maps allowed to rediscovery the agricultural vocation of the famous hunting Park of the Royal Villa, the largest fenced park in Europe, where the Gran Prix of Formula1 takes place. In some cases, the network derived from Lambro, such as Pelucca canal, which passes from north to south through the Park of Royal Villa, are still present but there is no water, and the riverbed is surrounded by vegetation. In other cases, such as Canal Pescapera in Gorgonzola (Milan), derived from Martesana canal coming from the Adda River, the track only partially exists, while a large part was filled with earth and turned into soil (Fig.7).

4.2 Integration of Multisource and Multispectral Data Using WMSs

The integration of information derived from remote sensing with historical maps can constitute an added value for different applications, allowing a deeper investigation of the history of our territory, enhancing the comprehension of the present and the dynamics of territorial transformation during the centuries. A richness bibliography has already shown that remote sensing, aerial and satellite images are a valuable tool for the characterization of the archaeological landscape and location of underground structure [18], [19], and for environmental analysis of complex water basin ecosystem [20]. Suitable remote sensing image interpretation techniques can give precious information on the water content, wet areas, river beds line detection, secondary channel recognition, morphological modifications: it has been giving the first examples of how remotely sensed digital imagery could have the potential to allow panoptic mapping of river hydro-morphology and human impacts[21], as well as the feasibility for fluvial morphology assessment in the San Francisco river [22]. Furthermore, automatic classification algorithms and feature extraction [23], could be experimented in recognition of ancient riverbed traces.

The objective here is to demonstrate the potential of the functionalities of the Atl@s and of the services available: tools for an integrated multi-source data analysis have been provided, enabling remote sensing images and products, distributed on the web by WMSs, to be added as new layers and visualized on basic historic maps. The following example underlines the necessity of a multisource service availability to get a systematic integration with historical and current maps and it highlights the fact that for targeted analysis, new information can be added exploiting existing external WMS, such as useful services distributed and provided by CNR-IREA (Italian National Research Council-Institute of the Electromagnetic Survey of Environment), or by the National Cartographic Service or other services (ex. GMES services).

The geoportal Atl@s has provided to the users the functionality 'Add WMS' in order to support any kind of useful spatial data already been available from other subject and source of spatial data. As the Atl@s is aligned to the most cartographic sites providing the users with possibility to add an existing WMS service distributed from other subjects (item Service WMS) like WMS of the European Environmental Agency (http://www.eea.europa.eu/data-and-maps) or the national environmental site, (http://www.pcn.minambiente.it/viewer/viewer.htm?service=progetto_natura&). As shown in Fig. 8 an application has been simulated on the area of the Adda River at the confluence of the derived Canal of Martesana: geo-referenced small scale ancient topographic maps of census areas of Vaprio D'adda, Concesa and Trezzo, were related to satellite images. The image of the Adda water courses in this area is derived by a Landsat TM image acquired on 13th September 1999 (pixel terrain resolution 30 m), and will be provided in the next future through WMS by CNR-IREA, (temporary hosted by the Institute of Ecosystem Study, http://ows.ise.cnr.it/demo/map.phtml), and visualized under the ancient maps within the Atl@nte geo-portal. More specific products derived from other remote sensors are going to be developed through the IREA WMS. In particular, it is been planned to exploit Quickbird and Geoeye spatial and spectral resolution integrated with historical data to verify the presence and the status of ancient riverbed segments.

Remote sensing images acquired from space satellite at different wavelengths of the electromagnetic spectrum, from visible to near infrared and thermal infrared, can be used to decode non visible traces, partially lying underground. A clear evidence of macro transformation of the river bed can be detected (Fig. 8) at scale 1:100,000 on the global view of the TM image classified for water content extraction.

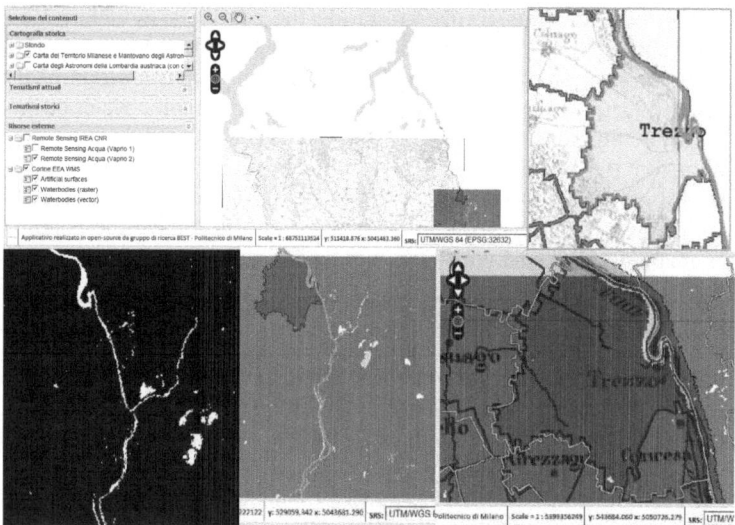

Fig. 8. Integration of different layers within Atl@ante: overlap of 'Map of Milano and Mantova Territories' with CORINE water bodies (EEA WMS), it is possible to extract cadastral maps belonging to the old census unit from the current municipalities boundaries in yellow (above). The Landsat TM image (CNR-IREA) WMS on the Chorographic map (1836) evidenced the changes of the Adda river over time and wetlands by water content classification (below).

4.3 UAV Flight Camp: A Test Filed for on-site Multispectral Data Collection

In case of lack of High Resolution (HR) images and data on the large scale, or due to high cost of HR satellite images, elaboration at a larger level of detail is required, as shown in a detail of the Census Section of Trezzo and Concesa (Figure 9, right). The ancient archaeological area of Trezzo, with roman and Lombard settlements along the river Adda and its riparian area, require further investigation of higher resolution thermal images analysis and classification. Comparing the state of the art recorded by historic map series during the centuries and the modified current situation of the territory, several hidden signs and traces of such transformations apparently lost can already be rediscovered and further analysed. To this aim, data and products distributed on the web as multispectral WMS are added as new layers and visualized on basic maps of the Atl@nte (Fig. 7, 8). As seen in the Fig. 9, the ancient maps are cut along the course of river Adda, the old borderline between Lombardy and Venetian Kingdom. Overlap of historical and current cartographic layers and transparency tool consents to identify the transformations of the course of the river Adda, today apparently lost, and hence the changes of territorial conformation around the river over time [17].

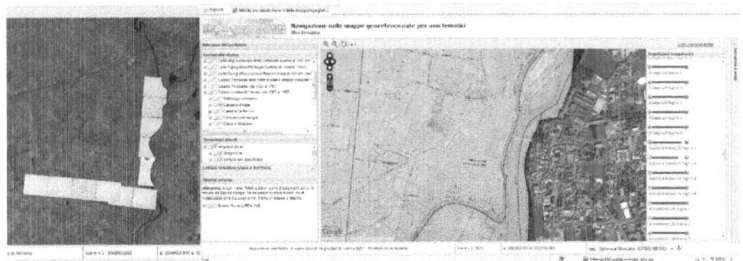

Fig. 9. The transformation of the secondary network channel of Martesana canals derived from Adda river related to the current Topographic DB (right). Multispectral image (IREA-CNR, 1:100,000) related to geo-referenced Cadastral Maps of Lombardy-Venetian Kingdom along the course of Adda river and Martesana canal on Google Maps (town of Vaprio d'Adda) (left)

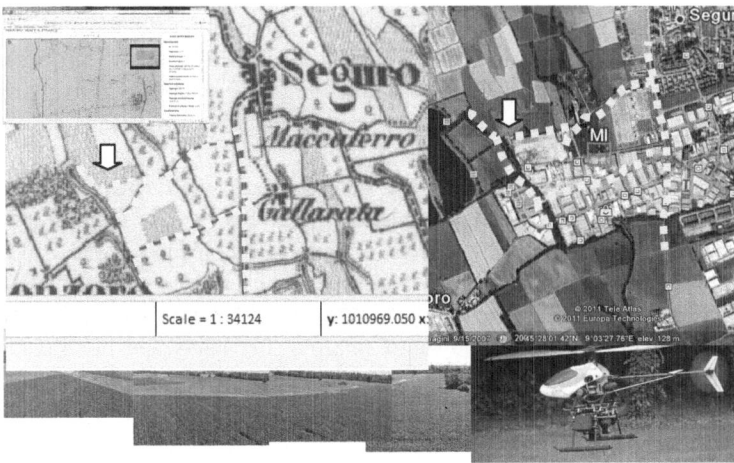

Fig. 10. ITM Topographic Map of Lombardy-Venetia Kingdom (1833) and the Cadastre of Lombardy-Venetia (1865-1887) map of the Census parcel unit of Seguro, extracted by WFS from the digital catalogue of National Archives in Milan (Divenire© project). Transformations of the wetland area is evidenced and will be investigated using low altitude UAV flights.

On one side, the systematic use of multispectral Quick Bird images related to the historic cadastres, within regional geo-portal, can achieve interesting potentials in downstream services distribution, following Global Monitoring for Environment and

Security (GMES) indications, to be further elaborated by strengthening automatic classification algorithms and feature extraction from the two different maps. On the other side, multispectral HR on site data collection and UAVs thermal imagery potentials can play an important role in the recognition and rediscovery of the secondary side canals for issues such as landscape and habitat preservation, biodiversity, flooding limitation thorough the re-use of secondary artificial network created over centuries to irrigate parcels and reduce water pressures.

UAV low altitude flight can be integrated to fill the gap of little scale satellite map to detect object dimension minor then 5 meters and details to supplement the thematic data available, allowing to observe the territory in a closer range (Fig.10) [25]. Research group of Politecnico di Milano has been developing a photogrammetric radio controlled (RC)/UAV platform with visible sensors on board for environmental and architectural purposes [24]. Multispectral sensors from low altitude platforms can introduce great potentials towards canals' reactivation and reconnection, along the river corridor and riparian area for mitigation and flooding monitoring. Acquisition with nearly infrared camera will be experimented (spring 2012) on a test area characterized by secondary canals along Lambro and Adda rivers registered on ancient cadastral maps (XVIII cent.). This action should help to address the investigation to detect, and where possible to recover, the permanence of canals hidden by cumulated detritus filling the banks or partially destroyed by man action.

5 Conclusions

The research described highlights the potentials of multi-scale and multi-temporal data integration of historic maps with current GeoDB, obtained from open access portals. Geospatial services can play an important role in decoding many territorial traces, addressing sustainable intervention and mitigation in the planning operations (Fig.11). Basic functions of Atl@nte allow to use WMS, allowing to interact with external data sources and providing the georeferenced map series within an open source GIS to external users. The increasing availability of data sources has become mandatory to support specific analysis driven by the planning aims and by the complex landscape awareness where built environment, water sources, agriculture and other require to be related. Potentials of multi-spectral data service integration with historic map layers has been shown to improve the landscape investigation, like macroscopic changes of ancient riverbeds that need to be assessed by local data acquisition. However, the fact that multispectral images are difficult to obtain by non-expert users like PA and general public, represent a huge barrier in landscape analysis at the moment. WMS development of satellite images, lie an ongoing example here shown and conducted by CNR-IREA, will be useful to overcome this gap, together with WPS of thematic image classification chosen by the users. Furthermore, the high level of detail of cadastral maps requires HR multi-spectral images that could be enhanced with on-site IR data collection using ultra-light UAVs, in order to recognize physical elements disappeared from the current maps and the non-visible traces lying underground. Advanced service implementation for real time data processing are expected to be an important key for the future development to support clients' operations: a gap of WPS development is represented by the non-authorized export of geo-referenced images available via WMS that is a necessary step for implementation of classification algorithm between different data sources.

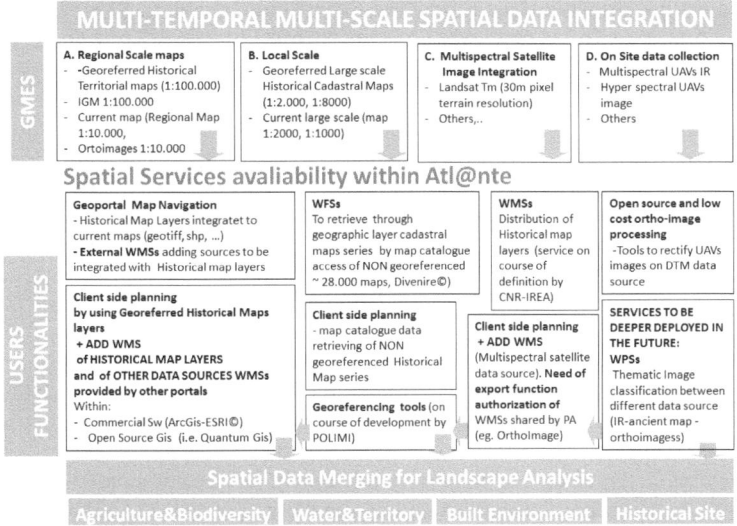

Fig. 11. Logical Scheme of Multi-Temporal Spatial Data Services for Landscape Analysis

Acknowledgments. Our acknowledgments to Fondazione Cariplo that funded the Atl@nte project together with ASMi (the National Archive of Milano) and Agenzia del Territorio (Cadastral Administration) for the availability of historical map series.

References

1. Council of Europe: European Landscape Convention. Chapter I, Article 5 Florence (2000)
2. Scottish Natural Heritage (2008), http://www.snh.gov.uk/planning-and-development/environmental-assessment/eia/
3. Craglia, M., Campagna, M.: Advanced Regional SDI in Europe: Comparative cost-benefit evaluation and impact assesment perspectives. IJSDIR 5, 145–167 (2010)
4. Cuca, B., Brumana, R., Scaioni, M., Oreni, D.: Spatial Data Management of Temporal Map Series for Cultural and Environmental Heritage. IJSDIR 6 (2011)
5. INSPIRE EU Directive 2007/2/EC of the EU Parliament and of the Council, http://eur-lex.europa.eu/ JOHtml.do?uri=OJ:L:2007:108:SOM:EN:HTML (last access February 27, 2012)
6. Scazzosi, L.: Landscape opportunities for territorial organisation. LOTO Project. Guidelines for landscape management of territorial transformation: methodological indications. Strasbourg Cedex. In: Vth WS of the Council of Europe for the Implementation of the European Landscape Convention, Girona, Spain, pp. 179–182 (2008)

7. Montaner, C., Roset, R.: Impact of the internet users on the Map Library of Catalonia access collections. In: 5th Int. Work. on Digital Approaches in Cartographic Heritage, Vienna, pp. 392–394 (2010)
8. Brumana, R., Prandi, F., Signori, M.: Local and global approaches for the integration of up-to-date geo-db and ancient maps within the Atl@s portal. In: Caglioni, M., Scarlatti, F. (eds.) Representation of Geographical Information for Planning, Esculapio, Bologna, pp. 48–58 (2009)
9. Oreni, D., Brumana, R., Scaioni, M., Prandi, F.: Navigating on the past, as a bird flight, at the territorial scale of historical topographic maps. WMS on the Corografie Delle Province Del Regno Lombardo-Veneto, for Accessing Cadastral Map Catalogue, E-Perimetron 5, 194–211 (2010) (ISSN:1790-3769)
10. Monti, C., Mussio, L.: L'attività geodetico astronomica, topografica, cartografica degli astronomi di Brera dal 1772 al 1860, atti dell'Osservatorio. Istituto Lombardo di Scienze e Lettere, Milano (1980)
11. Livieratos, E.: ICT in Heritage Research, GIM International, vol. 20(5) (2006), http://www.gim-international.com/issues/articles/id661-ICT_in_Heritage_Research.html
12. OpenGIS Web Map Server Implementation Specification. Open Geospatial Consortium (OGC) Inc. (2006), http://www.opengeospatial.org/standards/wms (last access February 28, 2012)
13. Lieberman, J.: OpenGIS Web Services Architecture, OGS specification, 03-025 (2003)
14. Henrie, D.: Ordnance Survey historic town plans of Scotland, 1847-95: Georeferencing and web delivery with ArcIMS and OpenLayers. e-Perimetron 4(1), 73–85 (2009)
15. Steiner, C., Marzorati, P., Schabl, A., Buchecker, M., Dobrovovodska, M., Kruse, A., Martinovic-Vukovic, B., Mavar, Z., Pungetti, G., Raguz-Lucic, E., Robbiati, C., Roth, M., Scazzosi, L., Spulerova, J., Velarde, M.D.: Cap V. Planning the future of European agricultural landscapes considering their cultural value and heritage. In: Pungetti, G., Kruse, A. (eds.) European Culture Expressed in Agricultural Landscapes. Perspectives from Eucaland Project, pp. 179–204 (2010)
16. Lüderitz, V., Zerbe, S., Jüpner, R., Arevalo, J.R.: Ecosystem restoration and sustainable management of rivers and wetlands – Introduction to the special issue. Forest Ecology, Landscape Research and Nature Conservation, Heft (2010)
17. Livieratos, E., Tsorlini, A., Pazarli, M., Boutoura, C., Myridis, M.: On the digital revival of historic cartography: treating two 18th century maps of the Danube in association with Google-provided imagery. In: 24th ICC, Santiago, Chile (2009), http://www.icaci.org/documents/ICC_proceedings/ICC2009/html/nonref/25_6.pdf
18. Brivio, P.A., Pepe, M., Tomasoni, R.: Multispectral and multiscale remote sensing data for archaeological prospecting in an alpine alluvial plain. Journal of Cultural Heritage 1(2), 155–164 (2000)
19. Kucukkaya, A.G.: Photogrammetry and remote sensing in archaeology. Journal of Quantitative Spectroscopy & Radiative Transfer, 83–88 (2004)
20. Fregonese, L., Monti, C., Brumana, R., Achille, C.: Usability of satellite high resolution images: extraction of geometrical, morphological and environmental information about the urban and coastal structure referred to the Venice lagoon. In: Proceeding CD, VII International Congress of Earth Sciences Chile (2002)
21. Gilvear, D.J., Davids, C., Tyler, A.N.: The use of remotely sensed data to detect channel hydromorphology; River Tummel, Scotland. River Research and Applications 20(7), 795–811 (2004)

22. Carvalho, O.A., Guimaraes, R.F., Santos, N.F., Martins, E.S., Gomes, R.A.T., Shimabukuro, Y.E.: Analysis of channel morphology of San Francisco River using remote sensing data. International Journal of Remote Sensing 31(8), 20, 1981–1994 (2010)
23. Prandi, F., Brumana, R., Fassi, F.: Semi-Automatic Objects Recognition in Urban Areas Based on Fuzzy Logic. Journal of Geographic Information System 2, 55–62 (2010) ISSN: 2151-1950
24. Barazzetti, L., Remondino, F., Scaioni, M., Brumana, R.: Fully automated UAV image-based sensor orientation. In: IAPRS&SIS, Calgary, Canada, on CD-ROM, vol. 38(1) (2010)
25. Brumana, R., Scaioni, M., Barazzetti, L., Cuca, B., Oreni, D., Alba, M.: Panoramic UAV views for landscape heritage analysis integrated with historical maps atlases. In: XXIIIth CIPA Symposium, Prague, Czech Republic, September 12-16. CD-Rom (2011)

From Concept to Implementation: Web-Based Cartographic Visualisation with CartoService

Hartmut Asche and Rita Engemaier

University of Potsdam, Geoinformation Group, Department of Geography
Karl-Liebknecht-Strasse 24/25, 14476 Potsdam, Germany
{hartmut.asche, rita.engemaier}@uni-potsdam.de
http://www.geographie.uni-potsdam.de

Abstract. This paper deals with the implementation and prototypical application of a concept of a web-based on-demand service for the modelling and visualisation of quality maps (CartoService, CS). Drawing on state-of-the-art web technology and software architecture CS can be considered a cartographic customisation of the standard visualisation system. The CS architecture is generic and completely based on the use and integration of web-services, more precisely, on a set of loosely coupled, self-contained methods and information units (services). Map modelling and visualisation components are sequentially ordered in a process chain which facilitates an automated production workflow. Interaction between web-service players (communicators) is based on a three-tier architecture. Tasks assigned to CS essential services are discussed and exemplified for one common application scenario: CS as map configurator. Presently, full implementation of CS is under way. As a fully operational product, CS will significantly contribute to the enhancement of web-based map visualisation in comparison to existing standards-based map dissemination.

Keywords: Geovisualisation, cartographic modelling, web cartography, quality web mapping, web services, quality maps.

1 Introduction

It is estimated that about 95 percent of all digital data have a geospatial component (Hamilton, in Perkins, 2010). Geospatial reference defines both the absolute and relative position of objects or phenomena on the earth's surface. The absolute position is generally determined by coordinates, the relative position refers to neighbourhood and spatial relations with other geo-objects. Traditionally, geospatial reference of data is visualised in map form, whether analogue or digital. In the pre-digital age production of maps was, in essence, a manual process executed with specialised map design and production tools. The lack of publicly accessible written sources on cartographic methods and the expertise required for their application restricted map production to a minority of qualified specialists. For this reason, the production, distribution, and, to a lesser extent, use of the essentially professional, adequately modelled maps was limited up to the 1980s. Ubiquitous access to digital information and communication technology (ICT) has, i.a., made techniques and media for the production and distribution of map graphics available to everyone. As a consequence, since the 1990s, there has been an ongoing explosion of map products from laypersons. Because the producers of theses map products have insufficient knowledge of cartographic visualisation methods, the content and cartographic quality of the vast majority of these maps can only be characterised as naive and inadequate.

Recent paradigms, such as the often-cited pictorial turn (Mitchell, 1994), visual studies and the ongoing debate on information visualisation (McCormick et al., 1987), have emphasised the importance of the proper visualisation of digital data and, in particular, of their visual analysis, in science, the economy, media and the public in general. Geovisual analytics, the geoinformation science variant of visual analytics, has been massively boosted in the aftermath of the 9/11 attack of 2001 on the New York World Trade Center in the context of US Homeland Security. Both R&D and applications in this field have established visual analysis of 2D/3D map graphics as a complementary access to geospatial information in comparison to non-graphic data analysis in geoinformation systems (GIS). Against this background this article outlines a web-based system (CartoService) that facilitates automated geovisualisation of geospatial data in cartographic modelling quality (Engemaier, 2011, Engemaier & Asche, 2011).

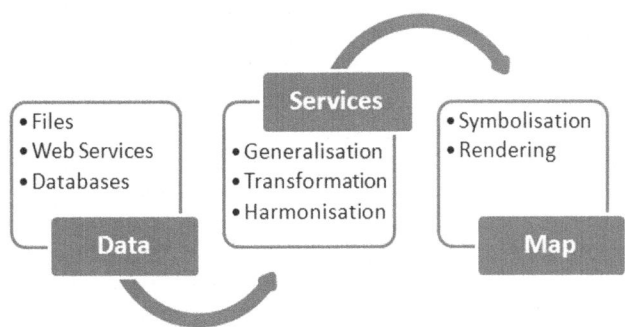

Fig. 1. CartoService rationale: Using web services to generate quality maps from geodata

2 Concept: Component-Based Cartographic Modelling Pipeline

CartoService (CS) designates a concept that aims to demonstrate how state-of-the-art web technology and software architecture can be employed to generate quality maps (fig. 1). In this respect CS can be considered a domain-specific, cartographic customisation of the standard visualisation system for the generation of web-based quality maps. The CS map modeling system uses an automated, rule-based workflow to create meaningful quality maps from web-based input data and visualisation requirements. CS thus maps the classical map design and editing process onto an ICT rule base. In compliance with the principles of (thematic) cartography, the modelling and visualisation process facilitates the mental processing of a non-graphic data model (primary model) from geospatial raw data to a graphic-oriented cartographic model (secondary model), i.e. a purpose- and user-defined quality map, which can be explored and analysed.

Classical map design and production is based on the following sequence of process steps: data acquistion, data processing, object-sign-reference, generalisation, map composition, production and distribution. Subject to map type, map topic, map users and map producers the application of these sets of methods is governed by the relevant map design guidelines. The above set of cartographic methods is incorporated into the following principle service components of CS (fig. 2).

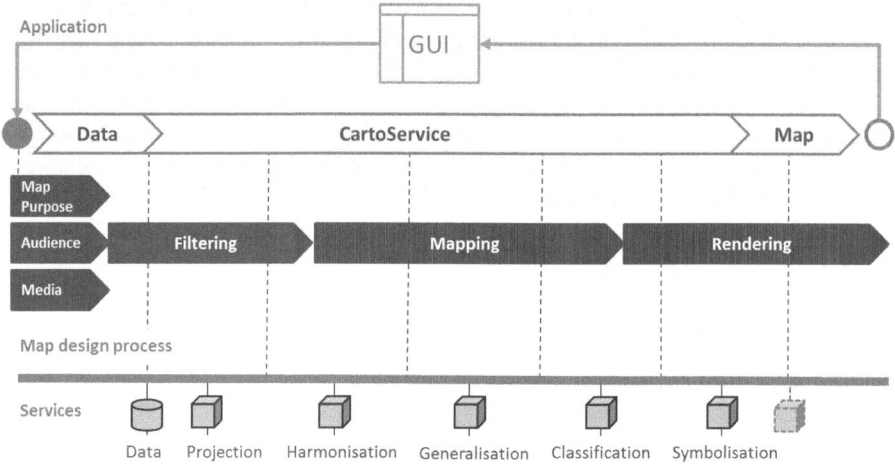

Fig. 2. Map generation process in CartoService

Preprocessing/Filtering
The initial modelling component performs evaluation and preprocessing operations on the geospatial data sets entered into the system. From a modelling perspective these data are raw data from diverse sources, e.g. data files, databases or web services. Processing of these data in CS includes operations, such as selection, reduction, assessment, classification (data generalisation) or error correction. The output of this step is a preprocessed non-graphic map dataset which represents a dedicated data model of a defined area on the earth's surface.

Graphics generation/Mapping
In this second component the all-important object-sign-reference of the preprocessed map dataset is executed. Operations performed include data-to-geometry-transformation (coordinate system, map scale), assignment of graphical primitives and map symbols, application of graphic variables as well as semantic generalisation conditions. The output is a geo-referenced, graphical map presentation typical of GIS map graphics. It is important to note that such graphic is a visual representation of the primary data model only which, by definition, lacks any graphic modelling. Graphical presentation *per se* thus does not constitute a secondary graphic model of the processed geodata.

Map Generation/Rendering
In this final and decisive component the presentation graphic is rendered into a proper map model in line with the initial media, distribution and map use specifications. Relevant transformation operations include cartographic generalisation (scale-/topic-oriented), annotation (lettering), media-specific map design as well as map/data interaction functions, where applicable. The end result is a professionally modelled quality map. Its modelled graphic structure facilitates effective communication of its visualised geospatial data as well as subsequent in-depth visual exploration and analysis.

The above components of map modelling and visualisation are arranged in a *map processing pipeline* which implements the process flow in an automated production sequence. Only when one component has been completed successfully can the subsequent component be activated. In this way professional modelling and visualisation of geospatial data can be guaranteed in accordance with minimum standards of (thematic) cartography. Each processing component can be enhanced or specialised in its functional range. Examples of such specialisations are the map configurator application discussed below or the related atlas toolbox (Asche & Engemaier, 2011).

3   Architecture: Service-Oriented Components and Layers

In principle, the ICT-based implementation of the CS concept into a technical architecture is built on the architecture patterns of the internet, the paradigms of service-oriented architecture (SOA) and object-oriented programming. More precisely, the outlined modelling and visualisation processes are modelled as a generic set of loosely coupled, self-contained methods and information units (services). The services are embedded and made available in an ICT environment in which web-service players (communicators) communicate in a three-tier architecture. By means of standardised XML-based document languages (e.g. XML, WSDL, BPEL) CS specifies type, communication format as well as the processing operations of the information units exchanged among the communicators.

For this purpose CS draws on the well-established technology of web-based communication. Hence, from a logical perspective, system components are organised in layers (cf. fig. 3). In accordance with the respective service definition the logical layers execute specific tasks within an individual layer and the adjacent logical layer. In the system architecture physical services are allocated to the logical layers (e.g. transformation, classification, generalisation). Integration and binding of these services is by standard communication interfaces.

To organise the selection (in compliance with the functions requested) and binding of external services the system architecture uses a service management component (service respository and matching). The management component not only controls the physical data stream but also logical control flows, such as the supervision and control of transactions and operations. In this way specific flows and the traffic associated to them are defined on the basis of the relevant input and process parameters. The flows can be output as a workflow description or saved to a connected database. This allows for repeated execution if required. The technical architecture of CS outlined here is completely based on the application and integration of web services. As has been show, these services are broken down into logical service layers according to different tasks and responsibilities.

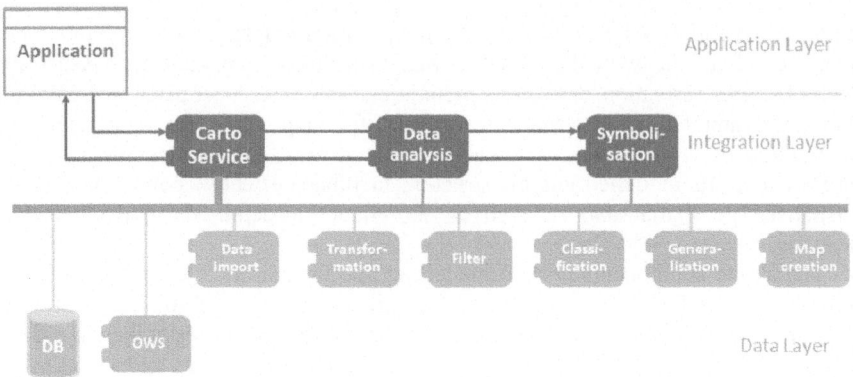

Fig. 3. CartoService architecture (i): Components and layers

The central layers of the CS system are the *data layer* (including data integration components), the *processing layer* (including components of non-graphic and graphic information processing) and the *application layer* (user interaction and monitoring). The complete process chain is controlled and monitored by the service management component as is the binding of external services. Both external and internal services are managed in a repository (cf. fig. 4). This archive contains services relevant to visualisation, their description and sources. It is constantly updated through a connection to corresponding external repositories.

Due to the modular nature of the service components the system architecture is generic and can be expanded in a flexible manner. The request-driven integration of services allows for the extension of the process flow, and, at the same time, for the skipping of single processing steps. In total, CS can be characterised as a component service which itself is composed of separate service units. Communication among the components is effected by messages exchanged in standard XML-based document languages.

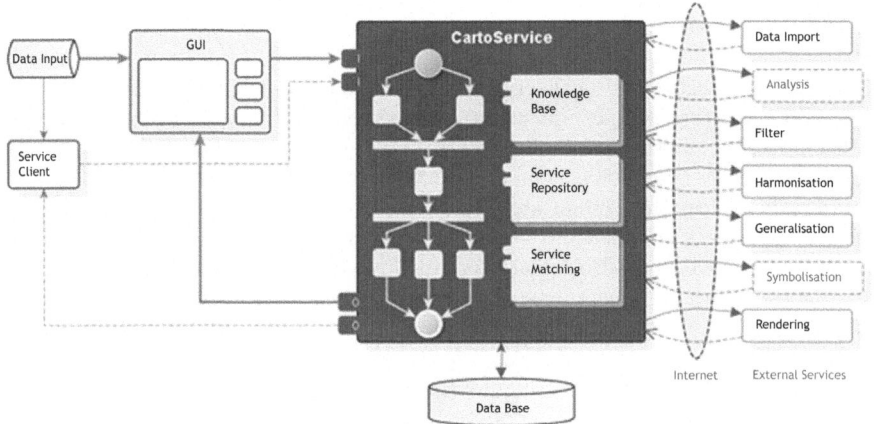

Fig. 4. CartoService architecture (ii): External web services and process flow

4 Modelling Process: Workflow and Reference Model

In the modelling and visualisation process detailed above the layers and components – the CS modelling services that turn geospatial data into quality maps – are assigned the following tasks (cf. fig. 5):

Data integration
First, requests are submitted to CS via a specified communication interface. Depending on the input data forwarded CS initiates the procedures of data collection, integration and data connection. Integration of input data is accomplished by suitable data connectors and adapters. Apart from this standard use case, data integration application scenarios in particular the map configurator scenario (cf. below, Engemaier, 2011), can also be carried out by an appropriately configured client application (e.g. user interface). Execution of this process provides an integrated input dataset for further processing in CS.

Data processing
In this process the input data are, in a first step, assessed for their geometric, thematic and temporal attributes. This assessment does not modify the data, but augments them with metadata which may be integrated in successive processing steps. Data evaluation is, in a second step, followed by the actual data preprocessing. For that purpose the augmented input data are initially filtered according to the criteria specified at the start of map generation. Only data meeting the parameters specified are passed on for further processing. If required, data are harmonised in a data harmonisation component integrated into the workflow (e.g. transformation into a uniform map projection). There is no predefined location where data harmonisation will have to be integrated into the workflow. Hence the order of data filtering and data harmonisation is variable and can be adapted to the respective application requirements. The output of this process is a filtered, harmonised, non-graphic map dataset.

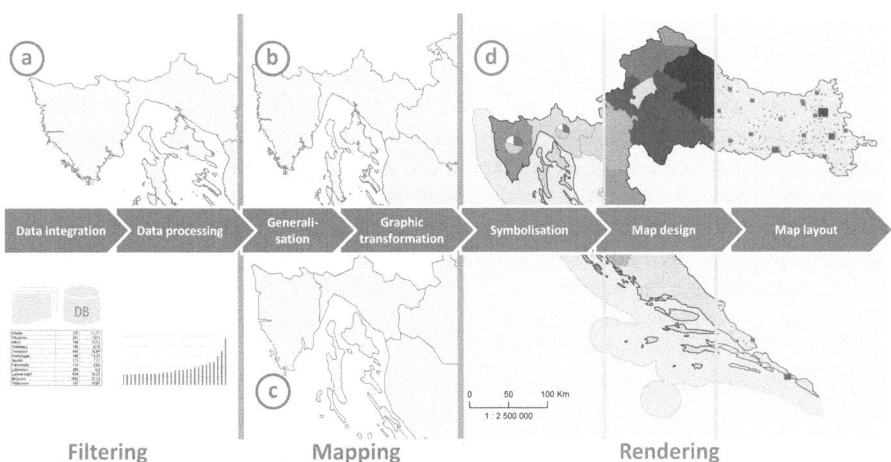

Fig. 5. CartoService process flow: (a) Raw data, (b) filtered data, (c) georeferenced, generalised map presentation, (d) rendered quality map model in 3 thematic versions

Data presentation
In this successive process the filtered, non-graphic map data are, in a first step, transformed into a presentation graphic of the map geometry. If required, this presentation includes a predefined graphical simplification of the map geometry by generalisation (cf. fig. 5). Should the framework requirements defined (e.g. map topic, application, distribution media) neccessitate additional abstraction of specific thematic attributes this is executed in this step as well. In either case, critera and parameters revelant to generalisation are derived from input data metadata and from cartographic modelling rules applicable to this task. As has been mentioned, specification of input conditions as well as selection and application of parameter sets is performed in the managment component. The results are graphical presentations of the non-graphic map data conforming to the predetermined specifications of geometric and thematic data quality. In a second step the map data are symbolised. For that purpose every geospatial object is assigned a map symbol representing the geometric, thematic and temporal object features by object-sign-reference. Symbolisation, however, does not imply creation of a proper cartographic map model. In fact the mapping result is a vector graphics representation of map objects with their graphic attributes. Subject to data features and prerequisites, map symbols and reference or localisation rules (symbol binding), respectively, are generated instead and returned in a defined description language.

Map creation
In this core process of image synthesis the presentation graphic is transformed into a quality map by graphics-oriented cartographic modelling and visualisation operations. For that purpose the presentation data are first put to cartographic generalisation. The graphic generalisation procedures aggregate, exaggerate or displace the graphic map objects in accordance with the data features, map scale, map purpose and presentation medium specified as well as geometric, topological and semantic context. In this way a complementary, fully generalised, symbolised graphic model is generated in professional map quality (secondary model) from the initial non-graphic data model (primary model). Subsequent map design operations then reshape the complex map graphic according to the relevant principles of visualisation and graphic design. In this second step additional graphical elements are integrated into the map graphic, such as map lettering (annotation, static or dynamic) and hill shading. The resulting map visualisation incorporates the full range of modelling features that make all the difference between a non-modelled presentation graphic and a graphically modelled map face. In a third step the layout component completes map generation by the design and placement of map title, map legend and other marginal information. From identical data this final step may yield different map compositions, depending on the specified presentation medium (e.g. screen map, paper map) as well as the intended map use and interaction. Thus, for example, the concept of smart legend (Sieber et al., 2005) can only be implemented in electronic maps.

The processing process outlined above, based on a project-specific, interacting set of CS components, can be considered the standard use case or reference model of map visualisation with CS. As has been demonstrated, the user is an integral part of the CS reference model, enabling her to control the majority of steps of the modelling process. User access to interaction and manipulation functions is, however, restricted wherever unprofessional user manipulation of essential modelling procedures affects the cartographic quality of the map to be generated. For that purpose the basic data processing and map visualisation procedures are initiated and sequentially executed in an automated workflow. These limitations help to ensure that user-centered processing of user data will, in any case, produce quality map graphics complying with the minimum requirements of cartographic modelling principles.

5 Application Scenario: CartoService as a Map Configurator

Based on the CS reference model of map visualisation an application scenario is presented as an example. The map configurator scenario is chosen to illustrate the use and interaction potential of the CS map visualisation system. The presentation focusses on those features and functions that distinguish CS from other web-based map visualisation systems currently available. It has been mentioned that, for the time being, CS is not fully implemented. Hence service functions and qualities as well as their integration in potential applications have been simulated under constructed system conditions. For that purpose an experimental setup has been configured equalling the process flow discussed above. Only the integration of external services conceptualised as an automatic operation is accomplished interactively.

The map configurator scenario can be considered a standard application of the CS visualisation system. In this scenario the use of CS services is governed by the following specifications:

1. A range of different geospatial datasets are available to the user on her computer as well as on the internet. Detailed information, however, on content and structure of this data (metadata) is lacking. In this case CS enables the user to get an overview of the data and gain a deeper understanding of the data-coded geospatial information by applying the available cartographic testing mechanisms and a graphical presentation of the input data. To this end, CS assesses both geometric and topological properties as well as thematic features of the selected datasets. Using these test criteria CS subsequently selects an appropriate cartographic visualisation method, applies it and produces a graphical preview.
2. Based on this preview the user is able to modify or specify her visualisation goal (e.g. topographic, thematic map style) by means of the respective service functions. She is also able to influence individual steps of the visualisation process (e.g data selection, level of generalisation). Modified settings can be saved to an individual user profile to be re-usable in other applications at a later stage. In this case CS serves as a back-end solution for cartographic visualisation. The user is able to access essential service features via a graphical user interface (GUI) specifically customised for that purpose (cf. fig. 6).
3. In this environment the user is able to access more extensive functions on her client through the GUI not considered in the server-based configuration outlined above. Additional features can be implemented by means of java or XML-based techniques (AJAX) which expand or reshape native CS operations in a transparent, user-specific way. Input data, for instance, can be filed to the GUI via drag & drop. From here the ensuing operations of evaluation, harmonisation, graphical presentation and map modelling are executed and controlled. Filter panels allow the user to track and control application of the visualisation parameters transparently and

interactively. Due to the GUI settings modifications of visualisation parameters will instantly affect the graphic structure of the map face of a particular application. Fig. 6 shows a GUI design with selected functions for the selection of input specifications, generalisation and symbolisation. The accompanying map examples were generated by manual integration of two external services relevant to map visualisation. Generalisation functions have been derived from the *MapShaper* application (Harrower & Bloch, 2006), whereas the well-known *ColorBrewer* (Harrower & Brewer, 2003) application has been used for the definition of colour scales.

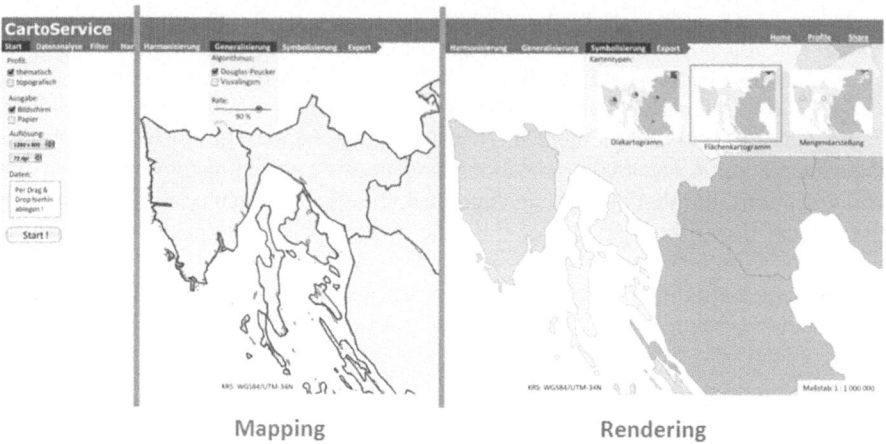

Mapping   Rendering

Fig. 6. CartoService graphical user interface: Map modelling and visualisation control options

Combining server-based CS functionality and client-sided interactivity CS provides the user with an application (e.g. the map configurator outlined above) that presents cartographic visualisation in a transparent, rule-based and attractive way. In contrast to other applications available on the internet CS makes it possible for experts and laypersons alike to visualise their own data in a professional cartographic way.

6 Conclusion

This article presents the concept and partial implementation of a web-based service (CartoService) for the generation of quality map graphics to the standards of cartographic modelling and visualisation. It is demonstrated that the relevant theoretical and methodical expertise required for professional map modelling can be operationalised and made available through web-based services. On the one hand the automated workflow guarantees professional visualisation of user-provided input data in the form of quality maps. On the other hand the user can influence the visualisation process individually and as the respective application requires, as has been shown with the common map configurator application scenario. At present, full implementation of CS is under way in the framework of a R&D map visualisation project. From experimental tests and applications it can, however, already be concluded that the application of a fully operational CS will improve the cartographic quality of web-based map visualisation compared to, for example, existing products or services based on Open Geospatial Consortium (OGC) standards. Thus CS is not only an important contribution to professional visualisation of qualitiy maps on the internet. CS-produced map graphics visualised in compliance with cartographic modelling principles will, at the same time, also support and promote effective visual communication, exploration and analysis of geospatial data in a broad range of applications, including the emerging field of geovisual analytics mentioned in the Introduction.

References

1. Alameh, C.: Chaining geographic information web services. IEEE Internet Computing, IEEE Educational Activities Department 7, 22–29 (2003)
2. Andrieko, G., Andrieko, N., Voss, H.: Organization of mapping services in the internet: Possible solutions. In: Ruiz, M., Gould, M., Ramon, J. (eds.) 5th AGILE Conference on Geographic Information Science, Palma, Balearic Islands, Spain, April 25-27, pp. 537–546 (2002)
3. Asche, H.: Der Atlasbaukasten: nachhaltiges Produktionskonzept im Geoinformationszeitalter? Eine_Bestandsaufnahme. Kartographische Nachrichten 59(1), 3–12 (2009)
4. Asche, H., Engemaier, R.: The Atlas Toolbox: Concept and development of a rule-based map component for a GIS-VIS production environment. In: Ruas, A. (ed.) Advances in Cartography and GIScience. Selection from ICC 2011 Paris. Lecture Notes in Geoinformation and Cartography, vol. 1, pp. 147–159. Springer, Heidelberg (2011)
5. Di, L.: Distributed geospatial information services – architectures, standards, and research issues. In: The International Archives of Photogrammetry, Remote Sensing, and Spatial Information Sciences, vol. XXXV(2) (2004)
6. Engemaier, R.: Webbasierte kartografische Visualisierung. Theoretisch-methodische Grundlegung und Ableitung eines Konzeptes zur Erzeugung von Qualitätskarten auf Basis von Webservices. Diss. Potsdam. Potsdam (2011)
7. Engemaier, R., Asche, H.: Carto Service: A Web Service Framework for Quality On-Demand Geovisualisation. In: Murgante, B., Gervasi, O., Iglesias, A., Taniar, D., Apduhan, B.O. (eds.) ICCSA 2011, Part I. LNCS, vol. 6782, pp. 329–341. Springer, Heidelberg (2011)
8. Friis-Christensen, A., Ostländer, N., Lutz, M., Bernhard, L.: Designing service architectures for distributed geoprocessing: Challenges and future directions. Transactions in GIS 11(6), 799–818 (2007)
9. Harrower, M., Bloch, M.: Mapshaper.org: A map generalization web service. IEEE Computer Graphics and Applications 26(4), 22–27 (2006)
10. Harrower, M.A., Brewer, C.A.: ColorBrewer.org: An online tool for selecting color schemes for maps. The Cartographic Journal 40(1), 27–37 (2003)
11. McCormick, B.H., DeFanti, T., Brown, M.D.: Visualization in scientific computing. Computer Graphics 21(6), 1–14 (1987)
12. Mitchell, W.J.T.: Picture theory, Chicago (1994)
13. Neun, M., Burghardt, D., Weibel, R.: Automated processing for map generalization using web services. GeoInformatica 13, 425–452 (2009)

14. Perkins, B.: Have you mapped your data today? Computer World online (2010), http://computerworld.com/s/article/350588/Have_You_Mapped_Your_Data_Today_ (access: February 2012)
15. Sahin, K., Gumusay, M.U.: Service oriented architecture (SOA) based web services for geographic information systems. In: The International Archives ofthe Photogrammetry, Remote Sensing and SpatialInformation Sciences, vol. XXXVII(B2), pp. 625–630 (2008)
16. Sieber, R., Schmid, C., Wiesmann, S.: Smart legend – smart atlas!_Mapping approaches into a changing world. In: Proc. 22nd Intern. Cartogr. Conference A Coruna [CD-ROM], July 9-16. International Cartographic Association, A Coruna (2005)
17. W3C, Web services architecture. W3C Working Group Note (February 11, 2004), http://www.w3.org/TR/ws-arch/ (access: March 2012)

Multiagent Systems for the Governance of Spatial Environments: Some Modelling Approaches

Domenico Camarda

Technical University of Bari, Italy
d.camarda@poliba.it

Abstract. The concept of community governance is generally intended as the management of complex processes underlying the inherent complexity of the environment.

As a matter of facts, this complex approach provides the diffused, delocalized and multiagent management processes taking place on settled territories with a multiscalar, multisectoral and transdisciplinary vision.

The aspects of interest of this study about the governance theme deal with the definition of an ICT-based, multiple-agent systems (MAS) for the representation of knowledge, roles, relationships, tasks and operational levels involved in governance processes. This model is oriented toward the construction of MAS-based system architectures to support development policymaking, managed through process models of community governance .

The paper starts with a first general introduction on the research background of the paper, particularly governance and multi-agent systems. A second chapter discusses multi-agent modelling and the practical case-study of Foggia (Italy). Concluding remarks end up the paper.

Keywords: Environmental governance, Multi-agent systems, Decision support, Knowledge interaction management, Spatial planning.

1 Introduction: A Research Framework on Governance

The concept of community governance today can be reasonably ascribed to that group of 'nomadic concepts' which crosses the environmental lexicon through different domains, yet assuming ontologically consistent meanings [1][2]. In fact, scientific literature looks at a multiplicity of different sectors, with related critical theoretical, methodological and at times operational and implementing discussions. In this sense, explicit knowledge contents range from economics to sociology to environmental planning, to knowledge engineering and technology. Here, governance is generally intended as the management of complex processes that underlie the inherent complexity of the environment, its phenomena and its relations in the physical, cognitive, social, economic sectors of life [3][4][5][6].

It is important to draw out the importance of governance concept from a common thread stemming out from a number of often interrelated sectoral interpretations. First of all, the concept of *institutional governance* has been largely dealt with in scientific literature. It is used with reference to the formalized mechanisms of collective action, aimed at attaining general benefits. In such typology, particularly public institutions and formal organizations play essential roles by structurally interacting with other agents and other organizations. In this sense, the *governance* term is used as a generic concept indicating abilities, rather than specific governing modalities. The attempt is to turn rigid administrative practices into abilities of performing decentralized and distributed administration, and, therefore, into institutional governance. The process of urban regeneration in Great Britain is one of the most representative examples in this concern. The model of the municipal administration as a unique board to develop public policies is replaced by a network of agencies working in partnership, with much closer relationships among central and local public boards, industries, commerce and community [7][8][9].

Another context which is significantly investigated by literature, particularly in recent times, is the one related to the relationships between governance and knowledge. Such relation stems from the growing awareness of the negativities, in development terms, that the communities have to suffer because of the presence of information or knowledge which is insufficient or asymmetrically distributed toward power centers. In the mean time, the need of managing the complex processes of environmental planning by drawing on commonsense knowledge pushes public administrations toward the involvement of citizens' distributed knowledge. Then, the role of knowledge becomes critical in governance processes and represents a linking foundation toward decision-support models connected to the management of large cognitive databases [10].

A further foundation of this research is the exploration of the possibilities of conceptual and operational modelling around the ontological articulations of community governance. In this concern, there is now an established research approach that tends to bring spatially (and sometimes temporally) decentralized organizations to agent-based structuring. This is often made with the aim of simulating roles, behaviours, relationships, drawing out logical and basic operating instructions for multi-agent decision support. A multiagent model can contain multiple human but also artificial, automatic agents, or a hybrid mix of human/artificial agents. A modelling of this type can rather easily be addressed to the management of a system of formal agents. Yet it shows its great potential in the environmental domain, allowing a significant reproduction of the ontological-phenomenological richness implicit in the complexity of the environment itself, and therefore allowing the maintaining of the necessary knowledge for decisionmaking. Baseline studies for multi-agent systems in the environment are not very widespread, but the different reflections, especially in terms of social simulation, are of great importance for the orientation of our research [11][12][13][14][15][16].

Moreover, multiagent modelling contains references to one of the structural aspects of environmental and land management processes, namely the hierarchical levels of tasks and reciprocal behaviors. An example in a more formal context is represented by economic supply chains, where agents spread along the production chain can vary from simple routine tasks to coordination and supervision. This happens commonly but in more complex modalities in environmental, urban and/or regional, contexts. In them, relations between community (human, natural, artificial) agents typically occur through between operational levels that are often hierarchically very different from one another. Multi-agent models are intrinsically suitable to follow these dimensionally complex organizational structures, thus providing potentials of management support within the so-called multi-level governance [17][18][19][20].

Consistently with these premises, the research moves in a complex-systems approach toward the formulation and implementation of development policies in settlement organizations, managed through governance processes. As often confirmed by scientific literature, a complex approach better provides a multiscalar, multisectoral and transdisciplinary vision to the processes of diffused, decentralized and multiagent government taking place in settled communities [21][22].

The aspects of interest in the present research essentially stem from a domain of urban engineering and environmental planning, interpreted through model-oriented, cognition-oriented perspective, that draw on the potentials expressed by the Information and Communication Technology (ICT) [23][24][25]. On the specific issue of building up ICT-based organizational/management models of governance, oriented toward efficient policies of sustainable development and management of environmental resources (the object of this research), literarature is growing only quite recently. In particular, MAS-based models have been developed and will constitute the area of interest of the present paper.

This field is of considerable interest for the scientific investigation carried out by the present research, oriented to the definition of ICT-based prototypical system architectures to support processes of community governance [26][27][28].

The research framework in which the contribution of the present unit is located, is oriented toward the definition of new models of governance and action that are based on an ontological and phenomenological formulation of the concept of sustainability. In particular, this concept is expressed through socio-cultural, institutional, economic and environmental views and meanings, that are integrated and based mainly on giving adequate value to the so-called "collective institutions" [7][8][9]. In this context, these institutional bodies are intended in a broader sense as self-organizing systems of participatory democracy, in which members share consciously and rationally common interests, as well as the conscious intention of using competitive-cooperative oriented behavioural models of action.

Such approach is of dual importance, both from a methodological-theoretical point of view and from a strictly operational point of view. In the latter, however, the approach may come out in both logical and practical processes involving a plurality of agents, together public and private, who are involved in some organizational networks for the management of common projects and for the solution of problems that produce collective impacts.

The role of knowledge becomes crucial in the processes of governance, and provides a connection toward decisionmaking support models related to the management of large cognitive databases. In this context, the research makes explicit reference to the need to investigate the potential of process modelling in support of community governance, through the formal use of the prerogatives ensured by information and communication technology (ICT). ICT-based models allow the management of both decision-support operational outputs (for example in policymaking), and the knowledge-languages inputs (for example through algorithmic formalizations and/or ontologies). in particular, inputs management is an activity of critical importance, because the database reference used stem from a dynamic interaction with/among community agents, whose structural involvement is a fundamental part of the process of community governance [23][24][25][28].

In this concern, the structure of the paper is organized as follows. A first chapter is a general introduction oriented to introduce the research background of the paper. In particular, it focuses on the concept of governance and its linkages with the so-called multi-agent systems (MAS) organization approach. A second chapter then follows, subdivided into two sections, that discusses aspects and potentials of multi-agent modelling. In particular, the first subsection highlights some features of agents and relationships occurring in MASs, whereas the second subsection looks at the organization of a governance-oriented case-study of MAS-based modelling in the provincial plan of Foggia (Italy). A final chapter carries out some concluding remarks on the modelling approach as emerged from its practical application, together with some prospects for future developments.

2 Modelling Governance in Environmental Planning

2.1 Features for an Agent-Based Approach

Our research group has been carrying out a number of experimental studies in the last decade or so, dealing with supporting multi-agent interactions and decision making in real planning contexts. In particular, a 5-year EU-funded Concerted Action has been developed in Mediterranean Countries, oriented to enhancing the sustainable planning and management of environmental resources till 2003 [29]. Other national and local funds have allowed the further development of agent-based planning research in Italian contexts up to today [13]. Although governance as such was not always an explicit aim of the various research efforts, it was nonetheless the main interest framework in terms of expected outcomes of the financing board and/or supporting public administration. As a matter of facts, both central and local administrations did express the need of enhancing the effectiveness of managing physical and social environments, basing on a stable participated knowledge-exchanging approach among stakeholders. At the root of such expectation, there was an increasing awareness of the critical roles of multifarious agents at different institutional scales, in terms of knowledge, transformation, use, management of environmental resources in everyday life. The concept of governance as a concerted, multiscale, multi-domain, multi-agent activity management then became evident as a natural real-life framework for a multi-agent modelling approach.

Basing on the experimentations carried out in the above research arena, we will now try to sketch out some essential aspects that we have found useful for the building up of governance-oriented multi-agent systems (MAS) for environmental planning and management. For this purpose, a number of features is singled out and investigated, basically and broadly following a social-simulation MAS approach [11], such as the ones shown below.

Substance. Agents can be natural actors or environmental life (humans, animals), or artificial entities created for cognitively high-level or low-level (routinary) activities – such as machines or sensors. For example, in human agents, a coordination activity is generally considered as being higher-level in comparison with a routinary activity.

Agent levels. Actions and interactions occurring with different cognition levels may be prerogatives of different agents operating at different levels. Yet, different-level activities may be concentrated in one single agent, e.g., when circumstances induce particular agents to start high-level functions in addition to routinary activities.

Collective agent. Agent's relations may generate a community of agents oriented (consciously or unconsciously) to interact as one single entity with one or more other agents in performing given activities. A typical example is a hierarchically rigid organization, that provides bottom-up information to a top-level agent who uses that information to interact with external agents, in turn receiving feedback that is transmitted down along the chains for subsequent information updates. Rather than an agent, the collectivity is an *agency*, in that case.

Relations, rules & algorithms. Agents interactions may occur in different modalities, often (but not exclusively) basing on the type of substance of agents. For example, human-human interactions may occur through ICT-based media or simply through socio-physical contacts, whereas human-artificial or artificial-artificial contacts may be ensured by software-based routines. In formal terms, different relations can be quali-quantitatively supported by rules of different kinds. A typical approach to formalizing agent relations is based on game theory, particularly when dealing with human agents with different decision behaviours [30]. The implementation of formal relations can be based on logical rules pivoted on cause-effect (e.g., if-then-else) connections among agents [31]. In a more aggregate way, numerical analysis and algorithms can provide rules to connect agents, usually when synthetic representations of linkages are needed [32][33]. As a matter of facts, methodological and rule-based approaches can be present in an unsorted mixture in real life, so generating a hybrid set of formal relations that mirrors actual hybrid agent types and relations.

Agent roles. Agents may have different official roles during their interaction activities. In particular, human agents may have different institutional roles. This is important because it may involve different power prerogatives in the interaction process, that may affect a number of features of knowledge exchanged. For example, some past experiences of our group show that the quality and quantity of information exchanged during scenario-building activities change substantially depending on the presence or absence of 'incumbent' powerful agents during interactions [34]. Such kinds of situation may suggest ad-hoc interaction-supporting approaches able to preserve democratic involvement (and, therefore, knowledge exchange quality), e.g., by using ICT-based system architectures [35].

Agent types. According to Jacques Ferber a classification of agents can be carried out according to two criteria: typological (i.e., cognitive vs. reactive agents) or behavioural (teleonomic vs. reflected behaviour). The typological distinction basically concerns the agent's representation of the world: a cognitive agent is able to perform reasoning from her/his/its symbolic representation, whereas a reactive agent can merely have world perceptions, i.e., subsimbolic representations. The behavioural distinction discriminates among agents' ways of acting: a teleonomic behavior is connected to intentional actions toward explicit aims, whereas a reflected behavior is connected to perceptive tendencies coming from agents themselves or from the external environment [11]. Human, artificial, hybrid agents can be all represented by such classification. In particular, governance-oriented systems typically present a mixture of agent types and behaviours, also subsuming institutional models of relations that need to be implemented in a multi-agent-based model with ad-hoc approaches [36][37][38].

Environment-agent proxy. The environment may play different roles in a MAS model. Either intended as a formal infrastructure for computer-based agent relations, or as a natural living framework for agent interactions, the environment is an essential part of the system. Traditionally, the environment is a static playground with null or merely reactive attitudes to external stimuli. However, having reactive attitudes allow its categorization as a type of agent in a MAS model, with relations with external agents that need to be explored and formalized explicitly [39]. Furthermore, in recent times the environment has been even considered as a proactive agent in some situations, with interesting attempts to model inter-agent transactions by logic rules and theories [16][40]. Particularly when dealing with transformation processes impacting on natural resources, environmental features tend to be valorized toward achieving sustainable development. Governance processes are naturally aimed at supporting decisions and policies in this framework, and are today increasingly interested in environment-including MAS approaches.

2.2 Hints from a Provincial Planning Process

As put down by recent literature, environmental planning is essentially a process of building up and managing knowledge [41]. Governance networks are particularly concerned with knowledge-oriented processes, that are structurally embedded in decisionmaking and planning. The plan of Foggia province was actually conceived as a testing ground where to draw out and test model architectures to support interactions and decisions. Not explicitly aimed at institutional governance, then, but pivoted on a system of governance *in potentia*. The model built for the plan of Foggia is a multi-agent system whose agents and relations are expandable to a greater governance system of the entire provincial administrative prerogatives.

The Foggia case study concerns a planning drawing effort for the province of Foggia, north of Apulia (southern Italy), where a participatory interaction process was set up among expert and non-expert knowledge agents of the area. A first draft plan prepared by the provincial planning agency concerned a set of future visions of provincial strategic ambits, basing on experts' studies and researches, as canvases to draw the final plan on. Such complex knowledge base ranged from environmental to managerial issues, and was represented by thematic reports, as well as geographical and conceptual maps. Subsequently, a planning process architecture to support the sharing, enrichment and evaluation of this set of visions was set up, in order to let decision-makers issue the final plan with related policy programs.

The main process is sketched out in figure 1 An initial activity by expert agents (i.e., scholars and researchers working in different scientific sectors) leads to a draft set of future visions that is subsequently modified by non-expert local stakeholders (such as local citizens, entrepreneurs, activists etc.). This second phase starts at local levels (basically in various municipal halls of the province) and then continues as a web-based forum interaction among remote stakeholders. The last phase leads to the final plan, after integrating the local-based and web-based interactions, also validated by possible further expert control (figure 1). The process actually starts from a knowledge base generally available to agents, then it constantly builds up further knowledge by agents' integration, enrichment, modification activity. This knowledge-building activity is a multi-scale process, occurring at municipal, provincial and (virtually) global scales, with different agents, roles, behaviours and inter-agent relation types.

An explicitation of the general model of action and interaction of involved agents can represent an initial framework for a governance-support system architecture, at least in planning-oriented activities. The generation of the original set of future visions is an expert-based knowledge-building activity and its management develops at a high cognitive level (figure 2).

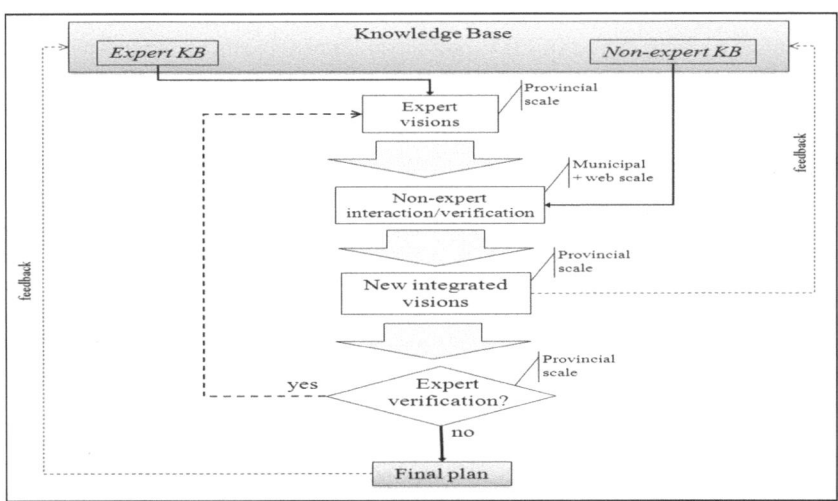

Fig. 1. The main planning process

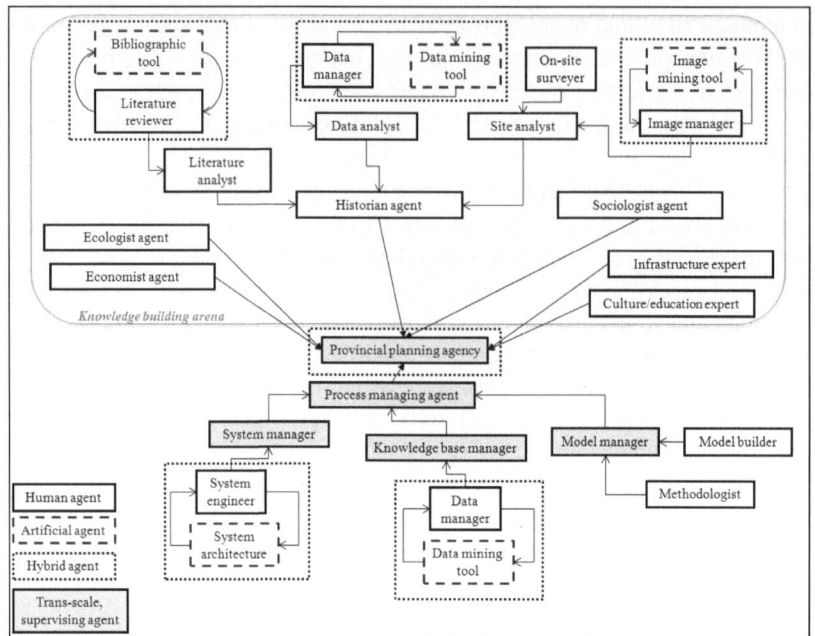

Fig. 2. The expert-knowledge multi-agent model at the provincial scale

That means it is a level in which a coordination role is essentially played by a complex agent, namely the provincial planning agency (PPA), over a number of operational knowledge-building agents. The former is typically oriented to build a much lower amount of substantial knowledge than the latter, yet it is a higher-level agent, with supervising prerogatives [11][12]. The consideration of PPA as an agency derives from its internal multi-agent mechanism, that in turn ensures its external behaviour as a single entity. In PPA, provincial human agents are connected to other human agents, as well as to supporting data management tools -as low level routinary agents. As a whole, PPA may be considered as a hybrid (human-artificial) agent, whose internal relations are supportable by a mixture of behavioural and logical formal rules.

At the same provincial scale, expert knowledge agents operate on a number of scientific domains, often relying on further, lower level (routinary) agents, both human and artificial as depicted above. An example is given by the historical agent, who typically exchanges knowledge with other literature and data management agents, involving diverse cognitive and operational levels. The connecting functions ensuring the necessary flow of knowledge do also operate differently, i.e., with different degree of complexity, ranging from logical or behavioural (human-human) rules to algorithmic functions when interfacing artificial tools.

Once implemented and fully operational, the entire multi-agent module of expert knowledge-building arena could be replaced by an expert system, so working as an proper artificial agent. In that case, controlling agents should be needed, so as to maintain and update the system itself when needed and avoiding black-box behaviours [15]. In that case, the controlling activity might be carried out by the

Process managing agency, i.e., an agent unit in charge of maintaining and expediting the entire multi-agent system, working at a higher coordination level (figure 2).

Once completed by the expert-based model, future visions are modified by the interaction with non-expert knowledge agents, at municipal levels (local stakeholders and, later, remote web-based stakeholders), so as to obtain new integrated visions. Such visions are then sent back to PPA for control phases and for the issuing of the final plan. This flow of knowledge interactions between PPA and lower level agents and agencies (such as the municipal boards) represents a formal model that can be structured as a proper system architecture to support planning efforts in governance processes at a infra-provincial scale (figures 3 & 4)

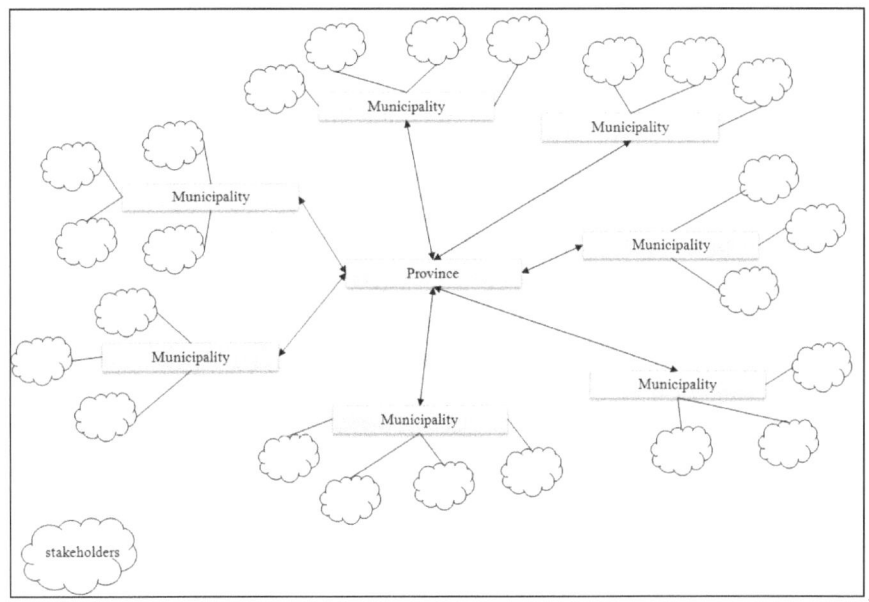

Fig. 3. Face-to-face knowledge building interactions

Institutional agents of the municipal planning agency can act as merely passive document-transmitting or arena-preparing agents (i.e., with very low cognitive level of activity). Yet, they can act as data collectors, or stimulators or even facilitators of the knowledge-building interaction among proper knowledge agents. In those cases, their cognitive level of interaction may range from routine activity to higher coordinating or supervising activities. In the local-based interaction, knowledge agents are local stakeholders in different domains of interest for local communities. The interaction arena is typically an institutional space of a municipal house, where the knowledge-building process develops vis-à-vis. Here, agent types, roles and relations are more complex than in the expert arena, due to some structural problems as non-formal languages, fuzzy data, non-scientific interests, trans-domain knowledge, etc., that need ad-hoc tools and platforms to solicit and formalize knowledge exchanged and resulting. Figure 5 shows a sketch of some major agents involved, with relevant interacting relations and types.

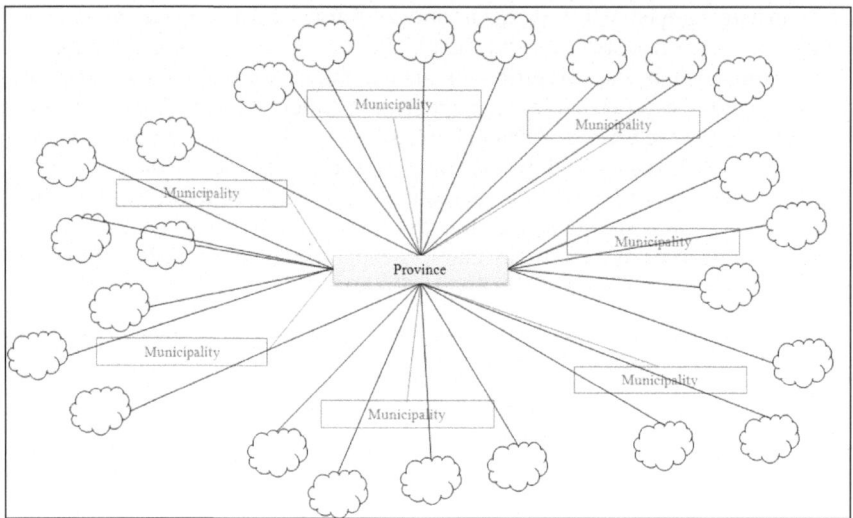

Fig. 4. Web-based Knowledge building interactions

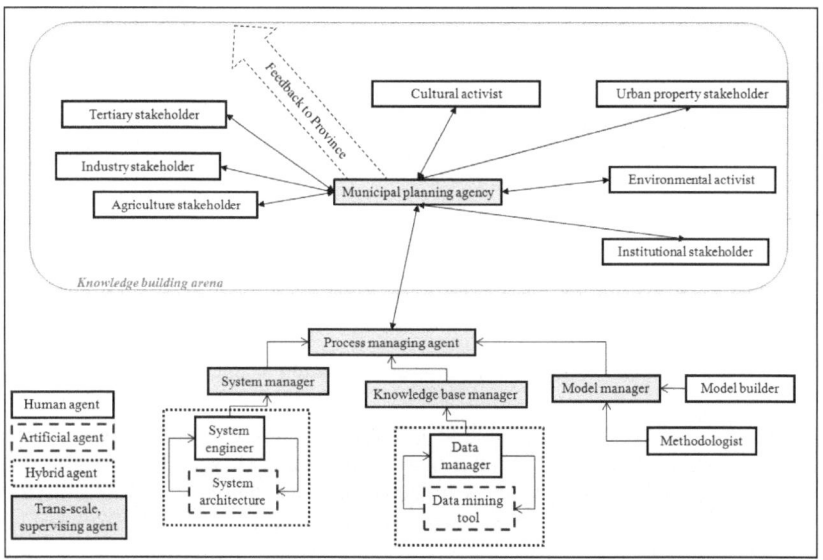

Fig. 5. A gents in knowledge base building (vis à vis)

After the local based interaction, the system architecture contains a subsequent web-based interaction, where agents and relations substantially change their types and scale of operativity, due to the change of environment, scale and outreach (figure 5).
An important aspect of multi-agent systems is the setting up of proper relation functions among agents, so as to ensure effective knowledge exchanges. Although this subject goes beyond the aims of the present paper, it is worth dealing with some issues able to complement the general framework model. The involvement of

artificial agents, with either cognitive (i.e., in robots) or routinary (i.e., in software tools) contribution, allows the setting up of basic deterministic IF-THEN causal rules, or even algorithmic functions, that can be easily formalized and performed in artificial/artificial or human/artificial agents' interactions [42]. Such rules can be multiplied, combined and integrated in different ways, so as to realize integrated modules, as subsystems of the general system architecture. Web-based interaction platforms, for example, do embed such resources in their virtual environment, so as to allow rich interactions and grasp multifaceted knowledge contents. In vis-à-vis stages of knowledge building, on the other side, this is slightly more complex, in that the real environment often prevents human agents from being clearly sincere and/or proactive because of community behavioural/psychological problems [34]. In this case, the use of simple deterministic rules can lead to the combinatorial explosions of relation functions, so leading to unreliable results. Therefore, particularly in human/human agents' interactions, this situation suggests the use of more open approaches for knowledge building, aiming at creating a longer, iterative, multi-modal (i.e., learning from oral, gestural, written, behavioural expressions) multimedia environment to support interactions. However, the complexity of such environments for real data management does remain generally backed by formal rules and interaction theories (such as game theory or prospect theories), in order to stay under a model-based approach that can allow the stable management of data within a general system architecture [30][43].

3   Concluding Remarks

This paper deals with the process of modelling and setting up a governance-oriented multi-agent system architecture to support a provincial planning initiative. The concept of governance is increasingly diffused in planning literature, yet little advancements and case studies do make reference to modelling governance in the environmental domain –that is intrinsically complex and difficult to be managed. Following a research thread started years ago in the Technical University of Bari [10], this study has dealt with a planning-drawing effort in the Italian province of Foggia. Through the analysis of a multi-agent architecture in public interaction forums, strategic future visions were envisioned, compared and fine-tuned, in the real context of the provincial region.

After an expert-based phase of initial knowledge setting up, interactive sessions were carried out, aiming at enriching the expert layout with non-expert, commonsense-based knowledge. The use of a multi-agent model was justified by the need to grasp/exchange the multiform knowledge involved and the different task expertizes. In this framework, the level of knowledge elicited, as well as the understanding of the types and behaviours of involved agents were main objectives of the study, in order to better support public decision-making in a governance-oriented model. From a substantial viewpoint, the knowledge-building interaction shows both strengths and limits. A strength feature was the not-experts knowledge raised, with contributions collected strategically and closer to local needs. A limitation of the study was the low level of coherence and depth of the issues raised, due to the heterogeneous qualifications and languages involved. This was particularly evident when compared to the intrinsic linearity and organization of the original expert visions. As a matter of facts, external process agents needed to be involved, so as to "formally" adapt each contribution and maintain the planning orientation and effectively adhere to governance aims. An unavoidable filtering action is a known disadvantage in this case.

As far as the decision-support effectiveness of the multi-agent system is concerned, the model organization is rather interesting. The MAS-based approach made the knowledge exchange easier and richer, coherently with the rich, complex nature of environmental features and stakeholders' attitudes. In order to facilitate and retain complex exchanges among human agents, conceptual mapping proved to be a fairly helpful tool, both in the generation and transmission of complex statements and in supporting the various process agents involved. In general terms, the study seems to show the feasibility and contextual coherence of a MAS-based modelling approach for community governance, in planning-oriented tasks. The extension of this approach to wider and differently-oriented tasks of public administrations is a stimulating follow-up for future researches in governance-oriented tasks.

References

1. Adams, W.A.: A transdisciplinary ontology of innovation governance. Artificial Intelligence and Law 16, 147–174 (2008)
2. Scandurra, E., Macchi, S. (eds.): Ambiente e Pianificazione. Concetti Nomadi nelle Scienze Urbane e Territoriali. Etaslibri, Milano (1996)
3. Healey, P.: Collaborative planning in perspective. Planning Theory 2, 101–123 (2003)
4. Nilsson, M., Persson, A.: Can Earth system interactions be governed? Governance functions for linking climate change mitigation with land use, freshwater and biodiversity protection. Ecological Economics 75, 61–71 (2012)
5. Robertson, M.: Performing environmental governance. Geoforum 41, 7–10 (2010)
6. Sharif, M.N.: Technological innovation governance for winning the future. Technological Forecasting & Social Change 79, 595–604 (2012)
7. Healey, P.: Creativity and urban governance. Policy Studies 25, 87–102 (2004)
8. Cars, G., Healey, P., Madanipour, A., De Magalhaes, C. (eds.): Urban Governance, Institutional Capacity and Social Milieux. Ashgate, Aldershot (2002)
9. Booth, P.: Partnerships and networks: The governance of urban regeneration in Britain. Journal of Housing and the Built Environment 20, 257–269 (2005)
10. Borri, D., Camarda, D., De Liddo, A.: Envisioning Environmental Futures: Multi-agent Knowledge Generation, Frame Problem, Cognitive Mapping. In: Luo, Y. (ed.) CDVE 2004. LNCS, vol. 3190, pp. 230–237. Springer, Heidelberg (2004)
11. Ferber, J.: Multi-Agent Systems: An Introduction to Distributed Artificial Intelligence. Addison-Wesley, London (1999)
12. Wooldridge, M.: An Introduction to Multi-Agent Systems. Wiley, London (2002)
13. Camarda, D.: Beyond citizen participation in planning: Multi-agent systems for complex decision making. In: Nunes Silva, C. (ed.) Handbook of Research on E-planning: ICTs for Urban Development and Monitoring, pp. 195–217. IGI Global, Hershey (2010)

14. Borri, D., Camarda, D., Grassini, L.: Learning and sharing technology in informal contexts: A multiagent-based supporting approach. In: Proceedings - IEEE International Conference on Mobile Data Management, Lulea, Sweden, pp. 98–105 (2011)
15. Arentze, T., Timmermans, H.: Multi-agent models of spatial cognition, Learning and complex choice behavior in urban environments. In: Portugali, J. (ed.) Complex Artificial Environments. Springer, Berlin (2006)
16. Weyns, D., Omicini, A., Odell, J.: Environment as a first class abstraction in multiagent systems. Autonomous Agent and Multi-Agent Systems 14, 5–30 (2007)
17. Baud, I., Dhanalakshmi, R.: Governance in urban environmental management: Comparing accountability and performance in multi-stakeholder arrangements in South India. Cities 24, 133–147 (2007)
18. Gertler, M.S., Wolfe, D.A.: Local social knowledge management: Community actors, institutions and multilevel governance in regional foresight exercises. Futures 36, 45–65 (2004)
19. Corfee-Morlot, J., Cochran, I., Hallegatte, S., Teasdale, P.-J.: Multilevel risk governance and urban adaptation policy. Climatic Change 104, 169–197 (2011)
20. Kern, K., Bulkeley, H.: Cities, Europeanization and Multi-level Governance: Governing Climate Change through Transnational Municipal Networks. JCMS: Journal of Common Market Studies 47, 309–332 (2009)
21. Batty, M.: Cities and Complexity: Understanding Cities with Cellular Automata, Agent-Based Models, and Fractals. The MIT Press, Cambridge (2007)
22. Bossomaier, T.R.J., Green, D.G.: Complex Systems. Cambridge University Press (2000)
23. Power, D.J., Sharda, R.: Decision Support Systems. In: Nof, S.Y. (ed.) Springer Handbook of Automation, pp. 1539–1548. Springer, Berlin (2009)
24. Turban, E., Aronson, J.E., Liang, T.P.: Decision Support Systems and Intelligent Systems. Pearson/Prentice Hall, New York (2005)
25. Allen, G.L. (ed.): Applied Spatial Cognition: From Research to Cognitive Technology. Erlbaum, New York (2007)
26. Egenhofer, M.J., Golledge, R.G.: Spatial and Temporal Reasoning in Geographic Information Systems. OUP, Oxford (1998)
27. Boroushaki, S., Malczewski, J.: Measuring consensus for collaborative decision-making: A GIS-based approach. Computers, Environment and Urban Systems 34, 322–332 (2010)
28. Borri, D., Camarda, D.: Spatial ontologies in multi-agent environmental planning. In: Yearwood, J., Stranieri, A. (eds.) Technologies for Supporting Reasoning Communities and Collaborative Decision Making: Cooperative Approaches, pp. 272–295. IGI Global Information Science, Hershey (2011)
29. Borri, D., Camarda, D., Grassini, L. (eds.): Sustainable planning for Soil and Water: The Mediterranean. L'Harmattan, Paris (2002)
30. Parsons, S., Gmytrasiewicz, P.J., Wooldridge, M.J.: Game Theory and Decision Theory in Agent-Based Systems. Kluwer Academic Publishers, Dordrecht (2002)
31. Mohammadian, M.: Intelligent Agents for Data Mining and Information Retrieval. Idea Group Pub., Hershey (2004)
32. Stankovic, M.: Control and Estimation Algorithms for Multiple-Agent Systems. BiblioBazaar, Charleston (2011)
33. Zinkevich, M.: Theoretical Guarantees for Algorithms in Multi-Agent Settings. School of Computer Science. Carnegie Mellon University, Pittsburgh (2004)
34. Khakee, A., Barbanente, A., Camarda, D., Puglisi, M.: With or without? Comparative study of preparing participatory scenarios using computer-aided and traditional brainstorming. Journal of Future Research 6, 45–64 (2002)

35. Barbanente, A., Camarda, D., Grassini, L., Khakee, A.: Visioning the regional future: Globalisation and regional transformation of Rabat/Casablanca. Technological Forecasting and Social Change 74, 763–778 (2007)
36. Sierra, C., Thangarajah, J., Padgham, L., Winikoff, M.: Designing Institutional Multi-Agent Systems. In: Padgham, L., Zambonelli, F. (eds.) AOSE VII / AOSE 2006. LNCS, vol. 4405, pp. 84–103. Springer, Heidelberg (2007)
37. Ferber, J., Stratulat, T., Tranier, J.: Towards an integral approach of organizations in multi-agent systems. In: Dignum, V. (ed.) Handbook of Research on MultiAgent Systems: Semantics and Dynamics of Organizational Models, pp. 51–75. IGI, Hershey (2009)
38. Searle, J.R.: The Construction of Social Reality. Free Press, New York (1997)
39. Ferber, J., Muller, J.: Influences and reaction: A model of situated multiagent systems. In: Tokoro, M. (ed.) Proceedings of the Second International Conference on Multi-Agent Systems, pp. 72–79. AAAI, Kyoto (1996)
40. Le Page, C., Becu, N., Bommel, P., Bousquet, F.: Participatory agent-based simulation for renewable resource management: The role of the cormas simulation platform to nurture a community of practice. Journal of Artificial Societies and Social Simulation 15, 10 (2012)
41. Maciocco, G. (ed.): The Territorial Future of the City. Springer, Berlin (2008)
42. Scerri, P., Vincent, R., Mailler, R.: Coordination of Large-Scale Multiagent Systems. Springer, Berlin (2006)
43. Briggeman, J.: Governance as a strategy in state-of-nature games. Public Choice 141, 481–491 (2009)

A Data Fusion System for Spatial Data Mining, Analysis and Improvement

Silvija Stankute and Hartmut Asche

University of Potsdam, Department of Geography,
Karl-Liebknecht-Strasse 24/25, 14476 Potsdam, Germany
{silvija.stankute,gislab}@uni-potsdam.de
http://www.geographie.uni-potsdam.de

Abstract. The availability of multiple spatial data and differences in their geometric and thematic qualities require the development of a new system with a new approach to data mining. This should include the processes of the analysis and improvement of heterogeneous spatial data structures. This paper describes the concept, architecture and functionality of such a system: the spatial Data Fusion System (DAFU). DAFU allows efficient use of heterogneous spatial information and creates the possibility to individual use of geo-data.

Keywords: data fusion, data integration, vector data, homogenization, heterogeneous geospatial data, data conflation.

1 Introduction

Real world object (streets, roads, homes etc.) acquisition is carried out by various companies or institutions. Each company or institution collects spatial information for different purposes. For this reason the rules for object acquisition are different in different companies. The rules specify which method is used for the acquisition of geo-data, in what kind of a scale the objects are to be collected and which thematic attributes must be determined.

The consequence is the development of a variety of heterogeneous geo-spatial datasets, which represent the same area of the real world and are different in their geometrical and thematic accuracy. The heterogeneous geo-data have different quality. This problem describe [1], [2], [3]. Insufficient geometrical and thematic accuracy leads to the need for a new method of geo-spatial information acquisition. This solution is not effective and is often connected with high costs. In addition there is the decisive factor: time, which must be captured for the new data set. These problems require better solutions.

The university Postdam in the Geoinformation Research Group, in the earlier works [4], [5] and [6] has developed a method, which increases the thematic and geometrical quality of the available spatial datasets. The new method incorporates the data fusion process.

Data fusion is the process of combining the information from two or more geo-spatial data sets. The objective of this process is to make an improved data set, which is more superior to the source/input data sets in respect of either geo-spatial and/or attributive properties [5]. This process was implemented as a Data Fusion System (DAFU). Against this background this article outlines a Data Fusion System that improves the quality of available geo-spatial information.

2   A General Overview of DAFU

DAFU merges the available geo-spatial data sets from different origins to a user-defined individual data set. The final data set consists of geometrical and thematic components of the input data sets. Attributes not required are rejected by the user-defined fusion of the attributes. Therefore DAFU can be called be considered a filter of geo-spatial data (see Fig. 1).

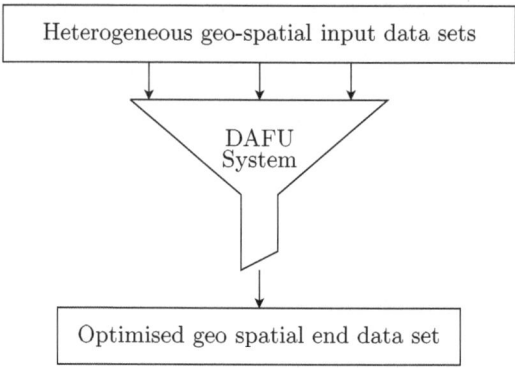

Fig. 1. The main idea of DAFU

DAFU has three components: 1) preprocessing of input data sets, 2) fusion/filtering of input data sets, 3) post-processing of end data set.

2.1   Pre-processing

The first DAFU component pre-processing has the following process steps: Analysis of the input data sets, determination of the data quality and data preparation (see Fig. 2). The subject of this process step is the analysis of the geo-spatial input information. Here the vector model of every input data set is analysed.

In the first pre-processing step each input data set must be converted in a unified coordinate system. This is important for the future geo-data merge. The next step is the transfer of the input data into the same data format.

In the third analysis step the uniqueness and completeness of spatial input data sets must be verified. All the data sets used in DAFU systems are vector data. Any spatial object contains geometric and semantic information. Objects in the real world are represented in geo-spatial data sets as points, lines or

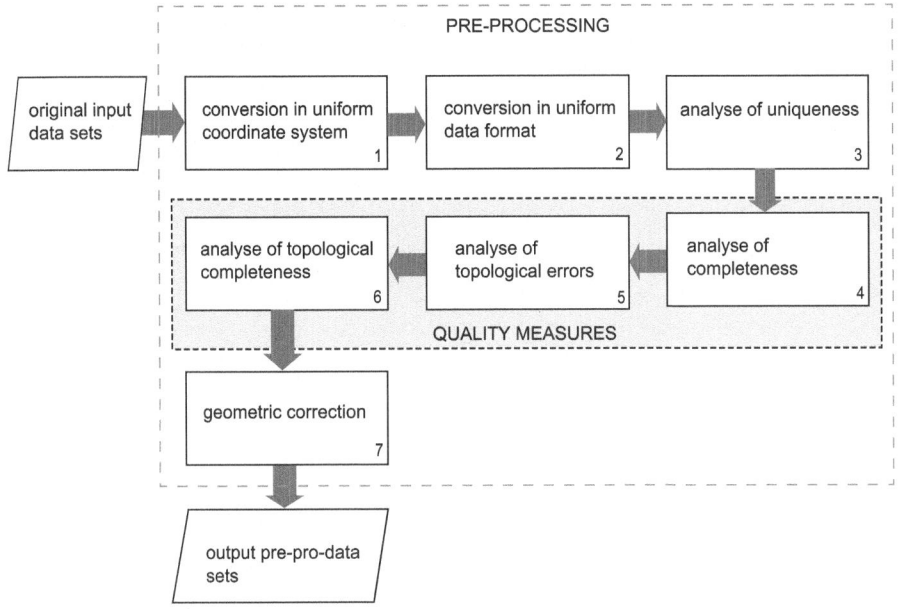

Fig. 2. Pre-processing component of DAFU: (1), (2), (7) - input data preparation; (3), (4), (5), (6) - analysis of input data; (4), (5), (6) - determination of the data quality

polygons. Lines and polygons consist of sequential points. All vector structures in vector data are based on points, which characterise all the geometrical information. Each point is described by x- and y-coordinate. Only one point or node having unique coordinates may occur at the same location with respect to topology. Geometrical primitives in vector data have to meet specific consistency conditions i.e. uniqueness and completeness. Object uniqueness means that one object can only be represented by one object in a data set. Completeness requires that all known relations are explicitly included in that data set [4].

Moreover, quality measures (or possibly a quality measure) will be computed. This defines the quality of the available input data sets. To be able to define the quality measure, two characteristics of each input data set have to be examined. The first characteristic is the degree of topological correctness. The second characteristic is the measurement of the thematic completeness.

To merge the geo-data with each other, so that they are correct from a topological point of view, it is necessary to carry out geometrical correction of the geo-data. This includes removing duplicate geometries.

The result of pre-processing component is *output-pre-pro-data sets*. These data will be used as input data for the fusion/filtering component.

2.2 Fusion/Filtering of Input Data Sets

After the successful analysis and preparation of the input data sets, the input data sets are merged (see Fig. 3). The DAFU system executes the algorithms

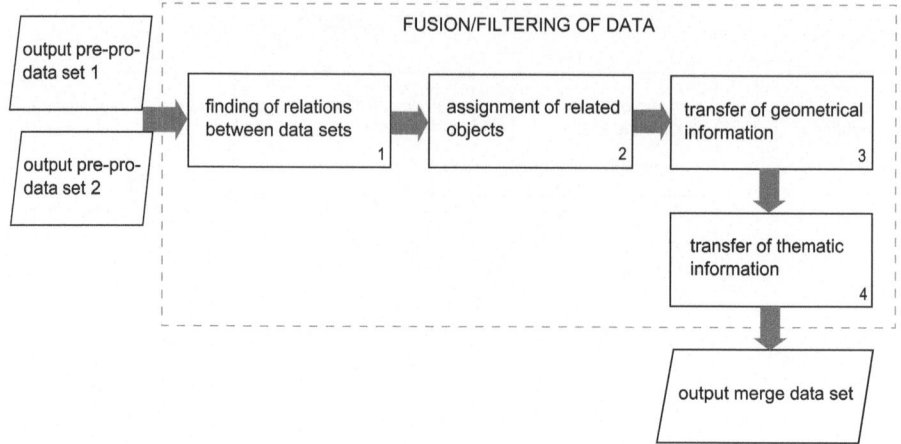

Fig. 3. Fusion/filtering component of DAFU in four steps

based on direct compar-ison of coordinates. For the fusion of the vector data with different geometrical types, separate algorithms were developed. The DCC algorithm (Direct Coordinate Comparison), which is responsible for the merge process of simple polygons, has been presented in previous work [5]. The algorithms for merging linear objects were presented in [4] and in some cases [6]. The above subroutines were developed and extended to cope with more complex geometries.

The requirements for the successful realisation of the subroutines are alike for all geometrical types. The input data sets must have the same coordinate system and the same data format. The next important requirement is redundancy-free input data sets. This means that each object of the space may be represented only with one geometry. All algorithms are based on the direct comparison of the coordinates. The relationship between objects in the various input data sets and objects in the real world are determined by using pairs of coordinates. Using pairs of coordinates are determined relationships between the objects that are present in the various input data sets and represent the same object in the real world.

By creating a relation between two objects, the transfer of attributes (thematic information) is ensured. The user-defined set of attributes ensures that thematic information is transmitted over an object from two or more input data sets. This transmission (or cross-referencing) means the extension of the attribute table and generation of new geometrical features. In one implementation of DAFU only one input data set with the other input data set can be extended. The user decides which data set will be extended. DAFU offers another possibility: After calculating the quality measures in the preprocessing component, DAFU displays the message with information about the quality of input data sets and a recommendation to use one of input data set as the target data set for following extension. On the basis of this information, the user can use the data set proposed by DAFU or he can decide without DAFU help what data he wants to expand.

If the user wants to enrich a particular data set with additional information from other data sets, it is called an extension of the record. If the user wants to create a new data set to include only some information from the input data sets, this is called a data filtering.

The result of fusion/filtering component is the *output-merge-data set*.

2.3 Post-processing Component

In the post-processing component (see Fig. 4) the *output-merge-data set* must be verified. The quality of fusion process is calculated and evaluated. This is

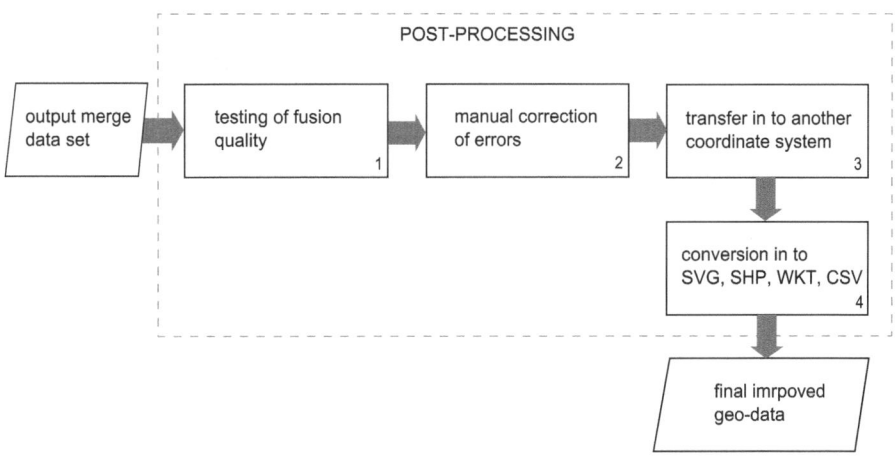

Fig. 4. Post-processing component of DAFU

followed by manual correction of the possible errors, which are usually below five percent for line-like objects and less than 10-15 percent for polygonal objects. The transfer of geo-data in to different coordinate systems follows after manual correction of errors. The next and final step of the post-processing component is to convert merged data set into other data formats. The last two steps are performed only by user request.

3 Architecture

The developed algorithms were implemented in the widely used interpreted, dynamic programming language PERL. One of the main strengths of this scripting language, and the reason for chosing it, is its high degree of portability. PERL runs on over 100 platforms, including Microsoft Windows and UNIX derivatives. This opens up the possibility of using the implemented algorithms on servers and clusters. This is important when using large volume input data sets. PERL is

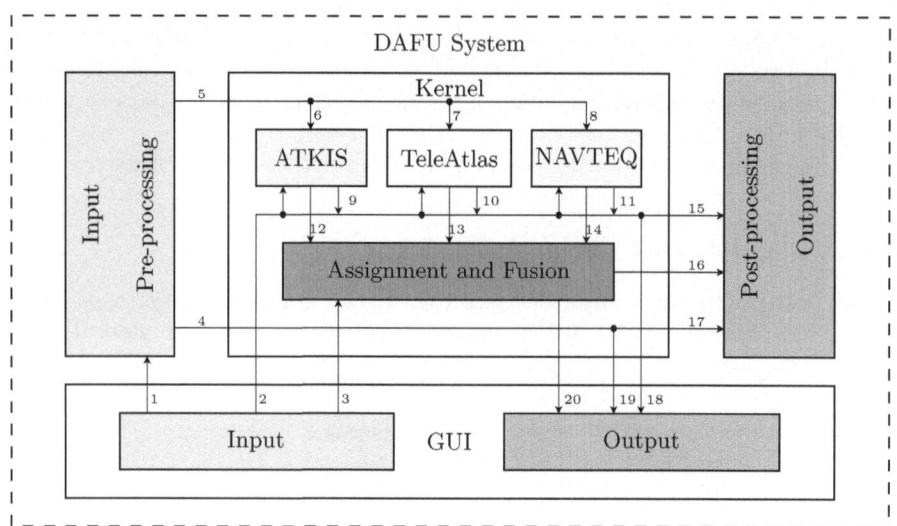

Fig. 5. Data dependency diagram of DAFU: 1 - control of input and pre-processing module; 2,3 - control of kernel module; 4 - delivers data for output and post-processing module; 5,6,7,8 - delivers data for ATKIS, TeleAtlas and Navteq module; 9,10,11,15,16,17 - delivers data for output module; 12,13,14 - supplying fusion-module with data; 18,19,20 - data for visualisation in GUI (modified Figure [4])

particularly suitable for the processing and/or manipulation of large ASCII data sets (DAFU works basically with ASCII files). Furthermore, it allows different programming paradigms, including modular and object-oriented programming, which makes the programming easier. The graphical user interface for easy control of the core routines were developed using the Tk library.

During the implementation of DAFU, modular construction of the software was particularly important. This allows generic programming and simplifies considerably the subsequent extension. DAFU contains a core with five modules (see Fig. 5). Three of them (ATKIS-, TeleAtlas- and Navteq-module) cover the input datasets. The objective is a conversion of input data in to the same internal data structure. If the input data sets during the pre-processing step are converted into the same data structure, the next step calls the assignment module. The assignment module relates individual objects of different data sets. This is neces-sary condition for the fusion module. The fusion module processes merges two different data structures into one.. This merging takes place according to certain rules, which are given by the user over the graphical user interface (GUI). The GUI module controls not only the core of DAFU, but also the periphery. The periphery includes the input and output modules, and the pre-processing and post-processing modules. The input module supplies the data for the core module, which provides data for the output module. Furthermore the GUI module is responsible also for the visualisation of the input data sets and the output data.

3.1 Call of Internal Routines

The modular structure shown in the Figure 5 is reflected in the programmed subroutines and their calls (see Fig. 6). In the context of the data input in the

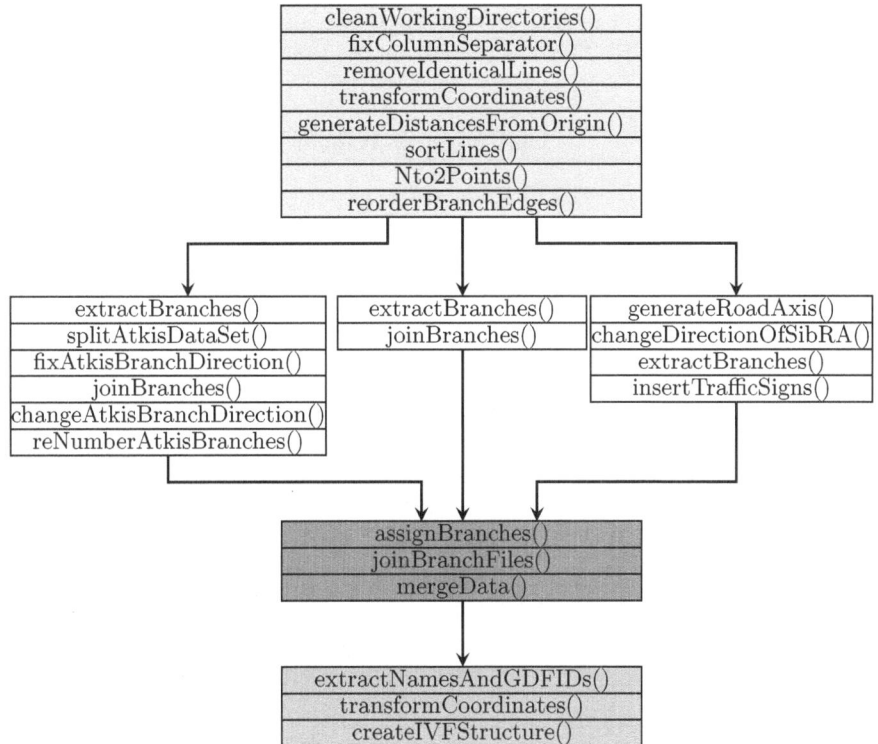

Fig. 6. Call of internal routines

preprocessing different routines were called. The routine *CleanWorkingDirectories()* deletes all temporary files that are created from previous steps. Now routines for coordinate transformation and sorting are called. A goal of these routines is to produce uniform data structures, which are processed in later steps. Routines for the processing of the used data records are called after the call of the input routines. These calls depend on the way the program has been configured for the fusion process. The subroutine as *signBranches()* assigns each branch of the target data to a corresponding branch in the source data set. With this information, the subroutine *fusionData()* is called. The output data of this subroutine are processed afterwards through calls of the output and post-processing module. For example the call of subroutine *transformCoordinates()* executes a coordinate transformation. After this transformation the output data set must be converted by calling *createSHP()* into the *.shp-format. Figure 6 represents the expiration of the routine calls; the colors of the individual blocks correspond to the colors of the modules from Figure 5.

3.2 Graphical User Interface of DAFU

DAFU was implemented in such a way that it can also be run without a graphical user interface (GUI). This has the advantage that DAFU can also be imple-

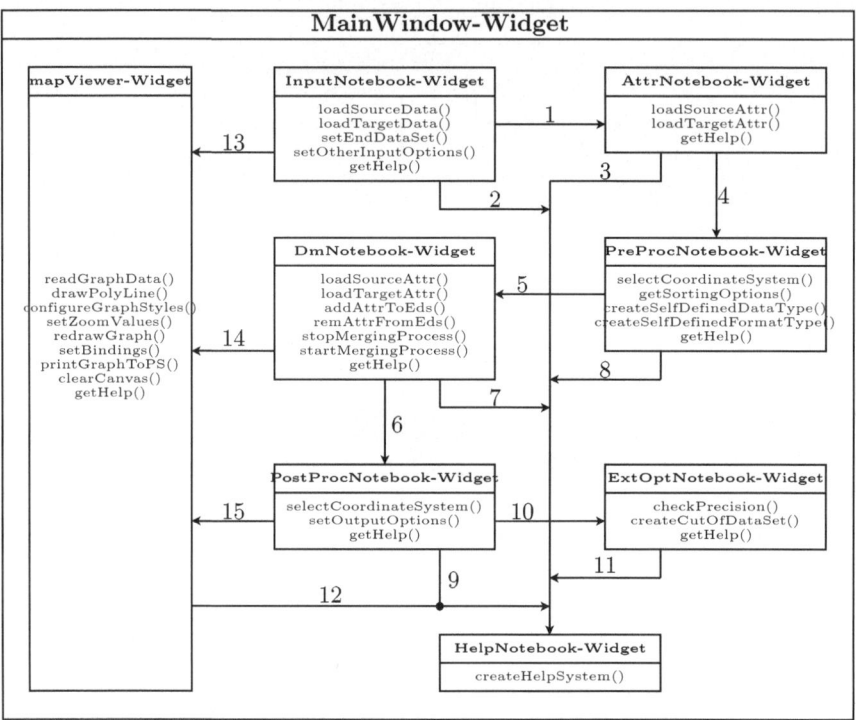

Fig. 7. Structure and dependencies in the GUI

mented on remote systems (eg. servers, clusters etc.), which have only command line interfaces. Furthermore, it can also be used in other scripts and programs with minimal adjustments.

A graphical user interface (front-end) has, on the one hand the advantage that the core (back-end) is tested in depth and on the other hand that the operation for users is simplified.

Figure 7 shows the modular structure and dependencies in the GUI. The entire GUI consists of 8 widgets, which communicate with each other via data exchange and signal exchange (bindings). All functions of the core of DAFU are covered by the GUI widgets and further implemented for data visualisation. A

A Data Fusion System for Spatial Data Mining, Analysis and Improvement   447

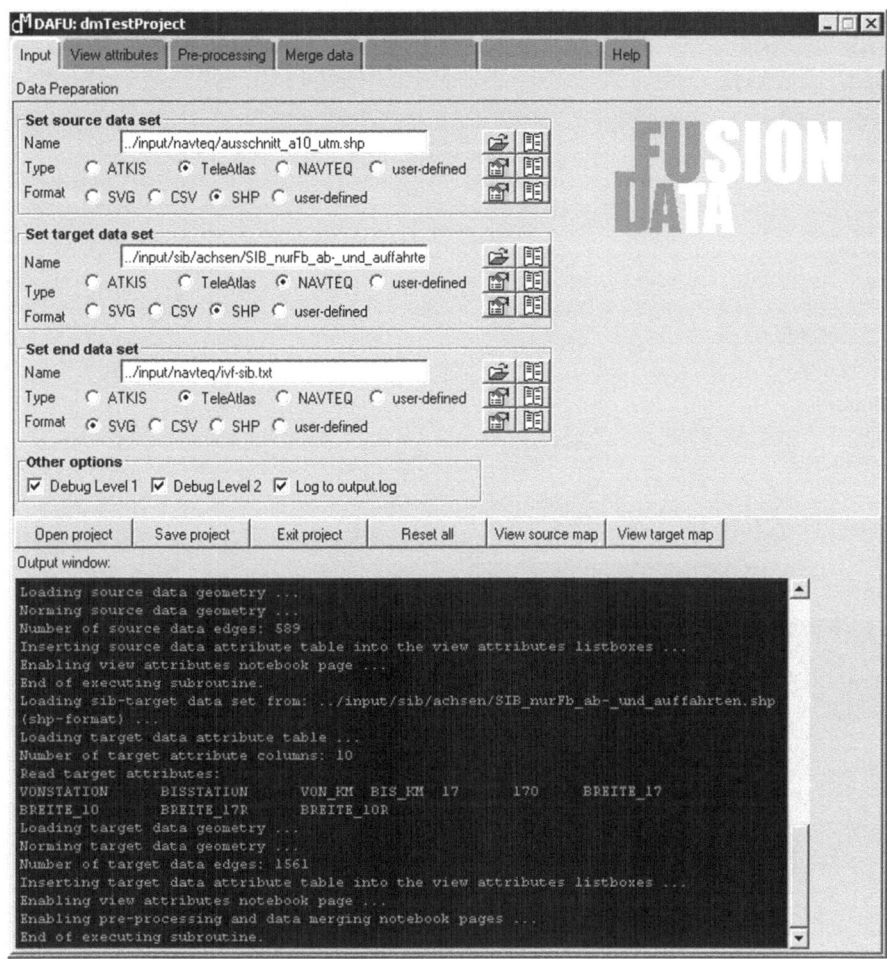

Fig. 8. DAFU-GUI: the input-section

simple and flexible support system provides the user with valuable information for the understanding and use of DAFU.

DAFU can be used in two ways. Firstly, by command input (console) without GUI and secondly, with the help of a graphical user interface. The configurations of DAFU is done on the command line via a configuration file. The configuration file contains, for example, information about the source data set, the target data set and the end data set and in particular which attributes should be united with each other. This configuration file is called on the GUI-level project file. It contains all the necessary settings DAFU requires for a successful implementation and can be created with the GUI. The project file can be reloaded into the GUI later, but it can also be used as a configuration file from the command line DAFU.

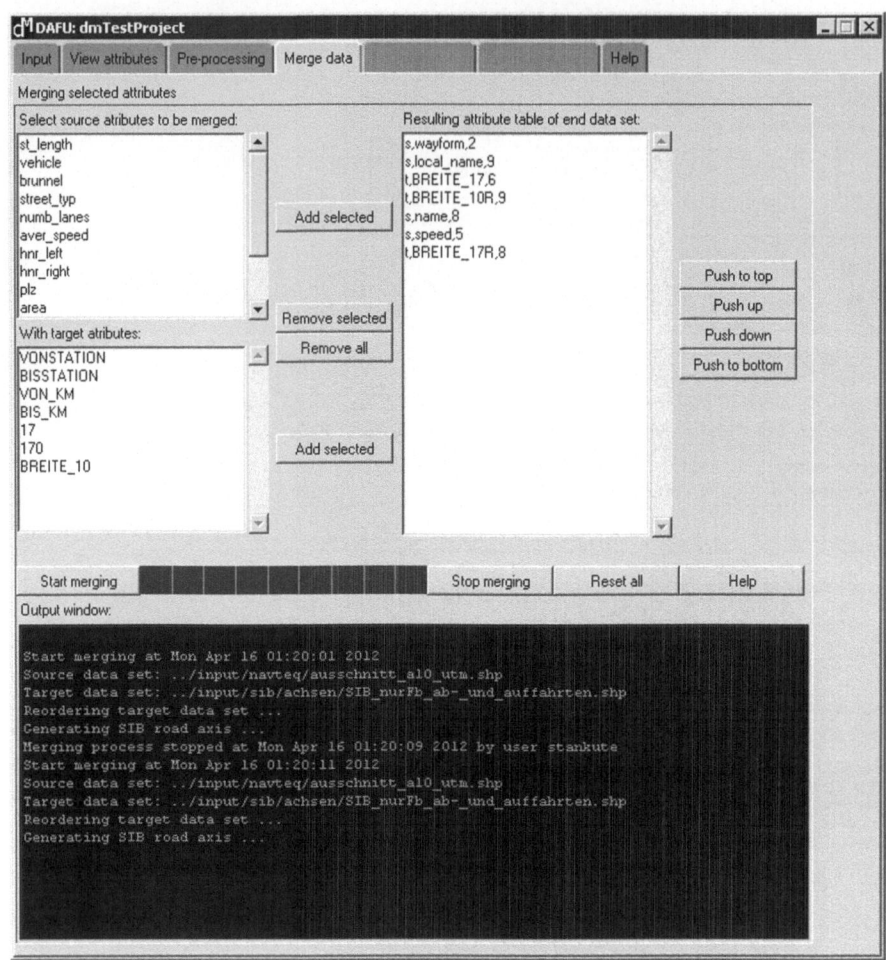

Fig. 9. DAFU-GUI: DataMerge-section

Figure 8 shows the input elements of the DAFU GUI. Here information about the source, target and final data sets is entered. It is important to set the type of the data set (ATKIS, TeleAtlas, Navteq or user-defined) and the file format (SVG, SHP, WKT, or user-defined). DAFU implements different sub-routines based on these settings. It is possible to set four different debugging levels (0 to 3). Once the input data has been entered successfully in the input section, the user can begin to analyse the input datasets. This is carried out in the attribute-section. The analysis of the various attributes and the values of the source and target data sets is important in order to determine which attributes should be included in the final data set.

The attributes that the final dataset must contain are clearly defined in the Data-Merge section (see Figure 9). Any combination of attributes from the source and target data set is possible. It should be noted that the basic geometry of the final data set is always based on the geometry of the target data set. After defining the attributes the merge process is started. The result of data fusion can be saved in different file formats (see input-section). It can also be visualised in the Map-Viewer in the post-processing section.

DAFU System allows efficient use of heterogneous spatial information and creates the possibility to individual use of geo-data.

4 Conclusion

This article presents the DAFU system for merging spatial geo-data. DAFU software allows the use of available data within a wider range. Also, during the fusion process the quality of the data increases.

References

1. Devillers, R., Gervais, M., Bedard, Y., Jeansoulin, R.: Spatial Data Quality: From Metadata to Quality Indicators and Contextual End-User Manual. In: Proceedings of the OEEPE/ISPRS Joint Workshop on Spatial Data Quality Management, Istanbul, March 21-22, pp. 45–55 (2002)
2. Hunter, G.J., Bregt, A.K., Heuvelink, G.B.M., De Bruin, S., Virrantaus, K.: Spatial Data Quality: Problems and Prospects. In: Navratil, G. (ed.) Research Trends in Geographic Informations Science. Lecture Notes in Geoinformation and Cartography, pp. 101–121. Springer, Heidelberg (2009)
3. Onchaga, R., Morales, J., Widya, I., Lambert, J.M.: An Ontology Framework for Quality of Geographical Information Services. In: Proceedings of the 16th International Symposium on Advances in Geographic Information Systems, ACM GIS 2008, Irvine, California, USA, November 5-7 (2008)
4. Stankutė, S., Asche, H.: An Integrative Approach to Geospatial Data Fusion. In: Gervasi, O., Taniar, D., Murgante, B., Laganà, A., Mun, Y., Gavrilova, M.L. (eds.) ICCSA 2009. LNCS, vol. 5592, pp. 490–504. Springer, Heidelberg (2009)
5. Stankute, S., Asche, H.: Geometrical DCC-Algorithm for Merging Polygonal Geospatial Data. In: Taniar, D., Gervasi, O., Murgante, B., Pardede, E., Apduhan, B.O. (eds.) ICCSA 2010. LNCS, vol. 6016, pp. 515–527. Springer, Heidelberg (2010)
6. Stankutė, S., Asche, H.: Improvement of Spatial Data Quality Using the Data Conflation. In: Murgante, B., Gervasi, O., Iglesias, A., Taniar, D., Apduhan, B.O. (eds.) ICCSA 2011, Part I. LNCS, vol. 6782, pp. 492–500. Springer, Heidelberg (2011)

Dealing with Multiple Source Spatio-temporal Data in Urban Dynamics Analysis

João Peixoto[1] and Adriano Moreira[2]

Mobile and Ubiquitous Systems Group, Centro Algoritmi,
Universidade do Minho, Guimarães, Portugal
peixoto@kanguru.pt, adriano.moreira@algoritmi.uminho.pt

Abstract. Capturing, representing, modelling and visualizing the dynamics of urban mobility have been attracting the interest of the research community recently. One of the drivers for recent work in this area is the availability of large datasets representing many aspects of the urban dynamics. Applications for these studies are diverse and include urban planning, security, intelligent transportation systems and many others. Quite often, the proposed approaches are highly dependent on the data type. This paper describes the definition of a set of basic concepts for the representation and processing of spatio-temporal data, sufficiently flexible to deal with various types of mobility data and to support multiple forms of processing and visualization of the urban mobility. A place learning algorithm is also described to illustrate the flexibility of the proposed framework. Available results obtained by the integration of geometric and symbolic data reveal the adequacy of the proposed concepts, and uncover new possibilities for the fusion of heterogeneous datasets.

Keywords: Urban modelling, space-time dynamics, data fusion.

1 Introduction

The mobility of citizens in an urban area is the source of various problems: traffic congestion, environmental impacts, inadequacy of public transport, and spreading of diseases, among others. For this reason it is important to understand the mobility behaviour of individuals in space, understand space itself, and understand the use people make of the urban space as a way to reduce and possibly eliminate these difficulties.

The dynamics associated with the mobility in urban areas always has two components, Time and Space, rising new challenges on how to capture, represent and visualize these dynamics. While capturing the presence and mobility of people in urban spaces has evolved enormously in recent years, movement representation and visualization still faces many challenges. Actually, the huge size of datasets being collected these days is creating more challenges to representation and visualization rather than solutions (in spite of their great potential for mobility analysis).

As referred by Yu and Shaw [1] the current Geographic Information Systems (GIS) are structured to represent the spatial component of data but lack good support for the temporal component. For this reason, some authors have developed platforms that provide support to spatio-temporal data, such as SECONDO [2], with the aim of representing the time dependence of mobile artefacts through the use of abstract definitions to represent the positions or shapes of objects over time and space, respectively [3]. On the other hand there are also challenges in how this information could or should be presented for human movement analysis. Thus, several studies presented recently explore different forms of visual representation of movement data, such as the iSPOTS project [4] which illustrates the occupation of space or, for example, the visualization of human travel behaviour based on the trajectory of money bills [5].

However, regardless of how the dynamics of an urban space is represented, most of these works focus only on one type of mobility data. Although our research work is focused on the analysis of the dynamics of urban space, our initial aim is to create a flexible and comprehensive conceptual framework for the representation of movement processes that allows the same concepts to be applied to different types of data from different sensors, such as GPS, Wi-Fi, GSM, ticketing systems, as well to the different modalities of urban mobility. Our approach to capture the dynamics of the urban space is based on merging the individual mobility profiles of people. This approach aims to benefit from the current capability of smartphones and other personal devices to be used as proxies to observe human spatio-temporal behaviour. The first step in the analysis of the urban dynamics is, then, the automatic creation of personal mobility profiles from multi-sensor data.

The next section in this paper describes some of the work developed in the field of analysis and visualization of urban mobility. Section 3 describes the concepts that are the basis of our proposed framework for the representation of spatial-temporal data. In Section 4, data from three types of sensors is mapped into the proposed concepts and three of the major transformation processes are described. Finally, in Section 5, some conclusions and open questions are discussed.

2 Related Work

Recently, several studies have been presented in the area of visualization of urban mobility dynamics, using different techniques. One of these techniques analyse urban mobility using the temporal variation of the occupation that individuals make of the urban space [4] [6]. This type of representation is based on the creation of temporal snapshots of space occupation. However, due the dynamics of the urban space, this approach may not be the most appropriated for the analysis of pattern changes [7]. Another problem is the definition of mobility in these approaches, because they represent the mobility through the variation of space occupation over the time and not the real movement of individuals. The analysis of mobility based on the use of space does not allow the extraction of more depth conclusions about the urban mobility. Thus, the application of this approach may be useful for the planning of urban space based on the detection of concentration areas of individuals, but not adequate for the detection of problems caused by mobility itself, such as traffic congestion.

Deep understanding of the phenomena associated with urban mobility involves the visualization of trajectories and flows, which truly reflect the movements of individuals. In this area one of the approaches to represent trajectories is with vectors [8]. Through this approach it is possible to have some sense of mobility, since it allows the representation of an individual according to a spatio-temporal reference. However, as this representation is based on the observation of the instantaneous movement, the simultaneous perception of the origin and destination of the movement is not easily transmitted. Alternatively, some researchers are using a different technique for representing paths through interconnected source-destination pairs [5]. Although with this approach it is possible to visualize the trajectories, several questions arise regarding the outcome of the visualization. First, if the interval between samples is large, intermediate movements are lost and then trajectories become twisted, as such some behaviours are not represented. Second, to connect the source to destination we may have to affect the Time component, since the analysis is not done continuously, but by time intervals, consequently losing this representation the notion of space change over the time and creating the same issues associated with representation by snapshots. In our research work we intend to study these and other issues related to the visualization of mobility through the representation of trajectories so as to explore new paradigms for the representation of mobility. Our approach is based on abandoning the snapshot representation of artefacts (individual or object), and create personal and global maps of mobility. To achieve this goal it is important to, first, properly structure the information in order to have a conceptual framework for the representation of mobility in an urban area and verify what type of data match with this structure.

3 Concepts for Movement Representation

The work that we have done so far defines eight general concepts that characterize our conceptual framework for the representation of mobility of an individual artefact (Figure 1). These concepts are designed to fit the data since its acquisition stage, until we get homogeneous representations of the major movement processes, be they of a single individual or of a group of individuals.

3.1 Raw Data

It all starts with the data collected by a multitude of sensors about the movement of an artefact. Until recently, GPS receivers have played a major role in data collection about movement. Fortunately, recent advances in mobile devices created the possibility to collect large amounts of data about the mobility of their users. These devices not only support the collection of geo-referenced data through their integrated GPS receivers, they also enable the collection of data about the use of Wi-Fi and GSM networks, the detection of nearby devices (persons), the logging of data generated from accelerators, and much more. Urban infrastructures are also contributing to these increased sensorial capabilities. Among others, public transportation operators often make use of ticketing systems that collect data about people entering or leaving buses and metro stations. Public authorities also collect data about the intensity of traffic flows across a particular street segment.

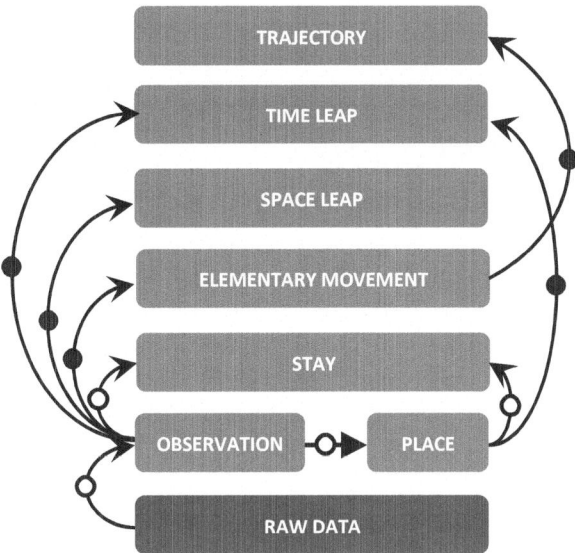

Fig. 1. Layered Structure of the Information Concepts

The result of this technological equipped urban environment is, from the mobility analysis point of view, a giant sensor network producing data in huge quantities. These datasets are, though, very heterogeneous: on the formats used to represent the mobility data; on the position precision and accuracy; on the time stamping accuracy and synchronization; on the sampling rate; on the spatial reference system being used to represent points or places in space.

The heterogeneity of the spatial reference model is of major importance when fusing data coming from different sensors. The two major spatial referentials are those based on the WGS-84 datum, and those based on addresses or names of places. The first one is a geographic space model, while the second is a symbolic one. It is on the second type where we find the greatest diversity: postal addresses, postal codes, networks cells identifications, bus stops identifications, etc.

Throughout the remaining of this section we propose a set of concepts for the representation of the mobility processes, in an attempt to merge most of the above mentioned heterogeneity into a more generic and homogeneous reference model.

3.2 Observation

A mobile artefact can be observed from different perspectives through different sensors. Each of these sensors can collect information on the artefacts with different attributes or features. As these raw datasets include different attributes, some form of normalization is required. The *Observation* concept aims to realize the first step in this normalization process, defining a basic set of information necessary to characterize the observation of an artefact from the mobility point of view.

Regardless the sensor used to acquire the mobility data, these data reflect the observation of a phenomenon. The *Observation*, and can be described abstractly as:

The observation of an artefact in a specific point of a spatio-temporal space

and is represented by the following attributes:

(Id_Observation, Artefact, Location, Timestamp)

The first attribute allows distinguishing the observation of the same artefact in the same place and time instant by multiple sensors. The Artefact attribute uniquely identifies the observed artefact. The Timestamp attribute records the time instant of the observation. The Location attribute describes, with variable levels of detail, the location where the artefact was observed. Since the raw data obtained from different sensors can define the location of an artefact according to a geometric or symbolic referential, the Location concept combines both geometric and symbolic representations. Geometric, when the representation of Location is based on cartographic coordinates. Symbolic, when the Location is characterized by the description of its location, for example a building name, an address, or the ID of a cell in a GSM network.

3.3 Location vs. Place

Regardless the type of sensor used to characterize an *Observation*, each *Observation* describes an artefact as a point in spatio-temporal referential (as shown in Figure 2a), i.e., the artefact is described by the instant time and space dimension.

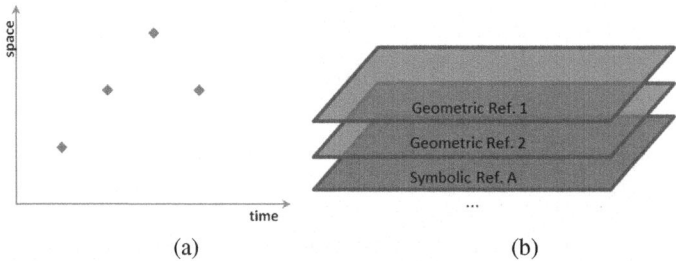

Fig. 2. Observations in a spatio-temporal referential (a), and the multitude of the spatial dimension (b)

For dimension time there is an almost universal definition and referential and, as such, the *Timestamp* attribute definition of an *Observation* does not cause major difficulties. The same does not apply with the spatial dimension, where such uniformity does not exist. As discussed earlier, space may be described in a geographic or symbolic manner and using several independent referentials (Figure 2b).

We define *Location* as a point in an arbitrary spatial referential, geometric or symbolic, or both, and represent it as: *location = {symbolic name, position}*, where position is represented by a pair of coordinates such as:

position = (latitude, longitude) or: *position = (X,Y)*

In turn, the aggregation of one or more nearby *Locations* defines a *Place*. The *Place* is not so much a specific point in one space dimension, but one region of space defined by the aggregation of *Locations*, whether they are symbolic or geometric.

The necessity to aggregate *Locations* emerges from the following reality. Considering that the location of an artefact in a sequence of *Observations* is described geometrically (for example, by pairs of GPS coordinates), an immobile artefact might be reported to be observed in different points of the space domain, in consecutive *Observations*, due to the sensor noise. Similar phenomena are observed in the symbolic domain. However, the symbolic space is usually a discrete one, while the geometric space is continuous, making this last one more prone to sensor noise. This way, even if there are slight changes of position when we conduct the process of aggregation and the subsequent detection of a *Place*, we get a better geographical characterization of the artefact and consequently a smaller number of distinct places visited by the artefacts. Extracting *Places* from sets of *Observations* might require different processes for geometric and symbolic data. Although many approaches have been described in the literature for both domains, we assume that combined approaches are still an open research challenge.

3.4 Suspension of Movement

Based on consecutive observations $\{o_{i-1}, o_i\}$ of an artefact over time, the artefact can be in one of two possible states: a moving state or a stationary state. The artefact is in stationary state when in the most recent observations the *Place* (P) is the same, no matter the *Location* (l).

$$\left(l_i \in P_j \right) \wedge \left(l_{i-1} \in P_j \right) \tag{1}$$

Stays vs. Time Leaps
Associated with the stationary state of an artefact, we define the concept of *Stay* as:

Time interval between the first and last observation of an artefact in the same place

and represent it by the following attributes:

(Id_Stay, Artefact, Place, Timestamp_Initial, Timestamp_Final)

Thus, we assume a *Stay* when for several consecutive observations of the artefact, the *Place* is the same. In the case of symbolic representations of *Place*, the transformation process that extracts *Stays* from *Observations* is simpler, as opposed to geometric representations, due the fluctuations of the positioning information. Several authors have already addressed this problem of detecting stays (also referred as stops, or regions of suspension of movement), in particular for the case of GPS traces.

It should be noted, however, that there are two formal restrictions to the definition of *Stay*. First, the time interval must be longer than zero. Second, the largest time interval between two consecutive *Observations* (from where the *Stay* emerged) should be short enough so that we can assume that the artefact did not leave the corresponding *Place* between *Observations*. Otherwise, if there is a long time gap between *Observations*, we cannot derive a *Stay* since it is not reasonable to infer intermediate observations (along a straight line connecting the two observations in the spatiotemporal space). These situations occur when, for example, a GPS tracked artefact enters a building. In order to address these situations we defined a concept of *Time Leap*, described as:

Long time period between two sequential observations of an artefact in the same place

and represented by the following attributes:

(Id_TimeLeap, Artefact, Place, Timestamp_Initial, Timestamp_Final)

Note that, the representation of a *Time Leap* and that of a *Stay* are identical, while representing different concepts with different semantic meanings.

3.5 Movement

The artefact is in the moving state when there is a change of *Location* (l) and the two most recent observations are not in the same *Place* (P), (one or both observations might even not be associated to a *Place*).

$$(l_i \neq l_{i-1}) \wedge \left[\forall_j (l_i, l_{i-1}) \notin P_j \right] \qquad (2)$$

Elementary Movement vs. Space Leap

In the movement state, we expect to observe variations in the *Location* attribute over the time, so there must be a concept that represents this variation. We name it *Elementary Movement*, and describe it as:

A Change of Location of an artefact occurred over time

represented by the following attributes:

(Id_Movement, Artefact, Location_Start, Location_End, Timestap_Initial, Timestap_Final)

We only consider that an *Elementary Movement* exists when, for a pair of observations at different locations in consecutive time instants, it is reasonable to interpolate the intermediate positions at intermediate time instants, i.e., only in situations where we can assume that the artefact moved from the initial *Location* to the final *Location* along a straight line at constant speed. This is often assumed when transforming a set of consecutive GPS records into a line representing the artefact trajectory.

For example, if a pedestrian is observed based on its GPS trace and these observations have been acquired at time intervals of one second, it is reasonable to infer that the artefact has realized elementary movements, as it is also reasonable to infer the intermediate position between two observations. On the other hand, if the time interval between observations is in the order of one minute, one cannot easily derive *Elementary Movements* from pairs of consecutive observations, because the space that can be travelled in such time interval is significant and we cannot assume that the artefact has actually travelled along a straight line. On the other hand, sampling the geographic position of a flying airliner at one minute intervals might be enough to derive *Elementary Movements*.

In situations where the time interval between consecutive observations of the artefact does not allow inferring with precision the intermediate positions of the artefact due to the large space of possibilities, we cannot consider that an *Elementary Movement* occurred, but instead a *Space Leap*. The *Space Leap* is described as:

A Change of Location of an artefact occurred over a long time period

represented by the following attributes:

(Id_SpaceLeap, Artefact, Location _Start, Location _End, Timestap_Initial, Timestap_Final)

Note that the representations of a *Space Leap* and of an *Elementary Movement* are similar, but their semantic meanings are different.

Trajectory

Finally, at the top layer of our conceptual framework we have the concept of *Trajectory* that represents a set of *Elementary Movements* ordered in time for the same artefact. The concept of *Trajectory* is described as:

Time-ordered list of Elementary Movements of an artefact over the space

and represented by the following attributes:

(Id_Trajectory, Artefact, List of Elementary Movements)

In turn, the set of existing trajectories at a given spatio-temporal interval for a given artefact or group of artefacts might lead to the representation of flows that exist in the urban space, the final goal that we aim to achieve with our study.

4 Mapping Real Data into the Proposed Framework

In order to validate the concepts of our proposed framework for the representation of spatio-temporal data, it is important to realize how well real data obtained through different sensors (different raw data) match the concepts in the framework. This task also triggers the design of a set of transformation processes responsible for the selection and adaptation of existing information, and also, to infer new information based on existing data.

In this section we take a data set comprising a set of RAW records collected by one single user, and describe how these data is mapped into our conceptual framework. Our focus is on the concepts of Observation, Place and Stay, and on the processes used to derive Places from Observations, and Stays from Observations and Places.

This data set includes records obtained from three types of sensors: GPS, Wi-Fi and GSM. By describing the transformation processes, we propose an approach to data fusion, where the three types of records are processed simultaneously to extract Places and Stays.

4.1 The Data Set

Many of the previous works presented in the area of visualization of urban mobility rely on collections of readings obtained from GPS receivers. These records include data beyond the position, such as the information of the temporal moment in which the position was acquired (a timestamp), the speed, the orientation (bearing), and other attributes. Since this type of mobility information has frequently been used in the study of urban mobility and it is easy to obtain, our framework should also adapt to this type of mobility data. On the other hand, resorting to GPS data to represent human mobility raises some concerns. The first one is about temporal coverage since it is difficult to collect GPS data indoors, and humans spend more than 80% of their time indoors. Therefore, most of the observation period is not actually observed. The second problem is that collecting GPS data in real world situations is technically difficult. One common approach is to give people a GPS receiver and a data logger to carry on during a specially prepared experiment (eventually disturbing the real daily live routines). The limitations of this approach are the limited number of persons observed, and the short observation period. An alternative is to rely on people's smartphones. Most of the recent smartphones integrate multiple sensors, and these sensors can be used to observe people spatio-temporal behaviours in many ways.

In our experiments, we resorted to the smartphone approach, aiming to overcome the limitations of the GPS only solution. By asking a group of people to install a small application on their Android smartphones we have been able to collect data during long periods of time and from multiple sensors: GPS coordinates, the nearby Wi-Fi Access Points, and the nearby GSM cells. After a few days, people using this application forget its use, and do not constrain their movement behaviour – they do not feel being involved in an experiment. The second advantage is that we collect data both outdoors (the three types) and indoors (Wi-Fi and GSM).

The data used throughout this section, for illustration purposes, were collected for a single user and over several months. Many other users were involved, but their data is not used here. The data reflect the mobility of a person in his daily normal activities. The following tables illustrate the raw data that were collected.

Table 1. Raw data collected from the GPS sensor

Timestamp	Latitude	Longitude	Altitude	Speed	Accuracy	Bearing
2011/06/29 15:25:07	1,297077	103,7808	93,5	0,75	17,88854	65
2011/06/29 15:25:18	1,297077	103,7808	108,2	0,75	26,83282	162,4
2011/06/29 15:25:31	1,297213	103,7806	134,4	1	40	283,8

Table 2. Raw data collected from the Wi-Fi sensor

Timestamp	BSSID	RSSI	SSID
2011/06/29 15:25:08	00:27:0d:07:d6:c0	-90	NUS
2011/06/29 15:25:11	00:27:0d:07:d6:c0	-88	NUS
2011/06/29 15:25:12	00:27:0d:07:d6:c0	-88	NUS

Table 3. Raw data collected from the GSM sensor

Timestamp	CID	LAC	MNC	SIGNAL_STRENGTH
2011/06/29 15:25:08	962335	441	3	9
2011/06/29 15:25:10	962335	441	3	8
2011/06/29 15:25:11	962335	441	3	8

The Wi-Fi raw data include a timestamp, the BSSID (MAC address that identifies the AP to which the artefact is currently connected), the RSSI (received signal strength indicator), and the SSID to identify the network to which the artefact is connected (other attributes were collected but are not represented here). The GSM raw data includes a timestamp, the cellID, the Location Area Code, the Mobile Network Code, the received signal strength, and other attributes not represented here.

4.2 Mapping RAW Data into Observations

Mapping GPS, Wi-Fi and GSM raw data into a set of *Observations* is straightforward. However, the transformation processes must be specific for each type of raw data and, in this example, different for GPS and Wi-Fi/GSM. For GPS data, we mapped the raw timestamp into the *Observation* timestamp, and the pair of coordinates into the *Observation* Position. For the Wi-Fi data, we mapped the raw timestamp into the *Observation* timestamp, and the BSSID into the Location. For GSM, we mapped the raw timestamp into the *Observation* timestamp, and the cellID into the Location. No data cleaning was performed. The table 4 illustrates the resulting set of *Observations* (the identification of the artefact was not included, and the timestamp was simplified).

4.3 Extracting Places from Observations

Automatically detecting Places that are relevant for one single person (e.g. the workplace) or for a group of persons (e.g. a popular place at a certain urban location) is an activity known as "place learning". Many approaches for place learning are described in the literature, most of them dealing with a single type of observations at a time, like GPS or GSM. Recent work in this field is addressing place learning by integrating observations from multiple sensors, such as GPS, Wi-Fi, and accelerometers, collected using smartphones [9]. In this work we propose a method for place learning based on a probabilistic model applied to observations obtained from GPS, Wi-Fi and GSM sensors. The novelty of this method comes from the simultaneous processing of the three types of observations, thus performing data fusion while clustering the observations to identify places. The proposed approach is an alternative to density-based spatial clustering algorithms.

Table 4. A set of *Observations* obtained from GPS, GSM and Wi-Fi raw data sets (the *Observations* are sorted chronologically)

Timestamp	Location			Optional Attributes
	Position		Symbolic Name	Sensor_type
	Latitude	Longitude		
15:25:07	1,297077	103,7808		GPS
15:25:08			00:27:0d:07:d6:c0	WIFI
15:25:08			962335	GSM
15:25:10			962335	GSM
15:25:11			00:27:0d:07:d6:c0	WIFI
15:25:11			962335	GSM
15:25:18	1,297077	103,7808		GPS

Density-based spatial clustering algorithms, such as DBSCAN [10] or SNN [11] define the similarity between multidimensional points using a distance function. For datasets where each point is described by a pair of coordinates, the most popular distance function is the Euclidean distance (2-norm). In our case, while GPS observations include a position (pair of coordinates) in their description, Wi-Fi and GSM observations do not. Therefore, a different distance function must be defined. Our approach to define such a distance function relies on the basic concept of *Observation*: one Observation describes what a particular sensor measured at a particular time instant and location, and not the location itself. Consider the example of a GSM based observation: the observation states that, at a particular time instant, the strongest GSM cell in the neighbourhood was C, and not that the observation was taken at the location of the C cell tower. Since GSM cells are often quite large, the consequence is that two samples of the GSM sensor taken at two far apart locations might be similar. Therefore, when trying to learn places from sets of observations, these two samples (observations) might end up being part of different places, meaning that cell C was "visible" from these two places.

We model the above described concept through a distance function that describes not the Euclidean distance between points in the dataset but the probability that two points (observations o_i and o_j) having been taken at the same place:

$$P_{sameplace}(o_i, o_j) \tag{3}$$

Dealing with three different types of observations simultaneously requires the use of 6 different probability functions:

Table 5. Probabilities that two observations have been taken at the same place

Prob. function	GPS	Wi-Fi	GSM
GPS	P_1	P_2	P_3
Wi-Fi	P_2	P_4	P_5
GSM	P_3	P_5	P_6

For the probability P_1, between two GPS observations, the Euclidean distance is a good indicator. If two observations are geometrically close, then they probably refer to the same place, and, therefore, P_1 can be described as:

$$P_1(o_i, o_j) = e^{-ED(o_i, o_j)/R_1} \tag{4}$$

where $ED()$ is the Euclidean distance between observations o_i and o_j, and R_1 is a parameter that relates the Euclidean distance to the closeness of the two observations. Note that P_1 takes a value of 1 for two observations taken at exactly the same position, and tends to 0 as the Euclidean distance goes to infinity.

For P_2 and P_3, the geometric distance cannot be used since Wi-Fi and GSM observations are described by symbolic locations. The same applies for P5 (Wi-Fi - GSM), since both observations are described by symbolically. In these cases we rely on the time difference between the observations: two samples taken within the same short time interval must refer to the same place. Therefore, P_2, P_3 and P_5 are defined as:

$$P_k(o_i, o_j) = e^{-|t_i-t_j|/R_k}, k = 2, 3, 5 \qquad (5)$$

where R_2, R_3 and R_5 are parameters that relates the time difference between the observations and the closeness of the two observations.

While estimating the closeness of two Wi-Fi observations, the characteristics of the Wi-Fi networks must be taken into account. The coverage area of one single Access Point (AP) is typically small and assumed to be a circle with a radius of 50 meters or less. Therefore, if two observations refer to the same AP, one can assume that they were taken from the same place. If they refer to different APs, then the time proximity can be used as in (5). So, for pair of Wi-Fi observations, P_4 is defined as:

$$P_4(o_i, o_j) = \begin{cases} P_{sameAP}, & AP_i = AP_j \\ e^{-|t_i-t_j|/R_4}, & AP_i \neq AP_j \end{cases} \qquad (6)$$

where P_{sameAP} is the probability of two samples referring to the same place given that the observed AP is the same. We do not set this probability to one because in places with poor coverage of Wi-Fi networks, the same AP might be detected from different nearby places. By setting this probability to a lower value (we are using 0.975), we do not limit the place size to the typical Wi-Fi cell size.

For the closeness of two GSM observations, also symbolic, the model used for Wi-Fi cannot be used since GSM cells are typically much larger in coverage area. Here we resort to the temporal proximity, but weighting differently depending if the two cells are the same or different:

$$P_6(o_i, o_j) = \begin{cases} P_{sameCell} \times e^{-|t_i-t_j|/R_6}, & cell_i = cell_j \\ P_{difCell} \times e^{-|t_i-t_j|/R_6}, & cell_i \neq cell_j \end{cases} \qquad (7)$$

the parameters $P_{sameCell}$ and $P_{difCell}$ being used to weight the probability.

For building the places, an iterative algorithm is used, where each new observations is added to one of the existing places if the probability of being taken at one place is higher that a predefined threshold (P_{min}). Otherwise, the observation is used to create a new candidate place. If P_{min} is exceeded, the observation is added to the place with higher probability.

One place is described by its GPS part, the Wi-Fi part, and the GSM part. The GPS part is represented by the centroid of all the GPS observations that have been added to the place, and the timestamp of the most recent GPS observation added to the place. The Wi-Fi part is described by the BSSIDs of all Wi-Fi observations that have been added to the place, and the timestamp of the most recent Wi-Fi observation added to the place for each BSSID. A similar representation is used for the GSM part.

The probability that an observation has been taken at a given place is the highest of the three probabilities that compare that observation with the three parts describing a place. Since each new observation that do not exceed P_{min} is used to start a new candidate place, and since most of the observations collected while the person is moving do not exceed P_{min}, a large number of candidate places are created by the algorithm. A candidate place is assumed to be a real relevant one if the total accumulated time spent at that place is longer than a minimum amount of time (e.g. two minutes).

Figure 3 illustrates the results of using the above described algorithm to detect the places visited by one single person during a one month period. A total of 13 places, with more than 2 minutes of total staying time have been detected. The place in red is the most relevant one, with a total staying time of 402,4 hours (in one month). Note that the algorithm has been able to distinguish between different places in very close locations (inset in Figure 3).

Fig. 3. Places detected from one month of data (159380 observations, from which 92991 are GPS, 60427 are Wi-Fi, and 5962 are GSM)

The detailed assessment of the quality of the place learning approach here described is out of the scope of this paper due to space limitations. However, a validation has been performed by comparing the detected places with a diary of the person under observation. All the 13 detected places are actually places that have been visited. Therefore, the false positives rate is 0%. On the other hand, not all the relevant places registered on the diary have been detected. In a few cases, this was due to the lack of observations (failure in the acquisition process). Other cases were due to the fact that only GSM observations were collected during the stay at those places, and those observations were too sparse in time to be grouped into one place.

4.4 Extracting Stays from Places and Observations

A place, as described in the previous section, is completely described by the set of observations that were clustered to create it. Therefore, computing the stays at each place from the place description is straightforward. In the following analysis, we assume that a stay occurred whenever the time elapsed between consecutive observations in a place do not exceed a given threshold (T_{max}). By concatenating all the consecutive time intervals that do not exceed T_{max}, one detects the stays at a given place. Figure 4 represents the stays (black lines) extracted from the set of 13 places shown in Figure 3. In Figure 4, the blue dots represent the GPS observations (y represents the distance from a reference point), the red dots represent the Wi-Fi observations (y represents different BSSIDs), and the green dots represent the GSM observations (y represents different cells). The inset in Figure 4 shows the details for one single day. This example shows that most of the time (stays) is assigned to one of the 13 places (56,5% of the total time in one month). Figure 4 also shows that there are temporal gaps between stays. These gaps represent the periods of movement, the periods where there is no data (inset in Figure 4), and the periods where the observations are too sparse to be grouped into a place.

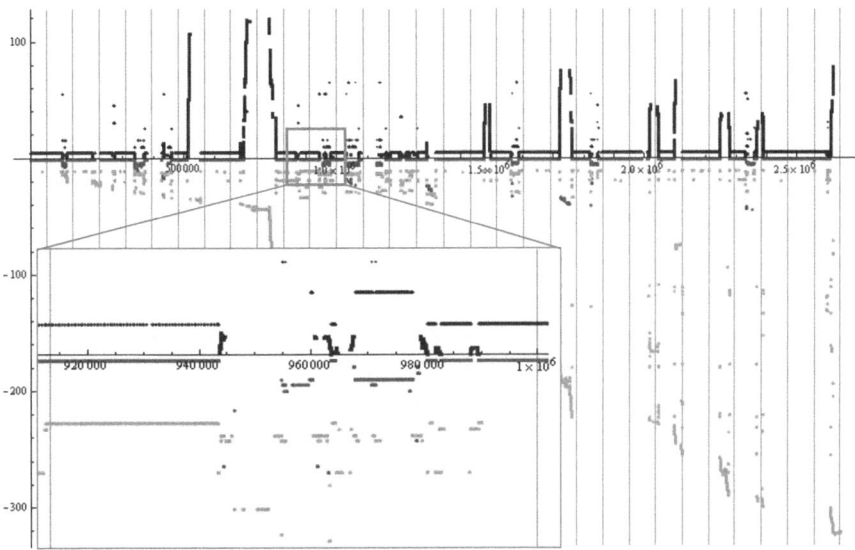

Fig. 4. Stays extracted from the observations in each place (T_{max}=60 seconds)

Stays represent a fundamental concept for the characterization of places. The results in Figure 4 uncover relevant information about the time profile of each place and, consequently, about the importance and relevance of that place for its tenant. Similar information about public places can be used to understand how particular urban areas are used by persons, and how the use relates to local events or availability of infrastructures. In our particular case, stays are of major importance for the characterization of transportation requirements in urban spaces, as the time profiles of places can be used to distinguish residential from commercial or industrial areas.

Stays are also the basis for the detection of origin-destination trajectories. A description of that process, as well as the description of the other mapping processes identified in Figure 1 is left for future articles.

5 Discussion and Conclusions

The mapping of multi-sensor data performed in this study allowed us to verify that the proposed concepts are appropriate to represent the three types of records used. It also supported the identification and design of various transformation processes required to map the data into the proposed concepts.

As illustrated in section 4.2, the proposed structure for an *Observation* can be used to represent both symbolic and geometric positions. The unified representation for the observations enabled the design of a generic process to infer *Places* while performing fusion of multi-sensor data. Extension of the proposed place learning algorithm to accommodate other sources of data only requires the definition of the probability functions that measure the probability of pairs of observations being taken at the same place. The results in section 4.4 reveal another benefit of using a unified representation for observations by illustrating how simple is the process that extracts *Stays* from the set of *Observations* that describe each *Place*.

Mapping *Places* and *Observations* into the remaining concepts identified in Figure 1 demands the design of other transformation processes. Besides these transformation processes, additional concepts might also need to be defined. Among them is a generalization of the *Trajectory* concept as our notion of *Trajectory* is only linked to the concept of *Elementary Movement*, i.e., we only consider a *Trajectory* exists when it is possible to infer intermediate positions between observations. As such, we are not covering the situations where *Space Leaps* occur while going from one place to another. One possible solution to this problem might be to define a new type of trajectory as a sequence of *Space Leaps* between places (eventually merged into an origin-destination trajectory). Currently our work is focused on the validation of the proposed concepts and transformation processes using a variety of datasets, including data from transportation systems (e.g. ticketing data used in buses).

Other challenges in this context are related to processing massive datasets. One of the difficulties already identified with the mapping we have been conducting is related to the space occupied at the level of storage in the database system of the observations data. Because we are working with individual data without any kind of aggregation, transformation processes need to deal with a large number of records (for example, the dataset used in section 4, representing one single user for a period of one month, is made of more than 150k records). Dealing with these large datasets requires efficient processing algorithms. In this context, the proposed place learning algorithm is an interesting contribution since the clustering process is quite efficient.

Acknowledgements. Research group supported by FEDER Funds through the COMPETE and National Funds through FCT – Fundação para a Ciência e a Tecnologia under the Project: FCOMP-01-FEDER-0124-022674.

References

1. Yu, H., Shaw, S.-l.: Representing and Visualizing Travel Diary Data: A Spatio-temporal GIS Approach. In: 2004 ESRI International User Conference, pp. 1–13 (2004)
2. De Almeida, V.T., Güting, R.H., Behr, T.: Querying Moving Objects in SECONDO. In: 7th International Conference on Mobile Data Management MDM 2006, p. 47. IEEE Computer Society (2006)

3. Erwig, M., Güting, R.H., Schneider, M., Vazirgiannis, M.: Abstract and discrete modeling of spatio-temporal data types. In: Proceedings of the Sixth ACM International Symposium on Advances in Geographic Information Systems, GIS 1998, vol. 3, pp. 131–136 (1998)
4. Sevtsuk, A., Ratti, C.: iSPOTS. How Wireless Technology is Changing Life on the MIT Campus. In: Proceedings of the 9th International Conference on Computers in Urban Planning and Urban Management, CUPUM 2005 (2005)
5. Brockmann, D., Theis, F.: Money Circulation, Trackable Items, and the Emergence of Universal Human Mobility Patterns. IEEE Pervasive Computing 7, 28–35 (2008)
6. Reades, J., Calabrese, F., Sevtsuk, A., Ratti, C.: Cellular census: Explorations in urban data collection. IEEE Pervasive Computing 6, 30–38 (2007)
7. Hagen-Zanker, A., Timmermans, H.: A Metric of Compactness of Urban Change Illustrated to 22 European Countries. In: Bernard, L., Friis-Christensen, A., Pundt, H. (eds.). Lecture Notes in Geoinformation and Cartography, pp. 181–200. Springer, Heidelberg (2008)
8. Moreira, A., Santos, M., Wachowicz, M., Orellana, D.: The impact of data quality in the context of pedestrian movement analysis. In: Painho, M., Santos, M., Pundt, H. (eds.) Geospatial Thinking, pp. 61–78. Springer, Heidelberg (2010)
9. Chon, Y., Cha, H.: LifeMap: A Smartphone-Based Context Provider for Location-Based Services. IEEE Pervasive Computing 10, 58–67 (2011)
10. Ester, M., Kriegel, H.-P., Sander, J., Xu, X.: A Density-Based Algorithm for Discovering Clusters in Large Spatial Databases with Noise. Computer, 226–231 (1996)
11. Ertoz, L., Steinbach, M., Kumar, V.: Finding Clusters of Different Sizes, Shapes, and Densities in Noisy, High Dimensional Data. In: Second SIAM International Conference on Data Mining, pp. 47–58 (2003)

Public Decision Processes: The Interaction Space Supporting Planner's Activity

Giuseppe B. Las Casas[1], Lucia Tilio[1], and Alexis Tsoukiàs[2]

[1] Università degli Studi della Basilicata, Viale dell'Ateneo Lucano, 10, 85100, Potenza, Italy
[1] LAMSADE, Université Paris Dauphine,
Place de LPlace du Maréchal de Lattre de Tassigny,
75775 PARIS Cedex 16
luctil82@gmail.com

Abstract. The aim of research is to test the model of interaction space as a tool to support the plan conception, in the context of a public decision process.

Interaction space model allows to analyze the interaction mechanisms that a public process generally activates, and help the planner, or more generally the analyst, to understand what kind of development the process could have, in order to address its progress.

The model has been tested during the planning process in Laurenzana, small village in the South of Italy. The paper describes the implementation and carries out some criticisms related to the not well structured relation between the space interaction model and the public decision process: once the interaction space is completely developed, the public process is generally at the beginning of implementation phase.

Keywords: Interaction space, decision process, planning process.

1 Introduction: Planning As a Decision Process

Spatial planning pertains to space transformation, when some problems are recognized, and a strategy needs to be defined in order to solve problems and produce improvement in conditions of territory and its inhabitants, taking into account several variables, several courses of actions, several stakeholders interacting eachother. Spatial planning, therefore, can be considered a decision process, and inherit all its characteristics.

According to Simon [1], decision process can be modeled as a sequence of five phases, three of which, intelligence, design and choice, are the core of the process, and the two last, implementation and ex-post analysis, concern more what happens once a decision is taken. A classic scheme of planning process takes into account some more detailed phases, considering also technical and bureaucratic moments of the process itself, as shown in figure 1, adapted from [2].

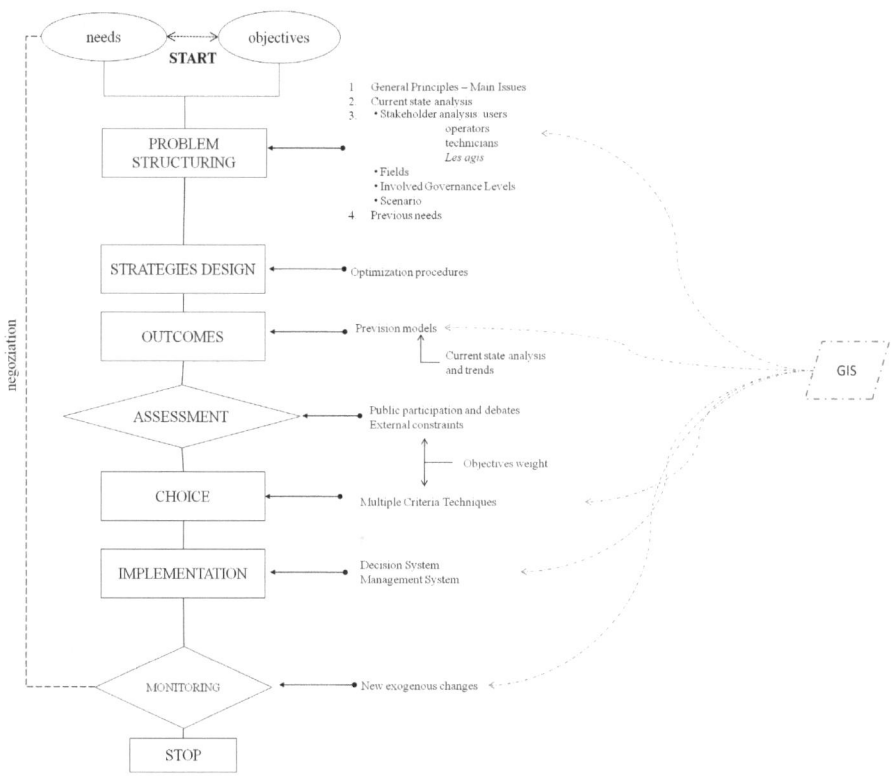

Fig. 1. Planning process, translated and adapted from [2]

Planning design, starting from recognizing inefficiency in cities and territories functioning [3], and searching reasons in inability to reach defined objectives and to respect planning principles [4], passes through an important phase of problem structuring, where analysis of current state is carried out, in order to clearly define needs and objectives, and then strategies. For each strategy, outcomes need to be forecast, in order to allow an assessment between alternatives and identify the optimal choice, to implement and monitor. Monitoring is important as it allows to identify gaps between forecasting and reality, and to adopt corrective actions.

GIS can support the whole process [5], not only in geographic information representation, but, moreover, in analysis, producing scenario and helping in choice [6].

In a comparison between the two schemes, it is interesting retrieve some analogies, as shown in figure 2.

The scheme on the left, adapted from [2], takes into account an important aspect of decision processes: negotiation. It characterizes several phases during the process; in planning procedures, there exist some well defined steps, with protocols and procedures, to allow a comparison between decision makers, analysts, citizens and groups. In spite of this, sometimes negotiation can happen also out of structured protocols, implicitly, without rules, but with the strength to influence the process itself.

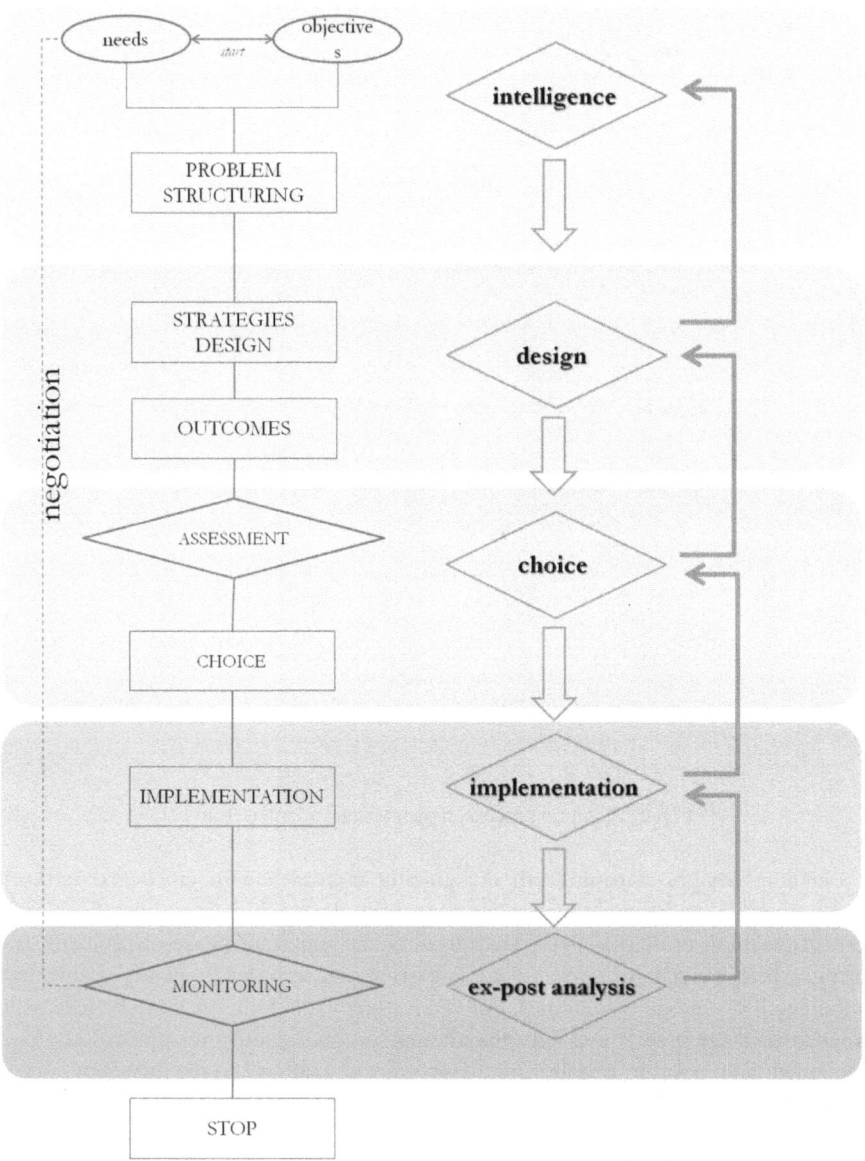

Fig. 2. Comparison between [2] scheme and Simon's scheme, in [1]

Therefore, in order to support planning process, analysts must be able to recognize all possible forms of negotiation in order to intervene, and try to control them. An useful tool to is the Ostanello-Tsoukiàs model [7], described in the following paragraph.

After a brief description of the model itself, it will be presented an application case.

2   The Ostanello-Tsoukiàs Model of Interaction Space

An analyst called to give its support in a problematic situation will define a representation of problematic situation [8], analyzing interaction mechanism between subjects involved in negotiation [9], identifying in a first moment all those elements related to the decision object.

In fact, the interaction space represents the virtual space where subjects interested in decision object transformation interact, trying to affirm their interests, using their own resources to influence the decision process. Interaction space is the place of negotiation mechanisms, both formal and structured ones and implicit ones.

A problematic situation can be modeled as a triplet of three components [8].

$$\mathcal{P} = \langle \mathcal{A}, \mathcal{O}, S \rangle \quad (1)$$

where A represents the set of subjects intervening in the process, O the set of interests that each subjects introduces into the process, and S the set of resources used by participants to protect their interests. The IS is characterized by this triplet and relations between them.

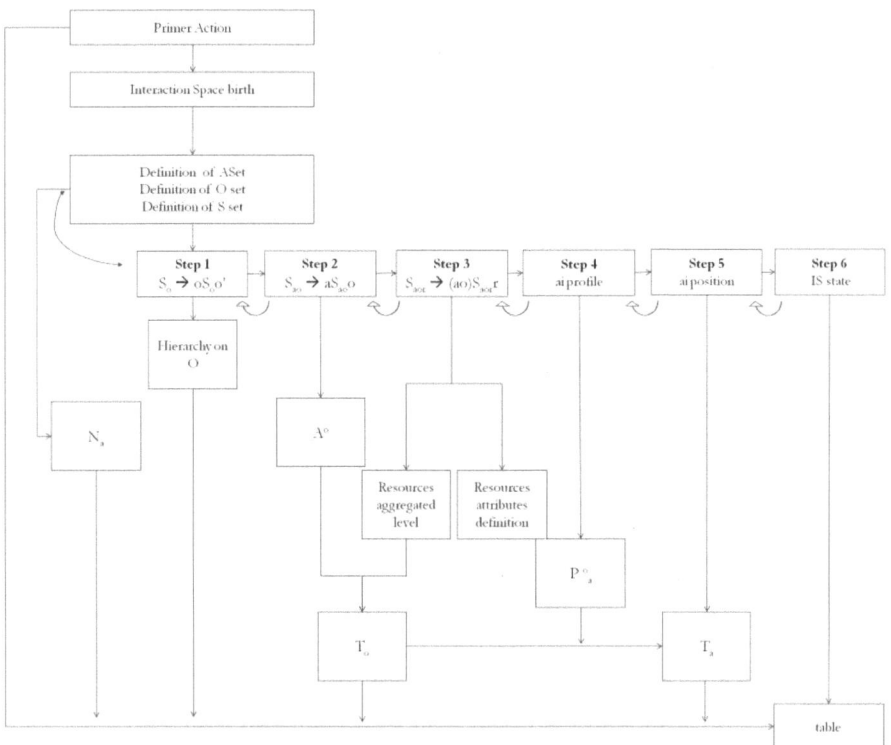

Fig. 3. The Interaction Space model, adapted from [7]

When a subject, that can be called promoter, promotes a transformation action on an object concerning its main interests and competences, he starts a primer action, implying birth of relation with other objects and other subjects: the Interaction Space is set. Then, the analyst can model all kind of relations between the three components of the triplet, and, through them, identify the state of interaction space.

Clearly, due to dynamicity in decision process, the state can change during the process itself, and it is important to define it in order to adopt strategies and actions to influence its development, mainly to avoid conflicts and negative situations.

Modeling consists in following some steps to collect all elements to identify Interaction Space state. Steps are schematically presented in figure 3. As the process is dynamic, steps can be iterated.

3 The Study Case: Laurenzana

Interaction space model has been tested during a real planning process, lead in 2007, in order to define the master plan of Laurenzana municipality.

Laurenzana, in Potenza province, South of Italy, is a small village, characterized by membership to a National Parc, Parco Nazionale dell'Appennino Lucano Val d'Agri – Lagonegrese, for almost 40% of its territorial surface, and at the same time, involved in petroleum extraction.

In the last decades nor the significant environment components, nor the economic importance of petroleum extraction became an element of development, and Laurenzana, as most of contiguous municipalities, is more and more depopulated, as the following graph synthesizes, and young people prefer to migrate in other bigger municipalities in the region or, more frequently, in northern Italian regions, where to find a job is more simple.

Despite these un-development conditions, municipality must accomplish to regional laws in the field of urban and spatial planning, and in 2007 started the process of masterplan definition.

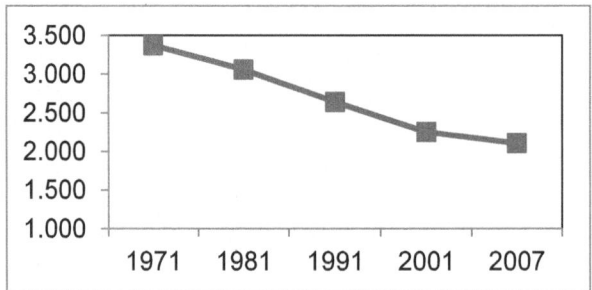

Fig. 4. Population trends during last decades in Laurenzana. Source: ISTAT

Between several strategies, one of them arose from indications of Regional Touristic Plan [10], focused on importance of traditions and cultural, historical heritage, strengthening the role of this kind of elements. In Laurenzana, a castle is present, and tourism can be developed around it, implying some decision at municipal level.

One other strategy arisen from Regional Touristic Plan is related to the concept of spreading receptivity as a development tool for tourism, called in Italian language, "borgo albergo". The document sustains the creation of new forms of receptivity, not concentrated in buildings uniquely devoted to this activity, but using empty houses, and financially sustaining holders in renovation actions.

These two issues are related to masterplan definition, but they require also a more enlarged debate to community, as they can influence the territory vocation, and, in order to become effective, need the cooperation of several stakeholders, not last the holders of empty houses. For these reasons, analysts involved in masterplan definition recognize the existence of a sub-process inside the main planning process, and retain it useful to study the interaction space model.

Fig. 5. The area interested by transformation: the castle and the area of empty houses in Laurenzana historical centre

4 Interaction Space

The need to open a debate concerning the future role of the castle and the possibility of making Laurenzana a "borgo-albero" can be considered as the Primer Action, determining the birth of the Interaction Space. Promoter is represented by analysts, in agreement with Local Administrators.

As the first step in order to model the Interaction Space is the representation of problematic situation, analysts identify the main useful elements to, answering to simple questions, as shown in the scheme in figure 6.

Answers are presented later in the paper, after a brief discussion.

As mentioned, primer action is represented by the need to find a role for the castle, also defining normative and management aspects, in order to make sustainable and efficient the choice of building a touristic development round about it.

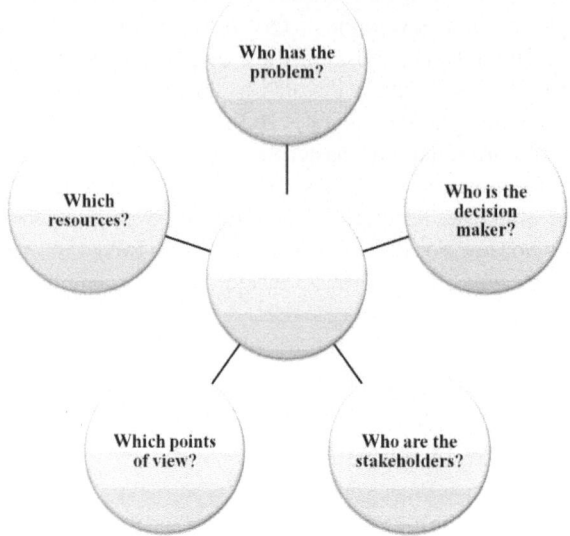

Fig. 6. Problematic situation: the fundamental questions to answer

Moreover, in 2007 the castle needed some renovation actions, implying also ground stabilization, as GNDCI group (National Group for Hydrogeological catastrophic events prevention) recognized some active processes of destabilization, and forbid access to the area. Each intervention on the castle, therefore, needs the advice of Sovrintendenza dei Beni Culturali (that is the Italian Institution with competences and veto power concerning cultural heritage).

As municipality resources were not strong, an important issue related to the castle is its management, once renovation is concluded: an interesting proposal was the possibility to identify some local actors interested in castle management and able to find resources, as, for instance, the local Pro Loco (Pro Loco are associations, diffused on the whole Italian territory, composed by volunteers interested in maintain and promote local traditions)

The other issue is related to the presence of empty houses, particularly in the closest neighborhood of the castle. These houses, moreover, were not only empty, but also abandoned, in a decay condition, determining a negative impact on landscape. In several cases, houses were reduced to ruins, and needed a strong intervention. During masterplan definition, moreover, some other areas have been recognized, as contiguous to castle area, but not yet urbanized, representing a possibility in order to localize some attractive activities, as an open-air amphitheatre or a falconry school, renewing an old tradition of the region. In the following, all these areas, including the empty houses, and the not yet urbanized areas are called *NU*. In order to take into account these areas, their owners must be involved in the decision process.

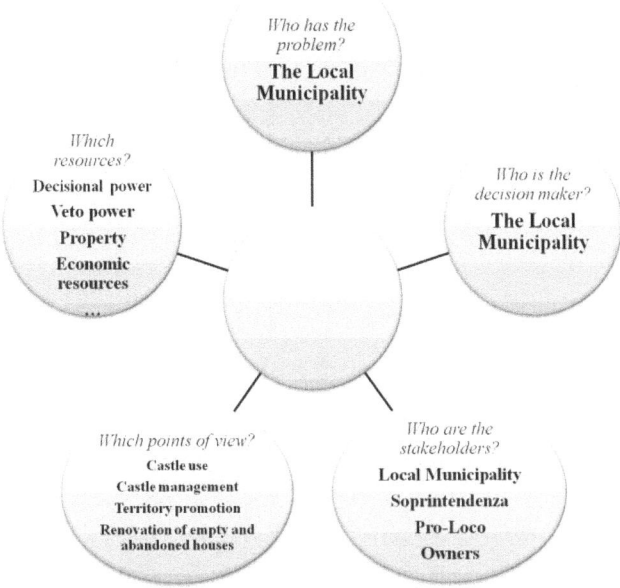

Fig. 7. Problematic situation: answers to fundamental questions

4.1 The Interaction Space Step-by-Step

Participants

As mentioned in the previous paragraph, a problematic situation can be modeled as a triplet composed by the set of participants, the set of objects and the set of resources, and model of Interaction space consists in identification of relation on these three set, following a step-by-step procedure.

At each step, some information need to be collected, in order to model relation. Concerning the set of participants, they are identified by their governance level and the role in the interaction space. This kind of information is useful to understand the possible dynamics, as it is a symptom of the strength of a certain participant to influence the process. In table 1 is presented a synthesis of identified participants.

Objects

In a more detailed analysis, the transformation concerning use and role of castle implies some other transformation also concerning other objects, as the mentioned empty houses. Table 2 presents a synthesis of identified objects.

Resources

Each participant enters in the Interaction Space with some resources, useful to promote its interests in the negotiation with other participants. Resources have been identified through interviews, but, even if a set has been defined, there is no certainty that the set is complete and represents all the resources really used during the process. For this reason, we can call identified resource as *possible*. They are presented in table 3.

Table 1. Identified Participants

Participant Set: A= { $a_1, a_2, a_3, a_4, a_5, a_6$ }

Participant	Notes	Governance Level	Role
Municipality Administration (a_1)	The Administration starts a debate concerning castle transformation, during masterplan definition. In reason of impossibility to manage the castle, due to scarce resources, it looks for someone, as Pro Loco, interested in.	Municipal	Promoter
Pro Loco (a_2)	It is a volunteer association, with the aim to promote traditions of its territory. It is composed by members and by an elected executive council.	Municipal	Invited
Owners (a_3)	Owners of NU are invited to enter into debate, in order to involve them into renovation of historical centre, through financial measures to transform their empty houses in a spreading receptivity elements. Therefore, they are invited to participate, but then, they are legitimate to participate to discussion, in reason of their property right on some soils. It was not simple to contact them.	Local	Invited/ legitimated
Soprintendenza dei Beni Culturali (a_4)	Soprintendenza, as the Institution with competences and veto power on cultural heritage, has the right to enter in the Interaction Space.	Regional	Legitimated
Analysts (a_5)	The analysts implied in masterplan design are involved also in modeling the interaction space, but they can be considered as participants, because they interact with other stakeholders, giving some technical support in choices	-	Legitimated
Region (a_6)	The Region is involved as the municipal planning tools must be coherent with regional planning tools.	Regional	Legitimated

Table 2. Identified objects

Objects Set: $O = \{ o_1, o_2, o_3, o_4, o_5, o_6, o_7, o_8, o_9, o_{10} \}$

Object	Notes
Castle use (o_1)	Decision concerning castle use is the primer action determining Interaction Space birth.
Urban planning tools (o_2)	Castle use is therefore related to urban planning tools, as use must be defined by masterplan.
Territory promotion (o_3)	To decide that castle becomes relevant in order to tourism development in the area means to adopt a strategy of territory promotion.
Castle valorization (o_4)	Territory promotion through castle means enhance the value of castle itself.
NU (empty houses and other soils in the neighborhood) (o_5)	In order to adopt strategy concerning spreading receptivity, empty houses need to be taken into account.
Castle management (o_6)	As Local Administration has scarce resources, it is needed a solution for castle management, in order to optimize resource and results.
Economic development (o_7)	Castle and NU renovation, in order to promote territory can mean to promote the economic development on tourism.
NU soils renovation (o_8)	In order to adopt strategy concerning spreading receptivity, empty houses need to be not only taken into account, but also renovate in order to promote territory.
Soils tax removal (o_9)	A tool to promote actions on NU soils can be the removal of tax.
Intervention coherence and homogeneity (o_{10})	Each decision must be taken with reference to a general coherence of masterplan, and in order to give homogeneity to interventions.

Table 3. Identified resources

Resource Set: $R = \{ r_1, r_2, r_3, r_4, r_5, r_6, r_7 \}$

Resource	Notes	Type
Decisional power (r_1)	This resource is available, in different way and different contexts, for Municipality Administration, Soprintendenza dei Beni Culturali, Region	Behavioural resource
Persuasion capacity (r_2)	Generally speaking, this kind of re source can be attributed to carismatic stakeholders. In this case, it characterizes the analysts, as facilitator of the process	Not quantifiable resource
Veto power (r_3)	Soprintendenza dei Beni Culturali can express its veto concerning castle transformation. Region has a more weak power concerning relation between municipal masterplan and regional planning tools.	Behavioural resource
Expert knowledge (r_4)	This is a resource that analysts can use supporting the process.	Not quantifiable resource
Economic resources (r_5)	Municipal Administration can use some economic resources in order to program activities, sometimes with help of Region, through programs aiming at sustain local actions for promoting territory. Generally, for this last kind of fundings, Pro Loco can be a beneficiary.	Quantifiable resource
Property right (r_6)	NU owners have property right on their soils.	Not quantifiable resource
Support policies (r_7)	Other kind of resources, generally promoted by Region, in order to support local territories.	Quantifiable resource

Relations on the set A, O, R.

Once elements in the Set A, O, R are identified, hierarchy of object must be built. This implies to make explicit relation between objects, considering projection and evocation. The obtained hierarchy is schematically presented as a graph, in the following figure.

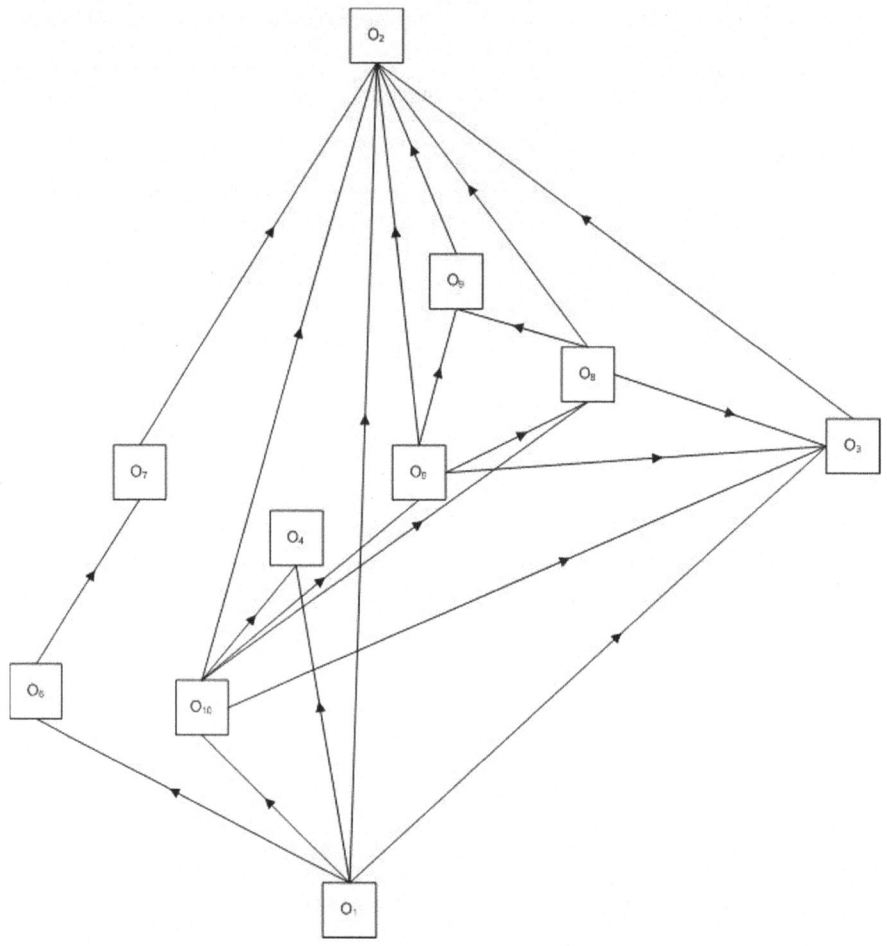

Fig. 8. Hierarchy of objects

In the second step, each object is related to one or more participants. This step imposes some revision of the first formulation of A,O,R sets. Figure 9 shows the identified relation for each participant.

Third step implies the assignment of resources to each identified relation between a participant and an object, that is to each arc of the previous scheme. This relation has been built considering also that resources are not only considered if they are available, but also if they are used, they are necessary, they are searched. Therefore, table 4 represents an excerpt of the table synthesizing this information; for each resource is

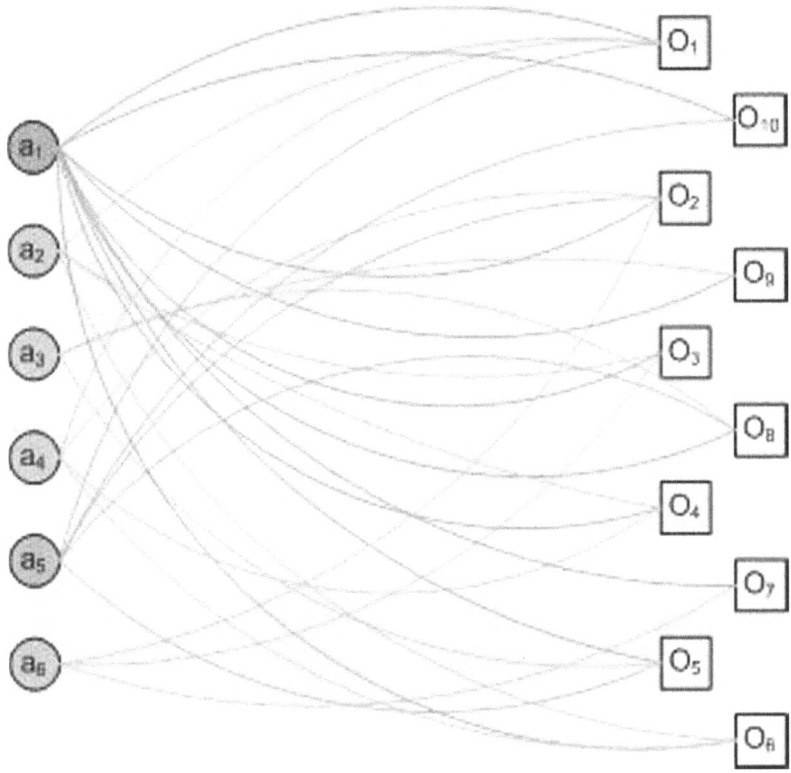

Fig. 9. Assignment of objects to participants

indicated with "D", "U", "N","R" if it is available, used, necessary or searched. In each row is indicated the relation between participant and object, and in each column a resource. The cells present a value (D,U,N,R) if that resource is used for the relation between participant and object.

Considering these three first steps, it is possible to define some information necessary to identification of state of interaction space, as the participants profile, and their position in the interaction space, and continue the process until the sixth step, that consists into identification of IS state.

Interaction Space state

At the moment t-1, the Interaction Space is characterized by Primer Action: Municipal Administration, during masterplan definition, decided to open a debate concerning the role of the castle in the touristic development of the village.

At the moment t, a limited number of participants is entered into the IS, the meta-object, identified in the castle in reason of relations of projection, has a strong meaning, and each interest in the IS is strongly dependent by the course of decisions. Adopting [7] model, the IS is characterized as showed in table 5.

Table 4. Excerpt of table of assignment of resources to relation between objects and participants

	r_1	r_2	r_3	r_4	r_5	r_6	r_7
Sa_1o_1	U						
Sa_2o_1		D					
Sa_4o_1			U				
Sa_5o_1		D		U			
Sa_1o_2	U				U		
Sa_4o_2			U				
Sa_6o_7		D			D		D
Sa_1o_8	U						
Sa_3o_8		D				U	
Sa_5o_8		D		U			
Sa_1o_9	U						
Sa_3o_9		D				U	
Sa_1o_{10}	U				D		
Sa_5o_{10}		D		U			

Table 5. Characteristics of Interaction Space

IS characteristics	
Participants number N_a	Several
Participants type T_a	Any kind but Opportunist
Objects type T_o	Any kind
Meta-Object meaning M	Strong, Stable Meaning
Interaction Space State at t-1, I_{t-1}	Primer Action

The recognized IS state is the Controlled Expansion (CE). In this state, the Meta-Object is used by the principal actor as a way to control the IS. Participants act in the space in order to find a solution, in a cooperation and negotiation climate.

5 Results

5.1 Model Results

Knowledge of Interaction Space state allows to understand the possible evolution of process and what kind of actions can be undertaken, helping the analyst to influence the process itself, avoiding some negative dynamics.

As the role of analysts was related to masterplan definition, the Interaction Space was dissolved when design was concluded. Decisions concerning castle and neighborhood were not yet mature, but they need to find formalization also in the masterplan. Therefore, masterplan assigns to castle and NU areas the urban regime of "areas for collective uses", and imposes different way to operate on the NU areas, considering different role for private owners.

5.2 Process Conclusion and Remarks

This decision must represent the Institutionalization for Interaction Space, and its conclusion. The definition of rules is the normal closure for a masterplan,

Therefore, some remarks are necessary: even if the planning process seems closed, as shown in the figure 1, it can be considered characterized also by implementation and monitoring, in order to follow the realization of forecast strategies and actions. In the reality, the plan is made effective, so strategies and actions are adopted and realized, but generally this happens without an analyst following the process, and generally the possibility to adjust some choices, if outcomes are different from the foreseen ones, is not considered.

Concerning the interactions space, moreover, the absence of an analyst means that no one can takes care of possible interactions, and if a conflict overcomes, it can be not controlled.

Therefore, in order to make effective the interaction space model, it is important to define a longer path for analysts, to make them active also when the decision is taken and the actions must be implemented.

References

1. Simon, H.A.: Administrative behaviour: a study of decision making processes in administrative organizations. Mac Millan, New York (1947)
2. Las Casas, G.: Processo di piano ed esigenze informative. In: Clemente, F. (ed.) Pianificazione del Territorio e Sistema Informativo. FrancoAngeli, Milano (1984)
3. Pontrandolfi, P., Lanza, V., Tilio, L.: I Laboratori di Urbanistica Partecipata a Potenza: sperimentazione di tecniche e strumenti. In: Las Casas, G., Pontrandolfi, P., Murgante, B. (eds.) Informatica e Pianificazione Urbana e Territoriale, Atti della Sesta Conferenza Nazionale, INPUT 2010, Libria Edizioni, vol. 3, pp. 393–402 (2010) ISBN 9788896067475

4. Las Casas, G.: Una cultura della pianificazione in un approccio rinnovato alla razionalità del piano. In: Francini, M. (ed.) Modelli di Sviluppo di Aree Interne ad Alta Ruralità, Scuola Estiva 2006. Centro Editoriale e Librario Università della Calabria (2006)
5. Goodchild, M.F.: The state of GIS for environmental problem solving. In: Goodchild, M.F., Parks, B.O., Steyaert, L.T. (eds.) Environmental Modeling with GIS, pp. 8–15. Oxford University Press, New York (1993)
6. Chakhar, S., Mousseau, V.: An algebra for multicriteria spatial modelling. Computers, Environment and Urban Systems 31(5), 572–593 (2007)
7. Ostanello, A., Tsoukiàs, A.: An explicative model of public interorganizational interactions. European Journal Of Operational Research 70, 67–82 (1993)
8. Tsoukiàs, A.: On the concept of decision aiding process: an operational perspective. Annals of Operations Research 154(1), 3–27 (2007)
9. Mazri, C.: Apport méthodologique pour la structuration de processus de décision publique en contexte participatif. Le cas des risques industriels majeurs en France. Ph. D Thesis, not published. Université Paris Dauphine (2007)
10. Regione Basilicata, Piano Regionale Turistico, L.R. 34/2006 (2001)

Selection and Scheduling Problem in Continuous Time with Pairwise-Interdependencies

Ivan Blecic, Arnaldo Cecchini and Giuseppe A. Trunfio

Laboratory of Analysis and Models for Planning (LAMP)
Department of Architecture and Planning - University of Sassari,
Palazzo Pou Salit, Piazza Duomo 6, 07041 Alghero, Italy
`{ivan,cecchini,trunfio}@uniss.it`

Abstract. We propose a general framework for modelling selection-and-scheduling problem with interdependencies in continuous time. Such problem may frequently arise in many real-world evaluation and decision-making contexts, such as project portfolio selection and scheduling in organisations, urban planning, and scheduling of public policies. For the purpose of conducting computational experiments, we further formulate a specific example model to optimise, whose benefit is to require a relatively limited number of input data. Given the NP-hardness of the problem, we employ the Covariance Matrix Adaptation Evolutionary strategy for solving it, and discuss few results.

Keywords: project portfolio selection, scheduling, interdependencies, continuous time optimisation, evolutionary algorithms, decision-making.

1 Introduction

The selection and scheduling of actions and projects under constraints is a class of problems with many real-world evaluation and decision-making applications. Project portfolio selection in organisation, implementation of urban projects, scheduling of public policies are just few examples where this type of problem might arise.

In its simplest form, the sole problem of selection of actions under constraints may be formulated as a typical optimisation and ranking problem, be it single- or multiple-criteria [1,2], depending on the nature and the number of constraints and objective criteria. The problem grows in complexity if there are specific supplementary constraints or combinatorial aspects and interactions among actions [3], such as interdependencies and logical or technical constraints, among others.

The scheduling is in itself an additional problem, since it adds a temporal dimension to the problem of selection of the bundle of actions to implement. In fact, the addition of the time variable requires a search for the optimal times to undertake each action within a given time horizon. This, in turn, brings about the need to provide a formal representation of time and to intertwine it with a model of interaction and interdependency among actions mentioned before.

In this work, we propose an approach for modelling a family of selection-and-scheduling problems with interdependencies whose novelty is to offer a more realistic treatment of time as a continuous variable, while possibly not imposing excessive burden in terms of the required input data.

To keep things simple, the performance of actions is evaluated on a single objective, since an extension to a multi-objective case, though straightforward, was unnecessary given the purpose and the focus of this paper.

2 Background

In the context of our discussion, interdependencies among projects and actions may be of very different nature, but can generally be classified in benefit, resources and technical interdependencies [4,5]. There are various approaches to modelling interdependencies (e.g. [5-13]), but very few handle jointly both interdependencies and scheduling [13,14], while being at the same time not too demanding in terms of the necessary information as input data.

Few proposals are bounded to modelling interdependencies between only two or at most three projects (e.g. [5-8]). Few other models are not confined to that limitation, like in [9]. This latter contribution is interesting because is suggests a generalised way to model interdependencies: for each interdependency it is necessary to declare the subset of interdependent projects together with the conditions of applicability of that interdependency among the projects in that subset. Subsequently, the evaluation procedure examines project portfolios for existence of projects and conditions for every specific subset. Altogether, the approach proposed in [9] is quite comprehensive and in the line of principle a more general model for describing interdependencies, possibly extendable to a continuous-time situation. But, given this generality, it is of difficult operational applicability to real-world cases with relations and interdependencies slightly above trivial, for it puts a significant burden on experts who need to evaluate and provide parameters exponentially growing in combinatorial complexity with the number of projects.

To tackle the problem of growing number of values to estimate and to feed the evaluation model with, an interesting proposal was made in [15], where a binary project-on-project "dependency matrix" is used as the basis for calculating the benefits of a given portfolio at given time. An interesting feature of such an approach is to base the evaluation of interdependencies in larger project portfolios and sequences only on binary relationship among projects. This is a clear advantage in terms of the amount of input information required by the model, but also in terms of simplicity and comprehensibility of the model's mechanics to non-expert and non-technical users. Similar approach was followed in [16].

In this vein, in this paper we make an extension of such pairwise-based approach, extending it to a model with continuous time.

Continuous time is of course a common and a natural assumption in many portfolio optimisation problems, such as financial analysis of alternative financial investments and portfolio allocation. But in those cases no intrinsic interdependencies among the performance of alternatives arise, and therefore they do not require specific modelling of such interdependencies in continuous time.

There have been attempts to address the scheduling with interdependencies in discrete-time frameworks [17,14]. However, besides being based on such pseudo-continuous time approach, they fail short in one or another aspect of the requirements we pose to our model for a realistic though practically tractable approach. For instance in [17] an overly-simplified approach to modelling intrinsic interdependencies is adopted (based on defining mandatory precursor projects and mutually exclusive projects), while in [14] a more sophisticated model of interdependencies is exploited, essentially based on [9], but this further increases the original combinatorial complexity, not only of computational nature, but foremost importantly, in terms of the input data required.

The paper is organised as follows. In section 3. we present the general framework of modelling interdependencies in continuous time. A specific optimisation problem within this general framework is defined in section 4. The optimisation search heuristic used for its solution is presented in section 5., while in section 6. we show and briefly discuss few computational experiments. Some conclusions and directions for further research are drawn in the final section 7.

3 Modelling Interdependencies in Continuous Time

When no other action has been previously undertaken, the performance (Π) of an action, measured against some performance criterion or objective, is assumed to be a function of the time passed since the beginning of its implementation. Formally, given a set of actions a_i ($i = 1, \ldots n$), we define the *stand-alone performance function* $\Pi_i(\tau; t_i)$, which gives the performance at the time τ generated by the action a_i implemented at the time t_i, provided that *no other* action is being implemented.

When two actions are performing jointly, the possibility of their mutual interaction may bring about interdependencies between their performances. We model these interdependencies by making an assumption about how the performance of an action changes over time, given the fact that the other action started to be implemented at some specific point in time. Therefore, for modelling performance interdependencies between actions, we define *pairwise-interdependency performance function* $\Pi_{i|j}(\tau; (t_i, t_j))$, representing the performance at the time τ generated by the action a_i implemented at the time t_i, provided that the action a_j is implemented at the time t_j.

Extending such notation, we write $\Pi_{i|\{1,\ldots,m\}}(\tau; t_i; (t_1, \ldots, t_m))$ to designate the performance at the time τ of the action a_i implemented at the time t_i, when a subset of actions $\{a_1, \ldots, a_m\}$ are implemented respectively at the times $\{t_1, \ldots, t_m\}$.

We define this *multi-interdependency performance function* of the action a_i only in terms of its stand-alone performance function and in terms of the relevant pairwise performance functions:

$$\Pi_{i|\{1,\ldots,m\}}(\tau; t_i; (t_1, \ldots, t_m)) = \Pi_i(\tau; t_i) + \sum_{j=1}^{m} \left(\Pi_{i|j}(\tau; (t_i, t_j)) - \Pi_i(\tau; t_i) \right) =$$
$$= (1-m)\Pi_i(\tau; t_i) + \sum_{j=1}^{m} \Pi_{i|j}(\tau; (t_i, t_j)) \quad (1)$$

Thereafter, the *total performance* of a subset of actions $\{a_1, \ldots, a_s\}$ implemented at the times $\{t_1, \ldots, t_s\}$ is the sum the multi-interdependency performance functions for all the actions in the subset. Therefore, we can express it in all know terms:

$$\Pi_{\{1,\ldots,s\}}\left(\tau; (t_1, \ldots, t_s)\right) = \sum_{i=1}^{s} \Pi_{i|\{1,\ldots,s\}}\left(\tau; (t_i, t_1, \ldots, t_s)\right) = \\ = \sum_{i=1}^{s}\left[(1-s)\Pi_i\left(\tau; t_i\right) + \sum_{j=1}^{s}\Pi_{i|j}\left(\tau; (t_i, t_j)\right)\right] \quad (2)$$

This *total performance* function yields the *instantaneous* performance generated by all the actions in the subset at any particular time τ. The *overall* performance generated by the actions in a given time interval $[0, \tau^*]$ should therefore be calculated as the defined integral of total performance function under that interval. This overall performance, thus defined, will be our objective function to maximise.

4 Specifying Performance Functions and Constraints

In the previous section we described the general form of the modelling approach hereby proposed. For the purpose of presenting few numerical exemplifications, we henceforth devise a fully specified model of actions' performance and interaction, using specific formulations of the stand-alone and pairwise-interdependency performance functions, together with all the relevant parameters.

It is important to emphasise that while the we previously described the model in its general form, the following discussion uses particular specifications of the *stand-alone performance functions* and of the *pairwise-interdependency performance functions* (based on simple linear relation with time). We picked those specifications for the sake of simplicity and for the purpose of providing an exemplification of the modelling approach proposed by us. By doing this, we are not making any claims that the real-world performances and interaction among actions are necessarily of that form. That, in fact, is the matter of specific applications in question, and should be established from case to case.

4.1 Stand-alone (Instantaneous) Performance Functions

In the case when only one action is implemented along the entire timeline, we assume that there is an initial time interval during which the action is *progressively* being implemented and therefore reaches its full performance only after such an implementation process. This is a realistic assumption in many situations where it takes time to fully develop and implement action.

In particular, for such stand-alone action's performance, we use functions of the general form:

$$\Pi_i(\tau; t_i) = \begin{cases} 0 & \tau < t_i \\ \dfrac{\pi_i}{\epsilon_i}(\tau - t_i) & \tau \geq t_i \wedge \tau < t_i + \epsilon_i \\ \pi_i & \tau \geq t_i + \epsilon_i \end{cases} \quad (3)$$

where the constant ε_i is the time required for the action a_i to reach its full performance, and π_i is that full performance.

The form of Eq. (3) shows that we assume that the action's performance is increasing linearly during its implementation/development period, and that once fully developed, the performance remains constant with time. To be sure, this is by no mean the only possible, nor it is probably the most plausible form. Indeed, it may be argued that the development phase has a non-linear, and even not necessarily monotonic form (e.g. many projects of transportation network improvement usually interfere negatively with the quality of the transportation network, before eventually bringing about improvements). Furthermore, it would be plausible to assume that once the action has been developed, its performance would have decreasing returns with time without investments in maintenance, due to consumption and agedness. However, for the sake of simplicity, we do not handle these arguments in this specification of the model.

4.2 Pairwise (instantaneous) Performance Functions

To define the general form of the pairwise-interdependency performance function of an action a_i given the action a_j, we make one further assumptions. We posit that the influence at the time τ of the action a_j on the performance of the action a_i is proportional to the fraction of the full performance reached by the action a_j at the time τ. Consequently, the pairwise-interdependency performance function of the action a_i given the action a_j is:

$$\Pi_{i|j}(\tau; (t_i, t_j)) = \begin{cases} 0 & \tau < t_i \\ \Pi_i(\tau; t_i) & t_i \leq \tau < t_j \\ \Pi_i(\tau; t_i)\left(1 + \beta_{i|j}\dfrac{\tau - t_j}{\varepsilon_j}\right) & \tau \geq t_i \wedge t_j \leq \tau < t_j + \varepsilon_j \\ \Pi_i(\tau; t_i)\left(1 + \beta_{i|j}\right) & \tau \geq t_i \wedge \tau \geq t_j + \varepsilon_j \end{cases} \quad (4)$$

where ε_j is the time required for the action a_j to reach its full performance, and $\beta_{i|j}\dfrac{\tau - t_j}{\varepsilon_j}$ ($\beta_{i|j}$ is assumed by us to be constant) is the marginal performance of the action a_i due to the interdependency from the action a_j with respect to time.

Substituting Eq. (3) into Eq. (4) we obtain $\Pi_{i|j}(\tau; (t_i, t_j))$ in terms of ε_i, ε_j, π_i and $\beta_{i|j}$. The noteworthy behaviour of that function is substantially determined by the time distance between the implementation of the two actions. To illustrate this, in Figure 1. we show two cases (one with positive and one with negative $\beta_{i|j}$) of an action's performance in terms of the time distance from an interacting actions.

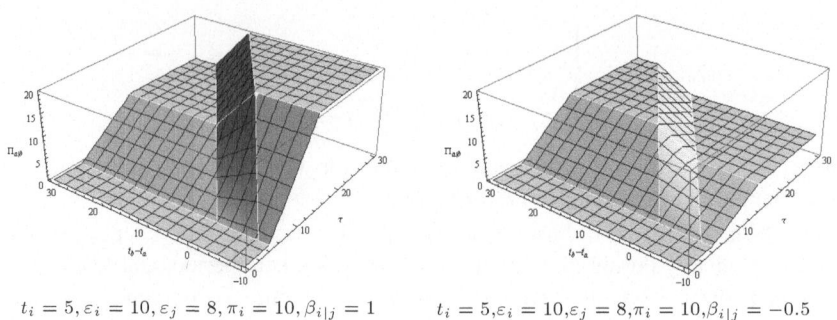

$t_i = 5, \varepsilon_i = 10, \varepsilon_j = 8, \pi_i = 10, \beta_{i|j} = 1$ $t_i = 5, \varepsilon_i = 10, \varepsilon_j = 8, \pi_i = 10, \beta_{i|j} = -0.5$

Fig. 1. Instantaneous performance of an action (axis Π) over time (axis τ) in relation to the time difference (axis $t_b - t_a$) of implementation from an interacting action.

4.3 Total (Instantaneous) Performance of a Subset of Actions

The explicit expression of the total performance of a subset of actions $\{a_1, \ldots, a_s\}$ implemented at the times $\{t_1, \ldots, t_s\}$ would be too extensive to write, as well as unnecessary for it stems quite straightforwardly by substituting the appropriate stand-alone and pairwise-interdependency performance functions in Eq. (3) and Eq. (4) into the general form of the total performance Eq. (2). As discussed, the *overall performance* generated by a subset of actions in a given time interval $[0, \tau^*]$ is the area under the total performance function, which is our objective function to maximise.

4.4 Budget Constraint

For the purpose of this numerical experimentation, we assume that each action has a cost a_i attached to it, and that the acting entity (planning authority, organisation) faces a budget constraint. In particular, we model that actions' costs have to be entirely paid for upfront, while there is an initial endowment of budget resources w_0 and a cash inflow with time at a constant rate of ω. Therefore, given a time-ordered bundle of actions $\{a_1, \ldots, a_m\}$ implemented respectively at the times $\{t_1, \ldots, t_m\}$ (such that $i < j \Leftrightarrow t_i \leq t_j$), we have the following set of m constraints:

$$w_0 + \omega\, t_k - \sum_{i=1}^{k} c_i \geq 0 \qquad k = 1, \ldots m \qquad (5)$$

5 Search Heuristic

The selection-and-scheduling problem with interdependencies is know to be NP-hard [18,19]. We therefore attempt to solve the above specified continuous optimisation problem with an Evolution Strategy (ES), in which a population of individuals is evolved through an iterative application of the recombination, mutation and selection operators. In the evolved population, each individual, representing an actions' scheduling, is composed of a real-valued vector $\{t_1, ..., t_n\}$ with $t_i \in [0, t^*]$. In order to address the possible exclusion of actions from the optimum scheduling, a value $|t_i - t^*|$ below a suitable small tolerance is assumed to imply that the i-th action is not part of the bundle represented by the individual.

In particular, we employed the Covariance Matrix Adaptation ES (CMA-ES) [20] for evolving the population of candidate solutions, which is considered one of the most successful optimisation algorithms for solving continuous optimisation problems. The adopted CMA-ES search heuristics is based on two variation operators: intermediate recombination and additive Gaussian mutation. The former recombines individuals through a weighted sum in which weighs are attributed to parent individuals according to a ranking criterion based on their fitness (i.e. it corresponds to computing the centre of the mass of individuals in the parent population). Mutation is then realised by adding a vector given by a multivariate random distribution with zero mean. In order to improve the search strategy, the covariance matrix of the mutation distribution is updated during evolution through an adaptive process which tries to learn a second-order model of the underlying objective function. Here, an elitist version of the CMA-ES algorithm was used (i.e. the best individual in the population is always preserved by the selection operators).

In general, ESs yield results which might, to some extents, depend on the particular initialisation of the population. Thus, in order to obtain more robust solutions, each optimisation process was composed of different runs of the CMA-ES, each initialised with a different seed for the random number generator.

Two constraint handlers are considered in the evolutionary optimisation process:

- a "box" constraint handler aimed at assuring that each of the evolving real-valued vector belongs to the set $[0, t^*]^n$;
- a constraint handler that guarantees the feasibility of each individual with respect to the cost-constraints defined by Eq. (5).

In both cases, specific repairing strategies are adopted in order to transform an infeasible solution produced by the search operators into a feasible one. In particular, the box-constraint handler simply replaces the individual lying outside the feasible set with the closest solution on the box boundary. The cost-constraint handler is instead based on a repairing heuristics that, starting from an infeasible scheduling, tries to repeatedly move forward in time some actions that are scheduled for the instant of time in which a constraint is currently violated. The procedure is iterated until all the cost-constraints are fulfilled. Its is worth to note that both constraint handlers may lead to the exclusion of one or more actions from the final bundle.

The numerical experiments were executed using the evolutionary computation framework included in SHARK Machine Learning Library [21].

6   Computational Experiments

The experimental setting, the results of which are hereby presented, consisted of 10 projects, with their respective parameters ε and π and all the pairwise βs (90 values).

The optimisation algorithm was ran under four basic configurational settings, combining two cash inflow rates ($\omega = 0$ and $\omega = 20$), once without and once with the effect of interdependencies. These four configurations were executed under different initial budget endowments, and with the value 20 as the target time horizon. The overview of the results is shown in Figure 2.

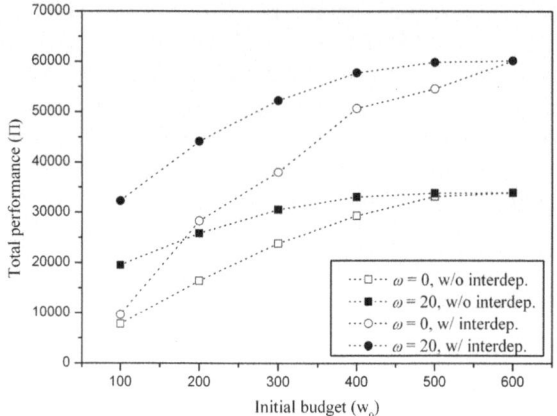

Fig. 2. Total performance for optimal action bundles calculated for 6 initial budget conditions (100, 200, 300, 400, 500, 600), two values of ω (0 and 20), with and w/o interdependencies.

Tables 1 and 2 show the actual bundles (i.e. the scheduling of the selected actions) for each computed solution. As it may be seen, notwithstanding the apparent regularity of the relation between the initial budget endowment and the total performance obtained by the optimal solutions (Fig. 2.), the related bundles yielding these performances vary quite remarkably, both among four basic configurational settings, and within each of them.

Table 1. Action bundles *without* interdependencies under different initial budget endowments (left $\omega = 0$, right $\omega = 20$), (rif. Fig. 2.). The empty dots along the timelines represent the point in time of the beginning of implementation of its respective action.

Constr.	Bundle (timeline length = 20)	Constr.	Bundle (timeline length = 20)
100 ☐	a4,a8	100 ■	a8, a9, a4, a2, a7 a6 a1 a3
200 ☐	a7,a8,a9	200 ■	a4,a6,a8 a9 a2 a7 a1 a3 a10
300 ☐	a6,a7,a8,a9	300 ■	a6,a7,a8,a9 a4 a2 a1 a3 a10 a5
400 ☐	a1,a4,a6,a7,a8,a9	400 ■	a1,a4,a6,a7,a8,a9 a4 a3 a5 a10
500 ☐	a1,a2,a3,a4,a6,a7,a8,a9,a10	500 ■	a1,a2,a3,a5,a6,a7,a8,a9 a4 a10
600 ☐	a1,a2,a3,a4,a5,a6,a7,a8,a9,a10	600 ■	a1,a2,a3,a4,a5,a6,a7,a8,a9,a10

Table 2. Action bundles *with* interdependencies under different initial budget endowments (left $\omega = 0$, right $\omega = 20$), (rif. Fig. 2.). The empty dots along the timelines represent the point in time of the beginning of implementation of its respective action.

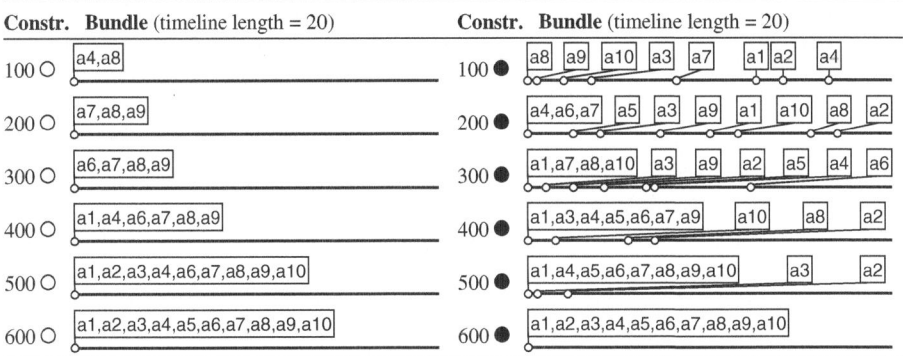

The effect of interdependencies can be appreciated by comparing the right-hand sides of tables 1 and 2. Indeed, the bundles in Table 2 (showing the cases with interdependencies turned on) present much greater irregularity than that in Table 1. While the solutions in Table 1. essentially crowd the same actions towards the beginning of the timeline progressively as the initial budget endowment increases, the solutions in Table 2. do that by significantly modifying the composition of the bundles. For example, in the solution "100●" the action "a8" is scheduled first, while the solution "200●" puts it in the second half of the timeline, after all but one action. Similar pattern can be observed in many other cases, for example in the case of the action "a6" when comparing solutions "300●" and "400●". These shifts in bundles are clearly due to the interdependencies, which add combinatorial complexity to the bundles.

7 Conclusions

The most relevant contribution of this paper should be seen in the proposal of a general framework for modelling continuous-time selection-and-scheduling problem with interdependencies. We further formulated a specific model for the purpose of conducting computational experiments. Though observations may be raised upon that *specific* model in the matter of its plausibility in various real-world situations, it is not entirely lacking of soundness, and has the fair benefit of not requiring excessive amount of input data, which often is an insurmountable difficulty in practice.

Considering the NP-hardness of the problem, it was necessary to use a search heuristic for its solution. Given the focus of this paper, we did not fully explore few other search strategies than the one hereby employed, nor did we further investigate whether an *ad hoc* devised strategy could possibly perform better. These questions are going to be addressed in future research.

Also, as a suggestion for further research, we will in future explore alternative specific models in relation to a real-world urban planning situation with possibly richer relations among performance, time and interdependencies among actions.

References

1. Roy, B.: Multicriteria Methodology for Decision Aiding Nonconvex Optimization and its Applications. Kluwer Academic Publishers, Dordrecht (1996)
2. Bouyssou, D., Marchant, T., Pirlot, M., Tsoukiàs, A., Vincke, P.: Evaluation and Decision Models with Multiple Criteria: Stepping stones for the Analyst. Springer, New York (2006)
3. Mavrotas, G., Diakoulaki, D., Caloghirou, Y.: Project prioritization under policy restrictions. A combination of MCDA with 0–1 programming. European Journal of Operational Research 171(1), 296–308 (2006)
4. Aaker, D.A., Tyebjee, T.T.: Model for the selection of interdependent R&D projects. IEEE Trans. on Engineering Management 25(2), 30–36 (1978)
5. Fox, G.E., Baker, N.R., Bryant, J.L.: Economic models for R and D project selection in the presence of project interactions. Management Science 30(7), 890–902 (1984)
6. Carraway, R.L., Schmidt, R.L.: An improved discrete dynamic programming algorithm for allocating resources among interdependent projects. Management Science 37(9), 1195–1200 (1991)
7. Kuei, C.-H., Lin, C., Aheto, J., Madu, C.N.: A strategic decision model for the selection of advanced technology. Intl. J. of Production Research, 32-v3232(9), 2117–2130 (1994)
8. Santhanam, R., Kyparisis, J.: A multiple criteria decision model for information system project selection. Computers and Operations Research 22(8), 807–818 (1995)
9. Stummer, C., Heidenberger, K.: Interactive R&D Portfolio Analysis With Project Interdependencies and Time Profiles of Multiple Objectives. IEEE Transactions on Engineering Management 50(2), 175–183 (2003)
10. Talias, M.A.: Optimal decision indices for R&D project evaluation in the pharmaceutical industry: Pearson index versus Gittins index. European Journal of Operational Research 177(2), 1105–1112 (2007)
11. Medaglia, A.L., Graves, S.B., Ringuest, L.J.: A multiobjective evolutionary approach for linearly constrained project selection under uncertainty. European Journal of Operational Research 179(3), 869–894 (2007)
12. Medaglia, A.L., Hueth, D., Mendieta, J.C., Sefair, J.A.: Multiobjective model for the selection and timing of public enterprise projects. Socio-Economic Planning Sciences 42(1), 31–45 (2008)
13. Zuluaga, A., Sefair, J.A., Medaglia, A.L.: Model for the Selection and Scheduling of Interdependent Projects. In: Systems and Information Engineering Design Symposium, SIEDS 2007, April 27-27, pp. 1–7. IEEE (2007)
14. Carazoa, A.F., Gómez, T., Molinab, J., Hernández-Díaza, A.G., Guerreroa, F.M., Caballero, R.: Solving a comprehensive model for multiobjective project portfolio selection. Computers & Operations Research 37, 630–639 (2010)
15. Dickinson, M.W., Thornton, A.C., Graves, S.: Technology Portfolio Management: Optimizing Interdependent Projects Over Multiple Time Periods. IEEE Transactions On Engineering Management 48(4), 518–527 (2001)
16. Blecic, I., Cecchini, A., Pusceddu, C.: Constructing strategies in strategic planning: a decision support evaluation model. Operational Research 8(2), 153–166 (2008)

17. Ghasemzadeh, F., Archer, N., Iyogun, P.: A Zero-One Model for Project Portfolio Selection and Scheduling. Journal of the Operational Research Society 50(7), 745–755 (1999)
18. Ehrgott, M., Gandibleux, X.: A survey and annotated bibliography of multiobjective combinatorial optimization. OR Spektrum 22, 425–460 (2000)
19. Roberts, D.L., Isbell Jr., C.L., Littman, M.L.: Optimization problems involving collections of dependent objects. Annals of Operations Research 163, 255–270 (2008)
20. Hansen, N., Ostermeier, A.: Completely derandomized self-adaptation in evolution strategies. Evol. Comp. 9(2), 159–195 (2001)
21. Igel, C., Heidrich-Meisner, V., Glasmachers, T.: Shark. J. Mach. Learn. Res. 9, 993–996 (2008)

Parallel Simulation of Urban Dynamics on the GPU

Ivan Blecic, Arnaldo Cecchini, and Giuseppe A. Trunfio

Department of Architecture, Planning and Design - University of Sassari,
Palazzo Pou Salit, Piazza Duomo 6, 07041 Alghero, Italy
{ivan,cecchini,trunfio}@uniss.it
http://www.lampnet.org

Abstract. In recent years, geosimulation models are becoming increasingly sophisticated and applied to real-world problems covering large geographical areas. As a result, they often require extended computing times. However, in spite of the improved availability of parallel computing facilities, the applications in the field of urban and regional dynamics modelling are almost always based on sequential algorithms. This paper makes a contribution towards a wider use of some high performance computing techniques, namely those based on General-Purpose computing on Graphics Processing Units (GPGPU), in the geosimulation applications. In particular, the relevant details of a parallel version of a typical Cellular Automata approach for simulating land-use dynamics are presented. Also, some computational results obtained on two typical GPU devices are discussed.

Keywords: urban cellular automata, land-use dynamics, model calibration, GPPU, CUDA.

1 Introduction

In recent years, a number of geosimulation models have been developed to better understand and predict urban growth and landscape changes. In particular, a clear trend that can be recognized from the analysis of the latest literature is the increasing size of the areas under study, which can often go beyond the traditional scale of a city [1-3], covering wider regional and nation territory [4-5] or even entire continent [6]. Furthermore, modern geosimulation models tend to be more and more sophisticated, also because they can take advantage of the increased availability of high resolution remote sensing data [7-9].

As a result, these models are often computationally intensive, since solving the real world application problems for which they have been developed can easily require long computing times.

In addition, an even more significant challenge arises from the calibration problem, which consists of adapting the model parameters to make the modelled phenomena matching the reality [10-13]. Such parameter tuning phase, which is usually carried out with automated search procedures, is often essential to improve the quality and accuracy of the model's results. Nevertheless, calibration usually involves large search spaces, the dimension of which grows remarkably with the number of parameters to be calibrated. Therefore, the calibration process is often highly computationally intensive and may require many thousands runs of the model [10]. For example, as reported in [14] the calibration of SLEUTH (a commonly used CA model for simulating urban growth [15]) for a medium sized data set can require up to 1200 CPU hours on a typical workstation. Considering that SLEUTH is based on only five parameters, it is clear that the calibration of a more complex model using a large high-resolution dataset can be infeasible using a standard computer [8].

However, while the computational needs of geosimulation models have been increasing, the same happened to the availability of high-performance computers. A variety of parallel computing systems are available today, including massive parallel computers, clusters, computational grids, multi-core CPU computers and the recently emerged devices enabling General-Purpose computing on Graphics Processing Units (GPGPU), which are multicore graphics processing units (GPU) that can perform computations traditionally carried out by the CPU. In spite of this increased availability of parallel computing facilities, as pointed out in [8] most current simulation models are still based on standard sequential algorithms. Indeed, few studies exist in the literature on the application of parallel computing to geosimulation models. Among these is the recent work by Guan and Clarke [8] where a general-purpose parallel library (pRPL) was developed and applied to speed up SLEUTH [15]. The pRPL library is based on the message-passing interface (MPI), a standard parallel programming library [16] which is usually available on massive parallel computers, computer clusters and computational grids.

This article offers an additional contribution to the research on high performance computing techniques for modelling and simulation of urban and territorial dynamics. In particular, GPGPU computing is applied to a widely used CA approach for land-use simulation based on the concept of transition potentials [1,9]. Such CA-based technique can provide future scenarios of land uses in the region of interest over a predefined period of time. By varying the inputs provided to the automaton (e.g. zoning status, transport networks, presence of facilities and services), the model can be used to explore the future development of the area of interest, under alternative spatial planning and policy scenarios. More specifically, the paper investigates the parallel speedup that can be achieved both in the case in which the land-use demand is exogenous to the cellular model (i.e. the so called constrained approach [1]) and in the unconstrained case, where the model parameters fully determines the land-use change dynamics. As a major contribution, the paper proposes a modified version of the original constrained approach [1] which allows for achieving a significant parallel speedup while maintaining all the relevant characteristics of the original model.

The paper is organised as follows. Section 2 offers an overview of the GPGPU, highlighting the specific approach adopted in this work. In section 3 the simulation models under investigation are outlined. Section 4 illustrates the relevant details of the parallel implementation that has been devised for the GPGPU platform described in section 2. Then, in section 5 the paper illustrates some results of the numerical experiments that have been carried out and section 6 concludes the paper outlining some possible future work in the field.

2 Parallel Computation Exploiting GPGPU

In the last six years, GPGPU has attracted the interest of many researchers in the field of high performance computing (e.g. [17-23]). This is mainly due to the following reasons: (*i*) the computational power of devices enabling GPGPU has exceeded that of the standard CPUs by more than one order of magnitude; (*ii*) the price of a typical high-end GPU is comparable to the price of a standard CPU; (*iii*) there has been a rapid increase in the programmability of these devices, which has facilitated the porting of many scientific applications leading to relevant parallel speedups [20,21].

Modern GPUs are multiprocessors with a highly efficient hardware-coded multi-threading support. The key capability of a GPU unit is thus to execute thousands of threads running the same function concurrently on different data (single-instruction, multiple-threads architecture). Hence, the computational power provided by such an architecture, which can easily reach a teraFLOP, can be fully exploited through a fine grained data-parallel approach when the same computation can be independently carried out on different elements of a dataset.

The particular GPGPU platform investigated in this paper is the one provided by nVidia, which consist of a group of Streaming Multiprocessors (SMs). Each SM can support a limited number of co-resident concurrent threads, which share the SM's limited memory resources. Furthermore, each SM consists of multiple Scalar Processor (SP) cores.

In order to program the GPU, in this paper we use the C-language Compute Unified Device Architecture (CUDA) [24], a popular programming model introduced in 2006 by nVidia Corporation for their GPUs. In a typical CUDA program, sequential host instructions are combined with parallel GPU code. The idea underlying this approach is that the CPU organizes the computation (e.g. in terms of data pre-processing), sends the data from the computer main memory to the GPU global memory and invokes the parallel computation on the GPU. After, and/or during the computation, the CPU invokes the copying of the computed results into the main memory for post-processing and output purposes. In some cases, the computing scheme outlined above can be enhanced including overlapping the CPU and GPU computation and overlapping memory copying with computation [24,25]. In CUDA, the GPU activation is obtained by writing device functions in C language, which are called *kernels*. When a kernel is invoked by the CPU, a number of threads (e.g. typically several thousands) execute the kernel code in parallel on different data. From the kernel code it is possible to distinguish the currently associated thread through some built-in variables (i.e. *threadIdx*, *blockIdx*, and *blockDim*). This allows to select from the device memory the data to associate to that thread (e.g. the cell of a CA). According to the nVidia approach to GPGPU, threads are grouped into blocks and executed on a SMs.

From a programmer's point of view, it is of a certain relevance to know that the GPU can access different types of memory. For example, to each thread block can be assigned a certain amount of fast *shared memory* (which can be used for some limited intra-block communication between threads). Also, all threads can access a slower but bigger *global memory* which is on the device board but outside the computing chip.

The device global memory is slower if compared with the shared memory but it can deliver significantly (e.g. one order of magnitude) higher memory bandwidth than traditional host memory (i.e. the main computer memory).

The latter is typically linked to the GPU card through a relatively slow bus. For example, in most hardware configurations accessing the host memory from the GPU can be more that 20 times slower in terms of bandwidth (i.e. Gb/s) than accessing the global memory. As a results, the parallel computation should be organised in such a way to minimize data transfers between the host and the device. For example, in some cases it is preferable to execute on the device the code which is inefficient (e.g. because that specific part of the whole computation does not fit well with the GPGPU model of parallelism) instead of running it on the CPU, if this allows to avoid large amount of CPU-GPU data transfers.

Although a simple porting of a sequential CA code on the GPU is often relatively straightforward, in order to achieve high speedups many optimization strategy can play a significant role. However, they would be outside the scope of this paper and for this reason the reader is referred to more specific literature [24-26].

3 A CA Model for Simulating Land-Use Dynamics

In order to empirically investigate the different levels of parallel speedup that can be achieved through the use of a GPGPU-based parallel implementation, two versions of a typical CA model for land use dynamics have been adopted: a constrained cellular automata model (CCA) [1] and the corresponding unconstrained version (UCA). Both models are based on the well known concept of *transition potential* [1],[9], which is a scalar value representing the propensity (or the probability in some application [9]) of each cell to acquire a specific land use. However, while in the CCA the aggregate level of demand for every land use is fixed by an exogenous constraint at each time step, in the UCA the amount of cells that are in a certain state at each time step depends on the internal model parameters (for this reason calibrating the UCA against real data can be much more difficult). As discussed later, the original CCA model [1] requires a ranking of all cells at each time step and this makes more difficult to obtain a high parallel efficiency. On the other hand, in the UCA the next state of each cell can be computed independently from the next state of the other cells. As a straightforward result, its computation can be parallelized with higher speedup.

In what follows, we first provide an outline of all the relevant computational characteristics of the standard versions (i.e. sequential) of adopted models, and then we illustrate their counterparts in GPGPU-based parallel implementations.

3.1 Model Overview

The cellular space consists of a rectangular grid of square cells, corresponding to the resolution of the data which are used as the source of land cover. Each cell of the automaton is primarily characterised by its *current land use/function*, which reflects the CORINE land cover data [27].

As proposed in [4],[5], the model includes three land use categories: *static, passive* and *active*. The cells having one of the static land uses will not change during the simulation. However, they can influence the land use dynamics in their neighbourhood exerting an attractive or repulsive effect on the active land uses. The dynamics of the latter is driven by a demand which is generated externally to the model. Instead, the passive land uses represent land available for being transformed into active land uses during the simulation.

In the present application, the actively modelled land uses include: *continuous urban fabric, discontinuous urban fabric, industrial areas, and commercial areas*. Passive land uses include most of the CORINE agricultural and natural land cover classes plus the *abandoned* state. As explained below, the transition from an active land use to the abandoned state may happen with mechanisms that are different in the constrained or unconstrained version of the model. Finally, static land uses include among others: *road and rail networks, subways, airports, wetlands, water bodies*.

To summarize, three types of state transitions are admitted by the model, namely: (*i*) from a passive to an active land use (i.e. land consumption or reusing of an abandoned area); (*ii*) from an active to a different active land use; (*iii*) from an active and use to the abandoned state.

Besides its current land use, each cell of the automaton is characterised by the following relevant properties:

- a *suitability factor* $S_j \in [0, 1]$ for each active land use. The suitability represents the "propensity" of a cell to support a particular activity or land use (e.g. it can be computed as a normalised weighted sum or product of relevant physical and environmental factors characterising each cell). Also, the suitability can be either pre-calculated in a GIS environment, and in this case it remains constant during the simulation, or dynamically computed by the CCA model itself during the simulation;
- an *accessibility factor* $A_j \in [0, 1]$ for each land use, reflecting the importance of access to the transportation networks for the various land uses or activities (e.g. commerce generally requires better accessibility than residence). These quantities are computed by the CCA module itself before starting the simulation using the street network provides as input;
- a value $Z_j \in [0, 1]$, defining the degree of legal or planning permissibility of the *j*-th land use (for example due to zoning regulations).

As mentioned above, the state of a cell also includes the *transition potential* P_j for each active land use *j*, which is computed by the transition function and expresses the land propensity level to acquire the *j*-th use.

As in every CA, each cell is characterised by a neighbourhood, namely the set of cells the state of which can influence the dynamics of the cell itself. In other words, the change of a cell state at each time-step depends on the states of its neighbouring cells. In the model adopted herein, the neighbourhood is defined as the circular region around the cell with an assigned radius ρ. The latter should be sufficient to allow local-scale spatial processes to be captured in the CA transition rules (e.g. 1 Km).

For both the UCA and the CCA models, the first phase of the transition function, executed at each step by all cells, consists of the computation of the transition potentials (one for each actively modelled land use) on the basis of the suitabilities, accessibilities, zoning, and states of the cells in the neighbourhood. In particular, the following equation was used [5]:

$$P_j = \begin{cases} A_j S_j Z_j N_j & \text{if } N_j \geq 0 \\ (1 - A_j)(1 - S_j)(1 - Z_j) N_j & \text{if } N_j < 0 \end{cases} \quad (1)$$

where N_j is the so called *neighbourhood effect*, which represents the sum of all the attractive and repulsive effects of land uses and land covers within the neighbourhood, on the j-th land use which the current cell may assume. Since, in general, more distant cells in the neighbourhood have smaller influence, in our version of the model the factor N_j is computed as:

$$N_j = I_k + \sum_{c \in V} f_{ij}(d_c) \quad (2)$$

where the summation is extended to all the cells of the neighbourhood V (which does not include the owner cell itself) and: i denotes the current land use of the cell $c \in V$, d_c is the distance between the central cell and the neighbouring cell c, and $f_{ij}(d)$ is a parameterised function expressing the influence of the i-th land-use at the distance d_c on the potential land use j. In addition, the positive term I_k, where k denotes the current land use of the cell, accounts for the effect of the cell on itself (zero-distance effect) and represents an inertia effect due to the transformation costs from one land use to another.

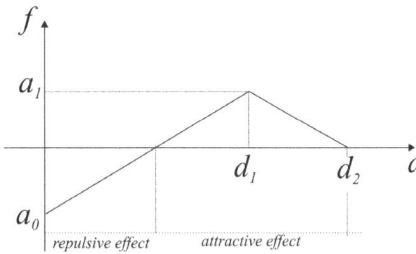

Fig. 1. The function $f_{ij}(d)$ expressing the influence of the i-th land-use at the distance d on the potential land use j

In this paper we assume that the $f_{ij}(d)$ are piecewise linear functions depending on four parameters as shown in Figure 1. Thus, if the model has n land uses and n_a active land uses, the number of model parameters is $(4n + 1)n_a$, including the parameters defining the functions $f_{ij}(d)$ and the inertial effects I_k.

Once all the transition potentials have been computed for each cell, a second phase of the transition function is carried out for the two different allocation schemes under investigation as follows.

Unconstrained transitions. In the UCA, of all the possible land uses, a cell is transformed into the one having the highest transition potential. Note that in this case the inertial term I_k of Equation (2) allows to avoid land use oscillations. Clearly, in the UCA case the transitions can be carried out locally, that is, as soon the potentials have been computed for a cell its land use for the next CA step can be assigned independently from the other cells. This also implies that the UCA scheme does not increase the asymptotic computational cost of the whole CA step.

Constrained Transitions. In the CCA, after the computation of all potentials, this phase takes place on a non-local basis and consists of transforming each cell into the state with the highest potential, given the constraint of the overall number of cells in each state imposed by the exogenous trend for that step. In practice, the cell transitions can be organised according to the procedure outlined in Algorithm 1.

Algorithm 1. sequential constrained allocation

1	$\mathcal{C} \leftarrow$ list of all candidate cells (i.e. those having a non fixed land use);
2	Sort $\mathcal{C}$ in decreasing order according to the highest potential of each cell;
3	**while** there are still cells in $\mathcal{C}$ **and** there is still demand for active land uses **do**
4	Extract the first cell $c \in \mathcal{C}$;
5	Assign to c the use corresponding to its highest potential;
6	Update the demand for active land uses accounting for the last transformation;
7	**if** (the requested number of cells of a land use has been achieved) **then**
8	Sort $\mathcal{C}$ in decreasing order according to the highest potential of each cell and ignoring the potentials for the satisfied land uses;
9	Assign the state "abandoned" to the remaining cells in $\mathcal{C}$ having an active land use

In the CCA transition strategy illustrated in Algorithm 1, the largest computational cost is due to the required scans of both the CA cells and the list $\mathcal{C}$ plus the n_a sorting algorithms.

4 GPGPU Implementation of CCA and UCA

The CA approach is known as one of the typical parallel paradigm of computation. In fact, the whole system is composed of a set of independent cells, which are influenced only by their neighbours. This allows for: (*i*) computing the next state of all the cells in parallel; (*ii*) accessing only the current neighbours' states during each cell's update, thereby giving the chance to increase the overall efficiency and making easier even the implementation in distributed memory machines. As in the sequential case, the typical CA parallel implementation involves two memory regions representing the *current* and *next* states for the cell. For each CA step, the neighbouring values from *current* are read by the transition function, which performs its computation and writes the new state value into the appropriate element of *next*.

In particular, the GPGPU parallel implementation of the CA illustrated in section 3, accordingly to the recent literature in the field [17-23], was based on the following design choices:

- one or more CUDA computational kernels (i.e. thread) are assigned to each cell of the automaton [20,21]. In case of particularly lightweight transition functions, also the allocation of multiple cells to each thread might be effective. However, in the present work this variation was not investigated;
- most of the automaton data (i.e. both the *current* and *next* memory areas mentioned above) is stored in the GPU global memory. This involves: (*i*) initialising the *current* state through a CPU-GPU memory copy operation (i.e. from host to device global memory) before the beginning of the simulation and (*ii*) retrieving the final state of the automaton at the end of the simulation through a GPU-CPU copy operation (i.e. from device global memory to host memory). Also, at the end of each CA step a device-to-device memory copy operation is used to re-initialise the *current* values with the *next* values.

In order to speed up the access to memory, the automaton data in device global memory should be organised in a way to allow coalescing. To this purpose, a best practice recommended by NVIDIA is to use of "structures of arrays" rather than "arrays of structures" in organising the memory storage of cell's properties [25]. Thus, while the typical sequential implementation of the model described in section 3 is based on structures or objects encapsulating all the cells properties, in the GPGPU implementation it is more convenient to use simple arrays to store the automaton. In particular, an array with the size corresponding to the total number of cells in the automaton has been allocated in the CPU memory for each of the properties mentioned in section 3.1. All of such arrays are then mirrored in the GPU together with some additional arrays for storing the neighbourhood structure and the model parameters. As mentioned below, some auxiliary arrays were also allocated only in device memory (e.g. the current transition potentials).

A key step in the parallelization of a sequential code for a GPGPU architecture according to the CUDA approach, consists of identifying all the sets of instructions that can be executed independently of each other on different elements of a dataset (e.g., on the different cells of the automaton). As mentioned in Section 2, such sequences of instructions are grouped in CUDA kernels, each transparently executed in parallel by the GPGPU threads.

As for the UCA model described in section 3, the computation performed at each step by each cell consists of two phases: (*i*) the computation of the transition potential according to Equations (1) and (2); (*ii*) the assignment of a new land use. Since both can be carried out independently for each cell, they were included in a single kernel, thus avoiding the overhead related to invocation of an additional kernel. Note that the source code of such kernel can be straightforwardly obtained from the corresponding sequential code, just removing the loop over the automaton cells and mapping the thread id (provided by the automatic CUDA local variables mentioned above) to the cell index for which the kernel instructions must be executed.

However, while in the CCA case the computation of transition potentials can be carried out independently for each cell, adapting the constrained allocation procedure to the GPGPU data-parallel scheme appears more difficult. For this reason, at each parallelised CCA step first the kernel *computePotentials_inParallel()* is invoked for each cell of the automaton, and then the specifically designed procedure described in the following section is carried out.

4.1 Parallel Constrained Transitions

In the constrained allocation procedure described by Algorithm 1, both the creation of the list $\mathcal{C}$ and its sorting could be carried out in parallel in the GPU global memory. However, the downwards scanning of the list (lines 4-5) must be carried out according to the list order, one cell at a time. This is because as soon as a land-use demand is satisfied, a new ranking of cells must be performed before any further cell transition. In addition, the constraints on the total number of cells to be assigned for each type of use represents a strong condition of dependency between the cells. Therefore, such sequential scan, even if carried out by the GPU would lead to a bottleneck for the whole CA step due to the low level of parallel efficiency.

For this reason, a different constrained allocation procedure has been devised, which is able to better exploit the GPU while maintaining the essential characteristics of the original constrained approach.

Algorithm 2. The proposed parallel constrained allocation. During the parallel potential computation phase all cells having an active land use are transformed into the "abandoned" state.

```
1   do
2       stillCellsAvailable ← false
3       stillLandUsesToSatisfy ← false
4       setMaxPotentialIndex_inParallel();
5       for all active land uses u do
6           ℒ_u ← createCellIndicesArray_inParallel (u) ;
7           ℒ_u ← sortArray_inParallel (ℒ_u) ;
8       for all active land uses u do
9           nToTransform ← min( #ℒ_u, reqNumOfCells[u] );
10          if ( nToTransform < #ℒ_u ) then
11              stillCellsAvailable ← true
12          if ( nToTransform > 0 ) then
13              assignLandUse_inParallel(ℒ_u, nToTransform);
14              reqNumOfCells[u] ← reqNumOfCells[u] - nToTransform;
15          if ( reqNumOfCells[u] > 0 ) then
16              stillLandUsesToSatisfy ← true
17  while ( stillCellsAvailable and stillLandUsesToSatisfy )
```

The current CUDA implementation of the constrained land use allocation, outlined in Algorithm 2, is an iterative process partly based on the *Thrust parallel algorithms library* (http://code.google.com/p/thrust/), which provides a collection of data parallel primitives for CUDA, useful as building blocks for implementing complex algorithms. As shown in Algorithm 2, the iterative procedure terminates when there is no demand for any active land use (i.e. *stillLandUsesToSatisfy* is false) or when there are not cells available for transformation (i.e. *stillCellsAvailable* is false). The first significant step of each iteration consists of computing for all the cells the index of the land use corresponding to the highest potential (line 4 of the algorithm), ignoring the land uses for which there is no demand. This is efficiently accomplished in parallel through the CUDA kernel function *setMaxPotentialIndex_inParallel()*, in which an independent thread is associated to each cell having a non-fixed land use. It is worth noting that, although omitted here for the sake of clarity, the first call to *setMaxPotentialIndex_inParallel()* could be avoided including its computation in the *computePotentials_inParallel()* kernel illustrated above (thus avoiding the overhead related to a CUDA kernel invocation). Subsequently, at line 6 for each active land use u, the function *createCellIndicesArray_inParallel(u)* creates an array $\mathcal{L}_u$ in the GPGPU global memory containing all the indices of the cells which are still to be updated and having the highest potential for the land use u. This function uses an auxiliary array including all the indices of non-fixed cells, which is pre-built by the CPU and moved to the GPGPU global memory at the beginning of the simulation. In practice, *createCellIndicesArray_inParallel()* is based on a function available in the above mentioned *Thrust* library, namely *copy_if()*, which is an implementation of the so-called stream compaction algorithm [28]. After its creation, each array $\mathcal{L}_u$ is sorted in decreasing order according to the highest potential of each cell. Also the sorting of such arrays is carried out in parallel thanks to the *Thrust* library. In the next stage of the iterative step (lines 8-16), for each active land use u the number *nToTransform* of cells that can be assigned to the land use u is computed as the minimum between the size of $\mathcal{L}_u$ and the current demand for the use u. Then, the CA cells corresponding to the first *nToTransform* indices of the array $\mathcal{L}_u$ are transformed into the land use u. The latter computation (line 13) is carried out in parallel (i.e. a GPU thread for each cell) through the CUDA kernel *assignLandUse_inParallel()*. Clearly, after each block of state transitions, the current demand for the land use u is updated accordingly (line 14). If there is still demand for at least one land use and if there are still cells to be updated, in the next iteration of the procedure the CUDA kernel function *setMaxPotentialIndex_inParallel()* computes the new indices of the highest potential for all the cells accounting for the already satisfied land uses (i.e. ignoring them). In this way, it is possible that a cell is assigned to one of the still unsatisfied uses even if the highest potential of that cell was for a different use (for which there is no demand). However, differently from the sequential constrained transition procedure described in section 3.1, in the case of Algorithm 2 this is only possible when the demand for a land use exceeds the number of cells having their highest potential for that use. This can be considered acceptable for most applications of urban dynamics modelling.

As a final remark in this section, it is worth noting that the Algorithm 2 could be formulated to express a higher level of parallelism exploiting the concept of CUDA streams. A CUDA stream represents a queue of GPGPU operations (e.g. memory copy or kernel executions) that get executed in a specific order. In particular, in some recent devices different kernels can be executed simultaneously when they are queued in different streams (up to sixteen streams are currently supported). This, in some cases allows for achieving an higher level of occupancy of the GPGPU (i.e. the actual number of threads running concurrently over the maximum number of threads admitted by the device). For example, in the Algorithm 2 a different stream could be associated to each different active land use. Therefore, the kernels at lines 6 and 7 could be queued together in each stream in such a way to execute simultaneously the computation for each land use. Analogously, the parallel assignments at line 13 could be concurrently executed in the different streams. Nevertheless, in the present application such a higher level of optimization has not been adopted.

5 Computational Results

Two CUDA graphic devices were used in the experiments: an NVIDIA Geforce GT430 and an NVIDIA Geforce GTX480. Note that both are consumer-level devices belonging to the GeForce 400 Series, which is the 11th generation of NVIDIA's GeForce graphics processing units. In Table 1 some of the relevant characteristics of the GPGPU devices are reported. The sequential UCA and CCA reference versions, implementing the same algorithms parallelised for the GPGPU, were run on a desktop computer equipped with a 2.66 Ghz Intel Core 2 Quad CPU

Table 1. Adopted GPU hardware for all carried out experiments

	GT430	GTX480
Streaming Multiprocessors	2	15
CUDA cores	96	480
Global memory [MB]	1024	1536
Core clock rate [MHz]	700	700
Bandwidth [GB/s]	28.8	177.4
GFLOPs	268.8	1344

The computational experiments were carried out on two different datasets, extracted from the CORINE land cover (CLC2000) inventory, which were already used within the European project BRIDGE for computing future land use scenarios [10]. The first dataset concerns the area of the city of Florence and is composed of 242 × 151 cells of size corresponding to 100 m. The second dataset represents the urban area of Athens and is composed of 321 × 391 cells of size 100 m. All the simulations were carried out for both the CCA and UCA models and consisted of 30 simulation steps, which correspond to 30 years of future land use projection. For the CCA, a constant 3% increment, referred to the initial number of cells, was adopted as constraint for each active land use.

In the CA models described in section 3, the effort involved in the computation of transition potentials is almost proportional to the number of neighbouring cells. For this reason three different neighbourhood radius were considered, namely $r = 10$ cells, $r = 15$ cells and $r = 20$ cells. Note that a too small neighbourhood may prevent important land use influences to be taken into account. Hence, in general it would be better to adopt bigger neighbourhoods even if this may lead to longer computational times.

Tables 2 and 3 show the elapsed times obtained for all the runs that were carried out. It is worth noting that all the results only accounts for the CA steps excluding the copying of the data between the CPU and the GPU at the beginning and at the end of the simulation, as well as any pre- and post-processing phase. In particular, the timings for the GPGPU codes were obtained using the accurate approach based on the CUDA concept of *event* which was suggested in [25].

Table 2. Elapsed times (in seconds) for the Florence test case and for different values of the neighbourhood radious r

	$r = 10$		$r = 15$		$r = 20$	
	CCA	UCA	CCA	UCA	CCA	UCA
CPU	6.94	6.71	11.59	11.37	16.90	16.70
GT430	1.40	0.62	1.67	0.94	2.00	1.28
GTX480	0.26	0.12	0.33	0.18	0.34	0.22

Table 3. Elapsed times (in seconds) for the Athens test case and for different values of the neighbourhood radious r

	$r = 10$		$r = 15$		$r = 20$	
	CCA	UCA	CCA	UCA	CCA	UCA
CPU	25.65	22.89	44.20	41.37	66.66	63.79
GT430	2.93	1.76	3.82	2.56	4.90	3.62
GTX480	0.55	0.34	0.75	0.50	0.96	0.72

Table 4. Achieved parallel speedup in different computational experiments that were carried out

		$r = 10$		$r = 15$		$r = 20$	
		CCA	UCA	CCA	UCA	CCA	UCA
Florence	GT430	5.0	10.8	6.9	12.1	8.5	13.0
	GTX480	26.8	57.6	35.5	61.7	49.9	76.7
Athens	GT430	8.8	13.0	11.6	16.2	13.6	17.6
	GTX480	46.5	67.9	59.0	82.1	69.4	89.0

In Table 4 the achieved parallel speedups (i.e. the ratio between the sequential execution time and the corresponding time took by the parallelized algorithm) are also shown.

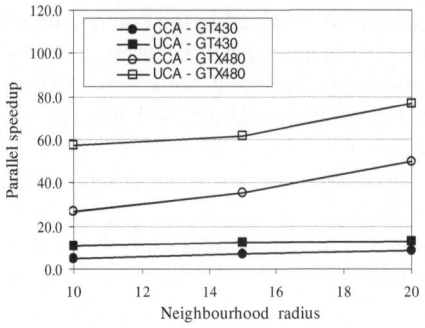
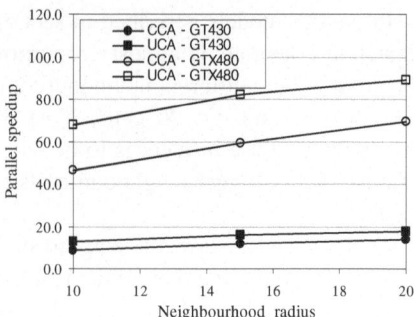

Fig. 2. The parallel speedup obtained for the Florence test case

Fig. 3. The parallel speedup obtained for the Athens test case

As it can be seen, the gain in terms of computing time is impressive. In particular, the most powerful GPU gave a speedup of 89.0 for the UCA and 69.4 for the CCA in the most computational intensive Athens test case.

As expected, the speedup of the UCA model was always superior to that achieved on the CCA model. In fact, apart from for the extended neighbourhood, the UCA can be considered as a standard CA since all the cells can be updated synchronously. This fits very well with the fine-grained data parallel approach offered by the GPGPU paradigm. On the contrary, in the CCA the constrained allocation phase, although specifically designed here, requires some parallel reduction operations (e.g. array sorting, array compaction) that notoriously can not lead to the same level of parallel efficiency.

Another important aspect concerns the different speedups obtained with different neighbourhood sizes. With the hardware used for the experiments, increasing the level of computation intensity of the kernel dedicated to the transition potentials, leads to a better speedup. As confirmed by the CUDA profiler (an analysis tool distributed with the CUDA toolkit) this can be related to the fact that by increasing the neighbourhood radius, the fraction of the total computational time used by the computation of the transition potentials (which is efficiently parallelized) grows more rapidly than the fraction of time used by the remaining computations. This is clearly shown in Table 5 for the case of the GT430 hardware.

Table 5. Percent of the time spent on the computation of transition potentials for the GT430 (results provided by the CUDA profiler)

	$r = 10$		$r = 20$	
	CCA	UCA	CCA	UCA
Florence	39.3	81.4	58.1	89.4
Athens	51.2	82.9	69,5	90.8

Interestingly, the maximum parallel speedup of 89.0 obtained for the UCA test case is comparable to the value of 98.5 obtained in [21] for a different CA model but using the same GPGPU device. The lower value obtained here can be easily explained. First, in the present application the main transition function kernel performs much more global memory accesses, due to the size of the neighbourhood, and less calculations. Second, the work presented in [21] was partly based on the *cudpp* CUDA library (http://code.google.com/p/cudpp/) which is known for offering more computational performance than the *Thrust* library used in the present work.

Overall, the results in terms of time savings are significant, also considering the relatively little effort required to develop the parallel versions of the models. It is also worth noting that, at the moment, the GT430 card costs less that 60 Euros while the GTX480 costs about 350 Euros.

6 Conclusions and Future Work

The entrance of the applications of CA models for real-world simulation of urban and spatial dynamics in its maturity requires not only theoretical advancements, better models and more accurate datasets. It also needs to take advantage of the recent innovations and developments offered by the computer science, to provide more usable, faster and easier to calibrate models and simulations.

In this sense, the hereby presented approach of parallel simulation on the GPGPU has proven promising and worth pursuing. The main advantage of such a parallelization lies in enabling an accurate calibration, which otherwise may not be possible in some cases involving models operating at regional or continental scale.

However, starting from the results presented above, ample margins of improvement are still possible and will be object of future work, since not all typical GPGPU optimization strategies have been implemented. Another important direction of research will be the transparent integration of the GPGPU approach to general purposes geo-simulation environments [29].

References

1. White, R., Engelen, G., Uljee, I.: The use of constrained cellular automata for high-resolution modelling of urban land use dynamics. Environment and Planning B 24, 323–343 (1997)
2. Clarke, K., Hoppen, S., Gaydos, L.: A self-modifying cellular automaton model of historical urbanization in the San Francisco bay area. Environment and Planning B 24, 247–261 (1997)
3. Clarke, K.C., Gaydos, L.J.: Loose-coupling a cellular automaton model and GIS: long-term urban growth predictions for San Francisco and Baltimore. International Journal of Geographic Information Science 12(7), 699–714 (1998)
4. White, R., Engelen, G.: High-resolution integrated modelling of the spatial dynamics of urban and regional systems. Computers, Environment and Urban Systems 28(24), 383–400 (2000)
5. Engelen, G., White, R., De Nijs, T.: Environment Explorer: Spatial Support System for the Integrated Assessment of Socio-Economic and Environmental Policies in the Netherlands. Integrated Assessment 4, 97–105 (2003)

6. Lavalle, C., Baranzelli, C., e Silva, F.B., Mubareka, S., Gomes, C.R., Koomen, E., Hilferink, M.: A High Resolution Land Use/Cover Modelling Framework for Europe: Introducing the EU-ClueScanner100 Model. In: Murgante, B., Gervasi, O., Iglesias, A., Taniar, D., Apduhan, B.O. (eds.) ICCSA 2011, Part I. LNCS, vol. 6782, pp. 60–75. Springer, Heidelberg (2011)
7. Armstrong, M.P.: Geography and computational science. Annals of the Association of American Geographers 90(1), 146–168 (2000)
8. Guan, Q., Clarke, K.C.: A general-purpose parallel raster processing programming library test application using a geographic cellular automata model. International Journal of Geographical Information Science 24(5), 695–722 (2010)
9. Santé, I., García, A.M., Miranda, D., Crecente, R.: Cellular automata models for the simulation of real-world urban processes: a review and analysis. Landscape and Urban Planning 96(2), 108–122 (2010)
10. Blecic, I., Cecchini, A., Trunfio, G.A.: A Comparison of Evolutionary Algorithms for Automatic Calibration of Constrained Cellular Automata. In: Taniar, D., Gervasi, O., Murgante, B., Pardede, E., Apduhan, B.O. (eds.) ICCSA 2010. LNCS, vol. 6016, pp. 166–181. Springer, Heidelberg (2010)
11. Rongo, R., Spataro, W., D'Ambrosio, D., Avolio, M.V., Trunfio, G.A., Di Gregorio, S.: Lava flow hazard evaluation through cellular automata and genetic algorithms: an application to Mt Etna volcano. Fundamenta Informaticae 8, 247–268 (2008)
12. Spataro, W., D'Ambrosio, D., Rongo, R., Trunfio, G.A.: An Evolutionary Approach for Modelling Lava Flows Through Cellular Automata. In: Sloot, P.M.A., Chopard, B., Hoekstra, A.G. (eds.) ACRI 2004. LNCS, vol. 3305, pp. 725–734. Springer, Heidelberg (2004)
13. Goldstein, N.C.: Brains vs. brawn comparative strategies for the calibration of a cellular automata based urban growth model. In: Proceedings of the 7th International Conference on GeoComputation (2003)
14. Clarke, K.C.: Geocomputation's future at the extremes: high performance computing and nanoclients. Parallel Computing 29(10), 1281–1295 (2003)
15. Project Gigalopolis, NCGIA (2003), http://www.ncgia.ucsb.edu/projects/gig/
16. Gropp, W., Huss-Lederman, S., Lumsdaine, A., Lusk, E., Nitzberg, B., Saphir, W., Snir, M.: MPI: the complete reference, vol. 2. The MIT Press, Cambridge (1998)
17. Preis, T.: GPU-computing in econophysics and statistical physics. European Physical Journal-special Topics 194, 7–119 (2011)
18. Roberts, M., Packer, J., Sousa, M.C., Mitchell, J.R.: A work-efficient GPU algorithm for level set segmentation. In: Proceedings of the Conference on High Performance Graphics, HPG 2010, pp. 123–132. Eurographics Association, Airela-Ville (2010)
19. Szerwinski, R., Güneysu, T.: Exploiting the Power of GPUs for Asymmetric Cryptography. In: Oswald, E., Rohatgi, P. (eds.) CHES 2008. LNCS, vol. 5154, pp. 79–99. Springer, Heidelberg (2008)
20. Filippone, G., Spataro, W., Spingola, G., D'Ambrosio, D., Rongo, R., Perna, G., Di Gregorio, S.: GPGPU Programming and Cellular Automata: Implementation of the SCIARA Lava Flow Simulation Code. In: Proceedings of the 23rd European Modeling and Simulation Symposium (EMSS), pp. 696–702 (2011)
21. Bilotta, G., Rustico, E., Hérault, A., Vicari, A., Russo, G., Del Negro, C., Gallo, G.: Porting and optimizing MAGFLOW on CUDA. Annals of Geophysics 54(5) (2011)

22. Pallipuram, V.K., Bhuiyan, M., Smith, M.C.: A comparative study of GPU programming models and architectures using neural networks. The Journal of Supercomputing, 1–46 (2011) (in press)
23. Krüger, F., Maitre, O., Jiménez, S., Baumes, L., Collet, P.: Speedups between ×70 and ×120 for a Generic Local Search (Memetic) Algorithm on a Single GPGPU Chip. In: Di Chio, C., Cagnoni, S., Cotta, C., Ebner, M., Ekárt, A., Esparcia-Alcazar, A.I., Goh, C.-K., Merelo, J.J., Neri, F., Preuß, M., Togelius, J., Yannakakis, G.N. (eds.) EvoApplicatons 2010. LNCS, vol. 6024, pp. 501–511. Springer, Heidelberg (2010)
24. NVIDIA CUDA C Programming Guide, version 3.2 (2010)
25. NVIDIA CUDA C Best Practices Guide, DG-05603-001_v4.1 (2012)
26. Reyes, R., de Sande, F.: Optimization strategies in different CUDA architectures using llCoMP. Microprocessors and Microsystems 36(2), 78–87 (2012)
27. EEA, CORINE Land Cover – Technical Guide, Office for Official Publications of European Communities, Luxembourg (1993)
28. Horn, D.: Stream reduction operations for GPGPU applications. In: Pharr, M. (ed.) GPU Gems 2. Addison Wesley (2005)
29. Blecic, I., Cecchini, A., Trunfio, G.A.: A General-Purpose Geosimulation Infrastructure for Spatial Decision Support. Transactions on Computational Science 6, 200–218 (2009)

Geolocalization as Wayfinding and User Experience Support in Cultural Heritage Locations

Letizia Bollini and Roberto Falcone

Department of Psychology, University of Milano-Bicocca
Piazza dell'Ateneo Nuovo 1, 20126 Milano, Italy
letizia.bollini@unimib.it

Abstract. People interact with a place by building mental models according to their priority and experience.

The geospatial and social dimensions with the massive introduction of mobile devices are changing the context use of the Web and strengthening the link between spatial, social and digital experiences in a even more *social-space-based* information architecture.

The paper –based on a case study– tries to investigate and propose by a conceptual and social-anthropological perspective, the design of a mobile application through Geo-location and Augmented Reality. The aim is to offer the *unusual* experience of visiting the *Cimitero Monumentale di Milano*, in real time & place, contextual information as biographical, architectural and artistic notes near gravestones, tombs and funerary monuments establishing a relational experience between people and the reale and digital space.

Keywords: Geo-localization, geo-localized interaction, digital territories, user experience design, mobile application design.

1 Knowledge Spaces: The Cultural Heritage Landscape

The contemporary exhibition design applied to the cultural heritage field and location is a strategic development act that means to evoke, to inform, to involve perceptually a new generation of visitors & tourists –or to better say– *cultural users*.

It defines an increasingly intertwining relational generating thick behavior layers, discoveries and reflections, which form its purpose and outcome. The contemporary exhibition design establishes an interior space relationship redefinition process that takes place with the identification of useful tools to fix and support narrative and performative practices recognized as messages carriers involved –more or less clearly– the cultural object itself. We can define the exhibition design as an *polyphonic action* project [1], whose gaze expands into space by redefining the time spent, so that relationships between container host, collected works and visitors defendants are realized in a virtuous convergence where, even through the synchronization of actions and dynamics of movement within the scene outlined, this one becomes a living environment of the narrative meanings represented.

This is *mise en scene* action based on harmony summary of three significant present together –the place, the cultural or artistic object, the user– intended to draw an interference and mutual attractions *map* which applies in all mental and physical activities interacting in exhibition enjoyment.

The *experience*, therefore, becomes a key role into the exhibit design configuration, building the exhibition event *assemblage* process between perceptual awareness and technology knowledge.

From the *reasoned deposit* (the taxonomic gallery, which meets density criteria, ordered by groups of histories and/or authors), through the *absent neutrality* (the white cube that chases the spatial disorientation of modern art domain, through the container perceptual removal) [2] and the *absent structure* (the black box that deals with space cancellation defined by the video-art), the exhibition action definitively move towards multi-modal practical coordination forms, searching for the possible development criteria of interaction interdisciplinary subsidiary processes, which transform the empty *cavea* interior space prepared for a generic entry of objects into a simultaneity of events place [3].

2 Multimodal Design: Reshaping the Space of the Interactive Experience

The contemporary exhibition design is of a particular communication medium where "consumers are encouraged to seek new information and to start up connections between different media contents" [4] The things you can *see* clearly, but especially things you can learn by your *sense*, change in exploration and understanding resources which can then be shared by exchange ways related to social interactions most appropriate to each visitor. Digital technologies are increasingly present in the contemporary exhibit world both as a *tool* in communication media both as *objects* itself. New perspectives thus open in the field of research into new languages, potentially capable to find a dimension that exploits the potentiality of information science support, not as a simple decantation of a consolidated culture towards a context still to explore (such as the many already diffuse stereotypical products that recall in a reassuring manner the morphemics of *similar* medias, such as multimedia that reduce the hypertextuality to a banal *defoliation* of screens) but as original research which experiments with new and adapted expressive solutions. As Papert [5] remarks, the main characteristic of the computer as a *medium* is its universality: i.e. the capacity to simulate other media, that is the capacity, in other words, to integrate in a single support more media channels. Furthermore the new mobile and smart devices are a *meta-medium* that can be seen – rather ambiguously - as an *instrument* and as a *mode*. The reflection of the terminological use of the word *medium* allows a noticeable change of perspective, implying to outline two poles upon which the design can be oriented: the slope of *support instruments*, that is of *media*, and that of *languages*, or modes of communication.

Then let's speak rather about a single *medium* that constitutes an integrated platform between a notational system and other channels of communication, or better: between modalities of communication. Multimodality, then, means to contemporaneously exploit various and complementary channels of communication.

This permits an interaction between the digital technologies and the users that mirrors the physical and cognitive richness of the interlocutor human [6]. In fact, also the perception, like the communication, is a cooperative and multimodal process: our senses are not *windows*, autonomous from the phenomena of the world that is around us, on the contrary, they constitute a collaborative system that allows us to realize a conscious coherence of reality. Similarly, also the semiological activity, as Donald Preziosi sustains, is multimodal because it involves the orchestration of signals originating from a variety of modalities existing together [7]. Therefore, the design of the communicative modality must correspond to this complexity of the semiological system, continually busy in the decodification and interpretation of percept, so that the perspectivity channels –auditory, tactile, visual– are aligned in the reception of information that precede from the communicative ambit. The multimodal approach to the exhibition space & design becomes then, not the non-exclusive point of view of the project, but the privileged one.

It could be argued that the contemporary exhibition design is an integral part of a generation system of *collective intelligence* [8] which is one of the organization forms, even more valuable because physiologically linked to the diffusion of a *high* knowledge. Contemporary exhibition design, to paraphrase Derrick De Kerckhove, could be define "The practice of multiple intelligences in relation to each other in real-time & place experience of a project […] gives immediately people the experience of their intelligence collectively in their [social] group. And it is pleasant to live because it is a new learning experience, or rather: a old cognitive experience which, however, is awareness accelerating and enriching." [9].

3 Locative Media and Cultural Heritage Scenario

Spaces with a cultural vocation, as a museum or open-air cultural parks, lives to communicate; more precisely, succeeds to be alive and vital if communicates effectively the collection of knowledge that has produced over time, in relation to historical and artistic recollections that compose his collection, pursuing what can certainly be defined his ultimate goal: contribute to the cultural growth of a society.

The communication, therefore, arises as instrumental action aimed to the activities, generation and dissemination of knowledge in favor of their audiences. Technologies, in this respect, gave a great impulse to the communication activities of the cultural heritage collection and location, offering a wide variety of channels through which to convey their information flows: recently, in fact, have multiplied the technological solutions able to improve the level and quality of interaction with this scenario using the Proximity Based Interaction, i.e. the interaction based on physical proximity of an individual with respect to an object.

New modes of interaction pragmatic-proxemics are defined for mobile devices of last generation, and exploiting the ability to know their position in space, have proved essential to interfere in the least intrusive manner possible with natural and proven habits of visit to a museum.

Using technologies such as NFC or QRCode, Augmented Reality, the visitor is able to receive, in addition, in a completely automated way, information flows also of multimodal type, as soon he is close to an POI.

Therefore, the visitor has no longer the need to activate a search procedure, more or less simple, of the information available in relation to the work that is observing. As the visitor approaches to the object an active dialogue between the device he is using and the object is activated, allowing the unique identification of the object ; at that point,the device triggers the system of automatic selection of information, which will forward the information contents required by the user.

The mobile device, permanent data recorder, becomes also heirloom, support and vehicle of a memory that exceeds the short time of the visit, an effect closely linked to that of personalization on the one hand and the collective re-appropriation of learning from the other.

Not only is the summation of documentary and encyclopedic basis on a global basis to build a network of interrelated issues and institutions, universally accessible but also the set of comments and annotations of visitors, as well as traces of their experiences within the museum space, forms a melting pot of knowledge that will not only upgrades the epistemological foundations of museum institutions, but it also redefines the shape using content that often go beyond the based on merely factual knowledge, presenting as true and own guides to action.

In the face of indisputable advantages arising from the use of mobile devices of last generation in the context of open air museums, the risk that appears more obvious, due to the splitting of the reading plans, can be recognized in that sort of swinging attentional, linked to the continuous jumping and the multiple references between the digital and physical plan, for which the subject is subjected to greater cognitive effort.

As a result turns out to be of fundamental importance the search for a proper balance, aimed to not distract excessively the visitor from poetry of exhibition spaces.

The logics, resulting from the discipline of the Context-Aware computing, adaptability-ability to recognize the user and remember the user preferences-and adaptivity -ability to update the interface interface and/or attributes of the context-necessarily must blend in an optimal performance.

4 Sentient Places and User Experencies

The increase of the computational power of the digital artifacts and the parallel miniaturization process of the computing devices, led to the integration of functions of, communication, sharing and perception of the surrounding reality exploiting devices of small dimension and of daily use.

Such process led to the accomplishment of design and production patterns that do not have any historical precedent, destined to create substantial novelty and to change radically the manner in which the people approach to the daily reality and to the use of the digital technology.

The point of beginning and the flywheel for the achievement f of this tecno-social revolution was the wide diffusion of the technologies of localization and of the relevant services Location Based to the mobile devices, where they met the world of the sharing of information on the net.

The supporting idea is that every object of this type, thanks to an identity and to a metrics that marks them in every moment in the time and in the space, is able to generate some stories - memories of the interactions. The memories of interaction are getting interwoven allowing unparalleled analysis of the data.

The process of interpenetration of atoms and bits, in addition to providing enhanced interaction with the surrounding environment, gives voice to the relationships between people and places and the objects that surround it, that become in this way stories, narrations that can be documented and worked out in real-time.

SpIme [10] artifacts arise from this perspective as mediators between the people and objects traditionally perceived as being inert, inanimate and ultimately providing them the chance to become sentient and to gain intelligence, able to convey information about itself and the context they are inserted and to access other aggregated information available on the net.

Devices of this type connect digital contents to geographic places, exploiting the character localized and geographically placed of the interaction. In consequence we are dealing with digital media applied and strongly connected to physical places, but technically independent from the position (of the device), in which the informative content has to to be georeferenced and produced from the space position (of the object). Associating objects and places of the physical world to their digital counterparts, objects are removed from their natural boundaries, ensuring an enhanced interaction between man and environment.

Taking into consideration the [11] media and the related practices from a sociological point of view, it is interesting to notice that in this scenario the places and physical environments, traditionally defined as real, are transforming into media-spaces in which forms and modes of social and technical mediation are pointing out (highlighting) the dynamical processes for the construction of places and communities located within the contemporary geomedial landscape.

A spatial context mapped and crossed through by practices that make use of locative media, therefore, appears both as the theatre both as the product of new forms of social interaction and cultural, a distributed agency where people converge, things and places, ever and ever mutually, self conscious. The current referencing and mapping systems are a new space for communication and support the narrative, acting as a link between virtual and real, placing the digital information in the physical world, and vice versa.

The locative media also play an important role as a tool for expression of the peoples and of the society, triggering dynamics related the recent phenomenon of neo-geography, a term which refers to a different set of practices that operate in a different way , or in the same way or similar to those of professional geographers. Rather than refer to scientific standards, methodologies of neo-geography are directed towards the intuitive, the expressive, the personal , the absurd and/or artistic, but may be simply the application of *real* geographical techniques.

The prefix *neo* has a lot to do with the technological and directly applicative aspects, in particular focusing on tools that allow neo-geographist to gather data and contents related to physical locations, allowing the achievement of an unprecedented democratization of geographic information promoted and carried out starting from the bottom. Using geo-tagging, namely. the manual entry of a spatial reference to the system, the neo-geographist are able to realize cartographic collaborative maps extremely detailed, pointing out geographically relevant content reporting and and registering the tracing of places and routes.

5 The Cimitero Monumentale of Milano: An Open-air Museum

The great monumental cemeteries of 19th-century European style, often become pantheons of illustrious passed peoples, has provided architects and sculptors a prestigious occasion to discuss and explore the wide range of feelings and emotions related to the solemn themes of death and Memorial, through the deepening of the design language, attention to proportions and balances, using different materials and techniques, which made over time these places real artistic parks of immeasurable value. The grandeur of the composition, the attention to detail and efficiency of functional solutions have made the Cimitero Monumentale di Milano as a model for the funerary architecture and one of the most famous and appreciated Italian eclecticism's accomplishments. The complex has been built over an area very wide, articulated and extremely varied, consisting of large squares, tiered galleries, arcades, advanced bodies, recesses, long corridors.

In the Monumental you can weave multiple readings: in it you can see the steps of various artistic seasons, but also the most self portraying history and image of the city through the famous names including buried here in different sections of the complex, which contribute to form a sort of collective identity, cultural, and historical of the metropolis of Milan. Versatile institution of culture and historical memory, the Monumentale di Milano needs to be enhanced and recovered, allowing visitors to walk the paths of memory that the site holds.

5.1 Features of the project

The application proposed aims to resolve some issues related to the experience of proxemics interaction within the Cimitero Monumentale di Milano.

Fig. 1. Erostrato: cover, UGC, navigation toolbar and narrative sequence based on bios

The mobile application using geolocation and Augmented Reality, will be able to:

• allow the visitor to search for the grave of a passed men or a monument
• show the path to the selected destination
• provide biographical information about famous passed men and for less famous too
• provide information on artistic monuments
• suggest other related points of interest
• allow the sharing of information about the visit

The POI discoverables by the application within the scenario, refers to to burials, architectural and artistic elements , specific places of complex (outputs, etc).

5.2   Interaction Modalities within the Scenario: User-Centered Approach

Starting from a logic focused on visitor interaction patterns [12] with Museum outdoor scenario and on analysis of the information needs and the user's point of view, the system allows to interact with the context through proxemics as follows:

• **biographical itinerary:** through this mode the system directs the user during the visit of the tombs of illustrious personalities buried within the cemetery (Fig. 1)
• **artistic journey:** through this mode the system directs the user during the visit of the tombs and monuments more interesting from an architectural or artistic point of view present inside the cemetery: usually organized as free path: through this mode you can view the most relevant items for biographical and artistic interest (Fig. 1)
• **map mode:** this mode the system displays on a map all points of interest, biographical and artistic related to the structure (Fig. 2)
• **search:** through this mode you can search for a specific point of interest: pointing the camera pointing device in the direction of the object, you can get contextual information about a point of interest: this mode is not restrained to a predetermined path. (Fig. 2)

The itineraries have been designed to meet the presence of structured criteria associated with the *Ant* visitor's own visit and can be limited for time interval, thus offering greater possibilities of customizing the visit. Map mode is suitable for the type of *grasshopper* visitor, that can be driven by previous knowledge of the scenario. Point and search mode fits the *butterfly* visitor needs, driven solely by the attractiveness of the objects and their own instincts.

5.3   Linking, Inserting and Modifying a POI: A User Generated Strategy

The inclusion of points of interest is carried out directly by the end user, using the same application, using the so-called shadow information. The user can easily create the biography of a dear and insert multimedia content, getting first-person witness and guarantor of memory associated with the passed people. By the same mean, the user can enter information relating to a monument or a artistic heirloom. The inclusion of a new point of interest occurs in the proximity of it and pointing the device in his direction; in this way the corresponding geographical coordinates will be registered.

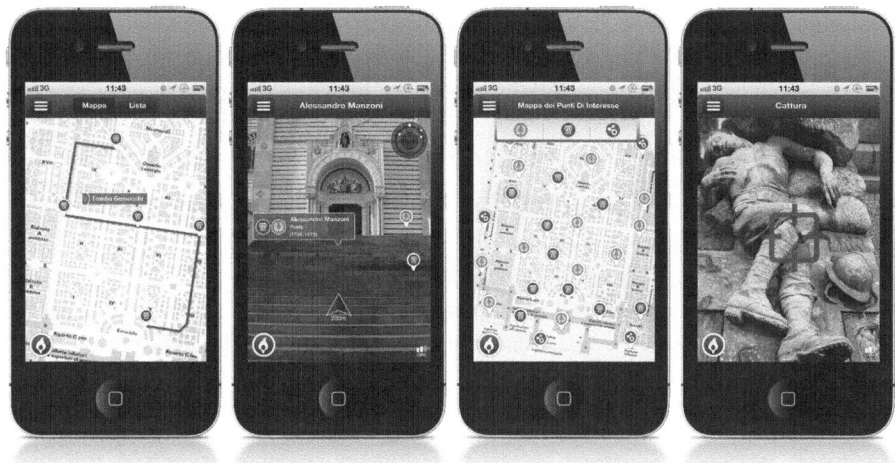

Fig. 2. Itinerarie, AR content, gelocated POI, spatial search mode

5.4 Geolocation and Wayfinding Support

The use of location technologies allows to brilliantly manage what, within a number of studies on the relationship between technology and cultural heritage location, has been identified as the first problem that tourists must face in preparing to visit a space within which is standing a widespread and spatially deployed quantity of points of interest: the findability of the works, their traceability within the overall scenario.

If the device is a valuable aid to the cognitive visitor orientation, his compass function has been enhanced even more as spatial tool, allowing not only the integration of topological maps of the structure, but offering in addition the opportunity to determine the users position within the map exploiting the sensors for localization. Thanks to which the user is able to navigate and to receive directions relative to the place where he will eventually go. A service of way-finding is of fundamental utility in the case of large structures, where it is not unusual to experience a sense of getting lost, physical or intellectual, that the eventual presence of signal systems is not always sufficient to dispel.

In addition, if the user has made a preparatory selection of works to observe, or the user has selected one of the exhibition proposed routes (Route planning), he will have the opportunity to be directed by the device during the visit, thus optimizing time spent inside the structure. This possibility open spaces of great interest from the point of view of behavior analysis and mode of use of the works exhibited by the visitor, as the device is of course capable of storing all the works that the user approached and is able to automatically generate the mapping of locations the user has visited.

6 Conclusions

The introduction of mobile devices will change the way we relate with the world and Internet supporting a more and more wide concept *contextual* experience. The possibilities offered by the geo-location, made accessible and usable also to a larger public, the different channels and modes of communication are enanching an experience more intuitive and *natural*. Locative media and user generated content, dissemination and condivision of local POI, content and –most important– emotion and personal commentes are buinding a social layer shared experience over the spatial and the individual one.

This living *plot* is one of the most promising for the design of applications that use the space and its location as a starting point for a complex esperiential path that leads within the here-and-now also of an *other* time-and-place typical of the artistical and historical sedimentation. Mobile devices will lead the third generation of IT tool to improve exhibition design and UX in cultural heritage and museum field.

Note: although the paper is a result of the joint work of both authors –Letizia Bollini and Roberto Falcone– Letizia Bollini is in particular author of part 1, 2, 3 and 6 and Roberto Falcone is author of part 4 and 5.

References

1. Borsotti, M.: Fruttiere: il progetto di allestimento tra luogo e arte. In: Borsotti, M., Satori, G. (eds.) Il Progetto di Allestimento Ela Sua Officina, pp. 36–57. Skirà, Milano (2009)
2. O'Doherty, B.: Inside the White Cube: the ideology of the gallery space. Lapis Press, Santa Monica (1986)
3. Bollini, L.: MUI: design of the HC Interfaces as a directing of communications modes targeted on human senses. In: Senses and Sensibility in Technology, Conference Proceeding, pp. 182–186. IADE, Lisbona (2003)
4. Jenkins, H.: Convergence culture. New York University Press, New York (2006)
5. Papert, S.: Mindstorms: Children, Computers and Powerful Ideas. Basic Book, New York (1980)
6. Bollini, L.: Registica Multimodale: il design dei New Media. Maggioli Editore, Santarcangelo di Romagna, RN (2008)
7. Preziosi, D.: Advantages and limitations of visual communication. In: Krampen, M. (ed.) Visuelle Kommunikation und/oder Verbalekommunikation?, pp. 25–35. Georg Holm Verlag, Hildesheim (1983)
8. Lévy, P.: L'intelligenza collettiva. Per un'antropologia del cyberspazio. Feltrinelli, Milano (1994)
9. De Kerckhove, D., Levy, P.: Due filosofi a confronto. Intelligenza collettiva e intelligenza connettiva: alcune riflessioni. Mediartech (1998), http://www.mediamente.rai.it/home/bibliote/Intervis/d/dekerc05.htm
10. Sterling, B.: La forma delle cose. Apogeo, Milano (2005)
11. Bleecker, J., Knowlton, J.: Locative Media: A Brief Bibliography And Taxonomy of Gps-Enabled Locative Media. Leonardo Electronic Almanac 3(14) (2006)
12. Levasseur, M., Véron, E.: Ethnographie de l'exposition. Bibliothèque Publique d'Information (1989)

Climate Alteration in the Metropolitan Area of Bari: Temperatures and Relationship with Characters of Urban Context

Pierangela Loconte, Claudia Ceppi, Giorgia Lubisco, Francesco Mancini, Claudia Piscitelli, and Francesco Selicato

Polytechnic of Bari, via Orabona, 4, 70124, Bari, Italy
(f.selicato@poliba.it)

Abstract. Urban planning exerts influences on environmental, social and economic system related with the city. Processes of land transformation and city growth determine radical changes in urban landscape morphology and as a consequence they affect air temperature and energy exchange. The urbanization can bear on local climate more intensively than the global warming does. This is due to the rapidity of human made changes related to natural ones. The present work analyzes the state of the art of studies involved in studying urban climate anomalies – Urban Heat Islands (UHI) – in order to explore relationship between urban planning morphology and urban climate. The research aim wants to indentify how urban geometry can be related with climate alterations in order to provide guidelines for planners to frame urban form planning and environmental quality. The case study is the city of Bari, located on the coast of the Mediterranean sea, in Apulia, one of the regions in Italy.

Keywords: Urban heat Island, urban morphology, GIS.

1 Introduction

The alterations of urban climate rely with the influence of urbanization on local climate itself and a careful urban planning could exert and influence on such phenomena and their oscillations. The growing of settled areas and the related land transformation processes determines radical changes in urban morphology and landscape. The influences of these processes is able to affect local climate in a more intense way than global warming does [1]. In order to consider the urban climate aspects in the planning phases, an effective management of the urban environment and a new approach to the governance of our urban areas are required [2]. Moreover, due to the complex heat exchange phenomena between the micro and meso scales, such an approach has to be necessarily designed at a varying scale. Local energy balances are also fundamental in the quality assessment of urban environment and the proneness of a city (or portion of it) to be an energy generator and/or an energy heat sink should be addressed. The alteration of urban climate (see for example the Urban Heat Island effect) could be investigated by a network of sensors able to monitor the climate parameters. Unfortunately such an infrastructure is rarely adopted by the municipalities and time series of meteo-data at a suitable spatial density very local scale are not commonly available. This paper is focused on the quantitative relationships among the urban heat, as derived from satellite survey, and several urban features, able to represent the main morphological features. A comprehensive study on alteration of thermal urban climate of a city should include the anthropogenic source of heat [3]. However, a quantitative assessment of such a source of energy is very difficult to achieve. On the other hand, the outgoing amount of thermal radiation sensed by the satellite includes the contribute of anthropogenic activities. The case study area is the city of Bari, Italy.

2 Literature Review

2.1 Background

Alterations of urban climate are ideally related to the Urban Heat Island (UHI) phenomenon. An increased sensibility by inhabitants towards the items of urban well-being has led to an increase of analysis and studies devoted to the understanding of processes able to influence the urban climate and implementation of mitigation strategies. The early studies on climate alterations of urban areas have been focused on the UHI phenomenon. Literature is available on the case study of London [4] and through several others international case studies [5], [6].

In this papers the phenomenon has been investigated in relation to urban-rural dichotomy, obtaining temperature differences between urban and rural areas. Some studies have been focused on the pre-urban temperature and its connection with the current ones. The differences of temperature with respect to the pre and post- anthropogenic actions have been therefore assessed [7]. However, the lack of data representing the urban shapes and a scarce dataset of temperatures did not allow

a complete understanding of the phenomenon and, nor, the spatial correlation between urban/rural features and thermal properties has been achieved. The concept of "rural" and "urban" remains very heterogeneous. Previous works, [8] have demonstrated that one-third of researches doesn't provide a description of the characteristics of sites, simply classifying them as urban or rural, while two-thirds provide only a qualitative descriptions. Under this perspective, researchers have theorized models able to distinguish the city from the countryside using satellite derived data.

With reference to the UHI development, various approaches have been proposed to understand the possible causes and find alternative mitigation strategies [9]. The driving mechanisms could be synthesized in the following main points [1]:

— Prevalence of the sensible heat flux on the flow of latent heat, caused by surfaces imperviousness and poor vegetation;
— More gradual decrease of latent heat flow with respect to trends related to rural areas;
— Release of heat by the structure of urban area occurring in the late evening and during the night;
— Contribution of anthropogenic heat as additional source of energy.

The attention has generally been focused on the individual parameters that influence the phenomenon from the macro-scale to micro-scale, like building geometry and morphology, surface materials, permeable and impervious horizontal surfaces, vegetational surfaces, electrical energy consumption, road traffic [1], colors, roughness, moisture content and canyoning effect .

In order to detect the genesis of temperature differences between one zone and the other, literature provides several methods of discretization of the landscape in homogeneous zones [10]. A possible territorial division is based on land use and vegetation covers [11]. Many researches has also shown the important role played by vegetation and urban green areas in the reduction of the use of energy for the daily needs and in the decrease of the UHI effect [12].

Several cases of study have faced the dependencies between land use changes and surface temperature. Experiments and statistical models have demonstrated that a reduction in temperatures in residential and urban areas is possible by a clever location of green areas [13].

In addition, the available literature addresses the typology of parameters used in the zoning of urban areas. One of the most relevant indicators is the urban morphology, because of its influence on many others parameters, including temperatures. A proposed classification scheme divides a generic urban area into three main types, based on the proximity of the buildings, which are further differentiated into 17 subtypes according to their function, location, height, method and age of construction [14][1]. This scheme is the base of an adopted classification of the Urban Climate Zones [15].

A successive classification divided the landscape into four series (City, Agricultural, Natural and Mixed), according to surface disturbance, caused by natural (trees, grass, soil) and anthropogenic elements (buildings, roads, crops) [16]. Each category is further divided into classes, according to the surface properties in micro-scale (10-100 mt): DRC Davenport Roughness Class [17], amount of impervious surface, Sky View Factor, thermal admittance, albedo, anthropogenic heat flux. The combination of the values defines the Local Climate Zones [18].

Moreover, literature provides many indicators to explain different morphologies and their relation with micro-climate conditions. The three-dimensional properties of different urban fabrics has been introduced into a set of indicators based on heterogeneous data (building density, absolute and relative rugosity, porosity, sinuosity, occlusivity, compacity, contiguity, solar admittance, mineralization) [19].

2.2 Urban Climate and Natural Systems

In the analysis of urban climate alteration as effect of changes in the natural systems three main perspectives could be identified. The first perspective focuses its scientific production on the relationship between thermal properties and urban issues; the second one explores the effects and the mitigation of green areas and natural city pattern on UHI; the third analyses the impact of climate alterations on biodiversity and human health. All studies are oriented on the evaluation of temperature with different technical and methodological approaches (i.e. GIS modelling, temperature sensing etc.).

The work focuses on the first two items, in order to understand how urban fabric and natural urban areas could be modeled and related to the climate alterations. The process of urban planning is related with climate being the central purpose of planning to create an environment suited to human need [3][20].

Both global and urban climates affect urban morphology, inhabitants' health, comfort, social life, and energy consumption, as climatic variables such as solar radiation, air temperature and wind are vital aspects of the functional and psychological components of a living place [21]. The green areas of a city determine trend and mitigation effects on climate and reduce the amount of pollutants in the environment.

Studies of applied ecology to urban areas shifted their focus and have been oriented to a better understanding of city/nature relationships. The knowledge of environmental system in the city is at the base for understanding urban ecological processes. Ecology in the city has been replaced by the perspective of the ecology of the city: in this way the urban system could be seen as sociological, and ecological frame in order to link urban planning and traditional ecology [23]. The positive impacts of green spaces on different climate issue such as air quality, hydrology, energy reduction and biodiversity has been proven [10] and the landscape planning has recognized the link between green space provision in the urban environment and environmental quality [24]. A green areas fragmentation can modify the thermal change capacity, reducing evapo-transpiration [22] whereas the high quantity of paved surfaces, with respect to green ones, alter the hydrological regime and the volume of superficial run off [24].

3 Methodology and Dataset

3.1 Case Study Area

Bari (41°7' N, 16°51' E), is the capital city of the Province of Bari, Apulia Region, Italy. Covers an area of 116,20 Km^2 and the resident population has been estimated in 326.915 habitants (up to 2006) with a density of 2813 ha/sqkm (PSTB, 2015). The climate is characterized as Mediterranean (after Koppen, 1936), continental with mild winter and hot dry summer. The presence of a long coastal area (12,79 km) determines a mitigation of the local climate. The area has been selected because of some interesting landscape features able to affect urban climate: the presence of the old core, the downtown area, the coast extension, the urban morphology and the presence of natural elements able to link agricultural systems with the city.

3.2 Available Data and Software Used

The ASTER-VNIR (Visible and Near Infrared) and TIR (Thermal Infrared) granules at 15 and 90 meters of ground resolution respectively have been used as dataset. NDVI and others indexes have been derived from the VNIR sensor whereas the thermal imagery have been retrieved from the 5 channels TIR unit. A dataset of diurnal

ASTER imageries spanning on a period from July 2001 and July 2006 has been selected for this study. The L1B Registered Radiance at Sensor level of processing was selected during the data purchasing. The transformation from the radiance levels to vegetation indices and temperature values were carried out by ENVI algorithm. In particular, the temperature-emissivity separation algorithm has been used to derive the temperature map from the thermal radiance at sensor.

3.3 The processing of Thermal Data

The interpretation of a thermal scanner image has to take into account several factors related to the ground physical properties and possible interactions between the thermal radiation and the atmospheric layer. Basically, the aim is the transition from the radiometry at sensor, expressed at an initial stage as Digital Number (DN), to the radiant temperature (the one that is remotely sensed by the thermal scanning) and, therefore, the kinetic temperature of the earth surface has to be derived. The latter is mostly related to the energy levels proper of the molecules constituting a body [25].

The ASTER Level-1B data are offered in terms of scaled radiance and, thus, the radiance at sensor can be obtained from DN values using the Unit Conversion Coefficient (Normal Gain) of each band [W/(m^2sr μm DN)] provided with the ASTER images. The processing chain proceeds with the atmospheric correction in the thermal infrared bands. Considering that meteo data across the area were not available at the epochs of acquisition, the ISAC (In-Scene Atmospheric Compensation) atmospheric correction algorithm (provided by ENVI software package by Research Systems Inc.)

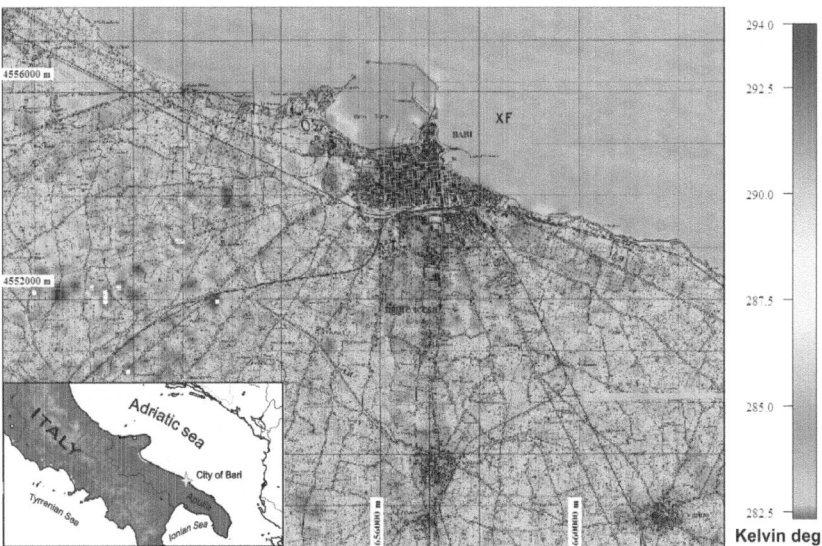

Fig. 1. Thermal image (K) of the city of Bari after the processing of thermal data acquired on 02.01.2006

was used. After correcting the atmospheric effects, the Emissivity Normalization algorithm was used to separate emissivity and temperature values from thermal infrared radiance data [26][27][28]. This technique infers the temperature for every pixel and band in the data using a fixed emissivity value (ability of a real body to emit in comparison to that of a blackbody). The highest temperature for each pixel is used to calculate the emissivity values using the Max Plank function after inversion. The Emissivity Normalization techniques is proven to be accurate, being possible to recover the temperature of the majority of the artificial radiance spectra to within 1.5K [27]. Results provided by the analysis of the ASTER thermal data over wintertime (February) and summertime (September) periods are reported in figures 1 and 2 respectively.

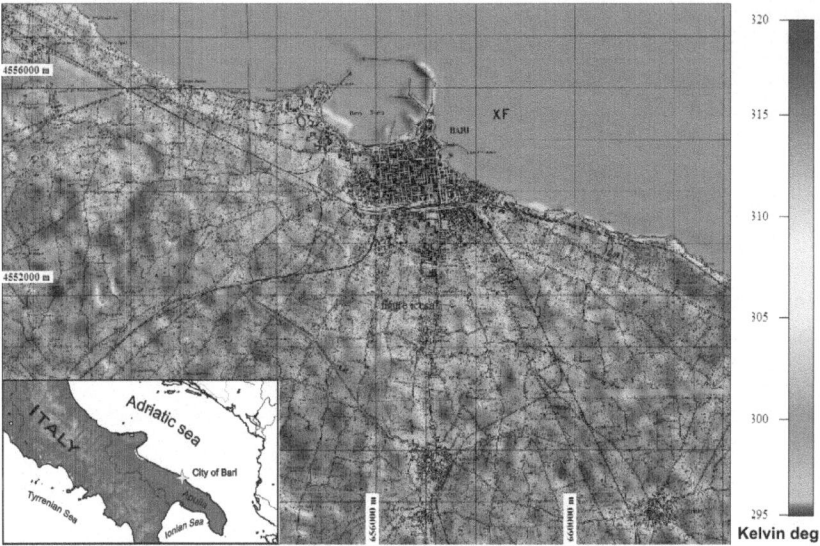

Fig. 2. Thermal image (K) of the city of Bari after the processing of thermal data acquired on 08.09.2001

3.4 Urban Morphology

The goal of this work is to understand the mechanisms able to explain the thermal climate alteration in the city of Bari and investigate the spatial relationships between satellite-based temperatures and the urban morphological features. However, the GIS approach used in this paper requires the definition of indices able to represents the urban forms and the way the area is settled. This task could be very difficult because the urban landscape arises from cultural traditions, and social and economic forces over time [29]. The following points related to the urban form could be summarized.

1. Urban form is defined by three fundamental physical elements: buildings and their related open spaces, plots or lots, and streets;

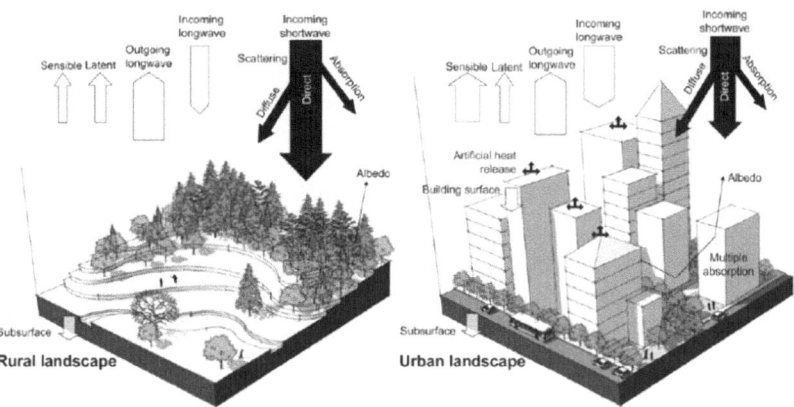

Fig. 3. Schematic depiction of radiation and energy fluxes over rural (left) and urban (right) landscapes on a clear day. The width of the arrows approximate the relative size of the flux [3].

2. Urban form can be understood at different levels of resolution. Commonly, four are recognized, corresponding to the building/lot, the street/block, the city and the region;
3. Urban form has to be considered as a dynamic object, since the elements of which it is comprised undergo continuous transformation and replacement [30].

Our attention is on the physical elements which compose the city and their features (geometry, density, typology, functions). These properties related to urban morphology have been derived by the official numerical cartography provided by the Apulia Region and National Department of Education and Scientific Research [30]. Different typologies of tissues have been recognized within the city of Bari. Tissues are group of buildings, open spaces, lots and streets, which form a cohesive whole either because they were all built at the same time or within the same constraints, or because they underwent a common process of transformation [30].

3.5 Relationship between Urban Environment and Temperatures: Factor Analysis

Two ASTER images acquired on Sept. 09 2001 and Feb. 01 2006 have been processed to produce thermal images. Images are based on 90 x 90 mt spaced grid and the analyses of the urban structure have been conducted using this value as reference parameter. Once the thermal images were produced by the ENVI package, all the dataset has been managed within the GIS. In order to characterize urban structures, a set of factors able to affects urban climate and UHI phenomenon have been recognized: Land Use, Urban Morphologies, Site Coverage and Absolute Rugosity.

Originally, the Land Use was derived from Geographic Information System provided by Puglia Region. It includes a categorized map of the land use where four classes are defined depending on the agricultural use they are dedicated to: arable land, permanent crops, meadowlands and mixed agricultural land. Successively, such classes were synthesized by grouping into classes with relevant features with respect to the thermal studies (i.e presence of tree canopy, predominance of herbaceous coverage). In this perspective arable land and meadowland have been fused into a single group. They are indicated as agricultural herbaceous coverage. Permanent crops and mixed agricultural lands are indicated as agricultural arboreal coverage.

The available natural land typologies, woodlands and lands with prevalent coverage by shrub and herbaceous, have been processed in the same way.

Green Urban Areas include all the different green typologies recognized in the urban tissues: cemitery, urban parks and sport fields.

The quantitative analysis of the Urban Morphology required the recognizing of the urban tissues by taking the 90 mt post spacing as reference spatial resolution. Thus, each cell is related to a reference tissue representing the area with its percentage of surface occupied. Beyond the urban tissue, industrial and mixed functional tissues have been recognized by the analysis. The following classes of residential tissues have been recognized:

— Old core: historical tissue of the city, characterized by high density, irregular texture of urban block, traditional building constructed until XIX century, located in the central core of the city;
— Regular high compact tissue: structured on regular grid, built in XIX century and characterized by high density, regular texture, regular urban block, located in the central core of the city, external to the old core;
— Medium compact tissue: structured on a semi-regular grid, built during XIX century and characterized by high density and bigger block dimensions, located near some main infrastructure of the city (principal street, railway);
— Low compact tissue: structured on a semi-regular grid, built during the XXth century and characterized by low density, bigger block dimensions and presence of open spaces (public and private spaces, gardens and residuals areas);
— Modern organic tissue: structured on a semi-regular grid, built during the XXth century and characterized by high density, medium dimension block, located in external and suburban areas;
— Open tissue: structured on a irregular grid, built during the XXth century and characterized by low density, high dimension block, located in suburban areas;
— Detached tissue: structured on isolated group of buildings, located in suburban/rural areas.

The surface temperatures exhibited by the urban tissues identified in this study are plotted in figure 4 for the dual analysis of satellite data.

The Industrial tissue structured on open and irregular grid near the principal infrastructures (railways, airport, highways) and was characterized by industrial building with extended surface, limited height (10-15 mt) and basically located in suburban areas. The Mixed Functional tissue is not associated to a well-defined spatial pattern and is characterized by public and private urban services.

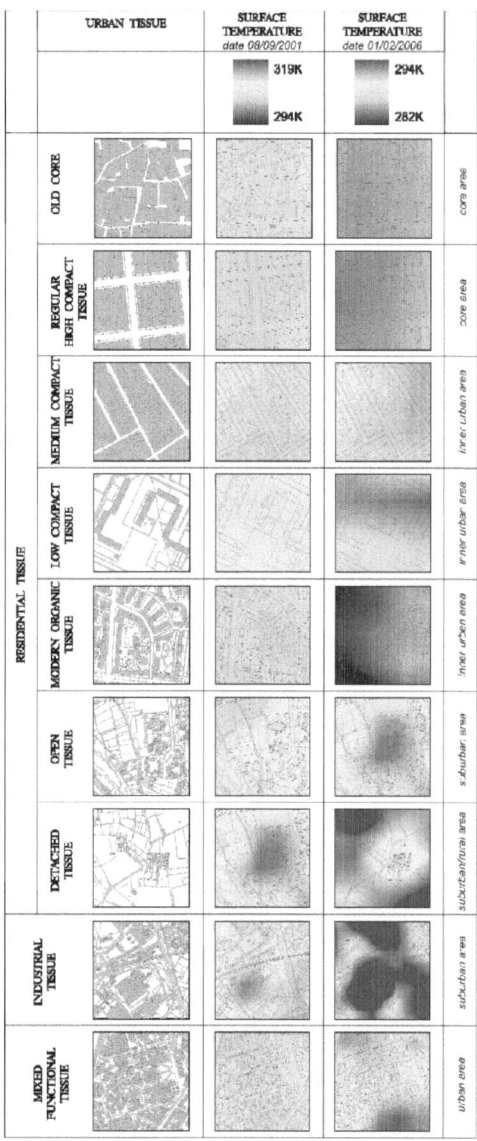

Fig. 4. Surface temperatures exhibited by the tissues identified in this study for both satellite imageries

To take into account the role of anthropogenic influence on temperatures, the Site Coverage (SC) factors has been designed for the analysis. It is the ratio of the total floor area of the building to the area of the cell. It is referred to the ground floor, so it also refers to impervious surfaces and represent an index of building density focused over the bi-dimensional cell (SC = Area $_{building}$ / Area $_{cell}$)

The Rugosity factor has been introduced to take the real urban fabric into account by evaluating the three-dimensional space. For each cell the volume of the building has been correlated to the total area of the cell. The building height map of the city of Bari has been elaborated through the GIS tools. Thus, for each cell, building volumes has been calculated. The Absolute Rugosity factor could explain the role of density of urban tissues on temperature (see also the ability of air to flow within the urban volume). The volume needed in the Rugosity take into account both the volume and empty spaces that lay over a reference cell and is derived from absolute rugosity formula introduced by Adolphe [19] where $R = V_{built} / A_{cell}$.

4 Analysis and Results

The interaction between the urban structures, as represented by the selected variables, and the surface temperature has been performed by the SPSS 17.0 statistical package through a cell by cell (90 mt) correlation analysis. The two images chosen are related to different periods and, in order to take the seasonality effect into account, both period have been investigated.

Table 1. The table summarizes the Pearson correlation coefficients obtained through correlation between the surface temperature and the land use classes

		02012006	09082001
SURFACE TEMPERATURE	Pearson Correlation	1	1
	Number of cases	7952	7973
AGRICOLTURE HERBACEOUS COVERAGE	Pearson Correlation	.256	.161
	Sig. (2-tailed)	.000	.000
AGRICOLTURE ARBOREOUS COVERAGE	Pearson Correlation	-.070	-.009
	Sig. (2-tailed)	.000	.482
NATURAL HERBACEOUS COVERAGE	Pearson Correlation	-.047	.145
	Sig. (2-tailed)	.518	.044
WOODLANDS	Pearson Correlation	.015	.114
	Sig. (2-tailed)	.560	.000
URBAN GREEN AREAS	Pearson Correlation	-.586	-.606
	Sig. (2-tailed)	.036	.022

The results of the bivariate correlation between surface temperature and land use classes, as described in the previous section, were estimated by Pearson's coefficient, as shown in the following table. The parameter for synthesizing the presence of the different classes has been represented by the percentage of area covered within each cell.

The areas with the types of land use called agriculture and natural herbaceous coverage, agriculture arborous coverage, and woodlands, of course, don't characterize inner parts of the city but the suburbs or the beginning of the rural border town.

The analysis of surface temperature distribution highlights how much land use classes are characterized by higher temperatures in suburban areas than in the urban ones: this result is confirmed by the correlation coefficients shown, while low value temperatures determine positive coefficients.

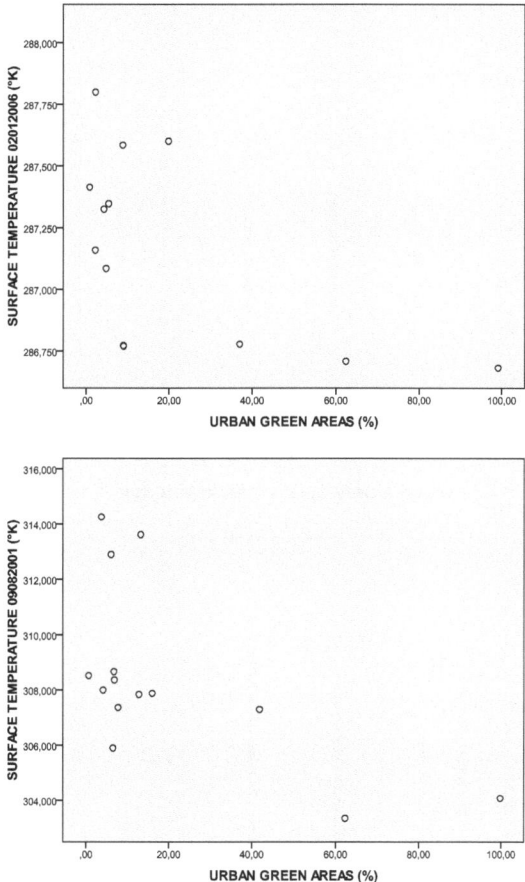

Fig. 5. Scatter plot representing the dispersion of the percentage of area occupied by urban green area for each cell, depending on temperature surface

We see that the correlation coefficient that takes a significant negative value for both thermal maps, is the percentage of area covered by urban green areas. Figure 5 shows the scatter plot that illustrates the relationship between the surface temperature, relative to the two detections, and the presence of urban green areas in the cell.

The distribution of the dispersion of the scatter plot seems to explain the ability by the sensor to recognize the surface temperatures for which the low presence of green areas does not produce a sensitive response. It colud be noted that from values greater than 20% of "green" presence, the temperature decreases in both thermal images.

Some papers, see for example [33], recognize such behavior and explained it by the consideration that the tree canopy intercepts the solar heat and provides cooler temperatures to the surface.

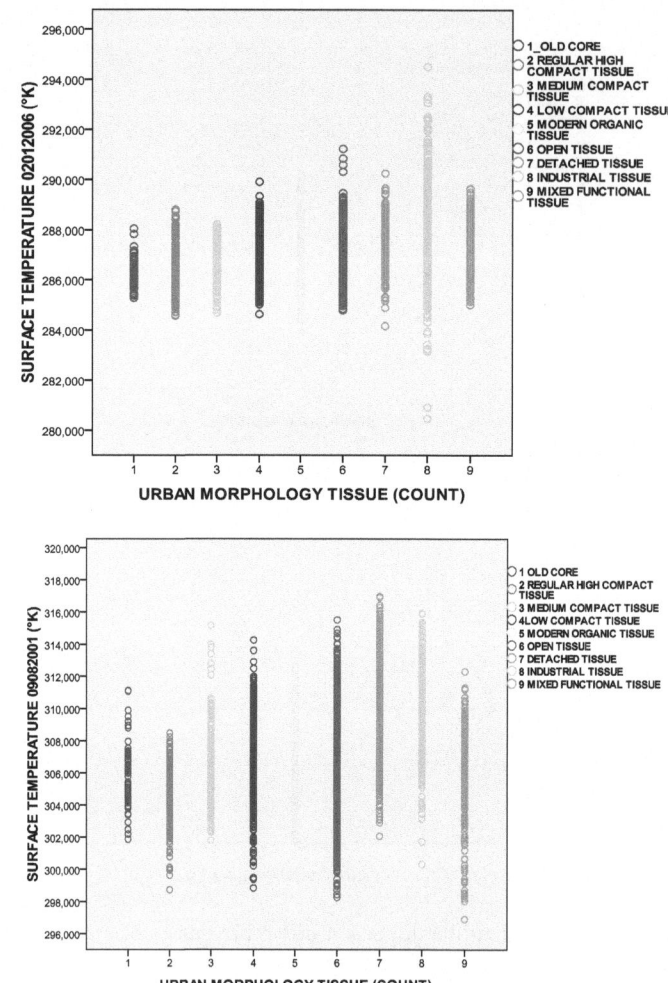

Fig. 6. Scatter plot representing the range of surface temperature assumed by the prevailing morphology

The observation of the distribution of types of urban morphologies with the surface temperatures (see fig. 6) seems to show that the compact tissue, and in particular the

Old core, abuts on lower average temperatures and it is characterized by a narrower range than in other types. It reads like the decrease of the compactness of the urban surface temperatures are higher. In this regard, for example, compare the behavior of the Old core and the types Open tissue and Detached tissue: the latter ones have, evidently, a broader range of temperatures, which higher average temperatures.

The behavior of the temperature decreases with the increasing of buildings dispersion: it is confirmed, in a quantitative way, by the correlations between two variables such as temperature and site coverage and absolute rugosity. In the following table, the Pearson coefficients are summarized and they are the result of the bivariate correlation between the surface temperature and the variables.

Table 2. Pearson correlation between the surface temperature and the site coverage and absolute rugosity

		02012006	09082001
SITE COVERAGE	Pearson Correlation	-.261	-.409
	Sig. (2-code)	.000	.000
ABSOLUTE_RUGOSITY	Pearson Correlation	-.318	-.435
	Sig. (2-code)	.000	.000

The site coverage of the buildings and the absolute rugosity show a negative weak correlation with the surface temperature.

By the analysis of thermal images, this trend is reinforced by the fact that areas outside the city and the open spaces in general show a higher temperature than the densely urbanized areas.

5 Limitations and Further Improvements

The correlation analysis shows that the phenomenon of UHI is reversed for the city of Bari. Densely urbanized zones reveal lower temperatures than the peripherals areas. The analysis of urban morphology provides an explanation about the spatial temperature trend even though the use of vegetation indices (such as the NDVI) would improve the analysis, especially for sensor like ASTER, able to acquire thermal and optical/infrared data simultaneously. This index provide an image to date description of the vegetation status at the ground, either for agricultural or natural areas. With regard to the urban fabric a more complete description of the fabric is required (within the index of roughness) to better summarize the range of possible morphologies and rates of porosities of the fabric. The difference in the elevation of buildings might constitute another relevant parameter to introduce in the analyses. By this way, the relationships between shaded/sunny areas and surface temperatures could be investigated. Moreover, the relationship between thermal data and urban features have to be assessed by considering the properties of satellite data. First of all, satellite data are related to a particular period and time of day (around 10.00 am for this study). Materials and surfaces with higher thermal capacity values didn't have enough time to accumulate and release after being illuminated by the Sun's radiation. Shaded areas were not warmed up by the solar radiation in the early morning period. According to Rinner [33], open areas are more exposed to solar radiations and exhibits a more pronounced heating of the ground surface.

The heterogeneous spatial resolutions of satellite and cartographic data could constitute a limiting factor of the analysis. In conclusion, the relationship between urban morphology and surface temperatures have been highlighted by this work. However, an effort towards a better definition of factor used in the correlation analysis and the widening of their number has to be fulfilled.

Acknowledgments. This work has been supported by the Apulia Region funds for Scientific Research (Project: PS-047 ECOURB: Analysing and modelling air and thermo pollution for urban ecolabelling systems, Responsible: Francesco Selicato and Dino Borri.

References

1. Arpa Emilia Romagna, http://www.arpa.emr.it/cms3/documenti/_cerca_doc/meteo/ambiente/bonafe_report_meteo_urbana.pdf
2. RCEP, The Urban Environment. TSO, Norwich (2007)
3. Zhao, C., Guobin, F., Xiaoming, L., Fan, F.: Urban Planning Indicators, Morphology and climate Indicators: A Case Study for a north-south transect of Beijing, China. Build. Env. 46, 1174–1183 (2011)
4. Howard, L.: The Climate of London. C. Baldwin, London (1833)
5. Arnfield, A.J.: Review two Decades of Urban Climate Research: A Review of Turbulence, Exchanges of Energy and Water, and the Urban Heat Island. Int. J. Clim. 23, 1–26 (2003)
6. Santamouris, M.: Heat island research in Europe – State of the art. Adv. Build. En. Res. 1, 123–150 (2007)
7. Lowry, W.P.: Empirical Estimation of the Urban Effects on Climate: A problem Analysis. J. Appl. Meteorol. 16, 129–135 (1977)
8. Stewart, I.D.: Landscape Representation and the Urban – Rural Dichotomy in Empirical Urban Heat Island Literature, 1950-2006. A. Clim. Chor. 40-41, 111–121 (2007)
9. Mirzaei, P.A., Fariborz, H.: Approaches to Study Urban Heat Island – Abilities and Limitations. Build. Env. 45, 2192–2201 (2010)
10. Gill, S.E., Handley, J.F., Ennos, R.A., Pauleit, S., Theuray, N., Lindley, S.J.: Characterizing the Urban Environment of UK Cities and Town: a Template for Landscape Planning. Land. Urb. Plann. 87, 2210–2222 (2008)
11. Auer, A.H.: Correlation of land use and cover with meteorological anomalies. J. Appl. Meteorol. 17, 636–643 (1978)
12. Akbari, H.: Shade trees reduce building energy use and CO_2 emissions from power plants. Environmental Pollution 116 (2002)

13. Kim, J.P.: Land use planning and the urban heat island effect. Dissertation presented in Partial Fulfillment of the Requirements for the Degree Doctor of Philosophy in the Graduate School of the Ohio State University (2009)
14. Ellefsen, R.: Mapping and measuring buildings in the urban canopy boundary Layer in ten US cities. Energy & Buildings 15-16, 1025–1049 (1990-1991)
15. Oke, T.R.: Initial guidance to obtain Representative Meteorological Observations at Urban Sites. IOM Report 81, World Meteorological Organization, Geneva (2004)
16. Stewart, I.D., Oke, T.: Classifying urban climate field sites by "local climate zones" the case of Nagano, Japan. In: The Seventh International Conference on Urban Climate, Yokohama, Japan (2009)
17. Davenport, A.G., Grimmond, S.B., Oke, T.R., Wieringa, J.: Estimating the roughness of cities and sheltered country. In: Proc. 12th AMS Conf. on Appl. Climatol., Asheville, North Carolina (2000)
18. Stewart, I.D., Oke, T.: Thermal differentiation of local climate zones using temperature observations from urban and rural filed sites (2010)
19. Adolphe, L.: A simplified model of urban morphology: application of the environmental performance of cities. Environment and Planning B: Planning and Design 28, 183–200 (2001)
20. Simonds, J.O.: Landscape Architecture: a Manual of Site Planning and Design. McGraw-Hill Professional, New York (2007)
21. Eliasson, I., Knez, I., Westerberg, U., Thorsson, S., Lindberg, F.: Climate and behavior in a Nordic City. Land. Urban. Plann. 82, 72–84 (2007)
22. Oke, T.R.: Street Design and Urban Canopy Layer Climate. Ener. Build. 11, 103–113 (1988)
23. Pickett, S.T.A., Cadenasso, M.L., Grove, J.M., Nilon, C.H., Pouyat, R.V., Zipperer, W.C., Costanza, R.: Urban Ecological Systems: Linking terrestrial ecological, physical, and socioeconomic components of metropolitan areas. Annu. Rev. Ecol. Syst. 32, 27–157 (2001)
24. Bridgman, H., Warner, R., Dodson, J.: Urban Biophysical Environments. Oxford University Press, Oxford (1995)
25. Lillesand, T.M., Kiefer, R.W.: Remote Sensing and Image Interpretation, 3rd edn. John Wiley & Sons (2000)
26. Hook, S.J., Gabell, A.R., Green, A.A., Kealy, P.S.: A comparison of techniques for extracting emissivity information from thermal infrared data for geologic study. Rem. Sens. of Environment 42, 123–135 (1992)
27. Kealy, P.S., Hook, S.J.: Separating Temperature and Emissivity in Thermal Infrared Multispectral Scanner Data: Implications for Recovering Land Surface Temperatures. IEEE Transactions on Geosciences and Remote Sensing 31, 1155–1164 (1993)
28. Gillespie, A., Matsunaga, T., Rokugawa, S., Hook, S.: A Temperature and Emissivity Separation Algorithm for Advanced Spaceborne Thermal Emission and Reflection Radiometer (ASTER) Images. IEEE Transactions on Geosciences and Remote Sensing 36, 1113–1126 (1998)
29. Levì-Strauss, C.: Tristes Tropiques, Terre Humaines, Paris (1955)
30. Moudon, A.V.: Urban Morphology as an Emerging Interdisciplinary Field. Urb. Morph. 1, 3–10 (1997)
31. Selicato, F.: Bari, Morfogenesi dello Spazio Urbano. Mario Adda Editore, Bari (2003)
32. Adolphe, L.: A Simplified Model of Urban Morphology: application to an analysis of the environmental performance of cities. Env. Planning 28, 183–200 (2001)
33. Rinner, C., Hussain, M.: Toronto's Urban Heat Island-Exploring the Relationship between Land Use and Surface Temperature. Remote Sens. 3, 1251–1265 (2011)

Study of Sustainability of Renewable Energy Sources through GIS Analysis Techniques

Emanuela Caiaffa[1], Alessandro Marucci[2], and Maurizio Pollino[1]

[1] ENEA - Italian National Agency for New Technologies, Energy and Sustainable Economic Development, "Energy and Environmental Modeling" Technical Unit, Rome, Italy
emanuela.caiaffa@enea.it, maurizio.pollino@enea.it
[2] University of L'Aquila, Natural Science Department, L'Aquila, Italy
marucci79@hotmail.it

Abstract. In an integrated vision of the problems concerning energy policies, the use of renewable energy sources should assume a significant role. The 2009/28/EC Directive of the European Parliament and Council has indicated ambitious energy and climate change objectives for 2020: greenhouse gas emissions reduction for 20%, renewable energy increase for 20%, improvement in energy efficiency for 20% [1].

The aim of this paper is to present a GIS based methodology able to support decision-making in energy supply from Renewable Energy Sources (RES). To decide what type of renewable energy font is the best choice for a specific territory, it's important to know the local energetic situation, exploring the potential renewable energy sources available in that specific area, deciding what is the more territory compatible/sustainable among them, and if it's exploitable by suitable environmental and economic point of view.

The methodology is largely directed towards the development of a tool to support siting decision.

Keywords: Renewable Energy Sources, GIS, sustainable development, solar radiation.

1 Introduction

In recent years, both by necessity and in response to international policies, interest towards alternative energy forms, i.e. not derived from conventional sources like fossil fuels, has considerably grown. Photovoltaic, solar thermal, wind, biogas, biomass, geothermal are commonly used terms and the associated technologies are reaching very high expertise standards, with considerable interest in world markets. The spirit that leads towards such technologies comes from to answer a real sustainable development need as well a rational resources use [2, 3].

One of the most interesting renewable sources features is their dispersion in the territory. This characteristic is on the one hand a strength, because potentially everywhere it is possible to exploit solar energy, wind power, etc., on the other is a limiting factor, because the energy concentration is reduced. Moreover, the diffuse nature of renewables can combine energy production with the fight against land depopulation and degradation phenomena, supporting the technological and economic development of small urban and rural reality. Renewable Energy Sources (RES) exploitation implies a more direct communities and local administrations involvement in finding the best solution for each energy source, use and location [4], promoting the concept of thinking globally and acting locally. In this perspective, and in a modern land management, it is also necessary to take into account potential and actual impacts due to installations for renewable sources production, privileging the *landscape ecology* assessing [5].

Changes in landscape take continuously place, with significant repercussions on quality of life and natural habitat ecosystems, mainly through their impacts on soil and ecosystems [6]. Landscape planning is strongly related to sustainable development issues, especially considering that landscape is both an influencing factor and a resulting product of the people-place relationship [7]. As pointed out in the European Landscape Convention [8], the analysis of landscape can allow to understand the wellbeing/discomfort condition of population in relation with their environment. A significant problem for some types of production plants, mainly those solar and wind, is related to their possible negative effects in terms of visual impact. A careful planning of single plant integration and the choice of devices less "visible" could reduce the problem, but certainly not eliminate it.

In this framework, the use of Geographic Information Systems (GIS) represents the most significant technological and conceptual approach to spatial data analysis, in order to provide reliable information for both planning and decision-making tasks. The GIS tool fits perfectly in this kind of survey and assessment, providing the mean to combine several features as ecological, territorial, socio-economic aspects, etc. useful to support landscape analysis to deal with issues concerning environmental impacts [9].

Moreover, projections for 2020 indicate that renewables could cover, for that date, from 20% to 30% of the world's energy needs [1]. A set of effective actions to achieve those objectives should include the use of new tools and methodologies to be implemented in a modern vision of territory governance [10, 11]. This requires the implementation of rules aimed at protecting the land primary productive function, so as to facilitate the development and exploitation of potential sources of energy therein. Therefore proposals for a judicious planning, land management, knowledge, location and accessibility of resources constitute the core for an efficient and sustainable governance.

2 Objectives

The main objective of this research is to find a methodology, territory related, able to support decision-making processes in choosing energy supply from renewable sources. The methodology, through a series of *informants*, will make the planner able to *read* the potential of a given area to house plants from renewables, and to identify what and how many renewable energy can be compatible and applicable to environmental, social, cultural, economic realities located on that territory [12].

The description of the potential energetic vocation of a place, requires a more detailed study composed of several variables and their combination. In fact, to precisely define an area potentially adapted to housing a plant from photovoltaic (PV), it is necessary to carry out a series of thematic maps quantifying the effective usable area. Several thematic maps like Parks, Sites of Community Importance (SIC), Special Protection Areas (SPA), areas of natural scenery constraint, urban, humid areas, particular appreciate habitat, and so on, are created and combined in order to define the remaining areas of possible production from RES. The types and the *forms* of intervention in the territory, for the production of energy, are multiple and similar to linear and areal infrastructures of anthropogenic nature. It follows that the *environment consumption* is one of the main phenomena found at different levels, which add to the already growing soil consumption. Unlike purely anthropic settlements, the energy production installations from RES, in a global vision, have a positive value because they are intended to produce energy in a sustainable way and in order to reduce the causes of deterioration and consumption of natural resources. Their inclusion in the territory should be considered in light of two fundamental aspects:

1. one hand, the environmental cost of their implementation, expressed in terms of consumption of resources, habitat loss, land use;
2. the other, the real energetic vocation of a territory in terms of potential energy obtainable.

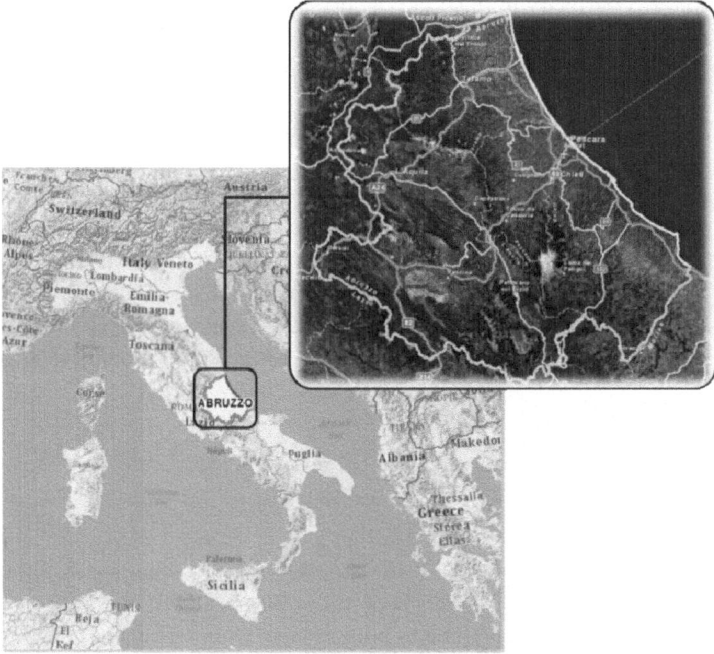

Fig. 1. Study area: Abruzzo region in Central Italy

This paper provides a framework for analysing the sustainability of renewable energy sources using GIS and presents a case-study for the Abruzzo region in Italy (Fig. 1). The region is situated at the centre of the Italian peninsula facing the Adriatic sea, with a 150 km long coastline. With an area of 10,763 km² and bordered on the east by the Adriatic and on the west by the Apennines, it is one of the most mountainous regions in Italy (the Corno Grande in the Gran Sasso massif, at 2,914 m, is the highest summit in the Apennines). The rivers, although numerous, are all seasonal except for the biggest, the Pescara and the Sangro. In the interior are the 500 km² of the Abruzzo National Park, where rare examples of Mediterranean flora and fauna survive (chamois, wolves, bears, golden eagles). The climate is varied: warm and dry on the coast, an alpine climate in the mountainous interior. Major roads and railway lines link the region to the south, west and north of Italy (Abruzzo western border lying less than 80 km due east of Rome).

3 GIS-Based Spatial Analysis

The production of electric energy from renewable source plants, being constituted by complex infrastructures that *need space*, can produce impacts on the natural environment. The clean energy quantification, that such technologies can produce, represents the value of sustainability that they have, in relation to interferences that they produce. The general methodology for analysis of solar radiation using GIS approach is summarised in Fig. 2. The time span, for which the average values of solar radiation have been calculated, goes from 1994 to 1999.

As a matter of fact, GIS techniques are efficiently exploited to analyse the effe of various factors, including the above mentioned thematic maps.

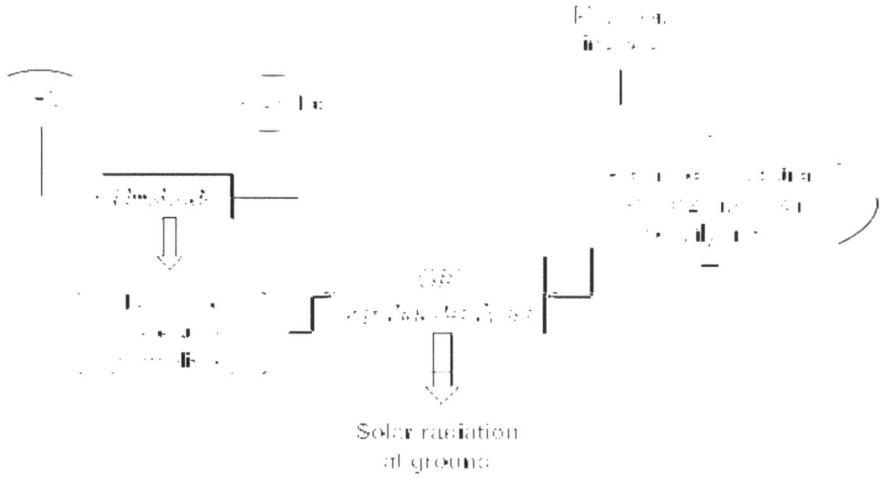

Fig. 2. General methodology schema

3.1 Spatial Distribution of Global Solar Radiation at Ground Level

The estimation method is based on the fact that the amount of solar radiation [13], reaching a certain portion of the land surface [14], is statistically correlated to the cloud cover on it. In addition, topographical characteristics such as latitude, slope, aspect and shadow cast, can affect solar radiation. In general, insolation is the result of the interactions between all these factors [15]. To determine solar radiation over an entire region, different approaches have been developed and described in literature [16, 17, 18, 19 and 20]. These studies create spatial databases of solar radiation using different interpolation techniques or calculate solar irradiance directly from meteorological geostationary satellites (e.g. the European based Meteosat).

In order to define the coverage of solar radiation on the ground, has been implemented a spatial analysis of land detected values by the program "The global solar radiation on the ground in Italy" compiled by ENEA [21], a specific database, extended at national level and offering free online access. The global solar radiation on Italy was estimated by processing images transmitted from the satellite Meteosat secondary band in the visible.

The monthly mean values of daily radiation on an horizontal plane thus estimated, were compared with those obtained from data measured by ground stations of the Central Office for the Italian Air Force Meteorological and of the agro meteorological National Network of Ministry of Agriculture. The average annual difference between the two sets of values is of 6-7%.

For this work are provided monthly data on Italy's radiation and monthly average daily values for the Abruzzo municipalities. The spatialisation of ground solar radiation data [2, 22] was a necessary step to implement the methodology discussed in this study.

The basic data required for the study are listed below:

- **DEM** (Digital Elevation Model), spatial resolution 20 m (Fig. 3).
- **Basic geographical layers:** vector databases, such as urban areas, road network, hydrography network, municipal boundaries, regional boundaries, etc.
- **Solar tables and charts:** data collection of the basic parameters of the solar azimuth angle and solar altitude angle, necessary to develop the basis of exposure.
- **Solar radiation:** data and results coming from the activity carried out by ENEA for the estimation of global solar radiation on a horizontal plane in Italy through the development of secondary images transmitted by satellite in the Meteosat visible band.

Digital elevation model has provided the structural basis for constructing thematic maps of solar radiation. The digital elevation model used in this work is derived from remote sensing data and is supplied in raster format: digital images in which each cell contains the altitude average value of the represented area.

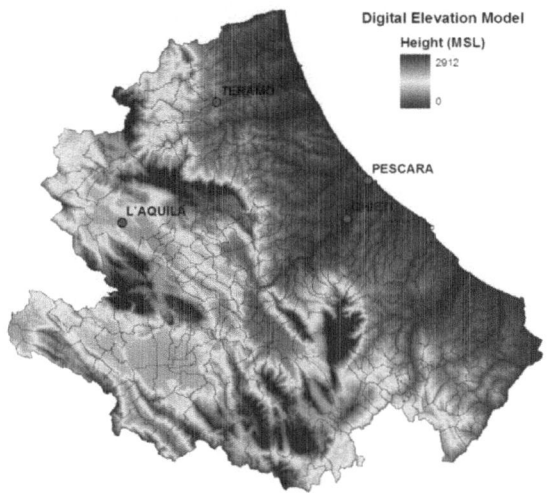

Fig. 3. Digital Elevation Model of Abruzzo region

3.2 Calculation of Average Slope

The *hillshade* GIS function [23], derived from Digital Elevation data Model, is a method to determine the hypothetical illumination of a surface, modelling the exposure surface using the shaded topography. On the setting of a hypothetical light source position, *hillshade* creates a hypothetical illumination of the topography and calculates the illumination level for each pixel. It's possible to compute the duration and intensity of insolation in a certain area, based on parameters of azimuth and altitude of the sun. In the table 1, 2 and 3 are shown the values used for the calculation.

Table 1. Solar Tables sunshine (Sshine) and sunset (Sset) values

Day	Sshine (CET)	Sset (CET)	Length of the day	Time equation	Eccentricity factor
Jan-17	7h 36'	16h 53'	9h 17'	-9'20"	10.340
Feb-16	7h 06'	17h 32'	10h 26'	-14'14"	10.251
Mar-16	6h 22'	18h 07'	11h 45'	-9'21"	10.108
Apr-15	5h 30'	18h 40'	13h 10'	-0'14"	6,897222
May-15	4h 49'	19h 13'	14h 24'	3'56"	6,790972
Jun-11	4h 33'	19h 35'	15h 02'	0'48"	6,729861
Jul-17	4h 47'	19h 34'	14h 47'	-6'01"	6,717361
Aug-16	5h 17'	19h 02'	13h 45'	-4'41"	6,76875
Sep-15	5h 48'	18h 13'	12h 24'	4'39"	6,865278
Oct-15	6h 21'	17h 20'	10h 59'	14'25"	10.059
Nov-14	6h 59'	16h 41'	9h 42'	15'20"	10.222
Dec-10	7h 28'	16h 27'	8h 59'	7'08"	10.319

Table 2. Solar altitude angle values

	Jan-17	Feb-16	Mar-16	Apr-15	May-15	Jun-11	Jul-17	Aug-16	Sep-15	Oct-15	Nov-14	Dec-10
03:00 CET												
04:00 CET												
05:00 CET					1°48'	4°21'	2°02'					
06:00 CET				5°25'	12°17'	14°34'	12°16'	7°38'	2°12'			
07:00 CET			7°01'	16°29'	23°12'	25°20'	23°03'	18°37'	13°15'	7°00'	0°13'	
08:00 CET	3°50'	9°12'	17°44'	27°27'	34°16'	36°22'	34°05'	29°40'	24°02'	17°10'	9°47'	4°51'
09:00 CET	12°30'	18°34'	27°41'	37°55'	45°07'	47°22'	45°05'	40°24'	34°04'	26°13'	18°04'	12°59'
10:00 CET	19°34'	26°26'	36°15'	47°12'	55°08'	57°49'	55°30'	50°12'	42°40'	33°29'	24°32'	19°22'
11:00 CET	24°27'	32°06'	42°32'	54°09'	63°01'	66°32'	64°16'	57°54'	48°44'	38°08'	28°32'	23°27'
12:00 CET	26°39'	34°52'	45°29'	57°06'	66°19'	70°40'	68°52'	61°33'	50°59'	39°23'	29°33'	24°48'
13:00 CET	25°52'	34°15'	44°25'	55°00'	63°13'	67°28'	66°37'	59°37'	48°47'	37°00'	27°28'	23°14'
14:00 CET	22°12'	30°22'	39°36'	48°37'	55°26'	59°10'	58°59'	52°59'	42°44'	31°26'	22°32'	18°58'
15:00 CET	16°06'	23°49'	31°59'	39°38'	45°28'	48°52'	49°00'	43°41'	34°10'	23°31'	15°23'	12°26'
16:00 CET	8°08'	15°20'	22°36'	29°19'	34°38'	37°54'	38°08'	33°10'	24°09'	14°04'	6°35'	4°12'
17:00 CET		5°35'	12°12'	18°23'	23°34'	26°50'	27°04'	22°10'	13°22'	3°39'		
18:00 CET			1°16'	7°19'	12°38'	16°01'	16°09'	11°08'	2°18'			
19:00 CET					2°09'	5°43'	5°41'	0°23'				
20:00 CET												
21:00 CET												

Table 3. Azimuthal angle values

	Jan-17	Feb-16	Mar-16	Apr-15	May-15	Jun-11	Jul-17	Aug-16	Sep-15	Oct-15	Nov-14	Dec-10
03:00 CET												
04:00 CET												
05:00 CET					113°52'	117°30'	117°25'					
06:00 CET				97°54'	104°11'	108°05'	107°49'	102°00'	92°31'			
07:00 CET			80°44'	87°52'	94°38'	98°53'	98°29'	92°11'	82°19'	72°07'	65°00'	
08:00 CET	56°58'	63°27'	69°59'	77°09'	84°26'	89°13'	88°43'	81°47'	71°20'	61°03'	54°23'	52°48'
09:00 CET	45°50'	51°50'	57°48'	64°41'	72°27'	77°58'	77°29'	69°44'	58°35'	48°21'	42°25'	41°29'
10:00 CET	33°15'	38°27'	43°16'	48°57'	56°36'	62°58'	62°48'	54°20'	42°52'	33°24'	28°45'	28°42'
11:00 CET	19°05'	22°58'	25°36'	28°12'	33°19'	39°39'	40°48'	33°07'	23°11'	15°58'	13°27'	14°31'
12:00 CET	3°44'	5°42'	5°06'	2°21'	0°36'	2°53'	7°05'	4°55'	0°07'	-3°02'	-2°50'	-0°33'
13:00 CET	-11°52'	-12°04'	-16°05'	-24°02'	-32°21'	-35°24'	-29°53'	-24°43'	-22°58'	-21°42'	-18°54'	-15°36'
14:00 CET	-26°41'	-28°47'	-35°15'	-45°46'	-55°58'	-60°24'	-55°54'	-48°17'	-42°42'	-38°23'	-33°41'	-29°41'
15:00 CET	-40°03'	-43°31'	-51°13'	-62°15'	-72°00'	-76°11'	-72°40'	-65°15'	-58°26'	-52°34'	-46°44'	-42°21'
16:00 CET	-51°50'	-56°12'	-64°25'	-75°09'	-84°04'	-87°47'	-84°52'	-78°10'	-71°12'	-64°41'	-58°11'	-53°35'
17:00 CET		-67°18'	-75°44'	-86°05'	-94°18'	-97°36'	-95°01'	-88°57'	-82°13'	-75°24'		
18:00 CET			-86°05'	-96°11'	-103°52'	-106°49'	-104°24'	-98°52'	-92°25'			
19:00 CET					-113°32'	-116°10'	-113°51'	-108°44'				
20:00 CET												
21:00 CET												

To each pair of values (altitude angle of the sun and solar azimuth angle) respectively in table 2 and table 3, corresponds a precise position of the sun across the sky, and so all surfaces are more or less illuminated, depending on their position relative to the sun. The solar radiation coming from space, for the purposes of our analysis, can be considered constant. The angle of incidence on the earth's surface, depurated of all those factors definable atmospheric and climatological, determines the effective power. So the objective of the study has been to develop surfaces of daily illumination where the exposure average values were expressed.

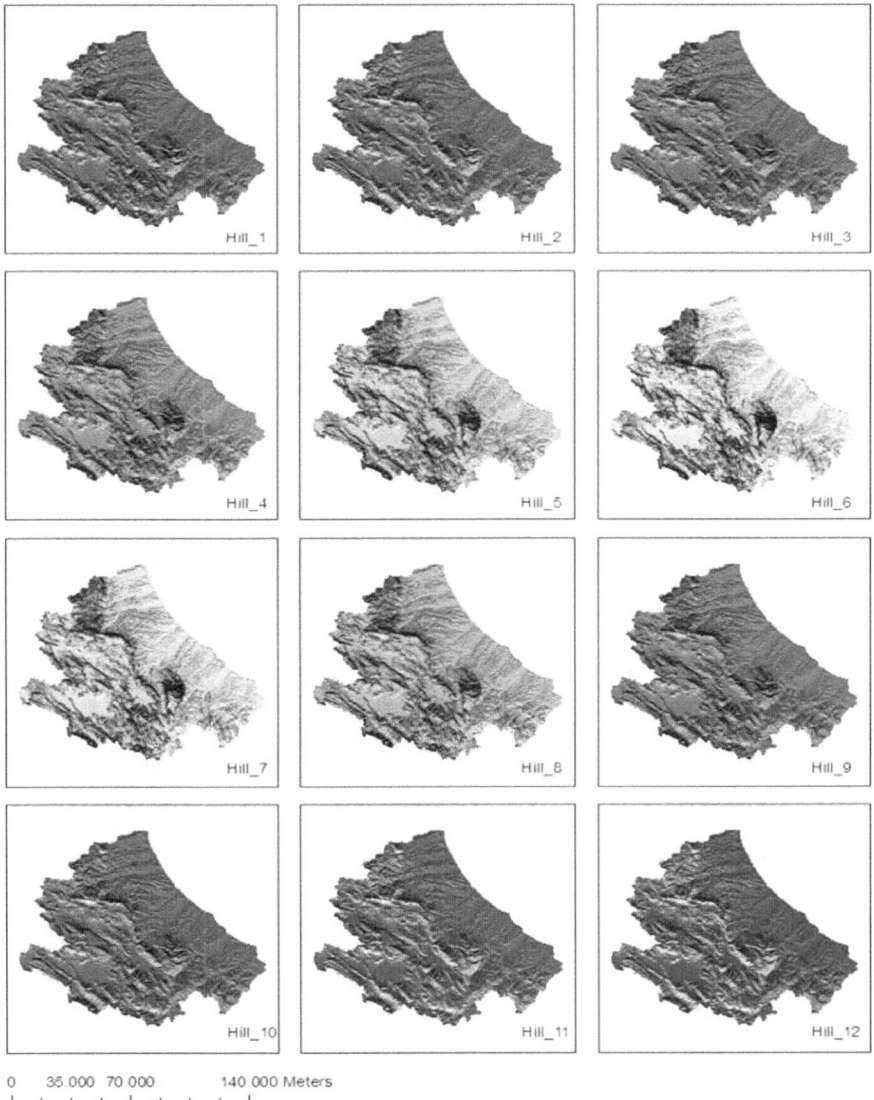

Fig. 4. Monthly exposure maps produced for a reference year

The thematic layers processing has been completed in ESRI ArcGIS environment [23], using spatial analysis and Map Algebra procedures: thematic maps on average daily values of radiation for the month they relate have been produced. In particular, the computing procedure used, has allowed the creation of 12 basic maps (Fig. 4) coming from the processing of 146 raster layers of exposure (according to the values reported in Tab. 1 and Tab. 2). These thematic maps have been obtained as result of the application of the *hillshade* function to DEM layer information.

To obtain the spatialisation of solar radiation values the maps of Fig. 4 were intersected with data of global solar radiation, that has been estimated processing secondary images transmitted from the satellite Meteosat [18] on visible band for the reference period from 1994 to 1999 [21]. Such data on solar radiation are available for all the national territory, for municipalities with at least 10,000 inhabitants and smaller municipalities with at least 5,000 inhabitants in sparsely populated places.

3.3 Estimation of Solar Electricity Potential

In order to estimate the potential solar electricity generation E [kWh], by PV configuration, the following formula has been used:

$$E = INS * SUP * \mu * v * K \quad [kWh] \tag{1}$$

This formula, derived from the one proposed by Šúri et al. [24], has been customised for our research purposes and with the intention to fit a real local situation. The following parameters have been made explicit:

- *INS* is the value of daily, monthly or annual insolation;
- *SUP* is the surface (expressed m^2) taken up by solar module (the values are derived from the dimensions declared by the manufacturer);
- μ is the average yield of photovoltaic cell (16%);
- v is the yield of the photovoltaic system (defined as a set of panels, inverters and electrical panels), usually ranging from 0.70 and 0.86;
- K is the reduction coefficient, defined to take into account possible shading effects (approximately ranging from 0.95 and 0.97) [25].

4 Results

The distribution and values of solar radiation on the ground are the results of analysis definable of fundamental importance for assessing the energetic potential of a territory. Knowledge of resources and their location represents a key step in planning for sustainable development [26].

Operatively several GIS layers (in raster format) have been produced, containing information about solar radiation at ground for the entire territory of Abruzzo region.

Starting from the processing of available data (1994-1999 interval), the relative thematic maps, reporting the average monthly values of solar radiation at ground, have been produced for each month of a reference year (Fig. 5).

Then, all the information obtained according the above described procedure have been synthesised in a single thematic map, reporting the values of solar radiation at ground for a whole year (Fig. 6). Once obtained the solar radiation values at ground, it has been possible to estimate the potential production from photovoltaic system, by applying the formula (1).

The map in Fig. 8 represents the potential solar electricity potential energy production from photovoltaic systems, assuming installations of 3 kW_p installed peak power. Moreover, from this map, it is possible to locate the areas potentially eligible to home an energy production plant from PV. To this end, the present study has evolved in defining where and how much it's possible to exploit this energy source, without constraints due to land use and local regulations.

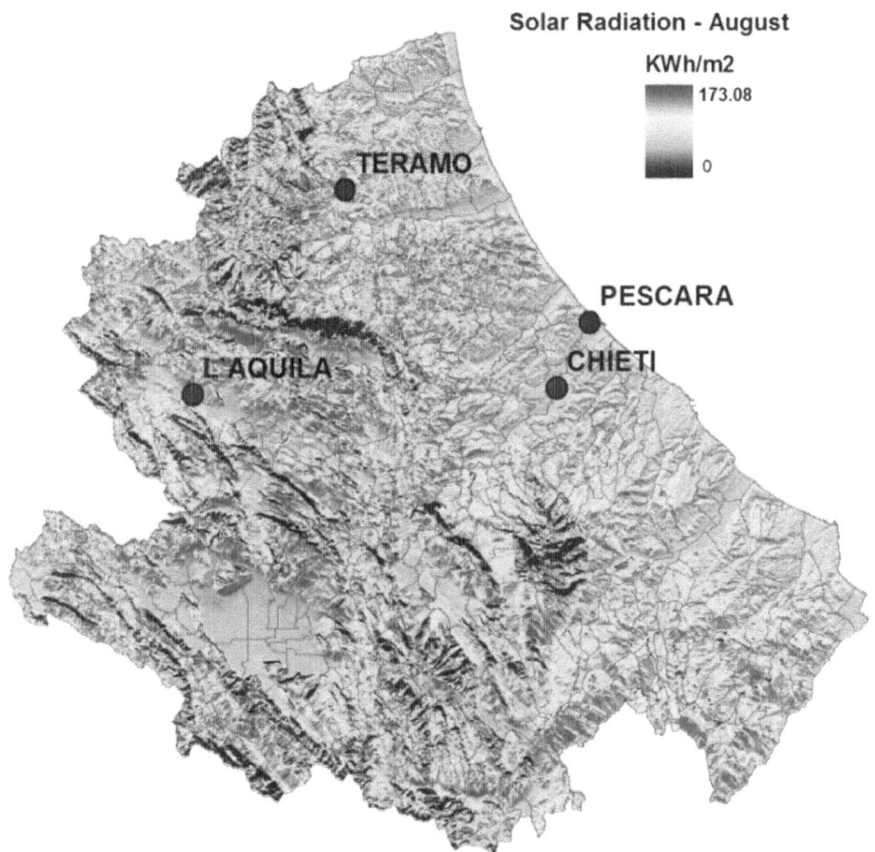

Fig. 5. Solar radiation at ground map for Abruzzo region: August month values

First of all, have been identified the places (*energy areas*) potentially suitable for the installation of a photovoltaic plant, net of all possible constraints. The constraints concern valuable areas such as parks, SPA, SIC, archaeological areas, landscaping, etc., and concern also areas of ascertained use like agricultural land areas, residential areas, infrastructures such as factories, highways, etc.. The remaining areas have been combined with the map of the solar radiation.

Subsequently, have been consulted planning tools like Urban Master Plans (at municipal level) and specific zoning laws, Landscape Plan, Energy Plan (at regional level) to produce the final potential energetic map. In fact, information about constraints reported in these documents, have been used inside GIS environment, in order to eliminate those areas not meeting the criteria of selection. The remaining areas (not subject to any kind of limitation) have been combined with the thematic map of the solar radiation at ground. After identifying these areas, it's necessary add another selection criteria, dictated by the terrain topography (northern exposure, steep slopes, etc.).

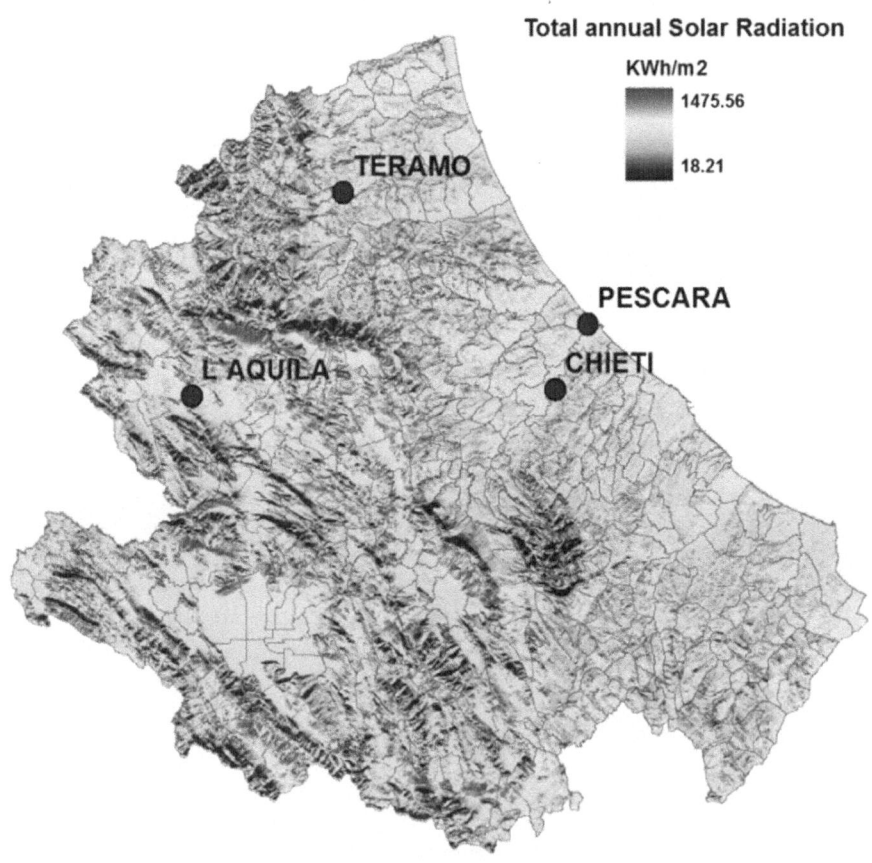

Fig. 6. Solar radiation at ground map for Abruzzo region: total annual values

To verify the real situation at the ground and to assess the feasibility of installations, the produced map has been checked using a set of thematic layers, such as DBPrior10K [27] GIS data (Administrative boundaries, Road network, Railways, etc., at 1:10,000 scale), Regional Technical Maps (RTM, at 1:10,000 scale) and digital aerial ortho-photos (available for consultation and visualization at the Italian National Geoportal [28], PCN). These layers have been used not only as reference and control, but also to take into account the presence of urbanized areas and other infrastructures (Fig. 7).

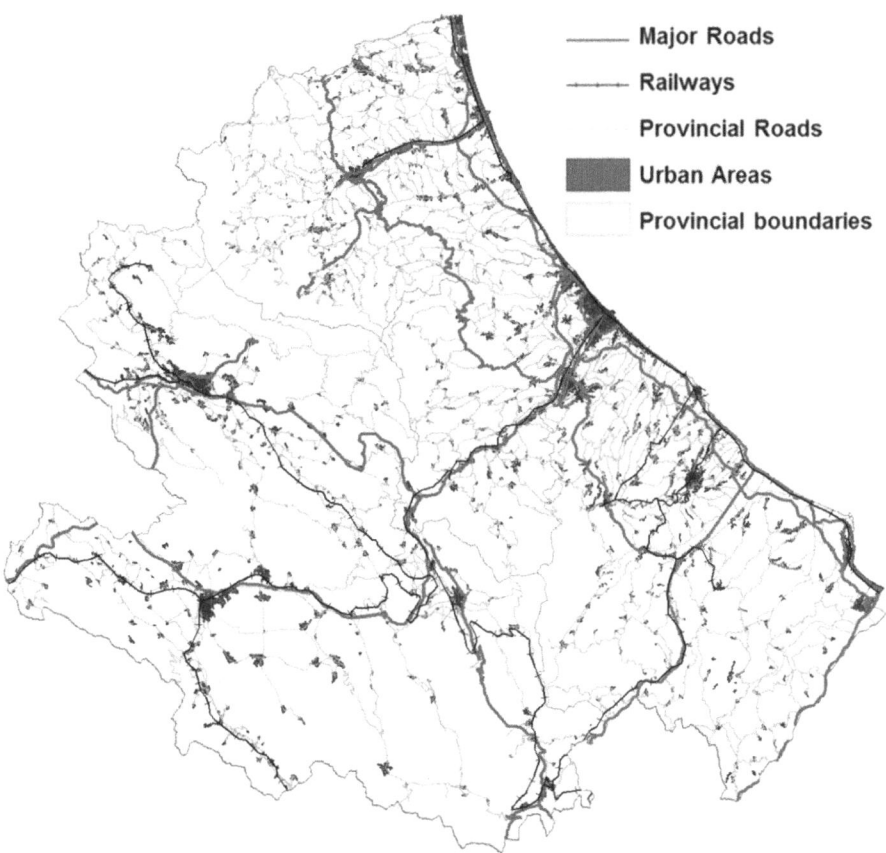

Fig. 7. Map of Urban areas and transportation network in the Abruzzo Region

An additional test was aimed to analyse the situation of the potential energy area both from the point of view of viability and of proximity to the electricity network (MV / LV network). The latter factor is fundamental to evaluate the capability to enter the energy produced in the national distribution network.

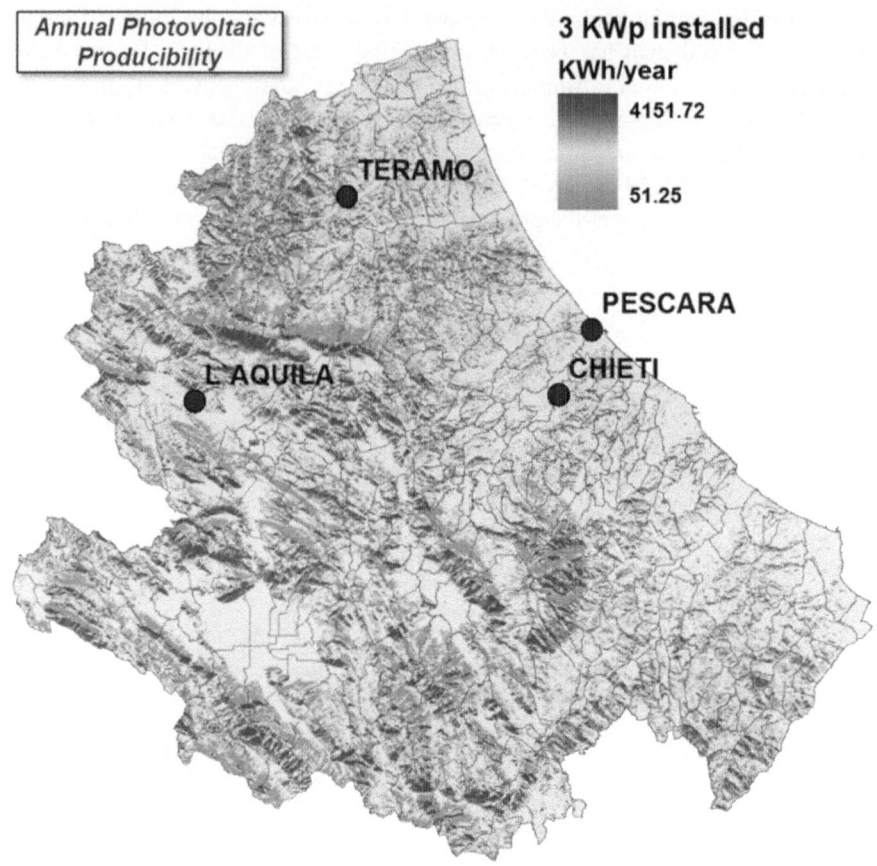

Fig. 8. PV potential production map

In particular, the distance analysis from existing energy production plants represents one of the future research directions, utilising - among others - VHR satellite images and digital cartography (RTM) in order to better evaluate the feasibility of the installations within the study area.

Finally, at the end of this further selection, the thematic layer enclosing only the areas potentially suitable for housing electricity production plants from PV has been obtained (Fig. 9).

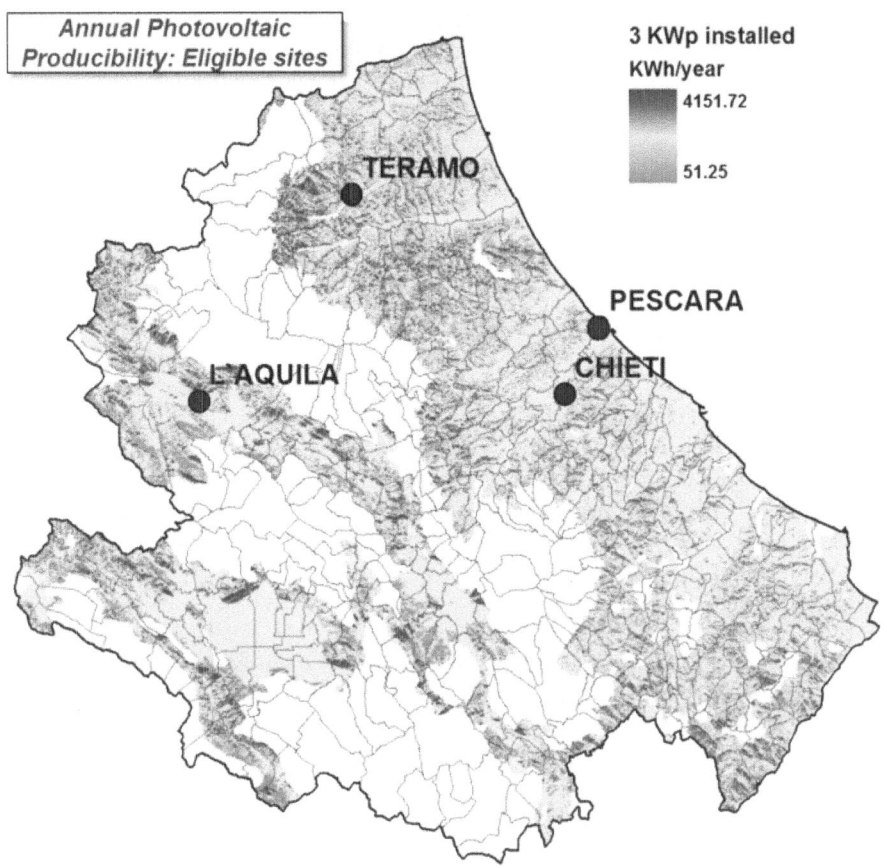

Fig. 9. Eligible sites for siting production plants from PV

5 Conclusions

In an overview of the issues that revolve around energy policy, the use of renewable energy sources has become a priority [29]. In fact, as already outlined by the 2009/28/EC Directive of the European Parliament and Council [1] (climate change objectives for 2020), all European countries are called to bring down the level of emissions of greenhouse gases in the atmosphere.

Many international bodies, as the UN Commission for Sustainable Development (UNCSD), the European Environmental Agency (EEA), the Organization for Economic Cooperation and Development (OECD), Eurostat (Statistical Office the European Community) and the IEA (International Energy Agency) are working on development plans in this direction [30].

The informatics layers realized for this study have proven a valuable tool for the analysis of the critical environmental purposes, and also at detailed scale a good result has been obtained. An accurate analysis of the environmental study has actually allowed to draw up clear and detailed entities and locations of possible with the ecosystem. The infrastructures insertion, both linear and areal, determines not only land consumption, due to the permanent occupation of the installations, but also produces influences on surroundings although minimum.

The greatest risk that could derive from insertion of the infrastructures, for the energy production from renewables, might be a micro-fragmentation and a micro-consume of the natural habitat. That circumstance doesn't show strong interference values, when analysed individually, but when considered in its entirety and cumulatively to all possible interventions in the area, could results in a high risk of degradation.

The development of information layers on the values and distribution of solar radiation is an important step for the evaluation of planning type both in energy and environmental field.

In conclusion we can say that through this study it was shown how GIS-based tools are able to analyse, interpolate and locate all the information necessary to make a complex environmental research [31]. Of course, the quality of input data determine the degree of survey accuracy and is a necessary condition to develop, in an appropriate way, environmental studies.

References

1. European Union: Directive 2009/28/EC of the European Parliament and of the Council of 23 April 2009 on the promotion of the use of energy from renewable sources and amending and subsequently repealing Directives 2001/77/EC and 2003/30/EC, http://eur-lex.europa.eu/LexUriServ/LexUriServ.do?uri=CELEX:32009L0028:EN:NOT
2. Šúri, M., Huld, T.A., Dunlop, E.D.: PV-GIS: A web-based solar radiation database for the calculation of PV potential in Europe. International Journal of Sustainable Energy 24(2), 55–67 (2005)
3. Pearce, J.M.: Photovoltaics: a path to sustainable futures. Futures 34(7), 663–674 (2002)
4. Hiremath, R.B., Shikha, S., Ravindranath, N.H.: Decentralized energy planning; modelling and application—a review. Renewable and Sustainable Energy Reviews 11(5), 729–752. Elsevier (2007)
5. Benson, J.F., Roe, M.H.: Landscape and sustainability, p. 423. Spon Press - Taylor & Francis Group, London (2000)
6. Antrop, M.: Changing patterns in the urbanized countryside of Western Europe. Landscape Ecology 15(3), 257–270 (2000)
7. Fichera, C.R., Modica, G., Pollino, M.: GIS and Remote Sensing to Study Urban-Rural Transformation During a Fifty-Year Period. In: Murgante, B., Gervasi, O., Iglesias, A., Taniar, D., Apduhan, B.O. (eds.) ICCSA 2011, Part I. LNCS, vol. 6782, pp. 237–252. Springer, Heidelberg (2011)
8. Council of Europe: European Landscape Convention, CETS No.176, Florence (2000), http://conventions.coe.int/Treaty/Commun/QueVoulezVous.asp?CL=ENG&NT=176 (retrieved on February 2012)
9. Šúri, M., Hofierka, J.: A new GIS-based solar radiation model and its application to photovoltaic assessments. Transactions in GIS 8(2), 175–190 (2004)
10. Caiaffa, E.: Geographic Information Science in Planning and in Forecasting. In: Institute for Prospective Technological Studies (eds.) Cooperation with the European S&T Observatory Network. The IPTS Report, vol. 76, pp. 36–41. European Commission JRC-Seville (2003)

11. Caiaffa, E.: Geographic Information Science for geo-knowledge-based governance. In: 8th AGILE Conference on Geographic Information Science, pp. 659–664. IGP Istituto Geografico Portugues, Estoril Portugal (2005)
12. Huld, T.A., Šúri, M., Kenny, R.P.: Estimating PV performance over large geographical regions. In: Conference Record of the IEEE Photovoltaic Specialists Conference (2005)
13. Zaksek, K., Podobnikar, T., Ostir, K.: Solar radiation modeling. Computers & Geosciences 31, 233–240 (2005)
14. Joint Research Centre (JRC) - Institute for Energy and Transport (IET): Photovoltaic Geographical Information System (PVGIS), http://re.jrc.ec.europa.eu/pvgis/ (retrieved on February 2012)
15. Duffie, J.A., Beckman, W.A.: Solar Engineering of Thermal processes, 2nd edn. John Wiley & Sons, USA (1991)
16. Hutchinson, M.F., Booth, T.H., McMahon, L.P., Nin, H.A.: Estimating monthly mean values of daily total solar radiation for Australia. Solar Energy 32, 277–290 (1984)
17. Beyer, H.G., Czeplak, G., Terzenbach, U., Wald, L.: Assessment of the method used to construct clearness index maps for the new European solar radiation atlas (ESRA). Solar Energy 61(6), 389–397 (1997)
18. Zelenka, A.: Combining METEOSAT and surface network data: a data fusion approach for mapping solar irradiation. In: Proceedings 10th Meteosat Scientific Users Conference, Switzerland, pp. 515–520 (1994)
19. Pellegrino, M., Caiaffa, E., Grassi, A., Pollino, M.: GIS as a tool for solar urban planning. In: Proceedings of 3rd International Solar Energy Society Conference-Asia Pacific Region (ISES-AP 2008), Sydney, Australia, November 25-28 (2008)
20. Cebecauer, T., Huld, T., Šúri, M.: Using high-resolution digital elevation model for improved PV yield estimates. In: Proceedings of the 22nd European Photovoltaic Solar Energy Conference, Italy, pp. 3553–3557 (2007)
21. Italian Atlas of solar radiation, http://www.solaritaly.enea.it/index.php (retrieved on February 2012)
22. Hofierka, J., Šúri, M.: The solar radiation model for Open source GIS: implementation and applications. In: Proc. of the Open Source GIS GRASS Users Conference, Italy (2002)
23. ESRI ArcGIS Resource Center, http://resources.arcgis.com/
24. Šúri, M., Huld, T.A., Dunlop, E.D., Ossenbrink, H.A.: Potential of solar electricity generation in the European Union member states and candidate countries. Solar Energy 81, 1295–1305 (2007)
25. Liu, B.Y.H., Jordan, R.C.: The Long-Term Average Performance of Flat-Plate Solar Energy Collectors. Solar Energy 7(2), 53–74 (1963)
26. Carriona, J.A., Estrelaa, A.E., Dolsa, F.A., Torob, M.Z., Rodriguez, M., Ridaob, A.R.: Environmental decision-support systems for evaluating the carrying capacity of land areas: Optimal site selection for grid-connected photovoltaic power plants. Renewable and Sustainable Energy Reviews 12, 2358–2380 (2008)
27. DBPrior10K GIS layers, http://www.centrointerregionale-gis.it/DBPrior/DBPrior.asp (retrieved on February 2012)
28. Italian National Geoportal, http://www.pcn.minambiente.it (retrieved on February 2012)
29. Sims, R.E.H.: Renewable energy: a response to climate change. Solar Energy 76(1-3), 9–17 (2003)
30. International Energy Agency (IEA): Key World Energy Statistics. International Energy Agency. OECD Publication Service, Paris (2008)
31. Muselli, M., Notton, G., Poggi, P., Louche, A.: Computer aided analysis of the integration of renewable energy systems in remote areas using a geographical information system. Applied Energy 63(3), 141–160 (1999)

The Comparative Analysis of Urban Development in Two Geographic Regions: The State of Rio de Janeiro and the Campania Region

Massimiliano Bencardino[1], Ilaria Greco [2], and Pitter Reis Ladeira

[1] Faculty of Political Science, University of Salerno, Via Ponte don Melillo,
I - 84084 Fisciano SA
mbencardino@unisa.it
[2] Faculty of Business and Economics, University of Sannio,
ilagreco@unisannio.it,
pitterladeira@hotmail.com

Abstract. In this paper we verify the possibility of implementing a regional comparative analysis between two systems, the State of Rio de Janeiro and the Campania Region. Primarily, we defined the boundaries of analysis area. Then, we will provide a comparison of the hierarchical structure of the two systems of cities, applying the rank-size distribution or Zipf's law on the top 15 cities of each one. Finally, we will analyze the specificity of each system and the metropolitan areas of Rio de Janeiro and Naples.

Keywords: Urban rank-size hierarchy, benchmarking, metropolitan areas, polycentrism.

1 Introduction

This paper analyzes two very different geographical regions: the region of Campania and the Federal State of Rio de Janeiro[1]. This comparison comes from the desire to produce a comparative evaluation of the metropolitan areas of Naples and Rio de Janeiro, in order to understand if the mechanisms used for the analytical formulation and articulation of the urban development of the first city can be used also for the another one, very distant and very different.

Naples is the ninth European city by population in the urban area and the third in Italy, while Rio de Janeiro is the third in South America (after São Paulo and Buenos Aires) and the second in Brazil.

For this reason, the characteristics of the functional region, especially in the metropolitan area, are analyzed with the purpose to comprise the aspects of the spatial distribution of urban functions.

---

[1] The paper is the result of a common reflection of the authors; however, the single sections can thus be attributed to: Massimiliano Bencardino paragraphs 1, 2, 3.1 and 5, Pitter Reis Ladeira paragraph 3.2 and Ilaria Greco paragraphs 4.1 and 4.2.

This comparison may be deficient or partial because not all factors were taken into consideration. But this paper represents only the first step of a more detailed study.

The first question we posed is to understand on what basis these two systems could be compared and what could be the local context that can make them comparable.

The second step was to analyze the system of cities located within the regions, trying to understand its evolution, the hierarchical structure and the functional relationships.

Finally, we analyzed the two metropolitan areas trying to understand the dynamics, the specificities, the government policies that will guide the development of them for the future and how the cities are able to meet the challenges of the globalization of economy, societies and cultures.

2   Basic Characteristics of the Considered Geographic Regions

The first question concerns the base of the comparison. Indeed, it's essential to understand what are and if there are equivalent territories on which to base the analysis, being the cities not only belonging to two different States, but also to two different continents.

The Italian State is organized into 22 regions, 110 provinces (in course of elimination) and about 8100 municipalities. By contrast, Brazil is a federal union of 26 states with the addition of a Federal District, which houses the capital Brasília. But, from 1889, the year of the proclamation of the Republic, to 1960, this District has coincided with the municipality of Rio de Janeiro, which was the capital of Brazil before 1960.

Brazilian States have a good self-government in legislative matters, public security and collection of taxes. At the head of the state government there is a Governor. The institution immediately below the State is the municipality. While, the legislative autonomy of the Italian Regions is notably increased beginning from the 2001 constitutional reform, but a full fiscal autonomy is not reached yet.

The State of Rio de Janeiro has a territory of 43.800 square kilometers, with 15 million inhabitants, it is the third most populous state in Brazil and the third last in extension, equal to 0.51% of Brazil. Campania is a region of about 13,6 thousand square kilometers with 5,8 million inhabitants, its capital is Naples and it's the second by population among the Italian regions (after Lombardia) and the first for density.

As mentioned, the analysis between the two regional contexts (Federal State of Rio and the Campania Region) seems fairly appropriate.

Dissolved the first question, the second step is to analyze the system of cities within the context of reference. In this study, which can be considered as initial, the demographic dimension of cities is the first parameter to do the analysis of the network of cities within the regional space and is one of the bases of the studies on urban hierarchy. In fact, we analyze it through rank-size distribution (Zipf's law) and this rule concerns the population of a city and its rank in the urban system.

In light of this simple calculation, we have a first compared image of the systems of city in the two geographical regions (Fig. 1).

From this comparison it's clear that the shape of the hierarchical distribution of cities is similar, despite the different sizes of the two areas for both spatial extent and number of inhabitants.

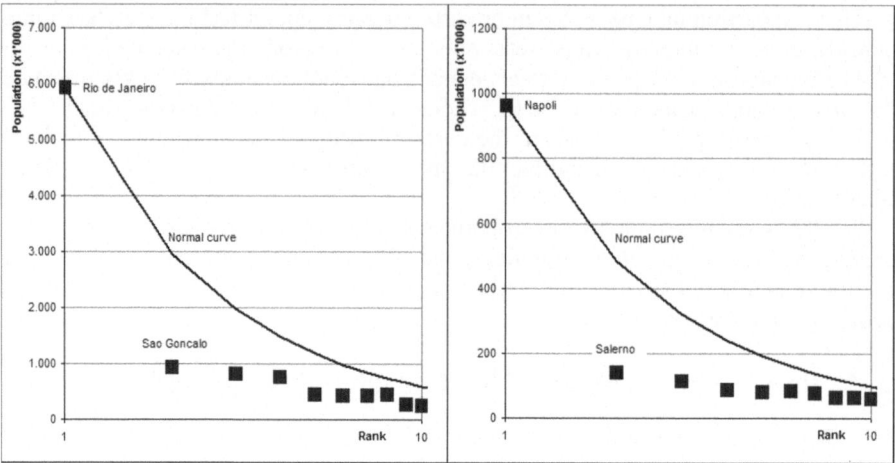

Fig. 1. Rank-size distribution applied to the two regions.

In fact, the two regions are both strongly polarized around their *primate*, Rio de Janeiro and Naples. The deviation of the other cities from the normal curve, in which the second most important city equal to 1/2 of first, the third equal to 1/3 of the first and so on, is quite similar. The size ratio of *Sao Gonçalo* with Rio de Janeiro is 1/6,3 and of *Salerno* with Naples is 1/6,9. Then, all the other cities are in the same proportions with the corresponding, to form a distribution absolutely equivalent, unless a factor of proportionality.

This result suggests that the two systems can be considered equivalent and that the analysis of the two contexts can be investigated to understand the different development and if there are cultural and economic conditions that can be analyzed in a comparative evaluation.

To continue the above analysis, we analyzed the spatial distribution and the urban dynamics of the top fifteen cities of the two geographic regions, as shown in Tables 1 and 2 and in Figures 2 and 3.

A first important item of discussion emerges from these tables: among the first fifteen cities of the State of Rio de Janeiro, 11 belong to the metropolitan region of Rio, and only the eighth city in the state hierarchy does not belong to it. Likewise for the Campania Region in the top 15 cities, 11 of them belong to the province of Naples, which essentially coincides with its metropolitan area.

The confirmation of the strong polarization of the two metropolitan areas around the city, on the one hand, confirms the comparability of the two urban systems and, on the other hand, indicates that the metropolitan area is the ideal scale for a precise comparability between the two urban systems.

Exactly on this scale the analysis of Chapter 3 and 4 of this paper will focus in-depth.

A second item that emerges clearly is the difference between the evolutionary dynamics of the two systems of cities. In fact, if the decline of Naples in Campania is accompanied by a substantial compensation growth in other provincial capitals, this is not done for the State of Rio de Janeiro.

In fact, in the State of Rio all cities show a substantial growth, both those belonging to the metropolitan system and those external to it.

Table 1. The first fifteen cities of the State of Rio de Janeiro

City	Population in 2010[2]	Population in 2000	±
Rio de Janeiro	5.940.224	5.857.904	+
Sao Gonçalo	945.752	891.119	+
Duque de Caxias	818.432	775.456	+
Nova Iguaçu	767.505	920.599	-
Belford Roxo	455.598	434.474	+
Niteròi	441.078	459.451	-
Sao Joao de Meriti	439.497	449.476	+
Campos dos Goytacases[3]	442.363	406.989	+
Petropolis	277.816	286.537	-
Volta Redonda	246.210	242.063	+
Magé	218.307	205.830	+
Itaborai	210.780	187.479	+
Mesquita	159.685	-	+
Nova Friburgo	173.989	173.418	+
Barra Mansa	172.484	170.753	+

Table 2. The first fifteen cities of the Campania region

City	Population in 2008[4]	Population in 2001	±
Napoli	963.661	1.004.577	-
Salerno	140.489	138.093	+
Giugliano in Campania	113.811	98.657	+
Torre del Greco	87.735	90.465	-
Casoria	80.028	81.887	-
Pozzuoli	83.335	78.938	+
Caserta	78.965	74.953	+
Castellammare di Stabia	64.866	66.706	-
Afragola	63.658	62.236	+
Benevento	62.507	61.773	+
Portici	54.743	60.068	-
Marano di Napoli	59.120	57.403	+
Ercolano	55.118	56.728	-
Aversa	51.947	53.219	-
Avellino	56.939	52.690	+

Therefore, the reasons for this strong differentiation may be searched in the different economic conditions of the two systems in question, in the different attractiveness of the two metropolitan areas compared to their countries or in different functional specialization, that both individual cities and the whole region express.

---

[2] Istituto Brasileiro de Geografia e Estatistica http://www.ibge.gov.br
[3] The underlined cities do not belong to the metropolitan area
[4] Atlante statistico dei comuni (ISTAT) http://www.istat.it/

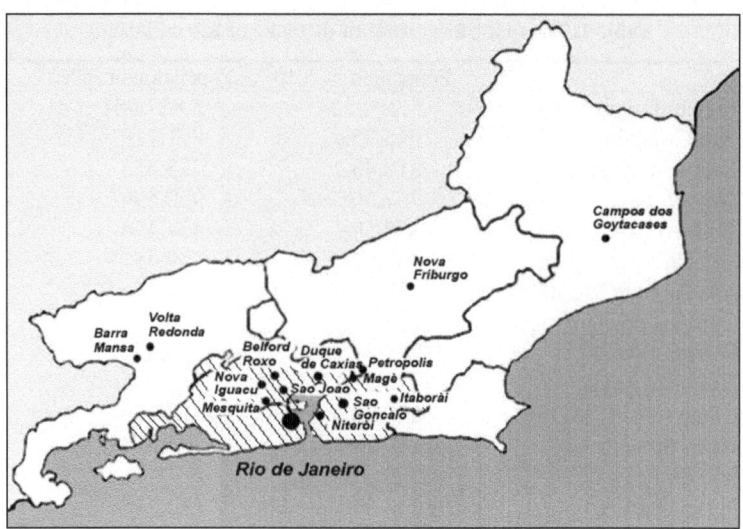

Fig. 2. The State of Rio de Janeiro

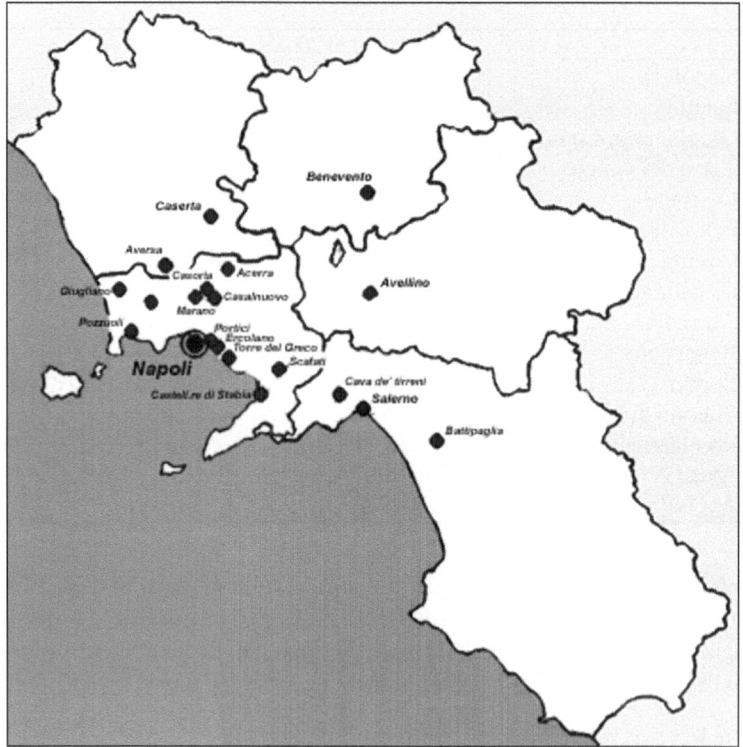

Fig. 3. Campania region

The State of Rio is located in South-east region, the most economically important in Brazil because it is responsible of over the 50% of the Pil of the Country together with the near states *St. Paul, Minas Gerais* and *Espírito Santo*.

The State of Brazil is strategically located with respect to Mercosur, the Southern Cone Common Market created with the signing of the Treaty of Assumption, signed by Argentina, Brazil, Paraguay and Uruguay.

The reserves of natural gas and oil are respectively 47.4% and 81% compared to the total reserves in the country. The State is the largest producer of oil and natural gas in the country with respectively 85% and 45% of national output and an output of 2 million barrels per day. In the whole State there are 23 thermoelettric plants operating and 10 planned. Among these working, 5 use as fuel the natural gas, 5 the diesel one, 1 the gases of refinery and 2 the enriched uranium.

The state of Rio de Janeiro has historically had a standard unbalanced development, due to heterogeneity between the economic and administrative capital and its peripheral zones. Its political separation had for more than 150 years but also other regional fragility have thus caused the concentration of production in its capital and therefore the separate growth of peripheral areas. This separation was reflected in a low degree of inter-regional integration, in a lack of a good network in urban and impoverished rural areas, which have been subjected to the negative effects of the polarization.

But, especially in recent years, in some cities, we can recognize various phenomenons of change, particularly in the cities belonging to the metropolitan region. One of these is *Sao Gonçalo*, a city with a diversified economy and that is investing in socio economic change, focusing on the agricultural production and on commercial enterprises, being also an obligatory passage for some tourist areas of the State.

Another important city is *Duque de Caxias*, whose growth has effects on the whole State of Rio, concentrating most of the specialized service industries. There are 1984 industries and 19'562 commercial establishments (SMF, 2009). The economic growth of the town seems to be coincident with the installation of the REDUC refinery, opened in 1961, the most complete oil refinery in the country.

Then, there is *Nova Iguaçu*, that joins the developed industry and a commerce in the alimentary sectors the birth, in 2003, of the *Center Federal de Educação Tecnológica* (CEFET), a ministerial institute for the superior and university education.

Finally, there are *Belford Roxo* with the chemical industry, *Niteroi*, a rich and commercial city and financial center of Rio, and other centers whose nature is purely commercial.

Campos dos Goytacazes, the sixth richest city of Brazil, stands among the centers of the State outside the metropolitan area. Its economy is based on petroleum royalties, but that has brought a greater diversity of investments, especially in services. It's a city of cultural and economic reference for the State.

Among other cities worthy of note, there are *Volta Redonda* and *Barra Mansa* which form a pole active mainly in the secondary, also benefiting from a strategic position lying on the main road Sao Paulo-Rio de Janeiro.

Beginning from the '80s, the "counterurbanization" appears in Italy in relation to the decline of population pressure and residential building in central cities and the growth of the move towards the first, the second and sometimes even to the most distant urban suburbs.

In Campania in particular, the economic and industrial situation of Naples deepens and, in the late '80s, unfavorable economic trends but also the closure of the industrial area of *Bagnoli*, in the west of Naples, that had characterized the previous years, set in motion a process that later will be called "deindustrialization".

A period of downsizing and closing of the production of entire departments that will also reflect the demographic data to a real turnaround begins. The city's population decreases, and the rate of unemployment increases.

Over the past 25 years, desertification continued in Campania especially in rural areas, but at the same time already since the '90s a process of rebalancing between the urban center of Naples and the other "strong" pole of a regional system is in progress, and these centers gradually begin to acquire a significant autonomy and to perform a function of territorial balance, such that for a long time geographers speak of polycentric development of the regional system, which has precisely the characteristics of a "metropolitan polycentrism" focused on Naples.

3   The Metropolitan Area of Rio de Janeiro

3.1   Territory, Urbanization, Activities and Functions

The Metropolitan Region of Rio de Janeiro is an area of approximately 5610 sq km, also known as Rio Grande; it was established by the Complementary Law No. 20, in July 1974, joining the states *ex-Rio de Janeiro* and *Guanabara*, which represented then the metropolitan areas of Rio Grande and Niterói (Figure 4). With 11,812,482 inhabitants (IBSG), it is the second largest metropolitan area in Brazil, South America and the third in the world (Census 2010).

In recent years, the average annual rates of increase of population, of 0.75% (1991-2000) and 0.82% (2000-2005) in the State capital and of 1,18% (1991-2000) and 1.05% (2000-2005) in the metropolitan area, indicate a slight decrease in growth rate of other municipalities and a small increase in the capital.

As considered by the IBGE, the Metropolitan Region (including Itaguai, Mangalore and Marica), has a GDP of $ 172'563'000, which constitutes the second largest center of the national wealth. It concentrates 70% of the economic power of the State of Rio and 8.04% of the all goods and services produced in Brazil. For many years it was the second largest industrial center of Brazil, with oil refineries, steel mills, shipbuilding, petrochemical, gas, chemical, steel, textile, printing, publishing, pharmaceuticals, beverages, cement and furniture. However, recent decades have witnessed a remarkable transformation in its economic profile, which is becoming more and more a great national center of services and businesses.

It brings together key national and international groups in the marine industry and the largest shipyards in the country and holds about 90% of the production of ships and offshore facilities in Brazil.

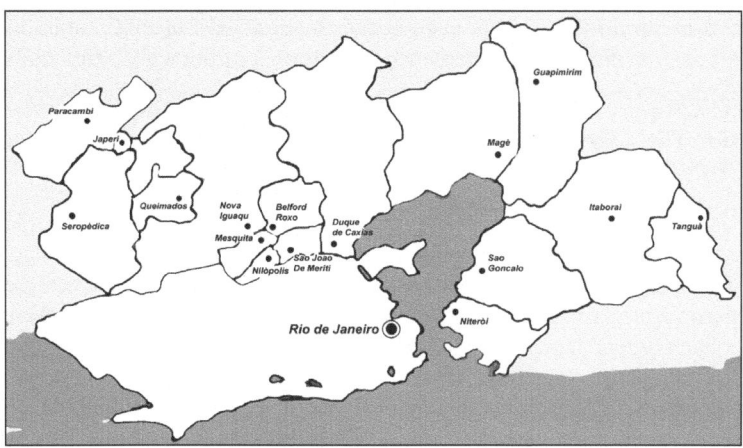

Fig. 4. Metropolitan Region of Rio de Janeiro

In the oil sector, there is a wide presence of over 700 companies, including the highest in Brazil (Shell, Esso, Ipiranga, Chevron Texaco, El Paso, Repsol, YPF). Many of these sustain research centers throughout the State and, together, produce more than four-fifths of the oil and fuel distributed to national service stations.

This situation was such that the state of Rio de Janeiro presented the highest level of polarization of its product and its people around its metropolitan region. After the 70s and 80s, who represented the worst moment of the region's economy, the decade of 1990 brought new expectations with the return of economic dynamism, coupled with the rising cost of living and helped the spread of the growth within the state, observing even a small redistribution of the population.

The territorial imbalance that marked the growth of the State, the decay of the primary sector and the impossibility of forming a modern agricultural system, combined with the factors of historical development and the scarcity of secondary sector, have stimulated a premature expansion of the tertiary sector.

Rio is a city of international events that began with the Conference on Environment and Development United Nations in 1992 and will culminate with the creation of the World Cup in 2014 and the 2016 Olympics, but it is also a city of constant insecurity due to ongoing conflicts and organized crime.

The agglomeration process together with the strong migration to the city of Rio has brought the phenomenon and the problem of the *favelas*. The phenomenon of the expansion of the favelas goes back to many decades ago, and has affected all the most populated centers in Brazil, from the poor northeastern states to those of the industrialized south, from Sao Paulo to Rio de Janeiro. The urban sprawl of the city is inextricably linked to the difficult realities of the countryside, where the large estates and monoculture have isolated many farmers, leaving them at the mercy of price fluctuations. In view of a stable employment, since the '30s, many people decided to leave their places and to start a new life in the big cities, where the opportunities were better for a living. In those years have begun the processes of mass migration to the metropolis and the consequent emergence of "informal settlements", called favelas.

3.2 Rio: The City of *favelas*. Issues and Programs for the Revitalization of Critical Areas

In the "Carioca" city the problem of the *favelas* occurred earlier than in other places, and has assumed quite specific contours. For a variety of reasons, Rio de Janeiro has seen the birth of the favelas earlier and more quickly. Rio was the capital of Brazil before the construction of Brasilia, and has attracted many people from all states of the federation.
For its political importance, also, Rio de Janeiro has been the center of several public works. Tashner (Tashner, 2003) recalls, for example, the urban reform of Pereira Passos, which has distorted the city and led to the construction of the great thoroughfare of Avenida Central (now Avenida Presidente Vargas). This work has forced thousands of persons to leave their homes and to build new illegal homes on the hills above the Zona Norte.

Moreover, the urban development and the construction of major public works required an enormous mass of workforce and many people were forced to emigrate in Rio de Janeiro just to get a job in the construction sector since from '50s.

Even the massive expansion of the city, which in recent decades has focused on the south area, led to the birth of the rich neighborhoods of Copacabana, Ipanema, Leblon and Barra da Tijuca and made use of cheap labor coming mainly from the Northeast. In this way, parallel to the major hotels and elegant condominiums in to the south, even the favelas have expanded beyond measure. So in a sense, the tourist vocation of the city was another factor which contributed to immigration in Rio, and then increasing *favelada* population, which required the construction of very large hotel complexes and that they needed a great use of workers.

If the birth of the first *favela carioca* dates back to 1897, when soldiers returning from the campaign Canudos found themselves homeless and occupied the area of the *Morro da Providencia* in the north area of the city, starting in the 30's we see the spread of the phenomenon, which was fed by the economic crisis and the collapse in coffee prices, which bankrupted many people (the term derives its name from the favela of a leguminous plant widespread in the region of Canudos). The soldiers were scattered on the hill where the current rises Morro da Providencia, as the plant on the heights of Canudos, and the Morro was renamed *Morro da Favela*. The name was used for all the illegal settlements that were built later.

The great public works, and building expansion, as was said, did the rest, and, in 1970, 10% of the population of Rio lived the favelas (1% in São Paulo), and in twenty years, 1973-1993, the percentage increased to 20%.

An another key element for understanding the specificity of the phenomenon of the favelas of Rio de Janeiro, is related to the particular physical and geographical city. This city is built in the middle of some mountain ranges covered with forest, and its favelas are laid on a steep rocky ridges. This feature meant that the favelas were perceived as free zones, distant and totally different from the rest of the city, and led to the familiar distinction between Morro and asphalt. The favela is isolated on the hill and the respectable citizens living in neighbourhoods where there are the roads.

This distinction has greatly influenced the relationship between population of favela and rest of the city, increasing distance and increasing prejudice.

The problem of the favelas is again returned to the phenomenon of violence. This has led the Prefecture to implement various programs for upgrading and urbanization.

The most important and famous is certainly the *Favela-Bairro program*. The program, started in 1994, is still in progress and it in the beginning focused on the favelas of medium size and in the first phase have benefited about 220,000 people. Its objectives are: 1) to build and to complement the urban structure with rehabilitation programs and democratization of access to services; 2) to produce the conditions for the reading of the *favela* as a neighbourhood of the city; 3) the participation of the inhabitants; 4) the introduction of urban values of the city as a sign of its identification as a neighbourhood.

Today Rio de Janeiro has more than six hundred favelas spread in every part of the city, but according to some, would be many more, since the definition of favela dall'IBGE provided and used for the official count, excluding the settlements are too small (Preteceille, Valladares 2000).

Of the 5,800,000 inhabitants of the metropolis, more than one million living in favelas, while about 880,000 were in 1991: during the decade of the nineties the population is increased by 3%, those living in favelas is increased by 24%. This especially is heading towards new areas of expansion of the city, the west and the area of Barra da Tijuca, where there is ample availability of land, even in the area and the north, however, the population of the favelas continues to grow at high rates (respectively 5% and 13%) (Cavallieri 2003).

Although public policies try to stop the growth of the phenomena of exclusion and support the integration of communities, the old and new problems continue to be many. In settlements that continue to emerge in the west of the city and the metropolitan area, urbanization and rehabilitation interventions are inadequate, and indeed even many central favelas have to live with the problems of hygiene and health risks.

Another program - Morar Carioca - for social inclusion through urban integration was launched in July 2010 by the City of Rio de Janeiro. It has as its objective the full and final social and urban integration of all favelas by 2020. In October 2010, in fact, it was sponsored a competition for the urbanization of favelas concerning the construction of infrastructure, facilities and services, with particular attention to aspects of social inclusion and environmental protection. An important innovation is the implementation of a system for maintenance and conservation of the structures, than the control, monitoring and employment law and land use. Have already been fixed, also the guidelines for action in the field of transport, health and environment for all disadvantaged communities. Sustainable practices are used in construction, waste recycling and reorganization of urban space. All of the favelas will have mini-centres for the recycling of solid wastes including organic. These could be used as fertilizer for reforestation in the communities where native vegetation has been demolished.

The activities of urbanization of Morar Carioca were divided into several phases and will be implemented according to the size and conditions of each community. In areas classified as "urban" are provided for the development of water supply, sewerage, storm water drainage, street lighting and flooring. In the case of non-urban communities, diagnosed by the City as "high-risk or not suitable for residential use", families will be recorded and resettled in a housing unit.

The second phase of the program, the more operational, will concentrate its work in communities within a radius of four kilometres from the future Olympic facilities in areas South, North and West.

In this phase, will be urbanized 216 favelas divided in 91 groups in which there are about 312,000 people. In the Southern area will be urbanized favelas as Chácara do Céu a Leblon, Vila Parque da Cidade a Gávea e Ladeira dos Tabajaras, between Copacabana and Botafogo. These communities are near to the core "Copacabana das Olimpíadas" which will host beach volleyball competitions, marathon swimming, triathlon, canoeing and sailing. Also the favela Morro dos Macacos a Vila Isabel and Engenho Novo, make the list with others. The prediction is that the second phase is finished before the 2014 World Cup.

Finally, Rio de Janeiro is a great laboratory for the testing of innovative projects in the non-profit. The city is full of community associations, non-governmental organizations and local, international organizations, cooperatives, foundations, research centres and universities. With the contribution of the highest human depth of characters in Rio were born and have developed important social movements for democratization and social inclusion.

In June 2012 in Rio de Janeiro will host a historic event. Representatives of governments of the whole world, international organizations, the most important research centers and civil society will gather together to discuss the future of Planet Earth conference in Rio + 20, namely the United Nations Conference on Sustainable Development. Twenty years after the historic conference of 1992, Rio de Janeiro returns leading development choices of our planet.

Exactly twenty years after the first edition of the Conference on Environment and Development of United Nations, will discuss new issues facing the world in which we live: the challenges of sustainable development, the problems of global warming, the need to preserve the natural environment in which we live, the need to change from the root the adopted development model and the base mechanisms of industrial production of our society, the lifestyles in favor of a more equitable, inclusive and respectful of nature approach.

The Earth Summit in 1992 represented a historic opportunity in which the foundations were laid for a radical change in thinking, acting and planning policies of many countries. For a big emerging city as Rio de Janeiro, the city of a thousand favelas, it was twenty years ago, as it is today, a very important opportunity to affirm the importance of emerging economies in guiding the new processes of development.

4 Metropolitan Area of the City of Naples

4.1 From Metropolitan Area to Metropolitan City of Naples

The urban settlement system in Campania has resisted for years on a centre-periphery model based on the city of Naples, hyper-congested inside with almost one million inhabitants (959,614 pop.) and uncontrolled growth on the outside, has exercised a physical, political and functional domain on the other cities of the Region.

Since the late Seventies, when for the first time defined the *Metropolitan Area* of Naples - considered at the time, for size and frequency of relationships among its components, the only metropolitan area in the Mezzogiorno (South of Italy) - the urban structure of the Campania region appears centred, more than in other southern regions, on a "mono-centric system" with a strong concentration of population and economic activities in the metropolitan area of Naples, embracing a territory of 800 square km with 4,4 million people, form an area with a strong «over urbanization in quantitative terms and under urbanization in terms of quality» (Mazzetti, Talia, 1977, p. 157).

The same city of Naples, on the basis of a plant radiocentric, has expanded over the years "to oil spot": the small scale of the urban coastal area has forced the city, on one side, to gradually embrace the surrounding hills, now the site of major districts with high residential as Capodimonte, Vomero and, secondly, to invade the valley of Fuorigrotta-Coroglio to west, the alluvial plain of Sebeto up the slopes of Vesuvius, the north side of Piscinola-Miano and, finally, to spread beyond the municipal boundaries by uniting the continuous band of neighbouring countries.

Today, this endemic absence of directionality in the development, of course with the political-economic and ethical-cultural factors, is among the main causes of the weakness of the urban structure of the Campania and unformed growth of the conurbation of the city of Naples which covers, now, the whole province of Naples.

The Province, in fact, without a cohesive pattern of development continues to grow in an undifferentiated way in every direction, presenting itself as a high concentration without development. It is, with 92 municipalities, currently, the third province in Italy by population, has a population density of 2632 inhabitants sq/km, almost 4 times that of Rome (705 inhab./Sq km) and receives, in addition to the capital city of the region (Naples), as many as 9 of the 15 most populous cities of the Campania region. The city of Naples alone receives more than one-sixth of the entire regional population and almost a third of that of his province; it is currently 18 th in Europe by population.

The start of a process of re-balanced and the definition of a system of planning for large areas able to govern the growth of a conurbation in its complexity has over four million inhabitants[5], the first in the Mediterranean, seems to pass for the establishment, than 22 years after the institution of the Metropolitan Area (MA) of Naples with the ex Law 142/90 on the structure of local authorities, the Metropolitan City of Naples as a new institution the territorial government.

The start of a process of re-balanced and the definition of a system of planning for large areas able to govern the growth of a conurbation in its complexity has over four million inhabitants, the first in the Mediterranean, seems to pass for the establishment, than 22 years after the institution of the Metropolitan Area (MA) of Naples with the

---

[5] In reality, the size of the Metropolitan Area of Naples varies greatly depending on the source: the 2005 UN data assign Neapolitan conurbation with a population of approximately 2,200,000 inhabitants., Taking into account only a small area to the municipalities bordering the urban area of Naples. According to OECD estimates would come to about 3,100,000 inhabitants, behind Milan and Rome. The U.S. Census Bureau and Times Atlas of the World estimates a population of about 3 million, Eurostat data about 4 million, while sources SVIMEZ 4,434,136 distributed over an area of 2,300 km², making it the second Italian metropolitan area by population. For CENSIS, however, the Naples metropolitan area is second only to Mega Lombardy region, with 4,996,084 inhabitants.

ex Law 142/90 on the structure of local authorities, the *Metropolitan City* of Naples as a new institution the territorial government.

The Metropolitan City, required by Law 142 and revived the reform of Title V of the Constitution of 2001 and the Delegated Law n.42/2009, is an authority over municipal comparable to the Province, but with more administrative functions currently owed to the municipalities, and shall replace the province of Naples within 36 months from the date of entry into force of the law (by May 2012).

The establishment of Metropolitan City of Naples presupposes, therefore, the physical demarcation of the Metropolitan Area. In reality, the current legislation do not provide specific criteria for the delimitation of metropolitan areas, but simply provides that the identification should be between parts of the territory consisting of a central city and a number of smaller towns united to it by contiguity and «relations of close integration with regard to economic activities, essential services in social and cultural relations and territorial characteristics» (Article 17 of Law 142/90).

Abandoned the hypothesis of a metropolitan area restricted within the same pro-vince of Naples, which would lead to excessive fragmentation of an area that always has strong territorial imbalances with the rest of the region and according to recent changes applicable laws, would not enable the establishment of territories other than a new province of reference, the proposals for the delimitation of the metropolitan area of Naples can substantially attributable to two: the first assumes the coincidence of the current provincial boundaries; the second, knowing the widespread urbanization in recent decades, considers an integral part of the Metropolitan Area Neapolitan large areas of the provinces of Caserta (Agro Aversa), Salerno (Agro Nocerino Sarnese) and Avellino (Baianese).

The Metropolitan Area of Naples has lived, in fact, since the late Seventies continuous processes of urbanization and de-urbanization since the decline of the urban system of the city of Naples have involved, first, the neighbouring municipalities of the first crown, as Portici and Casoria, which in the eighties have experienced strong population growth, and decreases in the 90s in favour of municipalities in the second crown to the north and north-west of Naples to the periphery of Caserta. The same thing happened against the municipalities of third crown of Aversa, Nola, Sarno e Nocera and between Pompei and Scafati, which in recent years have absorbed the decrease of the municipalities of the second row, increasing the population with reduced rates for the greater distance from the capital[6].

The prevailing view is towards the definition of a metropolitan area that coincides with the present Province of Naples, as encountered fewer difficulties in the modification of existing administrative areas (filed by municipalities) and has advantages in relation to a range of geographical and of spatial organization that have sustained long-term active life of the province. Before the constitution of the Metropolitan City, the 30 neighborhoods of city of Naples, who until 2006 formed 21 districts, were grouped into 10 *Municipal District* of about one hundred thousand inhabitants (Figure 5).

---

[6] Among the major studies concerning the delimitation of Metropolitan Area of Naples and its expansion, see: Svimez (1958, 1970), Cecchini (1983); Bencardino (1980), Coppola, Viganoni (1994), Forte (2003), Amato (2007), Viganoni (2007), Sommella, (2009).

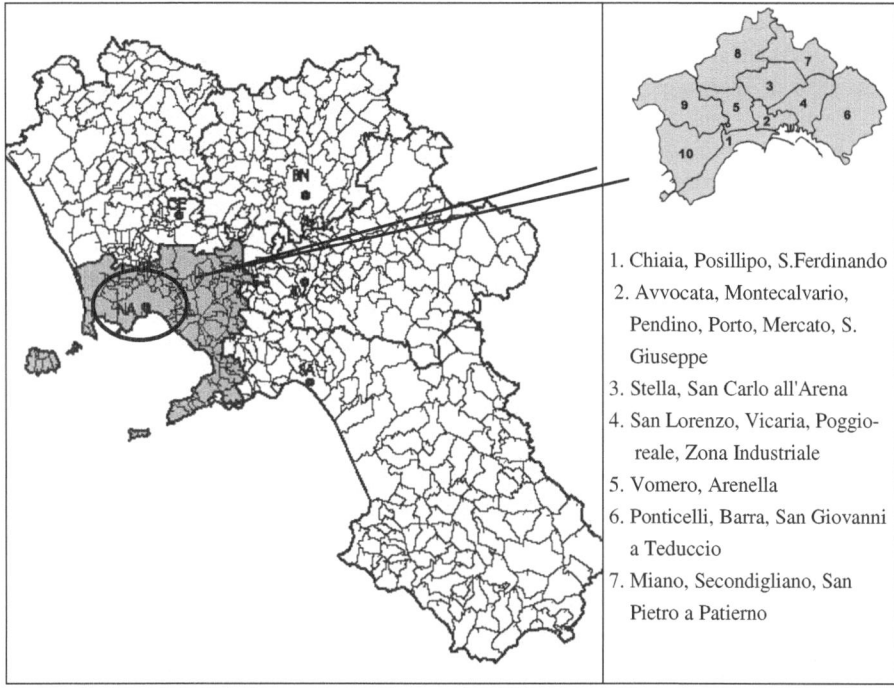

Fig. 5. Hypothesis of delimitation of Metropolitan City of Naples and New *Municipal District*

In Italy, the process of establishment of the metropolitan cities is, however, a impasse is due to the expectation of legislative decrees that the government should issue, according to L. 131/2003, for adapting legislation (Legislative Decree 267/2000), both a socio-cultural difficults: of the 14 Italian metropolitan cities only seven (Venice, Genoa, Bologna, Florence, Catania, Palermo and Messina) have defined the metropolitan area, while seven others including Naples (Turin, Milan, Trieste, Rome, Naples, Bari and Cagliari) did not proceed to formally identify the area, although some of these studies were still carried out and proposed perimeter.

4.2 Problems and Programs for the Revitalization of Critical Areas

More than twenty years after the institution of the Metropolitan Area of Naples, the vast and complex conurbation of Naples presents a series of problems, contradictions and inconsistencies, which now take the form of real "emergencies" to face in the process of territorial, urban, physical, environmental, administrative and economic development.

The establishment of the Metropolitan Area was intended, from the central city of Naples full of values and culture, heart of the Mediterranean, and able to guide the development of a wide area, the rationalization of the physical and functional relationships between the same "mother city" and its hinterland, both in terms of socio economic and political system and morphological and infrastructure relations. In reality, compared with these intentions, choices and decisions in terms of governance have only a small part contributed to the goal, leaving space for spread of forms of social and economic disintegration and an indiscriminate use of land (Sommella, 2008).

A view of short-sighted and fundamentally localist has, in fact, produced during the last thirty years many plans and programs characterized by being substantially independent and free of an integrated system of choices and evaluation although the obvious repercussions on the whole of the conurbation, represents essentially a set of choices in many cases contradictory. Some of the same ongoing operations (complex programs, Neighbourhood Contracts, Project Financing Management Centre) which in many cases have had all the necessary steps deliberative and approbation, are unfortunately in a significant delay for administrative inertia, internal conflicts to the same authorities and agencies with overlapping deficiencies, severe limitations and inconsistencies, although the need for such interventions by major and obvious physical and economic impacts on cities.

Particularly worrying is the future of the historic center, declared World Heritage Site that, unique in Europe, is for over 50 years without significant requalification or architectural restoration, in a vision of development and enhancement. The historic center, and the "old" in particular, was essentially left to its fate of decay and dilapidation because of a sterile and unjustifiable inaction that damages the very people, the social and economic system, and the important cultural and historical heritage of which the city is a privileged witness and known around the world. That of the suburbs is another problem to be addressed in the context of a metropolitan area, overcoming the barriers that prevent integration and eliminating the margin of which they are afflicted. To degradation of many areas and suburban of urban districts is added the absence of "active policies" with commercial building projects, urban regeneration and the distribution network both in town centers and in suburbs.

The absence of adequate multi-level governance is reflected, of course, the entire provincial conurbation in which, against a very intense urban development there is severe shortage of supply of services and infrastructure and employment, which feeds a widespread condition unemployment and crime, including young people, and important forms of environmental degradation.

Finally, although Napoli confirms the fourth Italian city for economic movement, after Milan, Rome and Turin, the industrial sector crisis that has hit major production companies makes its effects felt throughout the Province.

5 Conclusions

As we hypothesize from the beginning, this work was an interesting moment of reflection on the development of two important urban regions: Rio and Naples are respectively the twenty-second and hundredth Capital of the World (World Atlas, 2010).

These cities showed many similarities both in the hierarchical structure of their regions and for the problems that they had to deal during the centuries, but also some important differences that inevitably marked the development of them. It's very clear that the socio-economic contexts of Europe and South America are completely different and they inevitably have influenced the urban development but at the same time, some phenomena seem to be repeated identically and in parallel and others seem to be repeated in a differential time.

First of all, both cities are characterized by forms of social exclusion and deprivation, that diffusedly interest whole districts of the cities (*Favelas*, Scampia), unresolved issues and at the center of urban policies. These issues and these forms of development not regulated is similarly contextualized in a specific historical period in which liberal economic models have dominated and have drawn strong expansion of modern industrial cities. The agglomeration and conurbation urban sprawl were born in many cities of the world that had required a significant amount of workforce in a limited period.

Today, the greater attention to environmental problems, the deindustrialization and a less pressure on the cities gives us the opportunity to rethink them. And also this issue appear similar to both cities. In particular Rio in June 2012 will host the conference Rio + 20, the UN Conference on Sustainable Development, twenty years after the historic conference of 1992. And Naples in 2013 will be the site of the IV World Forum of Cultures: three months dedicated to the great themes like peace, sustainable development, multiculturalism, knowledge, urban identity. It's clear that both these opportunities can only be a moment of reflection also on the new urban identities.

Finally, a brief consideration should be made on the difference given by increased administrative complexity of the Neapolitan urban system than the Carioca urban system. The metropolitan region of Rio is composed of 20 municipalities and the metropolitan area of Naples, identified with its province, fewer than 92 municipalities. This should be a reason for reflection.

The results of this first analysis are then encouraging, and we suggest that a more detailed study can be done.

References

1. Amato, F.: Dall'area metropolitana di Napoli alla Campania plurale. In: Viganoni, L. (a cura di), pp. 175–211 (2007)
2. Bencardino, F.: L'armatura urbana nell'area metropolitana di Napoli. In: Bollettino della Società Geografica Italiana, Serie X, Roma, vol. 9, pp. 55–83 (1980)
3. Burgos, M.B.: Dos parques proletários ao Favela Bairro. As políticas públicas nas favelas do Rio de Janeiro. In: Um Seculo de Favela. FGV Editora, Rio de Janeiro (1998)
4. Cavalieri, F.: Favela–Bairro: integração de áreas informais no Rio de Janeiro. In: A Cidade da Informalidade. O Desafio Das Cidades Latino-Americanas. Livraria Sette Letras FAPERJ, Rio de Janeiro (2003)
5. Checchini, D.: Nota sulle aree urbane meridionali. Studi Svimez 36(11-12), 423–430 (1983)
6. Coppola, P., Viganoni, L.: Note sull'evoluzione recente dell'area metropolitana di Napoli. In: Citarella, F. (a cura di) Studi Geografici in Onore di Domenico Rocco, Napoli, Loffredo, pp. 471–486 (1994)
7. Forte, E.: Il ruolo delle aree metropolitane costiere del Mediterraneo. Area metropolitana di Napoli, Firenze, Alinea (2003)
8. Sommella, R.: Il contesto territoriale dell'indagine: l'area metropolitana di Napoli e le sue articolazioni. In: Amato, F., Coppola, P. (a cura di) Da Migranti ad Abitanti. Gli Spazi Insediativi Degli Stranieri Nell'area Metropolitana di Napoli, Napoli, Guida, pp. 147–174 (2009)

9. Sommella, R. (a cura di): Le città del Mezzogiorno. Politiche, dinamiche, attori, Franco Angeli, Milano (2008)
10. Preteceille, E., Valladares, L.: Favela, favelas: unidade u diversidade da favela carioca. In: Henriques, R. (ed.) Desigualdade e Pobreza no Brasil. Ipea, Rio de Janeiro (2000)
11. Taschner, S.P.: O Brasil e suas favelas. A Cidade da informalidade – O desafio das cidades latinoamericanas (2003)
12. Viganoni, L.: Il mezzogiorno delle città. Tra Europa e Mediterraneo. Franco Angeli, Milano (2007)
13. Secretaria Municipal de fazenda, http://www.rio.rj.gov.br/web/smf
14. World Atlas, http://www.worldatlas.com/capcitys.html

Land-Use Dynamics at the Micro Level: Constructing and Analyzing Historical Datasets for the Portuguese Census Tracts[*]

António M. Rodrigues, Teresa Santos, Raquel Faria de Deus, and Dulce Pimentel

e-GEO – Research Centre for Geography and Regional Planning
Faculdade de Ciências Sociais e Humanas, Universidade Nova de Lisboa
Av. de Berna 26-C 1069-061 Lisboa - Portugal
{amrodrigues,teresasantos,dpimentel,rdeus}@fcsh.unl.pt

Abstract. Historical census micro-data – data aggregated for small-areas – is of foremost importance as a tool for understanding detail patterns in the distribution of social phenomena. However, the non-coincidence of census tracts' geometries for different years hampers the dynamic analysis of such information. This article applies a methodology which uses auxiliary geographical data to build coherent historical datasets when asymmetric mapping occurs due to incoherent geometries. This data serves as *control zones* which are the source of the computation of a weighting scheme which allows the re-allocation of data for common spatial units. An application to a municipality in the Southern coast of Mainland Portugal – Portimão – helps to show the usefulness of this analysis. *Form*, *structure* and *functional* attributes are combined within a coherent framework. Proximity measures are used to help to identify local patterns. The final outcome highlight the potential of both the methodology used and the historical dataset produced.

Keywords: Census micro-data, asymmetric mapping, social dynamics.

1 Introduction

When analyzing the distribution of human activity over a given spatial surface, the availability of aggregated census data for very small units (census tracts) is the best source of information. This is so due to three reasons: first, data results not from samples, but from the total number of individuals (population); second, the level of geographical disaggregation is the highest generally available; third, the fact that (normally) it is commonly available allows cross-studies and validation. This highly disaggregated micro-data[1] is very sensitive to local variations; moreover, as is the case of Portugal, when geometries from different exercises are not consistent, issues

---

[*] This research is funded by the FCT - Fundação para a Ciência e Tecnologia (Grant SFRH/BPD/66012/2009).
[1] Strictly speaking, one should refer to census micro data as the lowest aggregation level, as data is collected at the individuals' level.

related to what is known as the Modifiable Areal Unit Problem (MAUP) become more acute. The particular characteristics of spatial data [1] imply that certain care should be taken and specific tools used which account for the effect of physical proximity.

The most common issue which results from the aggregation of data into groups (in the case of spatial data – regions or spatial units) is that behavior or distribution is considered homogeneous within each region. This is what is best known as *ecological fallacy* [2]. The opposite side of this phenomenon happens when the level of disaggregation is such that it becomes extremely hard to identify common traces (in the case of spatial data – spatial trends) [3]. In such cases, Exploratory Spatial Data Analysis (ESDA) techniques are of great use since they allow conditioning the behavior of an agent (or an aggregated number of agents) on proximity in relation to other agents [5].

Form, *structure* and *function* [6] are fundamental aspects of spatial units' characterization which define how successful is the exercise of building multi-temporal micro data from census tracts' information collected for different timestamps. Broadening the level of conceptual analysis, *form* refers to "a particular way in which a thing exists or appears" [7]; this implies attributes as shape, size, etc. The absolute or relative area occupied by a particular agent – or object, which serves as a control zone (ex: building) is one key element of that agent. The *structure* (distribution, height, etc.) of agents/objects within a common spatial geometry (the result of transforming an asymmetric map into a symmetric one) influences the weight of each object in the control zones' definition. Finally, the *function* of the object determines *how* that object should be taken into account; for example, a vacant/empty building may be understood as having no function (commercial, residential, etc.).[2] Geometries of spatial tracts (sub-sections) do not coincide in the last two Portuguese censuses, since their *form* differs [8]. This data issue, which is common to other countries [9, 10], implies that there is no error-free or automatic method to produce a coherent multi-temporal database. For any given variable, the value of each observation is highly dependent on shape and size. Figure 1 serves as an example of how different geometries lead to distinct datasets; the count dataset from both schemes represented by the point events on the left would result in the following counts:

$$Scheme\ 1 = \{8,4,2,4\}\ ;\ Scheme\ 2 = \{5,7,6\} \tag{1}$$

The resulting flows dataset is accordingly different between schemes. The quantification of inter-regional flows is dramatically different as some flows are simply ignored as their origin and destination is contained within the same spatial unit.

---

[2] The definition of *null values* in respect to a specific function (attribute) is naturally dependent on the classification scheme used – in particular when dealing with categorical variables.

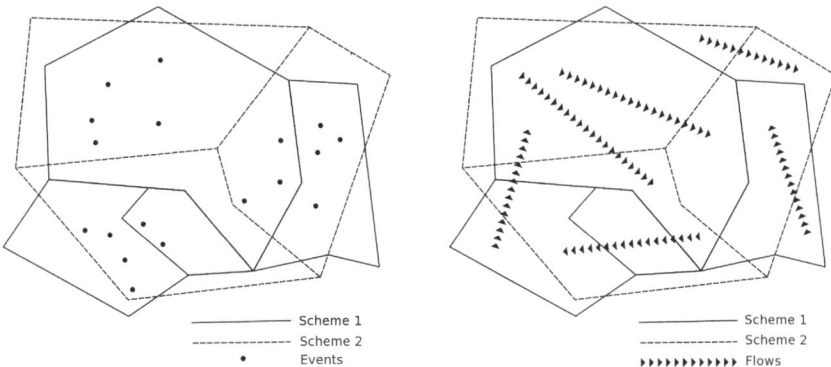

Fig. 1. Example of count and flow data aggregation according to distinct geometric schemes

Census data represent a static snapshot of a given reality. In order to perform cross-temporal analysis, there is a need to compare datasets from various census exercises. At the micro level, the number of possible applications of historical census datasets ranges from areas such as geo-demographics, geo-marketing, insurances and of course, local land planning [11]. Even if the disaggregation level of analysis is lower- the level of aggregation is higher (for example local authorities or municipalities) - census tract data allows for the analysis of intra-variability for each of these regions, minimizing to a great extent the problem of ecological fallacy. In such cases, these micro-data serves the purpose of structural analysis of spatial units; and *structure* can be represented by a probability density function representing intra-regional variability.

An *agent*, in the broadest of definitions within the scope of geographical sciences, operates in space. He (or it) has different characteristics – or *functions*. This implies that each agent makes a distinct contribution to the *functional characterization* of a set of agents. For example, if agents are individuals, and the function studied is literacy, each agent contributes differently to the description of the aggregate literacy level of a spatial unit - region. Accordingly, if agents are buildings, and the function studied is type of occupation (commercial, industrial, residential, etc.), then each agent contributes differently to the description of the type of land-use present in the same delimited area. *Function* allows the treatment of each agent, not as a homogeneous unit, but in an *object-oriented* way, as a unit with a common trace(s) (an individual, a building, etc.), but with distinct levels for distinct attributes.

If the starting point is data aggregated for different geometries and the objective is to build a coherent spatial-temporal database, some method must be used to overcome the problem of asymmetry in terms of spatial form. Asymmetric mapping techniques using control units/zones allow the creation of common geometries where data is re-allocated according to these common shapes [12, 13]. The definition and use of control zones is of the upmost importance since they are the source of the weighting scheme used to re-allocated data for a common geometric set of spatial units. Also, the shape of the resulting areas is important since *neighborhood effects* depend on the contact-area between regions.

The objective of the present study is two-folded: first, it is intended to build a coherent multi-temporal dataset for census data for 2001 and 2011. This is achieved through the use of control zones, which account for non-stationarity in the distribution of agents over census tracts. Using a land-use layer for the study-area, form, structure and function attributes are taken into account to produce a coherent result. The second objective is to demonstrate the potential of the created dataset; this is achieved through an analysis of urban and demographic dynamics for the municipality of Portimão, located in the southern coast of Continental Portugal

The rest of the article is organized as follows: the following section will describe and justify the choice of the study-area; also, data is described in detail, in particular the construction of the resident buildings' footprint layer (control zones layer). Section 3 discusses methodological issues, describing all steps necessary to build the multi-temporal dataset. The overlaying techniques used produce locally distinct outcomes, which are typified. Section 4 illustrates an application of the data for the analysis of urban and social dynamics. Section 5 presents the conclusions and the importance envisage in such study.

2 Case-Study and Data Considerations

This section presents the study-area, explaining its relevance as an appropriate location for testing the present methodology. Urban dynamics and demographic changes at the micro-level are ultimate themes of analysis, although the greatest focus of the article is on the nature of the methodology used; hence, the study-area chosen should present distinct realities and challenges from a purely technical perspective. These criteria were of paramount importance in the choice of a municipality in Southern Continental Portugal, as it represents a territory characterized by heterogeneous land-use, with both industrial and agricultural plots or residential and services buildings. Related with the latter, the tourist sector has an important weight.

2.1 Study-Area

Portugal is divided in administrative terms into several levels: according to the *Nomenclature of territorial units for statistics* (NUTS), the country corresponds to a NUTS 1 area, divided in seven NUTS2 regions, and 30 NUTS3 [14]. This division of the territory (created at the supranational level for statistical and administrative purposes) runs parallel to the national administrative areas which divide the same spatial surface into 18 districts ("distritos"), 308 municipalities ("concelhos") and 4260 local administrations ("freguesias") [15]; their origins date back to the division of the landscape into roughly natural regions ("províncias") - at the highest aggregation level, and local parishes ("freguesias") - at the lowest aggregated areas [16, 17].

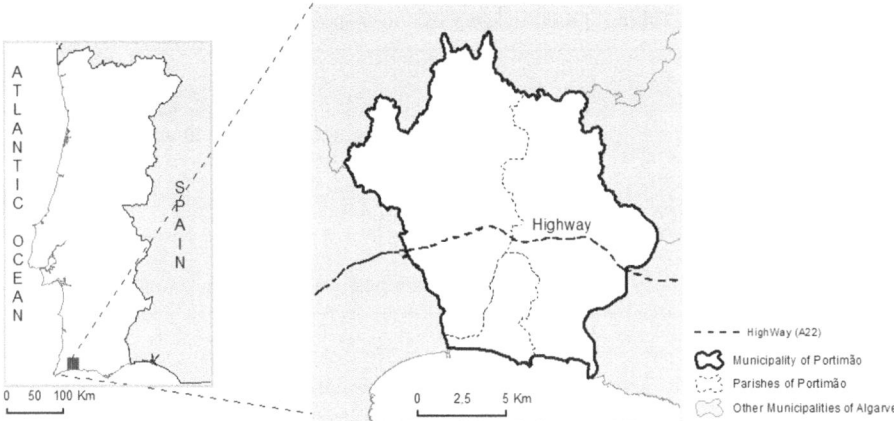

Fig. 2. Portimão municipality

The object of the present study is the municipality of Portimão, located in the southern coast of Mainland Portugal (figure 2). Its main node, Portimão, is one of the most important cities of the Algarve, second only to Faro; this is a region characterized by high rates of new construction, mainly related to the tourism sector. The municipality of Portimão, in particular, concentrates also important functions, as it is the largest urban node of the western part of the region. It is characterized by a densely populated area near the coast and sparse villages located inland, with 55 614 residents (2011 Census) and a surface of 182 km^2.

2.2 Data

The geographical data used in this study includes three planimetric sets: the census tracts (2001 and 2011) and resident buildings' footprint layer.

The census tracts in Portugal are available in vector format for the years 1991, 2001 and 2011. For 1991, the geographic base is in analogical format and is called "Geographic Spatial Referencing Base" (BGRE 1991). This map was then digitalized and made available in a Geographic Information System (GIS). For the following years, it is named "Geographic Information Referencing Base" (BGRI 2001, BGRI 2011) and is already based on a GIS [8]. When the BGRI 2011 was produced, drastic geometry changes meant that the National Statistical Institute (INE) was not able to maintain a coding system which would guarantee the existence of minimum common areas as was done in 2001 [18].

As mentioned above and it will be further explained in the methodology section, a resident buildings' footprint layer was used as the basis for control zones from which a weighting scheme was produced which allowed the re-allocation of 2011 data according to the 2001 geometries. It was derived from a land use /land cover map of the study-area, for the year 2010 (LULC 2010), developed within the doctoral project carried out by one of the authors. This layer is based on orthorectified digital aerial images of the year 2010. The ortho-imagery dataset was provided by a government institution - the Portuguese Geographic Institute (IGP). The technical specifications of the ortho-imagery are presented in Table 1.

Table 1. Ortho-imagery technical specifications

Technical Specifications	
Origin	digital photography obtained with digital aerial camera
Spatial resolution	0,50 m
Radiometric resolution	RGB and near-infrared
Format	raster
Flight plan	4 km * 5 km

The LULC 2010 is a vector map, with high thematic detail, ensured both by the minimum mapping unit (MMU) of 1000 square meters and by the nomenclature. The methodology used for the production of the latter and the technical specifications (such as the generalization rules) of the LULC 2010 map were based on the guidelines provided in the Technical Specifications development for the Land Use and Land Cover Map of Continental Portugal for 2007- COS 2007 (available in www.igeo.pt), applied and adapted to the territorial reality of the municipality of Portimão.

Table 2 presents an overview of the main technical specifications of land use/ land cover map 2010.

Table 2. LULC 2010 technical specifications

Technical Specifications	
Minimum Mapping Unit (MMU)	1000 m^2
Minimum Distance between lines and polygons	2 m
Reference System	ETRS89/PT-TM06
Ellipsoid	GRS80
Datum	ETRS89
Projection	Transverse Mercator
Nomenclature	COS2007 (3-level hierarchical nomenclature with 45 classes at the most detailed level)

The LULC 2010 map was derived by visual image interpretation to delineate, at the most general level, the boundary between the connected artificial areas and the surrounding agricultural and agroforestry areas, forests and natural and semi-natural areas, wetlands and water bodies. It is important to mention that LULC 2010 contains tree levels of thematic detail, with 45 classes at the most detailed level. In order to obtain the resident buildings' footprint layer from the LULC 2010 map, only the following four main classes of the artificial areas were used and aggregated: i) continuous urban fabric predominantly vertical, ii) continuous urban fabric predominantly horizontal, iii) discontinuous urban fabric and iv) sparse discontinuous urban fabric.

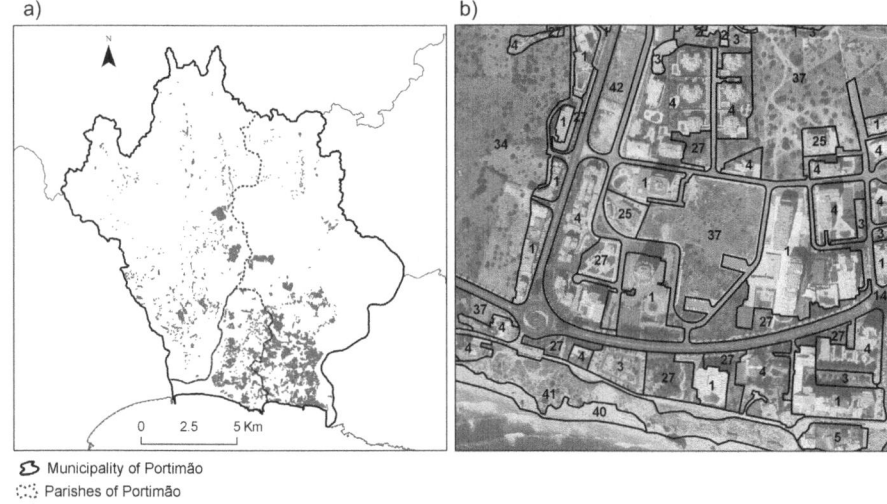

- ⌘ Municipality of Portimão
- ⁝⁝ Parishes of Portimão
- ⌒ Other municipalities of Algarve
- ⬛ Resident buildings' footprint layer

Fig. 3. Resident buildings' footprint layer (built from LULC2010 (a). Illustration of the land use/ land cover map production (b)

Urban class' accuracy was evaluated through visual analysis of images with higher resolution than those used to produced it. The sampling design included 100 random points, well distributed over the class area. The results indicate an overall accuracy of 95%, meaning that this class is well mapped, and strongly represents the land cover on the study area.

Figure 3 shows details of two thematic maps produced from the LULC 2010 map for the municipality of Portimão. Of these, figure 3a represents the derived resident buildings' footprint layer.

3 Methodology

Assuming agents, in a particular time-snapshot (timestamp), may be located using a pair of point coordinates (in a planimetric exercise), then ideally, point data would be the best source of information [10]. However, for distinct reasons, one of which is the need to maintain a certain degree of individuals' privacy, data is available, at best, for small area units (census tracts).

As mentioned above, any attempt to construct a multi-temporal census micro dataset is hampered by the non-coincidence of geometries, which originates the existence of asymmetric maps. In order to transform one of these into a symmetric map, two steps must be taken: first, a common geometry must be adopted; second, data must be re-allocated according to this new geometry. Thus, there are two sets of geographical structures: the original geometry layer(s) and the target geometry.

Three related methodologies can be used, which differ in terms of complexity and are totally dependent on the availability of ancillary data: first, if it is assumed that distribution of agents/events over the original spatial units is uniform, then the re-allocation exercise results from weighting data according to the size (area) of each unit – *form* attributes are necessary and sufficient since data is assumed to be stationary over the spatial surface.

Second, if we drop this assumption, then different weighting schemes must be designed. In general, this involves the use of control zones, representing detailed information on the distribution of some sort of objects which *control* and in fact determine the weighting scheme. Other than form, *structure* attributes are used; land-cover layers serve to discriminate between empty and non-empty spaces. Finally, if information is available in respect to land-use, stratification of objects according to their *function* becomes possible.

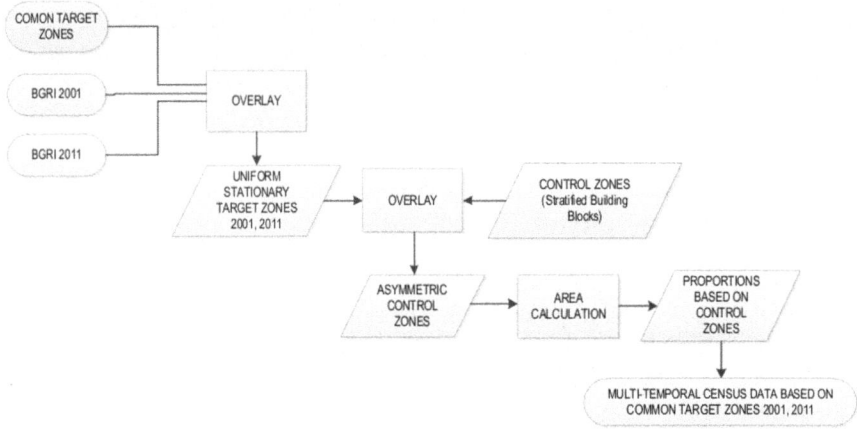

Fig. 4. Flowchart

Figure 4 represents the different stages of the exercise. It is important to note that in the present work, the common target zones correspond to the BGRI 2001. Although apparently redundant, the distinction between both in the flowchart is important to underline the role of the definition of a common geometry. Each unit's area represents the *form* attribute which would suffice if data was stationary. If building blocks were assumed to be of the same type (same function), then their area within each common target zone would represent the *structure* attribute. Since the LULC 2010 layer discriminates between land-use, then *functional* attributes were used. Building blocks were stratified resulting in the resident buildings' footprint layer, which overlaid with the stationary target zones, resulted in an asymmetric control zones layer. This was the basis for the computation of proportions which allowed a proper weighting scheme to be designed and data to be re-allocated for the common target zones.

When overlaying two layers from lattice datasets, distinct results may occur, simply in terms of geometry. Figure 5 illustrates these, starting with the simple case when form coincides. In these cases, census tracts from the two periods are the same, hence no re-allocation is necessary. Second, when in 2011 new census tracts were created from existing units without the compromise of borderline, new data could simply be aggregated to the original units. Third, when limits cross, the proportion of the geometrical subset of the 2011 unit which is inside the 2001 census tract serves to re-allocate 2011 data according to the weight duly computed.

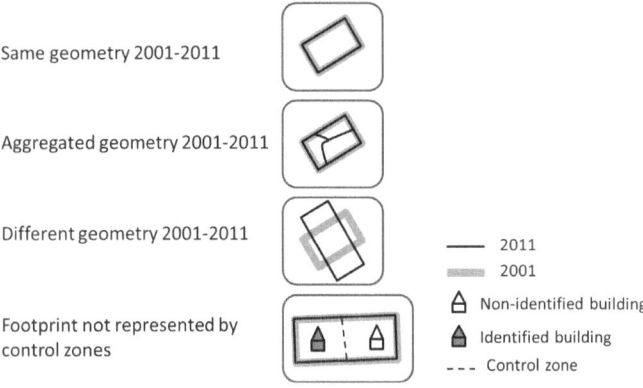

Fig. 5. Typology for the overlaying exercise

In the present study, data is assumed not to be stationary within census tracts; hence, control zones are used to account for non-uniform regional distributions. Moreover, building blocks (the chosen control zones) are stratified according to function. One problem occurs when, during the construction of the LULC 2010 dataset, blocks are omitted. This is the fourth and last case in the presented typology and results in omitted data. The probability of such occurrences is far greater in rural areas, where construction is scarce. In summary, since function is endogenous to the present methodology, error is minimized as control zones are chosen as only those buildings with a full or partial residential function.

4 Results

The use of the resident buildings' footprint as control layer allowed the creation of the multi-temporal micro-data information. In order to demonstrate its potential use, an exploratory analysis of the growth rate of three indicators from the census was performed. Figure 6 shows a representation of the number of buildings for the study-area focusing on 4 distinct dates: 1960, 1980, 2001 and 2011. For the first three dates, data was collected from the 2001 census. The latter period corresponds to data which results from the control zones mapping exercise. It is possible to observe great differences between 1960, 1980 and 2000, with growth concentrated around the old core of the city, spreading roughly in a concentric pattern outwards. Growth in the last decade (between 2001 and 2011) was sparse in rural areas, showing new types of occupation in rural areas. Important construction occurred within the city of Portimão, although growth rates were not as significant as before. The exploratory analysis of these results is conditioned by the fact that the later period represented is smaller than that of the first three figures (10 versus 20 years).

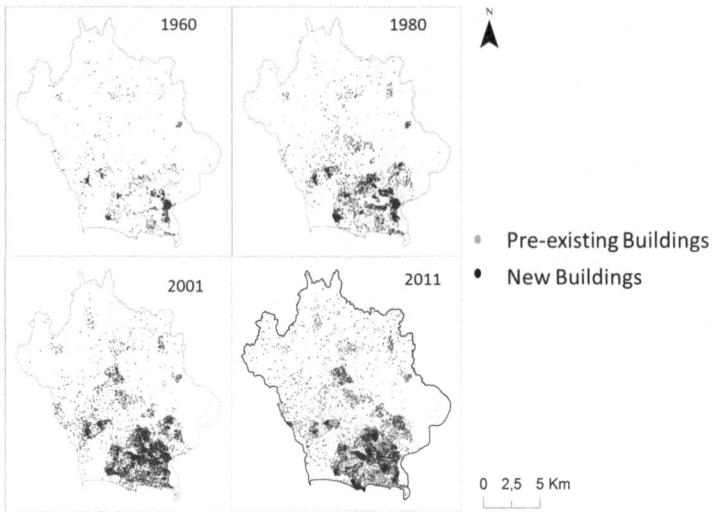

Fig. 6. Portimão municipality - expansion of number of buildings

The other two variables explored are resident population and the number of dwellings. The main goal was to analyze and compare growth rates for this set of indicators. Table 3 analyses the co-variation between the growth rates over the last decade. The non-parametric correlations (Spearman's coefficients) show some interesting patterns (all calculated coefficients are significant for the 99% confidence level): first, there is a positive correlation between the number of new buildings and new dwellings. Nonetheless, the fact that the coefficient is not higher than 0.766 indicates construction of distinct house types. The co-variation between the growth rate of population and buildings may reflect the phenomenon of second housing (common in the Algarve region) and urban regeneration with new residents occupying previously built-up areas.

Table 3. Non-parametric correlation matrix (growth rates - 2001-2011)

	R. Population	Buildings	Dwellings
R. Population	1		
Buildings	0.496	1	
Dwellings	0.614	0.766	1

The final exploratory exercise consisted in trying to identify particular spatial trends in terms of urban and demographic dynamics; a visual comparison of growth rates of resident population and the number of dwellings was performed. This exercise is more significant for the city of Portimão rather than the whole study-area since the smaller average size of census tracts for urban areas in comparison to the urban fringe allow a more detailed analysis. Spatial lags were computed using a binary contiguity matrix, suitable for large geographical scale exploratory exercises [5]. Spatial lags for variable X for observation i is given by the expression:

$$U_i^{lag} = \sum_{j=1}^{k} w_{ij} x_j, \qquad (2)$$

where w_{ij} corresponds to the neighborhood relation between i and j.

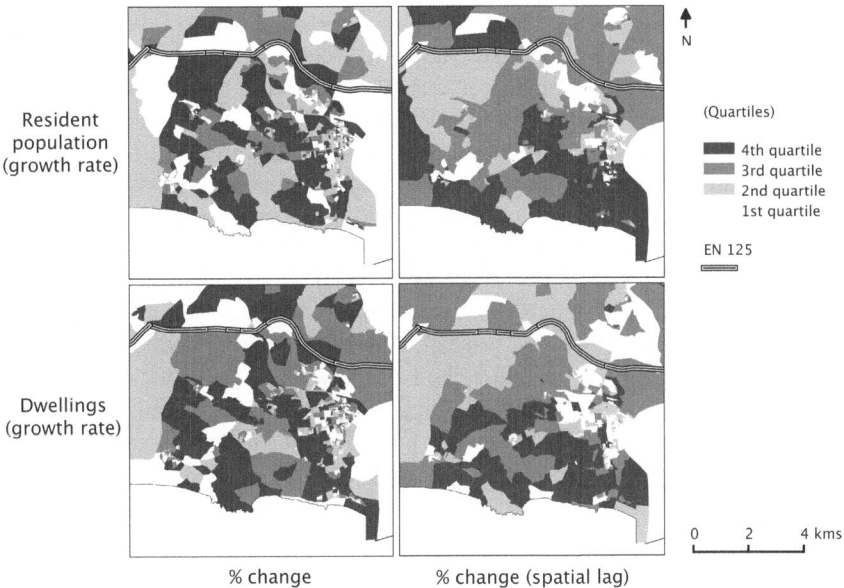

Fig. 7. Resident population and number of dwellings growth rates - 2001-2011 (values correspond to the observed rates and spatial lags computed with a binary spatial weights matrix)

Figure 7 permits the identification of census tracts where population and the number of dwellings grew at a higher rate. The spatial lags allow the analysis of common trends. Growth was still concentrated near the core, although some interesting differences exist. It is possible to identify areas where an increase in the number of dwellings did not result in a corresponding population growth. This exploratory analysis raises important questions related to the type of construction predominant in Portimão over the last 10 years. Answers to such questions, which fall outside the scope of the present article, will help understanding recent growth patterns over the recent past for the present study-area.

5   Concluding Remarks

The mapping exercise performed for the Portimão municipality served to identify the possible outcomes when combining data from conflicting geometries (figure 5). Results are totally dependent on the quality of control zones which, given the technical considerations of the LULC dataset, are assumed to meet high standards.

Although functional attributes were taken into account in the present study, future developments may include more complex weighting schemes which involved different proportions for the stratified building blocks. Yet, more complexity in the weighting scheme may introduce systematic noise which would cause final results to be biased.

The present work explored a new variation in an existing methodology used to build coherent geographical micro-datasets. Its application to a municipality in the southern coast of Mainland Portugal demonstrated one of the many uses such information may have. Yet, one should not forget that at such level of detail, trends are obscured by high local variations in the data. This fact emphasizes the importance of using techniques which help, through an endogenous use of neighborhood relations, the exploratory analysis of spatial trends. Although information at an even lower aggregation level would be necessary to infer on the quality of the data, it is believed that the deducted methodology is sufficiently pragmatic to guarantee to a large extent the quality of the end-product. Although the present methodology refers to a specific strand in the literature, both the scale and the study-area represent innovations in terms of potential accuracy and application to geographical areas which were not previously explored within this scientific domain.

References

1. Rogerson, P.: Statistical Methods for Geography. SAGE Publications (2001)
2. Freedman, D.A.: Ecological Inference and the Ecological Fallacy. In: Smelser, N.J., Baltes, P.B. (eds.) International Encyclopedia of the Social & Behavioral Sciences, vol. 6, pp. 4027–4030. Elsevier (2001),
   http://www.sciencedirect.com/science/article/pii/B0080430767004101
3. O'Sullivan, D.: Too Much of the Wrong Kind of Data: Implications for the Practice of Micro-Scale Spatial Modeling. In: Goodchild, M.F., Janelle, D.G. (eds.) Spatially Integrated Social Science. Oxford University Press (2004)
4. Anselin, L.: The Future of Spatial Analysis in the Social Sciences. Geographic Information Sciences 5(2), 67–76 (1999)
5. Rodrigues, A., Tenedório, J.A.: The Use of Spatial Weights Matrices and the Effect of Geometry and Geographical Scale. In: European Colloquium on Quantitative and Theoretical Geography - ECTQG 2011, Athens, September 2-5 (2011)
6. Santos, M.: Espaço e Método, 4th edn. Nobel (1997)
7. Abate, F., Jewell, E.J.: The New Oxford American Dictionary. Oxford University Press, New York (2001)

8. Santos, A.: Development and Use of Geocoding: Portugal Approach. United Nations Expert Group Meeting on Contemporary Practices. In: Census Mapping and Use of Geographical Information Systems, United Nations, New York (2007)
9. Walford, N., Hayles, K.: Thirty Years of Geographical (In)consistency in the British Population Census: Steps towards the Harmonization of Small-Area Census Geography. Population, Space and Place 10 (2011)
10. Martin, D.: Last of the censuses? The future of small area population data. Transactions of the Institute of British Geographers 31(1), 6–18 (2006)
11. United Nations - Department of Economic and Social Affairs Statistics Division; Handbook on Geographic Databases and Census Mapping (Draft version). In: Proceedings of the United Nations Expert Group Meeting on Measuring the Economically Active Population in Censuses, New York (2008)
12. Goodchild, M.F., Anselin, L., Deichmann, U.: A framework for the areal interpolation of socioeconomic data. Environment and Planning A 25(3), 383–397 (1993)
13. Freire, S., Aubrecht, C.: Towards improved risk assessment - Mapping spatio-temporal distribution of human exposure to earthquake hazard in the Lisbon Metropolitan Area. In: Boccardo, P., et al. (eds.) Remote Sensing and Geo- Information for Environmental Emergencies. International Symposium on Geo-information for Disaster Management (Gi4DM), February 2-4. CD-ROM, Torino (2010)
14. Nomenclature of Territorial Units for Statistics. European Commission, http://epp.eurostat.ec.europa.eu/portal/page/portal/nuts_nomenclature/introduction (accessed February 29, 2012)
15. Instituto Nacional de Estatística, I.P; Censos 2011 – Resultados Provisórios. INE, Lisboa (2011)
16. Santos, J.A.: A regionalização Portuguesa no Contexto Europeu. Instituto Fontes Pereira de Melo, Lisboa (1982)
17. ISCSP: Regionalização e Desenvolvimento. Forum 2000 Renovar a Administração. Instituto Superior de Ciências Sociais e Políticas/UTL (1996)
18. Geirinhas, J.: Conceitos e Metodologias BGRI - Base Geográfica de Referenciação de Informação. Direcção Regional de Lisboa e Vale do Tejo/INE (2001)

Using Hydrodynamic Modeling for Estimating Flooding and Water Depths in Grand Bay, Alabama

Vladimir J. Alarcon[*] and William H. McAnally

Geosystems Research Institute, 2 Research Blvd., Starkville MS, 39759, USA
{alarcon,mcanally}@gri.msstate.edu

Abstract. This paper presents a methodology for using hydrodynamic modeling to estimate inundation areas and water depths during a hurricane event. The Environmental Fluid Dynamic Code (EFDC) is used in this research. EFDC is one of the most commonly applied models to Gulf of Mexico estuaries. The event with which the hydrodynamic model was tested was hurricane Ivan. This hurricane made landfall at the Alabama Gulf Coast in September 16, 2004. Hurricane Ivan was the most severe hurricane to hit eastern Alabama. Results show that the EFDC model is able to generate instances of flooded areas before, during and after a hurricane event (Ivan hurricane). The model also estimated water depths and water surface elevation values consistent to measured data reported in the literature, and comparable to model-estimated data from a meso-scale Slosh model for the region (also reported in the literature).

Keywords: Grand Bay, Hydrodynamics, EFDC, modeling, grid generation, flooding, inundation.

1 Introduction

Floods and storms are intrinsic components of the natural climate system and climate variability and, as such, are a part of a natural disturbance regime, which is an important determinant of an ecosystem structure and function, particularly in the long run [1]. While floods are generally perceived as having negative effects (due to damage to the human environment), floods also contribute to enriching the flood plain soil with nutrients and humidity that are very important for agriculture or natural ecosystems. The most commons causes of flooding are the overflow of streams and rivers and abnormally high tides (resulting from severe storms) into the normally dry land area adjoining rivers, streams, lakes, bays, or oceans [2]. Those dry land areas in the nearby of water bodies are also called floodplain.

Coastal floods are usually primarily caused by storm surges but typically result from a combination of coastal tidal action, storm surge, rainfall, heavy surf, tidal

---

[*] Corresponding author.

piling, tidal cycles, topography, shoreline orientation, bathymetry, river stage, runoff and presence or absence of offshore reefs or other barriers [1].

The catastrophic loss of life and property by several hurricanes across Florida in 2004 and along the Gulf of Mexico coast during the summer of 2005 emphasizes the importance of accurate storm surge forecasts in highly populated coastal environments [3]. Among the weather events from that season, hurricane Ivan (the strongest hurricane of 2004) was composed by winds of more than 200 km per hour and devastated much of Gulf Coast in the southeastern US. The hurricane reached its peak strength on September 11, 2004, and made landfall at the Alabama Gulf Coast in the early morning hours of September 16. Hurricane Ivan was the most severe hurricane to strike eastern Alabama, western in many decades, approximating or exceeding design flood conditions along the Alabama shoreline, damaging severely the buildings closest to the coast [4].

The use of mathematical models is well established in the estimation of floodplains and inundation areas during hurricanes. Surge models and hydrological models are used by regulatory agencies (such as the US Federal Emergency Management Agency) to forecast potential floodplains for several types of storms. Hydrodynamic models of estuaries and bays are also used in conjunction to large scale surge models.

This paper presents a methodology for using the Environmental Fluid Dynamic Code (EFDC) for estimating flooded areas in the coast of Grand Bay, Mississippi (US) during the Ivan hurricane.

2 Methods

2.1 Study Area

Grand Bay is an estuary located in the northern Gulf of Mexico, at the border of the states of Mississippi and Alabama, USA (Fig. 1). It covers a geographical area of approximately 4300 hectares. The average depth is close to 1.1 m, although there are two areas where depths could reach 4.5 m (MLLW). The estuary receives waters from several small streams, being the two most important the Bayou Heron and the Crooked Bayou. The water depth at Bayou Heron ranges from 0.16 m to 1.69 m (depth readings represent the water depth above the depth sensor, which is located 0.5 m above the bottom). In terms of flooding, the effect of these small streams is negligible when compared to the impact of the ocean tides.

Grand Bay houses healthy estuarine salt marshes and fire-maintained pine savannas (some of the most biodiverse habitats in North America), environments that support many important species of fish and wildlife such as: finfish, shellfish, brown shrimp, speckled trout, oysters, sea turtles, bottlenose dolphin and, manatees and many species of carnivorous plants and orchids [5].

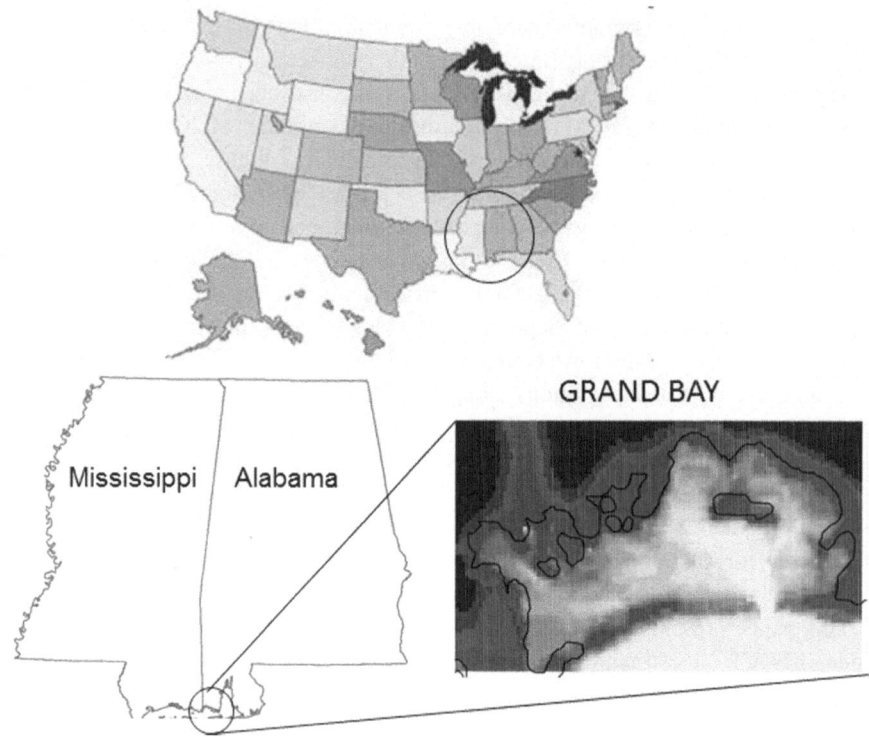

Fig. 1. Study area. The Grand Bay estuary is located at the border of Mississippi and Alabama

2.2 Computational Mesh

The bathymetry data used for producing the computational mesh for Grand Bay was downloaded from the NOAA-NOS Estuarine Bathymetry database. The dataset detailing the bottom topography of Grand bay provided by NOAA was referenced to the Mean Lower Low Water level (a tidal datum representing the average of the lower low water height of each tidal day observed over the National Tidal Datum Epoch). The grid generator for EFDC (GEFDC), capable of producing structured rectangular and curvilinear meshes [6], was applied for creating a structured grid of Grand Bay for hydrodynamics modeling.

The bathymetric information was downloaded in ASCII raster format. Manipulation of the raw ASCII bathymetry data was performed using ArcGIS and tailor-made C codes. This facilitated the production of one of the main input data required by GEFDC for the generation of the grid: the cell.inp file. This file specifies the interconnection between finite-difference cells, the type of cell (water, boundary, dry land, etc.), and whether it is a quadrilateral or triangular cell. Other files required by GEFDC such as dxdy.inp (specifying grid coordinates correspondence to bottom depths, cell dimensions, and bottom roughness), lxly.inp (cell center coordinates and orientation), etc., were also produced by those tailor made codes.

All pre-processing for the generation of the required GEFDC input files was performed in UTM Zone 16N coordinates. Figure 2 illustrates the process.

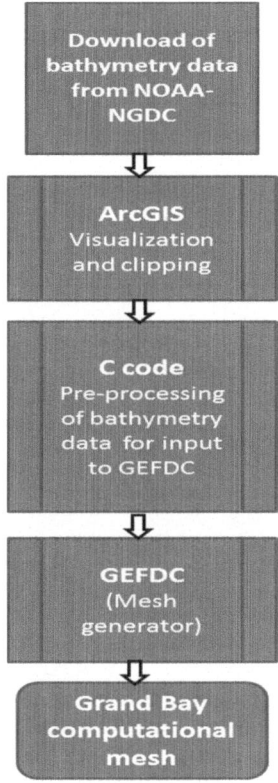

Fig. 2. Geo-processing of the bathymetry data for generation of the computational mesh for Grand Bay using the grid generator GEFDC and other computational tools

2.3 The Environmental Fluid Dynamic Code (EFDC)

The Environmental Fluid Dynamic Code (EFDC) is used in this research. EFDC is arguably the most commonly applied model to Gulf of Mexico estuaries for regulatory purposes. Ongoing or recent EFDC applications include the Back Bay of Biloxi, MS [7]; Bay St. Louis, MS [8] [9], Escatawpa and Pascagoula Rivers, MS [10]; Mobile Bay, AL [11][12]; Weeks Bay, AL (ongoing studies by GOMA, NGI), Tampa Bay [13], FL (and ongoing studies by GOMA) and many others.

The EFDC model is a public-domain surface water modeling system incorporating fully integrated hydrodynamics. It is used for 1D, 2D, or 3D simulations of rivers, lakes, reservoirs, estuaries, coastal seas, and wetlands [6]. It is currently used by federal, state and local agencies, consultants and universities [5].

2.4 Ivan Hurricane Data

Ivan was one of the strongest hurricanes of the first decade of this century. It landed at the Alabama Gulf Coast in September 16 of 2004. Figure 3 shows the Ivan hurricane path and the location of the study area.

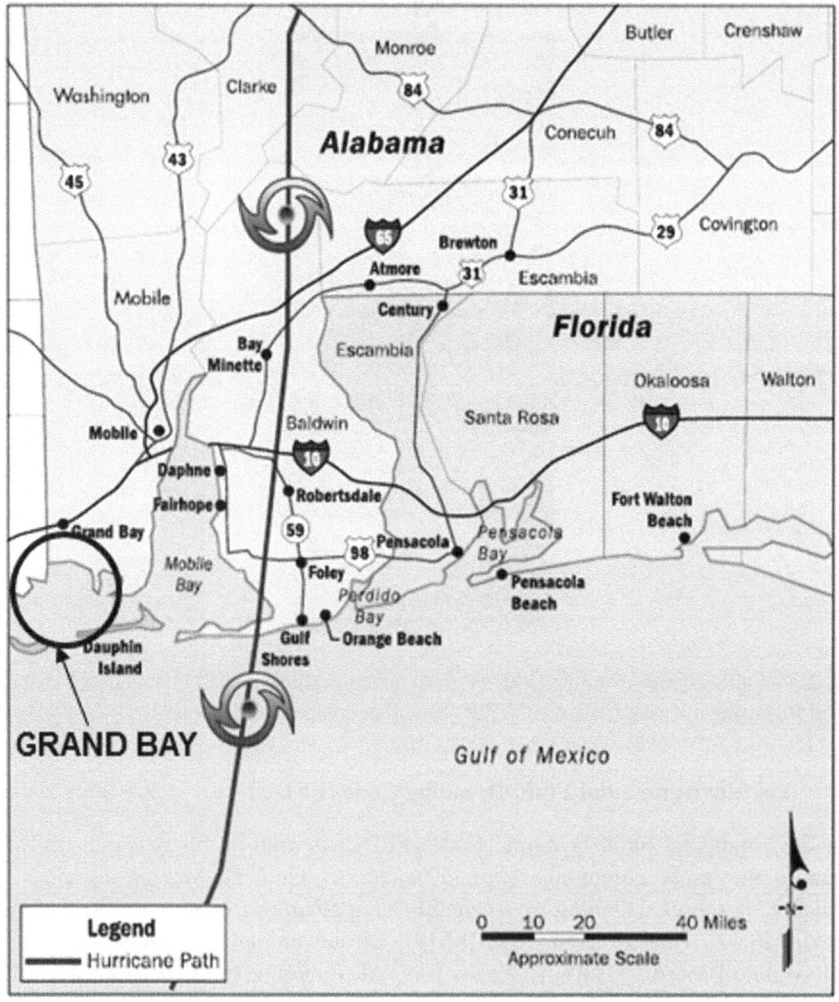

Fig. 3. Path of the Ivan hurricane and location of the study area (modified from [4])

A time series of tidal elevations was downloaded in ASCII format from the NOAA-NGDC website. The tidal data was referenced to the MLLW sea level in order for the vertical datum to be the same as the Grand Bay bathymetry. The closest station to Grand Bay that has records of water surface elevations occurring during the Ivan event is the Dauphin Island NOAA station. Since this station is approximately at the same latitude to Grand Bay, the data was used without further processing (spectral analysis or extrapolation) for hydrodynamic modeling simulation.

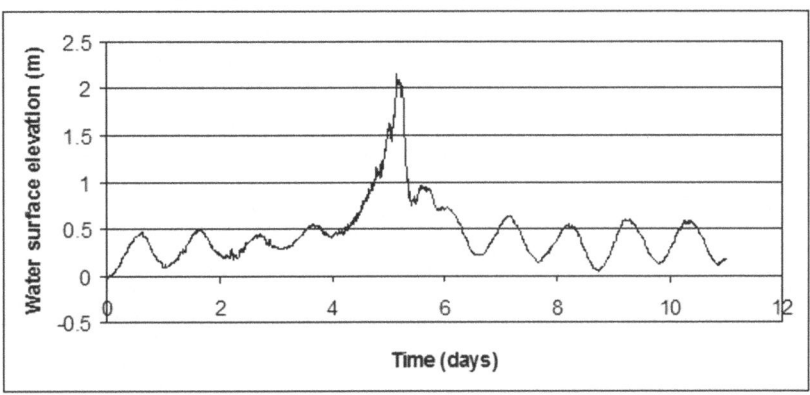

Fig. 4. Tidal heights recorded at the Dauphin Island NOAA station during Ivan hurricane. Water surface elevations are shown referenced to the Mean Lower Low Water level (MLLW).

3 Results

3.1 Computational Mesh for Grand Bay

Fig. 5 shows the structured grid developed for input into the EFDC model for simulating hydrodynamics in Grand Bay. The mesh consists of 8190 square cells (of 84.98 m per side) and covers not only Grand Bay but the dry land area surrounding the bay (floodplain). The figure also shows the location of ocean boundary conditions and freshwater boundary conditions. The model was set up with tidal ocean boundaries in the form of time-series of water surface elevation input at each cell shown as "ocean boundary" in Fig. 5. The fresh water boundary conditions corresponding to Bayou Heron (upper left boundary), and Crooked Bayou (lower left boundary) were input as time-series of stream flow occurring at those streams during the Ivan hurricane.

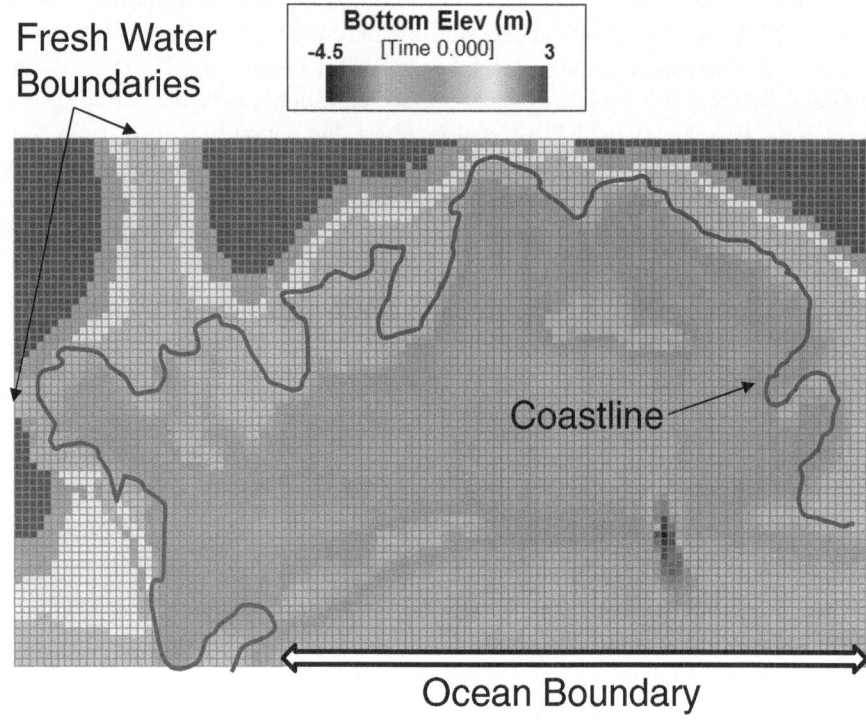

Fig. 5. Computational mesh for the EFDC hydrodynamic model of Grand Bay

3.2 Estimation of Flooded Areas during the Hurricane Peak

Fig.6 shows water depths occurring in Grand Bay at two instances: before the Ivan hurricane (top) and during the peak of the hurricane (bottom). The fact that square of known dimensions cells were used in the creation of the computational mesh for grand Bay greatly simplifies the estimation of flooded areas before and after the weather event. Raster operations for detecting active and inactive cells for the two instances shown in Fig. 6 were performed and the results are shown in Table 1.

Table 1. Estimation of flooded area

Instance	Number of active cells	Area in hectares
Before Ivan	6187	4467.84
At the peak of Ivan	8190	5914.28
Flooded Area		1446.44

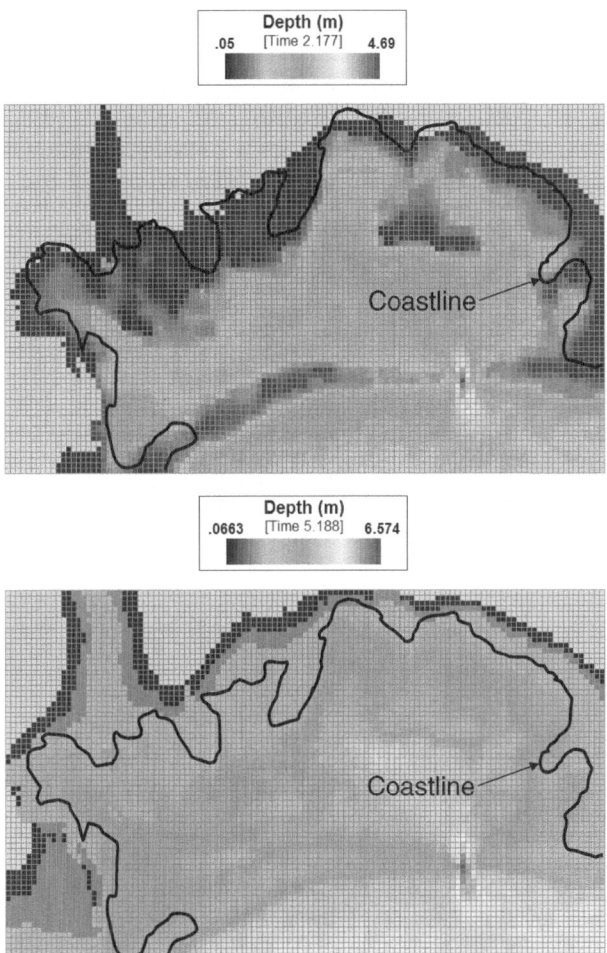

Fig. 6. Water surface elevations before hurricane Ivan (top) landing and at the peak of the hurricane (bottom)

A comparison of estimated water depths estimated using EFDC against measured water depths occurring during the hurricane is not feasible because of the lack of recorded data. However, and indirect validation of the results of this research is possible by comparing the water depths and inundated areas to meso-scale surge models. SLOSH (Sea, Lake and Overland Surges from Hurricanes) is a computerized model run by the National Hurricane Center (NHC) to estimate storm surge heights and winds resulting from historical, hypothetical, or predicted hurricanes by taking into account [14]. The US Army Corps of Engineers compared Slosh model results to high water marks observed after the Ivan event [15]. The report includes "inside high water marks" records inside of structures, which estimate storm tide elevation without the effect of waves, and "outside high water marks" which estimate the combined effect of storm tide and wave set up and run up. Outside high water mark elevations are generally higher than inside high watermarks because of the added wave effects [15]. The USACE report concluded that the comparison of observed storm surge hydrographs to the SLOSH model calculated storm surge hydrographs showed reasonable results. In this paper, the EFDC-estimated water depths are compared to Slosh model predictions of water depths for Grand Bay.

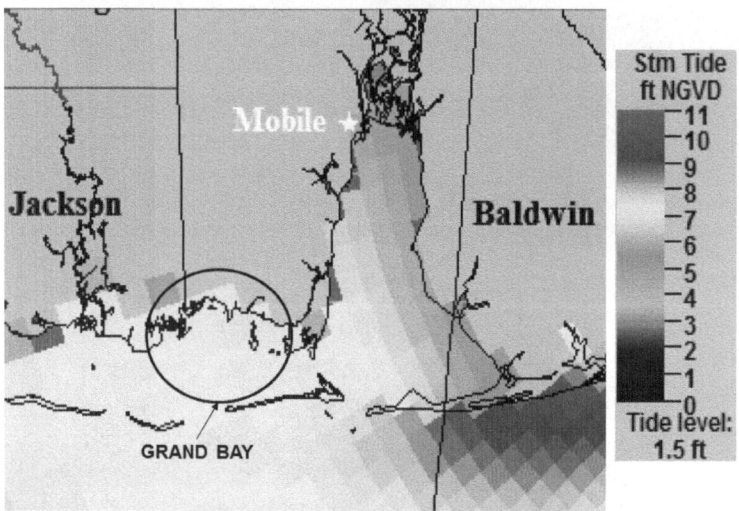

Fig. 7. The SLOSH model predictions of High Water (HW) tide levels showing the location of Grand Bay (modified from [15])

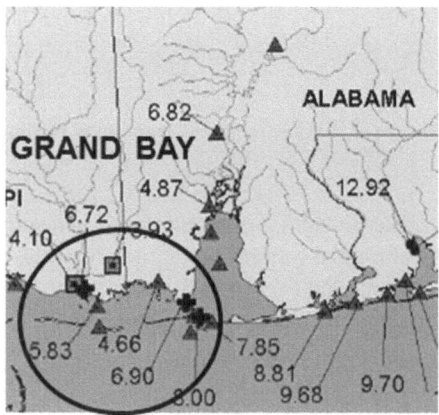

Fig. 8. USACE Mobile District tide gages peak elevation during Hurricane Ivan. In feet (modified from [16]).

As seen in Fig. 7, the Slosh model predictions for Grand Bay for tide levels (high water, HW) occurring during the Ivan hurricane average 2 m (6.5 ft.). Fig. 8 shows actual tidal elevations recorded by the USACE tidal stations in the Grand Bay area. Averaging the tidal elevations values for the Grand Bay study area results in a measured water depth of 1.8 m (5.9 ft.). The EFDC model for Grand Bay presented in this paper estimates water depths ranging from 2.07 m to 4.15.m (mean 2.7 m, median 2.5), as shown in Fig. 6.

4 Conclusions

This paper shows the potential of using shallow-water hydrodynamic models to estimate floodplain areas in coastal regions. The EFDC model is shown to be able to generate instances of flooded areas before, during and after a hurricane event (Ivan hurricane). Based on those calculations, the model estimates that the flooded area at the peak of the hurricane was 1446 hectares.

The model was also used to estimate water depths and water surface elevation values. The results of this computational exploration show that the EFDC model provides water depths estimations for the Grand Bay estuary (during the Ivan hurricane) consistent to measured data reported in the literature, and comparable to model-estimated data from a meso-scale Slosh model for the region (also reported in the literature). However, this indirect verification of results is insufficient for applying the technique to real inundation events. Actual validation of the methodology and flooded area estimates will be topic for future research.

Acknowledgments. This research was funded by the Northern Gulf Institute, and the NSF-EPSCOR grant number 1010578.

References

1. Ghimire, G., Hanson, R.: Flood and Storm Control. In: Ecosystems and Human Well-being: Policy Responses,
   http://www.maweb.org/documents/document.316.aspx.pdf
2. Mays, L.W.: Water Resources Engineering. John Wiley and Sons, New York (2001)
3. Colle, B.A., Buonaiuto, F., Bowman, M.J., Wilson, R.E., Flood, R., Hunter, R., Mintz, A., Hill, D.: New York city's vulnerability to coastal flooding Storm Surge Modeling of Past Cyclones, pp. 829–841. American Meteorological Society (June 2008)
4. Federal Emergency Management Agency (FEMA): Hurricane Ivan in Alabama and Florida Mitigation Assessment Team Report,
   http://www.fema.gov/library/viewRecord.do?fromSearch=fromsearch&id=1569
5. Peterson, M.S., Waggy, G.L., Woodrey, M.S.: Grand Bay National Estuarine Research Reserve: An Ecological Characterization (2007),
   http://www.nerrs.noaa.gov/Doc/PDF/Reserve/GRD_SiteProfile.pdf

6. Tetratech, Inc.: Draft user's manual for environmental fluid dynamics code Hydro Version (EFDC-Hydro) Release 1.00. Tetra Tech., Inc., Fairfax, Virginia (2002)
7. Mississippi Department of Environmental Quality: Fecal Coliform TMDL for the Back Bay of Biloxi and Biloxi Bay Coastal Streams Basin Harrison and Jackson Counties, Mississippi, Department of Environmental Quality, Office of Pollution Control, Jackson, Mississippi (2002)
8. Mississippi Department of Environmental Quality: Fecal Coliform TMDL for St. Louis Bay, Jourdan River (Phase Two), and Wolf River (Phase Two). Coastal Streams Basin: Hancock, Harrison, and Pearl River Counties, Department of Environmental Quality, Office of Pollution Control, Jackson, Mississippi (2001)
9. Huddleston, D.H., Liu, Z., Kingery, W.L.: Application of St. Louis Bay water quality model to develop TMDLs for tributaries, Prepared for Mississippi Department of Environmental Quality, Office of Pollution Control, Jackson, Mississippi (2007)
10. Rodriguez-Borrelli, H., Greenfield, J.M., Miller, J.: Using Biology and Hydrodynamics to Establish a Site Specific Dissolved Oxygen Criterion for the Escatawpa River And Estuary – Gulf of Mexico. In: XXX Congreso Interamericano De Ingeniería Sanitaria y Ambiental, Punta del Este - Uruguay (2006)
11. Wool, T.W.: TMDL Development for a Stratified Shallow Estuary: Mobile Bay Case Study, U.S. Environmental Protection Agency, Region 4 (2003)
12. Wool, T.A., Davie, S.R., Plis, Y.M., Hamrick, J.: The Development of a Hydrodynamic and Water Quality Model to Support TMDL Determinations and Water Quality Management of a Stratified Shallow Estuary: Mobile Bay, Alabama, U.S. Environmental Protection Agency, Region 4, Atlanta, GA (2004)
13. Grand Bay National Estuarine Research Reserve: An Ecological Characterization, http://www.nerrs.noaa.gov/Doc/PDF/Reserve/GRD_SiteProfile.pdf
14. National Hurricane Center (NHC); Hurricane preparedness: Slosh model, http://www.nhc.noaa.gov/HAW2/english/surge/slosh.shtml
15. US Army Corps of Engineers: Comparison of Observed and SLOSH model computed storm surge for hurricane Ivan (2004), along the north central Gulf of Mexico coastline, http://chps.sam.usace.army.mil/USHESdata/Assessments/2004Storms/PDFfiles/IVAN%20Slosh%20Report.pdf
16. US Army Corps of Engineers. Tide Gage Data for Hurricane Ivan. Mobile District Engineering Division Hydrology and Hydraulics Branch, USACE-ERDC, Vicksburg, http://water.sam.usace.army.mil/Hurricane_Ivan_Report.pdf

Comparison of Two Hydrodynamic Models of Weeks Bay, Alabama

Vladimir J. Alarcon[*], William H. McAnally, and Surendra Pathak

Geosystems Research Institute, 2 Research Blvd., Starkville MS, 39759, USA
{alarcon,mcanally,pathak}@gri.msstate.edu

Abstract. This paper presents a comparison of two hydrodynamic models of the Weeks Bay sub-estuary (Alabama, USA). One model was developed using the Environmental Fluid Dynamic Code (EFDC). The resulting model was compared to an existing hydrodynamic model (of the same water body) that was developed using the Adaptive Hydraulic modeling system (ADH). Comparisons were performed in terms of predicted water surface elevations in Weeks Bay. The computational grid was created using GEFDC (a mesh generator for EFDC) and NOAA's coastline and bathymetric data. The results showed that the EFDC model provides comparable water surface elevation (WSE) estimations for five out of seven control points located in the Weeks Bay study area. R^2 values for those points range between 0.88 and 0.99. Root mean square error values are shown to be lower than 0.15 m in those cases. For the rest of the control points, R^2 values range from 0.73 to 0.87 (RMSE range: 0.2 - 0.35), showing that the EFDC model provides acceptable estimations of WSE when compared to the ADH model WSE output. A finer computational mesh may improve EFDC WSE estimations for Weeks Bay as reported in the literature.

Keywords: Weeks Bay, Hydrodynamics, EFDC, ADH, modeling, grid generation, unstructured, structured.

1 Introduction

Estuaries and bays are water bodies semi-enclosed by land formations that cause the water to form a relatively shallow body. Although coastal waters such as estuaries and bays receive fresh water inputs from rivers draining to the water body, tidal influences (in the form of water surface elevation and water speed and direction) are felt miles upstream the incoming rivers.

The use of mathematical models by regulatory environmental agencies to set up waste load allocations, estimate Total Maximum Daily Loads (TMDLs), estimate impacts of remediation of contaminated sediments, and a variety of other purposes, is well established. Models of open waters and Gulf estuaries most commonly include

---

[*] Corresponding author.

both hydrodynamic and water quality models, due to the importance of transport on the fate of water quality constituents [1].

Measured water surface elevation (WSE) data are critical for establishing initial and boundary conditions for hydrodynamic models. The water surface elevations and water velocities calculated by hydrodynamic models for estuaries, bays and coastal rivers are strongly dependent on boundary conditions provided at open-ocean boundaries. However, the lack of tidal stations that collect WSE and other data, oftentimes forces modelers to interpolate or extrapolate these critical data, generating undesired uncertainty in the output of hydrodynamic models.

One other approach is to use calibrated predictions from existing models to compare the results of the new model. Nevertheless, this comparison is not straightforward. Existing models could be of bigger or smaller scale than the new model, or the new model could have been developed using a different numerical strategy.

This paper reports a comparison of two different hydrodynamic models of Weeks Bay, Alabama. A new model for Weeks Bay was developed using the Environmental Fluid Dynamic Code (EFDC). This code solves a 2-D finite-difference algorithm. Its estimations of water surface elevation values were compared to simulations of an existing model for Weeks Bay developed using the Adaptive Hydraulic Modeling system (ADH). ADH is a finite-element code. The specific objective of this exploration was to determine if the EFDC model for Weeks Bay, developed using the same datasets as the ADH model, could generate the same estimations of water surface elevations estimated by the ADH model.

2 Methods

2.1 Study Area

Weeks Bay is a sub-estuary located on the eastern shore of Mobile Bay (Alabama) in the northern Gulf of Mexico (Fig. 1). Its longitudinal axis is approximately 3.4 km long, running nearly north-south. Its widest point (3.1 km) is located near the center of the estuary [2][3]. The average depth is close to 1.5 m [2], although there are two areas where depths could reach 6 m. Tides are principally diurnal, and have a mean range of 0.4 m. The estuary receives waters from the Fish and Magnolia Rivers.

The Fish River watershed covers 14300 hectares and contributes approximately 73% to the total incoming freshwater flow, with the Magnolia River supplying the rest [4]. The combined discharge of both rivers averages approximately 9 cubic meters per second.

In 2009, Weeks Bay was listed on the Southern Environmental Law Center's 10 Most Endangered Places in the South due to development in the area and lack of zoning laws [3]. Weeks Bay was removed from the top 10 list in 2010 due to Magnolia Springs (AL) adopting runoff control ordinances that promote low-impact development laws which protect the Magnolia River.

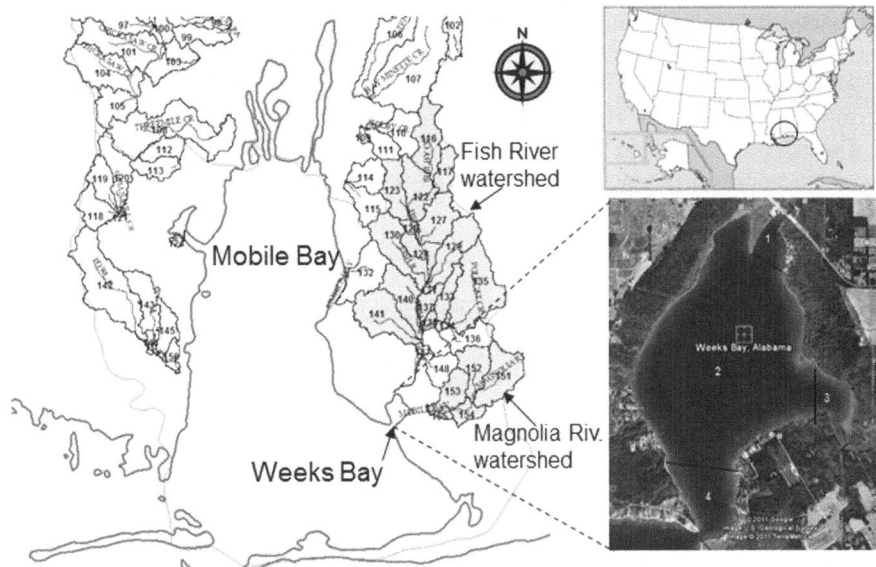

Fig. 1. Study area. Weeks Bay is a sub-estuary located within Mobile Bay, Alabama. Two main rivers drain waters into the sub-estuary: Fish River and Magnolia River. The watersheds for each river are shown.

2.2 Computational Meshes

The grid generator for EFDC (GEFDC), capable of producing structured rectangular and curvilinear meshes [5], was applied for creating a structured grid of Weeks Bay for hydrodynamics modeling. The bathymetry data (detailing the bottom topography with respect to mean sea level) was downloaded from the US National Oceanic and Atmospheric Administration National Geophysical Data Center (NOAA-NGDC) database. NGDC provides datasets of the U.S. coastal zone integrating offshore bathymetry with land topography into a seamless representation of the coast derived from NGDC's hydrographic surveys, multi-beam bathymetry, and track-line bathymetry; the U.S. Geological Survey (USGS); and other federal government agencies and academic institutions [6]. In this research, the US Coastal Relief Model bathymetric data was selected for its fine spatial resolution (84 m x 84 m raster cells), date of production (beginning date: 1999, ending date: present), and accuracy and reliability.

The bathymetric information was downloaded in ASCII raster format and was critical for producing one of the main input data required by GEFDC for the generation of the grid: the cell.inp file. This file specifies the interconnection between finite-difference cells, the type of cell (water, boundary, dry land, etc.), and whether it is a quadrilateral or triangular cell. Manipulation of the raw ASCII bathymetry data was performed using ArcGIS and tailor-made C codes. Other files required by GEFDC such as dxdy.inp (specifying grid coordinates correspondence to actual geographical coordinates), depdat.inp (bottom depths for each finite-difference cell corner), etc., were also produced by those tailor made codes.

All pre-processing for the generation of the required GEFDC input files was performed in UTM Zone 16N coordinates. Figure 2 illustrates the process.

The existing ADH model for Weeks Bay is composed by 9808 triangular elements and 4349 nodes. It is a non-structured finite-element mesh as required by the ADH solver. The mesh is shown in Fig. 3.

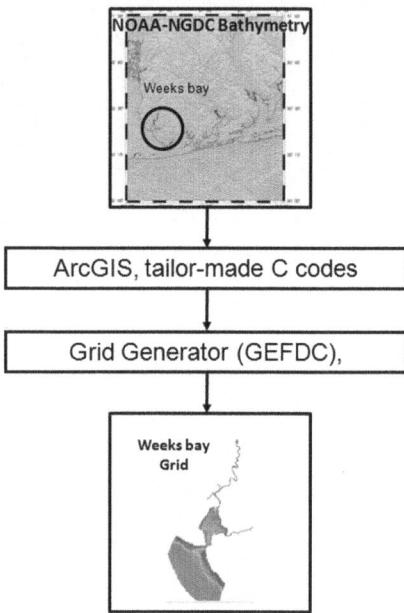

Fig. 2. Processing for the generation of the computational mesh for Weeks Bay for the EFDC solver

2.3 Solvers

The Adaptive Hydraulics (ADH) modeling system is a code that can describe both saturated and unsaturated groundwater, overland flow, 3D Navier-Stokes, and 2-D and 3D shallow-water problems. It is not on the public-domain and is designed to work in conjunction with the commercial software Surface Water Modeling System (SMS) (a Windows application) for building ADH application models, running simulations, and visualizing results [7]. It was developed at ERDC (USACE Vicksburg experimental station), and uses the finite-elements (triangular elements) strategy for solving the equations [7].

The EFDC Model is a public-domain surface water modeling system incorporating fully integrated hydrodynamics. It is used for 1D, 2D, or 3D simulations of rivers, lakes, reservoirs, estuaries, coastal seas, and wetlands [5]. EFDC was developed by

John Hamrick at Virginia Institute of Marine Science (VIMS) with primary support from the State of Virginia. Additional support has been provided by EPA and NOAA and it is presently maintained by Tetra Tech, Inc. It is currently used by federal, state and local agencies, consultants and universities [5].

3 Results

3.1 EFDC Computational Mesh

Fig. 3 (right-hand side) shows the EFDC grid developed for Weeks bay. The mesh consists of approximately 6300 square cells. Efforts were made to generate a grid as similar as possible to the ADH grid (left-hand side of Fig. 3).

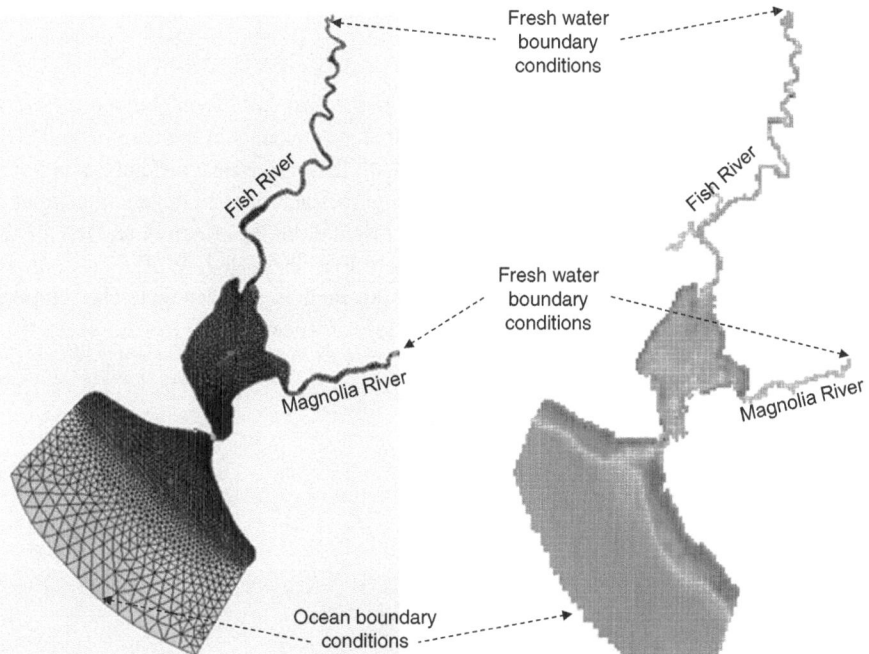

Fig. 3. EFDC and ADH computational meshes. A 6300-cell structured grid for EFDC input (right-hand side) was generated using the EFDC grid generator (GEFDC). The left-hand side of the figure shows the existing ADH grid.

Although EFDC works well with curvilinear (orthogonal) grids, the particular geometry of the coastline surrounding Weeks Bay did not allow generating a mesh with optimal orthogonality indicators (assessed by GEFDC). Fig. 4 shows one of the curvilinear meshes that were discarded due to abnormalities in the cells corresponding to the throat of Weeks Bay.

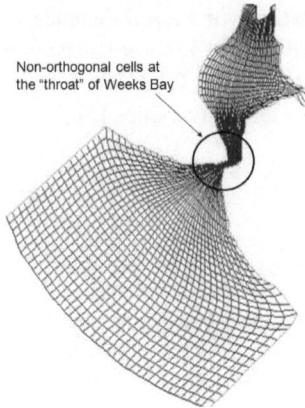

Fig. 4. Curvilinear grids for Weeks Bay. Cells at the "throat" of the estuary presented serious non-orthogonality issues

Ocean boundary conditions were imposed in the form of water surface elevation (WSE) time-series for all boundary cells (or boundary elements in the case of the ADH model) along the boundaries shown in Fig. 3. Due to the geometrical differences imposed by the geometry of EFDC's quadrilateral cells, and the fact that EFDC specifies input to boundary cells per East, North, South or West orientation, the ocean boundary for the EFDC Weeks Bay model was designed such that estimated WSEs at common geographical locations near the boundaries in both models are identical. The temporal resolution of the water surface elevation time-series was hourly.

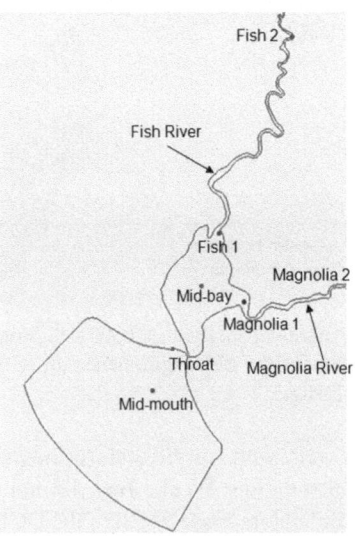

Fig. 5. Control points for which comparisons between EFDC and ADH output were performed

Fresh water boundary conditions were implemented at the most inland EFDC cells or ADH elements of the Fish and Magnolia river portions included in both computational grids (as shown in Fig. 3). Stream flow time-series constituted the fresh water forcings. Those time-series were produced by existing hydrological watershed models of Fish and Magnolia river watersheds from a previous research project [8]. In that research, several hydrological models of the Mobile Bay watershed and other watersheds located in the northern Gulf of Mexico, (Alabama, USA) were developed using the Hydrological Simulation Program Fortran (HSPF). Coastal watersheds of selected streams that drain directly to the Mobile estuary (namely: Fish River, Magnolia River, and Chickasaw Creek) were modeled and their corresponding HSPF models were calibrated. For this research, the existing hydrological models for Fish River and Magnolia River watersheds were updated with current precipitation and evapotranspiration time-series covering the period of hydrodynamic simulation. The temporal resolution of the HSPF-estimated stream flow time-series was daily.

3.2 EFDC vs. ADH Comparison

Fig. 5 shows the location of several control points, in the computational domain, for which water surface elevation estimations from the EFDC and ADH models were compared. These points were chosen because other research efforts for the area collected salinity and other water quality indicators at those points. The results shown in the following comparison charts use the location names specified in Fig. 5.

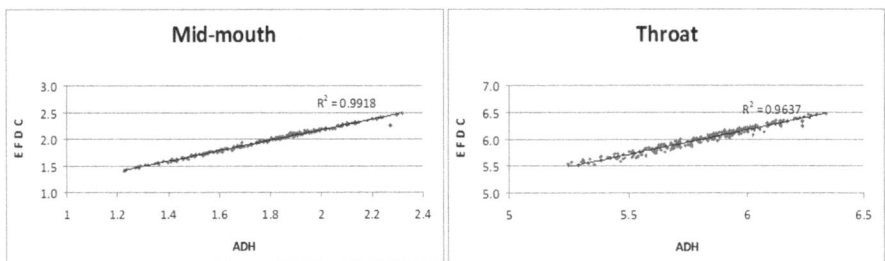

Fig. 6. Scatter plots of water surface elevations (WSE) calculated by the EFDC model of Weeks Bay compared to WSE estimated by the ADH model for the Mid-mouth and Throat control points. R^2 fitting coefficients are shown

A comparison of water surface elevations (WSE) estimated using EFDC against those WSE calculated by ADH are shown in figures 6 through 9. The scatter plots corresponding to the mid-mouth and throat control points (Fig. 6) show that both models estimate similar values of WSE as demonstrated by the high fitting coefficient (coefficient of determination R^2): 0.99 and 0.96, respectively. The plots show that EFDC tends to slightly overestimate WSE values with respect to those calculated by ADH. Overall, however, EFDC seems to produce comparable results.

Figure 7 shows a scatter plot comparing EFDC and ADH simulated output for the mid-bay control point. While the R^2 fitting coefficient is still acceptable (0.88), interestingly EFDC underestimates WSE values, and a spread of the scatter points is present.

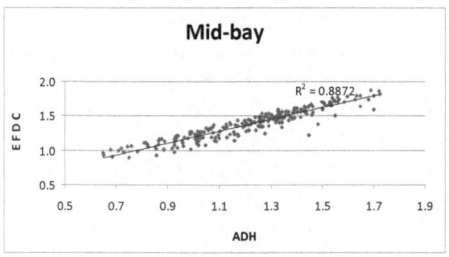

Fig. 7. Comparison scatter-plot of water surface elevations (WSE) calculated by the EFDC model of Weeks Bay versus WSE estimated by the ADH model for the Mid-bay control point

A similar quality in the EFDC simulation for two control points along the Magnolia River (Magnolia 1 and Magnolia 2) is evidenced by the scatter plots shown in Fig. 8. EFDC seems to slightly overestimate WSE values consistently for these control points. However, the R^2 values for both control points (0.88) do not decrease in comparison to the fitting coefficient values for the mid-bay control point, demonstrating that EFDC estimations of water surface elevations are consistent with the ADH estimations.

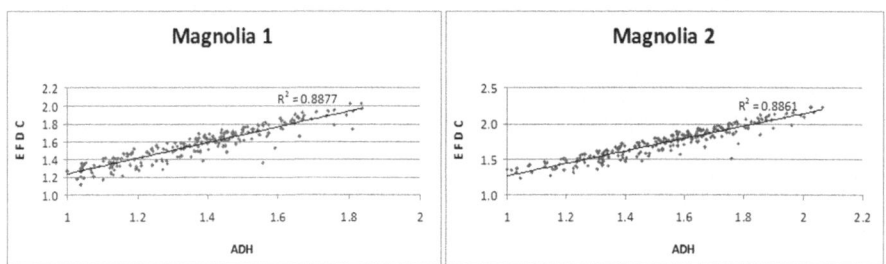

Fig. 8. Scatter-plots of water surface elevations (WSE) calculated by the EFDC model of Weeks Bay compared to WSE estimated by the ADH model for control points along Magnolia Rivers

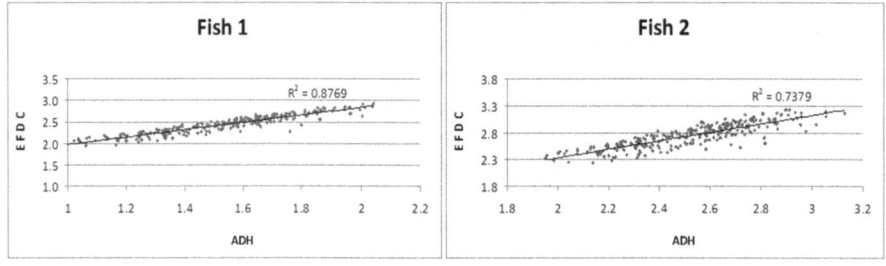

Fig. 9. Scatter plots of water surface elevations (WSE) calculated by the EFDC model of Weeks Bay compared to WSE estimated by the ADH model along Fish River

The comparison of EFDC water surface elevation estimations against ADH estimations for the two control points along the Fish River (Fish 1 and Fish 2 control points) is shown in Figure 9. R^2 fitting coefficients are 0.87.and 0.74, respectively. There exists a decrease in fitting quality directly proportional to the distance from the ocean boundary. The decrease, however, is not big, as evidenced by the still high R^2 value for the northern-most control point (R^2 value for Fish 2 is 0.74).

In order to further explore the actual quality of EFDC estimations, root-mean square-error values (RMSE) were computed. Also, since it seems that greater decrease in fitting quality is bigger with latitude than with longitude, RMSE and R^2 values are plotted against horizontal and vertical distances.

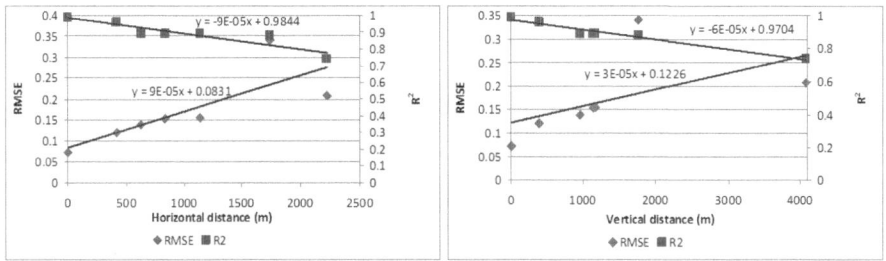

Fig. 10. R^2 and RMSE values against horizontal and vertical geographical distances, measured from the mid-mouth control point to the rest of the control points

Fig.10 presents a summary of fitting coefficients (R^2) and corresponding root-mean-square errors (RMSE) for all the control points. The values are plotted against horizontal and vertical spatial distances measured from the mid-mouth control point.

RMSE and R^2 values in Figure 10 show that the fitting inconsistencies increase with distances along the horizontal and vertical directions of the grid. The rate of error increase with respect to either direction is similar, probably due to the fact that the grid cells are square so the error propagates at the same rate vertically or horizontally. What is more important is that RMSE values range from 0.025 m to 0.15 m for most of the control points (Mid-mouth, Throat, Mid-bay, Magnolia 1, and Magnolia 2), showing that EFDC estimates water surface elevation values with small errors. For control points Fish 1 and Fish 2, although RMSE values are bigger (0.2 and o.35 respectively), EFDC estimations are still within acceptable error ranges.

4 Conclusions

The results of this computational exploration show that the EFDC model provides comparable water surface elevation (WSE) estimations for five of the seven control points located in the Weeks Bay study area. R^2 fitting coefficients for those points are greater than 0.88. Although EFDC slightly overestimates WSE for those points, the root mean square error is lower than 0.15 m. For the rest of the control points, R^2 values range from 0.73 to 0.87 (with RMSE values of 0.2 to 0.35 respectively), showing that the EFDC model also provide acceptable estimations when compared to the ADH model output. Providing EFDC of a finer computational mesh representing Weeks Bay seems may improve EFDC estimations for Weeks Bay. This conclusion is consistent with comparisons of finite-element and finite differences schemes reported in the literature [10] when applied to oceanographic modeling studies. However, the results show that the current grid will work well (RMSE < 0.2) for most of the cells within the computational domain.

Acknowledgments. This research was funded by the Northern Gulf Institute, and the NSF-EPSCOR grant number 1010578.

References

1. Martin, J.L., McCutcheon, S.C.: Hydrodynamics and Transport in Water Quality Modeling. CRC Press, Boca Raton (1999)
2. Riggs, A.A., Mortazavi, B.: Benthic Nitrogen Cycling in Weeks Bay, Alabama. In: Bays and Bayous Symposium, Science, Industry, Community: Building Bridges to Coastal Health, http://www.mobilebaynep.com/images/uploads/library/2010ProceedingsFinal.pdf
3. NOAA Ocean Service: Gulf of Mexico News. Office of Ocean and Coastal Resource Management, http://coastalmanagement.noaa.gov/news/docs/gomnews0109.pdf
4. McAnally, W.H., Evans, D., Martin, J.L., Sloan, J., Alarcon, V.J.: Sediment and Mercury Path and Fate Modeling: Progress Report. In: 2011 Northern Gulf Institute Annual Conference, Mobile, AL (2011)
5. Tetratech, Inc.: Draft user's manual for environmental fluid dynamics code Hydro Version (EFDC-Hydro) Release 1.00. Tetra Tech., Inc., Fairfax, Virginia (2002)
6. US National Oceanic and Atmospheric Administration National Geophysical Data Center, http://www.ngdc.noaa.gov/mgg/coastal/crm.html
7. Berger, R.C., Tate, J.N.: Guidelines for Solving Two Dimensional Shallow Water Problems with the ADaptive Hydraulics (ADH) Modeling System, USACE Research and Development Center (2008)
8. Alarcon, V.J., McAnally, W.H., Diaz-Ramirez, J., Martin, J., Cartwright, J.H.: A Hydrological Model of the Mobile River Watershed. In: Maroulis, G., Simos, T.E. (eds.) Computational Methods in Science and Engineering: Advances in Computational Science, vol. 1148, pp. 641–645. American Institute of Physics, Melville (2009)
9. United States Environmental Protection Agency, EPA. BASINS, Better Assessment Science Integrating point & non-point Sources, http://water.epa.gov/scitech/datait/models/basins/index.cfm
10. Thacker, W.C.: Comparison of Finite-Element and Finite-Difference Schemes. Part II: Two-Dimensional Gravity Wave Motion. Journal of Physical Oceanography 8, 680–689 (1978)

Connections between Urban Structure and Urban Heat Island Generation: An Analysis trough Remote Sensing and GIS

Marialuce Stanganelli[1] and Marco Soravia[2]

[1] University of Naples Department of urban and regional planning
piazzale Tecchio 80, 80125 Napoli
stangane@unina.it
[2] Province of Naples Geographical Information System
via Don Bosco 4f, 80141 Napoli
msoravia@provincia.napoli.it

Abstract. The phenomenon of Urban Heat Island is shown by an increase in temperature that mostly affects urban areas in comparison with the surrounding rural areas. This increase in temperature becomes problematic during the heat waves when it can give rise to problems of energy and health. The factors affecting this phenomenon are related to the morphology and location of the urban area, to the characteristics of building and roads materials, to the shape of urban structure. The paper investigates the phenomenon of UHI by analyzing in particular the influence of major urban planning features: the average height of buildings, the building density, the coverage ratio, the percentage of impermeable surface. The study was carried out starting from the analysis of a real case within the Province of Naples. The identification of areas liable to heat islands has been done by working out a thermal map of the Province of Naples through the creation of hyper-spectral satellite images using remote sensing techniques. This map has allowed to select some sample areas within which the main urban planning parameters have been detected through remote sensing techniques or spatial analysis. For each parameter, correlation curves "temperature - urban planning parameter" have been worked out. The main result is the development of an abacus that allows to estimate the expected temperature changes according to the decrease or increase of each urban parameter.

Keywords: Urban structure, energy consumption, climate change.

1 Urban Heat Island

Urban Heat Island -UHI- is a thermal anomaly affecting large urban areas that show temperatures which are higher than surrounding rural areas. The intensity of this phenomenon can be quantified as the maximum difference between the average temperature of urban air and that of the surrounding rural environment. Compared to the latter, the temperature increase is more pronounced at night than by day; in

daylight the temperature difference between urban and suburban areas can range from +1 ° C to +3 ° C, while at night it can reach values ranging from +7 to +12 ° C [1].
This phenomenon occurs especially in big cities where there is an extensive use of materials that retain heat.

The widespread overbuilding, the prevalence of paved surfaces on green areas, the use of building materials with low ability to dissipate heat are among the main causes of UHI. To these causes other factors are added due to the location of the urban area (local morphology, microclimate features, presence of huge waterbodies) and to human activities (emissions from motor vehicles, industrial plants, heating and air conditioning systems for household use).

In summer, the presence of this phenomenon leads to numerous problems, ranging from the peak demands for energy consumption, to air-conditioning costs, pollution and greenhouse emissions, health problems [2]. The phenomenon becomes dangerous during the increasingly frequent summer heat waves, which can cause power blackouts in metropolitan areas and a significant increase in mortality.

The scientific approach to the problem has so far focused mainly on how to detect the temperature within the heat islands and on the thermal properties of building materials. This paper investigates how urban structure can contribute to the occurrence of this phenomenon .

The incidence of urban shape in the determination of UHI is widely recognized in scientific field, though not much investigated in parametric form [3]. Until now, the spatial correlations between temperature, building density and vegetation have been investigated [4] [5] [6], this paper widen the number of urban structure indicators investigated in order to obtain a set of guideline for urban planning in future.

The proposed study examines specifically the relationship between some indicators of urban shape and the temperatures reached inside UHI and identifies the impact of selected urban structure indicators on the phenomenon. The field of application was the province of Naples.

The study was divided into four phases:

1. UHI identification
2. Identification of test areas subject to UHI phenomenon.
3. Recognition of the main indicators of the urban structure in the test areas.
4. Comparison among the test areas and correlation analysis for evaluating the incidence of each feature of the urban structure on the temperature variation.

First step developed has been Heat Islands identification.

There are two types of UHI: *surface* and *atmosferic* UHI. Surface Urban Heat Island refers to the temperature of urban surfaces exposed to the sun that is hotter than air. In summertime urban surfaces temperature could reach 50°C during the day, the difference with air temperature is smaller during nighttime. Atmosferic UHI refers to warmer air in urban areas compared to cooler air in rural surroundings, this phenomenoun is weak throughout the day and becomes more pronounced during nighttime due to the slow release of heat from urban surfaces.

To identify UHI scientists use direct and indirect methods.Direct methods such as fixed weather stations and mobile traverses are used to identify atmosfpheric UHI

[1,3]. İn this paper, remote sensing, an indirect measurement technique, has been used to estimate surface UHI.

Remote sensing data has been used to identify green areas too [7]. Trees and vegetation help reduce air temperatures trough the evapotranspiration process, in which plants release water to the surrounding air, dissipating ambient heat. In addition, the remaining rate of solar radiation not used for evapotranspiration and photosynthesis is reflected and so is neither absorbed nor emitted back later. This means that, when a vegetation surface is present, temperature values are lower.

The vegetation is easily detectable by satellite observations because it produces a specific spectral resolution characterized by a specific pattern of the reflectance coefficient.

Fig. 1. The localization of test areas

The temperature map has allowed the identification of test areas resulting warmer than surrounding environment. The selected areas have the same size (1 km^2) similar land use and building materials but different characteristics of urban geometry. The indicators used in each test area are the following:

- Non-permeable Surfaces Index: expressed by the ratio between paved open areas (i.e. streets, parking areas, courtyard) and the total amount of open areas;
- Percentage of Green Areas i.e. the area occupied by vegetation;

- Land cover ratio: expressed by the ratio between the built surface and the land area. The total built surface is the sum of the areas resulting from the horizontal plane projections of the buildings shapes;
- Geographical localization: location of each test area in relation to the surroundings, i.e. whether it is an inland or a coastal area;
- Building density: expressed by the ratio between the sum of the buildings volumes and the surface of the entire area;
- Mean altitude at sea level;
- Mean height of buildings.

The comparison between the value of each indicator in any test area and the average temperature of each area has allowed the correlation analysis among urban structure and temperature increase.

2   The Temperature and Vegetation Maps

The data used for this study were acquired by the air transported sensor MIVIS (Multispectral Infrared and Visible Imaging Spectrometer) made available by the Province of Naples with the scientific and technical cooperation of the Air Laboratory for Environmental Research, LARA, of the National Council for Research (CNR).

MIVIS is a sensor that operates with high spectral and spatial resolution which registers the radiation issuing from the earth's surface. The high spectral resolution consists in the high number of acquisition channels: in fact, the radiation from the earth's surface is divided into 102 channels, each with a small range of wavelengths. The high spatial resolution of the images obtained by the MIVIS sensor, however, offers a great detail in the number and geometrical characteristics of the elements that compose the images thanks to the 3x3m pixel size, which allows more precise and detailed analysis of phenomena. Data processing has been performed by the software ENVI 4.7 (Environment for Visualizing Image of the "Research System Inc"), which allows the visualization and analysis of data in different formats.

The realization of the temperature map has included the whole province of Naples, which consists of 92 municipalities. The area analyzed, given its size, was not detected in a single flight but was completed in a period from 28th of june 2005 to 27th of july 2005. This resulted in a marked difference in temperature between group of strips realized in different days, and that was taken into account in the selection of test areas. The entire detection consist of 116 strips which cover 1170 km^2 of the province of Naples.

The temperature map realized is a two-dimensional image that shows the temperature of the bodies derived by measuring the intensity of infrared radiation emitted by the concerned bodies.

The temperature map has been realized analyzing the channel 93 of each strip as it provides temperature values, which in so low altitude flight (about 1500 meters) is very similar to the temperature measured at ground level [8]. Channel 93 of each strip includes the range of wavelengths ranging from 8.2 μm to 8.6 μm which is not detectable by the human eye and detects the temperature of the bodies Therefore in the range of channel 93 the detected electromagnetic energy is only the radiance of the bodies. In fact, the amount of energy emitted per unit of area and per unit of wavelength range is not affected by the reflected solar radiation.

In addition the range 8.2 ÷ 8.6 μm of the channel 93 lies in the atmospheric window ranging from 8 μm to 14 microns where the interposition of the atmosphere between the satellite and the Earth's surface is almost meaningless. In this range the particles making up the atmosphere are crossed by the electromagnetic radiation and the one that reaches the sensor is exactly the radiance emitted by terrestrial bodies, without the interference of the atmosphere through the absorption or diffusion phenomena.

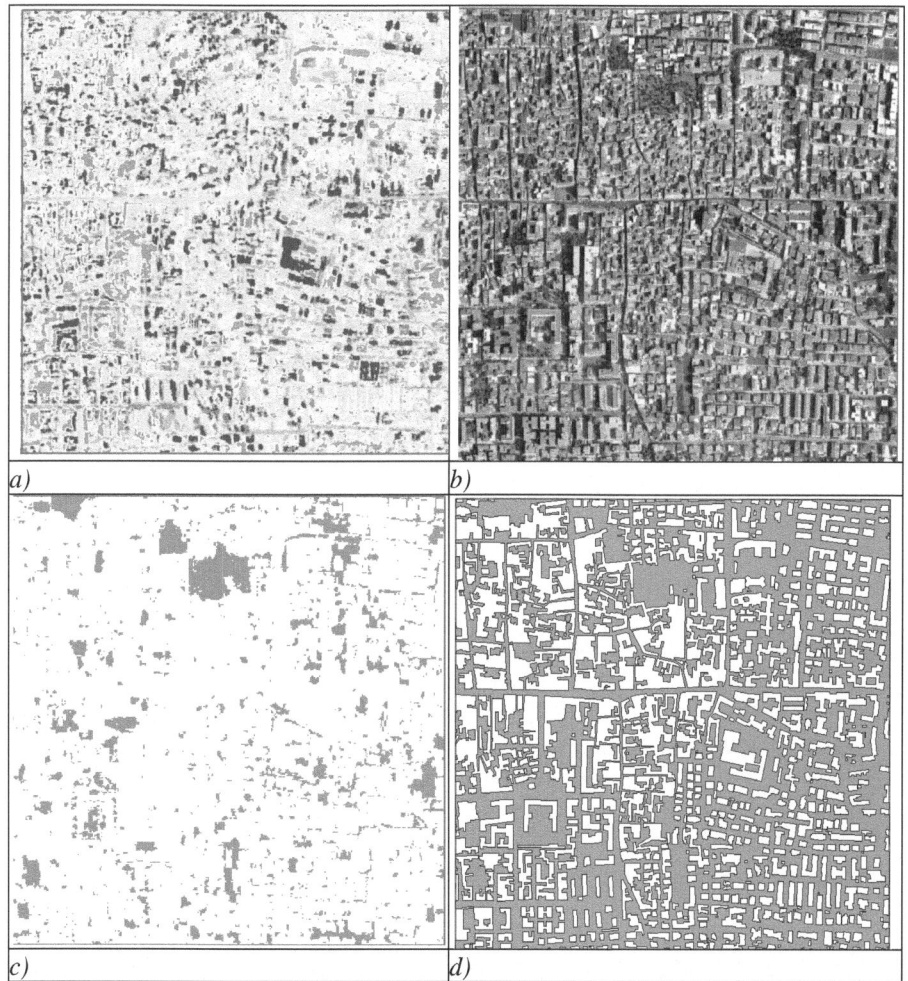

Fig. 2. The first test area: a) the temperature map; b) the orthophoto; c) the NDVI map; d) open spaces map

A preliminary operation to the analysis of such a vast territory as the province of Naples was to make a single mosaic image from the many strips issued from MIVIS. The obtained mosaic image, with datum Gauss Boaga Roma40 (in use in the period of the survey realization), was converted to UTM WGS-84.

The representation of the mosaic image in classes of temperature has assured an immediate reading of values, showing very high values, above 50 ° C, which are evidence of thermal anomalies that are the basis for the development of the Urban Heat Island phenomenon.

Thanks to the high geometric resolution of data it is possible to identify in the temperature map the single existing buildings that reach these anomalous peaks. The high geometric resolution, therefore, plays a key role in making a more detailed analysis allowing to derive the thermal properties of the different natural or anthropogenic elements present in the temperature map [9].

The comparison between orthophoto and temperature map clearly shows the role of materials on temperature, in the same area we find surfaces exceeding 50° C and surfaces below 35° C. The black bituminous built-up roofs and the paved wide open spaces (parks, squares) reach the highest temperatures.

Remote sensing data has been used to realize the vegetation map too.

Most vegetation indices are based on the fact that there are significant differences of reflectance in the electromagnetic spectrum and are based on the analysis of the relationship between defined wawelenghts, where there are different behaviours of reflection and absorption [10].

A widely used index is the NDVI (Normalized Difference Vegetation Index) which is based on the normalized difference between BV (brightness value) of pixels in the Near Infrared range (NIR) and those of the Red Infrared (RED):

$$NDVI = (NIR - RED) / (NIR + RED)$$

Typical values of NDVI -which is limited in the range (-1, 1)- are: 0.2 to 0.6 for vegetation; $0.1 \div 0.1$ for soils and rocks; 0, 2 for water.

Even the vegetation map through NDVI has been produced from the data of MIVIS sensor. For each strip two acquisition channels were considered: Channel 13 for the red infrared with a range of wavelengths ranging from 0.673 μm to 0.693 μm and the channel 20 for the Near Infrared ranging from 0.813 μm to 0.833 μm. To obtain a thematic map showing the vegetation it was necessary to change the decimal values of NDVI in integer values, by performing a linear stretching consisting in converting a 32 bit picture in a 8bit one where the BV of pixels have values ranging from 0 to 255 (ie, there is a passage from 2^{32} to 2^8 colour graduation).

The map shows all the vegetation of the Naples province. The high geometric resolution characterized by a 9m^2 pixel allows to make a detailed analysis identifying every single tree that has a meaningful influence on the temperature value as it interacts with the solar radiation. In fact, in green areas the recorded temperature is lower since urban vegetation has a substantial effect on reducing urban temperature, then, its increase has to be considered as one of the most effective measures against the UHI phenomenon.

3 Urban Structure in Test Areas

The values of the indicators describing urban structure have been derived from the official map of the Province of Naples. The land cover ratio has been calculated using the shapefile representing the building surfaces of the Province, this one contained the building heights too. The calculation of the percentage of green areas has been obtained counting the pixel belonging to BV 123 – 255 in the NDVI map and relating them to the overall number of pixel. Non permeable Surfaces Index has been calculated considering all the open spaces of the areas except the green spaces identified by NDVI map. Mean temperature has been obtained by the temperature map.

The table 1 shows the values calculated for all the investigated areas.

For each indicator the correlation with the mean temperature of the area has been studied. There is a direct correlation between land cover ratio and mean temperature: in flat inland areas, an increase of about 0.26 of the land cover ratio creates a temperature gradient of about 4.4 ° C. Indeed, from a coverage ratio equal to 0.18 and a temperature of 38.1 ° C, representing the area n. 3, we move to values of 0.44 and 42.5 ° C in the area n.1.

In coastal areas, with similar land cover ratio, temperature is lower, the decrease is between 3 ° and 6 ° C to confirm the beneficial effect of sea on temperature.

The percentage of green areas has a positive impact on decreasing temperature. In coastal areas the temperature moves from 39.75 ° C to 34.5 ° C at the respective green rates of 2.35% and 73.83%. The inverse correlation is not linear as the temperature is significantly affected by the exposure to the sea.

Table 1. Indicator values for test areas

Test Area	Land cover ratio	Green Areas %	NonPermeable Surfaces Index	Mean Temperature C°	Geographic Localization	Building Mean height (m)	Building density (m3/m2)	Mean altitude (meters on sea)
1	0.441	13.05	42.88	42.88	Inland Plain	7,49	4,17	97
2	0.25	31.55	43.4	39	Inland Plain	5.01	2.016	111
3	0.18	51.94	30.53	3801	Inland Plain	4.92	1.23	112
4	0.425	18.51	39	40.48	Inland Plain	5.34	2.67	102
5	0.43	15	41,61	36	Coastal Area	12.35	6.45	49
6	0,192	49,98	30,81	35,5	Coastal Area	6,94	1,71	62
7	0,5	2,35	47,806	39,75	Coastal Area	21,65	12,25	19
8	0,11	73,83	14,92	34,5	Coastal Area	6,87	1,14	86

In inland areas, however, the decreasing trend is almost linear since the reduction in temperature is only influenced by the vegetation surface. For an average temperature of 42.5 ° C there is a area covered by vegetation amounting to 13.05%, while at a temperature of 38.1 ° C there is 51.98% of vegetation surface. When the vegetation in the area increases by about 40% there is a temperature decreases of about 4 units.

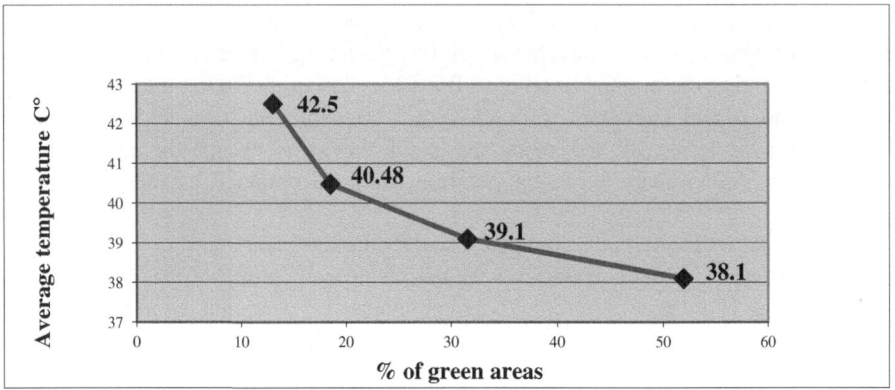

Fig. 3. Percentage of green areas - Temperature in inland plain areas

There is an increasing temperature trend according to the increase of the average height of the buildings. In fact, the more the buildings height is, the more it is difficult to dissipate heat in the atmosphere, since it remains trapped in building environments for a long time causing the "canyon effect".

In inland plain areas, we move from a temperature of 38.1 ° C with a mean height of buildings of 4.92 m to a temperature of 42.5 ° C at a mean height of 7.49 m. In coastal areas temperature moves from 34.5 ° C to 39.75 ° C in correspondence of the mean height of building equal to 6.87 m and 21.65 m respectively.

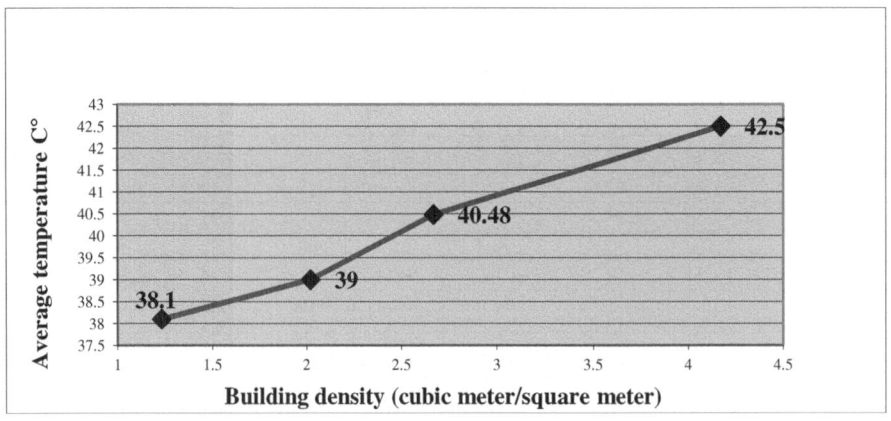

Fig. 4. Building density - temperature in inland plain areas

Also for this indicator the trend is not linear and making a comparison between inland and coastal areas, in the first case the temperature of 39 ° C is achieved at a 'height of 5.01 m, while in the second case that temperature is reached with a significantly elevated height of about 21m. This fact shows, once again, the dissipative action of the sea toward heat. The building density is a cross indicator that links building height to the covered surface. Of course, like the two primary indicators it issues from, it shows a rising trend as the temperature increases: almost linear for the inland plain areas, broken for coastal areas.

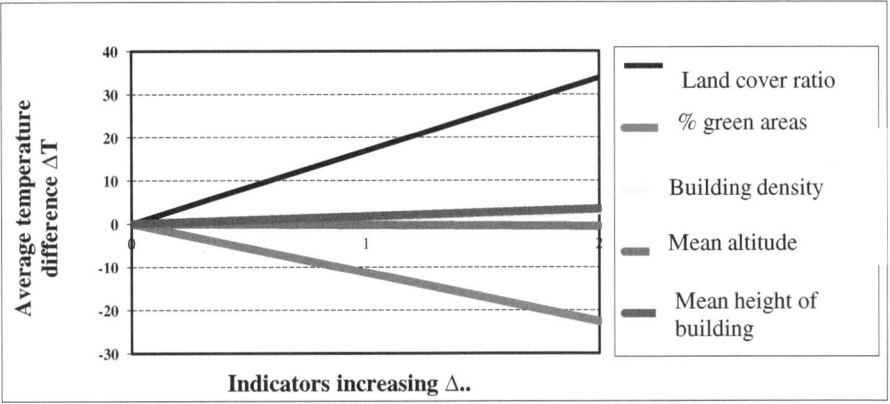

Fig. 5. Average temperature trend related to indicator increasing

4 Conclusions

The work proposed is still in progress, we are going to increase the number of test areas in order to have a sample survey more meaningful of different conditions (by geographical localization or building materials) present in an urban environment. What exposed until now has mainly a methodological value and is a first test of the procedure set up.

The definition of the relations that shows the increase or decrease in temperature ΔT (u), related to a unit increase of the examined indicators $\Delta..^u$, has been crucial for realizing an abacus that allows to estimate the temperature changes expected with the increase of each indicator connected with the increase of temperature inside an area of 1 km 2 according to the unit increase of the indices.

With the same morphological conditions, i.e. with a inland plain area with heights equal to 100 meters above sea level and building materials, this kind of abacus (implemented with the data still in progress) may be useful to estimate the increases in temperature deduced from some parameters of urban and regional planning.

Immediate actions to reduce UHI are the cooling of roofs and street pavements using lighter colored materials and the increasing of green areas and green roofs.

Aknowledgements. Paragraph 1 and 2 are due to Marco Soravia; paragraph 3 and 4 are due to Marialuce Stanganelli. A special thank to Fabio Migliaccio for his help in the realization of temperature map.

References

1. Goward, S.: The thermal behavior of urban landscapes and urban heat island. Physical Geography 2 (2009)
2. Akbari, H., et al.: Energy Saving Potentials and Air Quality Benefits of Urban Heat Island Mitigation. Solar Energy (2001)
3. Zhang, H.: Remote sensing evaluation of UHI and spatial pattern of the Shanghai metropolitan area, China. Ecological Complexity (2009)
4. Wu, P., Zhang, X.: Urban heat island effects study based on built-up index and vegetation index in Beijing city. In: Proceeding SPIE Remote Sensing and GIS Data Processing, China, vol. 7498 (2009)
5. Yang, F., Lau, S.S.Y., Qian, F.: Summertime heat island intensities in three high-rise housing quarters in inner-city Shanghai China: Building layout, density and greenery. Building and Environment (2009)
6. Qiu, J., Jia, L., Wang, Y.: Spacial correlation between Heat Island and green space in Qingdao City based on Remote Sensing. Journal of Southwest Jiaotong University 43(4) (2008)
7. Gallo, K.P.: The use of a vegetation index for assessment of the urban heat island effect. International Journal of Remote Sensing 14(11), 2223–2230 (1993)
8. Mauro, G., Di Lullo, A.: L'utilizzo dei dati iperspettrali MIVIS per un'analisi geografico – territoriale: la Media Valle del Tagliamento come caso studio. Bollettino dell'Associazione Italiana di Cartografia 123, 124, 125 (2005)
9. Akbari, H., Konopaki, S.: The impact of reflectivity and emissivity of roofs building cooling and heating energy use. In: Thermal Performance of the Exterior Envelopes of Building, VII Conference, Miami (1998)
10. Bresciani, M., Stroppiania, D., Fila, G., Montagna, M., Giardino, C.: Monitoring reed vegetation in environmentally sensitive areas in Italy. Rivista Italiana di Telerilevamento 41 (2009)

Taking the Leap: From Disparate Data to a Fully Interactive SEIS for the Maltese Islands

Saviour Formosa[1], Elaine Sciberras[2], and Janice Formosa Pace[1]

[1] Institute of Criminology, University of Malta, Humanities A (Laws, Theology, Criminology), Msida MSD 2080, Malta
{saviour.formosa,janice.formosa-pace}@um.edu.mt
[2] Malta Environment & Planning Authority, Floriana FRN 1230, Malta
elaine.sciberras@mepa.org.mt

Abstract. Taking environmental and spatial planning data to the masses has proven an arduous, expensive and barrier-strewn reality. Data costs, inaccessibility and lack of interactive sites have impeded the implementation of online analysis and knowledge building. The implementation of the Aarhus Convention, the transposition of the INSPIRE Directive and the launching of the SEIS initiative have enabled the Maltese Islands to take the next step into the dissemination of spatial and environmental data to the academic, scientific and public communities. This was made possible through the implementation of a project entitled "Developing national environmental infrastructure and capacity" co-financed under the 2007-2013 European Regional Development Funds (Structural Funds) Programme for Malta, which project was aimed to take national environmental monitoring capacity from a semi-analogue state to a fully interactive online system. The project aims at procuring hardware and software, taking innovative scans of the terrestrial and bathymetric domains, launching SEIS-based information management systems and disseminating the all the data for free.

Keywords: Aarhus, SEIS, INSPIRE, data interoperability, geoportal, LIDAR, spatial data, environmental data.

1 The Need for Speed

1.1 Identifying the Ills of Data Access and Inaccessibility

The state of affairs in information management and dissemination is peppered with both success stories and uphill struggles. Users plodding through the data-information-knowledge-action process experience various barriers to access to data, high costs, archaic mapping, baurocratic procedures and aged data the currency of which is unusable in a rapidly changing world. Various initiatives have enabled users to partake to the process with the onus placed on the national agencies. Such initiatives such as the Aarhus Convention [1], the INSPIRE Directive [2][1] and the Shared Environmental Information System [3] initiative have laid the proverbial carpet to ensure that the process is eased. Though the legislative aspects have been instrumental in enabling the academic and scientific field to be aware of the availability of data for further analysis, the general public lagged in the understanding of the use to which such data could be employed at governance and locality levels.

The Malta case study sought to analyse the situation on the ground as at 2006 most datasets were disparate, basemaps were dated as at 1988, environmental data capture was ad hoc and the dissemination, a mapserver was available with a date tag of 2000 and data were available to the public on a request basis [4].

The idea to bring all the aspects of baseline information, a comprehensive nation-wide digital terrain and bathymetric model, environmental and spatial data into an interactive medium was brought to the fore with the eventual application for ERDF funding. This was aimed at the eventual dissemination of all information for free through an INSPIRE compliant online tool. This sequence of events brought to the fore the need for a speedy approach to implement the changes to ensure compliance with the legislative requirements [5].

1.2 State of Play: The Spatial Irony

Malta has a strong GIS background with developments in the field dating back since the colonial period (pre-1964) with various progressive steps being taken that culminated in the most recent drive to digitize the legacy and to be prepared for the rapid changes that were brought about due to the improvements in visualization and rapid data transfer [6].

A brief overview of the history of geographic information systems (GIS) reveals a rapid (post-1985) pace of development in this arena. Figure 1 below depicts that the two main phases of the process were based on i) the Digital Mapping/ data collection phase and ii) the GI application phase pertaining to Phase i).

Whilst the process entailed the setting up of a national mapping agency in 1988 with a remit to digitize the basemap, the transition to a fully digital secenario required the implementation of GIS between 1994 and 1998, which was brought to the fore for public consumption in 2000 through the launching of a mapserver [7].

Whilst the first phase concentrated on the national implementation of hardware/software installations as infrastructural activities across the different social and physical agencies, phase two took on a legalistic and strategic trust aimed at ensuring that the diverse initiatives were pushed by the international scene. The latter called for integration, which is where the cracks started to appear in terms of dissemination of spatial data in the wider for a. The first pitfall referred to the lack of a appropriately projected base datasets which would allow integration to international cross-border datasets.

---

[1] The acronym INSPIRE refers to the Directive 2007/2/EC of the European parliament and the Council of 14 March 2007 with the aim to establish an Infrastructure for Spatial Information in the European Community. The directive entered into force on 15 May 2007 and will be fully implemented by member states in 2019.

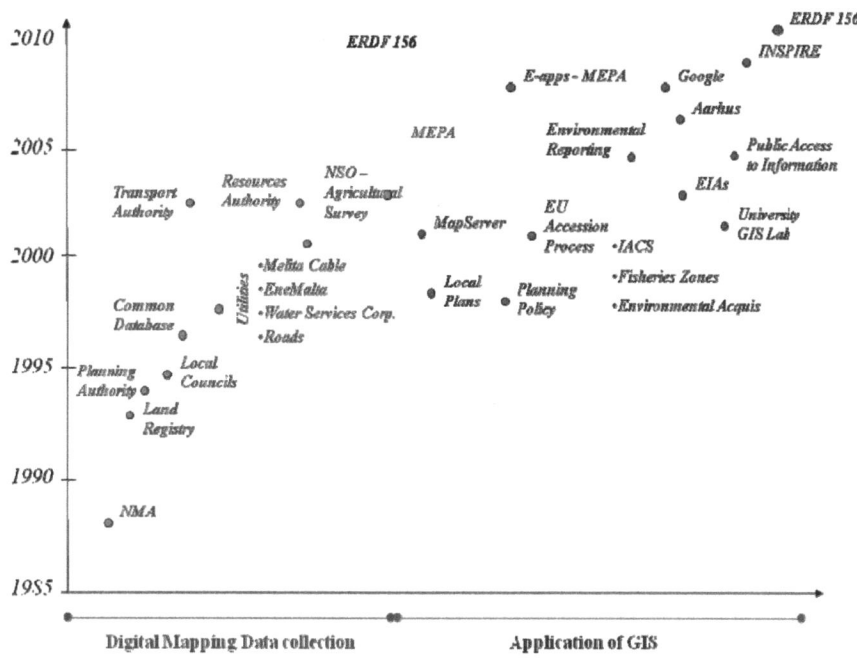

Fig. 1. The GI data process

Due to severe hardware restrictions and high storage costs, a decision was taken to truncate all the spatial data (Malta falls within 1 zone on the UTM projection). Such was done in order to save money on harddisk space, a decision that was aimed to ease the rolling-out costs but which is currently hindering all the EU/EEA data reporting to the various agencies, since any such data need to be reconverted to the original full UTM. This space and cost-saving measure is now estimated to incur EUR100,000s since the whole country's organisations have to reconvert all their data to fit within the international projection and which is causing severe restrictions on data convergence.

This process was further hindered by Malta reporting requirements to the EU which dealt several shocks to the system. Late deliveries due to reporting mishaps, missing or non calibrated data, spatial elements not supported by metadata and other hiccups resulted in negative outcomes for Malta's position in the EEA league [8]. However, the process had its bright moments with the creation of several high level outputs in conjunction with other agencies such as the Austrian Umwelbundesampt (UBA)[2]. Such datasets included the Corine Landcover 1990-2000-2006 and the relative change analysis, elevation maps, environmental protection maps (EEA

---

[2] The project was developed through an agreement between the European Environment Agency, the Malta Environment and Planning Authority and the Umweltbundesamt-Vienna, the latter providing interpretation and implementation expert support through the Twinning Project MT2002/IB/EN-01 "Establishing Institutional Capacity in the Environmental Sector".

CDDA, Natura 2000, SAC) [9], terrestrial and marine habitats as well as a Posidonia baseline survey and, in 2011, the Land use/cover area frame survey (LUCAS) field survey [10].

In conjunction with the above process, various research initiatives were initiated. These have ranged from organisational change initiatives [6], remote sensing [11], census web-mapping [12], environmental-landuse [13], GML-related dissertations [14], ethics in GIS [15], 3D GIS for spatial planning [16], environmental [17], quality improvement cycle [18], through to socio-technical approaches to GIS [19]. Each served a role to create and ultimately shape GI-related change in the Islands which lead to the preparation of the new foundation for national geographic information data cycle.

1.3 Taking the Plunge and the Availability of Resources

With the integration of the spatial planning and the environmental agencies, the need was felt to consolidate all the data gathering processes, create a comprehensive strategy for data reporting, integrate the initial data-related projects such as those emanating from the Structural Funds 2007-13, EAFRD 2007-13, Transition Facility Programme for Malta, Pre-Accession Funds, and other programmes, with the main project being that related to the establishment of institutional capacity in the environment sector [20]. The latter project that laid the ground for the current project was entitled "Establishing Institutional capacity in the Environment Sector" and was financed by the National Plan for the Adoption of the Acquis Programme for Malta 2002 which aimed to prepare the Maltese administration to face the challenges posed by EU accession. The overall objective of this project was to strengthen the institutional capacity within the environment sector in order to achieve compliance with the EU environmental Acquis. The project team closely examined the organisational structure of the institutions involved in the implementation and enforcement of the environmental Acquis and provided multi-disciplinary training to staff engaged in carrying out environment protection functions. Finally, the project also laid foundations for the development of further technical capability to manage new and existing infrastructure in various sectors, including waste, air quality, biodiversity and water, establishing technical standards and codes of practice for various functions related to environmental protection.

The outcome of the project which was concluded in 2004 set out the groundwork for the creation of a strategy to improve the data availability to the experts trained through the 2002 project. The groundwork laid out, the time was ripe to take a decision on the identification of a way forward for the integration of the environmental domain with the spatial development domain through the integration of information resources and information technology infrastructure in line with the Aarhus, INSPIRE and SEIS requirements as well as the outcomes of such projects as Plan4all [21].

1.4 The Integrative Project

This project is co-financed by the European Regional Development Fund, which provides 85% of the project's funding and the Government of Malta, which finances the rest under Operational Programme 1 - Cohesion Policy 2007-2013 - Investing in

Competitiveness for a Better Quality of Life. The authors are implementing the project through the Malta Environment and Planning Authority (MEPA) in collaboration with the Malta Resources Authority (MRA), the Department of Environmental Health, the National Statistics Office (NSO) and the University of Malta. The project has a budget of € 4,780,480, of which € 4,657,760 are eligible costs.

As a small densely populated and highly urbanised island economy, Malta faces wide-ranging environmental challenges in the areas of air, water, waste, nature protection, noise, chemicals, industrial pollution, sewerage, land use planning and control of urban sprawl, and other sectors. Various initiatives have already been undertaken to upgrade Malta's environmental regulatory capacity, including efforts to ensure full compliance with the relevant Community Directives as well as national legislation. However, environmental monitoring and reporting are hampered by lack of baseline environmental data on ambient conditions, lack of monitoring infrastructure, modern monitoring equipment and limited human resources. These factors result in ineffective monitoring and reporting on the state of the environment, increased risk of failure to detect, prioritise and address environmental degradation, as well as in an inefficient use of financial, human and natural resources.

The project focuses on radically improving the national environmental monitoring capacity in five environmental themes – air, water, radiation, noise, and soil. It will result in the procurement of equipment, information management systems, environmental baseline surveys, training of staff, and the enhancement of the national monitoring programmes in these five environmental themes. The purpose of the project will be achieved through a series of various service and supply contracts that will lead to the procurement of equipment, development of information management systems, gathering of critical baseline data, development of monitoring programmes and training of human resources in the operation of equipment and systems. The following outputs will be delivered:

- 100% of environmental monitoring requirements in the areas of air, water, radiation, noise and soil will be assessed, and an environmental monitoring strategy and detailed monitoring programmes will be designed and drawn up by Q2 of 2013 to cover all monitoring requirements. The strategy will be accompanied by detailed tender specifications for the procurement of equipment, systems, training and data collection requirements that could not be identified prior to the completion of the strategy;
- Air, noise and radiation equipment, information resources systems and infrastructure procured, installed, tested and commissioned, and relevant staff trained in their operation by the Q4 of 2012;
- Baseline studies with 100% scan coverage of the Maltese Islands conducted in the areas of water, radiation, noise and soil, together with terrestrial spatial surveys and bathymetric surveys of coastal waters within 1 nautical mile by Q2 of 2013;
- Shared Environmental Information System (SEIS) designed and implemented by Q2 of 2013;
- Results of the project will be disseminated throughout the project to a wide range of stakeholders and the public through an information campaign.

As a result, public awareness of issues pertaining to the environment will be improved, and better policy decisions can be taken in the field of the environment.

The above investment will contribute to achieving between 25% and 100% compliance with the monitoring requirements of around 17 legislative instruments (Directives, Regulations, Decisions) in the areas of air, water, soil, noise and radiation.

2 The Technologies and the Project Launch

The study, initiated in 2006, identified the situation that environmental monitoring and reporting is hampered by an incomplete monitoring strategy, a lack of baseline environmental data on ambient conditions, a lack of monitoring infrastructure & modern monitoring equipment and limited human resources. The process identified the need to enhance the national monitoring programmes in the five environmental themes[3] through the identification of information gaps in monitoring processes and filling data gaps, carrying out environmental baseline surveys and through the procurement of monitoring equipment & information management systems and finally the training of staff.

The project results are aimed to reach the following objectives:

(1) Environmental monitoring requirements in the areas of air, water, radiation, soil, noise and chemicals assessed, an environmental monitoring strategy and detailed monitoring programmes designed and drawn up. The strategy will be accompanied by detailed tender specifications for the procurement of equipment, systems, training and data collection requirements that could not be identified prior to the completion of the strategy;

(2) Air, water, noise and radiation equipment, information resources systems and infrastructure procured, installed, tested and commissioned, and relevant staff trained in their operation;

(3) Baseline studies conducted in the areas of water, radiation, noise and soil, together with terrestrial spatial surveys and bathymetric surveys of coastal waters within 1 nautical mile;

(4) Shared Environmental Information System (SEIS) designed and implemented;

(5) Results of the project will be disseminated throughout the project to a wide range of stakeholders and the public through an information campaign.

Therefore the project will result in the following deliverables which will enable the integration of the requirements for EU environmental reporting through the employment of the INSPIRE Directive for the spatial component, the use of the Aarhus Convention as the conveyor for the dataflow and ultimately the employment of a tool pertaining to the SEIS requirements for the eventual online dissemination[4]. This triumvirate of Directives and Initiatives were used as a fulcrum to ensure that the full spectrum of Maltese environmental and spatial information reaches the public at no expense, empowering both the individuals, non-governmental organisations and

---

[3] The 5 monitoring themes: air, water, noise, radiation and soil.
[4] Standards employed in the Malta-SEIS: OGC WMS, OGC WMS - T, OGC WFS, OGC WCS, ANSI SQL, INSPIRE, Z39.50 and CSW.

governing bodies to receive knowledge at first hand in order to debate on strategic and local plans as well as monitor the health of their environment and in turn give the social structure the power to react with readily available knowledge as against being faced with barriers to access. This process is the fruition of a process begun in 1995 by the authors to ensure that data are disseminated for free.

The project outcomes can be structured into 4 sectors: Environmental Acquisition, Spatial Constructs, Dissemination Media and Access as per Table 1.

Table 1. Project Outcomes

Environmental Acquisition	**Spatial Constructs**
(1) Air Strategy and Baseline Study (2) Water Strategy and Baseline Study (3) Noise and Radiation Strategy (4) Soil Strategy (degradation processes and contamination in diffuse sources)	(1) Full LIDAR Scan: Terrestrial and Bathymetric (2) Ground truthing for sea substrate type (3) Oblique aerial imagery & satellite imagery (4) An address point dataset
Dissemination Media	**Access**
(1) Online information service (2) Online mapservice - SEIS (3) Statistical backing for experts – inc. spatial stats	(1) ALL Data are to be disseminated for FREE

2.1 The GIS Components: Data Capture and Technologies

The requirements for data acquisition based on bathymetric and terrestrial 3D as well as aerial imagery were deemed essential to the analytical and decision-making processes related to environmental monitoring and the project since the state needed a new baseline from which to launch all its data capturing exercises across the different themes. Whilst some basic bathymetric data are available from legacy nautical charts, such data need to be updated to higher resolutions so as to be suitable for environmental modelling and EU reporting purposes. No comprehensive and detailed terrestrial 3D surveys have ever been carried out in Malta. The resulting lack of high quality 3D spatial data hinders land use planning, environmental monitoring and management processes that rely on such data. It was deemed useless to gather ad hoc datasets without an established reference baseline dataset and the decision was taken to take a series of scans using different technologies in order to provide a comprehensive seemless dataset that covers all the terrestrial area and the marine area up to 1 nautical mile from the baseline coast as designated by the Water Framework Directive which falls within the remit of the domains under study in this project. The aim was to address the identified data lacuna by delivering high resolution 3D terrestrial data coverage for the Maltese Islands using a combination of oblique aerial imagery and Light Detection and Ranging (LIDAR) data (Figures 2 a-b), as well as through a bathymetric survey of coastal waters within 1 nautical mile (nm) radius off the baseline coastline, using a combination of aerial LIDAR surveys, acoustic scans and a physical grab sampling survey. The use of this technology, as well as other fieldwork technologies, has equipped the researchers with a launching pad for the diverse physical, environmental and social studies that are undertaken in relation to social and environmental health.

The outputs from the project include the following services and supply (Table 2):

Table 2. Project Outputs

Services	**Supply**
(1) LIDAR Scan: Terrestrial (Topographic Light Detection and Ranging (LiDAR)) Digital Surface Model (DSM) and Digital Terrain Model (DTM) (316 km.sq) – Figure 3a (2) Bathymetric LIDAR aerial survey - depths of 0 m to 15m within 1 nautical mile from the Maltese coastline (38 km.sq) – Figure 3b (3) Bathymetric Scan: Acoustic (side scan sonar) Digital Surface Model and an acoustic information map of sea bed (361 km.sq) – Figure 3c (4) High resolution oblique aerial imagery and derived orthophoto mosaic and tiled imagery of the Maltese Islands (316 km.sq) (5) Satellite imagery (6) A complete address point database	(7) Remote GPS Cameras (Remote capture GPS receiver) (8) 3D scanner (9) GIS Handhelds (10) Global Navigation Satellite System Station and geodetic receivers

Fig. 2. a. Oblique Imagery Output 2011

Fig. 2. b. LIDAR Scanning Exercise 2012

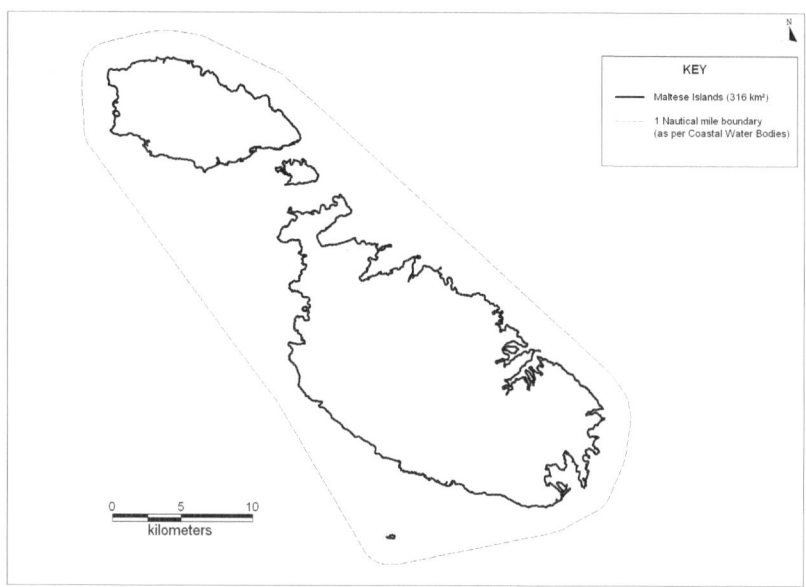

Fig. 3. a. Maltese Islands coast inclusive of 1 nautical mile boundary from the baseline coastline

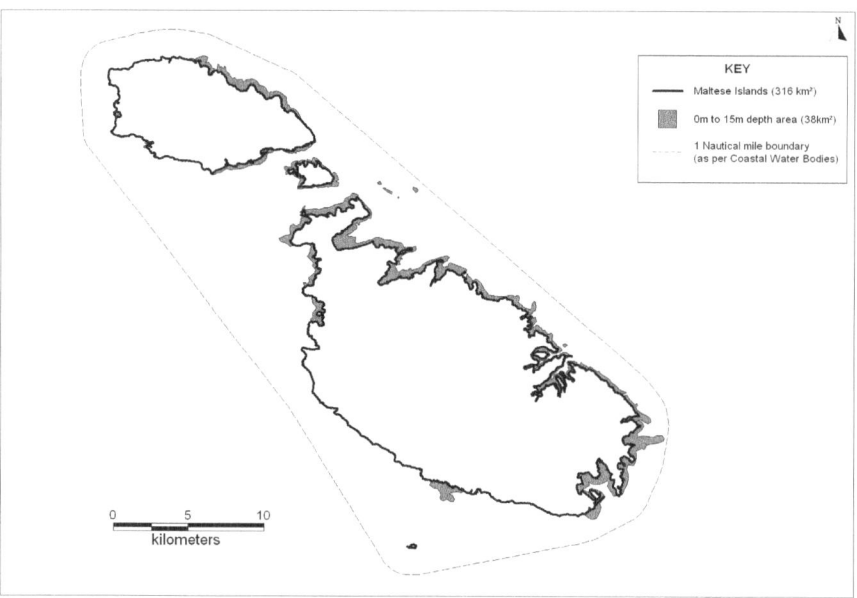

Fig. 3. b. Maltese Islands showing coastal water area with depths of 0m to 15 m within 1 nautical mile from the Maltese baseline coastline

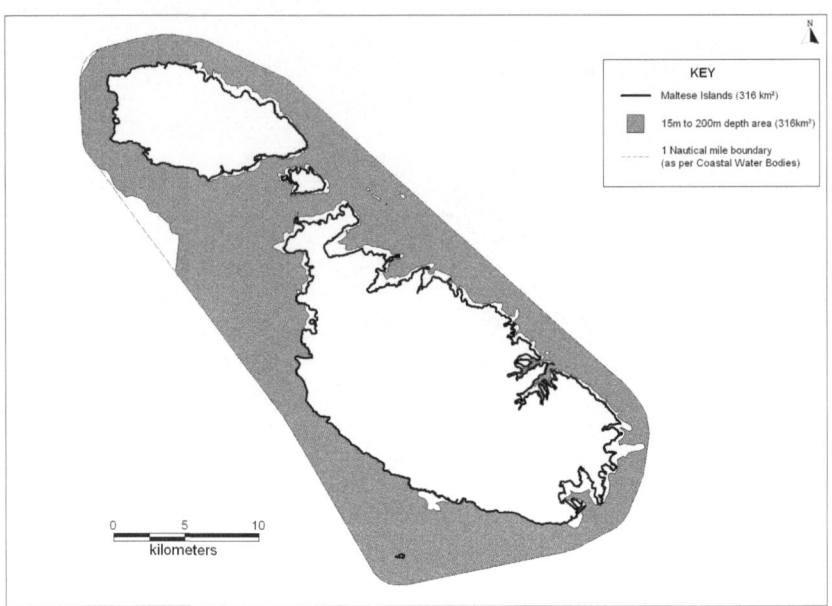

Fig. 3. c. Maltese Islands showing coastal water area with depths of 15 m to 200 m within 1 nautical mile from the Maltese baseline coastline.

3 Integration and Dissemination

The final phase but most crucial of the project entailed the implementation of an SEIS-based strategy and implementation of an online tool. This tool will build on the previous experiences with the MEPA Mapserver, the EEA CDR and the Aarhus capacity building project[5]. The objectives of this phase were set out to review and analyse the MEPA's, national and EU-level requirements for the development and operation of the SEIS, taking into consideration all relevant factors. On the basis of this review, design the Maltese component of the SEIS and in turn develop and implement the SEIS, including a dedicated geoportal based on ArcGIS geodatabase design as based on ArcGIS server architecture and finally train staff on the use, operation, data analysis, maintenance and customisation of the developed SEIS.

The phase is aimed to review the state of play of the current developments with respect to the Shared Environmental Information System (SEIS), including the following:

- EU Directives supporting the EU's SEIS initiative and any proposed recommendations of the EEA, JRC, EUROSTAT;
- Commission's Communication COM (2008) 46 Final "Towards a Shared Environmental Information System";

---

[5] www.ambjent.org.mt (a project implemented under a Austrian – Maltese Twinning Project MT/06/IB/EN/01 "Further Institution Building in the Environment Sector".)

- SEIS developments by the European Environment Agency (EEA);
- Overview and updates on the SEIS-BASIS (Shared Environmental Information System Baseline and Evolution Study) project which aims to provide guidance on how to improve the comparability and quality of environmental data, as required by SEIS;
- The outputs of the NESIS project and roadmap developments on how to move from the current information systems of EU's environment agencies towards an INPSIRE-SEIS based system. To include relevant results for the NESIS State of Play study on examples of best practice as a source of guidelines for MEPA' s proposed SEIS as informed by recent developments;
- Relevance of the INSPIRE Directive (Directive 2007/2/EC) and the Aarhus Convention to the EU's SEIS;
- Linking of an integrated reporting system is required in line with the EEA Reportnet initiative and its CDR (Common Data Repository) structure to SEIS;
- An analysis of the existing Maltese information management systems and platforms, as well as an assessment of the present institutional capacity necessary for the operation of the Maltese component of the SEIS;
- New or emerging reporting standards currently being adapted, such as XML-related standards, to which the SEIS should conform.

This process will involve the analysis of the current systems in place to process environmental monitoring data and data flows required, the design of the SEIS for Malta, and the development of such a web-based environmental information system.

The output will result in the creation of a web-based environmental information system, on the basis of existing platforms, as well as on the basis of any other additional platforms and components that may be required, to achieve full interoperability and functionality of the Maltese component of the SEIS, in line with the applicable guidelines and best practices in this field. This phase forms a key part of the ERDF-funded project and will integrate all the environmental monitoring data acquired through this project and through other environmental monitoring initiatives administered by MEPA and other Maltese entities.

The development of the SEIS will amongst other include the delivery of a web-based GIS dedicated to environmental monitoring data incorporating existing aerial orthophotos and basemaps as well as newly acquired satellite imagery, oblique aerial imagery, LiDAR terrain datasets and bathymetric data. Moreover, since the plans are in place to migrate from an ArcInfo database to an ArcGIS geodatabase structure, the SEIS will be developed using an ArcGIS Server platform. The SEIS is expected to be a modular and scalable system which is flexible to meet the varying demands of usage and applications with time.

The proposal was to design a geodatabase data model that is flexible, caters for potential expansion, easily adaptable by the environmental agency and supports migration from current data structures. The following steps were highlighted in order to improve the development of the geodatabase schema.

i) Identify the data sources and key data themes for the GIS and characterise each thematic layer (including symbology, annotation, map scale, accuracy, data use, integration with other datasets)
ii) Develop representation specifications and relationships of the geodatabases (modelling of feature classes, definition of tabular database structure, spatial behaviour and data integrity rules, were relevant)
iii) Define the data capture procedures, map display properties and assignment of procedures for geodatabase building, conversions, editing and maintenance.
iv) Document the proposed geodatabase design (such as schema diagrams, map layer examples, metadata documents)
v) Since an ArcGIS Server platform will be used for the development of the SEIS, ArcSDE must be employed to manage the underlying geospatial data that will be stored in Microsoft's SQL Server RDBMS (Figure 4) [22].

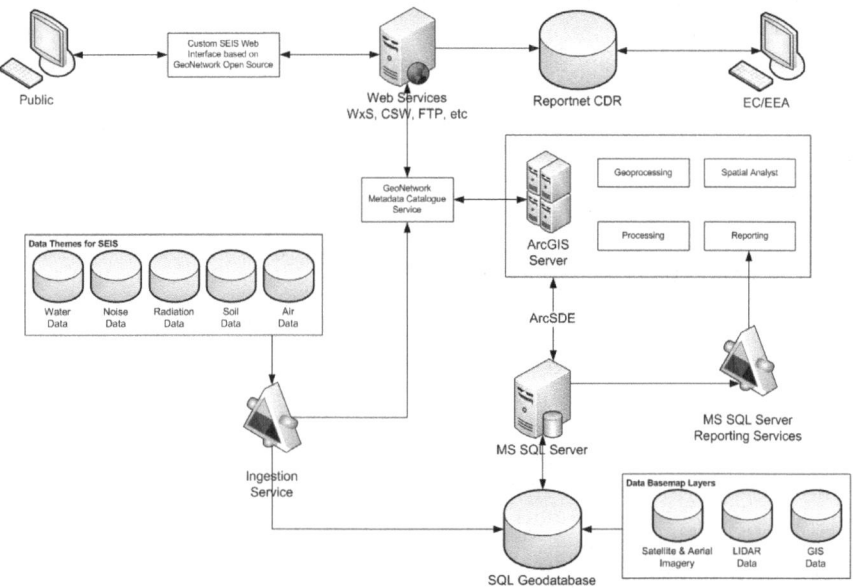

Fig. 4. Malta-SEIS Architecture: Source: Bonazountas M., and Karampourniotis I (2012, 17)

Finally, the project envisaged a structure that conforms to the Aarhus, INSPIRE and SEIS structures as follows:

Web Interfacing and Security
ArcGIS Server's .NET API will be used to interface with ArcSDE to present the data on the web. This system would allow the viewing, data querying, reporting and analysis of all spatial datasets identified in the five themes and the information resources datasets;

Front end Flexibility
Since it is expected that the SEIS will develop with time based on the increasing levels of geospatial data to be acquired, varying reporting demands and data analyses, the Malta SEIS will have an extensible front end that can be tailored to meet the new requirements of a changing and dynamic environment;

Discovery Services/Functions
The SEIS will include appropriate discovery services and functions since this is a requirement for users and such automated tools requesting data. Key spatial data sets should be published in ways that support the discovery of the data and provide access to these resources via product-neutral visualization and downloading services;

Geoportal
The geoportal is to be linked to the National Spatial Data Infrastructure (SDI) should this be established during the lifetime of the project and eventually linked to the EU SDI under development. All relative documentation relating to how spatial data and services can be interlinked respect the INSPIRE Directive's implementing rules as already outlined in the Plan4all project Malta test (Formosa, Magri, Neuschmid, Schrenk; 2011);

Social Dataset Integration
The system will ensure that other environmental domains not covered in the five themes will be covered as identified in the Plan4all Malta case study which reviewed the potential for such systems to integrate information from the physical, social, criminological, psychological and health domains. The studies of the impacts of environmental factors on health issues are seen as a major component that will lead to the public interest in the system and may eventually create a real scenario for volunteered geographic information in the Maltese Islands;

Spatial dataset standards
All spatial datasets will conform to the INSPIRE Directive (Directive 2007/2/EC) standards. The SEIS must conform to the requirements emanating from the various standards established by the environmental thematic areas. Therefore the project takes into account developments within the various ETCs (European Topic Centres) and to ascertain that the SEIS conforms to and seamlessly links to the European ones, where such protocols have been created. The geographic coordinate system for the SIES should be based on a Projection (UTM Zone 33 S), Datum (ED 1950) and Ellipsoid (Hayford International 1924);

Spatial data transfer
The project structure requires that the spatial data-transfer processes of the systems currently in place and new ones acquired through the project are thoroughly understood and depicted into the model with a target to establish a coherent data integration system that will allow the further transmission through the online portal.

In conclusion, the project, whilst an ambitious one, ensured that the whole process was an integrated one which takes into account the conceptualisation of a nation-wide strategy for access to data [23] employing various Directives and Conventions in order to ensure that data is transmitted for free to the general public whilst ensuring that the quality is assured. The project which was launched in December 2009 is expected to be concluded in 2013, with a planned maintenance procedure as established through such standards as ISO 9001: 2008 and those emanating from the INSPIRE and information related Directives.

References

1. NESIS, Ambjent.org.mt Environmental Portal (MT), http://www.nesis.eu/index.php?option=com_wrapper&view=wrapper&Itemid=158 (accessed on February 29, 2012)
2. Official Journal of the European Union: Directive 2007/2/EC of the European Parliament and of the Council of 14 March 2007, establishing an Infrastructure for Spatial Information in the European Community (INSPIRE), L108, vol. 50 (April 25, 2007)
3. Shared Environmental Information System, http://ec.europa.eu/environment/seis/ (accessed on February 29, 2012)
4. Conchin, S., Agius, C., Formosa, S., Rizzo Naudi, A.: Does visualisation of digital landscapes serve itself? How topographic, planning, environmental and other thematic information is integrated and disseminated via web GIS. In: Buhmann, Pietsch, Kretzler (eds.) Peer Reviewed Proceedings of Digital Landscape Architecture 2010. Anhalt University of Applied Sciences. Wichmann Verlag, Heidelberg (2010) ISBN 978-3-87907-491-4
5. Malta Environment & Planning Authority, Developing National Environmental Monitoring Infrastructure and Capacity. MEPA, Floriana, Malta (2009)
6. Gatt, M., Stothers, N.: The Implementation and Application of GIS in the Planning Authority of Malta, Geographical Information. In: Second Joint European Conference and Exhibition on Geographical Information. IOS Press, Barcelona (1996) ISBN 90 5199 268 8, ISBN 4 274 90098 3 (OHMSHA)
7. MEPA mapserver, http://www.mepa.org.mt/Planning/index.htm?MapServer.htm&1 (accessed on February 29, 2012)
8. European Environment Agency, Priority Dataflows, http://www.eionet.europa.eu/dataflows/pdf2011 (accessed on February 29, 2012)
9. European Environment Agency, Common Data Repository, http://cdr.eionet.europa.eu/mt (accessed on February 29, 2012)
10. European Commission, Land use/cover area frame survey (LUCAS) Decision 1445/2000/EC (LUCAS), http://epp.eurostat.ec.europa.eu/statistics_explained/index.php/Glossary:Land_use/cover_area_frame_survey_ (accessed on February 29, 2012)
11. Tabone Adami, E.: Corrections for the estimation of chlorophyll concentrations in coastal waters from remotely sensed data, MPhil thesis. University of Cambridge, Cambridge (1998)

12. Formosa, S.: Coming of Age: Investigating the Conception of a Census Web-Mapping Service for the Maltese Islands. Unpublished MSc thesis Geographical Information Systems. University of Huddersfield, United Kingdom (2000), http://www.tcnseurope.org/census/1995/index.html (accessed on February 29, 2012)
13. Tabone Adami, E.: Integrated modeling of nutrient transfers from land-based sources for eutrophication assessment of Maltese coastal waters, PhD thesis. University of Cambridge, Cambridge (2001)
14. Aguis, C.: Using GML to represent Spatial Environmental Information, unpublished MSc GIS thesis, University of Huddersfield, Huddersfield (2003)
15. Valentino, C.: Developing a Coherent Approach to Ethical Use of Geographical Information in Malta, unpublished MSc GIS & Management thesis, Manchester Metropolitan University, Manchester (2004)
16. Conchin, S.: Investigating the development of 3D GIS technologies for Spatial Planning: A Malta study, unpublished MSc GIS & Environment thesis, Manchester Metropolitan University, Manchester (2005)
17. Farrugia, A.: Implications of EU Accession on Environmental Spatial Data: a Malta Case Study, unpublished MSc GIS Network thesis, Manchester Metropolitan University, Manchester (2006)
18. Rizzo Naudi, A.: A Continuous Quality Improvement Cycle for Geographic Information Systems within the Malta Environment & Planning Authority (MEPA). Unpublished M.Sc. dissertation, Manchester Metropolitan University (2007)
19. Formosa, S., Magri, V., Neuschmid, J., Schrenk, M.: Sharing integrated spatial and thematic data: the CRISOLA case for Malta and the European project Plan4all process. Future Internet 3(4), 344–361 (2011), doi:10.3390/fi3040344, ISSN 1999-5903
20. Internationally Funded Projects at MEPA, http://www.mepa.org.mt/internationally-funded-projectsatmepa (accessed on February 29, 2012)
21. Beyer, C., Wasserburger, W.: Plan4all Deliverable 2.2, Analysis of innovative challenges (2009)
22. Bonozountas, M., Karampourniotis, I.: MALTA-SEIS: Deliverable D2.1Report of Analysis and Detailed Proposal for SEIS, CT3067/2010 – 02, Malta (2012)
23. Formosa, S.: Access to data in a small island state: The case for Malta, Islands and Small States Institute, University of Malta, Tal-Qroqq, Occasional Papers on Islands and Small States, vol. 5 (2010); ISSN: 1024 6282

Analyzing the Central Business District: The Case of Sassari in the Sardinia Island[*,**]

Silvia Battino[1], Giuseppe Borruso[2], and Carlo Donato[1]

[1] DiSEA– Department of Economic and Business Sciences, University of Sassari,
Via Muroni, 25 - 01700 Sassari, Italy
`{Sbattino,cadonato}@uniss.it`
[2] DEAMS – Department of Economic, Business, Mathematic and Statistical Sciences,
University of Trieste, Via A. Valerio, 4/1 – 34127 Trieste, Italy
`giuseppe.borruso@econ.units.it`

Abstract. The cities are places where people, goods and information flows concentrate and, even if they differ in position, dimension and functions, they remain the main players of local, national and international development. Every city, even if represented by its "historic centre" - space that not always has maintained its role of principal central position in the urban evolutionary processes - has experimented developing areas of specialization and concentrations of functions, as districts marked by one or more dominant functions: commercial, administrative, residential or cultural. Starting from the definition of the Central Business District (CBD) or "central place" of the city and keeping as reference the "urban environments" or "districts" of the most recent town-planning project of the city of Sassari, we performed a quantitative and distributive analysis of the *core* urban activities of the whole territory of the Municipality of Sassari. Such analysis was realized on a point pattern represented by the spatial distribution of 'high level' urban activities and their elaboration by means of spatial indexes and density estimation. The spatial indexes, referred also to urban roads' length and to the number of residents of every urban subunit considered, highlighted that such kinds of activities refer mainly to the city center area and concentrate in just four of the thirty-two districts. The GIS elaboration and processing allowed us to represent the central area (CBD), highlighting a *core*, as Piazza d'Italia, where a limited number of residents corresponds to a greater presence of rare activities.

Keywords: City, Urban Core, Centrality, Spatial Indices, Kernel Density Estimation, Sassari, Central Business District.

---

[*] The paper derives from the joint reflections of the three authors. Silvia Battino realized paragraphs 2.1, 3, 4.1, while Giuseppe Borruso wrote paragraphs 2.2 and 4.2. Carlo Donato wrote paragraphs 1 and 5.
[**] The geographical visualization and analysis, where not otherwise specified, have been realized using Intergraph Geomedia Professional and Geomedia GRID 6.1 under the RRL (Registered Research Laboratory) agreement between Intergraph and the University of Trieste (Italy).

1 Introduction

Cities are portions of space characterized by flows and movements of people, goods and information, different in terms of their locations and the functions they play over radii of gravitation[1]. Changes happen in time and affect both the physical characters of the city and its role and functions. For decades authors of different disciplines debated on this topic revealing what are the elements determining the importance of a city through the years, from population growth, to accessibility and type of urban planning. The most famous author focusing on urban system was Christaller [1] followed by Loesch [2] with the Central Place theory in which an explanation of the distribution patterns and hierarchy of locations in relation to the size of population served by tertiary activities is proposed[2].

The first theoretical studies on urban structures date back to the Twenties of the Twentieth Century (1925), when Burgess [16] proposed a concentric model, leading the way for further interpretation of a monocentric image of cities[3]. Other studies developed theories about the city where it is possible to distinguish a single central area or Central Business District (CBD), with low population density, followed by a neighboring core, where residential density initially increases and then decreases towards the peripheral areas as the distance from the CBD increases[4]. Noting that the population density decreases as one moves away from the central district, where the rare functional units tend to cluster, theories and models are developed to describe the land use, urban income and location of residences[5]. The studies range from monocentric to polycentric city ([6], [18]), developing around multiple cores of activities among them dissimilar and distinct from the CBD. A change in theory occurred mainly due to changes in the distribution model of acquisition of goods and services by consumers, policies of urban and regional planning and to implementation of new areas of business, commerce, residences in places different from central locations ([24], [25]). Actually to-date urban structures and functions are distributed, concentrated and transformed creating "neighborhoods" or "urban districts" characterized by the presence of dominant features, in terms of the main

---

[1] A city's range depends on the functions' characteristics: basic or city forming that determine the importance and the degree of centrality of a city and non-basic or city serving, aimed primarily at meeting the needs of those who live outside the urban area ([3], [4], [5], [6], [7], [8], [9]).

[2] Italian authors, developed also the theory adapting it to the national context. See also Bonetti ([10], [11], [12] and [13]), Dematteis [14] and Corna Pellegrini and Pagnini [15].

[3] For further interpretation of this model see Rodrigue, Comtois, Slack [7] and Fortuna [17].

[4] The exam of population density is the basis for further theories and models describing urban income, land use and residence's location ([19], [20], [21]).

[5] Alonso's Bid Rent Theory [19] is based on the higher cost of urban land and its decreasing function with distance. Central locations close to the CBD will be occupied by players able to derive a higher utility from being located in a "central position". Farther locations will benefit lower land prices but higher transport costs. See Iommi et al [22] and Alvarez de la Torre [23], on the evolution and value of land through a function of "satisfaction" tied to localization processes, to free time and to amount of goods and services in a given area.

administrative, commercial, industrial or residential functions played. The district presenting the highest degree and concentration of important functions for the urban socio-economic life is generally represented by the CBD, which often overlap with the geographic urban center: specialized retail activities, public buildings and offices, high land values and high level of traffic are generally its main characteristics[6] [26]. The other urban "places" develop around the CBD and usually identify transition areas, where all city functions gather[7] together with residential and industrial areas.

A part from such studies focusing on the development of urban areas in time there are also empirical evaluations and observations trying narrowing the center of business and identifying it. Both quantitative and qualitative models have been used considering the concept of distance to evaluate the centrality of a place compared to the surrounding area[8].

Actually, "old" cities changed and transformed over time due to the insurgence of other factors, such as the free choice of private and public entrepreneurs in locating their activity or operations imposed by zoning land-use planning [27]. Below we propose the definition and the cartographic representation of the Central Business District of the city of Sassari through quantitative and distributive analysis of higher rank activities taking into account some indices of intensity.

2 The Methods. Spatial Indexing, Point Pattern Analysis and the Definition of the CBD

The research carried out in this paper was focused on the development of spatial indexes and point pattern analysis, applied to the city of Sassari in North-western Sardinia to identify the area(s) of its CBD, starting from data concerning economic activities georeferenced at address-point level. Different methods were adopted to highlight the CBD area and compared in order to highlight the *core* of the city in terms of dominance of the economic activities located there.

2.1 Point Patterns and Spatial Indexes

The method adopted was based on collecting urban activities and classifying them in categories, followed by their georeferencing at address point level. From that point

---

[6] The functional importance of the CBD derives both from the city size and from the breadth of umland served from the same city: the difference between the total services for the city and all those related to its attraction area leads to understanding and quantifying the degree of centrality [12].

[7] As part of the transition zone we can meet some sectors of assimilation: "active" in the case of residential spaces that are built over time from the business, "passive", in the case of areas with features of high qualities that are converted into smaller activities because of the shift or the decline of the central area; "inactive" in the presence of static areas than other areas. Several activities nowadays locate along major traffic arteries, as shopping centers in suburban areas [11].

[8] Other quantitative models are based on population distribution, buildings' height, urban land values, etc. ([11], [26], [5] and [28]).

different methods were adopted. We started from the pure observation of the scatterplot of activities, also in terms of their distribution classified per groups of activities. The analysis followed with the realization of spatial indexes to understand the spatial distribution of rare tertiary activities in the urban centre referred to urban district area units. In particular among the rare tertiary activities considered we recalled businesses with high added value (jewelers, boutiques), financial activities (banks and credit institutions), insurance businesses, professional works (lawyers, accountants, architects, medical doctors) and public bodies (chambers of commerce, local authorities' offices)[9].

Rare economic activities were therefore geographically referred not only to address point but also to area features as census units - these latter containing also figures about population - urban districts and a unit area representing the boundary of the 'compact city' of Sassari[10]. A first flavor of the concentration of activities was therefore gained by attributing them to area units as urban districts and deriving their density. Two other indexes were then computed: the first relates the number of core activities with the length of the roads concerning them and is expressed as a ratio (ac/10m), and the second index compares the residents and the core activities (res/ac).

2.2 Kernel Density Estimation

The analysis was also performed over the point pattern without attributing activities to area or linear units as specified in the previous paragraph. *Kernel Density Estimation* (KDE) in particular was used to transform point events in space in a continuous density function over the study region, in order to visualize the phenomenon via a 3D surface expressing the variation of density of point events on the area examined.

The method is used for modeling point data over a grid on the study region, with cells attributed density values according to the events' distribution. KDE is used in several research areas, these including earth science, biology and epidemiology, with recent applications to social science, urban studies, these including population dynamics and central and recreational activities ([41], Boffi, 2004; [50], [51], [28], [29]) and seismic analysis ([52], [53], [54]), just to cite a few.

The kernel functions are three dimensional and weight events within their range of influence according to the distance from the intensity's estimation point [31], as in the following formula.

$$\hat{\lambda}(s) = \sum_{i=1}^{n} \frac{1}{\tau^2} k\left(\frac{s - s_i}{\tau}\right) \qquad (1)$$

with $\hat{\lambda}(s)$ the estimate of the density of the spatial point pattern measured at location s, s_i the observed i^{th} event, $k(\ )$ the kernel weighting function and τ as the bandwidth, or radius, centred in location s, within which events s_i are counted and the density function produced [31]. The searching function's extent is determined by the

---

[9] Other factors could have an influence on activities' location in urban areas: wholesale retail pushing for out of town shops and the emergence of new business areas outside the urban center ([28], [29]).

[10] Urban districts are those defined by in the latest version of the City Urban Plan, including 32 units in the compact city and 38 in the extra urban part [30].

searching radius (or bandwidth), τ, the unique arbitrary variable. In general, low bandwidth's values produce mainly local peaks in the density distribution while wider values tend to dilute the phenomenon and over smooth the observed phenomenon. For each grid cell a kernel function is calculated considering events within a certain distance, weighted according to the distance of the events from the same cell's centroid [32]. Cells present density values represented as a density surface approximating a continuum in the space, with a 'terrain-like' visualization of the events' distribution.

3   The Study Area: Sassari and Its *core*. The Data Used

Sassari is the capital of the homonymous province and hosts 130,658 residents on January 1, 2011 [33], being the second city of Sardinia in dimension after Cagliari and is located in North-western Sardinia. The demographic evolution slowed since the early Eighties due to some initial processes of counter-urbanization to the neighboring smaller size municipalities. About 110,000 inhabitants of the contiguous municipalities (Alghero, Muros, Olmedo, Osilo, Ossi, Porto Torres, Sennori, Stintino, Uri, Usini and Tissi) gravitate to the city of Sassari. At present 70% of the population lives in the compact city[11], 23% in the sprawled city and just 7% inhabits the agricultural area of the municipality.

We relied on Yellow Pages service ([34] accessed December 2010) to study and examine the Central Business District of the City of Sassari, defining and quantifying the economic activities located there. We included civic, administrative - political, economic - productive and representative activities – categorized and divided into categories and subcategories [35]. This led to the identification of nine categories such as Clothing and Accessories, Arts and Culture, Banks and Insurance, Commerce, Professions, Business Services, Property Services, Leisure and Public Offices which in turn were divided in 47 sub-categories for a total of 1,980 activities[12]. The more populated categories are Professions (48%), followed, by Clothing and Accessories (12%) and Leisure (10%). The remaining six categories presents values below 10%

---

[11] For more on compact and sprawled city see Neuman [36] and Indovina [37], [38].
[12] The subcategories are divided as follows: 1 for Clothing and Accessories (clothes and shoes); 6 for Art and Culture (public auctions, art restorers, theater, cinema, art galleries, museums); 1 for Banking and Insurance (bank branches and insurance companies); 6 for Commerce (trade agencies, travel agencies, delicatessens-wine shops, jewelers and goldsmiths, bakeries, fashion); 11 for Professions (architects, lawyers, administrative and fiscal-tax consulting, financial and commercial advisory, chartered accountants studies, surveyors, notaries, medical doctors, road accidents experts, industrial experts, accountants); 7 for Business Services (import-export, internet services, internet web design, marketing, market research, events organization, industrial engineers, translators and interpreters); 1 for Property Services (real estate); 5 for Leisure (hotels, bars and cafes, pubs, restaurants, pizzerias, catering), 9 for Public Offices (chambers of commerce, municipalities, consulates, international organizations, provincial offices, regional offices, court and judicial offices, post offices, universities).

(Table 1). Data were georeferenced at address point level using the GIS data provided by the Municipality of Sassari (Figure 1) and checked via on-line geocoding services.

Table 1. The core activities of Sassari and their percentage in the compact city in 2010. Source: Our elaboration from Yellow Pages (2010)

Category	Clothing and Accessories	Arts and culture	Banks and Insurance	Commerce	Professionals	Business Services	Property Services	Leisure	Local authorities	Total
Weight-%	12	2	7	8	48	4	4	10	5	100

4 Results and Discussion

4.1 Spatial Indexes

The exam of the data scatterplot can suggest some initial reflections on the spatial distribution of activities, showing that the area of greatest concentration of activity is represented by the compact city, where most of the rare services sector surveyed are located - 1,691 (85%) over 1,980 points (Figure 1).

Overlaying the spatial pattern of the compact city's urban districts we can notice that four of them are characterized by the highest number of high-value activities: Centro Storico, Piazza d'Italia, Viale Dante and Viale Amendola - Viale Italia, which together include 942 units - 55% of total. Among these, Piazza d'Italia emerges as the area hosting, alone, 392 units. In eight other neighboring districts (Via Porcellana, Acquedotto, Cappuccini, Via Napoli, Via Rockefeller, Prunizzedda, Luna and Sole e Monte Rosello Basso) the central functions present smaller number of units ranging from 40 to 100[13] (Figures 2; 3).

Such concentration of activities in the four urban districts described above can lead us to highlight the boundary of our CBD with their perimeter. In order to verify this we calculated two spatial indexes of intensity over the entire compact city.

The first relates the number of core activities with the length of the roads (ac/10m). The different degrees of intensity, thus obtained, were divided into three orders of magnitude (low, medium and high)[14].

The roads with the highest intensity index are collected in the Piazza d'Italia district, where in just three streets we find 41% of the surveyed activities[15]. In particular we can highlight as high intensity street segments, Via Vittorio Bellieni (1.6 activity every 10m), Via Roma (1/10m), Via Carlo Alberto (1/10m) and Via Cavour (0.9 / 10m)[16] (Figure 4).

---

[13] The remaining districts host less than forty units or lack economic activities.
[14] We calculated the median of values. In addition, for easy reading and interpretation of data this ratio has chosen a denominator equal to 10m.
[15] They are Viale Umberto (17%), Via Cavour (13%) and part of Via Roma (10%). Important here from the functional point of view is the presence of the Court.
[16] A mean value is found along the segments of Viale Umberto (0.6 / 10 m), Via Alghero (0.5 / 10 m) and Via Armando Diaz (0.5 / 10 m).

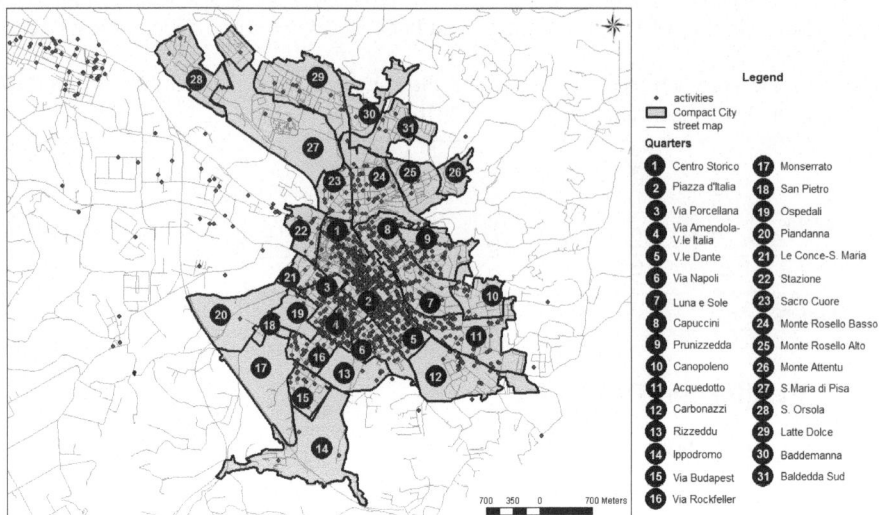

Fig. 1. Urban districts of the compact city of Sassari with a significant presence (> 40) of core activities in 2010 Source: our elaboration on GIS data from Municipality of Sassari [39]

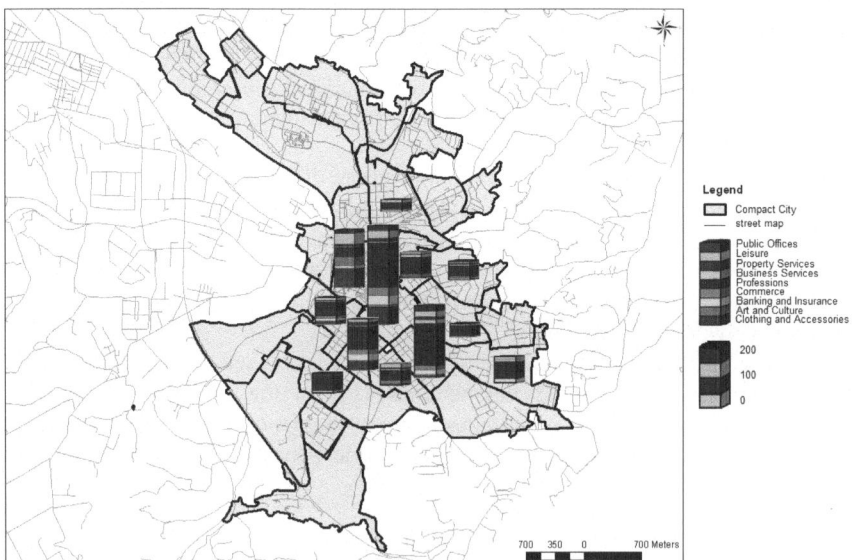

Fig. 2. Urban districts of the compact city of Sassari with (> 40) core activities in 2010 Source: our elaboration on GIS data from Municipality of Sassari [39]

Analyzing the Central Business District 631

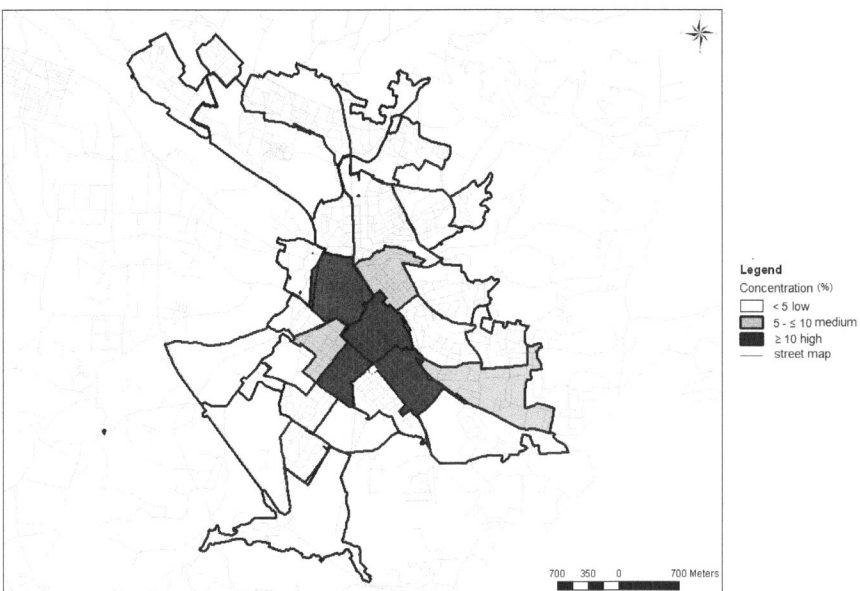

Fig. 3. Percentage weight of the core activities in the Urban districts of the compact city
Source: our elaboration on GIS data from Municipality of Sassari [39]

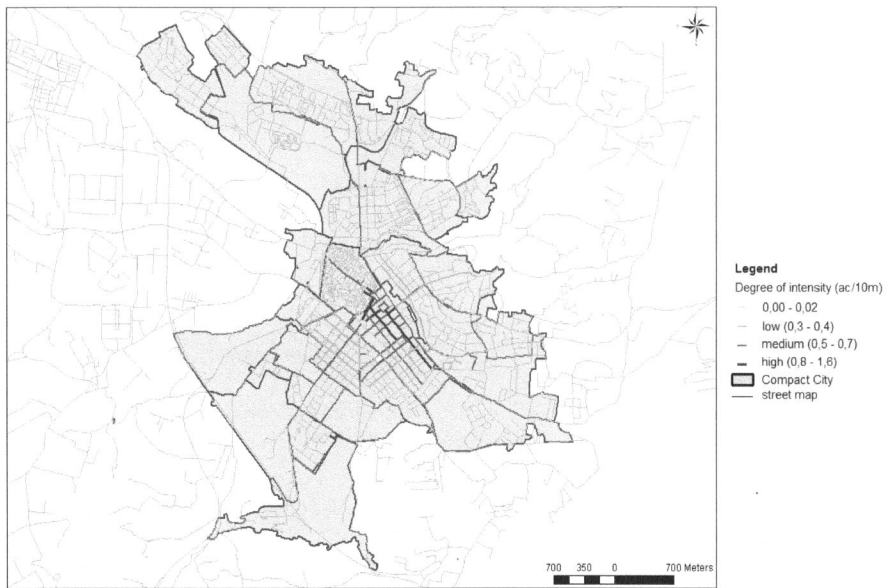

Fig. 4. Index of intensity "ac/10m" on the street network of the compact city of Sassari in 2010
Source: our elaboration on GIS data from Municipality of Sassari [39]

The second index compares residents and core activities (res/ac) in the different urban districts, identifying the excess and attractiveness of the functions surveyed[17]. At municipal level this index ranks over 65 units and holds a very low value (>110) both in areas characterized by sprawl, and in the compact city area identified by nineteen districts marked by core activities (<40ac). In the area in proximity of the Central Business District (CBD), the index shows an average value[18], but different in the eight areas that mark the area under investigation: high (≥20 - 50) in the districts surrounding our CBD such as Via Porcellana, Acquedotto, Cappuccini and Via Rockefeller, medium (≥50 - 80) in Via Napoli, low (≥80 - 110) in Prunizzedda and very low (>110) in Monte Rosello Basso. A very high degree of excess (<20) is proposed in the mentioned central area (CBD) and in particular in the Urban districts of Piazza d'Italia, Via Dante and Via Amendola - Viale Italia, and assumes a high value in Centro Storico (Figure 5).

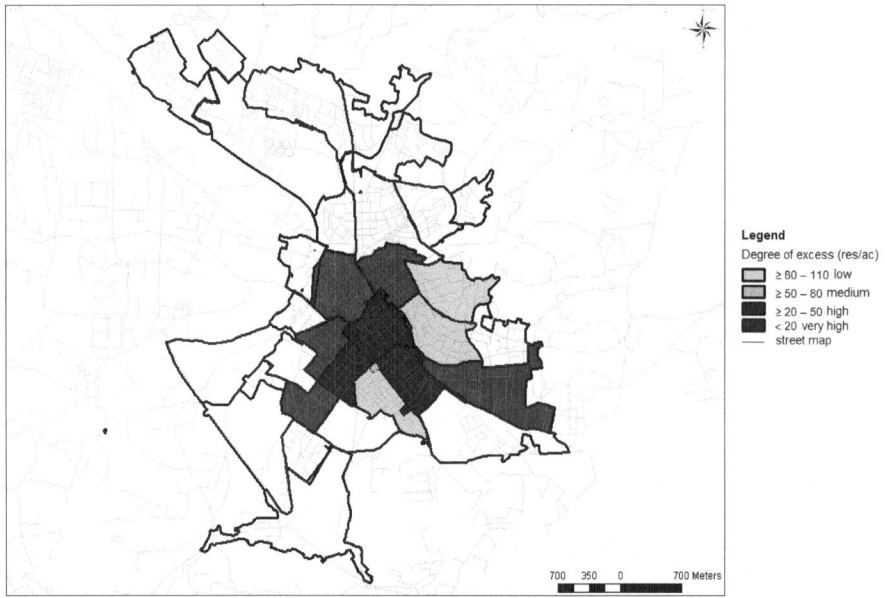

Fig. 5. Index of exceedeness "res / ac" in the compact city of Sassari in 2010 Source: our elaboration on GIS data from Municipality of Sassari [39], [55]

We need to specify however how the 'Centro Storico' district presents a slight growing population if compared to the overall central, compact city – increasing of 255 units between 2005 and 2010, while the central area decreased of 202 units – thus demonstrating some characteristics of 'transition district' around the true city centre

---

[17] By a close examination of the ratios obtained we identified the following five classes: very high <20, high ≥20 to 50, average ≥50 to 80, low ≥80 to 110 and very low >110.
[18] This mean value is also characterizes also the whole compact city.

and core. Non resident students, immigrants as well as non wealthy people tend to habit this area, often in building in bad shape. Some ghettoization processes involving immigrant population seem to be in place, while gentrification is just started and still not fully playing its effects [40]. The other three 'central' urban districts tend to reduce their inhabitants in the cited period and particularly the Piazza d'Italia district holds the lowest number of residents.

The results of this initial investigation confirm the existence of an urban Central Business District that substantially covers a compact area of the city, marked by the above-mentioned four contiguous districts. It presents an oblong shape, heading South from Centro Storico, and resembling a Greek letter 'lambda' (λ), where the upper and lower extensions identify, respectively, the same Centro Storico and Viale Dante, while the other leg leads to Via Amendola - Viale Italia; the bifurcation point of the letter, then, is the core, Piazza d'Italia. This design of the CBD, whose main route runs from North-West to South-East along Corso Vittorio Emanuele II and Via Roma for more than 1,300 meters, is due to the fact that it is identified by with these important roads flanked by others, mainly from the parallel Viale Dante and Viale Italia. Our CBD, thus, focuses on these roads and other ones define it, particularly Viale Umberto I, Piazza d'Armi and Via IV Novembre (Figures 2, 3, 4 and 5).

4.2 Density Estimation

Data obtained from the grouped Yellow Pages activities were also processed using Kernel Density Estimations. A fine 20m grid was overlaid over the study region of the Municipality of Sassari. As a kernel function a quartic one was chosen, with different bandwidth tested (122, 177, 269 and 355 meters, these corresponding to a K = 25, 50, 100 and 150 nearest neighbor activities within the searching radius), although the ones chosen for our analysis are the 122 and 355m bandwidth[19].

Higher values of the searching bandwidth can cause an excessive smoothing and dilution and provide a not so different message than that deriving from the 'simple' observation of a scatterplot. Such distances were considered as the most suitable for medium-size urban areas - as Sassari –, similarly to what implemented in other research on different cities ([41], [29]) particularly because the first one can allow highlighting local, proximity effects, while the second one is consistent with an average 5 minutes walking distance. The function provides in fact also a measure of accessibility, showing for each cell the events that can be reached within a certain distance, assigning a higher importance to closer events, as expressed by the weight inserted in the quartic function. In our case the 122 and 355 m distances are obtained also by means of a nearest neighbor computation using respectively K = 25 and K = 150 ranks, that implying a computation of the average of intra-events distances of different orders [42], thus linking the control of the variable to a k-nearest neighbor choice instead of an arbitrarily chosen radius.

---

[19] Although several functions can be used, a quartic function weights closer events to the cell more than those located apart.

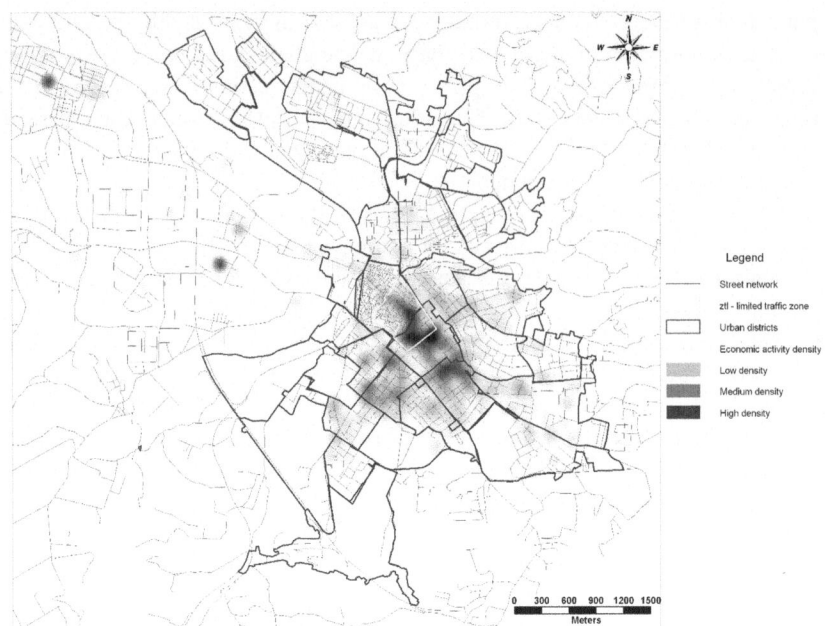

Fig. 6. Kernel Density Estimation over economic activities. 122m (K = 25) bandwidth. Source: our elaboration on GIS data from Municipality of Sassari [39]

We therefore obtained a density estimate for each grid cell over the study region, expressed as 'events per square kilometre', as the weighted number of events within, in the two analysis, 122 and 355 meters from each reference cell, counting and weighting events according to their distance from a cell's centroid according to a decreasing function and dividing the value by the area underneath the same function. Cells of relative density were mapped as pseudo-3D functions, obtaining isolines of homogeneous density values.

As the kernel density values can be expressed in (estimated) events per square kilometres, isolines were spaced of 100 activities per squared kilometre, to the peak in the city centre (Piazza d'Italia) with nearly 2,800 activities using the narrow bandwidth of 122 m (Figure 6) and around 1,500 activities per square kilometre using the wider 255 m one (Figure 7). Denser cells with a higher concentration of central activities are located in the city centre where a peak in the density surface can be noticed. This is also confirmed by the number of activities located in the area.

The density analysis confirms some of the results obtained using the other indexes and the use of the two different bandwidths allowed highlighting respectively more local and general effects. The narrower bandwidth as in the elaborations in Figure 6 can be easily compared to Figure 4, where activities were attributed to street segments. In Figure 6 we can notice several peaks in the distribution, representing clusters of central activities. The most important one is located in Piazza d'Italia, with other minor clusters in the neighboring urban districts, one of them located at the 'real' border between Piazza d'Italia and Centro storico districts. In the representation as in Figure 3 such evidence was split between the two districts given the design of the area units. However, peaks can be noticed in the four neighboring districts as already noticed using the other spatial indexes.

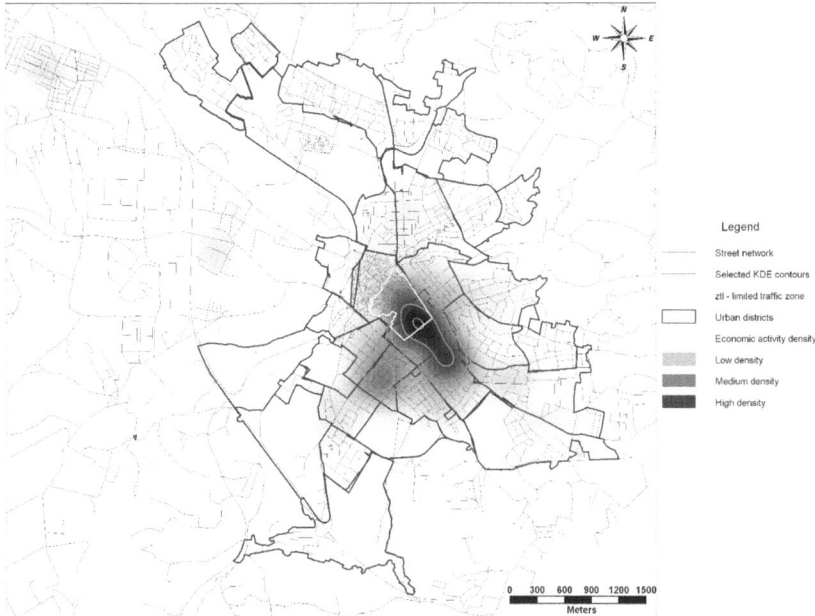

Fig. 7. Kernel Density Estimation over economic activities. 355m (K = 150) bandwidth. Source: our elaboration on GIS data from Municipality of Sassari [39]

The wider bandwidth is saying something more concerning a more general shape and orientation of the cluster and therefore of the CBD. It is confirmed that 'Piazza d'Italia', both in terms of the square name and of the urban district, represents the centre of the distribution as the area presenting the higher intensity value in all the elaborations. A 'proper' CBD could be highlighted in the narrow pale blue contour line – corresponding to a contour line of 1,100 events per square kilometre, following a North-west – South-east orientation, starting from the boundary of Centro storico district, crossing Piazza d'Italia and ending up in Viale Dante. Another, separate peak can be noticed in the urban district Via Amendola – Viale Italia. However, in considering a wider contour line (700 units per square kilometre) we can redraw the 'lambda' shape suggested in the previous indexes, therefore enlarging the margins of the CBD.

A final note can be added with reference to the recently introduced traffic restriction rules, that corresponding to the traffic-free area (ZTL) in Figure 6 and 7.

As traffic was intense and chaotic in the central area, at the border between the two districts of Piazza d'Italia and Centro Storico, a traffic-free area has been recently introduced. It does not match with the administrative division of urban districts as it 'cuts in two pieces' both the above mentioned ones. If we observe its shape with the clusters characterizing the CBD, we can notice that the denser part, located in the square 'Piazza d'Italia' is located at one of the margins of such central area. There, some streets very important in terms of their contribute to the centrality of the area, still lay in a trafficked area, and such cohabitation of urban central activities and mobility could produce conflicts in urban land uses and life. We must remind Appleyard [43] and Whitelegg [44] reporting a decrease in social interaction when traffic increases on urban road. As experimented in other cities, both in Italy and in the rest of the World, traffic limitation policies have allowed a higher liveability and use of the urban spaces, while the persisting coexistence of conflict in uses could deteriorate the quality and exclusiveness of some urban services.

5   Conclusions

Some final reflections can be done with reference to the methods adopted and the importance of delimiting central urban areas.

The spatial methods adopted to highlight the CBD in the city of Sassari offered interesting results. In particular using different methods, as spatial indexes referred to point, line and area features contributed to provide a consistent image of the spatial extent and characteristics of the central area. Also the results where confirmed by using density estimation, that allowed overcoming the traditional limits of human-drawn administrative subdivisions. The results were interesting as the methods allowed highlighting clusters of central activities in the urban areas, and also as their application, at different scales and using different parameters, can be easily repeated not only to other urban cases but also to highlight different characters and specialization of sub-areas in an urban environment, helping also in re-drawing, if needed, administrative subdivisions.

Central areas are the expression of the fact that the more an area is used and lived, the greater is its vitality and more important the functions a center plays also towards the external parts of the city, and the higher will be the push for renewal of the central offer ([45], [46]). We also need to remind that to-date urban balance have also changed pushed by social and spatial competition from consumption and marketing ([47], [48]). Urban central places do not therefore appear as spaces losing social cohesion, with ephemeral and repetitive characteristics losing local identities. The renewal and revitalization of various historical and cultural expressions, from production and trade related behavior, may be the solution regenerating the urban core and its functional areas [49].

References

1. Christaller, W.: Die Zentralen Orte in Suddeutschland. Fischer, Jena (1933)
2. Loesch, A.: The economics of locations. Yale University Press, New Haven (1954)
3. Conti, S.: Geografia dell'economia mondiale. Utet, Torino (1999)
4. Fano, P.L.: L'evoluzione dei sistemi urbani: un approccio probabilistico. Archivio di Studi Urbani e Regionali 1, 1–3 (1968)

5. Haggett, P.: Geography: A Global Synthesis. Pearson Education, Harlow (2000)
6. Hoyt, H.: The structure and Growth of Residential Neighborhoods in American Cities. U.S. Government Printing office, Washington D.C (1939)
7. Rodrigue, J.-P., Comtois, C., Slack, B.: The Geography of Transport Systems, 2nd edn. Routledge, New York (2009)
8. Toschi, U.: La città. UTET, Torino (1966)
9. Dematteis, G.: Le città come nodi di reti: la transizione urbana in una prospettiva spaziale. In: Dematteis, G., Bonavero, P. (eds.) Il Sistema Urbano Italiano Nello Spazio Unificato Europeo, pp. 15–35. Il Mulino, Bologna (1997)
10. Bonetti, E.: La teoria delle località centrali. Università di Trieste, Facoltà di Economia e commercio, Istituto di Geografia 6 (1964)
11. Bonetti, E.: La localizzazione delle attività al dettaglio. Giuffrè, Milano (1967)
12. Bonetti, E.: Le attività al dettaglio e la loro localizzazione. In: Atti del XX Congresso Geografico Italiano, Società Geografica Italiana, Roma (1967)
13. Bonetti, E.: Un riesame della teoria delle località centrali. Bollettino della Società Geografica Italiana 10.8, 475–487 (1979)
14. Dematteis, G.: Le località centrali nella geografia urbana di Torino. Pubblicazione del Laboratorio di Geografia Economica "P. Gribaudi" dell'Università. 2, Torino (1966)
15. Corna Pellegrini, G., Pagnini, M.P.: Recenti studi di geografia urbana. *Rivista Geografica Italiana 82, 489–509 (1975)*
16. Burgess, E.W.: The growth of the city. In: Park, R.E., Burgess, E.W., Mc Kenzie, R.D. (eds.) The City, pp. 47–62. University of Chicago Press, Chicago (1925)
17. Fortuna, G.: Modelli sull'uso del suolo urbano. Archivio di Studi Regionali 94, 165–182 (2009)
18. Harris, C.D., Ullman, E.L.: The nature of cities. Annals of the American Academy of Political and Social Science 242, 7–17 (1945)
19. Alonso, W.: Location and land use. Toward a general theory of land rent. Harvard University Press, Cambridge (1965)
20. Maarek, G.: Recherche sur l'urbanisation spontanèe, Paris (1964)
21. Mills, E.S.: An aggregative model of resource allocation in metropolitan area. American Economic Review 57, 197–210 (1967)
22. Iommi, S., Ferraina, G., Molinari, D.: Policentrismo e accessibilità della casa. Un abbinamento ottimale? In: 30° Conferenza Scientifica Annuale AISRe, Firenze, settembre 9-11 (2009)
23. Alvarez De La Torre, G.B.: El crecimiento urbano y estructura urbana en las ciudades medias Mexicanas. Quivera 12(2), 94–114 (2010)
24. Finocchiaro, E.: Città in trasformazione. Le logiche di sviluppo della metropoli contemporanea. Franco Angeli, Milano (1999)
25. Finocchiaro, E.: I nuovi luoghi del consumo nella città contemporanea. In: Cirelli, C. (ed.) Città e Commercio, pp. 61–80. Pàtron Editore, Bologna (2008)
26. Murphy, R.E.: The central business district. A study in urban geography. Aldine-Athertone, Chicago (1972)
27. George, P.: Geografia delle città. Edizioni Scientifiche Italiane, Napoli (1960)
28. Borruso, G.: Il ruolo della cartografia nella definizione del Central Business District. Prime note per un approccio metodologico. Bollettino dell'Associazione Italiana di Cartografia 126-127-128, 255–269 (2006)
29. Borruso, G., Porceddu, A.: A Tale of Two Cities. Density Analysis of CBD on Two Midsize Urban Areas in Northeastern Italy. In: Borruso, G., Lapucci, A., Murgante, B. (eds.) Geocomputational Analysis for Urban Planning. SCI, vol. 176, pp. 37–56 (2009)

30. Comune di Sassari: Piano Urbanistico Comunale (P.U.C.). Sassari (2009)
31. Gatrell, A.: Density Estimation and the Visualisation of Point Patterns. In: Hearnshaw, H.M., Unwin, D.J. (eds.) Visualisation in Geographical Information Systems. Wiley, Chichester (1994)
32. Levine, N.: CrimeStat III: A Spatial Statistics Program for the Analysis of Crime Incident Locations. Ned Levine & Associates, Houston, TX, and the National Institute of Justice, Washington, DC (2004)
33. Italian Institute of Statistics – ISTAT, http://demo.istat.it
34. Italian Yellow Pages, http://www.paginegialle.it
35. Cori, B.: Città, spazio urbano e territorio in. Franco Angeli, Milano (1993)
36. Neuman, M.: The Compact City Fallacy. Journal of Planning Education and Research 25, 11–26 (2005)
37. Indovina, F.: La città diffusa. Daest, Venezia (1990)
38. Indovina, F.: Dalla città diffusa all'arcipelago metropolitano. Franco Angeli, Milano (2009)
39. Comune di Sassari: Cartografia comunale. Sassari (2011)
40. Donato, C.: La "ghettizzazione" degli stranieri a Sassari. In: Krasna, F., Nodari, P. (eds.) L'immigrazione straniera in Italia. Casi, metodi e modelli. Geotema, 23, pp. 26–33 (2004)
41. Thurstain-Goodwin, M., Unwin, D.J.: Defining and Delimiting the Central Areas of Towns for Statistical Modelling Using Continuous Surface Representations. Transactions in GIS 4, 305–317 (2000)
42. Chainey, S., Reid, S., Stuart, N.: When is a hotspot a hotspot? A procedure for creating statistically robust hotspot maps of crime. In: Kidner, D., Higgs, G., White, S. (eds.) Socio-Economic Applications of Geographic Information Science, Innovations in GIS, vol. 9. Taylor and Francis, London (2002)
43. Appleyard, D.: Livable Streets. University of California Press, Berkley (1981)
44. Whitelegg, J.: Transport for a sustainable future. John Wiley & Sons, London (1993)
45. Monheim, R., Meini, M.: Le aree centrali urbane di Firenze e Norimberga tra potenzialità di sviluppo e rischio di decadenza. Il ruolo delle politiche di traffico e il comportamento dei city-users. In: Cori, B. (ed.) La città invivibile. Nuove ricerche sul traffico urbano. Pàtron Editore, Bologna, pp. 227–256 (1997)
46. Meini, M.: Mobilità urbana e sviluppo sostenibile: un problema irrisolto. In: Meini, M. (ed.) Mobilità e Territorio. Flussi, Attori, Strategie, pp. 238–251. Pàtron Editore, Bologna (2008)
47. Harvey, D.: L'esperienza urbana. Metropoli e trasformazioni sociali. Il Saggiatore, Milano (1998)
48. Bauman, Z.: Consumo, dunque sono. Editori Laterza, Roma (2008)
49. Rousseau, M.: Re-imaging the City Centre for the Middle Classes: Regeneration, Gentrification and Symbolic Policies in "Loser Cities". International Journal of Urban and Regional Research 33, 770–788 (2009)
50. Borruso, G.: Studio della popolazione e della sua evoluzione a scala urbana. Primi risultati di analisi di densità dei dati spaziali. In: Proceedings of the 7a ASITA Conference "L'Informazione Territoriale e la Dimensione Tempo", pp. 467–472 (2003)
51. Borruso, G.: Network Density Estimation: a GIS Approach for Analysing Point Patterns in a Network Space. Transactions in GIS 12, 377–402 (2008)

52. Danese, M., Lazzari, M., Murgante, B.: Kernel Density Estimation Methods for a Geostatistical Approach in Seismic Risk Analysis: The Case Study of Potenza Hilltop Town (Southern Italy). In: Gervasi, O., Murgante, B., Laganà, A., Taniar, D., Mun, Y., Gavrilova, M.L. (eds.) ICCSA 2008, Part I. LNCS, vol. 5072, pp. 415–429. Springer, Heidelberg (2008)
53. Danese, M., Lazzari, M., Murgante, B.: Geostatistics in historical macroseismic data analysis. Transactions on Computational Sciences 6(5730), 324–341 (2009)
54. Murgante, B., Danese, M.: "Urban versus Rural: the decrease of agricultural areas and the development of urban zones analyzed with spatial statistics" Special Issue on "Environmental and agricultural data processing for water and territory management". International Journal of Agricultural and Environmental Information Systems (IJAEIS) 2(2), 16–28 (2011)
55. Comune di Sassari: Popolazione residente al 2010. Sassari (2011)
56. Municipality of Sassari, http://www.comune.sassari.it
57. Sardinia Geoportal, http://www.sardegnageoportale.it

That's ReDO: Ontologies and Regional Development Planning

Francesco Scorza, Giuseppe B. Las Casas, and Beniamino Murgante

University of Basilicata, 10, Viale dell'Ateneo Lucano, 85100 Potenza, Italy
firstname.surname@unibas.it

Abstract. European Cohesion Policy generates several programs at territorial levels. An evident trend is the increasing of multi-level governance in the period 2007-2013, promoting a wider participation to programming processes. It is possible to affirm that new instances are coming out. We refer to problems generally connected with participation processes. The relation between problems in knowledge management and ineffective impacts of local development plans is confirmed. Therefore, the central role of communication determines relevant issues regarding the ability to understand the meaning of general and sectoral policies by stake holders, the awareness of citizens to manage technical instruments implementing such policies. Are they conscious of ex-ante comprehensive context analysis and/or can they share possible future scenarios? A way to tackle these problems is the use of ontologies. In this work we present the structural elements and an application of ReDO ontology (Regional Development Ontology) analyzing major steps of ontology design and nodal phases of ontology building (i.e. consensus on relations and restrictions, and switch from glossary to taxonomy).

Keywords: Regional Development Programs, Context Based Approach, Semantic Interoperability, Ontology.

1 Introduction

The planning process usually faces a complex and multidisciplinary dimension in which the knowledge management function increased its relevance. In planning activities, scientists and technicians develop their contributions on a multi-sectoral knowledge framework. The process always includes several active bodies, with different functions and responsibilities. Such an inclusion is mainly increased by the application of participative techniques (based on Internet and ICT e-government tools) and the role of communication, in planning process, has considerably increased during the last decades [20] [21] [22].

Ontologies assume a potential role in supporting and developing knowledge interchange issue dominating the process. Communication requires a sharing of ontologies between communicating parties [16] and it also needs new tools in order to facilitate a bottom-up participation process [9].

Research in ontology as the basis for the development of knowledge-interchange standards has expanded in recent years [7].

Within the complex framework of meaning concerning territorial classification and planning/programming specific contents, we agree on the assumption that a powerful tool to increase rationality of knowledge is the "ontology".

This paper suggests considerations connected to the issue of developing a "ready to use" ontology applied to the planning process. This approach implies a modelling activity and a knowledge engineering process in a multidisciplinary framework [10].

In this paper we describe the design and development processes of a sectoral ontology, the "Regional Development Ontology" (ReDO). In particular we propose the use of such tool for the representation of a sample of five Regional Operative Programs of the EU Programming Period 2007-2013.

2 How to Define It?

In order to discuss the approach and results of the research we have to start from a definition of ontology. In fact, the term 'ontology' can lead to misunderstandings connected to its adoption in different scientific or technical field of application.

Let's start from our definition of ontology: "Explicit and formal model of a domain". In our application we identified in the ontological approach a way to define a model concerning with the planning process. So we developed a representation of 'the plan' based on a specific ontology designed in order to accomplish the general objective of a deeper rationalization of the planning process aiming to achieve more equity, more effectiveness and more sustainability for the decision concerning territorial development. This representation has to be 'formal'; it means that it should be symbolic and mechanized (better computerized). In fact, we look at the complex process of sharing knowledge and information about 'the plan' among the community of stake holders directly or indirectly involved in the process. This aspect implies some specific instances: to use a shared interpretation model and so a shared symbology for the identification and understanding of the meaning of a concept included into ontology; to build a framework in order to develop functions and queries, to analyse and to validate the representation using ICT tools. Regarding the concept of 'domain' we assume the vision by Grüber [8] speaking of 'a subset of knowledge, dealt from a certain point of view".

This definition of ontology coexists with several other definitions depending on the field of application.

If the philosopher would define ontology as the "discipline dealing with theories of being", the informatics science significantly transformed the meaning of the term. A well-posed definition has been suggested by Ferraris [3]: "the theory of objects and their relations". Overcoming the traditional philosophical definition of ontology, we will use a slightly different notion (proposed, among others, by Grüber): a specific ontology seen as a model can be defined as "the explicit specification of an abstract, simplified view of a world we desire to represent" [8].

According with Genesereth and Nilsson [6], the base for representing knowledge is the process of conceptualization: objects, concepts and other entities that are assumed to exist in some area of interest and the relationships that hold among them. The term "ontology" describes the explicit specification of a conceptualization [7] of a 'part of reality'.

In information science ontologies describe a particular way to understand a part of the world [5]. Murgante et al. [14] refer to "ontology" as a meta-model of reality, where concepts and relations are used as boxes of the interpretative model, generating rules and bonds for relations.

For each data base it is possible (mainly necessary) to define a specific ontology [13]. This affirmation implies that we can have "n" local ontologies that should communicate each-others to build a shared knowledge. Laurini and Murgante [13] define the "domain ontology": an higher level ontology connecting different local level ontologies as "mediators" promoting the interoperability among different data bases. This represents an important field for recent researches and applications with many relevant results but with no general or standard solutions.

3 Towards 'Usability'

"In order to be useful, an ontology has to be shared" [2]. If we consider an international community, this concept strongly assumes the first priority of the research, but also in our "sectoral ready to use ontology" oriented to improve the planning process we need an agreement of stakeholders participating the process.

In order to minimize the effort (or, in other words, the cost) of adopting ontology in the planning process, we suggest to prefer a technical approach for developing ontology. We are in the case described by Corallo [1], where a limited group of experts defines the ontology and the community adopts it (or accepts it) as a tool of the process. The other case is that the ontology is collectively defined and developed, in order to immediately improve the collaborative definition of the world.

Another general issue to be faced are perspectives of the ontology. It has to be usable for future applications and perspective users (human beings or intelligent agents) and the usage (cataloguing, searching, exchanging information) has to be considered in the design of the ontological structure.

As a representation of real world is the result of a process of observation (or in other words a "building knowledge process" – see also [4]), such observation strongly depends on the observer point of view. His interpretation of the real world depends in turn on his cultural back-ground, his interests, his relation with the reality, etc. So we have to admit the presence of errors, imprecisions and uncertainties in results. There are various reasons for such limitations of the - physical, technical and cognitive - observation process, but they are fundamental and nearly nothing can be measured with absolute accuracy [4].

These considerations influence the process of building an ontology in the domain of planning. Indeed, planning processes are based on the interactions among politicians, technicians, stakeholders and context (intended not only in the physical dimensions but also in social, cultural and economic ones); therefore many points of views produce different visions, sometimes conflicting in terms of objectives, priorities, relevance, etc. The interaction of different actors on the scene of the plan generates problems connected with communication. A very important matter resides in the language and especially in the level of actors agreement on concepts and their definitions. It is the case of different databases containing complementary information but with no opportunity to "collaborate" in building a wider data-knowledge due to problems in meta-data, data-types, etc. It corresponds to a problem of interoperability.

This is a common situation in planning: different institutional (public) or private bodies build their own plans; they hold information systems (generally complex data infrastructures) containing general and specific data; each plan corresponds to a process of analysis and knowledge building, without opportunity of knowledge capitalization among different plans.

4 ReDO Design and Structure

After previous preliminary considerations, in this section we describe main stages of the operative research we called ReDO (Regional Development Ontology).

The first step regards the phase of "ontology design". It represents a crucial step in the procedure of applying ontologies to planning processes. Above all, attention should be paid to the structural elements of ontology: domain (or 'scope' of ontology), concepts ('classes'), hierarchy, attributes for concepts, restriction and relations between concepts, instances. The definition of such elements represents the 'ontology design'.

Our procedural scheme includes four steps [17] for ontology design:

- step 1: scope definition;
- step 2: class and slot design;
- step 3: constraints' enforcement;
- step 4: instances creation.

The domain is an abstraction of reality we want to represent. In the specific case study, the scope is represented by a complex reality: the program and its relationships with the context of implementation and with the community of actors and beneficiaries, the procedural scheme of implementation and management. It is composed by physical elements, relations among them, value systems, program actions, social issues, policy goals. In order to improve rationality process, the first issue is to circumscribe the domain. According to recent studies [21] [2], the fundamental questions to be answered in this phase are:

- Q1: Which is the portion of real world we want to describe through the ontology?
- Q2: Which are the answers we expect from our ontology?
- Q3: Which is the spatial dimension of the domain (in other words: "where does the ontology work")?
- Q4: Is the domain open or close?

Our objective is to represent European operative programs OPs(Q1) considered according to both strategic and operative/procedural components. In a general view, several European policies are implemented by OPs 2007-2013 at national, regional or interregional scale.

Answering Q3 might appear to be a consequence of the administrative border of each OP (Region, Country, aggregation of Regions). This choice might be an element of strong simplification of reality and therefore, it could imply errors in gathered evaluations. A way to control such errors is to consider the domain as open in space, time and objects (Q4).

Table 1. ReDO synoptic table [17]

Phases		Description	Output
1	Domain definition	Identification of ontology "scope". According to main questions described above, we defined the domain including the relevant aspects of EU OPs management and evaluation: components, actors, policies, tools, etc.	Domain
2	Concept identification	According to ontology structure, a team of experts (technicians and scientists) identified the relevant concepts for ReDO purposes after an analysis of 2007/2013 POs (PO ERDF Basilicata, PO ERDF Puglia, PO ERDF Campania - Italy).	Concept list (about 110 concepts)
3	Thesaurus	For each concept, the research team identified the pertinent definition using accredited sources. The result is a glossary (thesaurus) and it represents the first operative output of the process.	Thesaurus (about 110 concepts and definitions)
4	Extraction of ontology classes from thesaurus	Within the whole thesaurus, the research team defined the ontology classes through a pear to pear negotiation.	Classes (61 ReDO classes)
5	Taxonomy development	The 61 classes have been organized in a taxonomy: a hierarchical structure based on the taxonomic relation "IS_A"	Taxonomy
6	Application of attributes and restrictions to each class	Attributes and restrictions allow to realize an operative characterization of a class. The definition itself is an attribute of a class. Attributes correspond to data/information required for the individuals of the class. Restrictions are rules for class population.	Attributes Restrictions
7	Definition of relations among classes	Relations among classes allowed to represent procedures and functions connected to the management and the evaluation of OPs	Relations
8	Ontology population	After the construction of the ontological structure a very important step is the population of the ontology. It is the phase of operative representation of the domain in ReDO knowledge management tool.	Instances

The second methodological question (Q2) is probably the key of ontology design. What do we expect from our work? In a synthetic view, we intend to provide an operative tool for managing and control OPs, reinforcing the quality of interactions between each OP and the category of beneficiaries, also improving participation in local development processes. This ontological representation aims to obtain an improvement of rationality in policy making. This could be possible if contradictions and conflicts among different planning tools are removed or at least reduced. The activity (considered as a bottom-up and participated approach) leading to such an ambitious objective is evaluation, intended as a comprehensive and context based one [11]. The operative phases of ReDO build-up process are listed and commented in the synoptic table (table1) [18].

A brief description of each phase is provided in the synoptic table, but it is important to consider some crucial aspects: in the passage from thesaurus to taxonomy, the expert team agreed on a restriction of elements composing the ontology. This has happened out of any methodological prevision, and we can say it corresponds to a concrete pear to pear agreement process on conceptualization. Only the concept considered useful by the community of experts was included in the ontology. Probably, we could admit also the opposite case (the enlargement of thesaurus), but the relevant aspect rests in the agreement and sharing process as a necessary component of building an ontology.

It is important to underline that this representation is a report of a real process carried during ReDO research, and it has to be considered as a result of the methodological approach described in this work.

ReDO ontology is based on a simple structure of classes and relations. This simple model is oriented to usability.

We defined five main classes of ontology domain for our application:

1. Plan, defined as a "Written account of intended future course of action (scheme) aimed at achieving specific goal(s) or objective(s) within a specific timeframe. It explains in detail what needs to be done, when, how, and by whom, and often it includes best case, expected case, and worst case scenarios".
2. Project, defined as a "Planned set of interrelated tasks to be executed over a fixed period and within a certain cost and other limitations".
3. Policy, defined as "A specific statement of principles or of guiding actions implying clear but not mandatory commitment. A general direction that a governmental agency sets to follow, in order to meet its goals and objectives before undertaking an action program".
4. Tools, defined as "Financial, normative and methodological instruments for policies implementation".
5. Actors, defined as "Groups of private, public, no-profit bodies involved in development processes".

Among ReDO sets of relations, relevant ones are:

- Finances/Is_Financed_By: in the processes of planning and management of local development, financial tools represent a key variable. Through this relation, we make explicit the dependency between classes and financial aspects. This explanation has implications for operations related

to the management process which often presents problems of overlapping expertise and resources.
- Controls/Is_Controlled_By: responsibility, intended in terms of both ownership of programmatic function and process control (implementation and management of the program or of an intervention), is a key relationship in the design of the ontological model. In facts, OPs management structure does not allow easy attribution of such functions within the complex programming system. This leads to problems in connecting program and territory in terms of relationships between involved actors. In particular, the beneficiaries find it difficult to relate with the appropriate decision-making direction for specific issues.
- Implements/Is_Implemented_By: this relation expresses the ownership of the process of implementing policies, programs and interventions. This is a function given in different ways: for " hierarchical transfer", if a program directly implements one or more strategies (policies), for "competition", if policies are implemented by projects passing through a procedure of public competition (i.e. "Call for proposal").
- Evaluates/Is_Evaluated_By: the identification of the evaluation function within the ontological structure is one of the key results of ReDO. The evaluation function has always been unclear in UE Ops, for both periods 2000-2006 and 2007-2013. In order to clearly express fields (or classes) for which the evaluator (considered one of the key actors in the process) will exert his task is the basis for a proper comprehensive evaluation process [11].

The figure 1 shows a graph in which main classes are connected through the described relations.

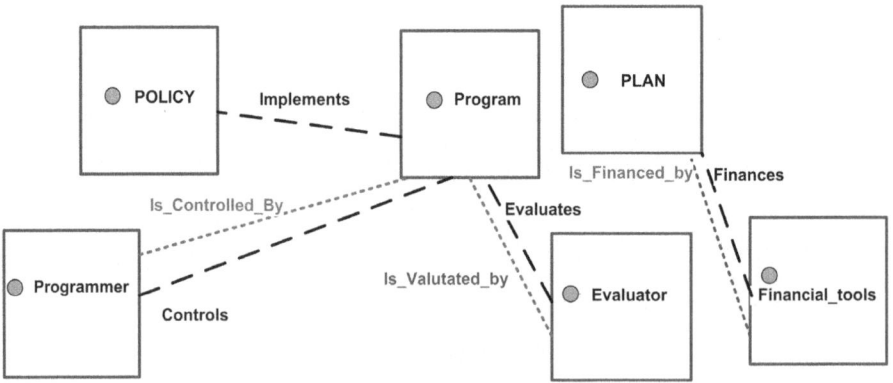

Fig. 1. ReDO relevant relations

5 Cognitive Structure VS Program Structure

The conceptual base of the application considers the analogy between what we call 'program structure' [11] and a 'cognitive structure'.

As described in previous works [18], program structure is the hierarchy between strategic and operative components of a plan, linked together by a logic nexus. In the following figure an example of program structure representation concerning the analysis of the POP FESR Basilicata 2000-2006 is proposed. In the figure 2, it is possible to identify the strategic component of the plan (overall and specific objectives) and the operative ones (results and activities). Through this analysis we obtained a graph – in particular a tree – in which nodes are components of the program and arcs are representative of cause-effect relations.

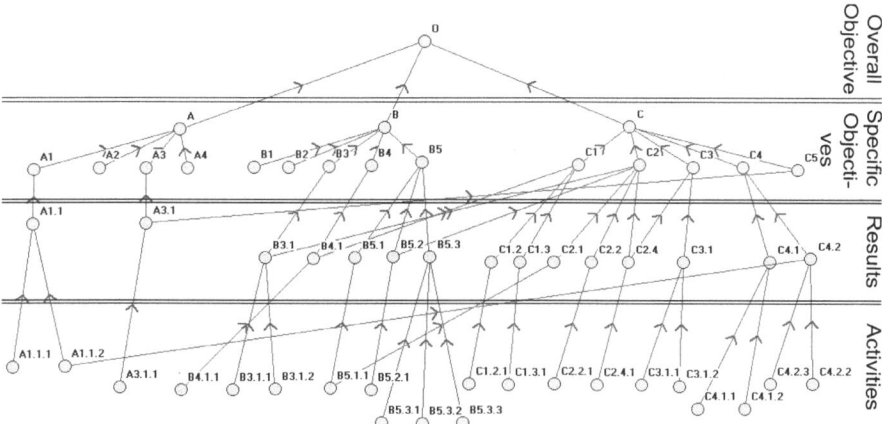

Fig. 2. OP Basilicata 2000-2006 – axis 4, Program structure

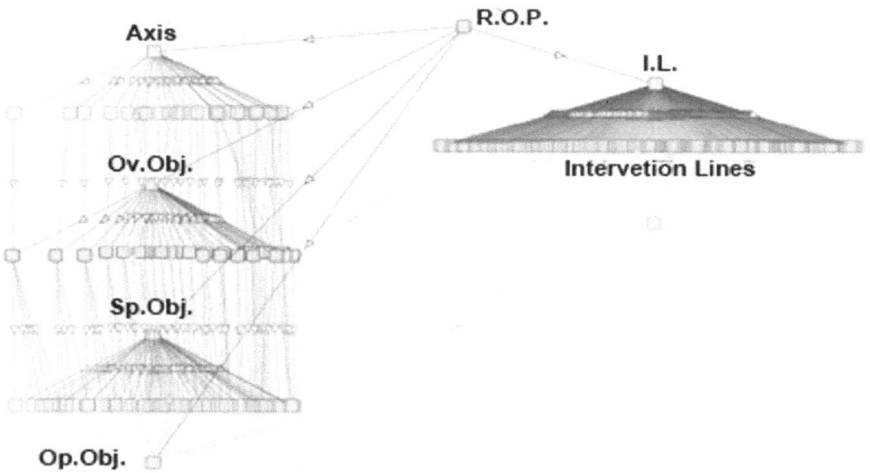

Fig. 3. ReDO ontology

On the other hand, cognitive structures are often arranged in a way that wide concepts are subdivided into narrower ones. At first, they seem to follow a hierarchical structure, where elements of the upper level are subdivided in smaller ones, such that a group of smaller ones makes up exactly one unit at a higher level. But this is not necessarily the case and in general a directed acyclic graph can be observed. There is an important parallelism between structure of an ontology and program structure.

For the aim of the research the static representation of program structure as an oriented graph does not verify the complex set of relationships connected to plan implementation phase. It refers to the functions of management and control, implementation of interventions, evaluation of impacts, etc..

Therefore, we identified ontology as a comprehensive knowledge management tool in planning field.

6 Five Regional Operative Programs and One Ontology

In order to test ontology as a model applied to local development planning, a wide application has been conducted on a sample of five Italian Regional Operative Programs 2007-2013 (R.O.P.) implementing EU regional policies.

The sample of experimentation has been selected in order to include a wide range of cases of Regions belonging to different UE mainstream objectives.

Table 2. Regional Operative Programs represented into ReDO ontology

Region		Program	Objective UE 2007-2013
1	Basilicata	O.P. ERDF Basilicata 2007-2013	Convergence (phasing out)
2	Puglia	O.P. ERDF Puglia 2007-2013	Convergence
3	Sardegna	O.P. ERDF Sardegna 2007-2013	Convergence (phasing in)
4	Emilia Romagna	O.P. ERDF Emilia Romagna 2007-2013	Regional Competitiveness and Employment
8	Toscana	O.P. ERDF Toscana 2007-2013	Regional Competitiveness and Employment

We observe how such programs are characterized by sectoral planning framework articulated in a program structure compatible with ReDO model.

The sample has been compared on two testing levels:

- first, we proceeded to analyse and represent single R.O.P., evaluating obtained results;
- in the second phase of the work, we developed an ontology containing all programs.

This extension of the representation has allowed to express comparisons and evaluations among operative programs.

Table 3. Individuals' encoding

Classes	encoding
Axis	A_1_'Region Name'
Overall Objective	OG_1_'Region Name'
Specific Objective	OS_1.1_'Region Name'
Operative Objective	OP_1.1.1_'Region Name'
Intervention Line	LI_1.1.1a_'Region Name'

In order to allow a proper management of this huge information system, a unique feature encoding in ReDO has been adopted. We proceeded, according to the description given in the below table, through the identification of elements belonging to R.O.P. hierarchical structure and specifying each R.O.P. by its name (ie. the name of the region).

At the end of the study we obtained an ontology with more than six hundred items: a very complex network. This is the dimension of the information baggage we used to deal with during the phases of plan implementation, management and evaluation.

At this step the useful ontological tools provided by Protégé helped to interrogate the network producing comparative results.

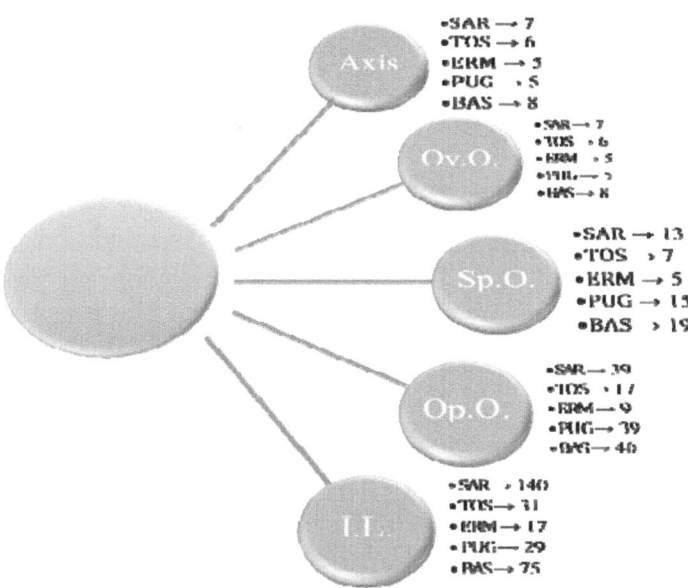

Fig. 4. ReDO in numbers

Some critical considerations emerged: if you are a planner, you should hold the knowledge about how to deal with a complex plan or program; if you are an applied technician you will understand and manage sectoral aspects of the plan connected to specific knowledge; but, if you are a politicians, a stake holder or a final beneficiary of a program, then which tools have you in order to understand and implement plan previsions, especially in a participative dimension? What we want to underline is that if we intend to implement a participative bottom up approach in local development, we have to provide not only technicians, but also common citizens of effective knowledge management tools in order to build-up a process of knowledge sharing.

Managing complexity is one of the permanent issues of planning theory, but such instance has increased in priority in incoming scenarios of shared planning in ICT environment. We think that ReDO assumes a relevant role as an applied knowledge management tool combining functions, queries, analytical and quantitative tools derived by other fields of application. In fact, the result of the research has allowed several interesting outputs: to investigate structures of programs, to compare contents of programs, to classify contents of each program within the framework of relations defined in the ontology, to allow semantic queries and navigation within the complex network.

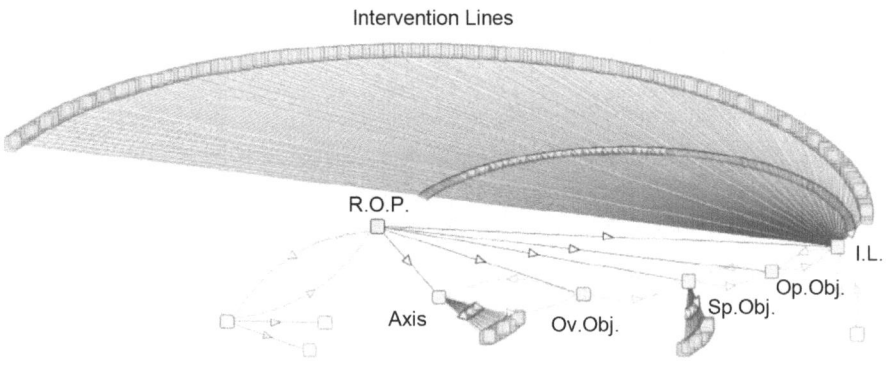

Fig. 5. ReDO R.O.P. network

7 Conclusions

ReDO ontology is the main output of the research. It is the result of a tested procedure for ontology design, methodological remarks regarding the role of users (or stakeholders) interaction in the process of building-up the ontology.

Indeed, the level of participation of technicians, scientists and potential users in the process of ontology development is directly proportional to usability of such knowledge management tools, especially in the field of planning [15]. This consideration identifies ReDO as a pure research output but it could bring to operative application starting from the ReDO model.

The current tools for managing ontologies (in this work we used the Protegé software) do not allow to integrate the spatial dimension within the ontological representation. Working in the field of territorial programming it is an important weakness and a perspective at the same time.

So, in order to assume ontology as a comprehensive DSS (Decision Support System) the problem of integration of Spatial Data Infrastructure should be faced.

To adopt effective knowledge management systems, responding to broad accessibility criteria, will allows 'program actors' (decision makers, citizens, stake holders) to have a complete information to the domain of interest.

As experimented in ReDO research, the ontological representation of the program gives important contribution to control and evaluate the program structure logic. Previous works [11] [12] show how logical weaknesses in program structures determine a lack of efficacy and effectiveness on the whole policy. Therefore, one of the most important applications of this tool concerns the field of program evaluation, intended as a comprehensive process [18].

References

1. Corallo, et al.: Enhancing communities of practice: an ontological approach. Paper Presented at 11th International Conference on Industrial Engineering and Engineering Management, Shenyang, China (2005)
2. Damiani, et al.: KIWI: A Framework for Enabling Semantic Knowledge Management. In: Damiani, E., et al. (eds.) Semantic Knowledge Management: An Ontology-Based Framework. New York Information science reference, Hershey (2009)
3. Ferraris, M.: Dove sei? Ontologia del telefonino, Bompiani Editore. Milano (2005)
4. Frank, A.U.: Ontology: a consumer's point of view, pp. 1–13 (1996)
5. Frank, A.U.: 'Ontology'. In: Kemp, K. (ed.) Encyclopedia of Geographic Information Science, Sage Publications, Thousand Oaks (2008)
6. Genesereth, Nilsson: Logical Foundations of Artificial Intelligence. Morgan Kaufmann Publishers, San Mateo (1987)
7. Gruber, T.: A translation approach to portable ontology specifications. Knowledge Acquisition 5(2), 199–220 (1993) doi: 10.1006/knac.1993.1008
8. Gruber, T.: Toward principles for the design of ontologies used for knowledge sharing? International Journal of Human-Computer Studies 43(5-6), 907–928 (1995), doi:10.1006/ijhc.1995.1081.
9. Knapp, S., Coors, V.: The use of eParticipation systems in public participation: the VEPs example. In: Coors, V., et al. (eds.) Urban and Regional Data Management, pp. 93–104. Taylor and Francis, London (2008)
10. Las Casas, G., Scardaccione, G.: Rappresentazione concettuale della conoscenza: ontologia del rischio sismico. In: Murgante, B. (ed.) L'informazione Geografica a Supporto Della Pianificazione Territoriale, pp. 279–299. Franco Angeli, Milano (2008)
11. Las Casas, G., Scorza, F.: Un approccio "contex based" e "valutazione integrata" per il futuro della programmazione operativa regionale in Europa. In: Bramanti, A., Salone, C. (eds.) "Lo Sviluppo Territoriale Nell'economia Della Conoscenza: Teorie, Attori Strategie" Collana AISRe – Scienze Regionali, vol. 41 (2009)

12. Las Casas, G., Scorza, F.: Comprehensive Evaluation: a renewed approach for the future of European Regional Convergence Policies. In: European Evaluation Society (EES) Biennial Conference 2008 'Building for the Future: Evaluation in Governance, Development and Progress', Lisbon, Portugal, October 1-3 (2008),
http://www.europeanevaluation.org/congressmedia&collectionId =7875192?page=7940885&collectionId=7875192&images7875192page =2 (accessed Dicember 12, 2008)
13. Laurini, R., Murgante, B.: Interoperabilità semantica e geometrica nelle basi di dati geografiche nella pianificazione urbana. In: Murgante, B. (ed.) L'informazione Geografica a Supporto Della Pianificazione Territoriale, pp. 229–244. FrancoAngeli, Milano (2008)
14. Murgante, B., Scardaccione, G., Las Casas, G.: Building ontologies for disaster management: seismic risk domain. In: Krek, A., Rumor, M., Zlatanova, S., Fendel, E.M. (eds.) Urban and Regional Data Management, pp. 259–269. CRC Press, Taylor & Francis, London (2009), ISBN: 978-0-415-055642-2, DOI:10.1201/9780203869352.ch23
15. Murgante, B., Scorza, F.: Ontology and Spatial Planning. In: Murgante, B., Gervasi, O., Iglesias, A., Taniar, D., Apduhan, B.O. (eds.) ICCSA 2011, Part II. LNCS, vol. 6783, pp. 255–264. Springer, Heidelberg (2011), ISSN: 0302-9743, doi:10.1007/978-3-642-21887-3_20
16. Mark, D.M., Smith, B., Tversky, B.: Ontology and Geographic Objects: An Empirical Study of Cognitive Categorization. Cognitive Science (1997)
17. Scorza, F., Casas, G.L., Murgante, B.: Overcoming Interoperability Weaknesses in e-Government Processes: Organizing and Sharing Knowledge in Regional Development Programs Using Ontologies. In: Lytras, M.D., Ordonez de Pablos, P., Ziderman, A., Roulstone, A., Maurer, H., Imber, J.B. (eds.) WSKS 2010. CCIS, vol. 112, pp. 243–253. Springer, Heidelberg (2010), ISBN 978-3-642-16323-4, doi:10.1007/978-3-642-16324-1_26
18. Scorza, F., Casas, G.L., Carlucci, A.: Onto-Planning: Innovation for Regional Development Planning within EU Convergence Framework. In: Murgante, B., Gervasi, O., Iglesias, A., Taniar, D., Apduhan, B.O. (eds.) ICCSA 2011, Part II. LNCS, vol. 6783, pp. 243–254. Springer, Heidelberg (2011), doi:10.1007/978-3-642-21887-3_19
19. Scorza, F. (ed.): Contributi all'innovazione degli strumenti per lo sviluppo locale – Edizioni Ermes, Potenza (2008)
20. Murgante, B., Tilio, L., Lanza, V., Scorza, F.: Using participative GIS and e-tools for involving citizens of Marmo Platano – Melandro area in European programming activities, special issue on "E-Participation in Southern Europe and the Balkans". Journal of Balkans and Near Eastern Studies 13(1), 97–115 (2011), doi:10.1080/19448953.2011.550809
21. Tilio, L., Lanza, V., Scorza, F., Murgante, B.: Open Source Resources and Web 2.0 Potentialities for a New Democratic Approach in Programming Practices. In: Lytras, M.D., Damiani, E., Carroll, J.M., Tennyson, R.D., Avison, D., Naeve, A., Dale, A., Lefrere, P., Tan, F., Sipior, J., Vossen, G. (eds.) WSKS 2009. LNCS, vol. 5736, pp. 228–237. Springer, Heidelberg (2009), ISSN: 0302-9743, doi:10.1007/978-3-642-04754-1_24
22. Murgante, B.: Interoperabilità semantica e pianificazione territoriale. Italian Journal of Regional Science 10(3) (2011), doi:10.3280/SCRE2011-003008

A Landscape Complex Values Map: Integration among *Soft* Values and *Hard* Values in a Spatial Decision Support System

Maria Cerreta and Roberta Mele

Department of Conservation of Architectural and Environmental Heritage,
University of Naples Federico II, via Roma 402, 80132 Naples, Italy
cerreta@unina.it, roberta_mele@hotmail.it

Abstract. The paper develops a Spatial Decision Support System (SDSS) for the identification and evaluation of the landscape complexity for the Massa Lubrense territory, in the South of Italy. Through the elaboration of a selection of spatial indicators and the combination of GIS and Analytic Hierarchy Process (AHP) method, it has been defined a decision-making process for the construction of a map of complex values, *soft* and *hard* values, that characterize the landscape of Massa Lubrense. The paper explores the potential of a Spatial Decision Support System (SDSS) in the field of land-use planning, recognizing different weights and priorities according to a complex definition of the landscape and its values.

Keywords: Complex values, Knowledge generation, Spatial indicators, Spatial Decision Support System, Multi-Criteria Analysis.

1 Introduction

In complex, uncertain and conflict-ridden planning contexts, different categories of values can be identified: direct-use, indirect-use, non-use and intrinsic values. The explicit recognition of the existence of multiple interdependent values establishes both the conceptual and empirical foundations to understand just how these value categories may be applied to the planning context. This means becoming aware of the "complex social value" of a context and its resources [1]. Thus, the explicit recognition of the existence of multiple interdependent values makes it possible to include instrumental and intrinsic values in evaluation process. In addition, by prioritizing values we can distinguish between them, highlight different perspectives and take into account various kinds of conflicts [2].

Intrinsic value allows us to move beyond the private sphere and reflect on collective benefits and externalities, explicating a clear ethical dimension. It expresses the "glue value", the system of immaterial relations, its specific character, its particular identity [3] [4]. It is a proactive value, able to create integration, to reduce marginalization, to overcome fragmentation, and stimulate vitality: a "catalyst" of material and immaterial energies, able to blend various value dimensions, helping to capture its deep unity [5].

Intrinsic value is consistent with the concept of value complex by Zeleny [6] [7] and conceived as a "meta-criterion", anchored and integrated in fundamental values that are broadly accepted and not subject to choice.

The value complex is the expression of a cognitive balance, characterized by candour and trust, based on principles, ethics and rules, mostly qualitative and expressible only in imprecise and fuzzy language, but rooted in specific contexts.

Recognizing all the different categories of values implies the recognition of the multiplicity and diversity of knowledge. Any representation of a complex system reflects only one subset of its possible representations [8] [9]. A consequence of these deep subjectivities is that, in any normative exercise connected to a public-decision problem, one has to choose an operational definition of "value". This despite the fact that social players with different interests, cultural identities, and goals may have different definitions of value. Consequently, to rank policy options it is first necessary to decide what is important for different social players and what is relevant for the representation of the real-world entity described in the model [10] [11]. Consistently with the complex-values approach, problem situations are closely related to the decision-making environment, which is strongly dependent on the interaction between knowledge and values. Complex values are, in fact, linked to the context and to the decision frame [12] and emerge from the cognitive frame shaping the physical, environmental, social and economic environment.

Complex values should be the driving force for any decision-making process: they help exploring the decision context and structuring the problem, by guiding information collection, uncovering hidden objectives, improving communication, facilitating involvement in multiple-stakeholder decisions, and interconnecting decisions. Complex values are "strategic values" able to guide strategic thinking, discover decision opportunities and create alternatives. They are embedded in a given problem context [2]. Full awareness of relevant values in a decision context depends on the different kinds of information and knowledge, *hard data* and *soft data* that characterize the decision contest. It is necessary to move beyond multiple and complex values, towards the realization that value judgments need to face multiple and often conflicting ways of valuing [13].

According to the above perspective, *thinking through complex value* [2] can be a useful approach to identify landscape values in a multidimensional perspective. Indeed landscape can be considered expression, at the same time, of diversity and identity, where changes in time are relevant and need to be analyzed studying processes relating to spatial patterns, extrapolating results in space and time, linking data of different qualities, and considering complex values as a driver of landscape change itself.

Over the past years many international programs were used for inventorying and assessing landscapes and for monitoring changes [14] [15] [16] [17] [18] [19], taking into account that many changes of the traditional landscape concept have been related to urbanisation, transportation, recreation, tourism, etc. New uses define new landscapes and identify *new values*. Moreover, landscape values depend on its heterogeneity, that can be related to biodiversity, cultural heritage and human appreciation [20]. Consistent with the European Landscape Convention, landscape character is defined as a distinct and recognizable pattern of elements that makes one landscape different from another, making an area unique [14] [21]. An increasing need for landscape inventorying, assessment, and monitoring demands also the development of landscape indicators, useful to evaluate the impact of policy measures and their effect on the landscape change [22], but also for sustainable conservation and protection of natural and cultural capital, spatial planning, and landscape management [23]. The many emerging needs carry many significant questions [24]: How many landscape changes can be tolerated? When does the landscape character change in irreversibly? What kind of development is acceptable where? How spatial planning can influence the landscape character?

In order to find a possible answer to these questions, this paper presents the elaboration of a methodological process where the recognition of different components of values is articulated into three main phases:

1. Recognizing the multiple dimensions of the landscape (spatial, geographic, economic, social, environmental, anthropologic, and cultural), identifying soft and hard data, and activating various forms of knowledge (explicit, systematized, experiential/practical-contextual, implicit) [25];
2. Specifying local complex values, identifying the values embedded in the activated knowledge through their spatial representation;
3. Exploring landscape opportunities within a broader decision-making context able to deal with shared knowledge and complex values in the light of local potential and criticalities.

According to the above perspective, through the empirical investigation in an operative case study related to the Massa Lubrense landscape, in the South of Italy, it has been possible to elaborate a "landscape complex values map", result of the integration of Analytic Hierarchy Process (AHP) [26] [27] [28] and Geographical Information Systems (GIS) in a Spatial Decision Support System.

2 Massa Lubrense Landscape: Soft Values and Hard Values Interplay

The town of Massa Lubrense[1] is in the local context of the Sorrento Peninsula, jutting out towards Capri and nestled between the gulfs of Naples and Salerno; with its extraordinary variety of views, it has the privilege to complete the charm of one of the most celebrated and beautiful corners of the world (Fig. 1).

The morphological complexity, the variety of landscape, the succession of natural vegetation and agricultural areas, and the close relationship between the natural and human components, the historic settlement pattern, and the presence of cultural and

---

[1] Massa Lubrense case study has been carried out within the elaboration of the degree thesis in Architecture of arch. Roberta Mele, on the subject "Landscape Chart of Massa Lubrense: an integrated approach between GIS and Multi-Criteria Analysis", tutor prof. Salvatore Sessa, co-tutor arch. Maria Cerreta, ing. Ferdinando Di Martino, arch. Barbara Cardone, University of Naples Federico II, december 2011.

environmental resources of high value give the area a strong landscaping specificity. This is underlined not only by the exceptional individual components, but also by the complex system of relationships between the different natural, cultural and social components. This system, over the centuries, has helped consolidating a specific landscape as a result of continuous and balanced relationship between *soft values* and *hard values*.

Fig. 1. Localization of Massa Lubrense

From the standpoint of environmental characteristics, the dominant feature of the landscape can be identified in the physiognomy of his "landscape-environment", which means the simultaneous presence in the territory of the elements of "great scenery" and elements of "landscape site", of individual characteristics such as geological, morphological, hydrological, flora and fauna that can be enjoyed up through the area. From the historical-cultural point of view, it is clear that many places of monumental and artistic interest insist on the area, making the cultural and historical heritage of the town one of the region's most important and connoting the historical landscape. Finally, from the socio-economic side, we can identify tourism as the main elements of the economy, that can grow up through a planning that provides the redistribution of sites for tourist, residential and commercial facilities. The territory of Massa Lubrense has a very heterogeneous network of receptivity, and a lots of businesses, from food and wine production to the entertainment and seasonal events, local and cultural activities, that take place throughout the year.

2.1 The Methodological Process

The evaluation of the Massa Lubrense landscape has been carried out through the application of a multi-methodological approach, where GIS support is relevant. In this work the GIS is essential to analyze the wide variety of geographic data necessary for the evaluation, while the multicriteria methodologies, and in particular the Analytic Hierarchy Process (AHP) [26] [27] [28], allow the explanation of weights to be attributed to the criteria and adopted for the analysis.

The growing importance of environmental and natural resources has led to the need of use methods for evaluating the effectiveness and the conservation of projects and enhancement involving the use of natural resources. Multi-criteria analysis is among the approaches that includes a rich variety of techniques that rely on the same pattern: making explicit the contributions of different options choosing among the various criteria or attributes. The criteria are the means by which the various alternatives are compared to each other with the objectives, taking into account simultaneous multiple quantitative and qualitative aspects. Evaluation is becoming an increasingly significant factor in structuring and processing of planning options for territorial changes. In practice it is clear the need to use integrated approaches that consider multidimensional techniques and tools able to promote dialogue and interaction between different knowledge [29] [30].

In this context, evaluation becomes a "tool" capable to integrate approaches, methodologies and models, adjusting to the many needs that the decision-making process itself reveals. Indeed, in spatial planning process it is hard to identify all the multidimensional effects of the proposed options, so, it is necessary to identify "new approaches", able to provide a framework for analysis and evaluation suitable to integrate the objectives and values in multidimensional decision processes [31]. At the same time, in decision making-process and decision management, GIS can be considered one of the most advanced tools available to address complex problems in a balanced mediation between economic, environmental and social objectives. It is an essential tool that, if properly used, can provide effective support for spatial planning [32] [33] [34] [35]. In this sense, geospatial technologies should be a driving force in the technical and socio-organizational work for the implementation of knowledge-based open platforms, integrated analysis and problem solving. Indeed, the various GIS applications can help improving the vertical and horizontal collaboration among all actors involved in sustainable development processes at all levels of government (national, regional and local); further growth of the availability of spatial data and developments within GIS allow us to continue an "information planning process" (analysis, design, evaluation, decision, management and communication) [33] [36]. Indeed, sustainable development planning, decision-making and management processes are dealing with multidimensional problems whose aim is to achieve a balance between economic development, environmental protection and social equity. Most of the problems related to the design and administration of land management methods require integrated multi-criteria decision making with GIS. Decision-making processes going with the planning and/or design need suitable tools for the identification of priorities, in order to pursue goals often in conflict. In these processes, complexity and uncertainty are involved, so it is necessary to include the use of evaluation as a tool for construction of choices, clarification of values, interests and needs, as well as exploration of the relevant factors. In Massa Lubrense case-study the whole approach of spatial multi-criteria analysis has been conceived as a dynamic process that combines and transforms the input of geographic data (maps of the attributes and criteria adopted for the analysis) to an output, also geographically, represented by the map of landscape complexity. The methodological steps necessary to identify and represent, through geo-referenced maps of spatial indicators, the complexity of landscape have been summarized as follows (Fig. 2):

- *Acquisition of data*: the information layers are selected for the landscape study and retrieved from various official data sources;
- *Processing*: raw data are processed in a GIS environment in order to create thematic maps in raster format related to the entire municipal area;
- *Standardization*: information resulting from the processing are standardized in order to classify all the criteria based on a common scale of value; established the scale value for each attribute, a spatial analysis was performed with buffer areas;
- *Calculation of weights*: standardized maps relating to the attributes are aggregated together as a function of the relative importance of attributes with respect to each criterion and with respect to the goals;
- *Construction of AHP hierarchy* by goal, criteria and spatial indicators: through processes of Map Algebra, managed in a GIS environment, the layers standardized information is aggregated by weighted sums, climbing up the hierarchy from the leaves at the top to provide maps of spatial indicators;
- *Results*: the calculation of the weights of each level of the hierarchy and the synthesis of the hierarchical reconstruction are expressed in the final map of the landscape complexity.

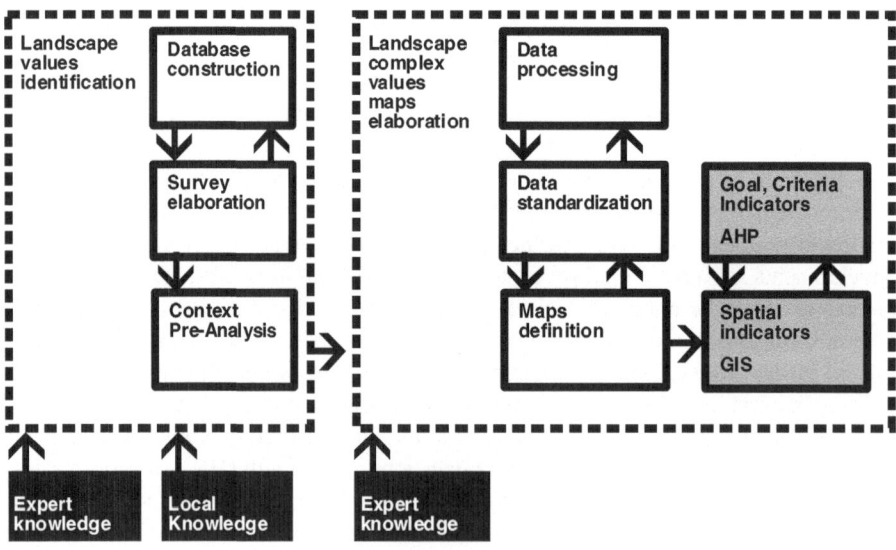

Fig. 2. The methodological steps

The construction of the cognitive framework of landscape components considered in the assessment phase was carried out by subdividing the landscape features into three thematic areas: physical-natural components, historical-cultural components and social-symbolic components (Fig. 3). Each of the thematic area expresses a category of values, able to identify the landscape complexity of Massa Lubrense territory. These thematic areas were, in turn, divided into sub-areas, which have been associated with certain characteristics. In particular, the physical-natural component has been divided into:

- Geo-physical characters: altimetry, geological sites, hydrogaphy and land use;
- Naturalistic characters: flora, marine and terrestrial fauna, natural protected areas (SCI), marine protected area, pathways, pinewoods, caves.

The historical-cultural component has been divided into:
- Urban character: historic centers, archaeological sites and water artificial system;
- Architectural character: private architectures, military architectures, religious architecture;

The socio-symbolic component has been divided into:
- Socio-economic characters: typical products, local events, local food and receptivity;
- Symbolic characters: viewpoints and socializing spaces.

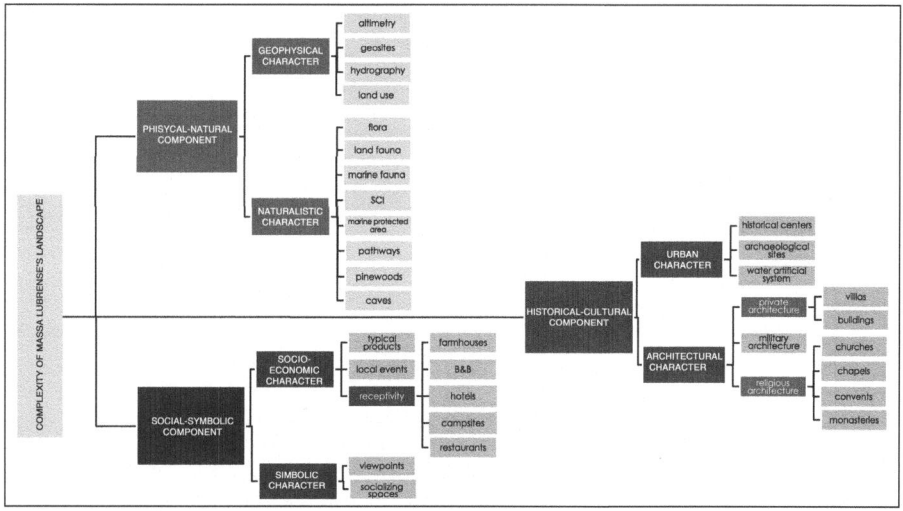

Fig. 3. The hierarchy of the main relevant components

All the thematic areas and their sub-areas can structure a hierarchy of main characters that compose the complexity of the landscape of Massa Lubrense territory.

2.2 Spatial Indicators as Expression of Local Identity

The different characters identified have been represented through selected spatial indicators. For the analysis of physical-natural components, historical-cultural components and social-symbolic components, a GIS was created, which incorporated data on natural and man-made elements of the territory. In the GIS all the components of the landscape were digitized, according to a hierarchy structure (Fig. 4).

Since the original data were not homogeneous, was added a new field ("value") in which a weight was assigned according to a range scale 1-5; in the scale, by convention, values close to 1 indicate less importance than those close to 5, which indicate the highest importance. Then, in order to obtain homogeneous classes in which the value represents the interval of belonging, the classification of "equal intervals" was carried out (Fig. 5).

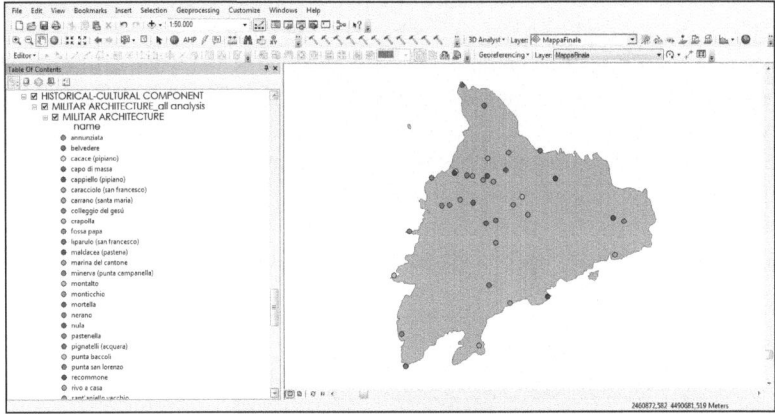

Fig. 4. Example of digitization: military architectures

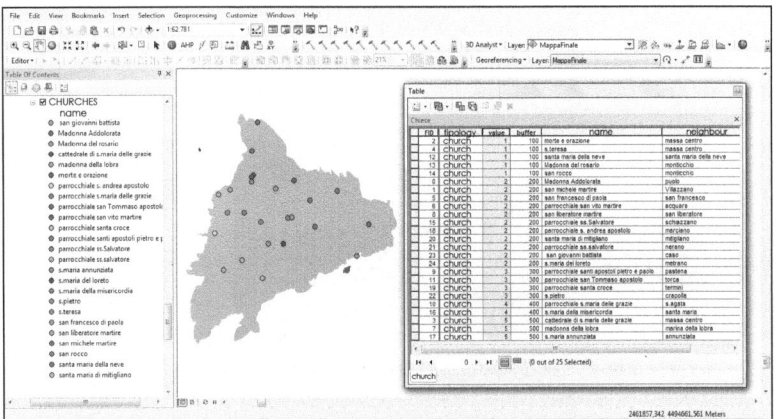

Fig. 5. Example of homogenization

In order to define the value for each element, an operation of spatial "buffer areas" was performed determining the impact areas by the radius of influence. The map was then converted from the buffer polygon shape-file in raster map, with a cell size of 10x10, according to the field "value". Through the "map algebra" function, a raster map was bounded only to the area of the district municipality. Subsequently, in the space analysis, we proceeded by selecting the function of the "raster calculator" (Fig. 6). Following the above process for each one of the selected indicator, it was possible to identify the main values of the Massa Lubrense territory related to the three selected thematic components.

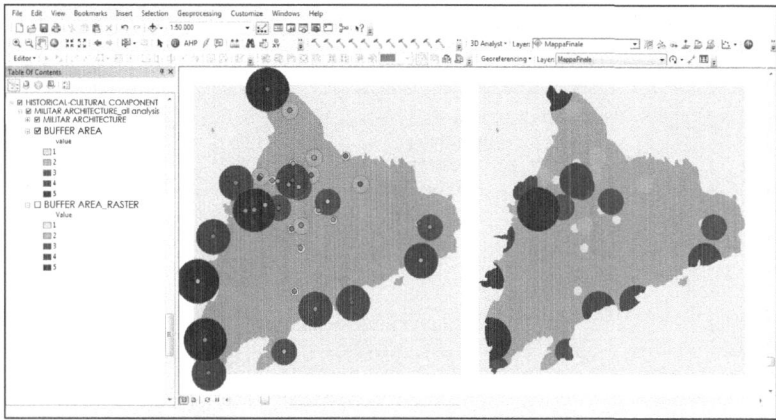

Fig. 6. Buffer and raster map: military architectures

2.3 The Identification of a Values Ranking

Defined the hierarchical decision tree and made all the raster maps of spatial indicators for each attribute, pairwise comparisons between elements of the hierarchy were made in order to determine which is the most important applying the extension of AHP, which implements the operations of "map algebra" through a weighted average between the weights of attributes (Fig. 7). The AHP structures the decision-making process in hierarchical form. The fundamental process of AHP involves perception, decomposition, and synthesis of a decision problem, in order to provide a methodology for modelling unstructured problems in the economic, social, and management sciences. The elaboration of a hierarchy is an abstraction of a system's structure to study the functional interactions of its components and their impacts on the entire system. This abstraction can take several related forms, all of which are essentially derived from an overall objective, down to sub-objectives, down further to a force that affects these sub–objectives, to the people who influence these forces, to the people's objectives and then to their policies, still further to the strategies, and finally the outcomes that result from these strategies [26]. From a procedural point of view, this approach consists of three main phases: 1. construct a suitable hierarchy; 2. establish priorities between elements of the hierarchy by means of pairwise comparisons; 3. check logical consistency of pairwise comparisons [27] [28].

In the present case-study, each thematic area (physical-natural components, historical-cultural components and social-symbolic components) has been organized according to a three levels hierarchical structure: 1. thematic components; 2. criteria; 3. values/characteristics. To the values/characteristics of the third hierarchical level have been associated some spatial indicators referred to the nature of the areas linked

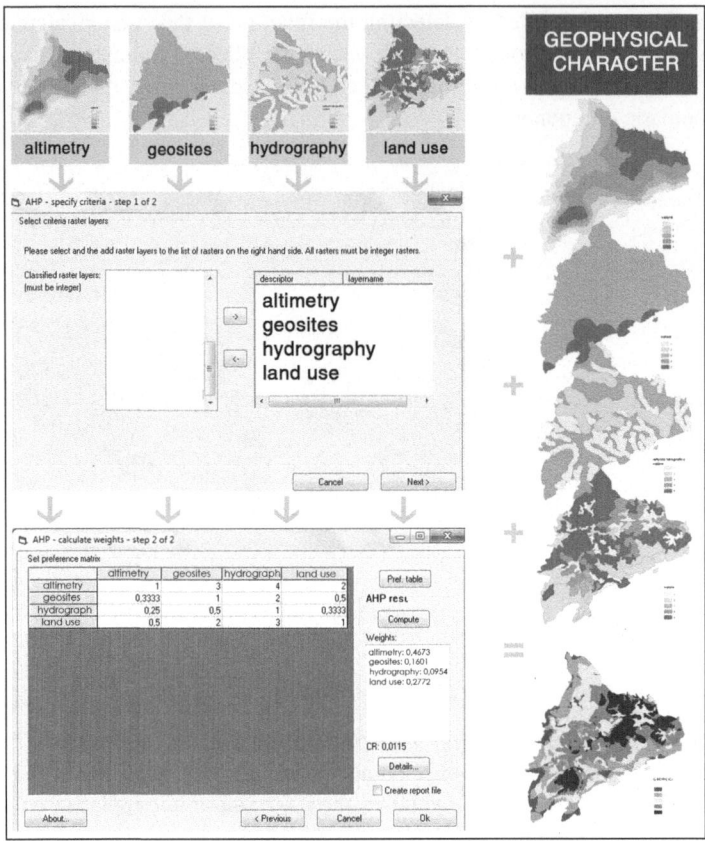

Fig. 7. AHP application in GIS framework

to a value judgement, expressed through a five points scale related to the complexity of landscape: high value; 2. medium–high value; 3. medium value; 4. medium–low value; 5. low value. To perform the "spatial assessment" it was used an extension of the AHP method within ArcGIS [37], obtaining "landscape complexity maps" of Massa Lubrense landscape. This has made it possible to make a pairwise comparison of the criteria of every hierarchical level. In order to apply the AHP method to each class of landscape complexity, a numerical value (score) and a chromatic scale have been associated to the five judgments. Indeed, in order to have a graphical representation of the results, to every score is related a colour, to be given to every pixel, according to the convention that goes from dark green to orange. The attribution of scores was made taking into account the results of the phase of the landscape values identification, where expert knowledge point of view and local knowledge have been combined in a dialogical process. The results of these comparisons identify the coefficients of dominance and constitute the elements of a matrix, called matrix of pairwise comparisons. After the phase of comparison the overall weight of the final level of the hierarchy has been determined, applying the principle of hierarchical composition which consists in multiplying the local weights of each element to those of the corresponding higher-level elements and sum the resulting products. In problems of multicriteria spatial analysis the phase of reconstruction is to perform hierarchical weighted sums of the first maps attribute obtaining maps of the criteria, then do the sums in weighted maps of the criteria resulting in a final map relative to the goal.

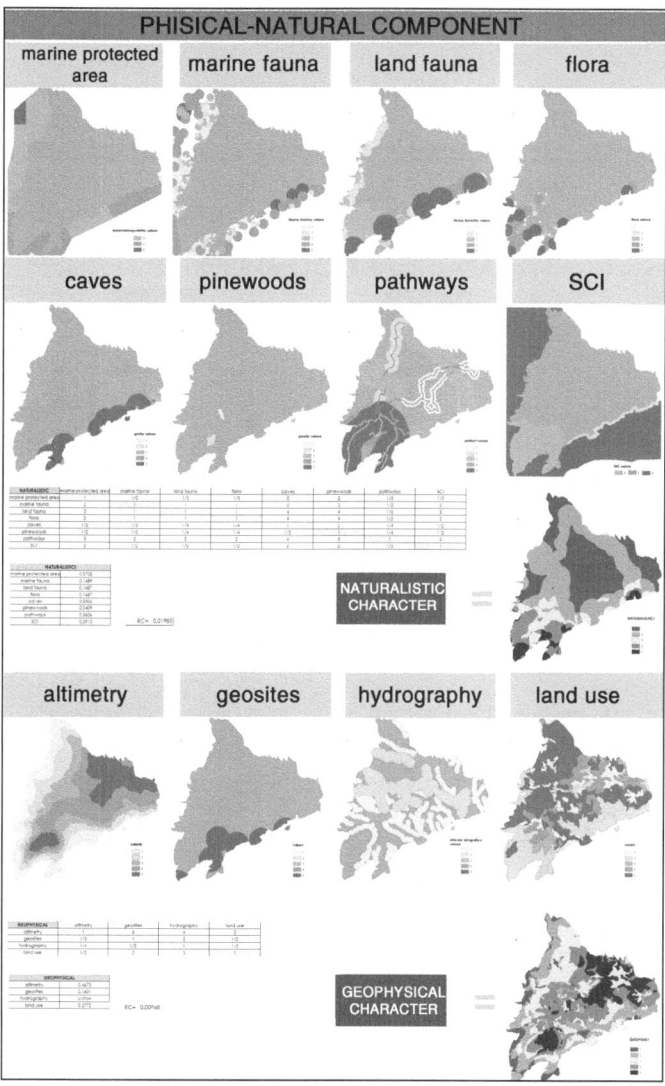

Fig. 8. Spatial indicators: physical-natural component

The maps of spatial indicators for each criterion was established for each component. It shows for example the physical-natural components (Fig. 8). Indeed, thanks to the AHP method application it is possible to combine the weights of criteria obtained through pairwise comparisons with the scores associated to the different classes of landscape complexity, obtaining, in synergy with GIS, the related "landscape complexity maps". For every pixel, it is possible to get a total value as a linear combination of weights of criteria by the score related to the landscape complexity taking into account the specific values/characteristics.

Considering all the thematic components and the related criteria and indicators of the hierarchy and putting together the data of all criteria belonging to the first hierarchical level, we can have the map of Figure 9, in which the colours from dark green to orange express the landscape complexity (from high to low) of Massa Lubrense territory. The dark green areas represent the most relevant expression of landscape complex values, where the highest hard values and the highest soft values are combined (Figg. 9-10).

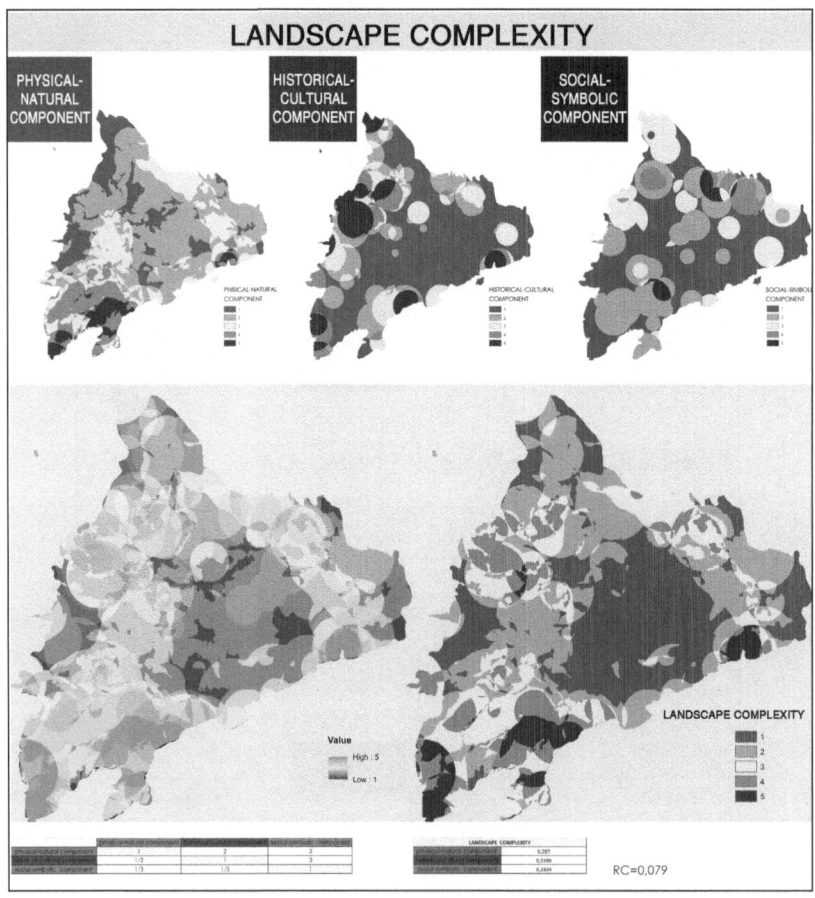

Fig. 9. Landscape complexity maps: synthesis of components

Fig. 10. A synthesis of the main steps

3 Conclusions

The spatial decision-making process elaborated for the interpretation of the main features of Massa Lubrense landscape points out how GIS combined with AHP can make full use of GIS functions such as space analysis, data processing and inquiry, where the complex data and their mutual influence can be included as relevant aspects, describing value and space position of each factor.

AHP makes more flexible the analysis; it is characterized by a relative ease of handling with multiple criteria, and it is simple to understand and to effectively deal with both qualitative and quantitative data. At the same time, GIS helps developing a good man-machine interactive interface. The methodological approach proposed shows that:

- It is possible to improve synthetic evaluation developing the GIS capability of spatial analysis and the AHP capability of multilayer analysis. The evaluation results and the distribution pattern obtained for each thematic component represent an effective way to communicate the complex values of the territory;
- it's useful to build a large and flexible multi-element system, where there are continuous and dynamic exchanges of information among various subsystems/themes, selecting comprehensive spatial indexes and/or indicators;
- GIS can be used to prepare spatial statistics and clustering processes to reveal the most suitable areas for sites selection, managing and analyzing large volumes of spatially resolved data from a variety of sources. Furthermore, it is able to deal with and simulate the necessary economic, environmental, social, technical, and political constraints;
- The AHP is a powerful tool to solve complex problems that may have interactions and correlations among multiple objectives;
- It allows decision-makers to clearly know current status of the integrated characteristics of local context and to help local community understand the interactions among different components in order to identify and implement shared actions.

The approach built during the case study may be considered as a basis to increase the level of integration of local and expert knowledge in a more extensive participatory GIS process, oriented to involve different expertise in order to improve the completeness of hard and soft data and the consistency of the whole evaluation. By means of the spatial decision-making process proposed, local communities can contribute actively to the implementation and updating of GIS data and to improve the evaluation of landscape complexity [38]. At the same time, the evaluation of intangible assets takes on a key role and makes it possible to analyze the social concept of value in a multi-dimensional approach, enabling the integration of cognitive and evaluative dimensions as well as technical and economic dimensions. At last, since an essential part of the work is based on the European Landscape Convention regarding the perception of the landscape from local community, the final value attributed to the complexity of the landscape can be considered dynamic and changeable. Indeed, one of the possible future developments of this work can be to improve the interaction between local and expert knowledge in order to give more relevance to the soft values in the definition of the landscape complex social value.

Acknowledgments. The authors would like to thank prof. Salvatore Sessa, ing. Ferdinando Di Martino, arch. Babara Cardone of Department of Constructions and Mathematical Methods in Architecture, University of Naples Federico II, for their valuable and collaborative support in GIS elaborations.

References

1. Fusco Girard, L.: Sustainability, Creativity, Resilience: Toward New Development Strategies of Port Areas through Evaluation Processes. International Journal of Sustainable Development 13(1-2), 161–184 (2010)
2. Cerreta, M.: Thinking Through Complex Values. In: Cerreta, M., Concilio, G., Monno, V. (eds.) Making Strategies in Spatial Planning. Knowledge and Values, pp. 381–404. Springer, Dordrecht (2010)
3. Fusco Girard, L., Nijkamp, P.: Le valutazioni per lo sviluppo sostenibile della città e del territorio. Franco Angeli, Milano (1997)
4. Fusco Girard, L., Nijkamp, P.: Energia, bellezza, partecipazione: la sfida della sostenibilità. Valutazioni integrate tra conservazione e sviluppo. Franco Angeli, Milano (2004)
5. Fusco Girard, L., Cerreta, M., De Toro, P., Forte, F.: The Human Sustainable City: Values, Approaches and Evaluative Tools. In: Deakin, M., Mitchell, G., Nijkamp, P., Vreeker, R. (eds.) Sustainable Urban Development. The Environmental Assessment Methods 2, pp. 65–93. Routledge, London (2007)
6. Zeleny, M.: Multiple Criteria Decision-Making: Eight Concepts of Optimality. Human Systems Management 17(2), 97–107 (1998)
7. Zeleny, M.: Human Systems Management: Integrating Knowledge, Management and Systems. World Scientific Publishers, Hackensack (2005)
8. Giampietro, M., Allen, T.F.H., Mayumi, K.: Science for Governance: the Implications of the Complexity Revolution. In: Guimaraes-Pereira, A., Guedes-Vaz, S., Tognetti, S. (eds.) Interfaces Between Science and Society, pp. 82–99. Greenleaf Publishing, Sheffield (2006)
9. Munda, G.: Social Multi-Criteria Evaluation for a Sustainable Economy. Springer, Heidelberg (2008)
10. Funtowicz, S.O., Martinez-Alier, J., Munda, G., Ravetz, J.: Multi-Criteria-Based Environmental Policy. In: Abaza, H., Baranzini, A. (eds.) Implementing Sustainable Development, pp. 53–77. UNEP/Edward Elgar, Cheltenham (2002)
11. Munda, G.: Social Multi-Criteria Evaluation: Methodological Foundations and Operational Consequences. European Journal of Operational Research 158(3), 662–677 (2004)
12. Strauss, K.: Re-Engaging with Rationality in Economic Geography: Behavioural Approaches and the Importance of Context in Decision-Making. Journal of Economic Geography 8(2), 137–156 (2008)
13. Richardson, T.: Environmental Assessment and Planning Theory: Four Short Stories about Power, Multiple Rationality, and Ethics. Environmental Impact Assessment Review 25(4), 341–365 (2005)

14. Swanwick C.: Landscape Character Assessment. Guidance for England and Scotland. The Countryside Agency, Scottish Natural Heritage (2002), http://www.ccnetwork.org.uk/lcatopic.htm
15. Pinto-Correia, T., Cancela d'Abreu, A., Oliveira, R.: Landscape evaluation: methodological considerations and application within the Portuguese national landscape assessment. In: Brandt, J., Vejre, H. (eds.) Multifunctional Landscapes. Theory, Values and History, vol. I, pp. 235–252. WIT Press, Southampton (2004)
16. Wrbka, T., Erb, K.-H., Schulz, N.B., Peterseil, J., Hahn, C., Haberl, H.: Linking pattern and process in cultural landscapes. An empirical study based on spatially explicit indicators. Land Use Policy 21, 289–306 (2004)
17. Vervloet, J.A.J., Spek, T.: Towards a Pan-European landscape map-a mid-term review. In: Unwin, T., Spek, T. (eds.) European Landscapes: From Mountain to Sea. Proceedings of the 19th Session of the Permanent European Conference for the Study of the Rural Landscape (PECSRL), pp. 8–19. Huma Publishers, Tallinn (2003)
18. Mücher, C.A., Wascher, D.M., Klijn, J.A., Koomen, A.J.M., Jongman, R.H.G.: A new European landscape map as in integrative framework for landscape character assessment. In: Bunce, R.G.H., Jongman, R.H.G. (eds.) Landscape Ecology in the Mediterranean: Inside and Outside Approaches. Proceedings of the European IALE Conference, Faro Portugal, March 29-April 2. IALE Publication Series, vol. 3, pp. 233–243 (2006)
19. Van Eetvelde, V., Antrop, M.: A stepwise multi-scaled landscape typology and characterisation for trans-regional integration, applied on the federal state of Belgium. Landscape and Urban Planning 91, 160–170 (2009)
20. Dramstad, W.E., Tveit, S.M., Fjellstad, W.J., Fry, G.L.A.: Relationships between visual landscape preferences and map-based indicators of landscape structure. Landscape and Urban Planning 78, 465–474 (2006)
21. Swanwick, C.: The assessment of countryside and landscape character in England: an overview. In: Bishop, K., Philipps, A. (eds.) Countryside Planning. New Approaches to Management and Conservation, pp. 109–124. Earthscan, London (2004)
22. Parris, K.: Measuring changes in agricultural landscapes as a tool for policy makers. In: Brandt, J., Vejre, H. (eds.) Multifunctional Landscapes. Theory, Values and History, vol. I, pp. 193–218. WIT Press, Southampton (2004)
23. Selman, P.: Planning at the Landscape Scale. Routledge, Oxon (2006)
24. Van Eetvelde, V., Marc Antrop, M.: Indicators for assessing changing landscape character of cultural landscapes in Flanders (Belgium). Land Use Policy 26, 901–910 (2009)
25. Healey, P.: Knowledge Flows, Spatial Strategy-Making, and the Roles of Academics. Environment and Planning C: Government and Policy 26(5), 861–881 (2008)
26. Saaty, T.L.: The Analytical Hierarchy Process. McGraw-Hill, New York (1980)
27. Saaty, T.L., Vargas, L.G.: Models, methods, concepts and applications of the Analytic Hierarchy Process. Kluwer Academic Publishers, Dordrecht (2001)
28. Saaty, T.L., Peniwati, K.: Group decision making: Drawing out and reconciling differences. RWS Publications, Pittsburgh (2007)
29. Abadi, H.N., Akbari, E., Etesami, H., Keshavarzi, A., Kohbanani, H.R.: Site selecting for dumping urban waste using MCDA methods and GIS techniques. World Applied Sciences Journal 7(5), 625–631 (2009)
30. Cerreta, M., De Toro, P.: Integrated Spatial Assessment for a Creative Decision-making Process: a Combined Methodological Approach to Strategic Environmental Assessment. International Journal of Sustainable Development 13(1/2), 17–30 (2010)

31. Fusco Girard, L., Cerreta, M., De Toro, P.: Integrated Spatial Assessment in Planning: Strategic Choices for Cava de' Tirreni Master Plan. In: Proceedings of the 11th International Symposium on the Analytic Hierarchy Process 2011, Sorrento (Italy), June 15-18, pp. 1–6 (2011)
32. Di Martino, F., Giordano, M.: I sistemi informativi territoriali, teoria e metodi. Aracne Roma (2005)
33. Campagna, M.: GIS for Sustainable Development. Taylor & Francis Group LLC, USA (2006)
34. Higgs, G.: GIS for Environmental Decision-Making. Andrew Lovett and Katy Appleton, USA (2008)
35. Murgante, B., Borruso, G., Lapucci, A.: Geocomputation, Sustainability and Environmental Planning. Springer, Berlin (2011)
36. Morse-McNabb, E., Sposito, V.: GIS-based modelling of regional conservation significance. Applied GIS 2(3) 2(3), 20.1–20.20 (2006)
37. Marinoni, O.: Implementation of Analytic Hierarchy Process with VBA in ArcGIS. Computational Geosciences 30, 637–646 (2004)
38. Bishop, I., Hossain, H., Sposito, V., Yingxin, W.: Using GIS in Landscape Visual Quality Assessment. Applied GIS 2(3), 18.1-18.20 (2007)

Analyzing Migration Phenomena with Spatial Autocorrelation Techniques

Beniamino Murgante[1] and Giuseppe Borruso[2]

[1] University of Basilicata, 10, Viale dell'Ateneo Lucano, 85100 Potenza, Italy
[2] University of Trieste, P. le Europa 1, 34127 Trieste, Italy
beniamino.murgante@unibas.it, giuseppe.borruso@econ.units.it

Abstract. In recent times a complete lack of attention to migration phenomena, in national and global policies, led to a huge concentration of foreigners in major cities of Europe and USA. This trend has been faced without effective policies and programs. Consequently, a great opportunity has been transformed in a great threat and the word immigration is generally associated with the term social security. In less than one century, Italy has been transformed from a country originating great migration flows to a country which is the destination of migration flows. The aim of this paper is to examine foreign immigration in Italy distinguishing according to nationality of foreigners. In order to analyze this phenomenon Shannon and Simpson Diversity Indices to measure the level of entropy in a distribution and the variation in categorical data have been used. The spatial dimension of migration flows has been analyzed in this paper using Spatial Autocorrelation techniques and more particularly Local Indicators of Spatial Association in order to analyze the highest values of a foreigner group considering the relationship with the surrounding municipalities.

Keywords: Migration Phenomena, Foreign Immigration analysis, Shannon Index, Simpson Index, Spatial Autocorrelation, Local Indicators of Spatial Association.

1 Introduction

During each political pre-electoral debate in Europe and in USA, one of the most discussed topics is migration phenomena and related policies. This situation arises due to a total lack of attention to such phenomena by national policies of these countries. Consequently, these phenomena have been completely unplanned and uncontrolled producing a huge social impact in major cities of these countries.

Foreigners' presence coupled with a careful integration of people with different demographic and social characteristics, cultural backgrounds, experiences and expectations may represent a great opportunity for destination areas of migration.

In order to avoid that such opportunities become threats, a continuous observation of the phenomenon is fundamental for programming measures and interventions suitable for an effective integration of immigrants and their families.

Migration has always been a natural process which produces other significant transformations in the environment as well as in everyday life, in economic systems, cultures, religions etc. The presence of foreigners is not easily detectable, because it is a particularly complex and rapidly evolving phenomenon.

Modern migrations are mainly characterized by two components, comparable in terms of absolute value: internal migration where part of the population moves within the country; external migration where part of the population reaches the country coming from another state. This work is completely concentrated on external component of migration towards Italy, distinguishing according nationalities of foreigners. An analysis of migration phenomenon in Italy which considers both the internal and the external components, without distinguishing between them, has been developed by Scardaccione et al. [6].

Several Indices have been used in this paper to analyze migration flows. Shannon and Simpson Diversity Indices, originally adopted in ecology to quantify habitat biodiversity, have been frequently used in several other fields and also in analyzing various demographic groups. These indices produce interesting results but they do not compute spatial dimension. Spatial autocorrelation techniques consider the intensity of a phenomenon inside a municipality, measuring at the same time the degree of influence over its surrounding municipalities. More particularly, for each foreigner nationality, Local Indicator of Spatial Association has been adopted in order to discover the highest values of the phenomenon coupled with the highest level of similarity with its neighbouring municipalities.

2 Methods

2.1 Shannon and Simpson Diversity Indices

Other than analyzing characteristics of single ethnic groups and comparing them, an element of interest lays in the analysis of the overall distribution and variability of foreign population in an area or in a single country, therefore considering both quantitative and categorical data concerning the presence of migrants. Indices of diversity can be useful when the focus is also on the different number and type of nationalities in an area, other than their figures.

Diversity indices originate from ecology and biology and are mainly aimed at measuring biodiversity of an ecosystem. They can be applied to measure diversity of a population in which each member belongs to a unique species. They have been used in studies concerning landscape [15] and social sciences [16], substituting the notion of 'species' with, for instance, land cover types rather than ethnic groups, and considering individual residents of an ethnic group instead of 'individuals belonging to species' [18], [19].

A commonly used index is Diversity Index or Index of Variability, frequently used to measure a variation in categorical data. It is known in the version of Shannon index [11] [12] that measures the level of entropy in a distribution, converting species in symbols and their population sizes in a measure of probability. Both number and evenness of species are considered and the values increase either by adding unique species or by means of a higher evenness of species. Shannon Diversity Index (SHDI) can be expressed by the following formula:

$$SHDI = -\sum_{i=1}^{N} p_i * \ln p_i \qquad (1)$$

where p is the proportion of individuals or objects in a category - the relative abundance of each type or the proportion of individuals of a given type to the total number of individuals in the category - and N is the number of categories. The index ranges from 0 to infinity, with 0 representing the case in which the analysed area is perfectly homogeneous in terms of population, while higher values represent a higher heterogeneity as the figures increase [10].

Simpson Diversity Index D (SIDI) is another index of Diversity - different from the Duncan and Duncan one [13] – and it is used for measuring the variation in categorical data.

$$SIDI = 1 - \sum_{i=1}^{N} p_i^2 \qquad (2)$$

p is the proportion of individuals or objects in a category and N is the number of categories. A perfectly homogeneous population would have a diversity index score of 0, while a perfectly heterogeneous population would have a diversity index score of 1 - assuming infinite categories with equal representation in each category.

As the number of categories increases, the maximum value of the diversity index score also increases.

The two indices are focused on different aspects concerning diversity. Shannon index particularly aims at highlighting the richness component, while Simpson index is more concerned with evenness and the analysis of dominant types [10]. According to some authors [14] Shannon diversity index is more sensitive to changes occurring in the importance of rarest elements, while Simpson index seems to respond to changes in the proportional abundance of most common community [10].

The indices are adapted in this study to the different ethnic groups and their weights in terms of resident individual, focusing the attention on different spatial levels.

2.2 Spatial Autocorrelation Techniques

Geographical objects are generally described by means of two different information categories: spatial location and related properties. In data analysis there is a huge literature concerning methods which separately compute attributes from spatial components.

The most interesting property of spatial autocorrelation is the capability to analyze at the same time locational and attribute information [3]. Consequently, spatial autocorrelation can be considered as a very effective technique in analyzing spatial distribution of objects assessing at the same time the degree of influence of neighbour objects. This concept is well synthesized in the first law of geography defined by

Waldo Tobler "All Things Are Related, But Nearby Things Are More Related Than Distant Things" [4]. Adopting Goodchild [3] approach, Lee and Wong [5] defined spatial autocorrelation as follows:

$$SAC = \frac{\sum_{i=1}^{n}\sum_{j=1}^{n} c_{ij} w_{ij}}{\sum_{i=1}^{n}\sum_{j=1}^{n} w_{ij}} \quad (3)$$

Where:

1. n is the number of objects;
2. i and j are two objects;
3. x_i is the value of object i attribute;
4. c_{ij} is a degree of similarity of attributes i and j;
5. w_{ij} is a degree of similarity of location i and j;

if $c_{ij}=(x_i-x_j)^2$ Geary C Ratio [7] can be defined as follows:

$$c = \frac{(N-1)(\sum_i \sum_j w_{ij}(x_i - x_j)^2)}{2(\sum_i \sum_j w_{ij})\sum_i (x_i - \overline{x})^2} \quad (4)$$

if $c_{ij} = (x_i - \overline{x})(x_j - \overline{x})$ Moran Index I [8] can be defined as follows:

$$I = \frac{N \sum_i \sum_j w_{ij}(x_i - \overline{x})(x_j - \overline{x})}{(\sum_i \sum_j w_{ij})\sum_i (x_i - \overline{x})^2} \quad (5)$$

These two indices are very similar, mainly differing in the cross-product term in the numerator, which in Moran is calculated using deviations from the mean, while in Geary is directly computed.

These two indices are global indicators of spatial autocorrelation. They provide an indication about the presence of autocorrelation. The precise location of elevated values of autocorrelation is provided by Local Indicators of Spatial Association. One of the most adopted indices of local autocorrelation is LISA-Local Indicator of Spatial Association [1] [2] considered as a local Moran index. The sum of all local indices is proportional to the value of Moran one:

$\sum_i I_i = \gamma * I$. The index is calculated as follows:

$$I_i = \frac{(X_i - \overline{X})}{S_X^2} \sum_{j=1}^{N} \left(w_{ij}(X_j - \overline{X}) \right) \quad (6)$$

It allows, for each location, to assess the similarity of each observation with its surrounding elements. Five scenarios emerge:

- locations with high values of the phenomenon and high level of similarity with its surroundings (high-high), defined as *hot spots*;
- locations with low values of the phenomenon and high level of similarity with its surroundings (low-low), defined as *cold spots*;
- locations with high values of the phenomenon and low level of similarity with its surroundings (high-low), defined as potentially *spatial outliers*;
- locations with low values of the phenomenon and low level of similarity with its surroundings (low-high), defined as potentially *spatial outliers*;
- locations completely lacking of significant autocorrelations.

LISA (Local Indicator of Spatial Association) provides an effective measure of the degree of relative spatial association between each territorial unit and its surrounding elements, allowing highlighting type of spatial concentration for the detection of spatial clusters.

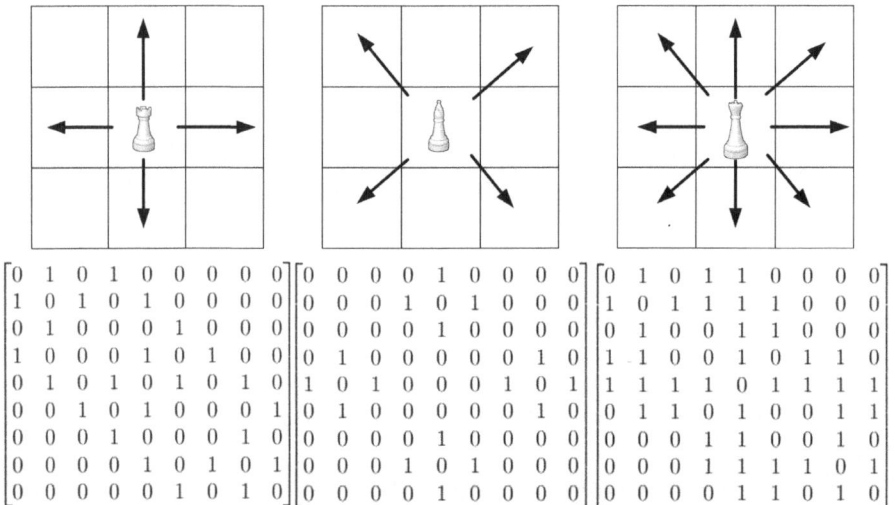

Fig. 1. Spatial weights matrix and the metaphor chess game

In equations 3, 4, 5, 6 the only term not well formalized is w_{ij} related to neighbourhood property. The most adopted approach in formalizing this property is spatial weights matrix, w_{ij} are elements of a matrix considered as spatial weights, equal to 1 if i and j are neighbours equal to 0 in the case of self-neighbour or if i and j are not neighbours.

This approach is based on the concept of contiguity, where elements share a common border of non-zero length. It is important to give a more detailed definition of contiguity and more particularly what does a border of non-zero length exactly mean.

Adopting chess game metaphor [9], contiguity can be considered as allowed by paths of *rook*, *bishop* and *queen* (figure 1). It is possible to adopt also the second or higher order of contiguity considering the crown of elements contiguous to first order neighbours. It is also possible to consider higher order of contiguity including the lowest order.

3 Italian Spatial Distribution of Foreign Immigration

In recent years Italy has tested a strong intensification of immigration. The biggest cities and metropolitan areas are major attraction centres for immigrants and even if less than in the past this phenomenon is mainly concentrated in Northern part of Italy. Despite today immigration phenomenon is seen as a crucial problem for the large number of immigrants who daily arrive on the Southern coast of the country, the dimension of the phenomenon is rather concerning movements from Italian regions where the economic crisis is more evident to those where there is more job offer.

In the past, migratory phenomenon in Italy was mainly characterized by abandonment of the nation to reach North and South America. Subsequently, migration originated in Southern Italy and it was mainly directed to Switzerland and Germany. Considering internal migration, an exodus directed to big industrial centres of Northern Italy occurred after the Second World War, when a lot of people left Southern Italy countryside to reach the big industrial cities of the North in search of a job.

Population growth observed recently in Italy is strongly determined by the foreign component. For this reason this paper is completely concentrated on this aspect. Data concerning foreign residents at municipality scale have been analyzed for years 2003 and 2009 using official data of Italian Institute of Statistics (ISTAT).

We considered 2003 because on 30 July 2002 Italian parliament approved a new law concerning immigration discipline and rules on conditions of foreigners. Consequently, a strong increase of residence permits has been registered in 2002 due to the regularization of foreign people having working permits. In fact, in 2003, working permits increased of about 355 thousand units for men and approximately 295 thousand units for women. While in subsequent years, the increase of residence permits was almost exclusively due to family reunification. We took into account 2009 because it is the last year of available data on foreign origin at municipality scale.

The analysis over data concerning foreign migration in Italy provided interesting results in terms of the pattern drawn by their spatial distribution in Italian municipalities. As a general note, we can recall that Italy experimented a dramatic increase in percentage of foreign population living in its territory, since the value more than doubled in less than a decade, from 3.5% at the beginning of the century to the current 7% of foreign population over the total. This value of course represents the average and local variations can be noticed in all administrative unit levels, these being regions, provinces and municipalities, as those analyzed here.

If we observe general data and compare the two years considered - 2003 and 2009 - we can notice that foreign residents more than doubled, from fairly 2 million people in 2003 to 4.2 million people in 2009. Such an increase of more than 2.2 million people was led by few national groups, since ten nationalities count for 73% of total immigrants in Italy in 2009 and a total of 20 nationalities explain the most of immigration process covering 88% of foreign residents.

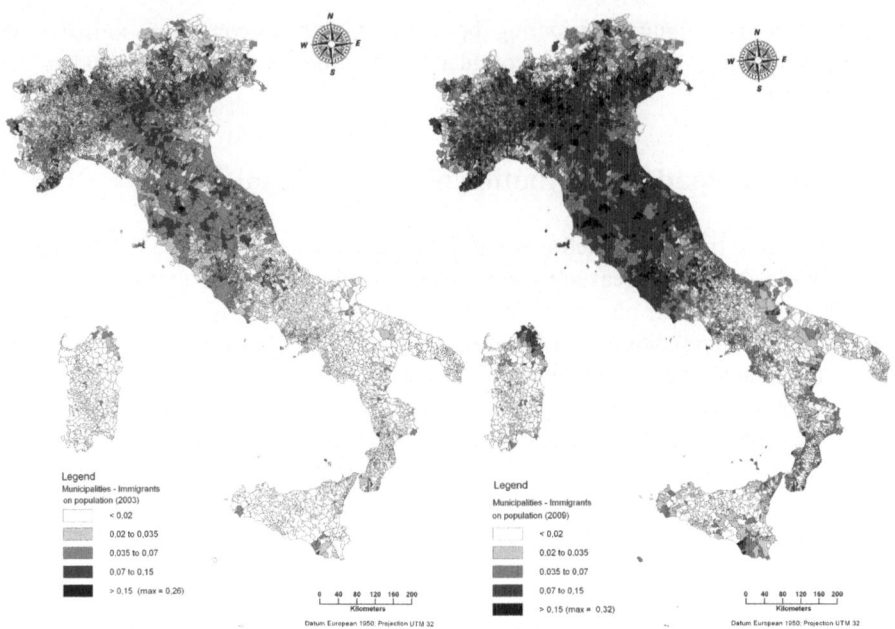

Fig. 2. Foreign residents in Italy. Percentage of foreign residents over population in Italy in 2003 and 2009. Source: our elaboration from Italian Institute of Statistics - Istat

In the top ten positions, in terms of absolute numbers – as well as absolute increase – we can find countries like Romania, Albania, Morocco, People's Republic of China, Ukraine, Philippine Islands, India, Poland, Moldova and Tunisia. In particular Romania, Albania, Morocco and China lead the way. Especially Romanians represent the first most numerous group which has experienced the most dramatic increase. A partial reason of such an increase can be possibly found in Romania accessing EU in 2004, which allowed an easier movement of people between the two States and therefore people to relocate to Italy.

Data show also that 'historical' groups as Albanians continue to choose Italy as a destination for migration, as well as other groups coming from North African countries, like Morocco, Tunisia, Egypt and Senegal. A most recent phenomenon is related to the immigration from South East Asia, particularly from China, India and Bangladesh, not to forget Philippine Islands, already 'settled' as a foreign group in Italy.

Italy represents a destination also for nationals from industrialized countries, as the rest of EU and USA. The number of people from these latter countries involved is not as high as those related to the countries already mentioned, but in any case they are interesting to understand some spatial patterns, as it will be more evident when observing local cases.

Table 1. Foreign residents in Italy. 2003 - 2009 comparison and absolute increases of single groups (absolute and percentage values). Source: our elaboration from Italian Institute of Statistics - Istat

Countries	Population 2003	Population 2009	Percentage increase	Absolute increase
Romania	177812	887763	399,27%	709951
Albania	270383	466684	72,60%	196301
Morocco	253362	431529	70,32%	178167
China	86738	188352	117,15%	101614
Ukraine	57971	174129	200,37%	116158
Philippine	72372	123584	70,76%	51212
India	0	105863		105863
Poland	40314	105608	161,96%	65294
Moldova	24645	105600	328,48%	80955
Tunisia	68630	103678	51,07%	35048
Macedonia	51208	92847	81,31%	41639
Peru	43009	87747	104,02%	44738
Ecuador	33506	85940	156,49%	52434
Egypt	40583	82064	102,21%	41481
Sri Lanka	39231	75343	92,05%	36112
Bangladesh	0	73965		73965
Senegal	46478	72618	56,24%	26140
Former-Yugoslavia	51708	57877	11,93%	6169
Nigeria	26383	48674	84,49%	22291
Total foreigners	1990159	4235059	112,80%	2244900
Total population	56890331	60320749	6,03%	3430418

3.1 Results and Comments with Shannon Diversity Index

Shannon and Simpson diversity indices have been calculated and applied to foreign immigration in Italian municipalities. In particular, they were tested for the presence of different nationalities and their figures.

The indices were computed over a subset of available data. In particular, SHDI and SIDI were calculated for years 2003 and 2009. Also, the indices considered both the whole dataset of foreign and Italian nationals in the territory and just the one of foreign nationals. It was thought that in the first case the indices could be useful to observe the presence of municipalities of high foreign immigration pressure on Italian nationals, while in the second case, if homogeneity occurred, it could be the subject for further investigations and it could highlight possible ghettoes and single-ethnic groups presence.

However the analysis and comments here are mainly focused on Shannon Diversity Index, as Sinpson's one seem a promising one but still providing results of difficult interpretation, a part from the fact that very high value occur, therefore showing a quite general and widespread co-presence of different nationals in same municipalities.

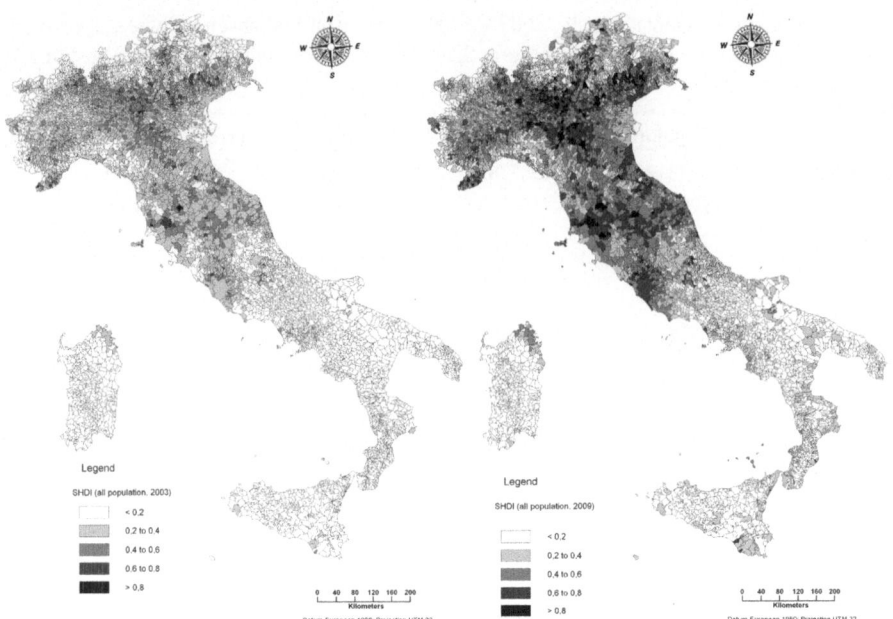

Fig. 3. Shannon Diversity Index considering the whole population 2003 (left) and 2009 (right)

The analysis of SHDI for the two years (2003 and 2009) and for the two dataset (considering for each year all national groups in one case and only foreign groups in the other one) is visible in figures 3 and 4.

We can notice a certain degree of matching with the picture drawn by spatial distribution of percentage of immigrants on overall population (figure 2). Maps could be easily overlapped and arising patterns appear quite similar.

A general note is that heterogeneity (i.e., higher values of SHDI) can be noticed in areas presenting higher percentage values of immigrants. Also, heterogeneity tends to increase as the weight of immigrations increases: comparing 2003 and 2009 figures and indices' values, we can notice higher values of heterogeneity in the same municipalities presenting higher percentage of immigrants over the entire population, and therefore highlighting also local variations in cases where immigration phenomenon is not so evident – as in Islands and Southern regions' locations.

SHDI applied to foreign residents alone offers a picture not 'affected' by the percentage values of immigrations over the entire population and therefore it allows observing the internal variation of foreign nationals. SHDI applied to 2003 data on foreign people alone confirms the 'image' of a concentration of people in urban areas in all regions, as well as in Italian industrial and agricultural districts. This is true for Northern regions, as well as for central and southern ones. Milan, Turin, Verona, Venice can easily be noticed, as well as central Italian cities as Florence and Rome, not to forget the 'linear' set of Emilia Romagna cities (Parma, Modena, Reggio Emilia, Bologna, etc.).

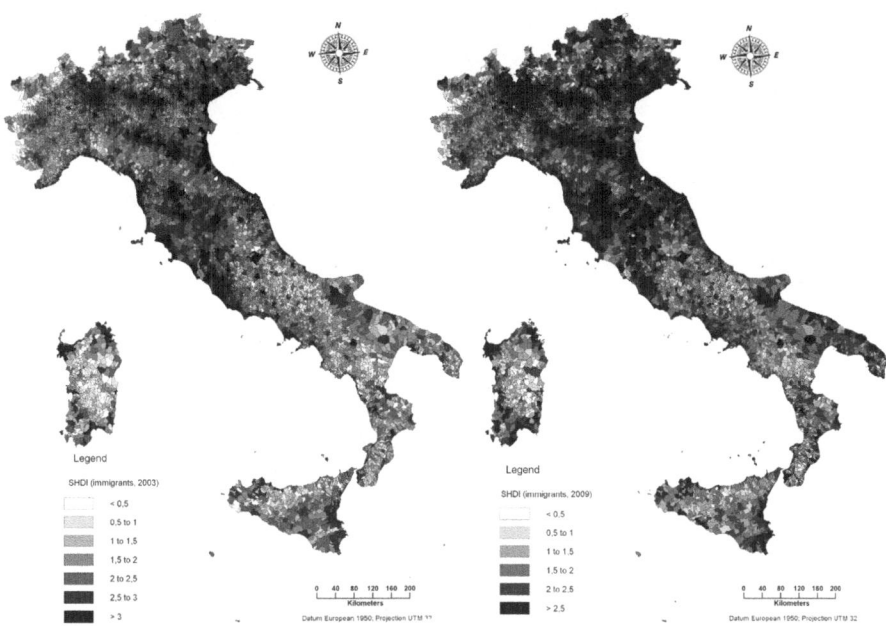

Fig. 4. Shannon Diversity Index considering only foreigners at 2003 (left) and 2009 (right)

Southern Italian cities also seem showing a high level of heterogeneity and therefore a high level of nationalities represented. This is true for cities like Bari, Foggia and Taranto in Puglia, Naples and Salerno in Campania, Sassari and Cagliari in Sardinia, just to mention a few of them. 2009 data analysis confirms what said for 2003, with a higher level of heterogeneity also in neighbouring municipalities with respect to those observed in 2003. This could imply a level of 'suburbanization' of immigrations, with an increase of the presence of immigrants in municipalities once less affected by the phenomenon, as well as an increase in the diversity of nationalities located there.

3.2 Results and comments with Local Indicators of Spatial Association Index

The application of LISA allows detecting clusters in a spatial distribution at local level; in this case the analysis was computed considering foreign population in Italian municipalities. The analysis was applied on some of the national groups.

Considering data at 2003, the Chinese group shows clusters particularly in some major metropolitan urban areas, as Milan and Rome and their hinterlands. Also, the phenomenon is interesting in its presence in Tuscany, between the provinces of Florence and Prato, as well as in the area crossing the three regions of Veneto, Lombardy and Emilia Romagna. In this latter region, there is a cluster of municipalities connecting urban areas of Parma, Reggio Emilia and Modena. In Veneto a cluster can be noticed around the city of Venice in the municipalities located in the mainland.

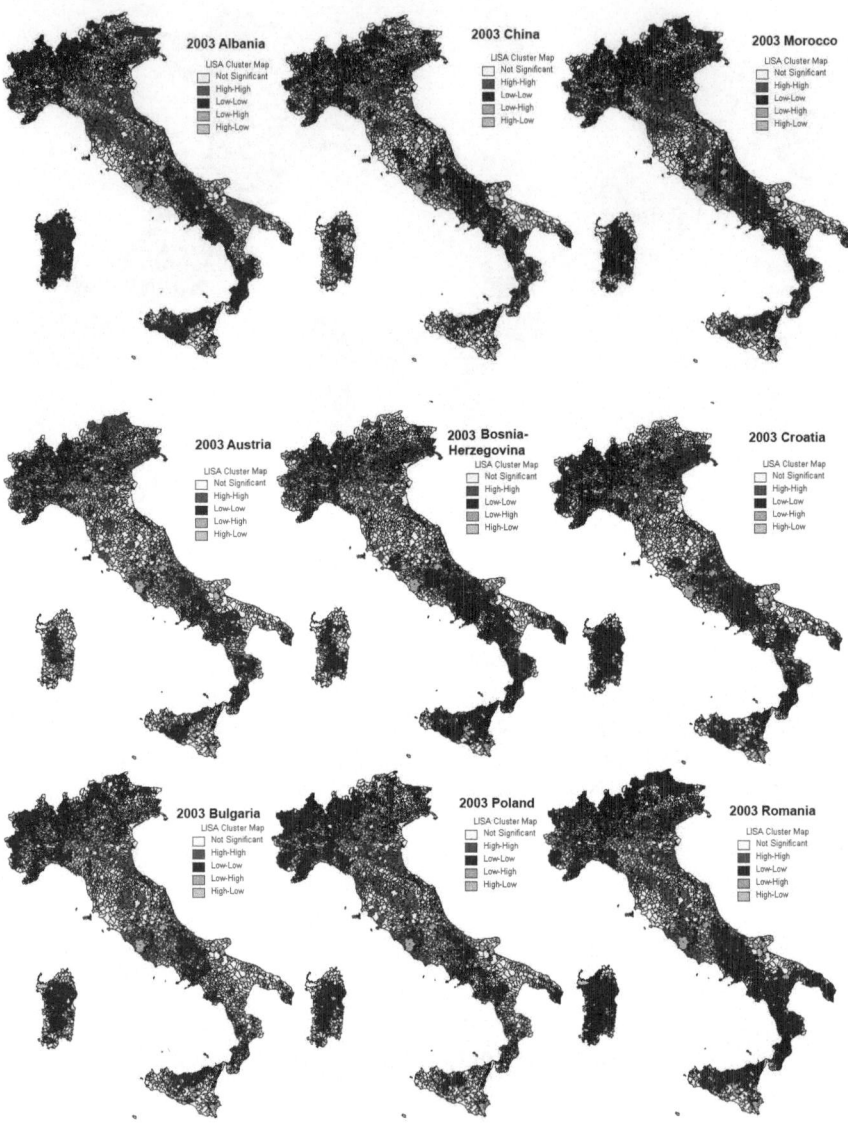

Fig. 5. LISA cluster map 2003. (our elaboration with GeoDa on ISTAT data)

A preference for location clustered in Northern – Central Italian regions seems to be noticed also in case of people from Morocco. We can fairly notice a single cluster of municipalities whose vertices can be observed in the areas around the cities of Venice and Milan and covering the area including Veneto, Lombardy and most of Emilia Romagna regions. Clusters can be also observed in northern Tuscany and in Umbria regions, as well as on the Adriatic Coast in Marche region.

'New' migrants from Poland and Bulgaria seem to prefer central Italian locations, being mainly clustered in Emilia Romagna, Tuscany and Lazio regions, here also preferring some urban and periurban locations (i.e., Rome and its hinterland). Former Yugoslavian states, as Yugoslavia (in 2003 grouping Serbia and Kosovo), Bosnia-Herzegovina and Croatia are present with groups mainly clustered in North-eastern Italy, starting from the North-eastern border between Italy and Slovenia and spreading westwards to Verona area and also (Croatians) clustering in the city of Milan.

Tunisian people are mainly concentrated in Emilia Romagna region, as well as in Milan area and in some Southern Italian locations, such as Naples area, Puglia region and Sicily (this latter possibly motivated by its geographical proximity to Tunisia and the integration of migrants in activities concentrated on fishery and agriculture).

With reference to the groups from industrialized countries, the algorithm used seems to be useful in highlighting some hot spots that can be quite easily explained.

Neighbouring Austrian people cluster in Alto Adige province in Northern Italy, located at the State Border with Austria, and presenting a strong Austrian-speaking community. Also, their presence is noticed in part of Friuli Venezia Giulia Region, here also close to the State Border, as well as in municipalities along the coastline. Such areas are renowned as tourist locations for Austrian people and, in some cases, once belonged to Austro-Hungarian Empire. Their presence can also be noticed in municipalities neighbouring important urban areas, as Venice, Verona, Milan, Florence and Rome.

Switzerland, UK and USA show also interesting settling pattern in Italian municipalities. For Swiss people some similar comments as for Austrian can be drawn, as the fact that clusters can be found in Lombardy and Alto Adige areas close to the State Border. Apart for that characteristic, all these groups tend to prefer also urban areas as Milan, Venice, Rome and Florence and particularly Tuscany as a region (among all, the "Chiantishire"), this latter therefore not to be considered just as a tourist destination, but also as a relocation site for these nationals. Some interesting patterns can be also noticed concerning USA people, as some clusters can be noticed close to important military installations, as Aviano and Sigonella Air Force Bases, respectively located in Friuli Venezia Giulia Region, close to Pordenone and in Sicily, in Catania province.

Some general conclusions can be drawn considering the different nationals. Foreign immigration appears as a phenomenon mainly characterizing Northern and Central Italy, in quantitative terms and with reference of its spatial distribution. Southern Italy and Islands appear less characterized by immigration, although, of course, important figures can be observed here (one for all, the presence of Tunisians in Sicily and Albanian people both in Sicily and Puglia).

Large urban areas tend to attract immigrants. This is visible both considering cities (i.e., Milan) and municipalities neighbouring urban areas (i.e., municipalities surrounding cities like Rome, Naples, Florence, Venice, Verona, etc.). Industrialized areas attract also immigrants, both in terms of 'traditionally' industrialized areas and also in the small-medium enterprises districts. This is visible particularly in North-eastern, North-west (Milan area) and Northern-Central Italy (Emilia Romagna and Tuscany Regions).

3.2.1 LISA 2009

The analysis of 2009 data can provide us with some information concerning the variation occurred in the years and new patterns of settlements.

We highlight here some major changes in some of the groups examined.

As a general remark, all groups analysed seem to maintain their spatial organization in the years, although clusters generally enlarge and new locations appear.

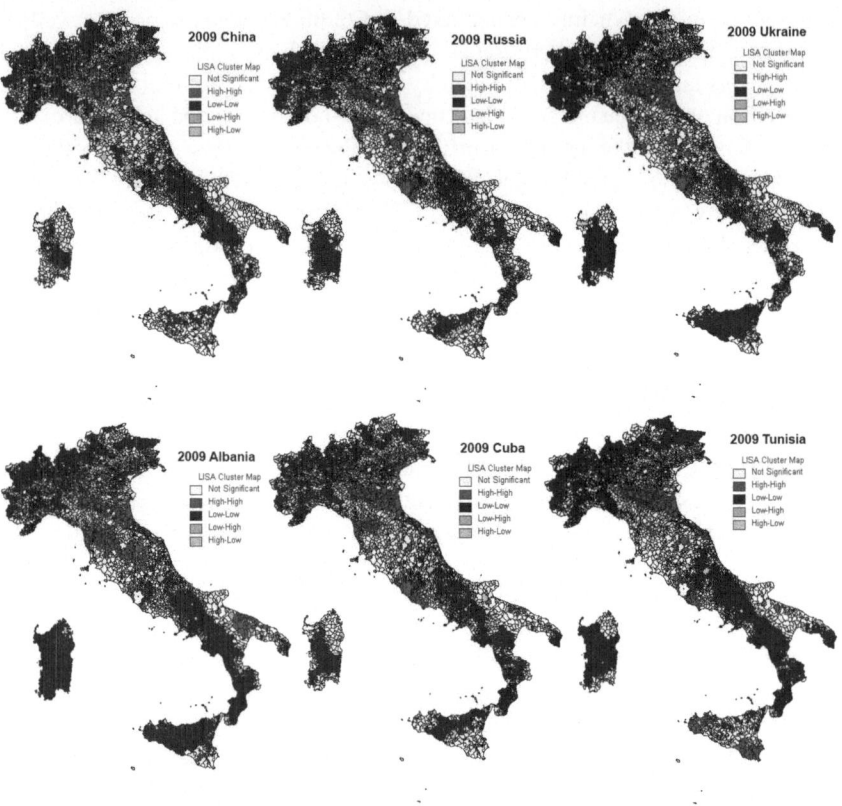

Fig. 6. LISA cluster map 2009. (our elaboration with GeoDa on ISTAT data)

In particular Bulgarian people reinforce their presence around the Italian Capital Rome, while a new cluster appears in Puglia region, centred in the city of Foggia and in its neighbouring municipalities. This seems to be due to activities of Bulgarian people in agricultural activities. Also news report of irregular immigration from Bulgaria being increasing in the area. Polish people confirm the same immigration pattern of 2003, although enlarging their clusters, reinforcing their presence in Lazio (especially Rome), and 'heading south', locating in Puglia region in Foggia area, as well as Bulgarian people, and also in Southern Sicily.

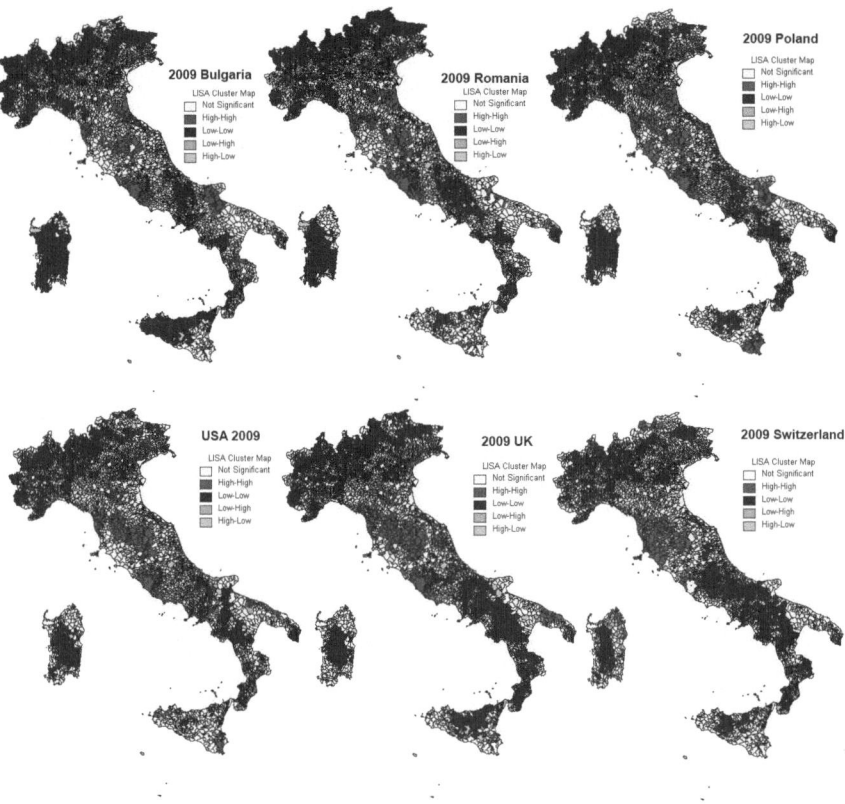

Fig. 7. LISA cluster map 2009. (our elaboration with GeoDa on ISTAT data)

Industrialized countries, UK and US confirm their spatial distribution, with new locations in Puglia and Lazio regions, particularly in the municipalities of Rome and neighbouring ones, and British people, in particular, tend also to move eastwards from the traditional location in Tuscany ("Chiantishire") to locate also in Umbria and Marche Regions. Similar patterns as those noticed can be detected for Switzerland. Here also people tend to settle in Tuscany and in neighbouring regions (Umbria and Marche) but also to experiment a 'spatial diffusion' towards the North-eastern coast of Sardinia.

Here also some general conclusions can be drawn. In general terms the considerations done with reference to 2003 can be confirmed, particularly in terms of North-South differences and polarization of migrants in (big) urban and industrialized areas. However, some interesting patterns seem to arise, as we can spot a trend of 'internal movements' of migrants, since clusters seem to grow in their dimensions and also to appear, with reference to some groups, in areas once not experimenting the phenomenon, in particular some other groups, part from Albanian and Tunisian people, settling in Southern Italy (i.e., Bulgarian and Polish people in Puglia, Campania and Calabria Regions).

4 Conclusions

The research carried out in this paper relied on the application of some spatial statistical techniques to immigration phenomena, focusing on the Italian case.

In particular, we tested some well known spatial analytical techniques for the immigration phenomena, thus trying to couple some quantitative methods with qualitative analysis and interpretation of the observed phenomena. We applied LISA algorithms and Entropy based indices to foreign residents in Italy, related to municipalities and diversified in terms of their nationality of origin.

As we run LISA methods on some of the major groups currently living in Italy, we were able to detect clusters in their spatial distribution, not limiting us to examin immigration in terms of their percentage weight, but also, and more importantly, in terms of spatial aggregation of foreign groups over the territory.

Entropy-based, diversity indices suggested the characters of heterogeneity versus homogeneity of the phenomenon. We could therefore observe some interesting clusters and trends, valid for immigration in wide terms and also in terms of national differences. In particular, we discovered some of the major characters of immigration in the two years, 2003 and 2009: in general terms it is a phenomenon especially characterizing Northern regions, cities – metropolitan areas and industrial districts and areas. Different nationals show differences in migration and settlement patterns. These can be explained by means of migration chains, geographical proximity and economic specialization in the country of origin that is proposed as expertise in the country/site of destination. Changes occurred in less than a decade, demonstrating a trend in internal movements inside Italy, visible in terms of enlargement of single nation clusters and creation of new ones, different from the original ones. It meant also suburbanization, since presence of foreign people in suburban municipalities around major cities and metropolitan areas increased. Also, Southern and insular Italy became areas of settlement for some migrant groups, if not with very large numbers, with interesting composition. In terms of heterogeneity, we noticed that the 'weight' – in percentage terms – of foreign population is in most of the cases characterized by an increase in number of nations in single municipalities, as well as of people. This means that in general, at least with reference to single municipalities, we do not observe ghettoization processes with dominance of single foreign groups – here we cannot say nothing related to dynamics *within* a municipality.

These same conclusions, of course, are compatible with other levels of data analysis and knowledge of the phenomenon. However, we must stress that the methods applied here were quite precise in highlighting some of the characters of immigration phenomenon that could be just imagined or hypothesized by means of other more traditional methods. It must be also stressed that this method can be interesting coupled with more qualitative local analysis. Indeed, it proved to be interesting in highlighting clusters and therefore local cases of some interest, therefore helping scholars to more precisely aim and deepen their research.

References

1. Anselin, L.: Spatial Econometrics: Methods and Models. Kluwer Academic, Boston (1988)
2. Anselin, L.: Local Indicators of Spatial Association-LISA. Geographical Analysis 27, 93–115 (1995)
3. Goodchild, M.F.: Spatial Autocorrelation. Catmog 47. Geo Books, Norwich (1986)
4. Tobler, W.R.: A computer movie simulating urban growth in the Detroit region. Economic Geography 46(2), 234–240 (1970)
5. Lee, J., Wong, D.W.S.: Statistical analysis with ArcView GIS, p. 192. John Wiley and Sons, New York (2001)
6. Scardaccione, G., Scorza, F., Casas, G.L., Murgante, B.: Spatial Autocorrelation Analysis for the Evaluation of Migration Flows: The Italian Case. In: Taniar, D., Gervasi, O., Murgante, B., Pardede, E., Apduhan, B.O. (eds.) ICCSA 2010. LNCS, vol. 6016, pp. 62–76. Springer, Heidelberg (2010), doi:10.1007/978-3-642-12156-2_5
7. Geary, R.: The contiguity ratio and statistical mapping. The Incorporated Statistician (5) (1954)
8. Moran, P.: The interpretation of statistical maps. Journal of the Royal Statistical Society (10) (1948)
9. O'Sullivan, D., Unwin, D.: Geographic Information Analysis. John Wiley & Sons (2002)
10. Nagendra, H.: Opposite trends in response for the Shannon and Simpson indices of landscape diversity. Applied Geography 22, 175–186 (2002)
11. Weaver, W., Shannon, C.E.: The Mathematical Theory of Communication. University of Illinois, Urbana (1949)
12. Shannon, C.E.: A mathematical theory of communication. Bell System Technical Journal 27, 379–423, 623–656 (1948)
13. Simpson, E.H.: Measurement of diversity. Nature 163, 688 (1949)
14. Peet, R.K.: The measurement of species diversity. Annual Review of Ecology and Systematics 5, 285–307 (1974)
15. Elden, G., Kayadjianian, M., Vidal, C.: Quantifying Landscape Structures: spatial and temporal dimensions, ch. 2. From Land Cover to Landscape Diversity in the European Union (2000), http://ec.europa.eu/agriculture/publi/landscape/ch2.htm
16. Gibbs, J.P., Martin, W.T.: Urbanization, technology and the division of labor. American Sociological Review 27, 667–677 (1962)
17. Borruso, G.: Geographical Analysis of Foreign Immigration and Spatial Patterns in Urban Areas: Density Estimation and Spatial Segregation. In: Gervasi, O., Murgante, B., Laganà, A., Taniar, D., Mun, Y., Gavrilova, M.L. (eds.) ICCSA 2008, Part I. LNCS, vol. 5072, pp. 459–474. Springer, Heidelberg (2008)
18. Borruso, G.: Geographical Analysis of Foreign Immigration and Spatial Patterns in Urban Areas: Density Estimation, Spatial Segregation and Diversity Analysis. In: Gavrilova, M.L., Tan, C.J.K. (eds.) Transactions on Computational Science VI. LNCS, vol. 5730, pp. 301–323. Springer, Heidelberg (2009)

From Urban Labs in the City to Urban Labs on the Web

Viviana Lanza, Lucia Tilio, Antonello Azzato, Giuseppe B. Las Casas,
and Piergiuseppe Pontrandolfi

Università degli Studi della Basilicata, Viale dell'Ateneo Lucano 10, 85100, Potenza, Italy
{lanza.viviana,lucti182}@gmail.com

Abstract. This paper reports an experience of planning participation, lead during 2010, with the objective to adopt traditional and innovative forms of participation, in the context of planning process simulation. The experience aimed at enhancing confidence in spatial planning processes, in a context where participation is not yet a custom. Some months later, a new attempt has been lead, to enlarge the set of adopted tools and test some electronic tools for e-valuation, asking citizens to involve other citizens, in order to enlarge the community.

Keywords: planning, e-participation, e-decision tools.

1 Introduction

The research presented in this paper is related to an experience of planning participation, lead in 2010 in the city of Potenza (in the south of Italy), where participation has not yet been strongly applied, with the aim of test both traditional and innovative tools for participation; in particular, here is presented the experimental and innovative component (for Potenza reality) of participation process, concerning the role of Internet and electronic tools in participation.

Considering that, according to [1], *Internet is the biggest public space that humanity has known [...], a place where all people can express themselves, gain knowledge, get ideas and not only information, deny, dialogue, participate to common life and so, build a different world where everyone could feel equally citizens*,], therefore it can be used to involve citizens in participation processes.

As affirmed in [2], e-tools seem to be attractive to promote participatory practices among citizens, because they are becoming more and more familiar to them, and help to overpass the traditional difficulties related to citizens' involvement. E-tools help to supply correct information, improve communication quality and wide interaction, so citizens who finally feel involved in the process, are stimulated to act and participate, and they succeed to propose and discuss, contributing to reinforce the bottom-up approach [3]. Several studies in literature highlight how the way in which information is presented influences the real perceived information and therefore the participation rank [4]; a pilot study in 2004 [5], for instance, shows some differences between traditional public meetings and other ones with the use of advanced technological tools: the latter had a higher popularity level, considering, for instance, that knowledge and learning benefit from, between several other aspects, a large use of visual systems as maps and imagines [6].

Nowadays, as observed by Thomas and Streib already in 2003 [7], citizens visit more and more websites, where they suppose to retrieve all needed information and where they try to interact with Public Administration. Moreover, several new tools are today available, contributing to enhance interaction: communities and social networks are the modern spaces where people can socialize, share contents and experiences, contribute in developing projects, and so on, and every day several new services are available, not only on the web, but also for smartphones, as the apps: exchange of information and social interaction are facilitated, and different types of groups of users interested in a particular field can born [8].

The *www-scenario* is generating changes in the way people communicate, transforming citizens in e-citizens, and determining also a different approach in the relationship between citizens and politicians and planners, and in the way in which they explore future. Public Administrations started using new technologies more and more often, with the aim of establishing a direct and transparent relationship with citizens, and building consensus in processes that are more and more democratic, due to citizens' involvement [9].

The paper is organized as follows: in the next section is presented a study case with an overview concerning e-participation approach and adopted tools. Some remarks about impacts of adopted tools on e-citizens are presented in third section, and the fourth introduces a second phase of electronic participation, with a stronger development of e-decision tools. Results analysis and future perspectives close the paper.

2   Electronic Participation Approach and Adopted Tools

In the light of the preliminary remarks, and with the idea that participation is something of objectively good for all community (mentioning Arnstein thinking [10]), in the city of Potenza a participative planning process has been simulated during 2010, testing both traditional and electronic participative models.

The entire process has been developed in four workshops in eight sub-areas of the town, referring to the approach of Consensus Solution, as one of several tools, developed in the United States with the goal of finding consensus solutions between stakeholders in conflict situations, through progressive reductions of disagreements. This approach recognizes that not all decisions can satisfy every stakeholder, and affirms that the real purpose of participative processes is not to find agreement, but to find a solution that stakeholders consider acceptable, because they feel that it has taken into account all points of view, in a fair and transparent process. In this context, therefore, an expert and neutral facilitator is fundamental in order to manage workshops; facilitators are super-partes actors, they can encourage participants to dialog and discuss, and they guarantee respect for everyone's opinions.

2.1 Electronic Participation Tools

Some e-participation tools have been developed to enrich traditional participative approach; analyzing the wide possibilities that the World Wide Web supplies, decision fell down on web 2.0 tools, which are becoming more and more familiar to users, with the aim of capitalizing on this familiarity as the key-element to stimulate interaction and participation. Therefore, we adopted the e-participation kit of Lanza, Prosperi [11] synthetically shown in next figure.

The e-participation kit is composed by four main kinds of tools covering the main kinds of interaction that can be achieved. The whole kit considers as the reference point the social tools, related to sharing, mapping and decision tools.

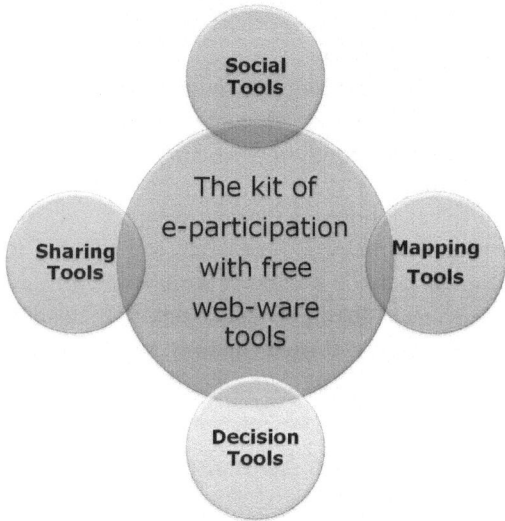

Fig. 1. The kit of e-participation adopted in the Urban Labs (our elaboration)

2.1.1 Social Tools

Social tools, as called in the e-participation kit, include all those tools that at now are generally indicated as social networks. Social tools, in fact, are developed to connect people with the same interests, creating some communities, where users have possibility to exchange opinions, discuss, comment, and keep informed, in synchronous and asynchronous way. In the World Wide Web, there are several services and platforms, from which we adopted two kinds of tools: first one is a Ning platform, and the second one is Facebook.

Ning is a very complete platform, born with the aim to build communities, including a blog, a personal page for each user, a chat, a mail service, some multimedia tools and so on. The account, available at www.lup-lisut.ning.com, has been created with the aim of keeping on activities of physical labs. The created community is composed by coordinators of labs, citizens who participated to physical workshops, and other interested citizens. The overall activities on Ning platform were not really high, and the number of reached users was low[1].

Fig. 2. The adopted social tools

Considering these data as a symptom of unsuccessful of Ning platform, a Facebook account was created, with the aim of enlarging community, taking advantage of the Facebook high popularity. Before to understand users difficulties with Ning platform, Facebook was not taken into account, supposing it could create confusion about lab mission, and retaining a threatening the possibility to reach rapidly a great community, that, in the testing phase, we were not able to manage. Facebook account (lup lisut) today counts about eight hundreds users and it has been used to publicize events and to enlarge community, maintaining ning platform as the reference point in the web.

The decision to "be on facebook" comes from the observation that it's one of the most popular social network, and in Italy, as it is the first social network available in Italian language, it connects a very high number of users, strongly higher than those of twitter, for instance. Considering data in [12], Facebook reaches about 16 millions of Italian users, while for twitter there are about 1,5 millions.

---

[1] At now, ning is no more a free platform, and for this reason, the Urban Labs promoter bought an account, in order to storage all materials retrieved during activities; the access to the platform is always free, and users can continue to register and participate for free, but unfortunately, even if the account is active, not all materials have been made available again.

2.1.2 Sharing Tools

Tools developed with the purpose of sharing information, as documents, images, other multimedia, ..., are grouped in this category. They are adopted in order to make transparent the process through spread diffusion of produced documents. We used several platforms and services, deepen described in the next list:

- Boxnet provides a solution to manage several kinds of documents online, sharing with other users simply by a link. Boxnet (luplisut) is the storage platform of urban labs, where each kind of documents is stored and a link is supplied and diffused to the relative users. For instance, some documents have been shared only among coordinators, other with citizens and so on.
- YouTube is the most known platform to upload and share videos.

Fig. 3. The adopted sharing tools

- Vimeo is another platform, less known and diffused, to upload and share videos. The exigency to have both, YouTube and Vimeo (account luplisut in both cases), depends on the service characteristics. In particular, YouTube allows uploading video up to 15 minutes, while in some cases, we need to upload longer videos, and we did this with Vimeo account.
- Flickr is a platform devoted to photosharing. It allows uploading freely a certain amount of image per week. On Flickr (account luplisut) we uploaded the first images of urban labs organization. In a second phase, we preferred to use the Ning platform also to share photo, taken during workshops.

- Slideshare is a web platform devoted to diffusion of slides and other similar stuff, where are stored slides showed during workshops (account luplisut).

2.1.3 Mapping Tools

Mapping tools are able to integrate Geographic Information System and public participatory tools (one of the latest innovation in the field of electronic participation). There is a large literature concerning the use of GIS in public and participatory planning (for instance, see Obermeyer [13]), also concerning the latest development, related to the diffusion of tools such as Google maps, making not expert users able to manage geographic information (consider the concept of Neogrography in Hudson-Smith et al. [14]) .

Between the several available tools, we chose to use Google Earth and Google Maps.

- Google earth is a free software enabling to surf on the earth surface, to define maps through markers, path, other shapes and content, and also to upload kml files. So, users can exchange their maps: the knowledge framework maps, created in a GIS environment, and used to support the planning process, have been exported by in kml format, and shared (through our box.com platform). Due to sensibility of some data, this kind of information has not been made available to everyone, but limited to citizens participating in workshops.

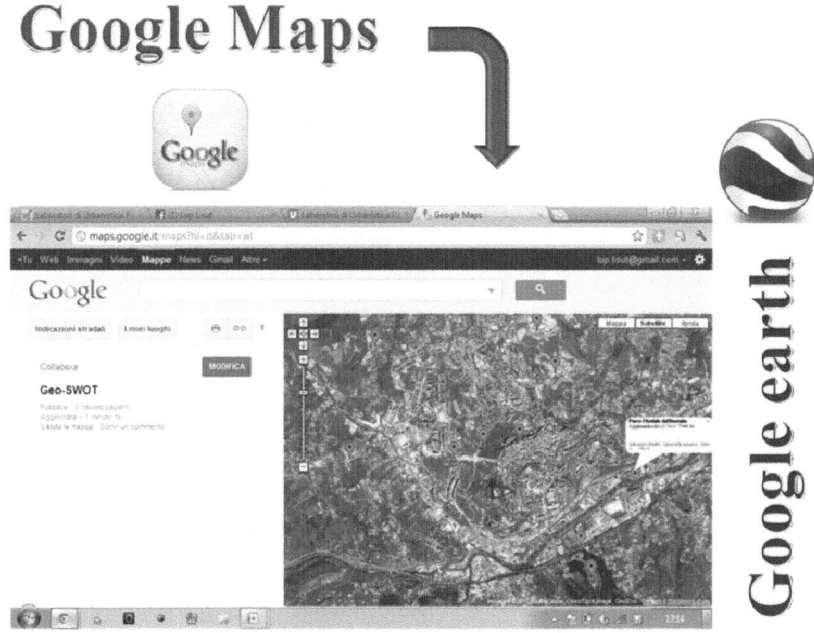

Fig. 4. The adopted mapping tools

- Google maps is a service offering powerful, user-friendly mapping technology. The GEO-swot analysis, built during workshops, has then been re-produced on Google maps: citizens have created some mashups [15], linking information, photos, and comments to a specific location.

2.1.4 Decision Tools

According to [11], in the decision tools category are grouped all those tools useful to register users preferences. Electronic vote is the best tool in order to take into account users' preferences, allowing considering preferences about several alternatives, but generally is difficult to translate these preferences during a decisional process; the main problem is related to vote legitimacy. Therefore, electronic vote make users feeling that their opinion is relevant.

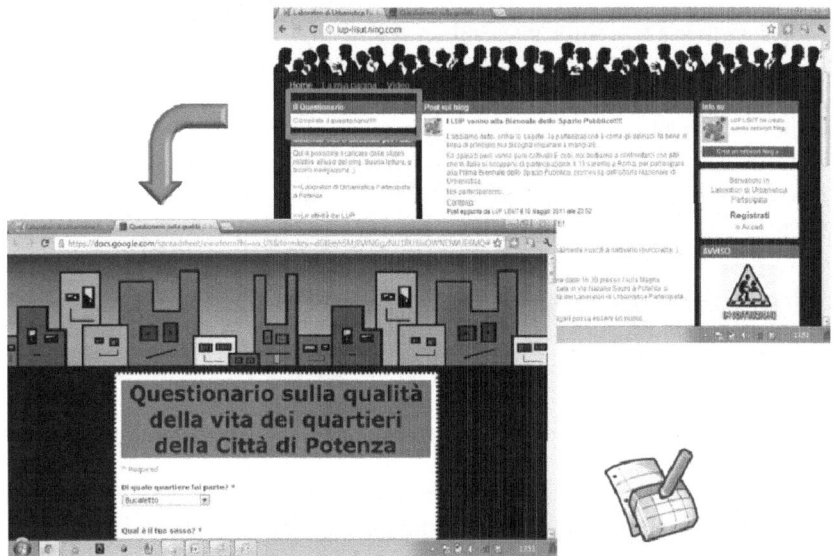

Fig. 5. The adopted decision tools

Another kind of tools in this category is the survey. Even if it is a softer tool than electronic vote, it supplies the possibility to register opinions and take into account users preferences in a not-direct way, using information from surveys. In Potenza case, considering that the planning process was just a simulation, a simple survey, powered by Google docs, has been created, diffused on the Ning platform and between citizens participating in workshop (https://docs.google.com/spreadsheet/viewform?hl=en_US&formkey=dGtEejh5MjBWNGgzNU1RU3lxOWNDWUE6MQ#gid=0), with the aim to collect more information about "life quality in the city", and using perceptions and opinions of city users. Citizens have been induced to reflect on several aspects by their own, and then results of surveys have been presented during workshops in order to collectively comment.

3 How Citizens Used Developed Tools

Experimentation assumptions rely on the idea that e-participation could be the innovative gymnasium to exercise the evaluation and the monitoring of citizens behavior concerning social daily life: research aims at producing some ideas about life quality, and ways to improve, to help decision makers, through citizens involvement in workshops, and supported by electronic tools. The experience cannot be considered completely successful. The traditional experience presented a certain degree of expected results achievement, with several citizens proposals, but at the same time, the electronic experience, considering that the web community activities were NON INTENSE, has been not so successful, not producing interesting results in terms of citizens contribution to planning process.

Therefore, it represents anyway an important step towards a new approach in participative planning, in particular concerning aspects related to knowledge framework sharing and aspects related to citizens' perception of re-thinking the town with their own perspective.

Main critical aspects, anyway, are related not on the adopted methodological approach, but mainly on the perception of the real utility of process: citizens felt that their contribute was not really taken into account during the decision process, and so their motivation was strongly weakened.

Concerning the electronic participation, some questions are yet open: which is the hitch that made difficult electronic participation? Are citizens, in this context, not yet ready to be e-citizens? Are there some technological aspects to consider, as the digital divide? A partial answer, in particular concerning mapping tools, could be found in Rocha words [16]: *when it is required to participate in public participatory GIS, the technological prowess and self initiative of the citizen makes a difference. And since the geospatial information is more difficult to manage, we will get less participation in the geospatial realm, with less skilled citizens.* This assumption reinforces the idea that, as Burby declares [17], *the key is for planners to work hard to both educate and learn from citizens,* and that citizens need to get use to participate. Our platform doesn't require any installation or other special requirements, but anyway the whole system has been percept by citizens as complex and requiring too much attention, and there was not enough time to make citizens confident with the available tools: the education role has not been completely carried out.

Anyway, for the success of a participatory process (whether electronic or not), it is believed helpful to a greater role to the "decision making" of citizens: it is for this reason that it has been carried out a new phase of participation based on decision tools, as described following.

4 From Strengthening Decision Tools Role towards e-Decision Tools

Considering some differences in the way citizens deal with the proposed e-tools, a preference was noticed referring to the survey on the life quality, disseminated through Google docs. Most of the citizens participating to workshops answered to the survey, but only few of them post comments or multimedia or used in other ways the available tools. Probably, citizens felt the need to express their opinions, even if in a closed way (questionnaire provided closes questions with a set of possible answers), and they choose to answer to the survey on the web. From this consideration, in 2011 *e-valuation* project is born, with the main aim of strengthen the role of decision tools in a participatory process, testing the kit of e-participation tools dedicated to decision and adopting an approach strictly similar to that one of survey. *e-valuation* project has been conceived as an iterative process, synthesized in the figure 6.

A new survey has been structured, as a decision tools, asking to e-citizens an evaluation of intervention proposals, defined by citizens during workshops in 2010. This *e-valuation* survey has been built and disseminated using Google Docs. As already tested with life quality survey, the tool is really simple to implement and use, and it enables to manage all the voting process, and to analyze answers.

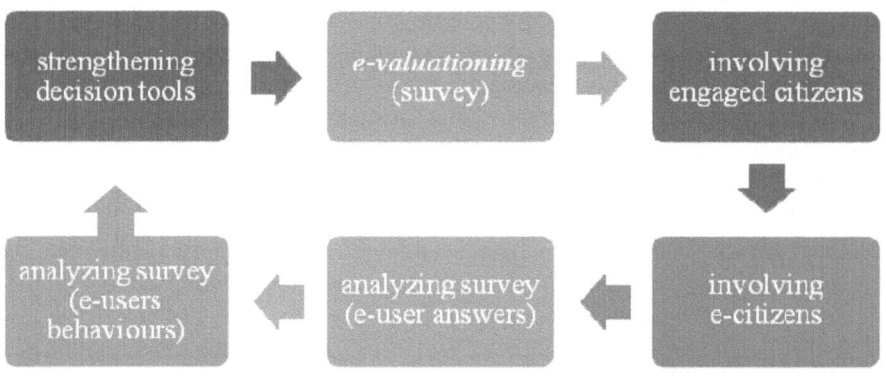

Fig. 6. e-decision process

The survey is divided into three sub-surveys, coherently with intervention groups: interventions have been categorized into three main groups, the strategic interventions, grouping those characterized by a strong investment and producing strong transformation on territory and strong effects on the town, the ordinary planning interventions, grouping those aiming at improving life quality through narrow transformations, and not-physical interventions, grouping mainly strategies and policies that not imply physical transformation. Some documents have been provided, containing details about interventions, as technical sketches and tables, rendering, information about objectives that the intervention can contribute to reach, stakeholders that can carry out the project (are they public, private or mixed?), financial resources availability and estimated costs, referred to sub-interventions to realize in order to implement the total intervention.

As during workshops, also in the case of *e-valuation* survey, citizens have been asked for order interventions according to priority, and considering that decisions about interventions depends on three main criteria: the strength of the relation between interventions and objectives, the costs and the priority identified by citizens. In order to avoid some possible influence, voters, both during workshops and in the electronic survey, have not been informed about scores that interventions obtained for the two first criteria. Priority assignment is mandatory, in order to avoid a partial survey completion.

Concerning dissemination phase, a chosen sample of citizens already involved in workshop has been contacted, in order to involve them in divulgation. Through emails, citizens have been asked for their availability to continue to help Urban Labs, in a new manner: help to involve other citizens! The *pass-the-word* approach has been the primer of a crowd sourcing [18] mechanism that we tried to start, since the users have voluntarily allowed to exchange information. In the future, this kind of mechanism can be used in order to contribute to proposals. At now, citizens were asked to answer to a survey, indicating some preferences about proposals. Tomorrow, they can produce some proposals, contributing to plan design. Moreover, they can supply information, and feed the knowledge framework.

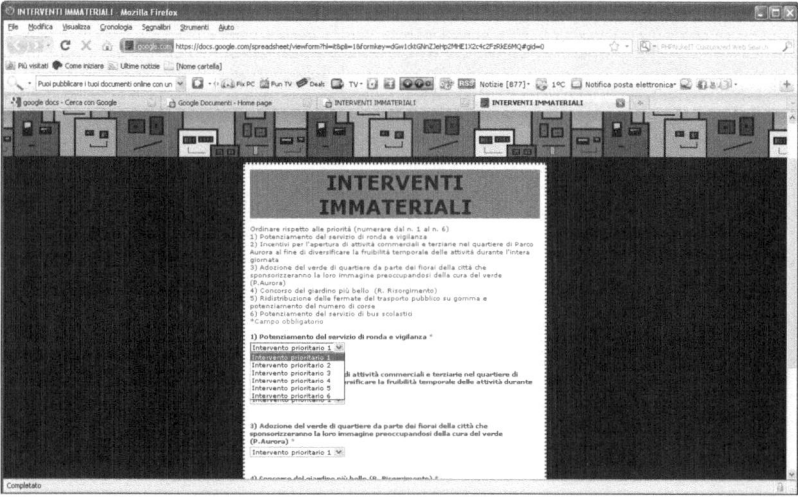

Fig. 7. Screenshot of the e-valuation survey, concerning in particular one of the three interventions categories: the no-materials interventions

In the e-mail, containing the link to the *e-valuation* survey and some instructions (strategic intervention: https://docs.google.com/spreadsheet/viewform?formkey= dG1TT0tybTd4eldPbEZXWjlkckdwR1E6MQ; ordinary planning interventions: https://docs.google.com/spreadsheet/viewform?formkey=dEdONzh1anlvTDg3RGg2 Ukw4UllVR3c6MQ; not physical interventions: https://docs.google.com/spreadsheet/

viewform?formkey=dGw1cktGNnZJeHp2MHE1X2c4c2FzRkE6MQ), they have been asked to contact friends, familiars, colleagues living in their same district, and invite them to express a vote to interventions conceived during workshops, as they already did. The basic idea is to stimulate involvement through citizens that have been involved, and that, accepting to help us, are strongly motivated into the achievement of their mission. Moreover, they were asked to help their friends, familiars and colleagues, in reason of their knowledge of interventions and of the voting procedure.

5 Analyzing Survey: Remarks on e-Citizens' Answers and Behavior

Looking for similarities or differences behavior in involved citizens and e-citizens, certain coherence can be revealed: both perceive interventions in the same manner, and both seem mainly interested in spaces organization, as is showed in figure 8.

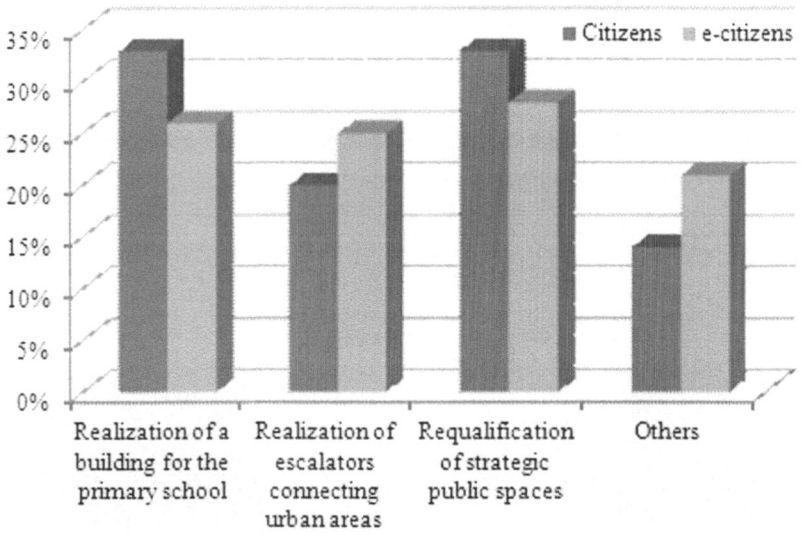

Fig. 8. Comparison between percentages of "high priority" rank in strategic works expressed by citizens and by e-citizens

Concerning results, voters express an implicit preference towards one of the three categories of interventions that is the ordinary planning interventions: answers are most concentrated on this category, revealing the implicit preference to involved e-citizens, that prefer some small projects, able to produce, in short period, improvement in life quality.

It is important point out that citizens consider as priority interventions that improve neighborhood quality and more generally life quality, with modest financial resources but easy to carry out. In fact, the interventions that have been considered as priority concern accessibility and usability of public spaces, both in terms of quantity and quality. Those interventions are as well related to improvement of social spaces, not enough and not well-equipped, and finally are related to safety augmentation.

Answers were surprising, because they do not reveal differences between a citizen and an e-citizen; often, people strongly using internet, develop more consciousness and a higher level of understanding, and he is generally more skilled etc, so that we expected to retrieve these differences in survey. Instead, they answer in the same way of citizens. This can mean that:

- Problems have been well identified, and proposed solutions are the only possible to solve, so that everybody recognize their utility.
- Survey is not deepening enough to allow a sensitive difference in responses.
- Citizens and e-citizens in Potenza cannot be distinguished.

Probably, these aspects are all present in our context, and in order to a deeper understanding some advanced inquiry is needed. Probably, a different way to involve e-citizens could produce different results.

Therefore, all efforts are not effective if there is not integration between citizens and decision makers. A citizen has the right to take part in the decision-making process by expressing his preference, and the decision maker has really the power to act and take the final decision for the development of strategies in the city; in a new perspective, citizens will not take the place of the decision maker, but they can contribute to the process and express their preferences as a part of the global preference expressed by decision maker himself. Moreover, if the evaluation or *e-valuation* process doesn't allow to find shared solutions, representing a global approval, but it emphasizes some irreconcilable positions, this doesn't mean that the process has failed, but that it has developed its full role in decision-making aid [19].

Last remark refers to the exigency that Administrators must be involved in such a process: *who have the power to act must be integrally involved in developing a strategy* [20]. If this does not happen, the efficacy and effectiveness of the process itself decrease, and citizens continue to feel not involved in choices. All kind of tools – and as showed there is a large amount of – can help into structuring participatory and e-participatory process, contributing brick after brick to a better planning process.

References

1. Rodotà, S.: The role of Parliaments in the development in the Information Society. Keynote speech-inter-parliamentary. In: Union International Conference, Genova (2007)
2. Murgante, B., Tilio, L., Lanza, V., Scorza, F.: Using participative GIS and e-tools for involving citizens of Marmo Platano – Melandro area in European programming activities. Journal of Balkans and Near Eastern Studies 13(1), 97–115 (2011), ISSN:1944-8953, doi:10.1080/19448953.2011.550809
3. Knapp, S., Coors, V.: The use of eParticipation systems in public participation: the VEPs example. In: Rumor, M., Coors, V., Fendel, E., Zlatanova, S. (eds.) Urban and Regional Data Management, pp. 93–104. Taylor and Francis, London (2007) ISBN: 978-0-415-44059-2

4. Simmons, D.A.: Communicating with the public: an examination of national park planning workbooks. Journal of Environmental Education 19, 9–167 (1987)
5. Conroy, M.M., Gordon, S.I.: Utility of interactive computer based materials for enhancing public participation. Environmental Planning and Management 47(1), 19–33 (2004)
6. Conroy, M.M., Evans-Cowley, J.: e-participation in planning: an analysis of cities adopting online citizen participation tools. Environment and Planning C: Government and Policy 24, 371–384 (2006)
7. Thomas, J.C., Streib, G.: The new face of government: citizen-initiated contact in the era of e-government. Journal of Public Administration Research and Theory 13(1), 83–102 (2003)
8. Apostol, I., Antoniadis, P., Banerjee, T.: From face-block to facebook or the other way around? In: Proceedings of Sustainable City and Creativity, Promoting Creative Urban Initiatives. Naples (2008)
9. Panopoulou, E., Tambouris, E., Tarabanis, K.: A framework for evaluating websites of public authorities. ASLIB Proceedings 60(5), 517–546 (2008), doi: http://dx.doi.org/10.1108/00012530810908229
10. Arnstein, S.R.: A ladder of citizen participation. Journal of the American Planning Association 35(4), 216–224 (1969)
11. Lanza, V., Prosperi, D.C.: Collaborative e-governance. Describing and pre-calibrating the digital milieu in urban and regional planning. In: Zlatanova, S., Fendel, E., Krek, A., Rumor, M. (eds.) UDMS 2009 Annual Urban and Regional Data Management, pp. 373–384. CRC Press (2009), ISBN: 978-0-203-86935-2, doi:10.1201/9780203869352.ch33
12. Cosenza, V.: Quanti italiani usano twitter? post on blog (2010), http://www.vincos.it, http://vincos.it/2010/10/10/quanti-italiani-usano-twitter/ (last access December 2011)
13. Obermeyer, N.J.: The evolution of public participation GIS. Cartography and Geographic Informatin Systems 25(2), 65–66 (1998)
14. Hudson-Smith, A., Crooks, A., Gibin, M., Milton, R., Batty, M.: NeoGeography and Web 2.0: concepts, tools and applications. Journal of Location Based Services 3(2), 118–145 (2009), doi:10.1080/17489720902950366
15. Hudson-smith, A., Batty, M., Crooks, A., Milton, R.: Mapping for the Masses: Accessing Web 2.0 Through Crowdsourcing. Social Science Computer Review 27(4), 524–538 (2009), doi:10.1177/0894439309332299
16. Rocha, J.G.: The Participation Loop: Helping Citizens to Get In. In: Murgante, B., Gervasi, O., Iglesias, A., Taniar, D., Apduhan, B.O. (eds.) ICCSA 2011, Part II. LNCS, vol. 6783, pp. 172–184. Springer, Heidelberg (2011), ISSN: 0302-9743, doi:10.1007/978-3-642-21887-3_14
17. Burby, R.J.: Making plans that matter. Citizen involvement and government action. Journal of the American Planning Association 69(1), 33–49 (2003), doi:10.1080/01944360308976292
18. Howe, J.: Crowdsorcing. How the power of the crowd is driving the future of business. Three Rivers Press, New York (2008)
19. Plottu, B., Plottu, E.: Modèle pour l'opérationnalité de l'évaluation démocratique (M.O.D.E.): quelques précisions et enrichissements. In: Proceedings of Outils Pour Décider Ensemble (2012)
20. Eden, C.: Strategic thinking with computers. Longe Range Planning 23(6), 35–43 (1990), doi:10.1016/0024-6301(90)90100-I

Bilayer Segmentation Augmented with Future Evidence

Silvio Ricardo Rodrigues Sanches[1], Valdinei Freire da Silva[2], and Romero Tori[1]

[1] Escola Politécnica da Universidade de São Paulo, Brasil, Av Professor Luciano Gualberto, Trav. 3, 158, Cidade Universitária, 05508-900, São Paulo-SP, Brasil
`silviorrs@usp.br,tori@acm.org`
[2] Escola de Artes, Ciências e Humanidades da Universidade de São Paulo, Av. Arlindo Béttio, 1000, Ermelino Matarazzo, 03828-000, São Paulo-SP, Brasil
`valdinei.freire@usp.br`

Abstract. This paper presents an algorithm that augments a previous model known in the literature for the automatic segmentation of monocular videos into foreground and background layers. The original model fuses visual cues such as color, contrast, motion and spatial priors within a Conditional Random Field. Our augmented model makes use of bidirectional motion priors by exploiting future evidence. Although our augmented model processes more data, it does so with the same time performance of the original model. We evaluate the augmented model within ground truth data and the results show that the augmented model produces better segmentation.

Keywords: bilayer segmentation, computer vision, image understanding.

1 Introduction

The image segmentation problem (the extraction of elements of interest in images or videos) has been under research since the beginning of the last century with the industry [24]. The industry of film and television productions traditionally use methods which extract one or more elements from an image or a video frame – most of these elements are people in the foreground – to create scenes from the combination of them with a new background [20,21]. Until the late 1970s such extraction was based on optical analog technology [5].

Traditional methods for image segmentation assume that the video frame was captured in a controlled environment, with a single color as background (usually blue or green) and with the environment lights configured to keep that color uniform [6,14,21]. By assuming a known background color, segmentation can be done in real-time and with low error.

Since the 1980s, new methods based on digital technology [6] have been developed and today there are methods able to extract elements not only in real-time but also from natural images (without a single color background) [2,4,18,22,26]. This possibility has boosted research in new areas of application, especially those in which the elements of interest are people in the foreground. Videoconferencing, videochat [4,18,26,25] and other systems [17] are examples of applications which can replace an original background before sending each video frame to remote users.

In this paper we address the efficient extraction of foreground layer (bilayer segmentation) to background substitution applications in which input streaming is captured in a natural environment by a monocular camera. The algorithm proposed here extends the segmentation model presented by Criminisi *et al.* [4]. The segmentation problem is modeled by a Conditional Random Field (CRF) which fuses color, temporal and spatial information. The temporal information is considered as a prior and it is conjugated with color and spatial models of observation in order to determine an *a posteriori* information. We improve on such a model by augmenting it with one future frame in the CRF model.

The rest of this paper is organized as follows: section 2 presents some previous researches related to bilayer segmentation and section 3 illustrates the methods which use temporal information. Section 4 introduces the basic model used in our solution, whereas the section 5 describes our augmented model. Section 6 discusses experimental results, and section 7 concludes the papper.

2 Related Work

Layer extraction [2,4,8,10,15,18,19,25] has long been an active area of research in computer vision. In recent approaches, a common characteristic is to treat the segmentation problem as an energy minimization problem. In a binary segmentation task each pixel of the processing frame is labeled as background or foreground (0 or 1), without considering fractional values to represent transparency. Briefly, from a set of pixels P and a set of labels L (in this case two labels), the goal is to find a label function $f : P \rightarrow L$ which minimizes a specific energy function [9][1].

In order to classify pixels, recent work in bilayer segmentation area has produced algorithms based on either depth [7,22] or motion [4,26]. Other algorithms require initialization in the form of a clean image of background [18].

Stereo-based segmentation [7] seems to achieve the most robust results for layer extraction with depth information. However, binocular video can be restrictive for some applications as well as approaches based in time-of-flight sensors for depth estimation [22]. In a teleconferencing or a videochat most users have only a single conventional web camera [18] and the necessity for calibration of two cameras for stereo is inconvenient [4].

Motion-based segmentation can be achieved by estimating optical flow [1], but, in the context of natural environments, the foreground motion cannot be described well by such rigid models. In addition, the optical flow computation is expensive [26].

---

[1] Determination of fractional transparency of pixel is necessary for precision in layer extraction. However, it can be determined after the binary segmentation, by techniques such as border matting [16].

Interactive color/contrast-based segmentation techniques have been demonstrated to be effective [3,16]. However, as demonstrated in [4], segmentation based on color/contrast alone is beyond the capability of fully automatic methods.

On the other hand, recent approaches show that fusing a variety of cues, for example, color, contrast and spatial priors [4,18,25,26] can produce segmentation computable in real-time with accuracy similar to the one obtained from stereo-based segmentation. Whereas some of those systems assume static background [4], others support distracting events and require no initialization [26].

In our model, visual cues such as color, contrast, motion and spatial priors are fused together within a CRF model, where the motion cue exploits bidirectional evidence in order to improve bilayer segmentation.

3 Temporal Information in Bilayer Segmentation Approaches

In applications where the bilayer segmentation is performed in controlled environments the misclassification of pixels can be avoided by user interventions. The environment light can be configured and the elements in the scene can be positioned so that the single background color remains constant. This can avoid the occurrence of shadows, reflections or noise on the background which are potential sources of error.

However, in natural environments, the background is arbitrary and any problematic situation must be handled by segmentation algorithm, avoiding user intervention. In such cases, as there is no prior knowledge about the background color, other information that can be obtained from the frame sequence become necessary, especially for segmentation methods based on monocular video.

Automatic and computationally efficient methods for real-time segmentation make use of these information as a set of *cuts* [4,26]. Color, contrast and motion are examples of cuts which are probabilistically combined by a framework of energy minimization.

As demonstrated in [4,18,25,26], information obtained from previous frames have proved important to aid the estimation current-frame labels. In those work, the temporal information is a cut in the energy minimization framework and it was used to identify pixels in movement. Due to the real-time restriction, the temporal information has been based only on the evidence from its past.

Although considering evidence from future (or bidirectional) may be prohibited in many applications, real-time ones in which some delay is already expected, the delay of one frame may be imperceptible. We demonstrate that this information is relevant for segmentation algorithms.

4 Notation and Basic Model

This section describes notation regarding frame observations and the probabilistic model for foreground/background segmentation proposed in [4] to which we refer as the basic model.

4.1 Notation

Given an input sequence of images, a frame is represented as a matrix

$$z = \begin{bmatrix} z_{1,1} & z_{1,2} & \cdots & z_{1,Y} \\ z_{2,1} & z_{2,2} & \cdots & z_{2,Y} \\ \vdots & \vdots & \ddots & \vdots \\ z_{X,1} & z_{X,2} & \cdots & z_{X,Y} \end{bmatrix} \qquad (1)$$

of pixels in the YUV color space. A frame at time t is denoted z^t. Temporal derivatives are denoted $\dot{z} = [\dot{z}_{x,y}]_{X \times Y}$ and are computed as

$$\dot{z}^t = |G(\mathbf{0}; \sigma_T) * z^t - G(\mathbf{0}; \sigma_T) * z^{t-1}| \qquad (2)$$

at each time t, where $G(\mathbf{0}; \sigma_T)$ is a 2D centralized Gaussian kernel with standard deviation σ_T and $*$ is the convolution operator. Spatial gradients $g = [g_{x,y}]_{X \times Y}$ are computed by convolving the frames with first-order derivative of Gaussian kernels with standard deviation σ_S, i.e.,

$$g^t = \sqrt{\left(\frac{\partial G(\mathbf{0}; \sigma_S)}{\partial x} * z^t\right)^2 + \left(\frac{\partial G(\mathbf{0}; \sigma_S)}{\partial y} * z^t\right)^2}. \qquad (3)$$

As in [4], we use $\sigma_S = \sigma_T = 0.8$. Spatio-temporal derivatives are computed on the Y channel only. Motion observation at time t is denoted $m^t = (g^t, \dot{z}^t)$. Given a sequence of image data $z^1, z^2, \ldots, z^t$ and a sequence of motion data $m^1, m^2, \ldots, m^t$, the segmentation task is to infer a binary label $\alpha_{x,y}^t \in \{F, B\}$ for every pixel in the current frame. F and B denote foreground and background, respectively.

4.2 Basic Model

The probabilistic model for layer extraction proposed in [4] uses an energy minimization framework and extends previous energy models for segmentation [3,7,16]. The model is a Conditional Random Field (CRF) [12] with independent terms that are set discriminatively, i.e., instead of working with join distributions, conditional distributions are considered [11]. The CRF models the conditional probability:

$$p(\alpha^1, \ldots, \alpha^t | z, \ldots, z^t, m^1, \ldots, m^t) \propto \exp-\left\{\sum_{t'=1}^{t} E^{t'}\right\} \qquad (4)$$

where $E^t = E(\alpha^t, \alpha^{t-1}, \alpha^{t-2}, z^t, m^t)$.

The energy E^t associated with time t is a sum of terms in which likelihood and prior are not entirely separated. The energy decomposes as a sum of four terms:

$$E(\alpha^t, \alpha^{t-1}, \alpha^{t-2}, z^t, m^t) = \qquad (5)$$
$$\eta V^T(\alpha^t, \alpha^{t-1}, \alpha^{t-2}) + \gamma V^S(\alpha^t, z^t)$$
$$+ \rho U^C(\alpha^t, z) + \phi U^M(\alpha^t, \alpha^{t-1}, m^t),$$

in which the first two terms are "prior-like" and the second two are observation likelihoods. η, γ, ρ and ϕ are normalizing parameters.

Temporal prior term $V^T(\cdot)$ imposes a tendency to temporal continuity of segmentation labels. Second-Order Markov chain is used in the energy minimization framework to incorporate the intuition that a pixel that was in the background at time $t-2$ and in the foreground at time $t-1$ is far more likely to remain in the foreground at time t than to go back to the background. The temporal transition priors are learned from labeled data. The temporal prior term is denoted

$$V^T(\alpha^t, \alpha^{t-1}, \alpha^{t-2}) = \sum_{m=1}^{X} \sum_{n=1}^{Y} [-\log p(\alpha_{x,y}^t | \alpha_{x,y}^{t-1}, \alpha_{x,y}^{t-2})]. \qquad (6)$$

Spatial prior term $V^S(\cdot)$ is an Ising term, imposing a tendency to spatial continuity of labels, and the term is inhibited by high contrast. Let C be the set of pairs of neighboring pixels in a frame[2]; and z_i and α_i be the values of pixel i in the YUV color space and the binary label to be attributed to pixel i respectively, i.e., i indexes a pixel in the matrices z and α. The Ising term is represented by an energy of the form

$$V^S(\alpha, z) = \sum_{i,j \in C} [\alpha_i \neq \alpha_j] \left(\frac{\epsilon + e^{-\mu ||z_i - z_j||^2}}{1 + \epsilon} \right). \qquad (7)$$

The contrast parameter μ is chosen to be $\mu = (2 \langle ||z_i - z_j||^2 \rangle)^{-1}$, where $\langle \cdot \rangle$ denotes expectation over all pairs of neighbors in an image sample.

The energy term $V^S(\alpha, z)$ represents a combination of an Ising prior for labeling coherence together with a contrast likelihood that acts to discount partially the coherence terms. The constant ϵ is a "dilution" constant for contrast. We set $\epsilon = 1$ as it was done in [7].

Color likelihood term $U^C(\cdot)$ evaluates the evidence for pixel labels using the color distributions in foreground and background. Likelihoods are modeled as histograms in the YUV color space. This term is defined as:

$$U^C(\alpha, z) = - \sum_{m=1}^{X} \sum_{n=1}^{Y} \log p(z_{x,y} | \alpha_{x,y}). \qquad (8)$$

In our experiments the foreground color likelihood model is learned from a first ground-truth segmented frame. The likelihoods are then stored in 3D look-up

---
[2] Here we work with a neighborhood of 4 neighbors, i.e., neighbors in each cardinal direction.

tables. The distribution is represented as a smoothed histogram to avoid overfitting within initialization.

Motion likelihood term $U^M(\cdot)$ uses spatial and temporal derivatives $m = (g, \dot{z})$ to capture the characteristics of the features under foreground and background conditions.

According to [4], the immediate history of the segmentation of a pixel falls into one of four classes: FF, BB, FB and BF. The observed image motion features $m_{x,y}^t = (g_{x,y}^t, \dot{z}_{x,y}^t)$ at time t is conditioned on those combinations of the segmentation labels $\alpha_{x,y}^{t-1}$ and $\alpha_{x,y}^t$. Temporal derivative $\dot{z}_{x,y}^t$ is computed from frames $t-1$ and t, so it should depend on segmentations of those frames.

The motion likelihood is learned from some labeled ground-truth data and then stored as 2D histograms to use in likelihood evaluation. The likelihoods are evaluated as part of the total energy, in the term

$$U^M(\alpha^t, \alpha^{t-1}, m^t) = - \sum_{x=1}^{X} \sum_{y=1}^{Y} \log p(m_{x,y}^t | \alpha_{x,y}^t, \alpha_{x,y}^{t-1}). \qquad (9)$$

4.3 Energy Minimization

Before minimizing energy E^t some considerations are done. First, parameters z^t and m^t are observations and, since they are extract from the current frame and the previous frame, they are not considered when minimizing energy E^t and we make it clear by writing $E^t = E(\alpha^t, \alpha^{t-1}, \alpha^{t-2} | z^t, m^t)$. Second, at time $t = 1$ parameters α^{t-1} and α^{t-2} are meaningless. Third, at time $t = 2$ parameter α^{t-2} is meaningless. Then, when labeling a frame at time $t = 1$, only terms V^S and U^C are meaningful and we have an energy $E^1 = E(\alpha^1 | z^t, m^t)$, and when labeling a frame at time $t = 2$, only terms V^S, U^C and U^M are meaningful and we have an energy $E^2 = E(\alpha^2, \alpha^1 | z^t, m^t)$.

Given the previous considerations, the labeling of pixels proceeds as follows:

- at time $t = 1$ minimizes energy $E^1 = E(\alpha^1 | z^t, m^t)$ and obtains $\hat{\alpha}^1$;
- by considering $\hat{\alpha}^1$ as an observation, at time $t = 2$ minimizes energy $E^2 = E(\alpha^2 | \hat{\alpha}^1, z^t, m^t)$; and
- by considering $\hat{\alpha}^{t-2}$ and $\hat{\alpha}^{t-1}$ as observations, at any time $t \geq 3$ minimizes energy $E^t = E(\alpha^t | \hat{\alpha}^{t-1}, \hat{\alpha}^{t-2}, z^t, m^t)$.

The simplification of considering $\hat{\alpha}^{t-2}$ and $\hat{\alpha}^{t-1}$ as observations decreases the neighborhood of a pixel when minimizing the energy function. Temporal dependences are solved by doing so and only spatial dependences are need to be solved. Since the energy E^t models a CRF, we can describe it as [11]:

$$E^t = \sum_{i \in S} \left[A_i(\alpha_i^t, \mathbf{o}^t) + \sum_{j \in \mathcal{N}_i} I_{ij}(\alpha_i^t, \alpha_j^t, \mathbf{o}^t) \right],$$

where S is the set of pixels in a frame, $\mathbf{o} = (\hat{\alpha}^{t-1}, \hat{\alpha}^{t-2}, z^t, m^t)$ is the observation at time t, $\mathcal{N}_i$ is the neighborhood of pixel i, A_i is the association potential and I_{ij} is the interaction potential.

Finally, to minimize energy we considered the implementation of graph cut done by Kolmogorov and Zabih [9].

5 Augmenting the Basic Model

Although Criminisi et al. [4] make use of motion characteristics in the motion term U^M, Criminisi et al. make use of motion characteristics only related to past frames. Since the labels α^t influences temporal derivatives, we can consider such derivatives backwards and forwards in time. In [4] only the first derivative is explored. Although considering evidence from future may be prohibited in many applications, if some delay is already expected in some real-time applications, the user would not perceive the delay of one more frame. Fig. 1 shows the relation between each variable considered in the CRF.

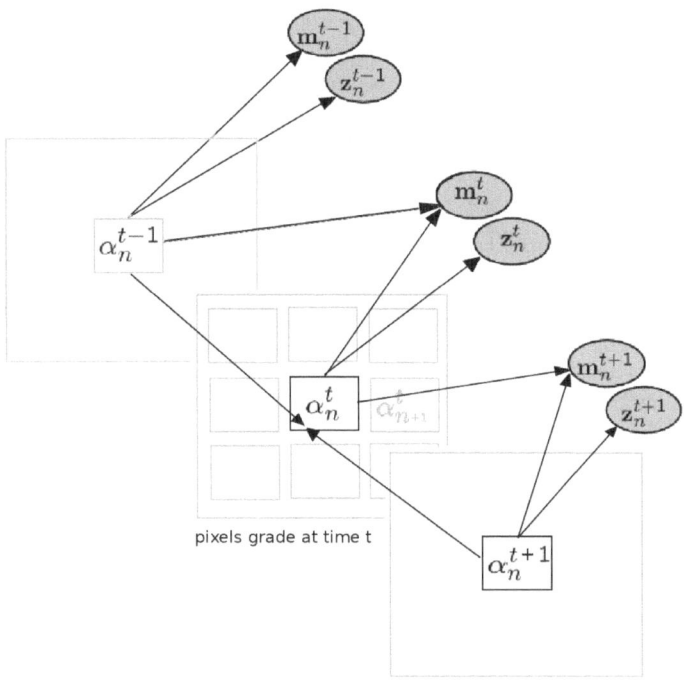

Fig. 1. Augmented CRF Model

The model by Criminisi et al. defines the observation $m^t = (g^t, \dot{z}^t)$ and minimizes the likelihood $p(g^t, \dot{z}^t | \alpha^t, \alpha^{t-1})$. However, if we wait for one more frame, another evidence $\dot{z}^{t+1}$ can be observed, given the opportunity of minimizing the likelihood $p(g^t, \dot{z}^t, \dot{z}^{t+1} | \alpha^t, \alpha^{t-1})$. If the delay in one frame is acceptable, such evidence may improve the segmentation of the foreground layer.

In order to evaluate the value of such evidence, we calculate the entropy when labeling sequence of images in videos which simulates videochat interaction. Given observations $\alpha^{t-1}, \dot{z}^t, \dot{z}^{t+1}$ and g^t, we combined them to classify labels α^t. In order to test the influence of each one of the observations, we grouped them into four groups and classified labels α^t with the maximum likelihood criteria. The likelihood was obtained from a grounded dataset and tested against itself. Table 1 shows our results: entropy and error rate for each group. Combining only two derivatives evidences ($\dot{z}^t, \dot{z}^{t+1}$ or g^t) gives similar values, but combining all of them together increases the classification based only on motion significantly if observation α^{t-1} is correct. Note that we cannot obtain the same results from table 1, since in our framework α^{t-1} is classified in previous steps and may present errors.

Table 1. Entropy Analysis

	Entropy	Error Rate
$g^t, \dot{z}^t, \alpha^{t-1}$	0.2100	0.033
$\dot{z}^t, \dot{z}^{t+1}, \alpha^{t-1}$	0.2013	0.031
$g^t, \dot{z}^{t+1}, \alpha^{t-1}$	0.2628	0.045
$g^t, \dot{z}^t, \dot{z}^{t+1}, \alpha^{t-1}$	0.1252	0.017

Since $p(g_n^t, \dot{z}_n^t, \dot{z}_n^{t+1} | \alpha_n^t, \alpha^{t-1})$ is kept in a look-up table and the future temporal derivative $\dot{z}_n^{t+1}$ should be calculated anyway at time $t+1$, our augmented algorithm calculates it in advance and no difference in computational time is observed. However, the size of the look-up table is multiplied by the spectrum of $\dot{z}^{t+1}$, increasing considerably the size of the look-up table.

If the size of the table is larger, we must also collect more data in order to learn the conditional probabilities $p(g_n^t, \dot{z}_n^t, \dot{z}_n^{t+1} | \alpha_n^t, \alpha^{t-1})$. Again, we analyze labeled videos in order to determine a better use of the look-up table, decreasing its size and improving its generalization when learning. Fig. 2 shows the histogram for the variable $\dot{z}$. It is clear the concentration around small values and the sparseness when values get large. Analyzing g proved to be similar.

Instead of using uniform discretization of the derivative spectrum, following our analysis, we divide the spectrum of derivatives in 39 bins. Around zero, bins are smaller, whereas bins increases as they are distant from zero. The size of bins from zero values to larger one are: five bins with size 1, six bins with size 2, twelve bins with size 4, eight bins with size 8, and eight bins with size 16.

6 Experiments

In order to experiment our ideas we compared both Criminisi et al. algorithm and our augmented version. We used a database of video images with 38 labeled video sequences which are available in [13]. In the experiments we derive our augmented version of Criminisi et al. from [23].

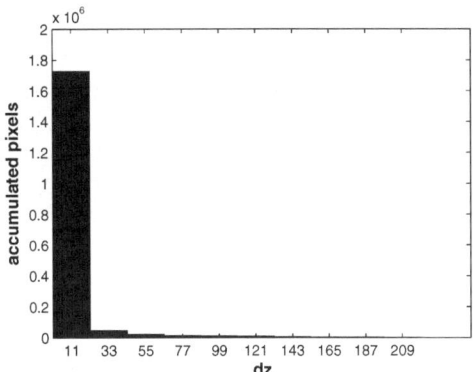

Fig. 2. Distribution of $\dot{z}^t$ values for a video

Since frames are labeled from 10-to-10 or 5-to-5 frames, we opted to use only such frames in our experiments. The labels of the first frames of each video is considered to be known in order to initialize the color likelihood model as well as initializing the prior regarding foreground layer. First, we obtained the temporal prior and the motion likelihood within all of the 38 labeled video sequences. In order to tune the parameters of segmentation models, we sampled randomly a hundred tuple of parameters $\langle \eta, \gamma, \rho, \phi \rangle$ and chose the tuple that obtained the best evaluation. Since the absolute values of parameters are irrelevant, we normalized parameter values by fixing $\gamma = 1$ and we sorted η, ρ and ϕ from interval $(0.0, 0.2)$ uniformly. Because γ proved to be more relevant in preliminary experiments, we sorted the parameters η, ρ and ϕ smaller than the parameter γ.

The evaluation of tuples was obtained by calculating the mean error among frames in every sequence. The error per frame was simply calculated as the error rate when compared to ground truth data, i.e., at frame t, given the ground truth α^t and the estimated label $\hat{\alpha}^t$, the error rate at frame t is given by

$$\epsilon^t = \epsilon(\alpha^t, \hat{\alpha}^t) = \frac{\sum_{x=1}^{X} \sum_{y=1}^{Y} |\alpha^t - \hat{\alpha}^t|}{XY}.$$

Remember that labels are binary, then $|\alpha^t - \hat{\alpha}^t| \in \{1, 0\}$. By considering all of the 38 video sequences we obtained the parameters $\eta = 0.0018$, $\gamma = 1$, $\rho = 0.0338$, $\phi = 0.0413$.

Because of our small database of video sequences, instead of using the same parameters to evaluate all of the video sequence, we opted by using leave-one-out method to training/testing video sequences. When evaluating the performance of each method regarding a video sequence, we took it out of the set of samples and chose the best parameters for the other 37 video sequences. In Fig. 3 foreground and background of a video sequence were separated automatically by our augmented model.

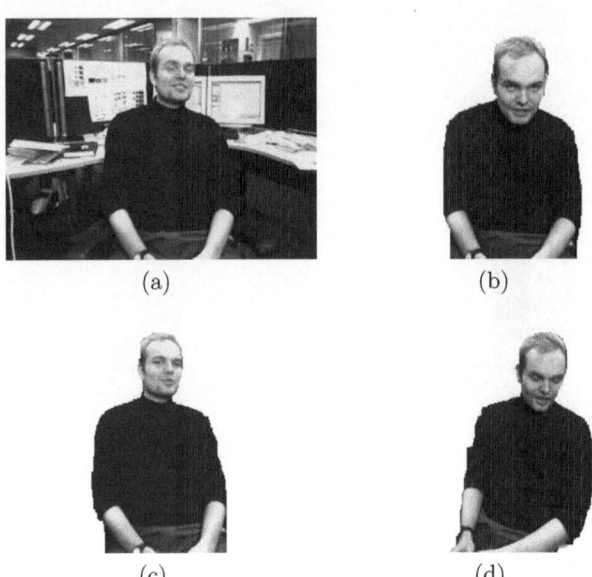

Fig. 3. Binary Segmentation. (a) A frame from the MS test sequence. (b,c and d) Automatic foreground extraction results for three frames by our augmented method.

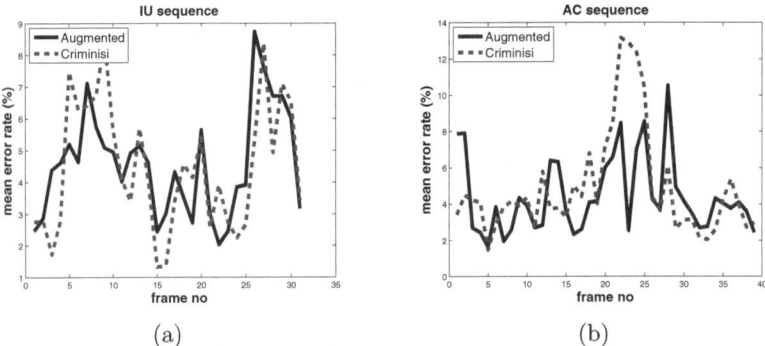

Fig. 4. Comparison between Criminisi et al. method and our augmented method. (a) IU test sequence and (b) AC test sequence. Note that the model parameters η, γ, ρ e ϕ were fully optimized for best performance.

We used the labeled video sequences for experimenting our augmented version against Criminisi et al. algorithm. Our approach had a mean error rate through the whole video of 0.041 whereas the basic approach presented a mean error rate of 0.054. Despite our small database, by applying t-student test we observed that our method was better than the original one with a significance of 0.18. Fig. 4 shows our results in IU and AC test sequences[3]. We found that the IU sequence

---

[3] The name of sequences IU, AC and MS are given in [13].

exhibit high illumination variation. Our best relative result were observed in this test sequence.

As a final example, Fig. 5 shows the results of our augmented method on a frame sequence. The original background was replaced with a new one. In this demonstration, the colour likelihood model was initialized manually.

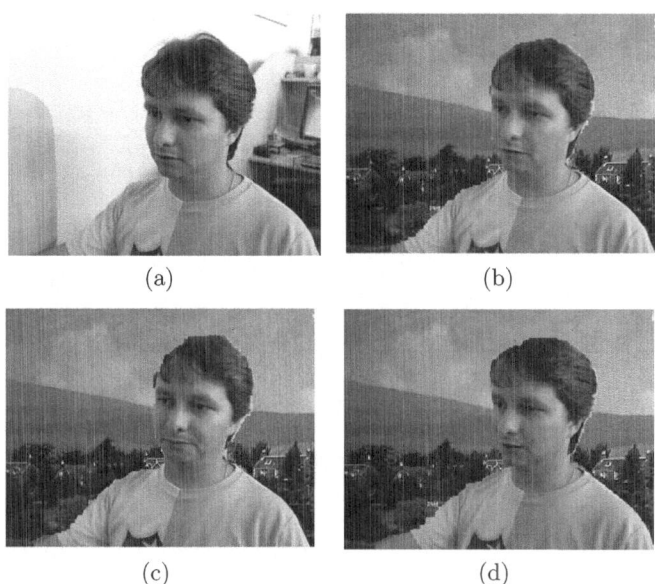

Fig. 5. A final example of Binary Segmentation and Background Substitution. (a) A frame with original background; (b, c and d) automatic background substitution for several frames by our augmented method. Note that no method for transparency of pixel was applied.

7 Conclusion

This paper has addressed the problem of bilayer segmentation of monocular video sequences. We extend the model presented in [4] to improve segmentation without incurring in more computational time. We accomplish it by extending the spatio-temporal coherence considering bidirectional evidence from a frame without losing real-time characteristic.

Although the delay of one frame can be prohibited in some high-performance application, most interactive applications accept small delays. If a high frame rate is considered, this delay can even be imperceptible.

The results show that an improvement was obtained, however more specific experiments must be done in order to detect in which kind of situation it is fruitful considering future frames.

Acknowledgment. Silvio Sanches would like to thank CAPES (Coordenação de Aperfeiçoamento de Pessoal de Nível Superior – scholarship) and Valdinei Silva would like to thank FAPESP (Fundação de Amparo à Pesquisa do Estado de São Paulo – proc. 11/19280-8) for their financial supports. The video sequences and the ground truth used in this research are courtesy of the Microsoft Corporation (Free research data [13]).

References

1. Barron, J.L., Fleet, D.J., Beauchemin, S.S.: Performance of optical flow techniques. International Journal of Computer Vision 12, 43–77 (1994)
2. Bergen, J., Burt, P., Hingorani, R., Peleg, S.: A three-frame algorithm for estimating two-component image motion. IEEE Transactions on Pattern Analysis and Machine Intelligence 14(9), 886–896 (1992)
3. Boykov, Y.Y., Jolly, M.P.: Interactive graph cuts for optimal boundary & region segmentation of objects in n-d images. In: IEEE International Conference on Computer Vision, vol. 1, pp. 105–112. IEEE Computer Society, Los Alamitos (2001)
4. Criminisi, A., Cross, G., Blake, A., Kolmogorov, V.: Bilayer segmentation of live video. In: CVPR 2006: Proceedings of IEEE Computer Society Conference on Computer Vision and Pattern Recognition, vol. 1, pp. 53–60. IEEE Computer Society, Washington, DC (2006)
5. Foster, J.: Mattes and Compositing Defined. In: The Green Screen Handbook: Real-World Production Techniques, pp. 3–15. John Wiley and Sons Ltd., Chichester (2010)
6. Gibbs, S., Arapis, C., Breiteneder, C., Lalioti, V., Mostafawy, S., Speier, J.: Virtual studios: an overview. IEEE Multimedia 5(1), 18–35 (1998)
7. Kolmogorov, V., Criminisi, A., Blake, A., Cross, G., Rother, C.: Bi-layer segmentation of binocular stereo video. In: CVPR 2005: Proceedings of the IEEE Computer Society Conference on Computer Vision and Pattern Recognition, vol. 2, pp. 407–414. IEEE Computer Society, Washington, DC (2005)
8. Kolmogorov, V., Criminisi, A., Blake, A., Cross, G., Rother, C.: Probabilistic fusion of stereo with color and contrast for bilayer segmentation. IEEE Transactions on Pattern Analysis and Machine Intelligence 28(9), 1480–1492 (2006)
9. Kolmogorov, V., Zabin, R.: What energy functions can be minimized via graph cuts? IEEE Transactions on Pattern Analysis and Machine Intelligence 26(2), 147–159 (2004)
10. Kumar, M., Torr, P., Zisserman, A.: Learning layered motion segmentations of video. In: Tenth IEEE International Conference on Computer Vision, ICCV 2005, vol. 1, pp. 33–40 (2005)
11. Kumar, S., Hebert, M.: Discriminative random fields: A discriminative framework for contextual interaction in classification. In: ICCV 2003: Proceedings of the Ninth IEEE International Conference on Computer Vision, p. 1150. IEEE Computer Society, Washington, DC (2003)
12. Lafferty, J.D., McCallum, A., Pereira, F.C.N.: Conditional random fields: Probabilistic models for segmenting and labeling sequence data. In: ICML 2001: Proceedings of the Eighteenth International Conference on Machine Learning, pp. 282–289. Morgan Kaufmann Publishers Inc., San Francisco (2001)
13. Microsoft research – free research data, http://research.microsoft.com/en-us/projects/i2i/data.aspx

14. Mishima, Y.: Soft edge chroma-key generation based upon hexoctahedral color space. U.S. Patent 5,355,174 (1994)
15. Parolin, A., Fickel, G.P., Jung, C.R., Malzbender, T., Samadani, R.: Bilayer video segmentation for videoconferencing applications. In: 2011 IEEE International Conference on Multimedia and Expo. (ICME), pp. 1–6 (2011)
16. Rother, C., Kolmogorov, V., Blake, A.: "Grabcut": interactive foreground extraction using iterated graph cuts. ACM Trans. Graph. 23(3), 309–314 (2004)
17. Sanches, S.R.R., Tokunaga, D.M., Silva, V.F., Sementille, A.C., Tori, R.: Mutual occlusion between real and virtual elements in augmented reality based on fiducial markers. In: 2012 IEEE Workshop on Applications of Computer Vision (WACV), pp. 49–54 (2012)
18. Sun, J., Zhang, W., Tang, X., Shum, H.Y.: Background Cut. In: Leonardis, A., Bischof, H., Pinz, A. (eds.) ECCV 2006. LNCS, vol. 3952, pp. 628–641. Springer, Heidelberg (2006)
19. Torr, P.H.S., Szeliski, R., Anandan, P.: An integrated bayesian approach to layer extraction from image sequences. IEEE Transactions on Pattern Analysis and Machine Intelligence 23(3), 297–303 (2001)
20. Vlahos, P.: Composite photography utilizing sodium vapor illumination. U.S. Patent 3,095,304 (1963)
21. Vlahos, P.: Comprehensive electronic compositing system. U.S. Patent 4,100,569 (1978)
22. Wang, L., Zhang, C., Yang, R., Zhang, C.: Tofcut: Towards robust real-time foreground extraction using a time-of-flight camera. In: Fifth International Symposium on 3D Data Processing, Visualization and Transmission (3DPVT) (2010)
23. Implementation of "bilayer segmentation of live video", http://vision.caltech.edu/projects/yiw/FgBgSegmentation
24. Williams, F.D.: Method of taking motion pictures. U.S. Patent 1,273,435 (1918)
25. Yin, P., Criminisi, A., Winn, J., Essa, I.: Tree-based classifiers for bilayer video segmentation. In: Proceedings of the IEEE Computer Society Conference on Computer Vision and Pattern Recognition, CVPR 2007, pp. 1–8. IEEE Computer Society, Los Alamitos (2007)
26. Yin, P., Criminisi, A., Winn, J., Essa, I.: Bilayer segmentation of webcam videos using tree-based classifiers. IEEE Transactions on Pattern Analysis and Machine Intelligence 33(1), 30–42 (2011)

A Viewer-dependent Tensor Field Visualization Using Multiresolution and Particle Tracing

José Luiz Ribeiro de Souza Filho, Marcelo Caniato Renhe,
Marcelo Bernardes Vieira, and Gildo de Almeida Leonel

Universidade Federal de Juiz de Fora, DCC/ICE,
Cidade Universitária, CEP: 36036-330, Juiz de Fora, MG, Brazil
{jsouzaf,marcelo.caniato,marcelo.bernardes,gildo.leonel}@ice.ufjf.br
http://www.gcg.ufjf.br

Abstract. This paper presents an adaptive method for visualization of tensor fields using multiresolution and viewer position and orientation. A particle tracing method is used in order to explore the benefits of motion to the human perceptual system. The particles are inserted and advected through the field based on a priority list which ranks tensors according to anisotropy measures and viewer parameters. Tensor fields representing colinear and coplanar structures are suitable for multiresolution analysis. Using multiple scales, we propose the use of anisotropic information in multiresolution, yielding an effective and simple method to compute priority values for particle creation. We also propose a new deterministic criterion for particle insertion in the field that balances their distribution in the tensor field domain. Our results show that our method enhances the visualization and reduces artifacts encountered in previous approaches.

Keywords: Tensor Field, Particle Tracing, Multiresolution, Scientific Visualization.

1 Introduction

Tensor field properties, such as curvatures and continuities, are sometimes hard to visualize. Particle-tracing methods using tensorlines provide a good way to observe these features. But the tensorlines could represent some inharmonious or even discontinuous paths present in tensor fields. Smoothness is an important factor to be analyzed. Being able to enhance this feature without changing the peculiarities of the field provides an opportunity to further explore these characteristics of tensor fields.

In this work, we propose an improvement of method presented in [1] which used particle tracing to generate a viewer-dependent visualization. This method used a particle creation criterion based on a priority list which sorted the tensors according to their importance. However, the choice of the tensor in the list was done through a normal distribution function, which sometimes resulted in creation of particles in less interesting sites. The previous approach also generated a flickering effect, due to particles reaching isotropic regions shortly after being created. In this paper, we propose a new approach using multiresolution, and we present a new criterion for the creation of particles. We decompose the original field in order to get smaller subsamples. With this approach, we were able to significantly reduce the amount of parameters necessary in the calculation of the scalar used in the priority list sorting.

One common problem in tensor field visualization is ambiguity. In glyph-based visualization, tensors with different forms may appear similar from a particular point of view. Tensors with linear anisotropy may be identified as an isotropic if the main eigenvector is aligned to the observer. To solve this problem, we follow previous works [2,1] in adopting a metric to evaluate the tensor orientation in regard to the observer. This strategy can be efficient not only to treat the degeneration problem, but also to improve other visualization methods. Aiding to that, we use the multiple scales of the field to enhance tensors based on their distance to the observer.

2 Related Work

Research in tensor field visualization is generally concerned with the problem of achieving a more intuitive visualization of the field. The large amount of information present in a field usually makes its analysis difficult for the observer. Thus, different approaches have been tried in past works. An overview about some of them is presented in this section.

In cases where punctual data is used to obtain information from the field, the discrete approach plays an important role. Shaw *et al*, in [3] and later in [4], proposed a glyph-based visualization of general multi-dimensional data using superquadrics, seeking to explore human perceptual system characteristics in order to obtain a meaningful display of the data. Kindlmann [5] later used superquadrics to specifically describe a tensor glyph that encodes the shape of the tensor and displays it in a consistent orientation. He used measures defined by Westin *et al* [6] to better adapt the geometry of the tensor, avoiding symmetry problems and ambiguity in the identification of its shape. These measures allow classification of diffusion tensors by its shape. They are useful in DT-MRI, since diffusion can be anisotropic or isotropic depending on the tissue characteristics.

Delmarcelle *et al* [7] used another approach, in which they produced a continuous representation of the data contained in the tensor field. They introduced the concept of hyperstreamline to define continuous paths along which the tensor field can be visualized. This method is, however, subject to degeneration [8] and more suitable to symmetric tensor fields. Thus, Weinstein *et al* [9] introduced the tensorlines method, in an attempt of stabilizing the propagation in regions of non-linear diffusion, where the hyperstreamlines method encountered difficulties. They proposed a combination of diffusion with advection vectors applied to DT-MRI. More information on the use of tensor glyphs and continuous methods, as well as a number of other DTI visualization techniques, can be found in [10].

Another approach has also been proposed by Kondratieva et al [2]. They provided a dynamic visualization, which aims at taking more advantage of the human perceptual system. A GPU particle tracing was used to produce motion in order to enhance the user perception. They advected particles along the directions of a generated vector field, while allowing the user to interactively visualize the tensor field. Leonel et al [1] used this same approach, but also taking into account the position and orientation of the observer. An adaptive visualization of the tensor field was provided, enhancing the features that are more likely to interest the viewer.

Some more recent works following [2] focused on improving fiber tracking algorithms. A stochastic method to determine connectivity in a fiber path was presented in [11]. A GPU implementation of the method is also presented. Köhn et al [12] and Evert et al [13] also made use of graphics hardware to achieve a better and faster fiber tracking, allowing for interactive visualization. Mittmann et al [14] presented a real-time interactive fiber tracking method, in which the user defined volumes of interest in the tensor field, and the algorithm calculated new fiber paths automatically based on the user choices. Finally, in [15] an interpolation method was introduced in order to avoid low-anisotropy regions in the trajectory calculation. When the algorithm reaches such a region, it interpolates the tensors in some neighborhood and continues the path along the main eigenvector of the interpolated tensor.

This work is focused on improving visualization of diffusion tensor images. Thus, we still used the tensorlines method [9] as the tracking algorithm. Our method was implemented in CPU, yielding good results and allowing a fast and real-time interactive visualization, even for a large amount of particles, as will be shown in the paper. We also adopted a multiresolution approach associated to the dynamic visualization employed by [2] and [1]. Multiresolution analysis of diffusion tensor images can be found in the literature. Rodrigues et al [16], for example, proposed a scale-space representation of a DTI image, using a multiresolution watershed segmentation method to separate coarse from fine data. They generated a hierarchical representation afterwards, through a cross scale linking of the segmented regions. In this paper, we present a different and much simpler multiresolution scheme, based on wavelet theory.

3 Fundamentals

3.1 Tensors

Second-order tensors can be defined as linear transformations between vector spaces. They are represented by 3x3 matrices. In this work, a tensor of particular interest is the one presented by Westin [6]. It is called a local orientation tensor, and it is a special case of a non-negative symmetric rank 2 tensor. This tensor can be used to estimate orientations in a field. Mathematically, it can be defined as following:

$$\mathbf{T} = \sum_{i=1}^{n} \lambda_i e_i e_i^T$$

where λ_i represent the eigenvalues and e_i the associated eigenvectors.

In $\mathbb{R}^3$, the equation above can be decomposed in such a way that $\mathbf{T}$ can be expressed in terms of its linear, planar and spherical intrinsic features [1]. So, the tensor definition becomes:

$$\mathbf{T} = (\lambda_1 - \lambda_2)\mathbf{T}_l + (\lambda_2 - \lambda_3)\mathbf{T}_p + \lambda_3 \mathbf{T}_s$$

This decomposition reveals an important geometric interpretation about the tensor. Assuming that $\lambda_1 \geq \lambda_2 \geq \lambda_3$, we can analyze the eigenvalues to identify the shape of the tensor, which is of much more use than its magnitude, for example. If, for instance, we have $\lambda_1 >> \lambda_2 \approx \lambda_3$, the tensor is approximately linear. If $\lambda_1 \approx \lambda_2 >> \lambda_3$, then the shape of the tensor is approximately planar. Finally, if all eigenvalues are almost equal, then the tensor is approximately isotropic. In this case, there is no main orientation present in the tensor.

Coefficients of Anisotropy. The tensor eigenvalues can be used to calculate coefficients of anisotropy. The eigenvalues are obtained by solving $\det(\lambda \mathbf{I} - \mathbf{D}) = 0$. We can define three of these coefficients: linear (c_l), planar (c_p) and spherical (c_s). These three coefficients must sum to 1.

$$c_l = \frac{\lambda_1 - \lambda_2}{\lambda_1 + \lambda_2 + \lambda_3} \tag{1}$$

$$c_p = \frac{2(\lambda_2 - \lambda_3)}{\lambda_1 + \lambda_2 + \lambda_3}$$

$$c_s = \frac{3\lambda_3}{\lambda_1 + \lambda_2 + \lambda_3}$$

It is also possible to calculate a number of coefficients which are insensitive to basis changing. These coefficients are called algebraic invariants. Among them, the ones presented below are helpful in the definition of a series of parameters that can be used to analyse the characteristics of the field.

$$J_1 = \mathrm{tr}(\mathbf{D})$$

$$J_2 = \frac{\mathrm{tr}(\mathbf{D})^2 - \mathrm{tr}(\mathbf{D}^2)}{2}$$

$$J_3 = \det(\mathbf{D})$$

$$J_4 = ||\mathbf{D}||^2$$

where $\mathrm{tr}(\mathbf{D})$ and $\det(\mathbf{D})$ are the trace and the determinant of $\mathbf{D}$, respectively [1].

Kindlmann [17] presents three more algebraic invariants, which are used not only to describe what is called the eigenvalue wheel, but also to define the central moments of the tensor. These invariants are defined as follows:

$$Q = \frac{3J_4 - 3J_1^2}{18}$$

$$R = \frac{-5J_1J_2 + 27J_3 + 2J_1J_4}{54}$$

$$\Theta = \frac{1}{3}\cos^{-1}\left(\frac{R}{\sqrt{Q^3}}\right)$$

The central moments are related to the geometric parameters of the wheel. The definition of the wheel, along with a detailed explanation of it, can be found in the work by Kindlmann [17]. Using the Kindlmann invariants, we can define the central moments as shown below:

$$\mu_1 = \frac{J_1}{3}$$

$$\mu_2 = 2Q$$

$$\mu_3 = 2R$$

The second central moment μ_2 represents eigenvalues variance. Taking its square root, we can obtain the standard deviation σ. This allows us to define an important parameter in this work, called the asymmetry of the eigenvalues. The asymmetry parameter varies from negative to positive as the tensor changes from planar to linear. It is calculated as follows [18]:

$$A_3 = \frac{\mu_3}{\sigma^3} = \frac{R}{\sqrt{2Q^3}} \qquad (2)$$

A more complete description of several anisotropy coefficients are found in [1]. In this paper, we use only the A_3 and c_l coefficients since they indicate linear and planar continuities suitable for particle tracing.

3.2 Tensorlines

Previous works intended for path tracing in tensor fields lacked stability in certain scenarios. The hyperstreamlines method [19] used just the main tensor eigenvector in order to obtain a smooth tracing, but it was subject to degeneration. Seeking to work around the inherent problems of this method, Weinstein et al [9] proposed an extension called tensorlines. Instead of only using the main eigenvector to determine the path, it applies the tensor to a vector corresponding to the propagation direction in the previous step.

$$\mathbf{v}_{out} = \mathbf{T}\mathbf{v}_{in} \qquad (3)$$

The new vector $\mathbf{v}_{out}$ produced by the transformation above is linearly combined with $\mathbf{v}_{in}$ and the main eigenvector e_1. Thus, we obtain the new propagation vector, which is dependent on the shape of the tensor. It is calculated as follows:

$$\mathbf{v}_{prop} = c_l e_1 + (1-c_l)((1-w_{punct})\mathbf{v}_{in} + w_{punct}\mathbf{v}_{out}) \qquad (4)$$

The parameter w_{punct} lies in the range $[0,1]$ and defines how much the propagation should penetrate planar tensors. This parameter is controlled by the user. The coefficient c_l is the linear anisotropy coefficient defined in the previous subsection.

The vector field produced by the tensorlines method can be applied to a particle tracing procedure to visualize the tensor field. The particles introduced in the field have no mass. At each time step, we update the position of each particle over time t, using a vector v from the generated field as the velocity.

3.3 Multiresolution

In this work, we used a multiresolution scheme based on the Daubechies analysing filters. The Daubechies low pass filter is applied to the tensor field, separately for each tensor component, with the purpose of generating lower resolution fields with half spectrum of the previous scale. Since the tensor fields we worked on have the maximum dimensions of $148 \times 190 \times 160$, two decimated scales seemed to be enough for our purposes.

A decimated tensor captures the anisotropy of a group of local tensors. It can be used, for example, to filter the particle paths during tensorlines computation. The anisotropic features of the scaled tensors are linear combinations of the underlying tensors shape in full resolution. As such, its anisotropic coefficents bring new information to form an improved priority list. Details about signal multiresolution can be found on [20].

In the previous work of Leonel [1], an extensive list of parameters were used to calculate the importance of a single tensor to the observer. This importance was determined by a scalar parameterized by the user. Here, we present a new formulation for calculating this scalar with a reduced number of parameters, taking into account the lower resolution fields obtained. The next section presents the equations for this calculation and other contributions of this work.

4 Proposed Method

The previous approach [1] was conceived to induce the human perceptual system to detect continuity using particle motion. Particles in motion represent the features of the tensor field. One critical point in visualization using particle tracing is to define the particle starting point. The easier approach to insert particles into the domain is to compute new positions randomly. A fixed distribution function, however, generally does not insert new particles in most interesting sites. Using tensorlines to indicate suitable particle paths, the idea was to carefully select the position where a new particle should start. It was based on tensor field anisotropic features and viewer-dependent relationships. The maximum number of particles at a time was fixed. A priority list determined which particle should be chosen. Several coefficients for particle sorting were presented.

In this work, we propose major modifications for the priority list calculation and new particles selection. Our approach is based on the use of multiresolution of the tensor field. Each scale of a tensor field in multiresolution combines the tensor of the previous, higher resolution, scale. We exploit the anisotropic features of the resulting tensors to provide a new scalar value for the priority list

(Eq. 5). Viewer dependent and independent coefficients in multiresolution are computed (Fig. 1).

4.1 Priority Features

Let $\mathbb{T}_{x \times y \times z}$ be a discrete and finite tensor field with lattice given by $x, y, z \in \mathbb{N}$, so that $\mathbb{T}^s = \{\mathbf{t}_1^s, \mathbf{t}_2^s, \mathbf{t}_3^s ... \mathbf{t}_n^s\}$ is composed by $|\mathbb{T}^s| = n$ tensors, where s is the scale index. For a given voxel (a, b, c), where $a, b, c \in \mathbb{N}$ and such that $a \leq x$, $b \leq y$ and $c \leq z$, we have the correspondent tensor $\mathbf{t}_i^s \in \mathbb{T}$. As explained in Section 3.3, the tensor fields are decomposed two times in this work, resulting in three scales: $s = 0$ is the original tensor field, $s = 1$ is the tensor field with half spectrum, $s = 2$ is the tensor field with a quarter of original spectrum.

The eigensystem of a tensor $\mathbf{t}_l^s$, $1 \leq l \leq n$ is represented by the eigenvectors $\vec{e}_1^s \perp \vec{e}_2^s \perp \vec{e}_3^s$ and the eigenvalues $\lambda_1^s \geq \lambda_2^s \geq \lambda_3^s \geq 0$.

The goal of the priority list is to define in which lattice location a new particle should be inserted. In this work we propose a new scalar $\Upsilon \in \mathbb{R}$ which defines the priority of a voxel having a particle created in it. This priority is calculated using multiresolution tensors characteristics and geometric features of the scene (Fig. 1).

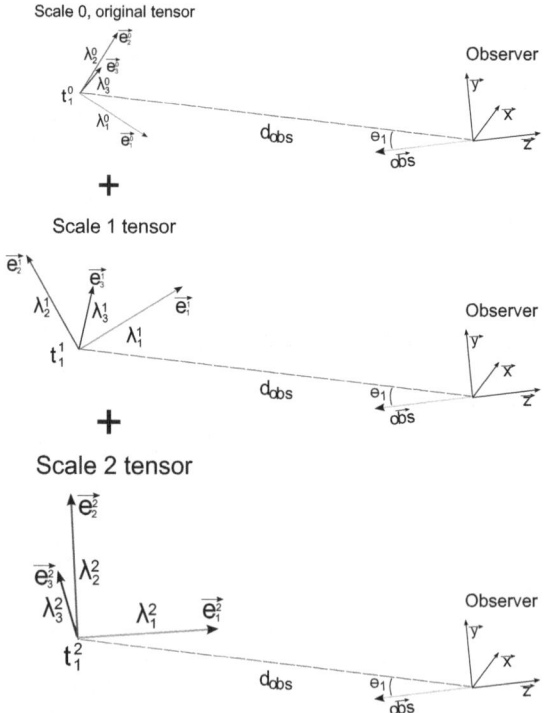

Fig. 1. Combination of tensors in multiple scales, related to the observer

In [1], the position and orientation of a tensor in relation to the observer are evaluated by three scalars k_1, k_2 and k_3. Using multiresolution with three scales $s = \{1, 2, 3\}$, the coefficients can be evaluated for each scaled tensor of a location. We propose the following scalars to capture the viewer-dependent orientation of the l-th multiresolution tensors $\mathbf{t}_l^1$, $\mathbf{t}_l^2$ and $\mathbf{t}_l^3$, all centered in the domain at position $\vec{x}_l$:

$$k_1^s = 1 - |\vec{e}_1^{\,s} \cdot \vec{obs}|$$
$$k_2^s = 1 - |\vec{e}_2^{\,s} \cdot \vec{obs}|$$
$$k_3^s = |\vec{e}_3^{\,s} \cdot \vec{obs}|,$$

where $\vec{e}_1^{\,s}$, $\vec{e}_2^{\,s}$ and $\vec{e}_3^{\,s}$ are the eigenvectors of the tensor and $\vec{obs}$ corresponds to the camera view vector (Fig. 1). Thus, we propose nine scalars to capture the orientation of the local tensor in relation to the observer, which means three values for each of the three scales. These nine values quantify the relative position of the observer with respect to the tensors eigensystems, so that we can prioritize tensors representing colinear or coplanar structures which are perpendicular to the observer.

We need to define the weight of each scale in the calculation of the priority value. A simple but effective approach is to fix the weights in 2.0 for the original tensor (scale 1), 1.0 for the intermediate tensor (scale 2), and 0.5 for the tensor of the maximum scale 3. The distance of the tensor to the observer d_{obs}:

$$d_{obs} = \frac{|\vec{x}_l - \vec{x}_{obs}|}{MAX(d_{obs})},$$

which is normalized by the greatest distance in the field $MAX(d_{obs})$, gives the the viewer-dependent weight:

$$w = 1 - d_{obs}.$$

The scalar Υ, that indicates the priority of a voxel to receive a particle, is defined as:

$$\begin{aligned}\Upsilon_t =& 2.0 \cdot w \cdot (A_3^1 + c_l^1 + k_1^1 + k_2^1 + k_3^1) + \\ & 1.0 \cdot w \cdot (A_3^2 + c_l^2 + k_1^2 + k_2^2 + k_3^2) + \\ & 0.5 \cdot w \cdot (A_3^3 + c_l^3 + k_1^3 + k_2^3 + k_3^3),\end{aligned} \quad (5)$$

which is a linear combination of the following terms:

- coefficient of linear anisotropy of the scaled tensor (c_l^s) (Eq. 1);
- asymmetry of the tensor eigenvalues (A_3^s): changes from negative to positive as the scaled tensor vary from planar to linear (Eq. 2);
- coefficients of orthogonality between the observer and the first eigenvector of each scaled tensor (k_1^s): bigger if the main direction of the tensor is perpendicular to the view vector;

- coefficients of orthogonality between the observer and the second eigenvector of each scaled tensor (k_2^s): bigger if the second main direction of the tensor is perpendicular to the view vector;
- coefficients of parallelism between the observer and the third eigenvector of each scaled tensor (k_3^s): bigger if the third eigenvector is aligned with the view vector, which implies that the other eigenvectors are perpendicular to the observer.

4.2 Particle Insertion

A maximum number of particles $N_p \in \mathbb{N}$ is fixed by the user. This value is generally small compared to the size of the tensor field. At most N_p particles exist and walk through the field at a given time. The priority list is used to achieve better visualization results by inserting particles in the more interesting sites.

When the simulation begins or the user changes its position or orientation, all tensors $\mathbf{t} \in \mathbb{T}^0$ have their priority value computed (Eq. 5). They are sorted in a list where the highest priorities are positioned on the top. Using the N_p topmost tensors, the total priority is computed:

$$m = \sum_{l=1}^{N_p} |\Upsilon_l|.$$

The topmost tensors $\mathbf{t}_l \in \mathbb{T}^0$, $1 \leq l \leq N_p$, are allowed to have

$$n_l = N_p \cdot \frac{|\Upsilon_l|}{m}$$

particles, which represents the proportion of new particles that can be assigned to the position of the tensor $\mathbf{t}_l$ along an insertion round. Note that some of the tensors will not have enough priority to receive a particle. Particles are thus created in less than N_p tensor positions.

Initially, there are N_p particles to be inserted into the domain at the beggining of an insertion round. If we insert n_l particles for the topmost tensor, there may be several particles walking together or very close to each other. This is not desired because multiple particles together are not visually salient. Thus, we propose to assign only one particle to each tensor of the list (with non-zero n_l) at a time, decrementing its n_l value upon insertion. If there are still particles left for insertion after visiting the position N_p of the list we return to its topmost tensor, running through the list in a circular way. When all particles are inserted, all n_l are zero, indicating the end of an insertion round. We then reestablish n_l and a new insertion round begins. This round-robin policy for particle insertion guarantees all sites with non-zero n_l have at least one particle inserted before any previously assigned tensor is visited again.

Note that only N_p particles are viewed in the domain. As the simulation runs, some particles are removed. Their reinsertion obeys the round-robin policy and the visual result are well distributed particle clouds. The topmost tensors are guaranteed to have more particles inserted during simulation.

The Υ_l scalar has viewer-dependent terms, so, it is necessary to reorder the priority list when the camera position or orientation changes. A merge sort algorithm is enough for having good response times with 100.000 particles.

The simulation and the particle removal steps are explained in [1]. Given the tensorlines, a simple advection step determines the next position of a particle. Due to isotropic regions in the tensor field, particles can get stuck. To reduce the creation of particles in an isotropic region, some tensors are flagged as bad places when particles inserted on them are removed after few advection iterations. Those tensors periodically receive particles that disappear rapidly, generating flickering regions. Their elimination from the particle insertion process resulted in a much better visualization.

5 Results

Here we present the results for the application of our method to three different tensor fields: the 3-point field, the helical flow and a diffusion tensor field of a brain. In all of the experiments, we used the color palette shown in Figure 2 to represent the importance of a given tensor to the observer. Each particle was represented as a pointer glyph, just as shown in [1].

Fig. 2. Color palette used for Υ [1]

For the helical field with 38x39x40 grid, we used 7000 particles spread through the sites according to the generated priority list. In this process, 3214 sites in isotropic regions were eliminated from the 12073 possible ones. Figure 3 shows the helical field. Figure 4 shows the visualization of the helical tensor field from different points of view. Notice that tensors nearer and perpendicular to the observer tend to have higher priority, thus their color being closer to red. As the camera orientation is changed, the priority list is recalculated and the new best ranked tensors in the list are then displayed with proper colors. This can be seen by looking at the density of particles. The amount of particles decreases as it gets far from the observer, since the nearest sites have higher priority.

Next, we present the results obtained for a diffusion tensor field of a brain [21] (Fig. 5). These tensor fields are usually generated by magnetic resonance

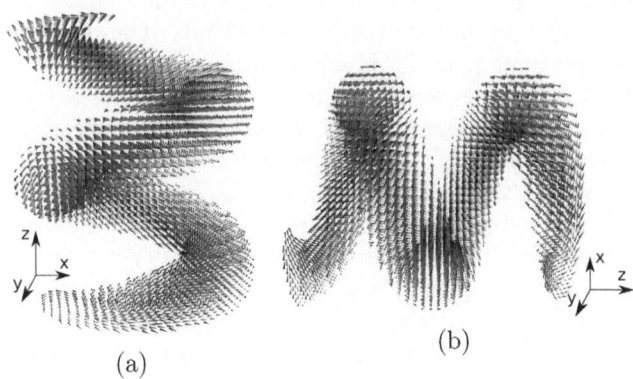

Fig. 3. The helical tensor field visualization from two different angles

imaging. They are very useful in detecting fibers, which are represented by regions of high linear and planar anisotropy. Similarly, the crossings of white matter tracts in the brain are also identified with higher planar anisotropy. Thus, we can use this knowledge to enhance the visualization of these regions of the brain DT image.

Figures 6 and 7 show examples of the field visualization under different camera orientations. In this simulation, the grid dimension was 74x95x80 and the number of particles created was 15000. Plus, 12366 sites were flagged for elimination from the 24582 initially available. In this field it was possible to see one of the main advantages of excluding sites from creation: significantly reduction of flickering effect. The removal criterion destroys particles which they could cause flickering by reaching isotropic regions. But particles that have a short time between their creation and destruction, like 1 or just 2 simulation steps, also result in flickering sensation. As mentioned in Section 4, we exclude sites in which created particles are soon destroyed, and that really presented a better view for the simulation.

Finally, we simulated a 3-point field (Fig. 8). It represents a 38x39x40 grid where there are three spherical charges at positions (0,0,0), (38,0,40) and (38,39,40). The tensor field is calculated as the geometric influence of all three charges at every position of the grid.

With this example it was possible to analyse some features of tensor fields that are of interest for visualization: continuities and curvatures. We could see the importance of the anisotropy factor on choosing where to create the particles (Fig. 9). This visualization used 30000 particles. This factor combined with the relative position of the observer creates a huge flow near the observer, allowing to follow the particles and to notice the smoothness of most of the field. There were 9115 eliminated sites from a total of 56495 initially possible.

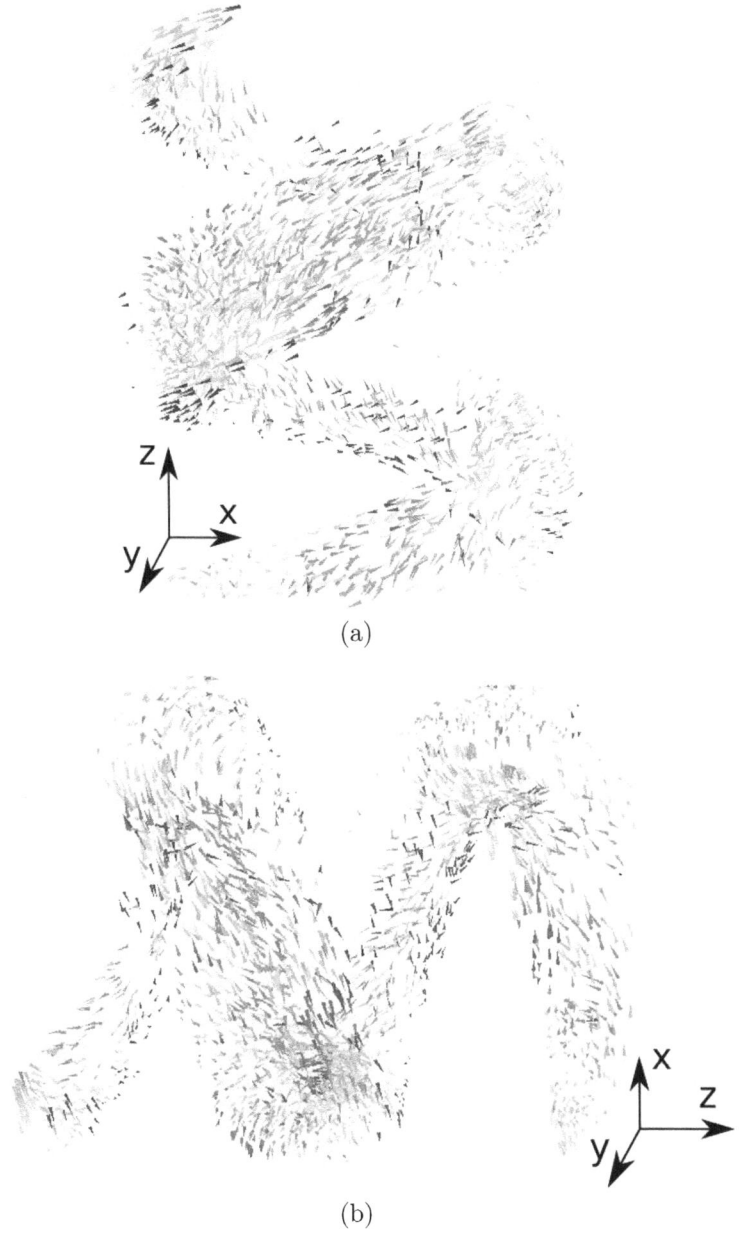

Fig. 4. Visualization of the helical tensor field with 7000 particles

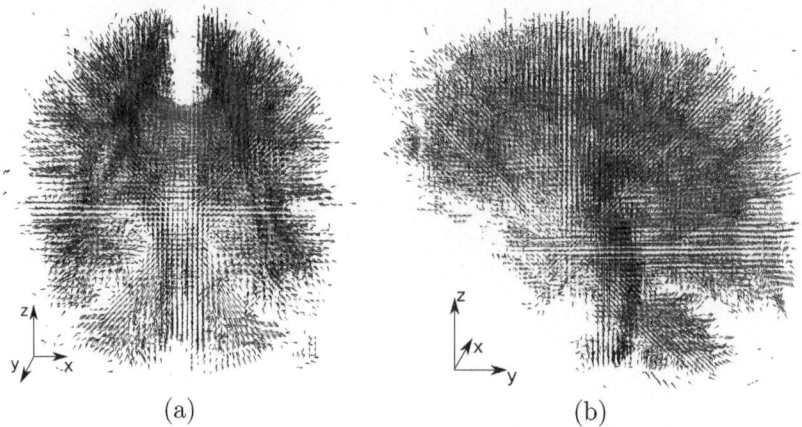

Fig. 5. The original brain tensor field visualization from two different angles

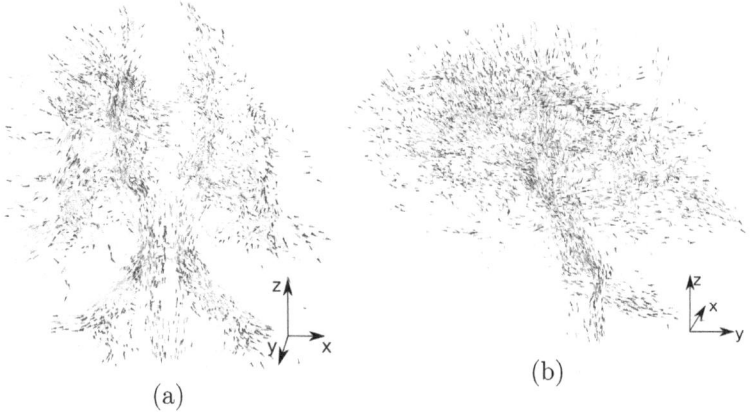

Fig. 6. Visualization of the brain field simulation associated to the viewing angles in Figure 5

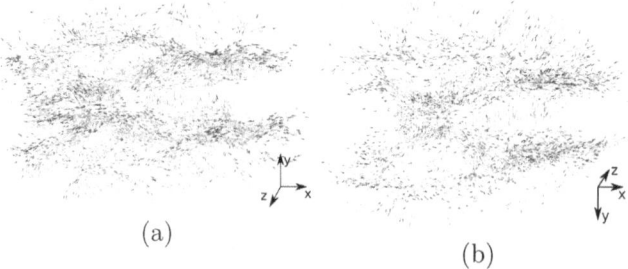

Fig. 7. Additional viewing angles from the brain simulation

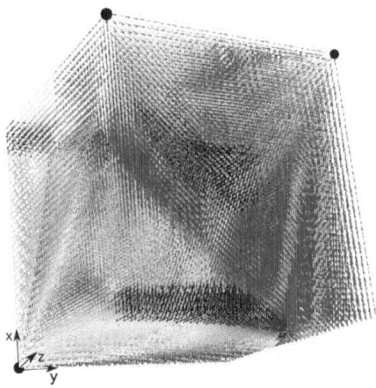

Fig. 8. Original 3-point field. The black circles represents charge positions

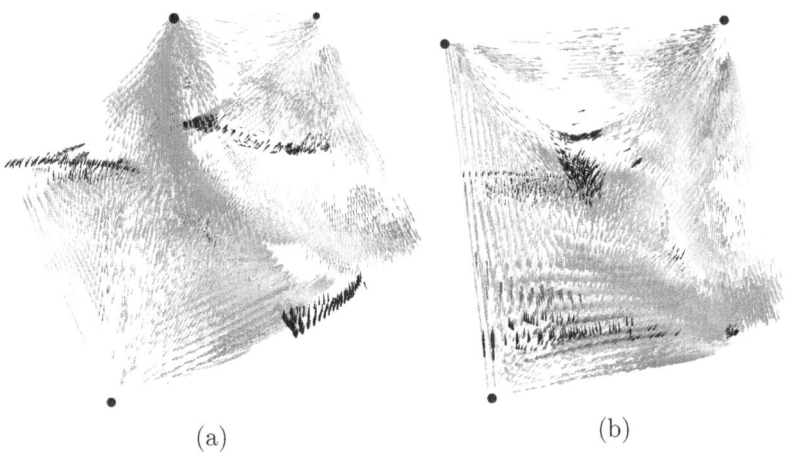

Fig. 9. Visualization of the 3-point field simulation at two different simulation steps

6 Conclusion

Choosing where to create particles in a tensor field for a good visualization is not an easy task. We have combined multiresolution coefficients of the field with viewer-dependent terms in order to evaluate the importance of each site of the grid at the current observer's position. The multiresolution coefficients allowed us to check the anisotropy of the field at different scales. With this information, it was possible to reduce the high amount of terms used on our last approach [1] for ranking each possible creation site. The priority list using multiresolution information and a deterministic algorithm for balanced particle insertion are the main contributions of this paper.

We have shown our results on three different tensor fields. Increasing the capacity for creating particles at higher priority sites did concentrate a large number of particles on the spots of most interest, near the observer. We have also determined rules to permanently remove sites from the priority list. By eliminating these sites from the list we could reduce the flickering problem we had to almost none.

The multiresolution terms of the priority value (Eq. 5) represent smoothed tensor structures. Our results show that these local and filtered anisotropy estimations have improved the particle tracing proposed in [1] for tensor field visualization, since tensors representing colinear and coplanar structures (anisotropic in several scales) tend to have more particles during simulation.

References

1. de Almeida Leonel, G., Peçanha, J.P., Vieira, M.B.: A Viewer-Dependent Tensor Field Visualization Using Particle Tracing. In: Murgante, B., Gervasi, O., Iglesias, A., Taniar, D., Apduhan, B.O. (eds.) ICCSA 2011, Part I. LNCS, vol. 6782, pp. 690–705. Springer, Heidelberg (2011)
2. Kondratieva, P., Krüger, J., Westermann, R.: The application of gpu particle tracing to diffusion tensor field visualization. In: Visualization, VIS 2005, pp. 73–78. IEEE (2005)
3. Shaw, C.D., Ebert, D.S., Kukla, J.M., Zwa, A., Soboroff, I., Roberts, D.A.: Data visualization using automatic, perceptually-motivated shapes. In: Proceeding of Visual Data Exploration and Analysis. SPIE (1998)
4. Shaw, C.D., Hall, J.A., Blahut, C., Ebert, D.S., Roberts, D.A.: Using shape to visualize multivariate data. In: NPIVM 1999: Proceedings of the 1999 Workshop on New Paradigms in Information Visualization and Manipulation in Conjunction with the Eighth ACM Internation Conference on Information and Knowledge Management, pp. 17–20. ACM, New York (1999)
5. Kindlmann, G.: Superquadric tensor glyphs. In: Proceedings of IEEE TVCG/EG Symposium on Visualization 2004, pp. 147–154 (May 2004)
6. Westin, C.F.: A Tensor Framework for Multidimensional Signal Processing. PhD thesis, Linköping University, Sweden, S-581 83 Linköping, Sweden (1994) Dissertation No. 348, ISBN 91-7871-421-4
7. Delmarcelle, T., Hesselink, L.: Visualization of second order tensor fields and matrix data. In: VIS 1992: Proceedings of the 3rd Conference on Visualization 1992, pp. 316–323. IEEE Computer Society Press, Los Alamitos (1992)
8. Delmarcelle, T., Hesselink, L.: Visualizing second-order tensor fields with hyper streamlines. IEEE Computer Graphics and Applications 13(4), 25–33 (1993)
9. Weinstein, D., Kindlmann, G., Lundberg, E.: Tensorlines: advection-diffusion based propagation through diffusion tensor fields. In: VIS 1999: Proceedings of the Conference on Visualization 1999, pp. 249–253. IEEE Computer Society Press, Los Alamitos (1999)
10. Vilanova, A., Zhang, S., Kindlmann, G., Laidlaw, D.: An introduction to visualization of diffusion tensor imaging and its applications. Visualization and Processing of Tensor Fields, 121–153 (2006)
11. McGraw, T., Nadar, M.: Stochastic dt-mri connectivity mapping on the gpu. IEEE Transactions on Visualization and Computer Graphics 13(6), 1504–1511 (2007)

12. Köhn, A., Klein, J., Weiler, F., Peitgen, H.: A gpu-based fiber tracking framework using geometry shaders. In: Proceedings of SPIE Medical Imaging, vol. 7261, p. 72611J (2009)
13. Evert, A., Neda, S., Andrei, J.: Cuda-accelerated geodesic ray-tracing for fiber tracking. International Journal of Biomedical Imaging (2011)
14. Mittmann, A., Nobrega, T., Comunello, E., Pinto, J., Dellani, P., Stoeter, P., von Wangenheim, A.: Performing real-time interactive fiber tracking. Journal of Digital Imaging 24(2), 339–351 (2011)
15. Crippa, A., Jalba, A., Roerdink, J.: Enhanced dti tracking with adaptive tensor interpolation. Visualization in Medicine and Life Sciences II, 175–192 (2012)
16. Rodrigues, P., Jalba, A., Fillard, P., Vilanova, A., ter Haar, B.: A multi-resolution watershed-based approach for the segmentation of diffusion tensor images. In: MICCAI Workshop on Diffusion Modelling, pp. 161–172 (2009)
17. Kindlmann, G.: Visualization and Analysis of Diffusion Tensor Fields. PhD thesis (September 2004)
18. Bahn, M.: Invariant and Orthonormal Scalar Measures Derived from Magnetic Resonance Diffusion Tensor Imaging. Journal of Magnetic Resonance 141(1), 68–77 (1999)
19. Delmarcelle, T., Hesselink, L.: Visualization of second order tensor fields and matrix data. In: Proceedings of IEEE Conference on Visualization 1992, pp. 316–323. IEEE (1992)
20. Mallat, S.: A Wavelet Tour of Signal Processing. The Sparse Way, 3rd edn. Academic Press (2008)
21. Kindlmann, G.: Diffusion tensor mri datasets, http://www.sci.utah.edu/~gk/DTI-data/

Abnormal Gastric Cell Segmentation Based on Shape Using Morphological Operations

Noor Elaiza Abdul Khalid[1], Nurnabilah Samsudin[1], and Rathiah Hashim[2]

[1] Faculty of Computer and Mathematical Sciences, UiTM,
40450 Shah Alam, Malaysia
elaiza@tmsk.edu.my, nurnabilahsam@gmail.com
[2] Faculty of Computer Science and Information Technology, UTHM,
86400 Parit Raja, Malaysia
radhiah@uthm.edu.my

Abstract. Cancer is the fourth leading cause of death among medically certified deaths in Malaysia. The most reliable diagnostic method to diagnose gastric adenocarcinoma is by inspecting the microscopic images of samples obtained through biopsy. These images are analyses by pathologist to identify the presence of cancer. However the process is time consuming and the interpretation varies with different pathologist. The application of image analysis techniques can assist pathologist towards a more efficient and faster diagnosis. Thus, this paper introduces an image analysis framework to automatically recognize and distinguished between normal gastric and gastric adenocarcinoma cells. The framework consist of the three phases of image analysis; preprocessing phase where the color tone issues are solved by component separation; processing phase which includes the thresholding and morphological techniques to segment the cells; post processing to identify the perimeter, area and roundness of the cells. This study shows that it is possible to automatically recognize and differentiate images with normal and abnormal cells.

Keywords: Segmentation, Stomach, Morphological Operation, Abnormal Cell detection, Roundness, Image Processing.

1 Introduction

In Malaysia cancer is the fourth leading occurrence of certified deaths and gastric adenocarcinoma is considered the 6th most common cancer affecting the local population [1]. Gastric adenocarcinoma is a type of gastric cancer which arises from epithelial cells of stomach/gastric glands [1], [2]. Abnormal gastric cells can be distinguished from the normal cells in terms of the shape (normal cells are usually round and abnormal cells are irregular in shape and often a fusion of several cells), alignment (normal cells are well aligned but abnormal cells are not aligned) and the nucleus of the mucosa cell (the size is almost the same but bigger in abnormal cells) [3]. Early diagnosis will enhance the prognosis for cancer patients. Currently images are processed digitally because it is fast, flexible, and precise [4]. The development of image processing algorithms is to mainly facilitate image manipulation that is suitable for different industries like medical [5], [6], engineering [7] and quality controls [8].

Recognition and feature extraction plays an important role in detecting various human anatomical structures and abnormalities [9] from medical images. Medical images are naturally complex, often noisy and have low in contrast [10].

Present method of diagnosing gastric adenocarcinoma is manual observation of the gastric cells pathological appearance in microscopy image [11]. This method is tedious, slow, tiresome, and may result in differential interpretation by different clinicians [12]. The distinct boundary of each cell makes it possible for a computer system to recognize the differences between normal and abnormal cell. Conversely this process involves several image processing and analysis techniques. Previous studies have shown that this kind of process should include the three basic phases in image analysis which includes the pre-processing, processing and post-processing phase [13], [14]. The preprocessing phase is usually the contrast enhancement and noise reduction phase which includes image smoothing [15], image enhancement [16] and color component separation [17]. On the other hand the processing phase usually involve feature and texture extraction which includes statistical [18], pattern recognition [19], clustering [20] and segmentation [21], [22], [23], [24] techniques. Finally the post processing techniques will include the recognition phase which engages the object identification [25] and shape descriptor [26], [27] method.

This paper proposes a framework of image processing techniques to distinguish between normal and abnormal gastric cells by computing the area, perimeter, and roundness. The framework consists of five steps namely color component separation, thresholding, image complement, opening morphology and finally calculations of the measurements of the cells.

2 Methods

2.1 Data Collections

The microscopic images of stomach epithelial cells were digitized and in JPG format (.jpg). In total, there were 30 images collected for both images of normal and abnormal cell; all images collected were used in this research. The microscopic images were obtained by clinicians through biopsy; where tissue sample were removed from the stomach [28] which followed by observation under the microscope where the abnormal cells can be detected through careful examination. These microscopic images were color images as clinicians dyed the images, producing RGB images.

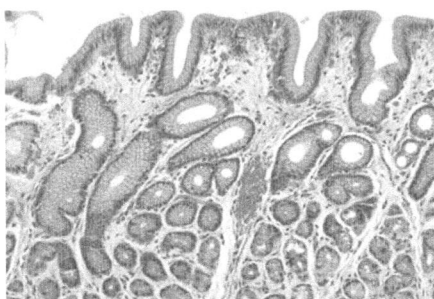 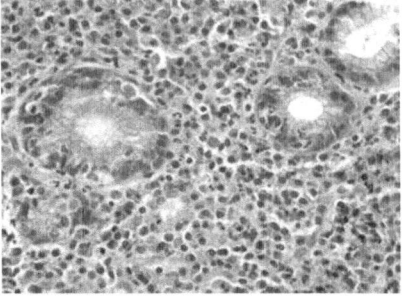

Fig. 1. Image of gastric cells where the cancerous cell shape is abnormal and not circle (left) [30] compared to normal gastric cell which have rounder shape (right) [31]

Figure 1 is the microscopic images of normal and abnormal cross section of gastric glands where the shape outline is obviously different. Gastric mucosa epithelial nuclei appear round or oval shaped unlike abnormal cell [29].

2.2 Proposed Framework Design

Table 1 is the research framework of the project; explained in details for three phases' image pre-processing, processing and post processing. The preprocessing stage consists of color channel separation. The processing stage involves the thresholding, image compliment, and morphological operation. Finally the post processing stage includes the recognition of the normal and abnormal cells.

Table 1. Research framework

Phase	Activity	Output images
Preprocessing	Color Channel separation; RGB image divided into channels	Red channel / Green channel / Blue channel
Processing (Segmentation)	Thresholding	
	Image Complement	

Table 1. (*Continued*)

	Morphological Operation	
Post processing	Roundness, Area, and Perimeter roundness	

(a) Pre Processing

Color channel means each channel express how bright the color for each pixel in the image [32]. Raw input image obtained were color image, which comprise of red, blue, and green plane (RGB). Due to the collection of original images which were mostly red, the red plane provides more info compared to the other two planes. However other suitable channel which is green can be chosen depending on which provides the best information on based gray scale distribution. Output of RGB sliced channel produced grayscale image output because the channels represent the intensity of the color and example can be seen in Figure 2.

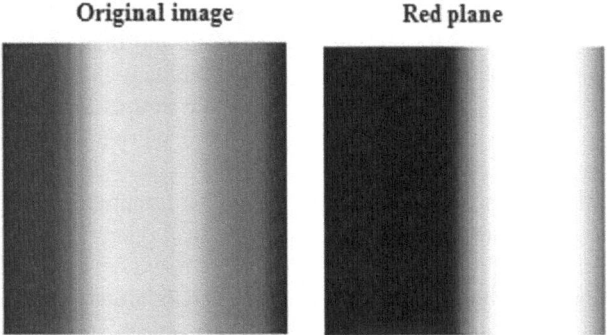

Fig. 2. Red plane (Right) represents higher intensity of red by producing white, while black area represents there is no red represented and for grey represent medium intensity of red intensity. Thus, plane slicing produced grey scale output images [33].

Color channel step is where user were able to choose either red or green channel based on which image outline is clearer and provide more suitable feature. The red plane were chosen because the dye used for the microscopic image is red, thus the red channel proves to provide more data compared to the other channels. However due to the variation of dyes and image background, sometime green channel can provide more data compared to red channel. However, the blue channel has the lowest intensity and provides less data thus is not chosen.

(b) Processing

Thresholding followed by segmentation using morphological operation may increase compact nature and roundness of cells [34]. The definition of thresholding is converting gray scale image to black and white image, where the values above threshold value were converted to white while values below the threshold becomes blacks [35].

The thresholding value varies for the images in the data collection due to the resource variation. However, the values used for abnormal cells vary from 90 to 250 and normal from 90 to 230. The variation relies manually to user where user may insert value they estimated to obtain the most preserved form of the shape outline. Thresholding aim in this research is to partially remove unwanted pixel while preserving the cell shape outline.

Image complement function is to change the binary color from black to white and white to black. This function is suitable for this research to make sure the object of interest (the cells) is white while the background is black. This step is essential for this research as it allows next steps to be carried out.

Morphological operations were originally developed for binary images but later were extended for the use of gray scale images. [36] used morphological operation on gray scale gastric tumour cell to detect the edges, however concluded that gray scale edge detection requires more complex computations compared to binary edge detection. The research also advocate the use of mathematical morphology for identifying shape of microscopic cell post to thresholding process and concluded that processing binary image for edge detection yield better result.

The operations involved include erosion and dilation. Erosion is to shrink an image object, depending to the structuring elements applied and dilation is to expand the object of an image. Morphological opening is the combination of erosion followed by dilation by using the same structuring elements, while morphological closing is dilation followed by erosion and the main purpose is to morphologically open binary images which were small objects.

Structuring element used is a set of 3 by 3 pixels at brightness value '1'. Morphological opening is combined by two basic morphological operators which are the erosion and the dilation (Refer to Equation 1) [37]. Both operators used same structuring element for shape smoothing, removal of smaller details around the border and also remove small bridges of pixel that connects two larger objects.

$$X \circ B = \left(\bigcap_{b \in B} X_{-b} \right) \oplus B \tag{1}$$

Where X is original image, B is the structuring element used for both operations.

(c) Post Processing

The outlines of the cell were obtained and to detect whether the cells are normal (round shape) or abnormal, it is evaluated through calculation of the area, perimeter, and most importantly the roundness measurement calculated using Equation 1, Equation 2 and Equation 3 respectively.

The area of the cell was calculated using the equation below;

$$A = \Pi r^2 \qquad (2)$$

Where A stands for area, while π value is 3.142 and r is radius of the circle. The perimeter of the cell was calculated using the equation below;

$$d = 2r \qquad (3)$$

Where d stands for diameter and r is radius of the circle. The roundness of each cell was calculated using the equation below;

$$Metric = 4\Pi \left(\frac{Area}{Perimeter} \right) \qquad (4)$$

The parameters previously mentioned were used to classify the cells into normal or abnormal. In order to evaluate the classification, the main source [38], [39] for measuring performance is through contingency table, or also been using terms like false positive, false negative, Type I error or Type II error. The contingency table is important to represent analysis and stating relationship [40] between two or more variables; this research variables' are the experimentally classified cells (Predicted Class) and the real state of cells; whether it is normal or abnormal (True Class). The table can be seen in Table 2.

Table 2. Contingency table

Predicted Class	True Class	
	Normal Cell	Abnormal Cell
Normal Cell	True Positive	False Positive (Type II Error)
Abnormal Cell	False Negative (Type I Error)	True Negative

If a normal cell (True Class) was experimentally classified as abnormal (Predicted Class), then it is False Negative, or also called Type I Error, meanwhile when an abnormal cell was experimentally classified as a normal cell, it is a False Positive or also called as Type II Error.

3 Result and Evaluation

The results will be discussed in to sections which are on visual appearance and evaluation using the perimeter, area and roundness measurements. The visual appearance of the normal cells can be seen in Figure 3 shows the original image on the left and the segmented image on the right. In the segmented image, there are colors like dark blue, light blue, very light blue, green, yellow, orange and red because after the image is labeled with its value, it is converted into RGB image to assist user eye to see the cells clearer and allow user to see and differentiate number of cells in the particular image segmented by the application. It is observed that most of the normal cells are of regular shaped which is almost round and oval apart from a few irregular shaped. Observing the segmented cells labeled as A, the cells are actually segmented incorrectly where two cells are segmented as one. This could be due to error during the segmentation process. On another note, not all the cells are detected and segmented. Furthermore, some non- celled area are also detected and segmented as the cell labeled B.

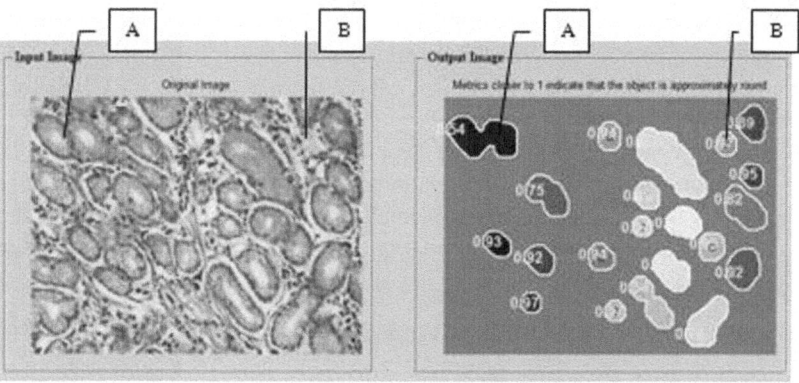

Fig. 3. Original image (Left) and the segmented normal cells (Right)

On another hand the abnormal cell images in Figure 4 are more easily detected due to is prominent size and less surrounding anatomical features.

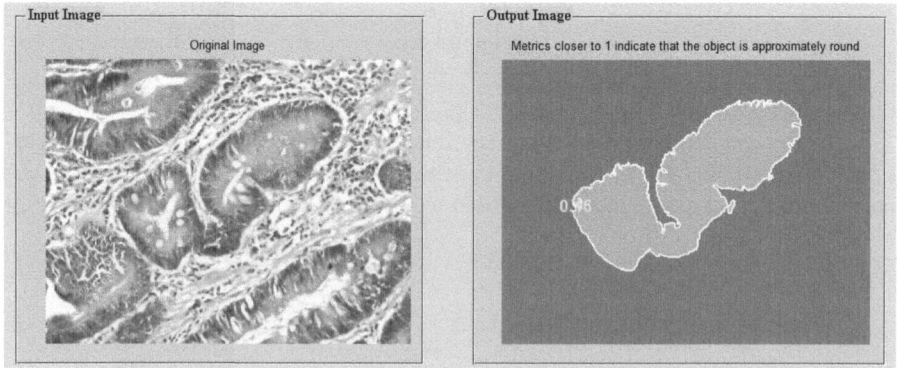

Fig. 4. Original image (Left) and the segmented abnormal cells (Right)

Table 3. Perimeter, Area and Roundness measurement comparison

Image	Perimeter	Area	Roundness
Normal Cells	350– 700	3 142 – 24 932	0.43 – 0.97
Abnormal Cells	1 000– 10 000	1 791 – 139 232	0.06 – 0.87

Table 3 shows the results for the post-processing phase which involves the extraction of the ranges for the perimeter, area and roundness.

Notice that the perimeter measurements for the abnormal cell images are significantly larger (between the range of 1000-10000) than the normal cell images (between the range of 350– 700).The area measurements for the images that contain the abnormal cells have a very large range (between the range of 1 791 – 139 232). This can be due to the detection of some normal and non-cell area. The area measurement range for the normal cells images are lower (between the range of 3 142 – 24 932) which could be due to the regularity and size of cells detected. Finally the roundness range for the normal cells ranges between 0.43 – 0.97. The lower value of roundness could be the outcome of the detection of non-cells area and some detected cells may not be well segmented. In the matter of the abnormal cells, the roundness are detected between 0.06 – 0.87. Lower range values should be the expected results due to the abnormal cells irregular shape but the upper range could be explained by the presence of some normal cells and abnormal cells that is round in shape.

Table 4. False positive (Type I Error) and false negative (Type II Error) values according to natural condition; normal gastric cells

Error Type	Perimeter (%)	Area (%)	Roundness (%)
Type I Error	50	33.33	50
Type II Error	-	-	-
No error	-	66.67	50
System error	50	-	-

Table 4 shows the error types occurred to all parameters used for cell classification, where the true state of the cell is a normal gastric cell. The research misidentified normal cell as abnormal (Type I Error) where there are 50% for both roundness and perimeter measurement while only 33.33% error by using area measurement to classify the segmented cells. The application proved to be able to classify the cells correctly where 66.67% correct by using area measurement and 50% for roundness measurement.

Table 5. False positive (Type I Error) and false negative (Type II Error) values according to natural condition; abnormal gastric cells

Error Type	Perimeter (%)	Area (%)	Roundness (%)
Type I Error	-	-	-
Type II Error	-	16.67	-
No error	66.67	33.33	83.33
System error	33.33	50	16.67

For classifying abnormal cells, Table 5 above shows that only 16.67% Type II Error was made by using area measurement. The application segmentation technique is reliable as it can classify abnormal cells where 83.33% by using roundness measurement, 66.67% by using perimeter, and 33.33% through area measurement.

4 Conclusion

Even though there are a number of discrepancies, image analysis techniques could be designed to automatically recognize and distinguish the difference between images with normal and abnormal cells. The proposed framework could be used to identify, recognize and distinguish between microscopic images that contain normal or abnormal cells. The framework sensitivity for segmentation and classification of abnormal gastric cell is quite high, 61.11% while for normal cell the sensitivity is 38.89%. The effectiveness, efficiency and accuracy can be further improved by using more accurate techniques such as segmentation and clustering image analysis.

Acknowledgement. The authors would like to thank Mara University of Technology (UiTM) and Universiti Tun Hussein Onn Malaysia (UTHM) for all supports.

References

1. Lim, G.: Overview of Cancer in Malaysia. Japanese Journal of Clinical Oncology 42, S37–S42 (2002)
2. Tadataka, Y.: Principles of Clinical Gastroenterology. Wiley-Blackwell, Massachusetts (2011)
3. Dicken, B.J., Bigam, D.L., Cass, C., Mackey, J.R., Joy, A.A., Hamilton, S.M.: Gastric Adenocarcinoma: Review and Considerations for Future Directions. Annals of Surgery 241, 27–39 (2005)
4. Rapp, C.: Image Processing and Image Enhancement (1996)
5. Ibrahim, S., Khalid, N., Manaf, M.: Seed-Based Region Growing (SBRG) vs Adaptive Network-Based Interface System (ANFIS) vs Fuzzy c-Means (FCM): Barin Abnormalities Segmentation. International Journal of Electrical and Computer Engineering 5(2), 94–104 (2010)

6. Khalid, N.: CR Images of Metacarpel Cortical Edge Detection-Bone Profile Histogram Approximation Method. In: Manaf, M., Aziz, M., Ali, M. (eds.) International Conference on Intelligent and Advanced Systems, ICIAS, Kuala Lumpur, pp. 702–708 (2007)
7. University, P.: Communications, Networking, Signal & Image Processing. Purdue University: School of Electrical and Computer Engineering,
   http://engineering.purdue.edu/ECE/Research/Areas/CommSigP.whtml
8. Spinola, C.: Image Processing for Surface Quality Control in Stainless Steel Production Line. In: Cañero-Nieto, J., Martin-Vazquez, M., Bonelo, J., Garcia-Vacas, F., Moreno-Aranda, G., Espejo, S., Hylander, G., Vizoso, J. (eds.) IEEE International Conference on Imaging Systems and Techniques (IST), Thessaloniki, pp. 192–197 (2010)
9. Bhattacharyya, D., Robles, R.J., Kim, T.-H., Bandyopadhyay, S.K.: Feature Extraction and Analysis of Breast Cancer Specimen. In: Chang, C.-C., Vasilakos, T., Das, P., Kim, T.-h., Kang, B.-H., Khurram Khan, M. (eds.) ACN 2010. CCIS, vol. 77, pp. 30–41. Springer, Heidelberg (2010)
10. Binh, N.T., Khare, A.: Adaptive Complex Wavelet Technique for Medical Image Denoising. In: Van Toi, V., Khoa, T.Q.D. (eds.) BME 2010. IFMBE Proceedings, vol. 27, pp. 196–199. Springer, Heidelberg (2010)
11. Madhloom, H., Kareem, S.: An Automated White Blood Cell Nucleus Localization and Segmentation using Arithmetic and Automatic Threshold, pp. 959–966 (2010)
12. Stomach Cancer Symptoms, Causes, Stages and Gastric Cancer Treatment. MedicineNet.com, http://www.medicinenet.com/stomach_cancer/article.htm
13. Rahnamayan, S., Mohamad, Z.: Tissue Segmentation in Medical Images Based on Image Processing Chain Optimization. In: International Workshop on Real Time Measurement, Instrumentation and Control, Toronto, pp. 1–9 (2010)
14. Ihalainen, T.: Image Pre-Processing and Post-Processing in CR and DR. HUS Helsinki Medical Imaging Center, Helsinki, Finland (2011)
15. Dawood, F., Rahmat, R., Dimon, M., Nurliyana, L., Kadiman, S.: Automatic Boundary Detection of Wall Motion in Two-dimensional Echocardiography Images, pp. 1261–1266 (2011)
16. Starovoitov, V., Samal, D., Briliuk, D.: Image Enhancement for Face Recognition. In: International Conference on Iconics, St. Peterburg, Russia (2003)
17. Abadi, M., Capelle-Laizé, A.-S., Khoudeir, M., Combes, D., Carré, S.: Grassland Species Characterization for Plant Family Discrimination by Image Processing. In: Elmoataz, A., Lezoray, O., Nouboud, F., Mammass, D., Meunier, J. (eds.) ICISP 2010. LNCS, vol. 6134, pp. 173–181. Springer, Heidelberg (2010)
18. Awad, A., Baba, K.: An Application for Singular Point Location in Fingerprint Classification. In: Snasel, V., Platos, J., El-Qawasmeh, E. (eds.) Communications in Computer and Information Science, Ostrava, Czech Republic, pp. 262–276 (2011)
19. Patil, B., Subbaraman, S.: Human Iric Pattern Recognition using Phase Components of Image. In: International Conference on Industrial and Information Systems, Sri Lanka, pp. 10–14 (2009)
20. Wu, Z., Leahy, R.: An Optimal Graph Theoretic Approach to Data Clustering: Theory and Its Application to Image Segmentation. IEEE Transactions on Pattern Analysis and Machine Intelligence (TPAMI) 15, 1101–1113 (1993)
21. El-Bendary, N., Hassanien, A., Corchado, E., Berwick, R.: ARIAS: Automated Retinal Image Analysis System. In: Proceedings Soft Computing Models in Industrial and Environmental Applications, 6th International Conference SOCO, Salamanca, Spain, vol. 87, pp. 67–76 (2011)

22. Poulos, M., Evangelou, A., Magkos, E., Papavlasopoulos, S.: Fingerprint Verification based on Image Processing Segmentation using Algortithm of Computational Geometry. World Scientific Publishing Company, WSPC (2004)
23. Ahn, S.: Digital Image Processing. Clayton School of Information Technology. Monash University Information Technology (2007)
24. Seemann, T.: Digital Image Processing Using Local Segmentation
25. Coffield, P.: Correlator Pre- and Post-Processing for Object Identification. In: SPIE Digital Library, vol. 317 (2000)
26. Pedronette, D., Torres, R.: Exploiting Contextual Information for Image re-ranking and Rank Aggregation. International Journal of Multimedia Information Retrieval CAIP (2011)
27. Lehmann, T.: Medical Image Processing and Management. Shape Analysis and Visualization, ch. 7
28. Egton Medical Information Systems, http://www.patient.co.uk/health/Biopsy.htm
29. Yin, G., Zhang, W., He, X., Chen, Y., Shen, X.: On the Classification of Chronic Gastritis at Molecular Biological Level. World Journal of Gastroenterology 9, 836–842
30. Normal Gastric Cells, http://www.hpylori.com.au/image/antrum_nml.jpg
31. Abnormal Gastric Cells, http://www.hpylori.com.au/image/antrum_chronic.jpg
32. Color Channel Definition. Pixmonix Gallery, http://www.pixmonix.com/scanning-glossary.php
33. Image Processing Toolbox. Mathworks Product Documentation, http://www.mathworks.com/help/toolbox/images/f14-13543.html
34. Ritter, N., Cooper, J.: Segmentation and Border Identification of Cells in Images of Peripheral Blood Smear Slides. In: The Thirtieth ACSC (2007)
35. Sachs, J.: Thresholding Definition. Digital Imaging Glossary, http://ftp2.bmtmicro.com/dlc/Glossary.pdf
36. Image Processing Toolbox for label2rgb. Mathworks Product Documentation, http://www.mathworks.com/help/toolbox/images/ref/label2rgb.html
37. Li, T., Wang, S., Zhao, N.: Gray-scale Edge Detection for Gastric Tumour Pathologic Cell Images by Morphological Analysis. Computers in Biology and Medicine (2009)
38. Olson, D., Delen, D.: Advanced Data Mining Techniques. Springer, Heidelberg (2008)
39. Handaga, B., Mat Deris, M.: Similarity Approach on Fuzzy Soft Set Based Numerical Data Classification. In: Zain, J.M., Wan Mohd, W.M.b., El-Qawasmeh, E. (eds.) ICSECS 2011, Part II. Communications in Computer and Information Science, vol. 180, pp. 575–589. Springer, Heidelberg (2011)
40. Eumetcal The European Virtual Organisation for Meteorological Training, http://www.eumetcal.org/resources/ukmeteocal/verification/www/english/msg/ver_categ_forec/uos1/uos1_ko1.htm

A Bio-inspired System for Boundary Detection in Color Natural Scenes

Karin S. Komati[1], Evandro O.T. Salles[2], and Mario Sarcinelli-Filho[2]

[1] Federal Institute of Education Science and Technology of Espirito Santo
Campus Serra
Rodovia ES-010, km 6.5, Manguinhos, 29173-087, Serra, ES, Brazil
karinkomati@gmail.com
[2] Federal University of Espirito Santo, Graduate Program on Electrical Engineering
Av. Fernando Ferrari, 514, Goiabeiras, 29075-910, Vitoria, ES, Brazil
{evandro,mario.sarcinelli}@ele.ufes.br

Abstract. This paper proposes a new unsupervised and fully automatic method to detect the boundaries in color natural images, inspired in the human visual model proposed by Grossberg. One of the hypotheses of Grossberg, the FACADE, admits complementary specialized streams at the bifurcation of the parvocellular pathway in the visual cortex: one of the branches performs edge processing and the other performs surface processing. In a similar way, this proposal has two parallel processes that are integrated at the end. The edge processing is implemented through a classical edge-detection method, whereas the surface processing is performed through a region growing method. The proposed integration scheme eliminates false contours resulted from the region growing guided by the result of edge detection, and eliminates the noise resulted from the edge detection as well, now guided by the result of the region growing, thus taking advantage of their complementary natures. Experiments on a large set of color images show that the results of the proposed system are closer to the human perception than the those correspondent to the individual methods (each branch), in quantitative and qualitative terms.

Keywords: digital image processing, boundary detection, multifractal descriptor, bio-inspired systems.

1 Introduction

Boundary detection is a crucial task in computer vision systems. As a matter of fact, high level procedures, such as object recognition, strongly rely on the quality of the boundary information.

To propose a fully automated and unsupervised boundary detection system is a complex task, since it is not known a priori what types of regions (uniform, with smooth color gradation and texture variations) exist in an image, or even how many regions a given scene contains. Examples of the variety and complexity of natural images are given in Fig. 1. Fig. 1(a) (a snake in a desert) the central element and the background have almost the same color, causing a ill-defined border. In Fig. 1(b), there is a mixture of artificial and natural textures (the bridge has geometric patterns, quite unlike the natural texture of the hill and the trail of smoke produced by the locomotive). In Fig. 1(c) (a koala in a tree), in spite of the complexity of the texture of the koala's fur, human perception recognizes it as a single element.

(a) (b) (c)

Fig. 1. Image examples (extracted from [1])

So far, the most efficient instrument to detect the boundaries of objects in a scene is the human eye. Generally, the human annotation is used as a reference for the accuracy tests of boundary detection algorithms. Thus, computer vision systems that are inspired by human vision system (HVS) represent a promising alternative [2,3,4,5].

However, the knowledge about the biological vision system, particularly the human one, is not complete and detailed [6,7]. Many researchers have proposed theories on how the human vision system processes information, and built computational systems based on such theories, for comparison of the visual behavior implemented with the human one. Such theories have evolved over the years, based on the increasing understanding of the human visual system, trying to describe how we "perceive" the positions of objects in a scene and their properties.

This work is inspired in the HVS model proposed by Grossberg [8,9,10,11,12]. Such theory has evolved since it was firstly proposed, in the 1960s, thus seeming to represent a good approximation of how the human vision system behaves. The model proposed by Grossberg is quite complete, with explanations about the human perception, 3D perception, binocular vision, ocular dominance, occlusion edges, learning and memory, among other challenges. Following the model of Grossberg, this work focuses on characteristics associated with boundary detection.

2 Human Visual System

Vision starts with two external sensors, the eyes. The retina, situated at the back of the eyeball, is where the image of the outside world is captured. The neurons of the retina transform light energy into electrical signals that are transmitted to the brain via the optic nerve. The retina ganglion cells, which drive the optical nerve, can be separated into two categories: parvocellular (small cells) and magnocellular (large cells). These two categories of cells are the starting point of parallel pathways that can be identified throughout most parts of the brain devoted to visual processing [6].

The optical nerve connects to the lateral geniculate nucleus (LGN) of the thalamus and this to the visual cortex (Fig. 2). The visual cortex is usually divided into 5 separated areas (V1, V2, V3, V4 and V5). This general arrangement can be subdivided into two parallel pathways, which start in the parvocellular (P) and magnocellular (M) cells of the retina. The dorsal pathway receives input from the M pathway but only the ventral pathway receives input from both the M and P pathways. These separated parallel pathways decode/extract the elementary properties of the objects in the visual field, such as color, form and orientation, producing a rather complex internal representation of what we see.

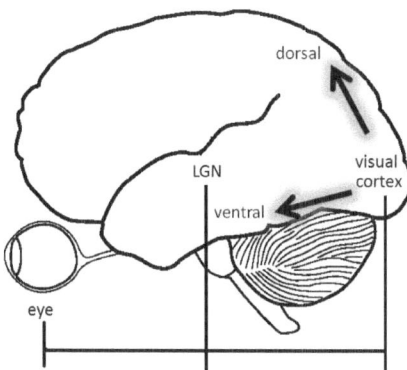

Fig. 2. The M and P pathways from the retina project through the lateral geniculate nucleus (LGN) to visual cortex

Retina's magnocellular ganglion neurons show a high sensitivity to contrast, low spatial resolution, and high temporal resolution or fast transient responses to visual stimuli. These characteristics make the magnocellular branch of the visual system especially suitable to quickly detect novel or moving stimuli.

Retina's parvocellular ganglion neurons show a low sensitivity to contrast, high spatial resolution, and low temporal resolution or sustained responses to visual stimuli. These cellular characteristics make the parvocellular visual pathways especially suitable for the analysis of details in the visual world, the perception of color and maintenance of color perception regardless of lighting (color constancy). The P pathway is responsible for complex form recognition, so that it is particularly interesting for boundary detection problems in 2D images.

2.1 The Grossberg Model

This section presents an overview of some assumptions of the model proposed by Grossberg that were used in the conception of the system proposed in this paper, which are presented below:

1. segmentations generated by the model are not the result of training on image exemples [9];
2. the Grossberg's model includes the "The Two Streams Hypothesis" concept, first proposed by [13]. Such hypothesis suggests that each of the visual pathways is projected to different areas, and thus each one treats different attributes. The dorsal stream, commonly referred to as the "where" stream, is involved in spatial attention, and communicates with regions that control eye movements and hand movements. The ventral stream, commonly referred to as the "what" stream, is involved in the recognition, identification and categorization of visual stimuli;
3. The "Complementary Computing" hypothesis suggests that brains are organized into parallel processing streams with complementary properties. Each of these streams exhibits complementary strengths and weaknesses. It is proposed that interactions between these processing streams overcome their individual deficiencies and generate behavioral properties that realize the unity of conscious experiences. In this sense, pairs of complementary streams are the functional units because only through their interactions key behavioral properties can be competently computed [11];
4. A neural theory, called FACADE (Form-And-Color-And-DEpth) theory [12], claims that boundaries and surfaces are the brain's perceptual units for the P stream. Boundaries and surfaces are computed in parallel processing streams that obey computationally complementary properties. These streams interact to overcome their complementary weaknesses and to transform their complementary properties into consistent percepts. The surface representation uses complementary properties provided by signals from boundaries representations.

The assumptions of the Grossberg's model (described, in a simplified way, in Fig. 3), can be resumed as: "The human's brains are organized into independent modules and independent modules should be able to fully compute their particular processes on their own. Brains are also organized into parallel processing streams with complementary properties. Interactions between streams are the key to create coherent behavioral representations that overcome the complementary deficiencies of each stream and support conscious experiences [10]."

3 The Proposed Method

The proposal is to model a system based on the Grossberg's model, whose goal is to detect edges in color images. In accordance with the assumption that segmentations generated by the model are not the result of training on image examples, this proposal is a fully unsupervised system, not demanding any training phase.

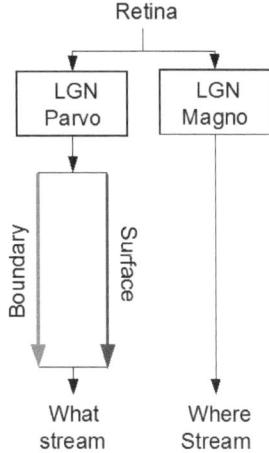

Fig. 3. Simplified view of the Groossberg's model

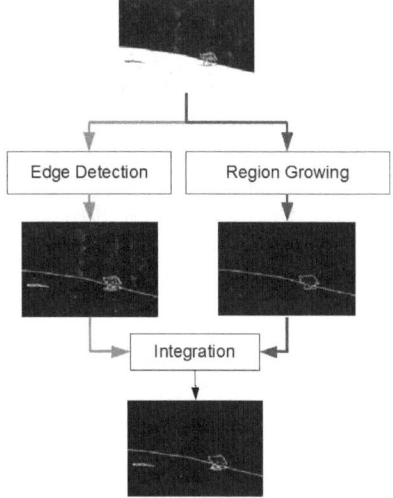

Fig. 4. Data flow for the image segmentation process. The integration step receives the edge and region maps as inputs.

The interest is to model the P pathway (the "what" stream) just as the FACADE theory suggests: the P pathway is divided in two complementary perceptual properties, boundary and surface. In our system, the data flow correspondent to the image segmentation process has two parallel processes that are integrated at the end (post integration, as shown in Fig. 4). The boundary processing is implemented through a classical edge-detection method, whereas the surface processing is performed through a region growing method.

Traditionally in image processing area, boundary-detection techniques are classified in two complementary approaches: region- or edge-based techniques. Region-based techniques rely on common patterns in intensity values within a cluster of neighboring pixels. The cluster is referred to as the region, and the goal of the segmentation algorithm is to group regions according to their anatomical or functional roles. Edge-based techniques rely on discontinuities in image values between distinct regions, and the goal of the segmentation algorithm is to accurately demarcate the boundary separating these regions. However, there may exist gaps and noisy edges in edge-approach results, whereas region-approach results tend to be over-segmented with inaccurate boundaries [14].

It is necessary that the edge-map be a soft map, for implementing the integration step. A soft map is a gray scale image where each pixel has a value from zero to one, where higher values mean greater confidence in the existence of a boundary. As for the edge map, some classical edge detectors (Sobel, Prewitt, Laplacian and morphological gradient) [15] generating an output known as a soft boundary map were tested, and the result is that the morphological gradient presents the quantitative measure (overall F-measure [21,23]) slightly better than the other detectors. Therefore, it was chosen as the edge detection method for this work. As for the usual image smoothing to eliminate noise before the edge detection, a classical non-linear edge-preserving smoothing filter, the Kuwahara filter [16], was selected, with 5×5 mask size.

We are considering that the region-map presents standard characteristics: the image is binary and all regions are bounded by closed contours. For the region growing method, we chose the MM-Frac approach proposed in [17], which is based on the fully-automatic JSEG method [18]. Essentially, such method merges the JSEG's homogeneity criterion with the Multifractal Measurement [19], in a merging process controlled by a global statistical property of the image being analyzed, the shape of its power spectrum [20].

3.1 The KSS Algorithm

The strategy integration, the KSS Algorithm [21], combines two complementary results: similarity (region) and discontinuity (edge) information. The region-growing result of MM-Frac and the morphological gradient information are separately extracted, in a parallel way.

The KSS algorithm is presented as a pseudo-code (see Algorithm 1). The post-processing integration method, the KSS system, is independent of how the edge-map and the region-map are got. However, it is necessary that the region-map be a binary image and the edge-map be a soft map.

In step 2, the sum operation will enhance all boundary pixels that match in the two different input maps. In the rest of the code, the logic is to eliminate or reduce false information.

In step 3 and 4, we should detect the weak edges from the edge map. The step 4 is basically a binarization process in the edge map, where each pixel with a low gray level value corresponds to a weak edge pixel. To automate the threshold

Algorithm 1. KSS

1: Inputs: edge-map and region-map
2: image-result = edge-map + region-map
3: computes $threshold_{weak}$
4: weak-edge-map = binarize(edge-map, $threshold_{weak}$)
5: **for** each pixel position **do**
6:    **if** (the pixel position is marked as boundary on region-map) and (the majority of the pixel neighborhood is marked as a weak on weak-edge-map) **then**
7:       set non-boundary at pixel position in image-result
8:    **end if**
9:    **if** (there is no pixel neighborhood marked as boundary on region-map) and (the pixel position is marked as a weak on weak-edge-map) **then**
10:      set non-boundary at pixel position in image-result
11:   **end if**
12: **end for**

value, in step 3, it was used an idea similar to the one in [22]: the value of the threshold is based on edge-map histogram h, and is given by

$$threshold_{weak} = \frac{\sum_{i=0}^{50} h_i}{\sum_{i=50}^{200} h_i}, \quad (1)$$

where $i = [0, 255]$ is the value of a pixel in the gray-scale edge map.

A noisy edge-map will result in low values of $threshold_{weak}$ while a strongly defined edge-map will result in higher values.

The higher the $threshold_{weak}$ value is, the more weak-edges are obtained, and thus more information will be eliminated in the process. The result presents less edge information from the region-map and less weak information compared to the edge-map. The result seems cleaner, preserving only the strong edges of both maps. However, when the image is noisy, all information about both maps is preserved, with the strong edges emphasized. In such images, we notice that region-map information is more valuable than edge-map information.

In step 5, the algorithm works with two binary images: the *weak-edge-map* (steps 3 and 4 result) and the region map (provided by MM-Frac). The first *if* statement (steps 6, 7 and 8) eliminates false boundaries provided by region-map. The second *if* statement (steps 9, 10 and 11) eliminates the noisy provided by region-map.

Thus, the KSS results erase some of the region-map's edge information and don't preserve weak information for the edge map. The result seems cleaner, preserving only the strong edges of both maps. Our strategy is to put the two maps together, eliminating the false boundaries in the region map, based on edge information, and eliminating the noisy edges in the edge map, based on region information.

4 Experimental Results and Discussion

We tested the proposed method with natural colored images provided by the BSDS ("The Berkeley Segmentation Dataset and Benchmark") image dataset [1], applying it to all 100 (one hundred) images of the test dataset. As discussed so far, our experiments do not include any parameter-tuning for individual images and there is not training phase, for it is a fully unsupervised system.

Quantitative performance comparison requires ground truth and well defined metrics. Both requirements can be found in BSDS. For each image in BSDS, there are at least five hand-labeled segmentations made by human beings, which constitute the ground truth. The standard metrics of BSDS are Precision, Recall and F-measure [23], which determines how well the boundary map obtained approximates the human ground truth boundaries. Precision is the probability that a machine-generated boundary pixel is a true boundary pixel and it is a measure of accuracy or reliability. Recall is the probability that a true boundary pixel is detected, thus being a measure of completeness. The F-measure metric is the harmonic mean of precision and recall.

Fig. 5 shows some results. There (a) shows the input image, (b) shows the human benchmark. Following, one has the segmentation result of the edge detection method (part (c)) and of the MM-Frac method (part (d)). Finally, the result of the KSS Algorithm is shown in part (e). Notice that each result has its F-measure metric computed.

In the third column, the results of edge detection are presented. Such results can be very noisy, mainly due to the fact that edge detection techniques rely entirely on the local information available in the image. The edge-map responds to all contrast variations over the texture regions, like in the sand area in the result of the image #196073. At the same time, the method of edge detection is responsible for highlighting details such as the stick in the left in the snowy area (results of image #167062), the leaf near to the snake (results of image #196073) and the stick in the hands of the aboriginal (results of image #101087).

One problem observed in the MM-Frac result is caused by the varying shades due to the illumination. For instance, the color of the sky can vary in a very smooth transition, as in the BSDS image #42049. Visually, there is no clear boundary. However, the MM-Frac result presents a circle region in the image. The human perception does not perceive this smooth varying of color as a different region. The result after KSS also does not present this false boundary. The smooth is not perceived by the edge detection, and then the boundary is erased by the KSS method.

Another example of false boundary is the contour in the sky over the smoke (MM-Frac result of image #182053). In the same area the edge detection result is quite clear, so that the false boundary is eliminated, reducing the oversegmentation associated to MM-Frac.

When the image is noisy (as images#196073,#134035 and#69015), most information from the region map is preserved and the information from edge detection is attenuated.

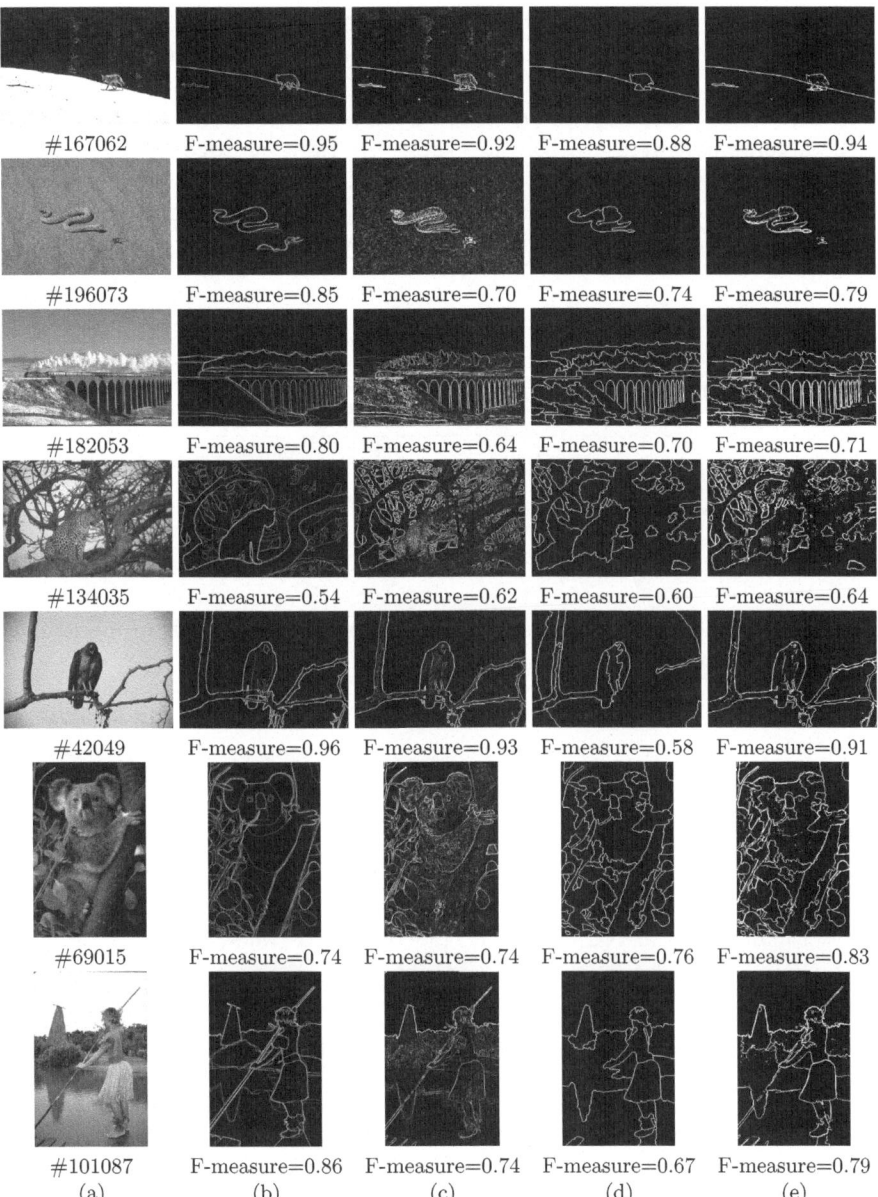

Fig. 5. (a) Original image (b) Human benchmark (c) Results of the edge detection method (d) Results of the MM-Frac method (e) Results of KSS algorithm

The results of the KSS method are presented in the rightmost column. Notice that when the edge detection result is reliable, the KSS algorithm works well and solves the problem of false boundaries provided by the region growing result. Details detected using edge detection method are kept, but the noise is attenuated. Now, the boundaries are more accurate and closer to the human perception.

Quantitatively speaking, the metrics recall, precision and F-measure of each method, computed by the BSDS, are shown in Table 1. Edge detection looses in terms of precision, because of the noisy pixels, while the MM-Frac approach shows the worst recall and good precision. After KSS integration step the F-measure increases to 0.61, which is the closest value, comparing to the human perception.

Table 1. Precision, Recall and F-measure metrics calculated by BSDS

	Recall	Precision	F-measure
Edge Detection	0.65	0.49	0.56
MM-Frac	0.63	0.56	0.59
KSS	0.69	0.54	0.61
Human	0.70	0.89	0.79

Comparing the integration result with each each individual KSS input, considering the F-measure metric:

- the result of KSS is better than the result of the MM-Frac in 82% of the images;
- the result of KSS is better than the result of the edge detection method in 81% of the images.

In addition, for 58 images the result of KSS showed F-measure metric greater than the greatest F-measure among the MM-Frac algorithm and the edge detector. Finally, only in two cases the KSS result was worse than the two KSS inputs.

4.1 Comparison with Mutilabel Graph-Cut

The boundary detection problem can be understood as an optimization problem, whose solution via multiway graph cut is the optimum labeling S [24]. We assume that the optimum labeling S is guided by two kinds of constraints: boundary and region, in our case represented by the information provided by the edge detection and region growing technique, respectively [25].

In terms of qualitative comparison, Figure 6 shows the same images of Figure 5. The column order is: (a) shows the input image, (b) the human benchmark (c) KSS results and (d) multilabel graph cut results. For each result we present its computed F-measure metric.

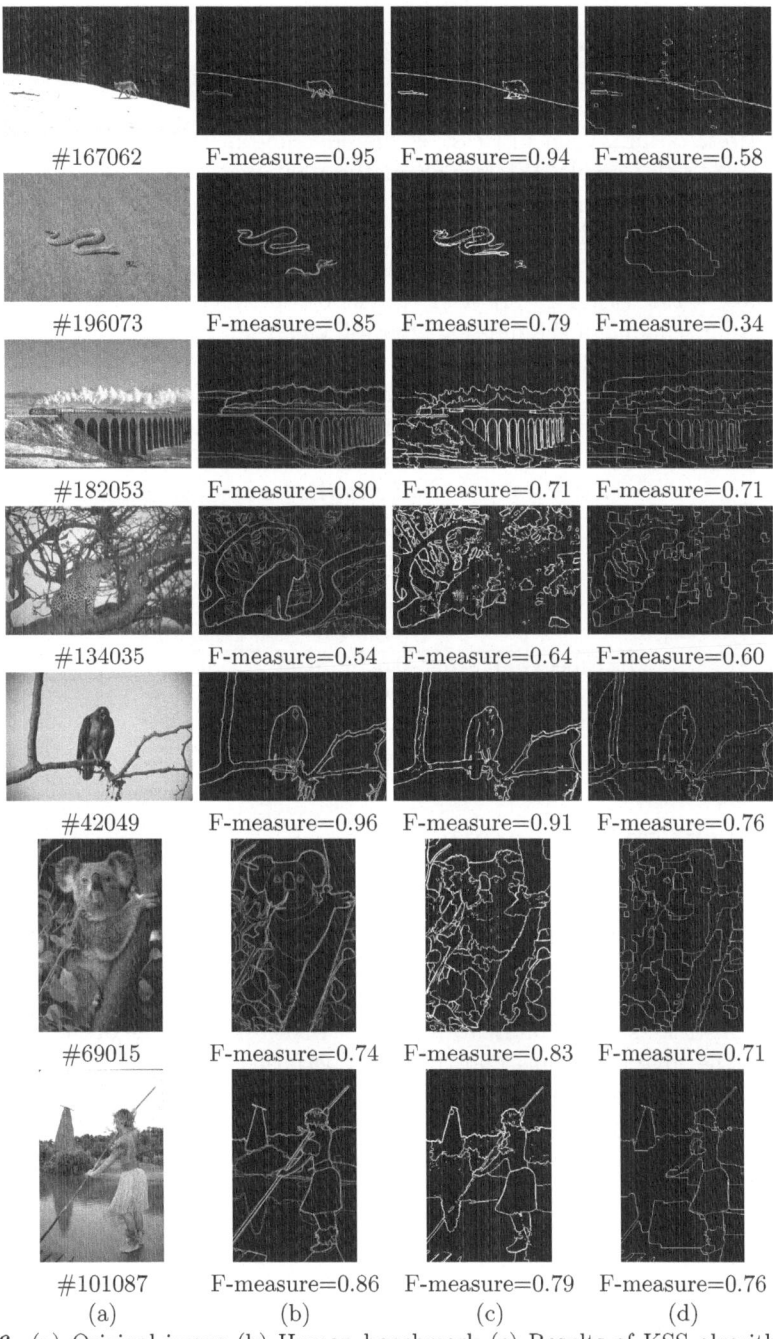

Fig. 6. (a) Original image (b) Human benchmark (c) Results of KSS algorithm (d) Results of Mutilabel Graph-Cut

In all multilabel graph cut results in Figure 6, it is noticeable blocking artifacts, because each image is represented by a 4-connected lattice [26].

An important characteristic of multilabel graph cut is its capacity to change the initial labeling considering the edge detection information. That is, the labeling generated by the region growing technique is the initial state of solution in a discrete optimization process, as the edge-line in the sky over the smoke of image #182053 result. While KSS method eliminates false information and enhances edge matching in both maps, but it does not change the boundaries positions, as multilabel graph cut does.

Sometimes, this feature is not well-suited to define the boundaries of objects (see image #196073). The problem caused by the varying shades due to the illumination remains in the multilabel graph cut result of image #42049.

Quantitatively speaking, the metrics recall, precision and F-measure of both methods computed by the BSDS method are shown in Table 2. KSS and multilabel graph cut improves the recall metric when comparing with MM-Frac and edge detection metrics. Moreover, KSS performs better F-measure then multilabel graph cut, which is the closest value to the human perception.

Table 2. Precision, Recall and F-measure metrics calculated by BSDS

	Recall	Precision	F-measure
Mutilabel Graph-Cut	0.65	0.56	0.60
KSS	0.69	0.54	0.61
Human	0.70	0.89	0.79

5 Conclusion and Future Work

This work proposes a new approach to boundary detection for natural color images, inspired in the human visual model of Grossberg. One hypotheses states that segmentation annotations made by humans are not the result of training on image exemplars. Another hypotheses, the FACADE, admits complementary specialized streams at the bifurcation in the "what" stream: one of the branches performs the edge processing and the other performs surface processing. These streams interact to overcome their complementary weaknesses and to transform their complementary properties into a better perception.

In a similar way, this proposal has two parallel processes that are integrated at the end. The edge processing is implemented through a classical edge-detection method, whereas the surface processing is performed through a region growing method. As the Grossberg model suggests, such system is fully unsupervised, without any parameter-tuning for individual images and any training step.

The integration step receives these two classical complementary information, the edge and region maps, as inputs. The main goal of such integration step is to combine the region-growing result from MM-Frac and edge information and overcome their complementary weaknesses.

Our strategy, called KSS, is to put together the two maps, and to eliminate the false boundaries in region-map, based on edge information, and the noisy edges in the edge-map, based on region information. Furthermore, all strong edges of both input maps are held, improving the boundary detection. Unfortunately, the KSS results present broken edges, not keeping the contour closed.

The conclusion is that the bio-inspired approach proposed here improves the boundary detection results, generating segmented images that match the human perception better than the results associated to the individual methods used in the architecture.

A future work is the implementation of other assumptions of the Grossberg model, using techniques of digital image processing and pattern recognition.

References

1. Martin, D.R., Fowlkes, C.C., Tal, D., Malik, J.: A database of human segmented natural images and its application to evaluating segmentation algorithms and measuring ecological statistics. In: Proc. 8th IEEE International Conference on Computer Vision, vol. 2, pp. 416–423 (2001)
2. Marr, D.: Vision. Freeman Publishers (1982)
3. Hu, D.-K., Li, J.-P., Yang, S.X., Gregori, S.: A Bio-inspired Model for Color Image Segmentation. In: Proc. of the International Conference on Apperceiving Computing and Intelligence Analysis (ICACIA 2009), pp. 317–320 (2009)
4. Komati, K.S., De Souza, A.F.: Using Weightless Neural Networks for Vergence Control in an Artificial Vision System. Journal of Applied Bionics and Biomechanics 1(1), 21–32 (2003)
5. Köppen, M., Ruiz-del-Solar, J., Soille, P.: Texture Segmentation by Biologically-Inspired Use of Neural Networks and Mathematical Morphology. In: Proc. of the International ICSC/IFAC Symposium on Neural Computation (NC), Vienna, Austria, pp. 267–272 (1998)
6. Kandel, E.R., Schwartz, J.H., Jessell, T.M.: Principles of Neural Science, 4th edn. McGraw-Hill, New York (2000)
7. Olshausen, B.A., Field, D.J.: How close are we to understanding V1? Neural Computation 17, 1665–1699 (2005)
8. Grossberg, S., Mingolla, E.: Neural dynamics of perceptual grouping: textures, boundaries, and emergent segmentations. Perception and Psychophysics 38(2), 141–171 (1985)
9. Grossberg, S.: 3-D vision and figure-ground separation by visual cortex. Perception Psychophysics 55(1), 48–121 (1994)
10. Grossberg, S.: Linking Mind To Brain: The Mathematics of Biological Intelligence. Notices of the American Society 47, 1361–1372 (2000)
11. Grossberg, S.: The complementary brain: unifying brain dynamics and modularity. Trends in Cognitive Sciences 4(6), 233–246 (2000)
12. Grossberg, S.: The art of seeing and painting. Spatial Vision 21(3-5), 463–486 (2008)
13. Mishkin, M., Ungerleider, L.G.: Contribution of striate inputs to the visuospatial functions of parieto-preoccipital cortex in monkeys. Behavioural Brain Research 6(1), 57–77 (1982)

14. Muñoz, X., Freixenet, J., Cufí, X., Martí, J.: Strategies for image segmentation combining region and boundary information. IEEE Pattern Recognition Letters 24(1-3), 375–392 (2003)
15. Gonzalez, R.C., Woods, R.E.: Digital Image Procesing, 2nd edn. Addison-Wesley Longman Publishing Co., Boston (2001)
16. Kuwahara, M., Hachimura, K., Eiho, S.: Processing of Riangiocardiographic Images. Digital Processing of Biomedical Images, 187–203 (1976)
17. Komati, K.S., Salles, E.O.T., Sarcinelli-Filho, M.: Two-level Strategy for Image Boundary Detection. In: Proc. of the International Conference on Computer Vision Theory and Applications (VISAPP 2011) Vilamoura, Algarve, Portugal, pp. 181–186 (2011)
18. Deng, Y., Manjunath, B.S.: Unsupervised segmentation of color-texture regions in images and video. IEEE Transactions on Pattern Analysis and Machine Intelligence 23(8), 800–810 (2001)
19. Chaudhuri, B.B., Sarkar, N.: Texture segmentation using fractal dimension. IEEE Transactions on Pattern Analysis and Machine Intelligence 17(1), 72–77 (1995)
20. Torralba, A., Oliva, A.: Statistics of Natural Image Categories. Institute of Physics Publishing: Computation in Neural Systems 14, 391–412 (2003)
21. Komati, K.S., Salles, E.O.T., Sarcinelli-Filho, M.: A Strategy for Image Boundary Detection Combining Region and Edge Maps. IEEE Computing in Science and Engineering 13, 46–52 (2011)
22. Rotem, O., Greenspan, H., Goldberger, J.: Combining Region and Edge Cues for Image Segmentation in a Probabilistic Gaussian Mixture Framework. In: Proceedings of the IEEE Conference on Computer Vision and Pattern Recognition (CVPR 2007), Minneapolis, MN, US, pp. 1–8 (2007)
23. Martin, D.R., Fowlkes, C.C., Malik, J.: Learning to detect natural image boundaries using local brightness, color, and texture cues. IEEE Transactions on Pattern Analysis and Machine Intelligence 26(5), 530–549 (2004)
24. Boykov, Y., Veksler, O., Zabih, R.: Fast approximate energy minimization via graph cuts. IEEE Transactions on Pattern Analysis and Machine Intelligence 23(11), 1222–1239 (2001)
25. Komati, K.S., Samatelo, J.L.A., Salles, E.O.T., Sarcinelli-Filho, M.: A strategy for Boundary Detection Combining Region and Edge Information. In: Proc. of the 24th Brazilian Symposium on Computer Graphics and Image Processing (SIBGRAPI 2011), Maceió, Alagoas, Brazil, pp. 305–312 (2011)
26. Boykov, Y., Kolmogorov, V.: Computing Geodesics and Minimal Surfaces via Graph Cuts. In: Proc. of the 9th IEEE International Conference on Computer Vision (ICCV 2003), vol. 2, pp. 26–33 (2003)

Author Index

Abdullah, Noryusliza IV-364
Abdullah, Nurul Azma IV-353
Abid, Hassan III-368
Adewumi, Adewole IV-248
Afonso, Vitor Monte IV-274, IV-302
Agarwal, Suneeta IV-147
Aguiar, Rui L. III-682
Aguilar, José Alfonso IV-116
Ahmadian, Kushan I-188
Alarcon, Vladimir J. II-578, II-589
Albuquerque, Caroline Oliveira III-576
Alfaro, Pablo IV-530
Ali, Salman III-352
Alizadeh, Hosein III-647
Almeida, Regina III-17
Alonso, Pedro I-29
Alves, Daniel S.F. I-101
Amandi, Analía A. III-698, III-730
Amaral, Paula III-159
Amjad, Jaweria III-368
Ammar, Reda A. I-161
Amorim, Elisa P. dos Santos I-635
An, Deukhyeon III-272
Anderson, Roger W. I-723
Angeloni, Marcus A. I-240
Antonino, Pablo Oliveira III-576
Aquilanti, Vincenzo I-723
Arefin, Ahmed Shamsul I-71
Armentano, Marcelo G. III-730
Arnaout, Arghad IV-392
Aromando, Angelo III-481
Asche, Hartmut II-347, II-386, II-414, II-439
Aydin, Ali Orhan IV-186
Ayres, Rodrigo Moura Juvenil III-667
Azad, Md. Abul Kalam III-72
Azzato, Antonello II-686

Bae, Sunwook III-238
Balena, Pasquale I-583, II-116
Balucani, Nadia I-331
Barbosa, Ciro I-707
Barbosa, Fernando Pires IV-404
Barbosa, Helio J.C. I-125

Baresi, Umberto II-331
Barreto, Marcos I-29
Baruque, Alexandre Or Cansian IV-302
Bastianini, Riccardo I-358
Batista, Augusto Herrmann III-631
Batista, Vitor A. IV-51
Battino, Silvia II-624
Beaver, Justin IV-646
Bencardino, Massimiliano II-548
Berdún, Luis III-698
Berenguel, José L. III-119
Bernardino, Heder S. I-125
Berretta, Regina I-71
Biehl, Matthias IV-40
Bimonte, Sandro II-373
Bisceglie, Roberto II-331
Bitencourt, Ana Carla P. I-723
Blecic, Ivan II-481, II-492
Boavida, Fernando II-234
Bollini, Letizia II-508
Boratto, Murilo I-29
Borg, Erik II-347, III-457
Borruso, Giuseppe II-624, II-670
Boulil, Kamal II-373
Braga, Ana Cristina I-665
Brumana, Raffaella II-397
Bugs, Geisa I-477
Burgarelli, Denise I-649
Bustos, Víctor III-607

Caiaffa, Emanuela II-532
Calazan, Rogério M. I-148
Caldas, Daniel Mendes I-675
Callejo, Miguel-Ángel Manso I-462
Camarda, Domenico II-425
Campobasso, Francesco II-71
Campos, Ricardo Silva I-635
Candori, Pietro I-316, I-432
Cannatella, Daniele II-54
Cano, Marcos Daniel III-743
Cansian, Adriano Mauro IV-286
Carbonara, Sebastiano II-128
Cardoso, João M.P. IV-217
Carmo, Rafael IV-444

Carneiro, Joubert C. Lima e Tiago G.S. II-302
Carpené, Michele I-345
Carvalho, Luis Paulo da Silva II-181
Carvalho, Maria Sameiro III-30, III-187
Casado, Leocadio G. III-119, III-159
Casado, Leocadio Gonzalez I-57
Casas, Giuseppe B. Las II-466, II-640, II-686
Casavecchia, Piergiorgio I-331
Castro, Patrícia F. IV-379
Cavalcante, Gabriel D. IV-314
Cecchi, Marco I-345
Cecchini, Arnaldo II-481, II-492
Ceppi, Claudia II-517
Cermignani, Matteo I-267
Cerreta, Maria II-54, II-168, II-653
Chanet, Jean-Pierre II-373
Charão, Andrea Schwertner IV-404
Cho, Yongyun IV-613, IV-622
Cho, Young-Hwa IV-543
Choe, Junseong III-324
Choi, Hong Jun IV-602
Choi, Jae-Young IV-543
Choi, Jongsun IV-613
Choi, Joonsoo I-214
Choo, Hyunseung III-259, III-283, III-324
Chung, Tai-Myoung III-376
Ci, Song III-297
Cicerone, Serafino I-267
Ciloglugil, Birol III-550
Cioquetta, Daniel Souza IV-16
Clarke, W.A. IV-157
Coelho, Leandro I-29
Coletti, Cecilia I-738
Conrado, Merley da Silva III-618
Corea, Federico-Vladimir Gutiérrez I-462
Coscia, José Luis Ordiales IV-29
Costa, M. Fernanda P. III-57, III-103
Costantini, Alessandro I-345, I-401, I-417
Crasso, Marco IV-29, IV-234, IV-484
Crawford, Broderick III-607
Crocchianti, Stefano I-417
Cuca, Branka II-397
Cui, Xiaohui IV-646
Cunha, Jácome IV-202

da Luz Rodrigues, Francisco Carlos III-657
Danese, Maria III-512
Dantas, Sócrates de Oliveira I-228
da Silva, Paulo Caetano II-181
Daskalakis, Vangelis I-304
de Almeida, Ricardo Aparecido Perez IV-470, IV-560
de Avila, Ana Maria H. III-743
de By, Rolf A. II-286
de Carvalho, Andre Carlos P.L.F. III-562
de Carvalho Jr., Osmar Abílio III-657
Decker, Hendrik IV-170
de Costa, Evandro Barros III-714
de Deus, Raquel Faria II-565
de Felice, Annunziata II-1
de Geus, Paulo Lício IV-274, IV-302, IV-314
Delgado del Hoyo, Francisco Javier I-529
Dell'Orco, Mauro II-44
de Macedo Mourelle, Luiza I-101, I-113, I-136, I-148
de Magalhães, Jonathas José III-714
De Mare, Gianluigi II-27
Dembogurski, Renan I-228
de Mendonça, Rafael Mathias I-136
de Miranda, Péricles B.C. III-562
de Oliveira, Isabela Liane IV-286
de Oliveira, Wellington Moreira I-561
de Paiva Oliveira, Alcione I-561
Deris, Mustafa Mat I-87, IV-340
De Santis, Fortunato III-481
Désidéri, Jean-Antoine IV-418
de Souza, Cleyton Caetano III-714
de Souza, Éder Martins III-657
de Souza, Renato Cesar Ferreira I-502
de Souza Filho, José Luiz Ribeiro I-228, II-712
De Toro, Pasquale II-168
Dias, Joana M. III-1
Dias, Luis III-133
do Nascimento, Gleison S. IV-67
Donato, Carlo II-624
do Prado, Hércules Antonio III-631, III-657
dos Anjos, Eudisley Gomes IV-132
dos Santos, Jefersson Alex I-620

dos Santos, Rafael Duarte Coelho
  IV-274, IV-302
dos Santos, Rodrigo Weber   I-635, I-649,
  I-691, I-707
dos Santos Soares, Michel   IV-1, IV-16

e Alvelos, Filipe Pereira   III-30
El-Attar, Mohamed   IV-258
Elish, Mahmoud O.   IV-258
El-Zawawy, Mohamed A.   III-592, IV-83
Engemaier, Rita   II-414
Eom, Young Ik   III-227, III-238, III-272
Epicoco, Italo   I-44
Esmael, Bilal   IV-392
Ezzatti, Pablo   IV-530

Falcinelli, Stefano   I-316, I-331, I-387, I-432
Falcone, Roberto   II-508
Fanizzi, Annarita   II-71
Farage, Michèle Cristina Resende   I-675
Farantos, C. Stavros   I-304
Farias, Matheus   IV-444
Fechine, Joseana Macêdo   III-714
Fedel, Gabriel de S.   I-620
Felino, António   I-665
Fernandes, Edite M.G.P.   III-57, III-72, III-103
Fernandes, Florbela P.   III-103
Fernandes, João P.   IV-202, IV-217
Ferneda, Edilson   III-631, III-657
Ferreira, Ana C.M.   III-147
Ferreira, Brigida C.   III-1
Ferreira, Manuel   III-174
Ferroni, Michele   I-358
Fichtelmann, Bernd   II-347, III-457
Fidêncio, Érika   II-302
Figueiredo, José   III-133
Filho, Dario Simões Fernandes   IV-274, IV-302
Filho, Jugurta Lisboa   I-561
Fiorese, Adriano   II-234
Fonseca, Leonardo G.   I-125
Formosa, Saviour   II-609
Formosa Pace, Janice   II-609
Fort, Marta   I-253
França, Felipe M.G.   I-101
Freitas, Douglas O.   IV-470
Fruhwirth, Rudolf K.   IV-392

García, I.   III-119
García, Immaculada   I-57
Garrigós, Irene   IV-116
Gasior, Wade   IV-646
Gavrilova, Marina   I-188
Gentili, Eleonora   III-539
Geraldes, Carla A.S.   III-187
Gervasi, Osvaldo   IV-457
Ghandehari, Mehran   II-194
Ghazali, Rozaida   I-87
Ghiselli, Antonia   I-345
Ghizoni, Maria Luísa Amarante   IV-588
Girard, Luigi Fusco   II-157
Gomes, Ruan Delgado   IV-132
Gomes, Tiago Costa   III-30
Gonschorek, Julia   II-208, II-220
Gonzaga de Oliveira, Sanderson Lincohn
  I-172, I-198, I-610
Görlich, Markus   I-15
Greco, Ilaria   II-548
Grégio, André Ricardo Abed   IV-274, IV-286, IV-302
Guardia, Hélio C.   IV-560
Gupta, Pankaj   III-87

Hahn, Kwang-Soo   I-214
Haijema, Rene   III-45
Han, Jikwang   III-217
Han, JungHyun   III-272
Han, Yanni   III-297
Handaga, Bana   IV-340
Hasan, Osman   III-419
Hashim, Rathiah   II-728
Hendrix, Eligius M.T.   I-57, III-45, III-119, III-159
Heo, Jaewee   I-214
Hong, Junguye   III-324
Huang, Lucheng   I-447

Ibrahim, Rosziati   IV-353, IV-364
Igounet, Pablo   IV-530
Ikhu-Omoregbe, Nicholas   IV-248
Im, Illkyun   IV-543
Imtiaz, Sahar   III-339
Inceoglu, Mustafa Murat   III-550
Iochpe, Cirano   IV-67
Ipbuker, Cengizhan   III-471
Ivánová, Ivana   II-286
Izkara, Jose Luis   I-529

Jeon, Jae Wook   III-311
Jeon, Woongryul   III-391
Jeong, Jongpil   IV-543
Jeong, Soonmook   III-311
Jino, Mario   IV-274, IV-302
Jorge, Eduardo   IV-444
Jung, Sung-Min   III-376

Kalsing, André C.   IV-67
Kang, Min-Jae   III-217
Karimipour, Farid   II-194
Kasprzak, Andrzej   IV-514, IV-576
Kaya, Sinasi   III-471
Khalid, Noor Elaiza Abdul   II-728
Khan, Salman H.   III-339
Khan, Yasser A.   IV-258
Khanh Ha, Nguyen Phan   III-324
Kim, Cheol Hong   IV-602
Kim, Hakhyun   III-391
Kim, Iksu   IV-622
Kim, Jeehong   III-227, III-238, III-272
Kim, Junho   I-214
Kim, Young-Hyuk   III-248
Kischinhevsky, Mauricio   I-610
Kluge, Mario   II-386
Knop, Igor   I-707
Komati, Karin S.   II-739
Kopeliovich, Sergey   I-280
Kosowski, Michał   IV-514
Koszalka, Leszek   IV-576
Koyuncu, Murat   IV-234
Kwak, Ho-Young   III-217
Kwon, Keyho   III-311
Kwon, Ki-Ryong   IV-434
Kwon, Seong-Geun   IV-434

Ladeira, Pitter Reis   II-548
Laganà, Antonio   I-292, I-345, I-358, I-371, I-387, I-401, I-417
Lago, Noelia Faginas   I-387
Laguna Gutiérrez, Víctor Antonio   III-618
Lanorte, Antonio   III-481, III-512
Lanza, Viviana   II-686
Lasaponara, Rosa   III-481, III-497, III-512
Le, Duc Tai   III-259
Lederer, Daniel   II-263
Le Duc, Thang   III-259
Lee, Dong-Young   III-368, III-376

Lee, Eung-Joo   IV-434
Lee, Hsien-Hsin   IV-602
Lee, Jae-Gwang   III-248
Lee, Jae-Kwang   III-248
Lee, Jae-Pil   III-248
Lee, Jongchan   IV-613
Lee, Junghoon   III-217
Lee, Kwangwoo   III-391
Lee, Sang Joon   III-217
Lee, Suk-Hwan   IV-434
Lee, Yunho   III-391
Leonel, Gildo de Almeida   II-712
Leonori, Francesca   I-331
Li, Yang   III-297
Lim, Il-Kown   III-248
Lima, Priscila M.V.   I-101
Lin, Tao   III-297
Liu, Yi   IV-100
Lobarinhas, Pedro   III-202
Lobosco, Marcelo   I-675, I-691, I-707
Loconte, Pierangela   II-517
Lomba, Ricardo   III-202
Lombardi, Andrea   I-387
Lopes, Maria do Carmo   III-1
Lopes, Paulo   IV-217
Lou, Yan   I-447
Lubisco, Giorgia   II-517
Luiz, Alfredo José Barreto   III-657

Ma, Zhiyi   IV-100
Macedo, Gilson C.   I-691, I-707
Maffioletti, Sergio   I-401
Maleki, Behzad   III-647
Mancini, Francesco   II-517
Mangialardi, Giovanna   II-116
Manuali, Carlo   I-345
Marcondes, Cesar A.C.   IV-470
Marghany, Maged   III-435, III-447
Marimbaldo, Francisco-Javier Moreno   I-462
Marinho, Euler Horta   IV-632
Martins, Luís B.   III-147
Martins, Pedro   IV-217
Martucci, Isabella   II-1
Marucci, Alessandro   II-532
Marwala, T.   IV-157
Marwedel, Peter   I-15
Marzuoli, Annalisa   I-723
Mashkoor, Atif   III-419
Masini, Nicola   III-497

Mateos, Cristian   IV-29, IV-234, IV-484
Mazhar, Aliya   III-368
Mazón, Jose-Norberto   IV-116
McAnally, William H.   II-578, II-589
Medeiros, Claudia Bauzer   I-620
Mehlawat, Mukesh Kumar   III-87
Meira Jr., Wagner   I-649
Mele, Roberta   II-653
Melo, Tarick   II-302
Messine, F.   III-119
Miao, Hong   I-447
Milani, Alfredo   III-528, III-539
Min, Changwoo   III-227, III-238
Min, Jae-Won   III-376
Misra, A.K.   IV-157
Misra, Sanjay   IV-29, IV-147, IV-234, IV-248
Miziołek, Marek   IV-514
Mocavero, Silvia   I-44
Mohamad, Kamaruddin Malik   IV-353
Mohamed Elsayed, Samir A.   I-161
Monfroy, Eric   III-607
Monteserin, Ariel   III-698
Montrone, Silvestro   II-102
Moreira, Adriano   II-450
Moreira, Álvaro   IV-67
Moscato, Pablo   I-71
Moschetto, Danilo A.   IV-470
Müller, Heinrich   I-15
Mundim, Kleber Carlos   I-432
Mundim, Maria Suelly Pedrosa   I-316
Mundim, Maria Suely Pedrosa   I-432
Munir, Ali   III-352, III-368
Murgante, Beniamino   II-640, II-670, III-512
Murri, Riccardo   I-401
Musaoglu, Nebiye   III-471

Nabwey, Hossam A.   II-316, II-358
Nakagawa, Elisa Yumi   III-576
Nalli, Danilo   I-292
Nawi, Nazri Mohd   I-87
Nedjah, Nadia   I-101, I-113, I-136, I-148
Nema, Jigyasu   III-528
Neri, Igor   IV-457
Nesticò, Antonio   II-27
Neves, Brayan   II-302
Nguyên, Toàn   IV-418
Niu, Wenjia   III-297
Niyogi, Rajdeep   III-528

Nolè, Gabriele   III-512
Nunes, Manuel L.   III-147

O'Kelly, Morton E.   II-249
Oliveira, José A.   III-133
Oliveira, Rafael S.   I-649
Oreni, Daniela   II-397
Ottomanelli, Michele   II-44

Pacifici, Leonardo   I-292, I-371
Pádua, Clarindo Isaías P.S.   IV-51
Pádua, Wilson   IV-51
Pallottelli, Simonetta   I-358
Panaro, Simona   II-54
Pandey, Kusum Lata   IV-147
Paolillo, Pier Luigi   II-331
Park, Changyong   III-283
Park, Gyung-Leen   III-217
Park, Junbeom   III-283
Park, Sangjoon   IV-613
Park, Young Jin   IV-602
Parvin, Hamid   III-647
Parvin, Sajad   III-647
Pathak, Surendra   II-589
Pauls-Worm, Karin G.J.   III-45
Peixoto, Daniela C.C.   IV-51
Peixoto, João   II-450
Pepe, Monica   II-397
Perchinunno, Paola   II-88, II-102
Pereira, Gilberto Corso   I-491
Pereira, Guilherme A.B.   III-133, III-187
Pereira, Óscar Mortágua   III-682
Pereira, Tiago F.   I-240
Pessanha, Fábio Gonçalves   I-113
Pigozzo, Alexandre B.   I-691, I-707
Pimentel, Dulce   II-565
Pingali, Keshav   I-1
Pinheiro, Marcello Sandi   III-631
Pirani, Fernando   I-316, I-387, I-432
Piscitelli, Claudia   II-517
Poggioni, Valentina   III-539
Pol, Maciej   IV-576
Pollino, Maurizio   II-532
Poma, Lourdes P.P.   IV-470
Pontrandolfi, Piergiuseppe   II-686
Poplin, Alenka   I-491
Pozniak-Koszalka, Iwona   IV-576
Pradel, Marilys   II-373
Prasad, Rajesh   IV-147
Prieto, Iñaki   I-529

Proma, Wojciech   IV-576
Prudêncio, Ricardo B.C.   III-562

Qadir, Junaid   III-352
Qaisar, Saad Bin   III-339, III-352, III-407
Quintela, Bárbara de Melo   I-675, I-691, I-707

Ragni, Mirco   I-723
Raja, Haroon   III-368, III-407
Rajasekaran, Sanguthevar   I-161
Rak, Jacek   IV-498
Ramiro, Carla   I-29
Rampini, Anna   II-397
Rasekh, Abolfazl   II-275
Re, Nazzareno   I-738
Renhe, Marcelo Caniato   II-712
Resende, Rodolfo Ferreira   IV-632
Rezende, José Francisco V.   II-302
Rezende, Solange Oliveira   III-618
Ribeiro, Hugo   IV-202
Ribeiro, Marcela Xavier   III-667, III-743
Riveros, Carlos   I-71
Rocha, Ana Maria A.C.   III-57, III-72, III-147
Rocha, Bernardo M.   I-649
Rocha, Humberto   III-1
Rocha, Jorge Gustavo   I-571
Rocha, Maria Célia Furtado   I-491
Rocha, Pedro Augusto F.   I-691
Rodrigues, António M.   II-565
Romani, Luciana A.S.   III-743
Rosi, Marzio   I-316, I-331
Rossi, Elda   I-345
Rossi, Roberto   III-45
Rotondo, Francesco   I-545
Ruiz, Linnyer Beatrys   IV-588

Sad, Dhiego Oliveira   I-228
Salles, Evandro O.T.   II-739
Salles, Ronaldo M.   IV-326
Salvatierra, Gonzalo   IV-484
Sampaio-Fernandes, João C.   I-665
Samsudin, Nurnabilah   II-728
Sanches, Silvio Ricardo Rodrigues   II-699
Sanjuan-Estrada, Juan Francisco   I-57
Santos, Adauto   IV-588
Santos, Maribel Yasmina   III-682

Santos, Marilde Terezinha Prado   III-667, III-743
Santos, Teresa   II-565
Sarafian, Haiduke   I-599
Saraiva, João   IV-202, IV-217
Sarcinelli-Filho, Mario   II-739
Schirone, Dario Antonio   II-1, II-17, II-88
Sciberras, Elaine   II-609
Scorza, Francesco   II-640
Selicato, Francesco   I-545, II-517
Selicato, Marco   II-144
Sellarès, J. Antoni   I-253
Sertel, Elif   III-471
Shaaban, Shaaban M.   II-316, II-358
Shah, Habib   I-87
Shukla, Mukul   IV-157
Shukla, Ruchi   IV-157
Silva, António   I-571
Silva, João Tácio C.   II-302
Silva, José Eduardo C.   I-240
Silva, Rodrigo   I-228
Silva, Rodrigo M.P.   IV-326
Silva, Roger Correia   I-228
Silva, Valdinei Freire da   II-699
Silva Jr., Luneque   I-113
Silvestre, Eduardo Augusto   IV-1
Simões, Flávio O.   I-240
Simões, Paulo   II-234
Singh, Gaurav   II-286
Skouteris, Dimitris   I-331
Soares, Carlos   III-562
Song, Hokwon   III-227, III-238
Song, Taehoun   III-311
Soravia, Marco   II-599
Soto, Ricardo   III-607
Souza, Cleber P.   IV-314
Stanganelli, Marialuce   II-599
Stankute, Silvija   II-439

Tajani, Francesco   II-27
Tasso, Sergio   I-358, IV-457
Tavares, Maria Purificação   I-665
Teixeira, Ana Paula   III-17
Teixeira, José Carlos   III-174, III-202
Teixeira, Senhorinha F.C.F.   III-147, III-202
Tentella, Giorgio   I-417
Thom, Lucinéia   IV-67

Thonhauser, Gerhard   IV-392
Tilio, Lucia   II-466, II-686
Timm, Constantin   I-15
Tori, Romero   II-699
Torkan, Germano   II-17
Torre, Carmelo Maria   I-583, II-116, II-144, II-157
Traina, Agma J.M.   III-743
Treadwell, Jim   IV-646
Tricaud, Sebastien   IV-314
Trofa, Giovanni La   II-144
Trunfio, Giuseppe A.   II-481, II-492
Tsoukiàs, Alexis   II-466
Tyrallová, Lucia   II-208, II-220

Usera, Gabriel   IV-530

Vafaeinezhad, Ali Reza   II-275
Vargas-Hernández, José G.   I-518
Varotsis, Constantinos   I-304
Vaz, Paula   I-665
Vecchiocattivi, Franco   I-316, I-432
Vella, Flavio   IV-457
Verdicchio, Marco   I-371
Verma, Shilpi   III-87
Versaci, Francesco   I-1
Vieira, Marcelo Bernardes   I-228, II-712

Vieira, Weslley   IV-444
Vyatkina, Kira   I-280

Walkowiak, Krzysztof   IV-498, IV-514
Wang, Hao   III-283
Wang, Kangkang   I-447
Weichert, Frank   I-15
Won, Dongho   III-391
Wu, Feifei   I-447

Xavier, Carolina R.   I-635
Xavier, Micael P.   I-691
Xexéo, Geraldo B.   IV-379
Xu, Yanmei   I-447
Xu, Yuemei   III-297

Yanalak, Mustafa   III-471

Zaldívar, Anibal   IV-116
Zenha-Rela, Mário   IV-132
Zhang, Tian   IV-100
Zhang, Xiaokun   IV-100
Zhang, Yan   IV-100
Zhao, Xuying   IV-100
Zito, Romina   I-583
Zunino, Alejandro   IV-29, IV-234, IV-484